MCAT® Organic Chemistry

2025–2026 Edition: An Illustrated Guide

Acknowledgments for the 2025–2026 Edition

Ensuring that the course materials in this book are accurate and up to date would not have been possible without the multifaceted contributions from our team of content experts, editors, illustrators, software developers, and other amazing support staff. UWorld's passion for education continues to be the driving force behind all our products, along with our focus on quality and dedication to student success.

About the MCAT Exam

Taking the MCAT is a significant milestone on your path to a rewarding career in medicine. Scan the QR codes below to learn crucial information about this exam as you take your next step before medical school.

Basic MCAT Exam Information

Scores and Percentiles

MCAT Sections

Registration Guide

Preparing for the MCAT with UWorld

The MCAT is a grueling exam spanning seven subjects that is designed to test your aptitude in areas essential for success in medicine. Preparing for the exam can be intimidating—so much so that in post-MCAT questionnaires conducted by the AAMC®, a majority of students report not feeling confident about their MCAT performance.

In response, UWorld set out to create premier learning tools to teach students the entire MCAT syllabus, both efficiently and effectively. Taking what we learned from helping over 90% of medical students prepare for their medical board exams (USMLE®), we launched the UWorld MCAT Qbank in 2017 and the UWorld MCAT UBooks in 2024. The MCAT UBooks are meticulously written and designed to provide you with the knowledge and strategies you need to meet your MCAT goals with confidence and to secure your future in medical school.

Below, we explain how to use the MCAT UBooks and MCAT Qbank together for a streamlined learning experience. By strategically integrating both resources into your study plan, you will improve your understanding of key MCAT content as well as build critical reasoning skills, giving you the best chance at achieving your target score.

MCAT UBooks: Illustrated and Annotated Guides

The MCAT UBooks include not only the printed editions for each MCAT subject but also provide digital access to interactive versions of the same books. There are eight printed MCAT UBooks in all, six comprehensive review books covering the science subjects and two specialized books for the Critical Analysis and Reasoning Skills (CARS) section of the exam:

- Biology
- Biochemistry
- General Chemistry
- Organic Chemistry
- Physics
- Behavioral Sciences
- CARS (Annotated Practice Book)
- CARS Passage Booklet (Annotated)

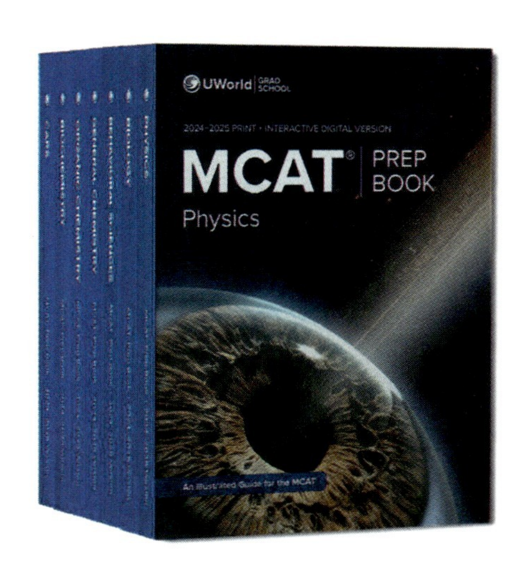

Each UBook is organized into Units, which are divided into Chapters. The Chapters are then split into Lessons, which are further subdivided into Concepts.

MCAT Sciences: Printed UBook Features

The MCAT UBooks bring difficult science concepts to life with thousands of engaging, high-impact visual aids that make topics easier to understand and retain. In addition, the printed UBooks present key terms in blue, indicating clickable illustration hyperlinks in the digital version that will help you learn more about a scientific concept.

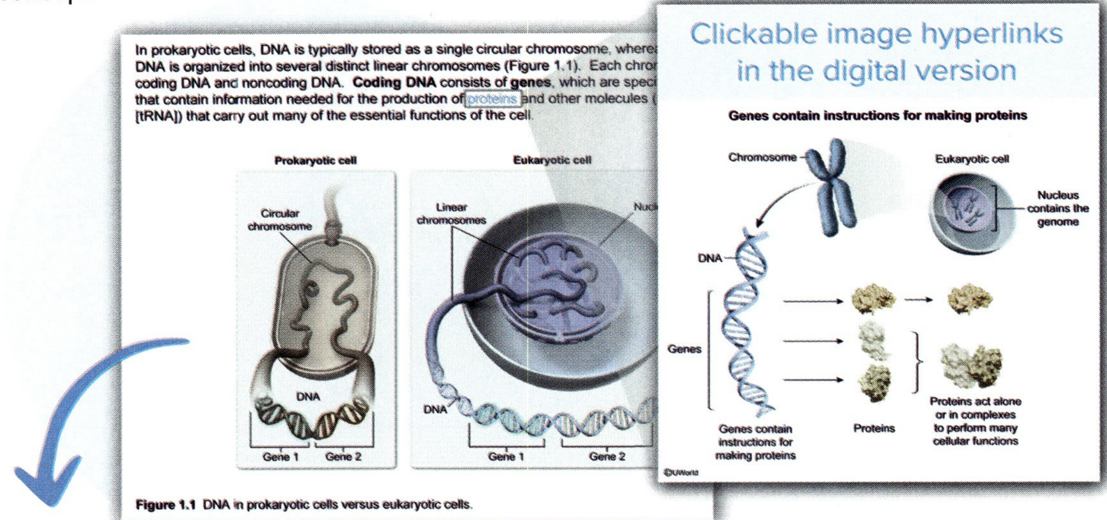

Test Your Basic Science Knowledge with Concept Check Questions

The printed UBooks also include 450 new questions—never before available in the UWorld Qbank—for Biology, General Chemistry, Organic Chemistry, Biochemistry, and Physics. These new questions, called Concept Checks, are interspersed throughout the entire book to enhance your learning experience. Concept Checks allow you to instantly test yourself on MCAT concepts you just learned from the UBook.

Short answers to the Concept Checks are found in the appendix at the end of each printed UBook. In addition, the digital version of the UBook provides an interactive learning experience by giving more detailed, illustrated, step-by-step explanations of each Concept Check. These enhanced explanations will help reinforce your learning and clarify any areas of uncertainty you may have.

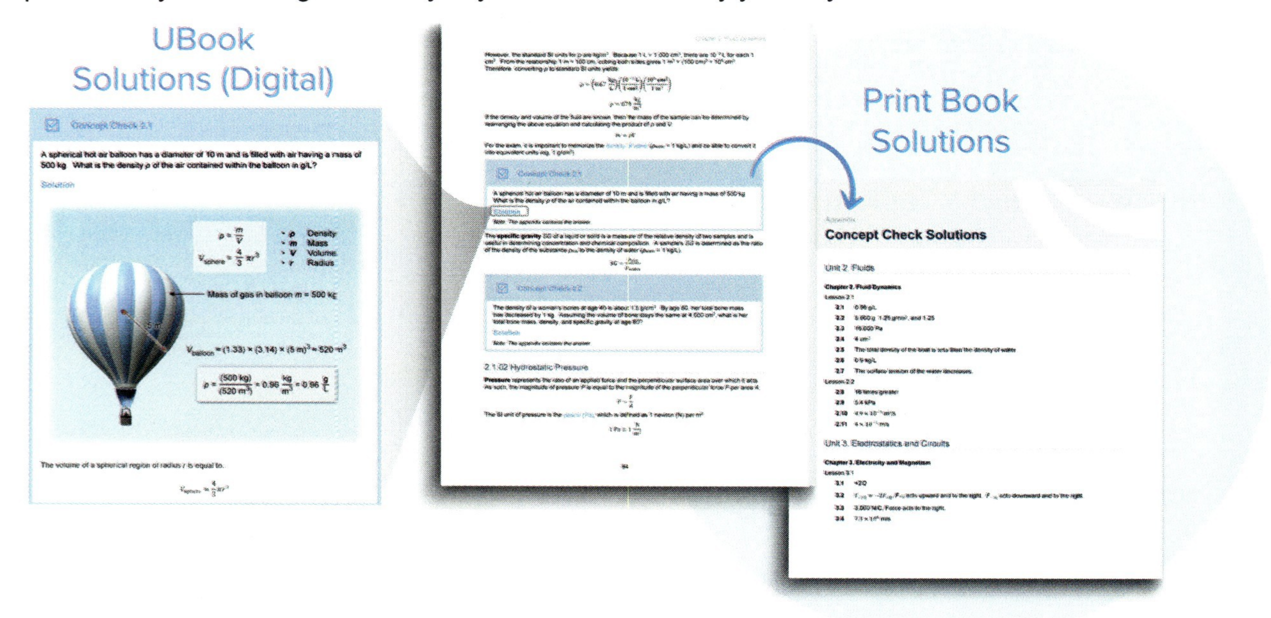

MCAT CARS Printed UBook Features

For CARS, the main book, or Annotated Practice Book, teaches you the specialized CARS skills and strategies you need to master and then follows up with multiple sets of MCAT-level practice questions.

Additionally, the CARS Passage Booklet includes annotated versions of the passages in the CARS Main Book. From these annotations, you will learn how to break down a CARS passage in a step-by-step manner to find the right answer to each CARS question.

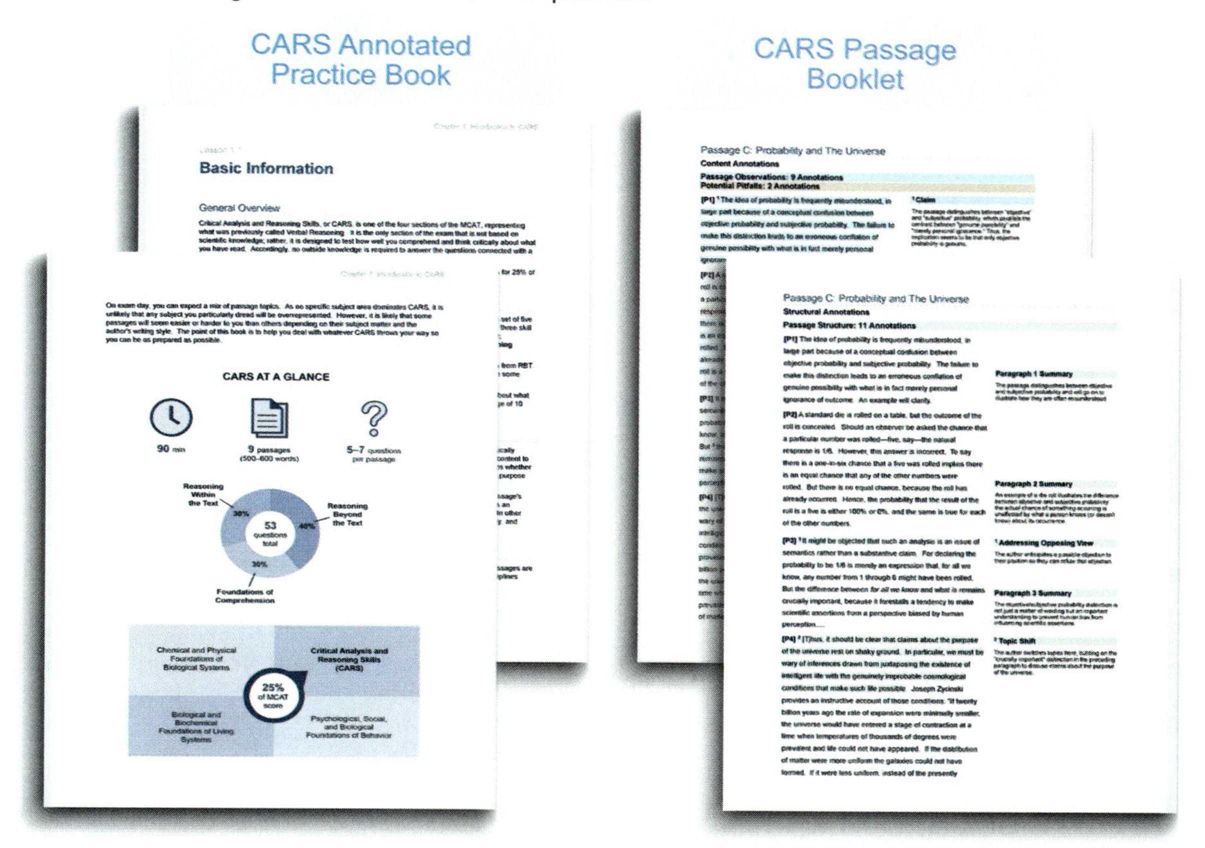

MCAT-Level Exam Practice with the UWorld Qbank

UWorld's MCAT UBooks and Qbank were designed to be used together for a comprehensive review experience. The UWorld Qbank provides an active learning approach to MCAT prep, with thousands of MCAT-level questions that align with each UBook.

The printed UBooks include a prompt at the end of each unit that explains how to access unit practice tests in the MCAT Qbank. In addition, the MCAT UBooks' digital platform enables you to easily create your own unit tests based on each MCAT subject.

To purchase MCAT Qbank access or to begin a free seven-day trial, visit gradschool.uworld.com/mcat.

Boost Your Score with the #1 MCAT Qbank

Why use the UWorld Qbank?

- Thousands of high-yield MCAT-level questions
- In-depth, visually engaging answer explanations
- Confidence-building user interface identical to the exam
- Data-driven performance and improvement tracking
- Fully featured mobile app for on-the-go review

Special Features Integrating Digital UBooks and the UWorld Qbank

The digital MCAT UBooks and the MCAT Qbank come with several integrated features that transform ordinary reading into an interactive study session. These time-saving tools enable you to personalize your MCAT test prep, get the most out of our detailed explanations, save valuable time, and know when you are ready for exam day.

My Notebook

My Notebook, a personalized note-taking tool, allows you to easily copy and organize content from the UBooks and the Qbank. Simplify your study routine by efficiently recording the MCAT content you will encounter in the exam, and streamline your review process by seamlessly retrieving high-yield concepts to boost your study performance—in less time.

Digital Flashcards

Our unique flashcard feature makes it easy for students to copy definitions and images from the MCAT UBooks and Qbank into digital flashcards. Each card makes use of spaced repetition, a research-supported learning methodology that improves information retention and recall. Based on how you rate your understanding of flashcard content, our algorithm will display the card more or less frequently.

Fully Featured Mobile App

Study for your MCAT exams anytime, anywhere, with our industry-leading mobile app that provides complete access to your MCAT prep materials and that syncs seamlessly across all devices. With the UWorld MCAT app, you can catch up on reading, flip through flashcards between classes, or take a practice quiz during lunch to make the most of your downtime and keep MCAT material top of mind.

Book and Qbank Progress Tracking

Track your progress while using the MCAT UBooks and Qbank, and review MCAT content at your own pace. Our learning tools are enhanced by advanced performance analytics that allow users to assess their preparedness over time. Hone in on specific subjects, foundations, and skills to iron out any weaknesses, and even compare your results with those of your peers.

Explore the Periodic Table

You will need to use the periodic table to answer questions on the MCAT for specific sections. Introductory general chemistry concepts constitute 30% of the material tested in the Chemical and Physical Foundations of Biological Systems section of the exam. In addition, General Chemistry constitutes 5% of the Biological and Biochemical Foundations of Living Systems section of the MCAT. Using and understanding the periodic table is a crucial skill needed for success in these sections.

1 **H** 1.0																	2 **He** 4.0
3 **Li** 6.9	4 **Be** 9.0											5 **B** 10.8	6 **C** 12.0	7 **N** 14.0	8 **O** 16.0	9 **F** 19.0	10 **Ne** 20.2
11 **Na** 23.0	12 **Mg** 24.3											13 **Al** 27.0	14 **Si** 28.1	15 **P** 31.0	16 **S** 32.1	17 **Cl** 35.5	18 **Ar** 39.9
19 **K** 39.1	20 **Ca** 40.1	21 **Sc** 45.0	22 **Ti** 47.9	23 **V** 50.9	24 **Cr** 52.0	25 **Mn** 54.9	26 **Fe** 55.8	27 **Co** 58.9	28 **Ni** 58.7	29 **Cu** 63.5	30 **Zn** 65.4	31 **Ga** 69.7	32 **Ge** 72.6	33 **As** 74.9	34 **Se** 79.0	35 **Br** 79.9	36 **Kr** 83.8
37 **Rb** 85.5	38 **Sr** 87.6	39 **Y** 88.9	40 **Zr** 91.2	41 **Nb** 92.9	42 **Mo** 95.9	43 **Tc** (98)	44 **Ru** 101.1	45 **Rh** 102.9	46 **Pd** 106.4	47 **Ag** 107.9	48 **Cd** 112.4	49 **In** 114.8	50 **Sn** 118.7	51 **Sb** 121.8	52 **Te** 127.6	53 **I** 126.9	54 **Xe** 131.3
55 **Cs** 132.9	56 **Ba** 137.3	57 **La*** 138.9	72 **Hf** 178.5	73 **Ta** 180.9	74 **W** 183.9	75 **Re** 186.2	76 **Os** 190.2	77 **Ir** 192.2	78 **Pt** 195.1	79 **Au** 197.0	80 **Hg** 200.6	81 **Tl** 204.4	82 **Pb** 207.2	83 **Bi** 209.0	84 **Po** (209)	85 **At** (210)	86 **Rn** (222)
87 **Fr** (223)	88 **Ra** (226)	89 **Ac+** (227)	104 **Rf** (261)	105 **Db** (262)	106 **Sg** (266)	107 **Bh** (264)	108 **Hs** (277)	109 **Mt** (268)	110 **Ds** (281)	111 **Rg** (280)	112 **Cn** (285)	113 **Uut** (284)	114 **Fl** (289)	115 **Uup** (288)	116 **Lv** (293)	117 **Uus** (294)	118 **Uuo** (294)

*	58 **Ce** 140.1	59 **Pr** 140.9	60 **Nd** 144.2	61 **Pm** (145)	62 **Sm** 150.4	63 **Eu** 152.0	64 **Gd** 157.3	65 **Tb** 158.9	66 **Dy** 162.5	67 **Ho** 164.9	68 **Er** 167.3	69 **Tm** 168.9	70 **Yb** 173.0	71 **Lu** 175.0
+	90 **Th** 232.0	91 **Pa** (231)	92 **U** 238.0	93 **Np** (237)	94 **Pu** (244)	95 **Am** (243)	96 **Cm** (247)	97 **Bk** (247)	98 **Cf** (251)	99 **Es** (252)	100 **Fm** (257)	101 **Md** (258)	102 **No** (259)	103 **Lr** (260)

Table of Contents

Unit 1 Introduction to Organic Chemistry

Chapter 1 Chemical Bonding

1.1 Covalent Bonds

1.1.01 Valence Electrons and the Octet Rule
1.1.02 Formation of Chemical Bonds
1.1.03 Lewis Structures
1.1.04 Formal Charge
1.1.05 Covalent Bond Trends

1.2 Hybridization

1.2.01 Atomic Orbitals
1.2.02 sp^3 Orbitals
1.2.03 sp^2 Orbitals
1.2.04 *sp* Orbitals

1.3 Bond Polarity

1.3.01 Electronegativity
1.3.02 Dipole Moments

1.4 Molecular Orbital Theory

1.4.01 Overview of Molecular Orbital Theory
1.4.02 Sigma Bonding
1.4.03 Pi Bonding
1.4.04 Molecular Orbitals and Multiple Bonds

Chapter 2 Structure of Organic Molecules

2.1 Molecular Shape

2.1.01 VSEPR Theory and Electron Domain Geometry
2.1.02 Relationship of Electron Domain Geometry to Hybridization
2.1.03 Molecular Geometry

2.2 Molecular Polarity

2.2.01 Molecular Polarity and Polar Bonds
2.2.02 Molecular Polarity and Lone Pairs

2.3 Intermolecular Forces

2.3.01 Ion-Associated Interactions
2.3.02 Dipole-Associated Interactions
2.3.03 Hydrogen Bonding
2.3.04 Temporary Dipole-Associated Interactions
2.3.05 Comparing the Intermolecular Forces

2.4 Overview of Physical Properties

2.4.01 Boiling Point
2.4.02 Solubility
2.4.03 Density

2.5 Drawing Organic Molecules

2.5.01 Condensed Structural Formulas
2.5.02 Line-Angle Formulas
2.5.03 Depicting Three-Dimensional Molecules

Chapter 3 Isomers

Chapter 4 Organic Nomenclature

Chapter 5 Overview of Organic Reactions

Lesson 1.1

Covalent Bonds

Introduction

Organic chemistry is broadly defined as the chemistry of carbon-containing compounds. Because the study of organic chemistry builds on the foundations covered in general chemistry coursework, Chapters 1 and 2 serve as a platform to bridge selected concepts from general chemistry to applications in organic chemistry.

1.1.01 Valence Electrons and the Octet Rule

Individual atoms will either transfer or share **valence electrons** to give each atom access to eight valence electrons, a feature commonly known as the **octet rule**. By fulfilling an octet, individual atoms become isoelectronic with the noble gases and gain increased stability (Table 1.1).

Table 1.1 The octet rule for elements in the first (n = 1) and second (n = 2) rows of the periodic table.

		n = 1			n = 2	
Group	**Element**	**Valence electrons**	**Needed for octet**[a]	**Element**	**Valence electrons**	**Needed for octet**
1A	H	1	1	Li	1	7
2A				Be	2	6
3A				B	3	5
4A				C	4	4
5A				N	5	3
6A				O	6	2
7A				F	7	1
8A	He	2	0	Ne	8	0

[a] First-row elements strive for a total of 2 valence electrons

Elements in the first row of the periodic table (ie, H, He) require only two valence electrons (the maximum number allowed by a lone 1*s* orbital). Elements in the third row of the periodic table (or greater) have access to *d* and *f* subshells that may allow more than eight valence electrons (an expanded octet). However, within organic chemistry most of the relevant elements appear in the first and second row.

1.1.02 Formation of Chemical Bonds

There are two primary ways that atoms gain valence electrons, complete their octet, and form a **chemical bond**.

In an **ionic bond**, one atom *fully* transfers one or more electrons to another, and **ions** are generated. The atom that gives up valence electrons becomes positively charged (a **cation**), while the atom that accepts valence electrons becomes negatively charged (an **anion**). **Salts** form when the oppositely charged ions become electrostatically attracted. Most ionic bonds are formed between a **metal** and a

nonmetal, as shown in Figure 1.1. Consequently, ionic bonds are relevant to organic chemistry, but they appear less frequently than covalent bonds.

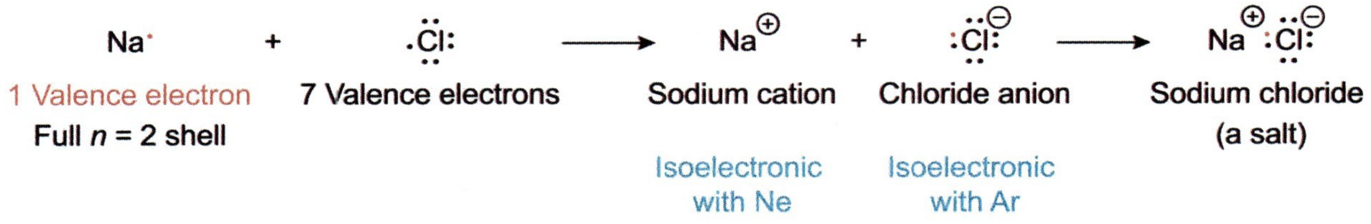

Na· + ·Cl: ⟶ Na⊕ + :Cl:⊖ ⟶ Na⊕ :Cl:⊖

1 Valence electron
Full n = 2 shell

7 Valence electrons

Sodium cation
Isoelectronic with Ne

Chloride anion
Isoelectronic with Ar

Sodium chloride (a salt)

Figure 1.1 Formation of an ionic bond.

A **covalent bond** is formed when atoms share electrons to achieve a full octet, as shown in Figure 1.2.

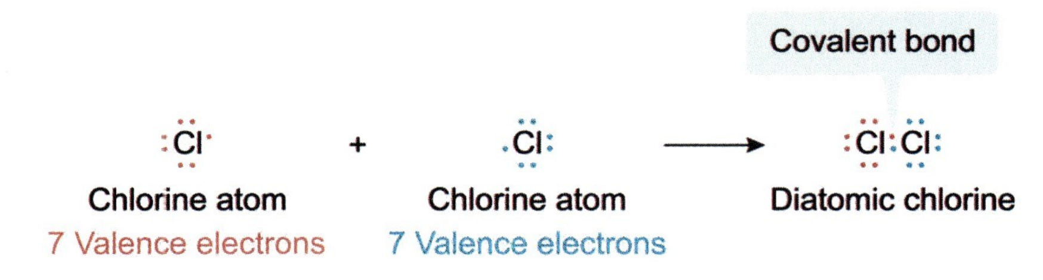

Figure 1.2 Formation of a covalent bond.

1.1.03 Lewis Structures

A **Lewis (dot) structure** represents chemical bonding between atoms in a molecule by depicting each valence electron as a single dot. Shared **bonding electrons** are represented by either a pair of dots or, more commonly in organic chemistry, a single line *between* the atoms (a **single bond**). **Nonbonding electrons** (ie, **lone pairs**) and are represented by dots *around* (but not between) atoms. Lone pairs are *very important* in determining the reactivity of organic compounds.

Sometimes atoms must share more than one pair of electrons for all atoms to achieve a full octet. A **double bond** forms when two pairs (four electrons) are shared, and a **triple bond** forms when three pairs (six electrons) are shared between atoms. Double and triple bonds are examples of **multiple bonds**, which are almost always represented by lines rather than dots (Figure 1.3).

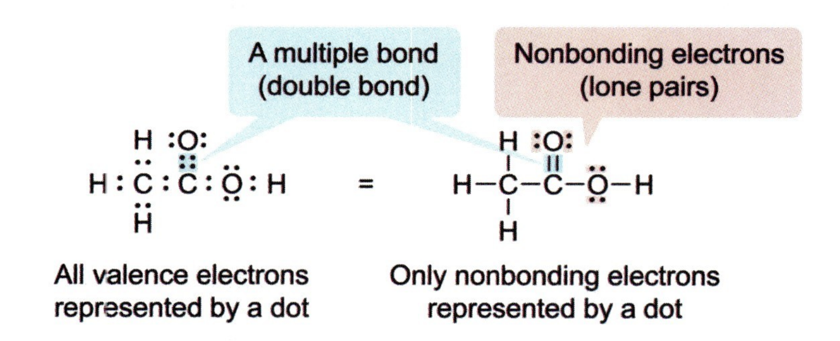

Figure 1.3 Multiple bonds and nonbonding electrons in Lewis structures.

1.1.04 Formal Charge

Formal charge quantitatively indicates the electron density of an atom in a molecule relative to the electron density of the atom in isolation according to the equation:

$$\text{Formal charge} = \text{Atomic valence electrons} - \text{Nonbonding electrons} - \frac{1}{2}\,(\text{Bonding electrons})$$

Formal charges often influence the reactivity and properties of a substance, and the sum of the formal charges of all atoms in a molecule is always equal to the overall charge of the molecule or ion (the **molecular net charge**). A general chemistry text can provide a review of the process for generating Lewis structures and assigning formal charges for polyatomic molecules.

Concept Check 1.1

What is the formal charge for each nonhydrogen atom in the following Lewis structure?

```
   H  H :O:
   |  |  ||  ..
H—N—C—C—O:
   |  |      ..
   H  H
```

Solution

Note: The appendix contains the answer.

Although the formal charge equation is useful, a recognition of how common bonding patterns affect formal charge is a great time-saver. Often, the number of valence electrons of an element indicates the total number of covalent bonds that the element prefers to form within molecules. However, there may be multiple different patterns of single, double, and triple bonds that result in the same formal charge.

For example, the element carbon has four valence electrons and prefers to form a total of four covalent bonds with other elements to fulfill its octet and have a formal charge of 0. Although there are several bond combinations (Figure 1.4) that meet this criterion, these variations create carbons with very different reactivity profiles. Deviations from these patterns result in a nonzero formal charge.

Four single bonds | One double bond, Two single bonds | Two double bonds | One triple bond, One single bond

Figure 1.4 Bonding scenarios for a carbon atom with a total of four covalent bonds.

The most common bonding patterns for the second-row elements relevant to organic chemistry are depicted in Figure 1.5. Committing these patterns to memory and understanding the relationship between variants are likely to be *very* helpful throughout organic chemistry.

Element	C Carbon	N Nitrogen	O Oxygen	X Halogens
Valence electrons	4	5	6	7
+1 Formal charge (cation)				
Total number of bonds	3	4	3	2
Number of lone pairs	0	0	1	2
Example	$-\overset{\oplus}{\underset{\vert}{C}}-$	$-\overset{\vert}{\underset{\vert}{N}}{}^{\oplus}-$	$-\underset{\vert}{\ddot{O}}{}^{\oplus}-$	$-\underset{\cdot\cdot}{\ddot{X}}{}^{\oplus}-$
No formal charge (neutral)				
Total number of bonds	4	3	2	1
Number of lone pairs	0	1	2	3
Example	$-\overset{\vert}{\underset{\vert}{C}}-$	$-\underset{\vert}{\ddot{N}}-$	$-\underset{\cdot\cdot}{\ddot{O}}-$	$-\underset{\cdot\cdot}{\ddot{X}}:$
−1 Formal charge (anion)				
Total number of bonds	3	2	1	0
Number of lone pairs	1	2	3	4
Example	$-\underset{\vert}{\ddot{C}}{}^{\ominus}-$	$-\underset{\cdot\cdot}{\ddot{N}}{}^{\ominus}-$	$-\underset{\cdot\cdot}{\ddot{O}}{}^{\ominus}:$	$:\underset{\cdot\cdot}{\ddot{X}}{}^{\ominus}:$

Figure 1.5 Relationship between valence electrons, covalent bonding, and formal charge.

1.1.05 Covalent Bond Trends

The distance between nuclei connected by a covalent bond (**bond length**) represents the optimal, lowest-energy distance with the least repulsion and greatest constructive interaction (Figure 1.6). If the nuclei are too close together, their electron clouds will strongly repel one another (ie, they will have high potential energy), and if the nuclei are too far apart, their valence electrons will be unable to interact.

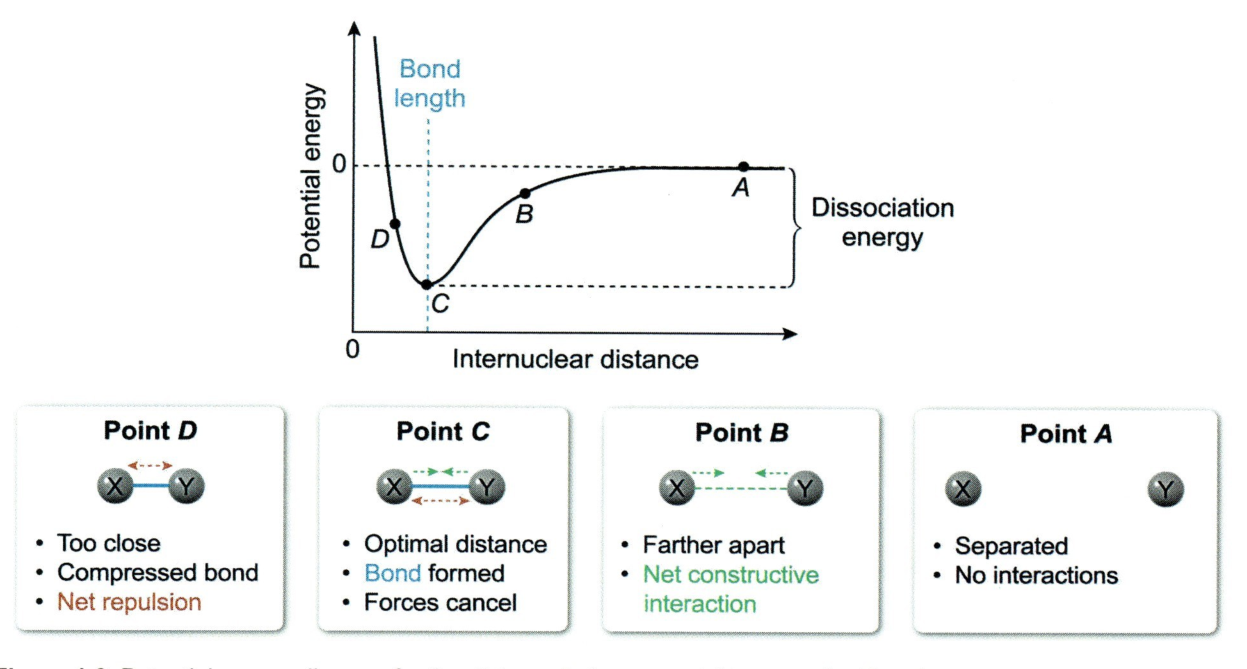

Figure 1.6 Potential energy diagram for the distance between nuclei in a covalent bond.

Atoms with valence electrons in higher shells farther from the nucleus have longer bond lengths. For example, the bond lengths of the hydrohalic acids (H–X) increase from fluorine to iodine.

Bond lengths also vary due to the multiplicity (**bond order**) of the covalent bond. Multiple bonds (eg, double and triple bonds) feature additional orbital interactions that bring the nuclei closer together and decrease the bond length (see Concept 1.4.04). For example, a carbon-carbon single bond has a larger bond length than a carbon-carbon double bond, which is larger than the bond length of a carbon-carbon triple bond.

The **bond dissociation energy** is the energy required to break a covalent bond and evenly divide the electrons in the covalent bond (Figure 1.7).

Energy

A : B ⟶ A· + ·B

Figure 1.7 Bond dissociation energy.

Consequently, the bond dissociation energy for a multiple bond is larger than that of a comparable single bond because more electrons must be disrupted (Figure 1.8).

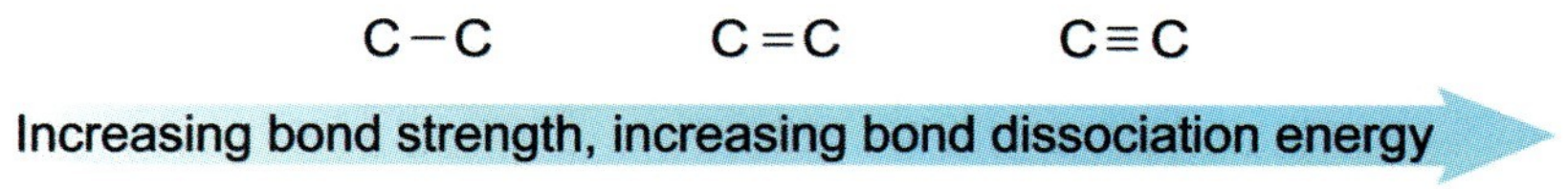

C−C   C=C   C≡C

Increasing bond strength, increasing bond dissociation energy

Figure 1.8 Comparison of bond dissociation energies by bond order.

Lesson 1.2

Hybridization

Introduction

Atomic orbitals are represented by mathematical expressions, which describe the behavior of electrons in an atom. The principal quantum number *n* (electron shell) describes the distance of an atomic orbital from the nucleus. The angular momentum quantum number ℓ describes an atomic orbital's shape. Quantum numbers are discussed in greater detail in General Chemistry Lesson 1.3. This lesson gives a review of *s* and *p* atomic orbitals and examines different types of associated hybrid orbitals.

1.2.01 Atomic Orbitals

In organic chemistry, the *s* and *p* atomic orbitals are most prevalent. An ***s* orbital** ($\ell = 0$) has a spherical shape with electron density distributed evenly around the nucleus; a ***p* orbital** ($\ell = 1$) has a dumbbell shape with two lobes that have opposite phases (ie, opposite signs [+ or −] within the mathematical expression) and are connected by a node (ie, a point where the mathematical expression changes phase and there is zero probability of electron density) (Figure 1.9).

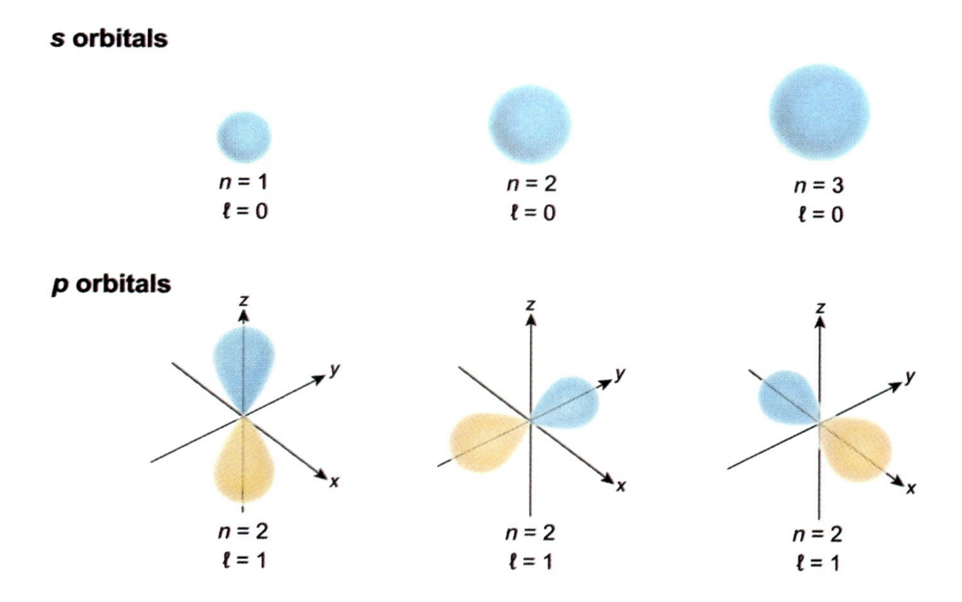

Figure 1.9 Shapes of *s* and *p* atomic orbitals.

Most organic molecules contain bonds made from **hybrid orbitals**, which are generated by the linear combination of *s* and *p* atomic orbitals on the same atom, yielding sp^3, sp^2, and *sp* hybrid orbitals. Because of the conservation of orbitals, the number of hybrid orbitals generated must be equal to the number of atomic orbitals that were hybridized. Each hybrid orbital has a distinct electron domain geometry that minimizes orbital repulsion and dictates the bond angles in the molecule (see Lesson 2.1).

1.2.02 sp^3 Orbitals

The combination of one *s* orbital and three *p* orbitals (four total atomic orbitals) results in four **sp^3 hybrid orbitals** (Figure 1.10), each having 25% *s* character and 75% *p* character.

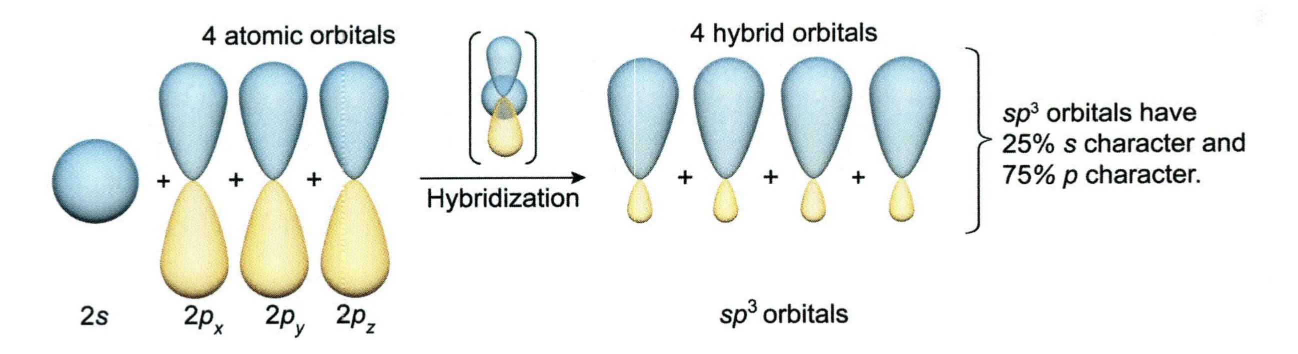

Figure 1.10 Formation of sp^3 hybrid orbitals.

The hybridization of an atom can be determined by examining the Lewis structure and counting the number of **electron domains** (ie, bonding regions and pairs of nonbonding electrons) around the atom. Each double or triple bond counts as one electron domain. The number of electron domains *must equal* the sum of the *s* and *p* superscripts in the hybrid orbital.

Atoms that are sp^3 hybridized have *four electron domains*. For example, the carbon in methanol (CH_3OH) has three bonds to hydrogen and one bond to oxygen (Figure 1.11) (see Concept 2.1.02 for more information).

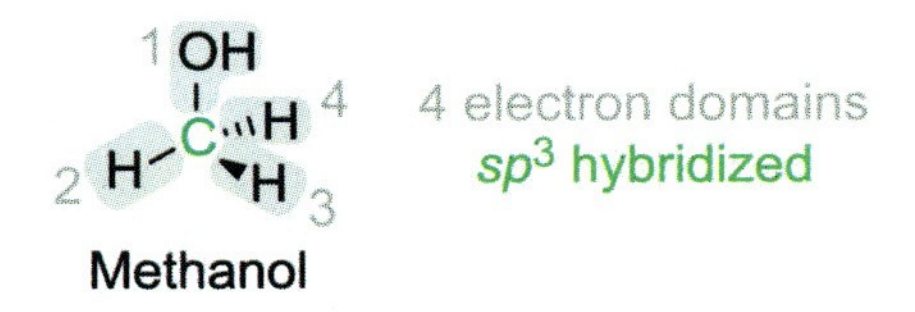

Figure 1.11 Electron domains and sp^3 hybridization.

1.2.03 sp^2 Orbitals

The combination of one *s* atomic orbital and two *p* atomic orbitals (three total atomic orbitals) form three **sp^2 hybrid orbitals** as shown in Figure 1.12. Each orbital has 33% *s* character and 67% *p* character. An **unhybridized *p* orbital** remains oriented perpendicular to the plane of the three sp^2 orbitals; unhybridized *p* orbitals can be vacant, contain a lone pair, or participate in a pi bond (see Concept 1.4.03).

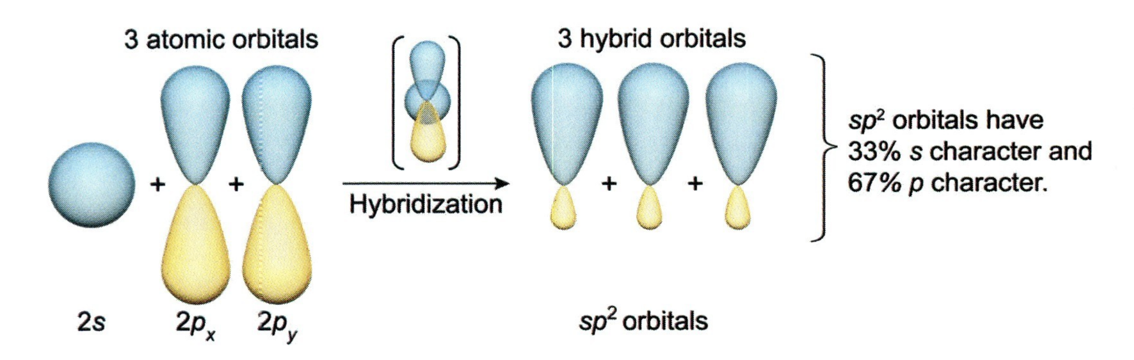

Figure 1.12 Formation of sp^2 hybrid orbitals.

Atoms that participate in double bonds are a common example of sp^2 hybridization. The carbon atoms in ethene (CH_2CH_2) each have two single bonds to hydrogen and one double bond to the other carbon; therefore, each carbon atom has *three electron domains* and is sp^2 hybridized (Figure 1.13).

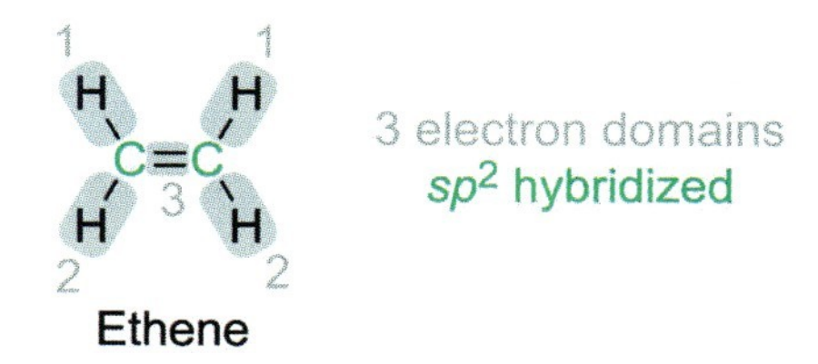

Figure 1.13 Electron domains and sp^2 hybridization.

1.2.04 *sp* Orbitals

Formation of two ***sp* hybrid orbitals** occurs by mixing one *s* atomic orbital and one *p* atomic orbital, leaving two unhybridized *p* orbitals (Figure 1.14). Because only one *p* orbital is used to form the two *sp* hybrid orbitals, the hybrid orbitals must be oriented along only the axis of the contributing *p* orbital. As such, *sp* hybrid orbitals have a **linear relationship**. Each *sp* orbital has 50% *s* character and 50% *p* character.

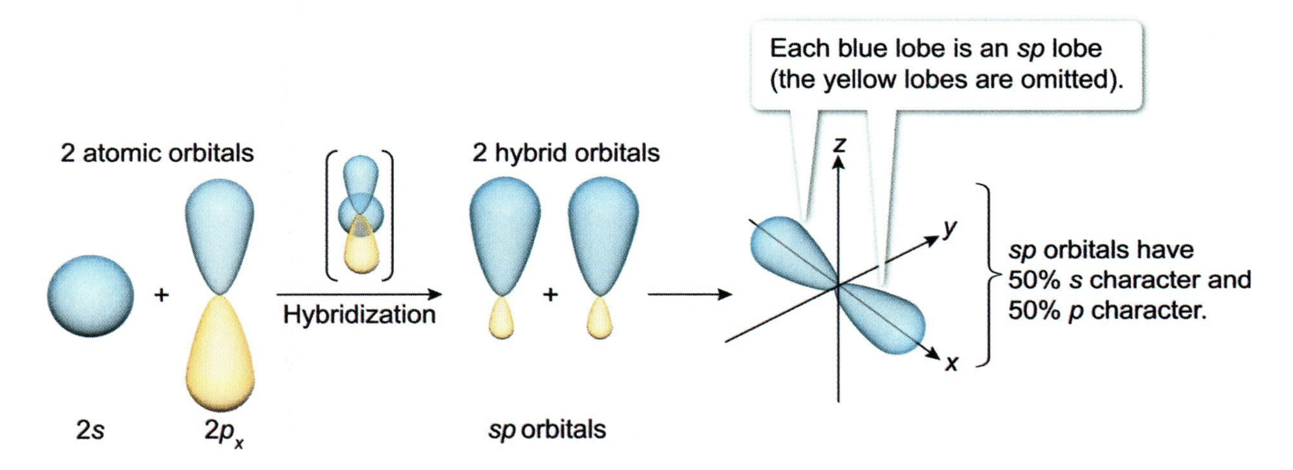

Figure 1.14 Formation of *sp* hybrid orbitals.

Atoms participating in one triple bond or two double bonds are *sp* hybridized. The carbon atoms in ethyne (C_2H_2) each have one bond to hydrogen and a triple bond to the other carbon; therefore, these carbons have *two electron domains* (Figure 1.15).

Figure 1.15 Electron domains and *sp* hybridization.

Concept Check 1.2

Identify the hybridization of the labeled atoms in the molecule in the following structure.

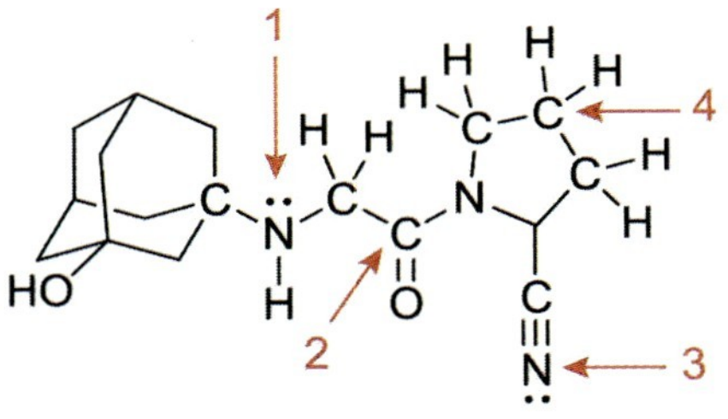

Solution

Note: The appendix contains the answer.

Lesson 1.3

Bond Polarity

Introduction

As discussed in Lesson 1.1, covalent bonds are formed when electrons are shared between two atoms, and ionic bonds are formed when an atom transfers one or more electrons to another atom. **Bond polarity** describes the nature of electron sharing in a covalent bond as a function of the **electronegativity** of the two atoms. This lesson provides a review of electronegativity and dipole moments.

1.3.01 Electronegativity

Electronegativity is the tendency of an atom to attract valence electrons in a bond. Ultimately, electronegativity is a direct result of an atom's effective nuclear charge. As a result, electronegativity displays periodic trends. Atoms that tend to accept electrons (ie, atoms with a higher electron affinity and a higher ionization energy) also tend to have a higher electronegativity. As shown on the periodic table in Figure 1.16, electronegativity tends to increase from left to right across a period and from the bottom to the top of a group.

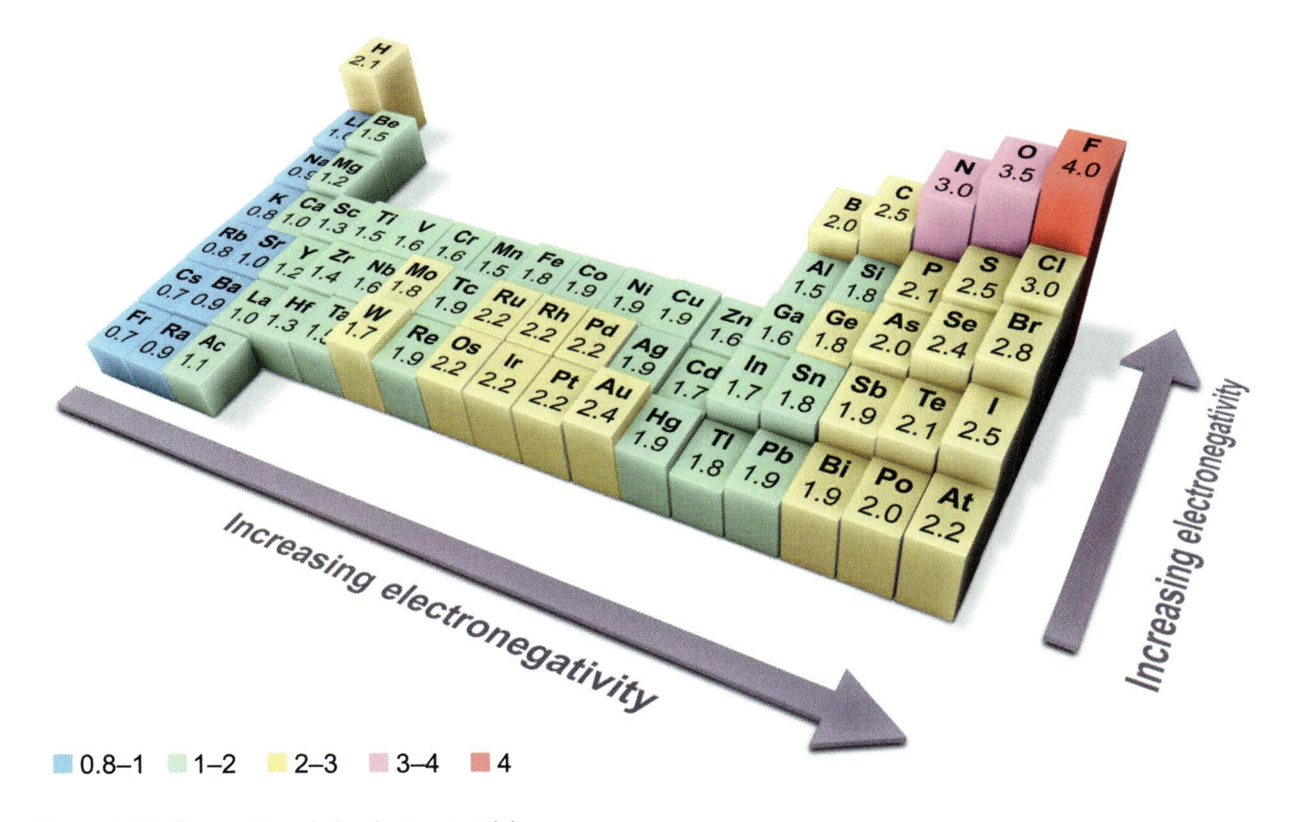

Figure 1.16 General trends in electronegativity.

Figure 1.17 shows how the difference in electronegativity between two atoms relates to the polarity of the bond. Atoms with an electronegativity difference of less than 0.5 essentially share electrons equally and form a **covalent nonpolar bond**. In contrast, atoms with an electronegativity difference between 0.5 and 1.8 do *not* share electrons equally and form a **covalent polar bond**. Atoms with large differences in electronegativity (greater than 1.8) *do not share* electrons; instead, one atom transfers electrons to the other atom, forming an **ionic bond**.

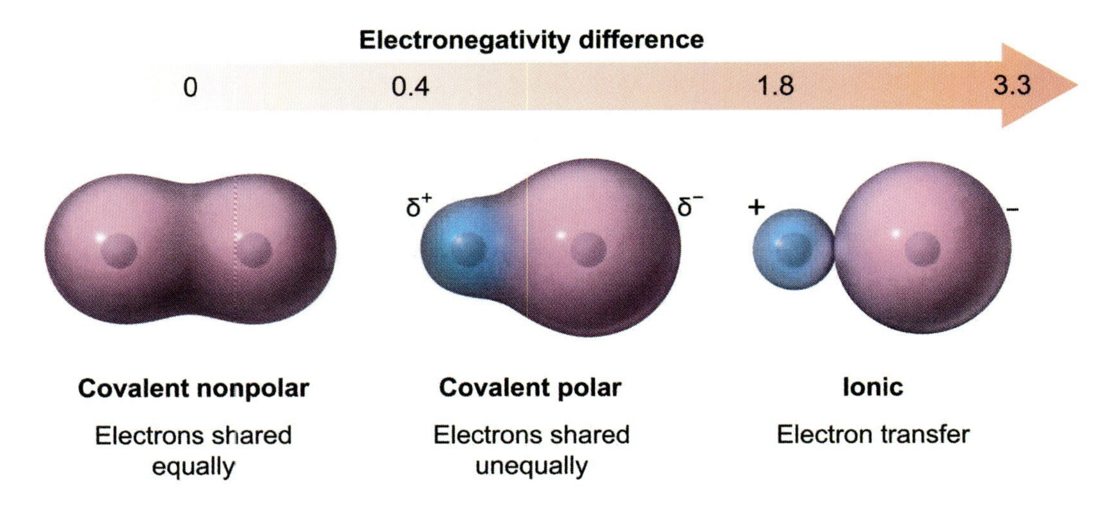

Figure 1.17 Electronegativity difference and bond polarity.

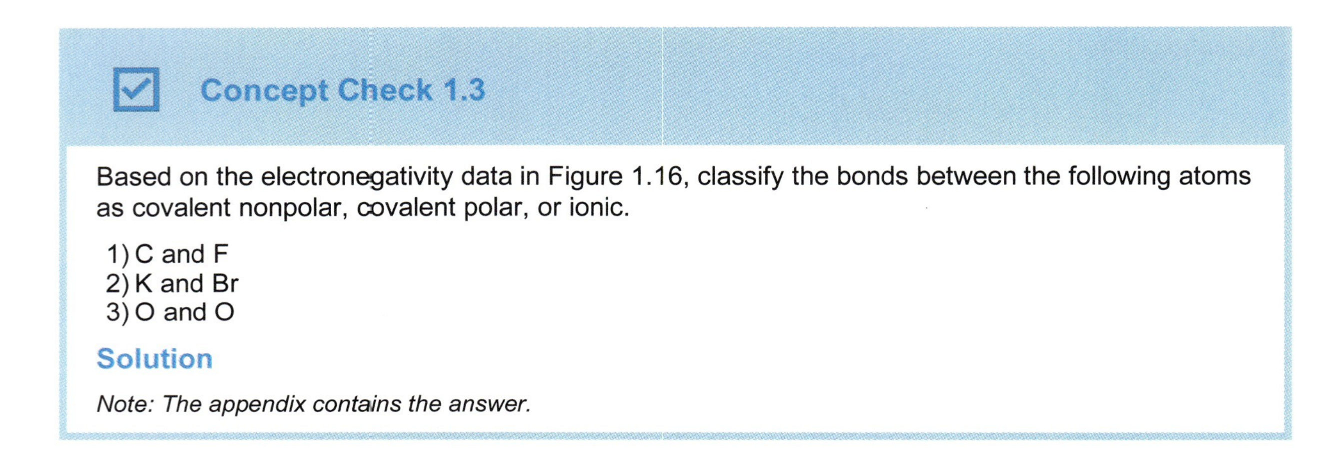

Concept Check 1.3

Based on the electronegativity data in Figure 1.16, classify the bonds between the following atoms as covalent nonpolar, covalent polar, or ionic.

1) C and F
2) K and Br
3) O and O

Solution

Note: The appendix contains the answer.

1.3.02 Dipole Moments

A dipole is created when an electronegative atom pulls electrons toward itself and away from an adjacent atom, resulting in a charge separation between two atoms in a bond. The more electronegative atom has a greater density of negatively charged electrons and carries a **partial negative charge (δ^-)**. In contrast, the less electronegative atom has a lower density of electrons and carries a **partial positive charge (δ^+).** The quantitative measure of a bond's polarity is the **dipole moment (μ)**, which is a vector quantity that depends on:

- Magnitude of charge separation (difference in electronegativity)
- Distance of charge separation (bond length)

The SI unit of measure for a dipole moment is the **debye (D)**, which is the dipole moment generated when one coulomb of charge is separated by one meter. A dipole moment is commonly denoted by an arrow with a cross at its tail; the arrow points toward the more electronegative atom (δ^-) and the cross is oriented toward the less electronegative atom (δ^+) (Figure 1.18).

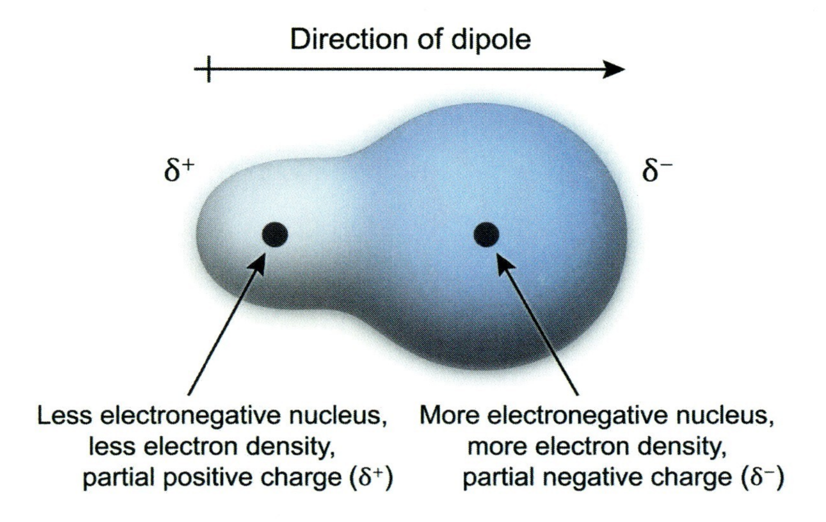

Figure 1.18 Electric dipole moments.

Individual bond dipole moments can be further influenced by neighboring atoms. Electron-donating groups are electron rich and can share electrons with nearby atoms through covalent bonds. In contrast, **electron-withdrawing groups** are electron deficient and pull electrons away from nearby atoms. These two scenarios lead to a shift in electron density known as the **inductive effect**.

Lesson 1.4

Molecular Orbital Theory

Introduction

Electron orbitals can be treated as mathematical expressions, and the linear combination of discrete atomic orbitals *on a single atom* leads to the generation of **hybrid orbitals** (Lesson 1.2). Similarly, the linear combination of orbitals from *different atoms in a molecule* leads to the generation of **molecular orbitals**. This lesson provides an overview of molecular orbital (MO) theory and introduces selected applications of MO theory in the study of organic chemistry.

1.4.01 Overview of Molecular Orbital Theory

An electron orbital is described by its **wavefunction**. As such, the associated wavefunction has a phase, which can be either positive (+) or negative (−) (Figure 1.19). Locations where the wavefunction changes phase (ie, crosses the *x*-axis) are called **nodes** and represent regions where the probability of electron density is zero.

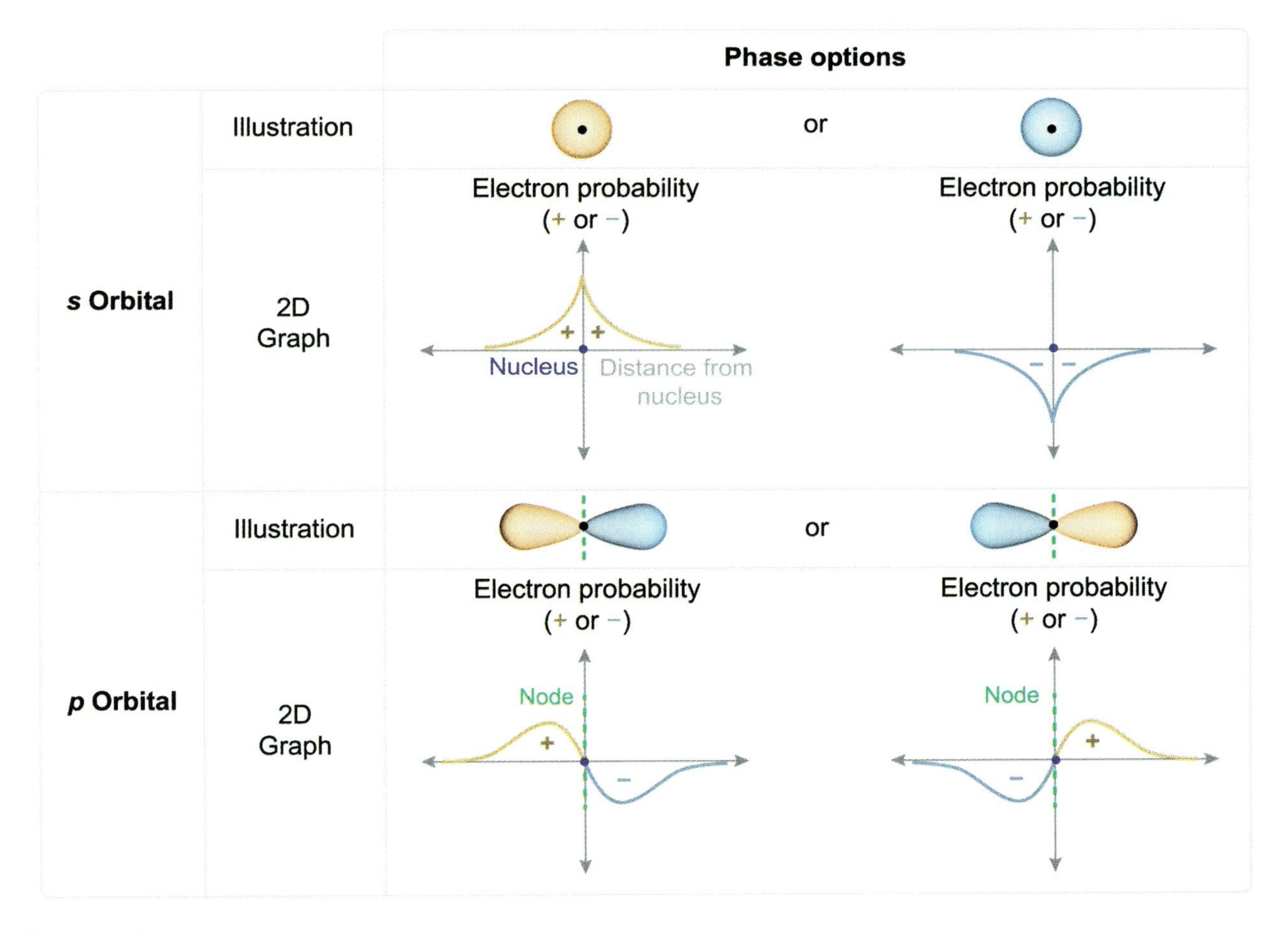

Figure 1.19 Wavefunctions of *s* and *p* atomic orbitals.

Wavefunctions that share the *same* phase (whether positive or negative) add together *constructively* and lead to an increase in amplitude (ie, probability of electron density). Wavefunctions that have *opposite* phases add together *destructively* to create new nodes. These interactions occur by a **linear combination of orbitals** and must demonstrate **conservation of energy** (ie, the total orbital energy remains unchanged) and **conservation of orbitals** (ie, the total number of orbitals is unchanged).

As a wavefunction oscillates between phases, there is an equal probability that the process of linear combination of orbitals will create a constructive or destructive interaction (Figure 1.20). Molecular orbitals that describe a constructive interaction of wavefunctions are called **bonding molecular orbitals** and have a region of shared electron density between the nuclei (ie, the chemical bond). Molecular orbitals that describe a destructive interaction of wavefunctions are called **antibonding molecular orbitals** (denoted by *) and have a node between the nuclei.

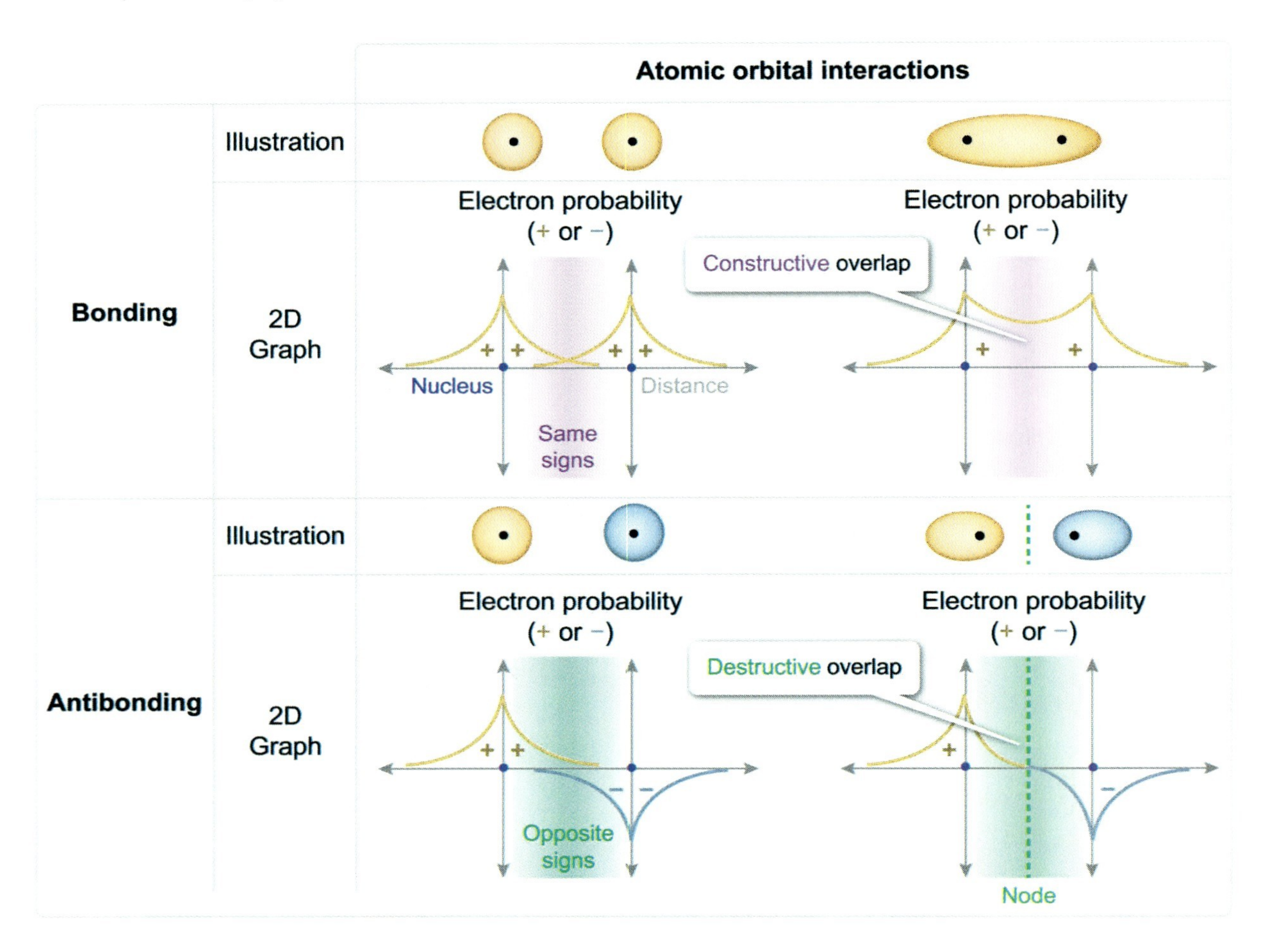

Figure 1.20 Bonding and antibonding interactions of *s* atomic orbitals.

As a result, a bonding molecular orbital has a *lower* energy and is more stable than the atomic orbitals prior to linear combination. In contrast, an antibonding molecular orbital has a *higher* energy and is less stable. The resulting molecular orbitals demonstrate conservation of energy, as the sum of the bonding and antibonding molecular orbitals equals the combined energy of the original orbitals. Conservation of orbitals is also observed as two orbitals generate two molecular orbitals (bonding and antibonding).

A **molecular orbital diagram** depicts all the features associated with the linear combination of orbitals. Figure 1.21 depicts the molecular orbital diagram for H_2, which results from the linear combination of the 1*s* orbital on each hydrogen atom. Although this is the simplest example of molecular orbital generation, it illustrates the major principles associated with MO theory.

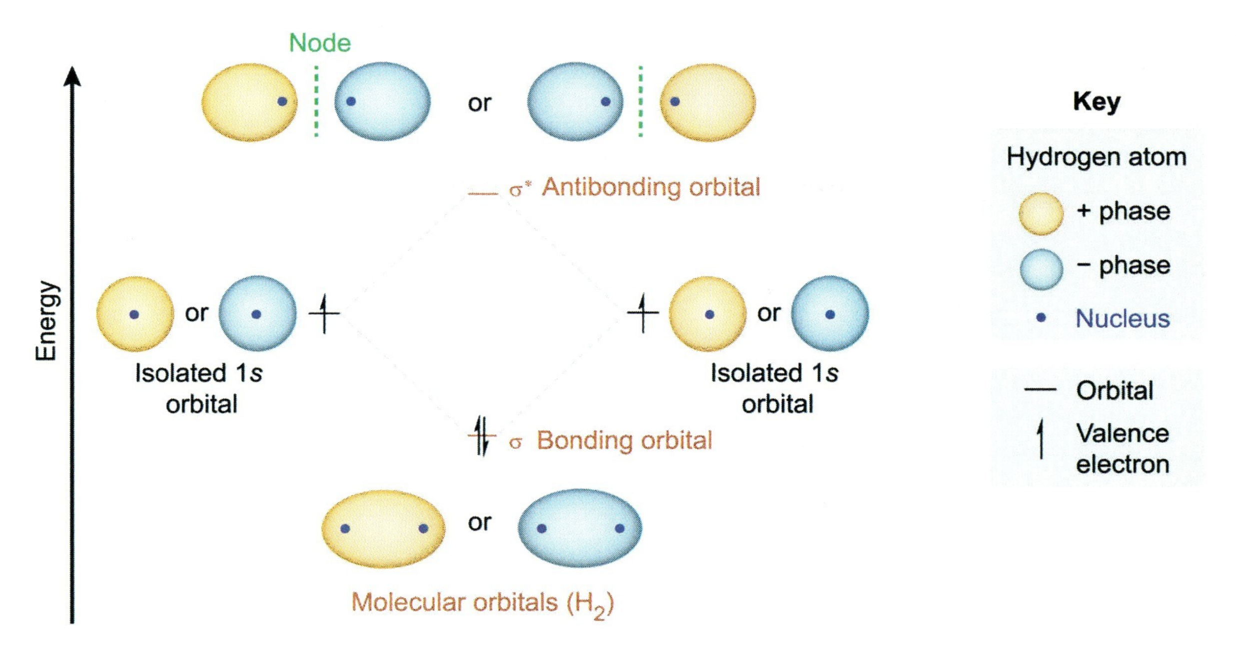

Figure 1.21 Molecular orbital diagram for hydrogen gas (H_2).

Once molecular orbitals have been generated, they are filled by valence electrons according to the same principles learned in general chemistry:

- According to the Pauli exclusion principle, an orbital can contain a maximum of two electrons (with opposite spin).
- Applying the **Aufbau principle**, electrons are placed in the lowest-energy (molecular) orbital available.
- If **degenerate orbitals** (ie, orbitals with the same relative energy) are present, electrons are first placed in separate degenerate orbitals before being paired.

In the example of hydrogen gas, there are two valence electrons (one from each hydrogen atom). Both electrons are placed in the lowest-energy bonding molecular orbital, and the antibonding molecular orbital is vacant.

1.4.02 Sigma Bonding

A **sigma (σ) bond** is a chemical bond that forms from the direct *head-to-head* overlap of orbitals of the participating atoms. The orbitals involved in formation of a σ bond may be unhybridized (eg, *s*, *p*) or hybridized (eg, *sp*, sp^2, sp^3) in any combination, but constructive, head-to-head overlap of electron density must occur *between the nuclei* (Figure 1.22).

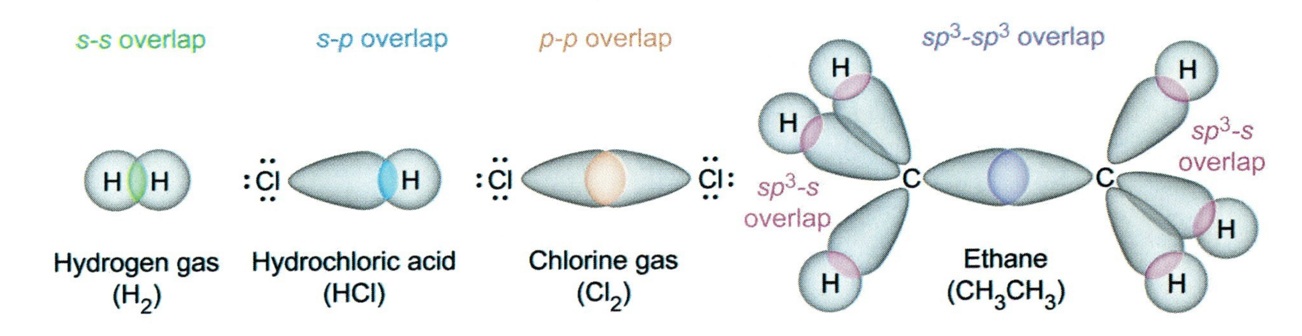

Figure 1.22 Types of sigma bonding.

The head-to-head orbital overlap of a σ bond allows rotation around the axis of the bond, which varies the **conformation** (ie, relative orientation) of any groups attached to the σ-bonded atoms. This feature is relevant to **conformational analysis**, which is discussed in Lesson 3.2.

1.4.03 Pi Bonding

A **pi (π) bond** is a chemical bond that forms from the *side-to-side* overlap of parallel *p* orbitals on the participating atoms (Figure 1.23). Unlike a σ bond, a π bond:

- Can be formed *only* from the overlap of unhybridized *p* orbitals (ie, hybrid orbitals and *s* orbitals *cannot* form a π bond).
- Has two regions of overlapping (constructive) electron density *above and below* the line connecting the nuclei, with minimal electron density directly between the nuclei.
- Provides a smaller total amount of constructive electron density and is a weaker type of bond.
- Cannot be the *sole* bond connecting two atoms.

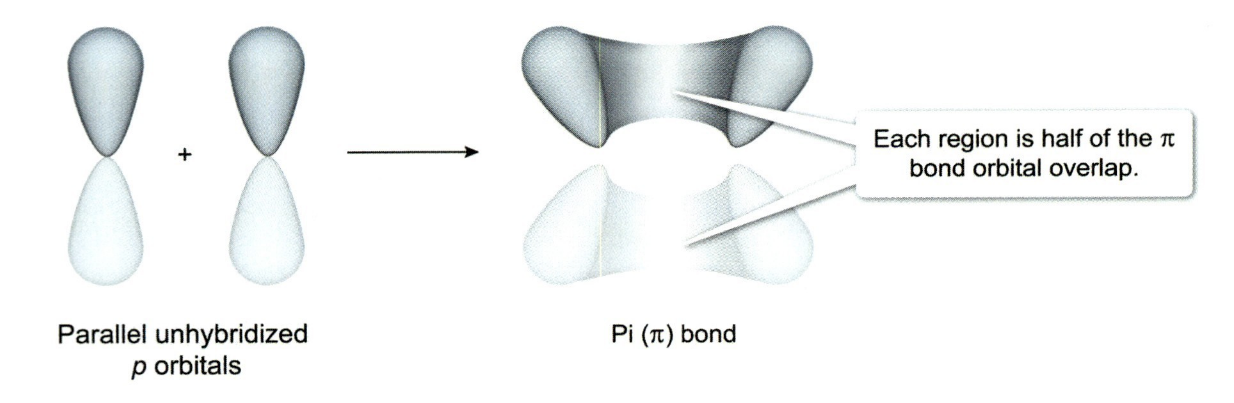

Figure 1.23 Formation of a pi bond.

1.4.04 Molecular Orbitals and Multiple Bonds

The general premises of σ bonding and π bonding allow multiple bonds to be redefined at an orbital level. Recall that a double bond occurs when four electrons are shared between atoms, and a triple bond occurs when six electrons are shared between atoms.

All single bonds are composed of a σ bond between atoms. Because σ bonds play an integral role in atom connectivity, the network of σ bonds in a molecule is sometimes referred to as the **σ bond skeleton**.

All **double bonds** are composed of one σ bond and one π bond (Figure 1.24). Because a π bond requires an unhybridized *p* orbital on each participating atom, the hybridization of atoms in a double bond can be either *sp* or sp^2 (atoms in a double bond *cannot* be sp^3 hybridized). The σ bond portion of a double bond lies between the nuclei and is buried between the two lobes of the π bond. Consequently, the π bond portion of a double bond, which is more accessible, is generally the reactive part of a double bond. Although an isolated π bond is weaker than a σ bond, a double bond consists of *both* a σ bond and a π bond and is therefore stronger than a single bond.

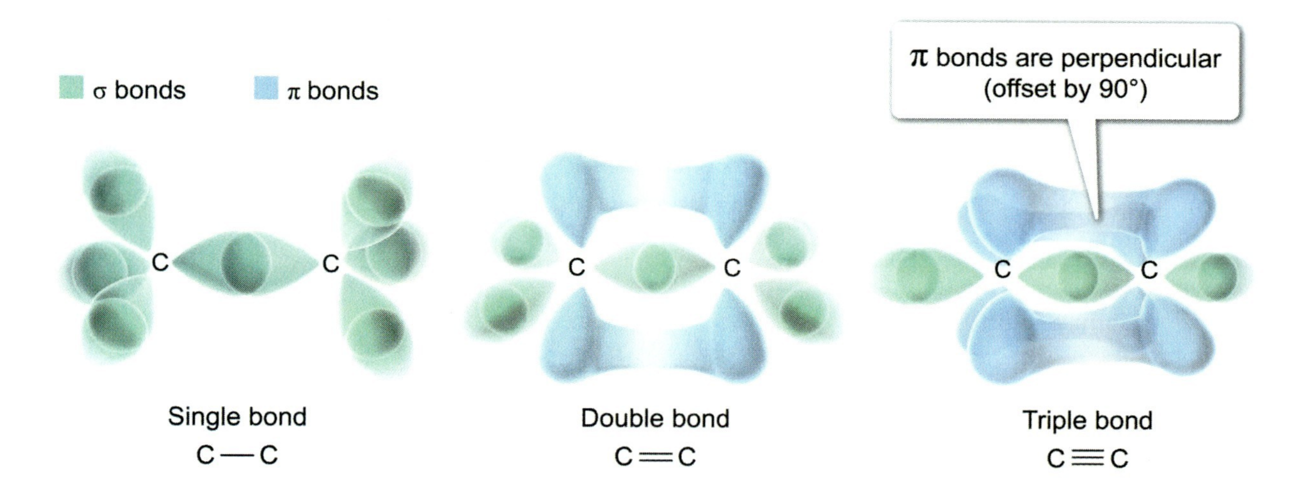

Figure 1.24 Orbital depiction of multiple bonds.

All **triple bonds** consist of one σ bond and two π bonds. Because *each* π bond requires an unhybridized *p* orbital, the hybridization of atoms within a triple bond *must* be *sp*. The π bonds in a triple bond are perpendicular to each other and form a cylinder of π electron density surrounding the axis of the *sp*-*sp* σ bond. As in double bonds, the π bonds in a triple bond are generally more reactive than the σ bond. Because a triple bond contains a σ bond and two π bonds, it is the strongest type of covalent bond observed in organic molecules.

Lesson 2.1

Molecular Shape

Introduction

Within general chemistry, student understanding of molecular structure evolves from the broad topics of atom stoichiometry and molecular formula to more-detailed Lewis structures. Lewis structures introduce the concept of covalent bonding, but two-dimensional Lewis structures *do not convey* the properties of molecules in *three-dimensional space*.

This lesson covers important concepts pertaining to the three-dimensional structure of molecules.

2.1.01 VSEPR Theory and Electron Domain Geometry

Lewis dot diagrams provide a snapshot of electron distribution in two-dimensional molecular structure and provide a platform to describe molecular shape in three dimensions. The **valence-shell electron-pair repulsion (VSEPR) model** is among the most useful approaches for predicting three-dimensional molecular structure from the information in a Lewis dot diagram. VSEPR is discussed in more detail in General Chemistry Concept 2.2.03.

As a brief review, the number of **electron domains** around an atom impacts their geometric distribution. In the VSEPR model, electron domains are either bonds or nonbonding electrons (lone pairs). Multiple bonds (eg, double bonds, triple bonds) are treated as one electron domain each, and lone pairs also count as one electron domain each.

In organic chemistry, which deals largely with second-row elements, most atoms with geometries of interest have two, three, or four electron domains.

- Atoms with two electron domains have their domains separated by 180° and therefore have a linear electron domain geometry.
- Atoms with three electron domains have their domains positioned in a single plane separated by 120° and therefore have a trigonal planar electron domain geometry.
- Atoms with four electron domains have their domains pointed toward the corners of a regular tetrahedron. With respect to the central atom, the angle between any two domains is 109.5°. Therefore, these atoms have a tetrahedral electron domain geometry.

Concept Check 2.1

What is the electron domain geometry for each of the nonhydrogen atoms in the following structure?

```
    H  H :O:
   ⊕|  |  ||  ..⊖
 H—N—C—C—O:
    |  |      ..
    H  H
```

Solution

Note: The appendix contains the answer.

2.1.02 Relationship of Electron Domain Geometry to Hybridization

As introduced in Lesson 1.2, the hybridization of atomic orbitals on an atom is directly related to the atom's electron domain geometry. For second-row elements:

- Atoms with a total of two electron domains have two *sp* hybrid orbitals in a linear relationship. These atoms will also have two unhybridized *p* orbitals oriented perpendicular (90°) to one another.
- Atoms with a total of three electron domains have three sp^2 hybrid orbitals in a trigonal planar relationship. These atoms will also have one unhybridized *p* orbital oriented perpendicular to the plane of the sp^2 orbitals.
- Atoms with a total of four electron domains have four sp^3 hybrid orbitals in a tetrahedral relationship. These atoms have no unhybridized *p* orbitals.

Nonbonding orbitals have different impacts on hybridization and electron domain geometry, depending on the role of the nonbonding orbital:

- Vacant nonbonding orbitals (as in carbocations, see Lesson 5.2) do not contain any electrons and are *not* counted toward the total number of electron domains. In most cases, vacant nonbonding orbitals will behave like unhybridized *p* orbitals and be oriented perpendicular to the plane of bonding orbitals.
- Nonbonding orbitals containing one electron (an unpaired electron called a radical) also *do not count* toward the total number of electron domains. In most cases, the orbital containing the radical electron is an unhybridized *p* orbital.
- Nonbonding orbitals containing two electrons (**a lone pair**) *do* count toward the total number of electron domains. Depending on the hybridization of the atom in question, the lone pair can be in an *sp*, sp^2, or sp^3 orbital.

2.1.03 Molecular Geometry

As discussed in Concept 2.1.02, the total number of electron domains of an atom is directly related to both its electron domain geometry and atomic orbital hybridization. While the full electronic structure (shape) of an atom is critically important for certain applications within organic chemistry, there are times when chemists are interested in the **molecular geometry** of an atom, the shape of only the *bonding regions*.

The molecular geometry of an atom builds upon the foundation of its electron domain geometry and retains both the general shape and angles between electron domains. However, electron domains containing nonbonding electrons (lone pairs) are factored out of consideration, as these inherently do not participate in bonding. The possible molecular geometries for each electron domain geometry are depicted in Figure 2.1.

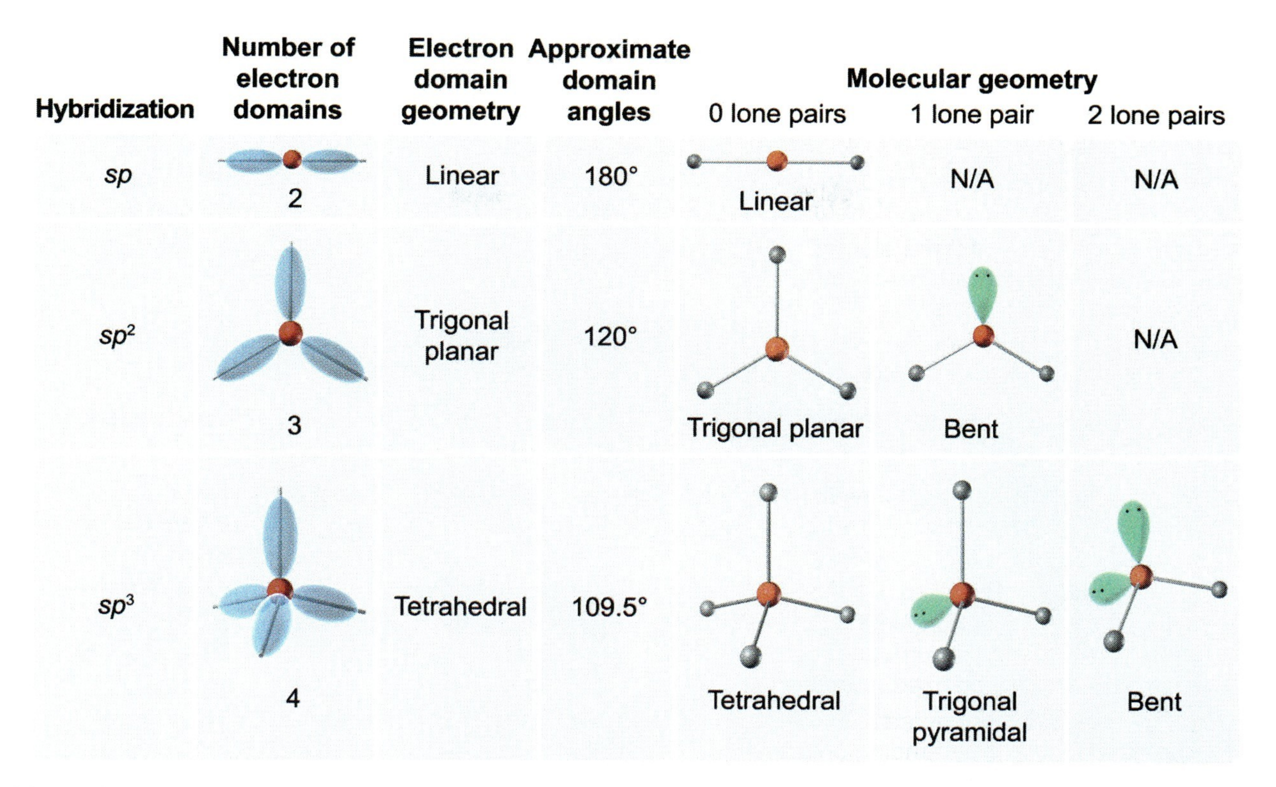

Hybridization	Number of electron domains	Electron domain geometry	Approximate domain angles	Molecular geometry: 0 lone pairs	1 lone pair	2 lone pairs
sp	2	Linear	180°	Linear	N/A	N/A
*sp*²	3	Trigonal planar	120°	Trigonal planar	Bent	N/A
*sp*³	4	Tetrahedral	109.5°	Tetrahedral	Trigonal pyramidal	Bent

Figure 2.1 The relationship between hybridization, electron domain geometry, and molecular geometry.

Molecular geometry is discussed in further detail in General Chemistry Concept 2.2.03.

Concept Check 2.2

What is the molecular geometry and what are the approximate bond angles for the indicated atoms in the following structure?

```
                 H
                 |
     H       H  :N   H
     |       |   |   |
  H—C—Ö—C—C—C—H
     |       |       |
     H       H       H
```

Solution

Note: The appendix contains the answer.

Lesson 2.2

Molecular Polarity

Introduction

Lesson 1.3 introduced the topics of bond polarity and bond dipole moment as a quantitative measure of bond polarity. In this lesson, these concepts will be applied to a molecule's three-dimensional structure (Lesson 2.1) to reveal the polarity of whole molecules.

2.2.01 Molecular Polarity and Polar Bonds

As introduced in Lesson 1.3, a polar covalent bond is a bond in which atoms do not equally share electrons in bonding orbitals due to a difference in **electronegativity**. A **bond dipole moment (μ)**, measured in units of debye (D), is a vector quantity that is mathematically defined as the product of an amount of opposing charge (q) times the distance of charge separation (r) (ie, $\mu = q \times r$). Increases in the magnitude of charge or distance of separation both increase a bond dipole moment.

The *molecular* dipole moment is the vector sum of all the individual bond dipole moments in a molecule and serves as an indicator of a molecule's overall polarity. As a vector quantity, the **molecular dipole moment** is dependent on:

- The magnitude of individual bond dipole moments
- The presence of lone pairs (nonbonding electrons) (see Concept 2.2.02)
- The geometric relationship (bond angles, symmetry) of the polar bonds and lone pairs in the molecule (see Lesson 2.1)

The role of the geometric relationship of polar bonds on molecular polarity can be illustrated by the two structural versions (isomers) of the molecule 1,2-dibromoethene. The most significant individual bond dipole moments in this molecule are the carbon-bromine bonds.

In the *cis* isomer, the bromine atoms are on the same side of the molecule, and the vector sum of the carbon-bromine bond dipole moments creates a molecular dipole moment of 1.9 D. However, in the *trans* isomer, the bromine atoms are on opposite sides of the molecule, and the vector sum of the carbon-bromine bond dipole moments is 0 D. In this case, the individual bond dipole moments fully cancel each other due to the symmetry of the molecule (Figure 2.2).

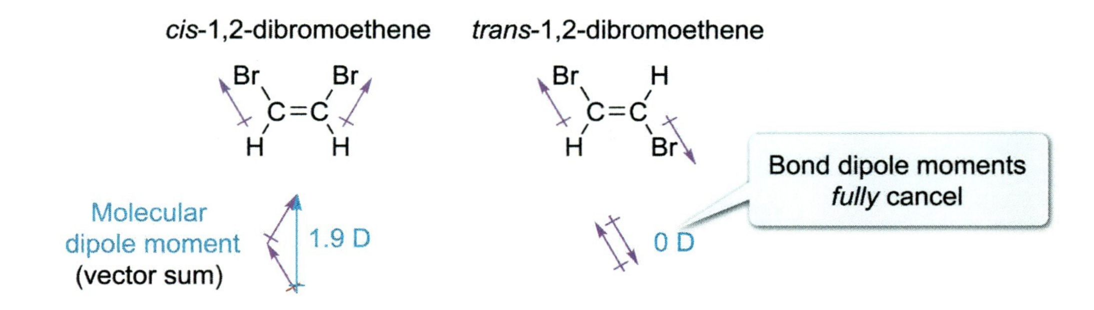

Figure 2.2 Impact of geometric relationship on molecular dipole moments.

Methane (CH_4) and its chlorine-substituted derivatives can also be compared (Table 2.1). Methane, which contains nonpolar carbon-hydrogen bonds oriented in a symmetrical relationship, has a molecular dipole moment of 0 D. In contrast, its derivative with a single polar carbon-chlorine bond (chloromethane) has a molecular dipole moment of 1.9 D.

Table 2.1 Effect of polar bonds on dipole moment.

Molecule	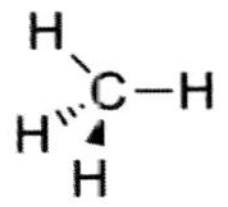	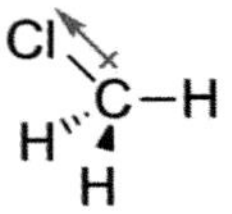	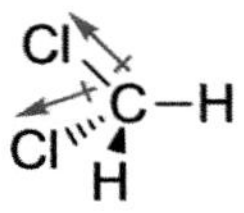	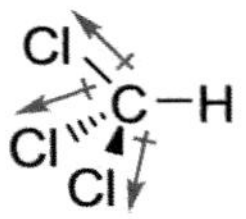	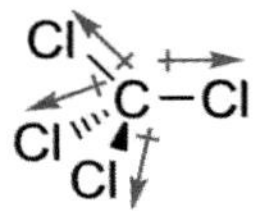
Name	Methane	Chloromethane	Dichloromethane	Chloroform	Carbon tetrachloride
Polar C–Cl bonds	0	1	2	3	4
Molecular dipole moment	0 D	1.9 D	1.6 D	1.0 D	0 D

Although students might think that adding a second polar carbon-chlorine bond (dichloromethane) would further increase the molecular dipole moment, this assumption is incorrect. While dichloromethane contains two polar carbon-chlorine bonds, these polar bonds are partially cancelled in the *vector sum* because they point in different directions. Therefore, the molecular dipole moment is *decreased* to 1.6 D. Likewise, the addition of a third polar carbon-chlorine bond (chloroform) leads to *further* decrease of the molecular dipole moment (1.0 D). The perfectly tetrahedral geometry of carbon tetrachloride *fully* cancels the dipole moments of the polar carbon-chlorine bonds, resulting in a molecular dipole moment of 0 D—the same as methane.

2.2.02 Molecular Polarity and Lone Pairs

The presence of a lone pair in a molecule also represents a form of charge separation because electrons are negatively charged and oriented a distance away from the positively charged nucleus. Consequently, each lone pair is treated like a bond dipole moment and represented by a vector arrow. Table 2.2 depicts the effects that lone pairs (and hybridization) have on molecular dipole moments.

Table 2.2 Impact of lone pair polarity on molecular dipole moments.

Molecule	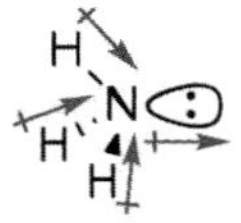	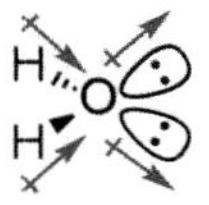	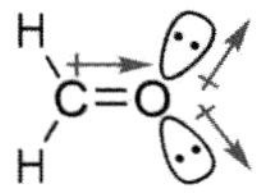	
Name	Ammonia	Water	Formaldehyde	Hydrocyanic acid
Hybridization of atom with lone pairs	sp^3	sp^3	sp^2	sp
Molecular dipole moment	1.5 D	1.9 D	2.3 D	2.8 D

The nitrogen atom in ammonia has one lone pair and three polar nitrogen-hydrogen bonds. The lone pair's vector adds to the dipole vectors of the three nitrogen-hydrogen bonds, resulting in a molecular dipole moment of 1.5 D. A similar situation occurs with water to generate an even more polar molecular dipole moment of 1.9 D. The oxygen atom in formaldehyde has a trigonal planar electron domain geometry. The additive effects of the very polar carbon-oxygen double bond and the two lone pairs result in a molecular dipole moment of 2.3 D. Lastly, the linear electron domain geometry of the nitrogen atom in hydrocyanic acid perfectly aligns the highly polar carbon-nitrogen triple bond with the vector of its lone pair, resulting in a molecular dipole moment of 2.8 D.

Lesson 2.3

Intermolecular Forces

Introduction

Intermolecular forces are foundational to nearly all aspects of chemistry. They are the attractions or repulsions that occur between individual molecules within both pure substances and mixtures. Intermolecular forces, described in more detail in General Chemistry Lesson 2.3, all relate conceptually to the electronic attraction or repulsion of charges. Opposite charges generate an attractive force, while similar charges generate a repulsive force. Charge-charge interactions can take place between full, formal charges (**ions**), partial charges (**dipoles**), temporary charges (**induced dipoles**), or any combination of these. This lesson provides a brief review of intermolecular forces with a particular emphasis on organic chemistry.

2.3.01 Ion-Associated Interactions

Ions are atoms or molecules that carry a charge. Positively charged species are **cations** and negatively charged species are **anions**.

The direct interactions between ion pairs are called **ion-ion forces**. If the ions are oppositely charged, the ion-ion force is attractive (an ionic bond); if the ions have the same charge, the ion-ion force is repulsive. Ion-ion interactions are among the strongest types of intermolecular forces.

There are two common ways that ions and ion interactions are involved within organic chemistry:

- As short-lived **reactive intermediates** formed during a chemical transformation (eg, carbocations, carbanions) (see Lesson 5.2).
- Through acid-base reactions (see Lesson 5.1).

More commonly, an ion interacts with a molecular region bearing a *partial* charge, an interaction called an **ion-dipole intermolecular force**. Figure 2.3 shows an example of ion-dipole intermolecular forces being responsible for the solvation of ions by the polar solvent water.

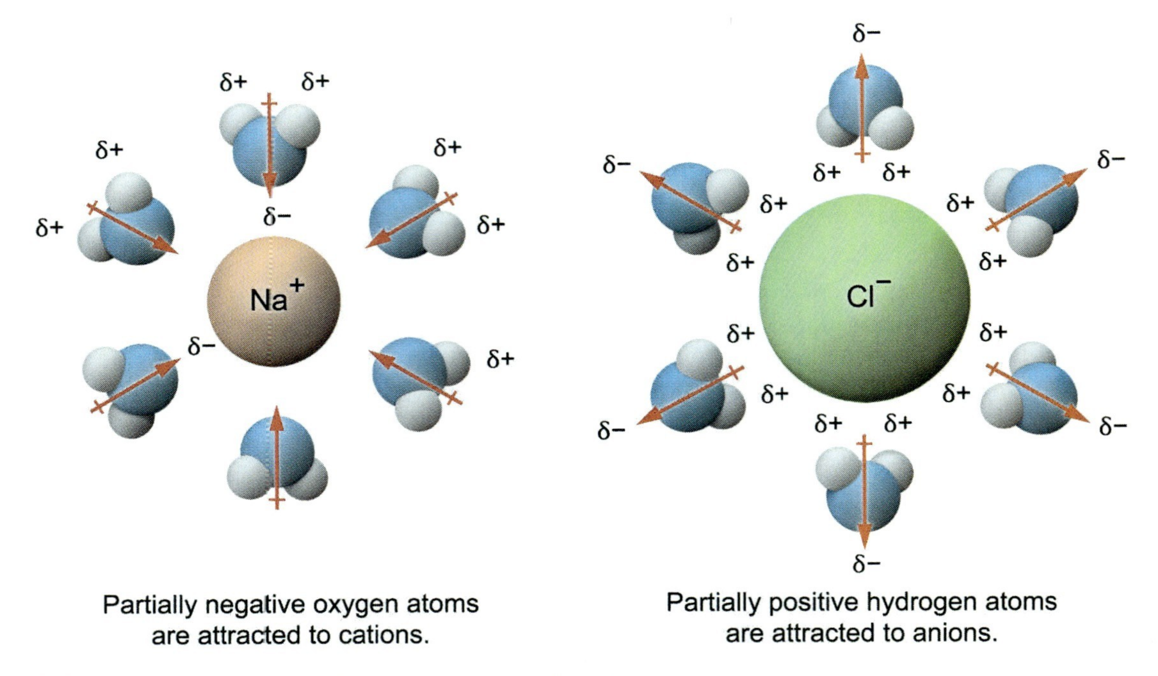

Figure 2.3 Ion-dipole interactions of sodium chloride with water.

Within organic chemistry, there are few examples of interactions between an ion and a nonpolar molecule to generate an **ion-induced dipole force**. A biochemical example is the interaction between an iron (II) cation and molecular oxygen (O_2), as observed in the proteins myoglobin and hemoglobin.

2.3.02 Dipole-Associated Interactions

As discussed in Lesson 1.3, a **covalent polar bond** occurs when atoms joined by a covalent bond have a difference in **electronegativity** between 0.5 and 1.8. The electrons in a covalent polar bond are not shared evenly:

- The more electronegative atom attracts more of the electron density and bears a **partial negative charge (δ^-)**.
- The less electronegative atom attracts less of the electron density and bears a **partial positive charge (δ^+)**.

The partial separation of charge in a covalent polar bond generates a **bond dipole moment**, which then contributes to the **molecular dipole moment** (see Lesson 2.2).

Dipole-dipole forces describe the interaction between the molecular dipole moments of different polar molecules. Like all other intermolecular forces, dipole-dipole forces can be either attractive or repulsive. In attractive dipole-dipole forces, an electron-rich (δ^-) end of one molecule interacts with an electron-deficient (δ^+) end of another molecule (Figure 2.4).

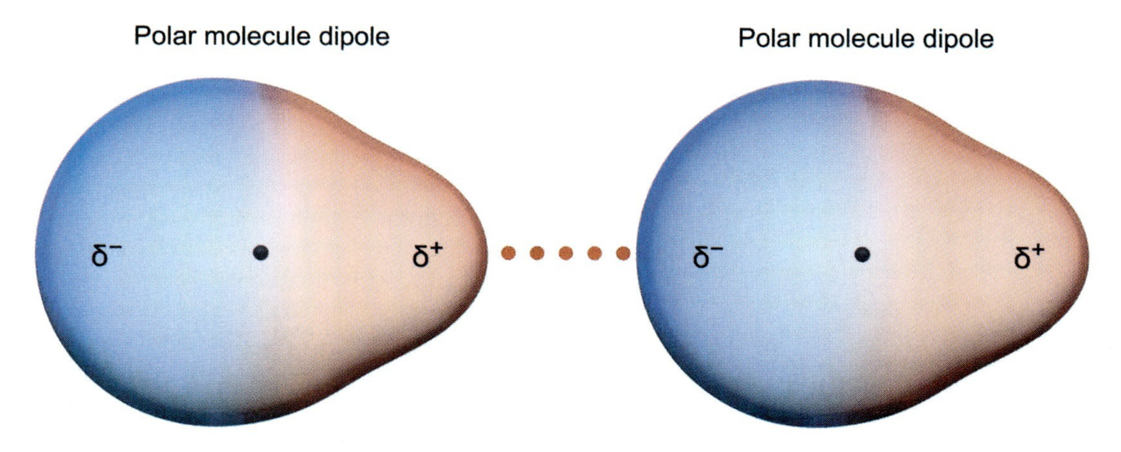

Figure 2.4 Dipole-dipole interaction.

Dipole-dipole forces are among the most common types of intermolecular force observed in organic molecules. Frequently, dipole-dipole forces are associated with the presence of certain bonding patterns of atoms (**functional groups**, see Lesson 2.8) that inherently contain polar bonds and result in a polar molecule. For example, molecules containing a polar carbon-oxygen double bond tend to exhibit dipole-dipole forces between individual molecules.

Given the prevalence of molecular dipole moments in organic molecules, intermolecular forces can also exist between a molecule with a permanent dipole moment and a nonpolar molecule capable of generating a temporary dipole moment (a dipole–induced dipole force). These types of interactions are frequently encountered in mixtures of polar and nonpolar organic molecules.

2.3.03 Hydrogen Bonding

A **hydrogen bond** is the interaction between an electron-deficient hydrogen atom and the lone pairs of a highly electronegative atom. The only three elements with an electronegativity high enough for hydrogen bonding are fluorine (F), oxygen (O), and nitrogen (N). Most of the applications of hydrogen bonding in organic chemistry involve *oxygen* and *nitrogen*.

A hydrogen bond requires both a hydrogen bond donor and hydrogen bond acceptor (Figure 2.5). A **hydrogen bond donor** is a hydrogen atom that is covalently bonded to F, O, or N. The large electronegativity difference between these atoms and hydrogen generates a large bond dipole moment and a highly electron-deficient (δ^+) hydrogen atom. A **hydrogen bond acceptor** is a lone pair attached to the electronegative elements F, O, or N, which are typically electron-rich (δ^-).

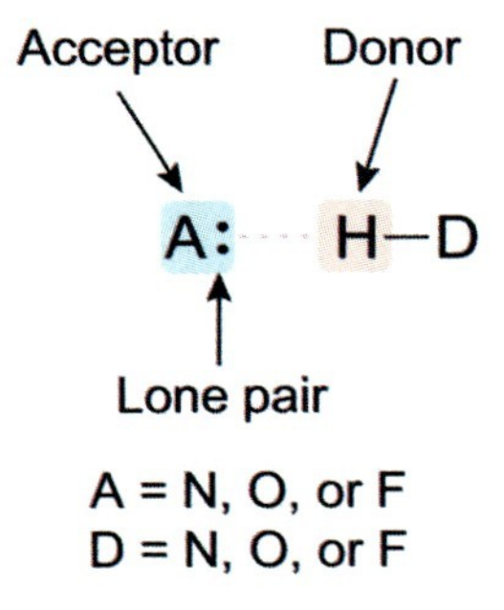

Figure 2.5 A generalized hydrogen bond.

Although historically called a bond, a **hydrogen bond** is conventionally classified instead as a very strong type of specialized dipole-dipole interaction.

Hydrogen bonding can occur between molecules of the same substance, molecules of different substances, or between atoms in a single molecule (Figure 2.6). In all scenarios, there must be *both* a hydrogen bond *donor* and a hydrogen bond *acceptor*.

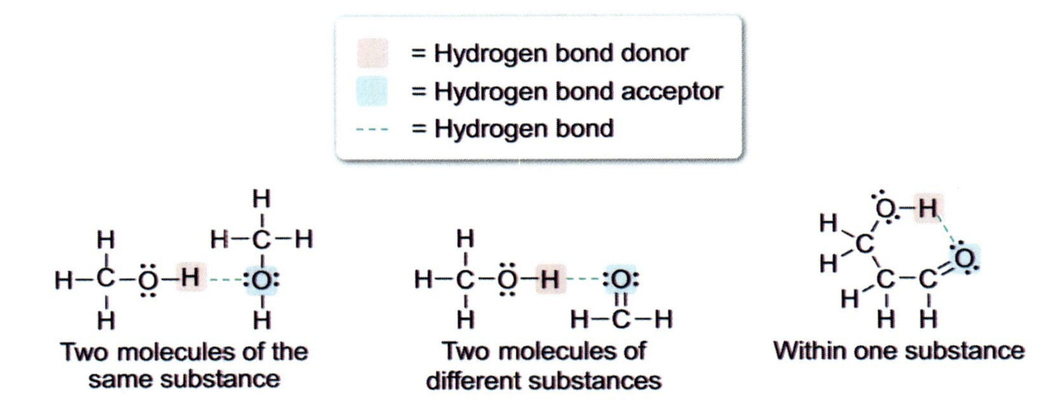

Figure 2.6 Examples of hydrogen bonding.

Several common bonding patterns of atoms (functional groups, see Lesson 2.8) have the required components for hydrogen bonding, a feature that significantly impacts the properties of these classes of materials. Functional group characteristics are discussed in greater detail in Unit 2.

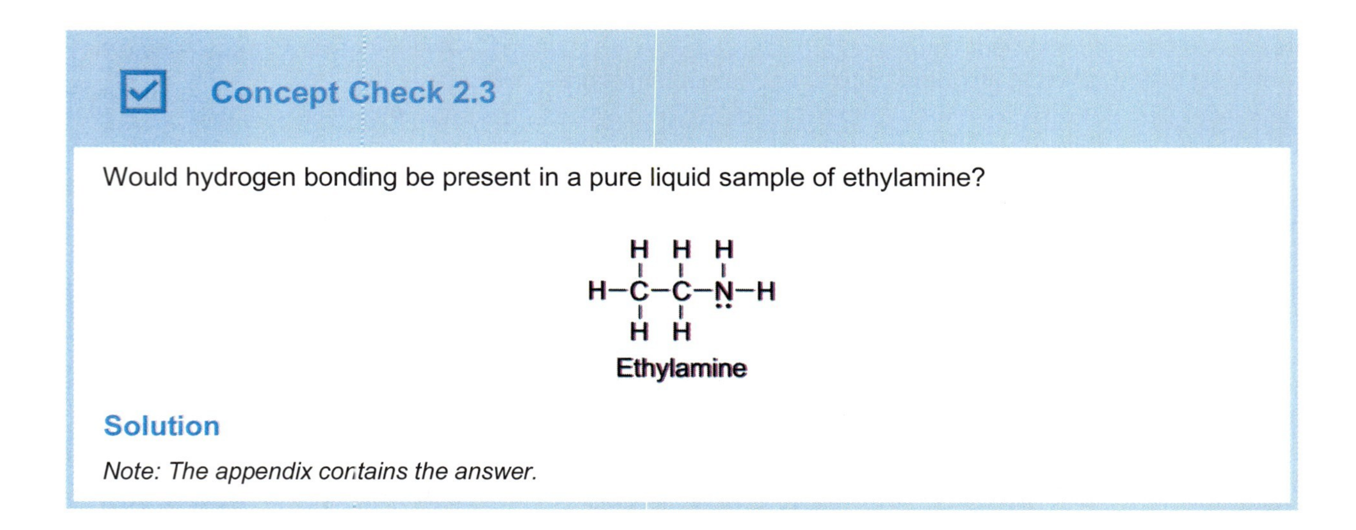

Concept Check 2.3

Would hydrogen bonding be present in a pure liquid sample of ethylamine?

Solution

Note: The appendix contains the answer.

2.3.04 Temporary Dipole-Associated Interactions

The principal intermolecular force for *nonpolar* molecules is the **London dispersion force** (Figure 2.7), which is generated through the interaction of *temporary* dipole moments in nonpolar molecules. While the electron cloud in a nonpolar molecule is typically described as evenly distributed, electron clouds become distorted (**polarized**) when molecules approach one another. This process causes the formation of short-lived dipole moments, which interact and cause an attractive force. As nearly all molecules have a cloud of electrons (with a notable exception of the hydrogen cation, H^+), *all molecules* demonstrate and are influenced by the London dispersion force.

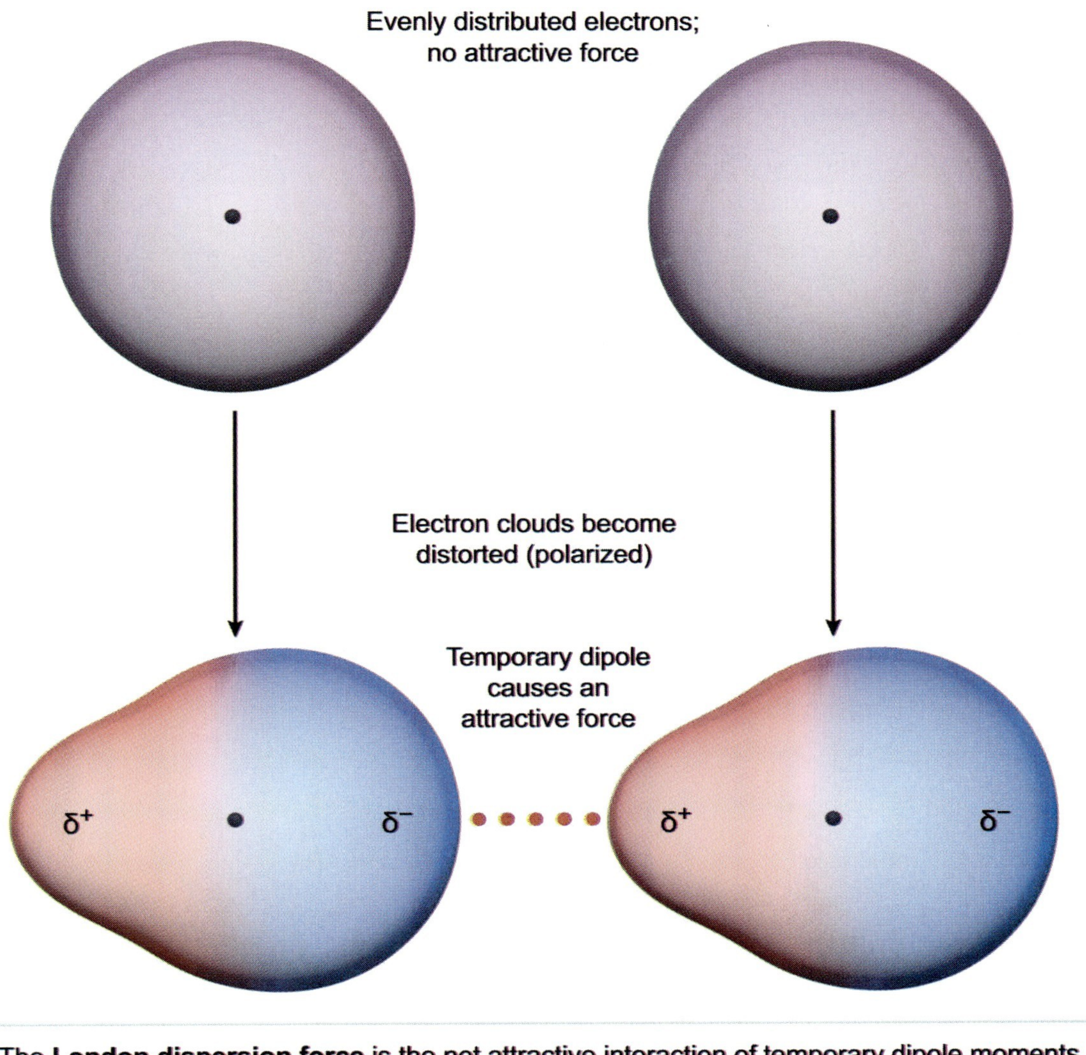

Figure 2.7 London dispersion force.

Unlike polar intermolecular forces, which are defined by static structural features (eg, dipoles), the London dispersion force is proportional to the *surface area* of the electron cloud for a molecule. Therefore, molecules with larger surface areas have proportionally larger London dispersion forces. As will be discussed in later lessons, molecular surface area is correlated to both the molecular weight of a substance and the amount of branching present in its structure.

2.3.05 Comparing the Intermolecular Forces

Because the electric charge of an atom or molecule can fall into one of three general categories (full, partial, or induced), a total of seven combinations of intermolecular forces have been discussed in this lesson (Figure 2.8).

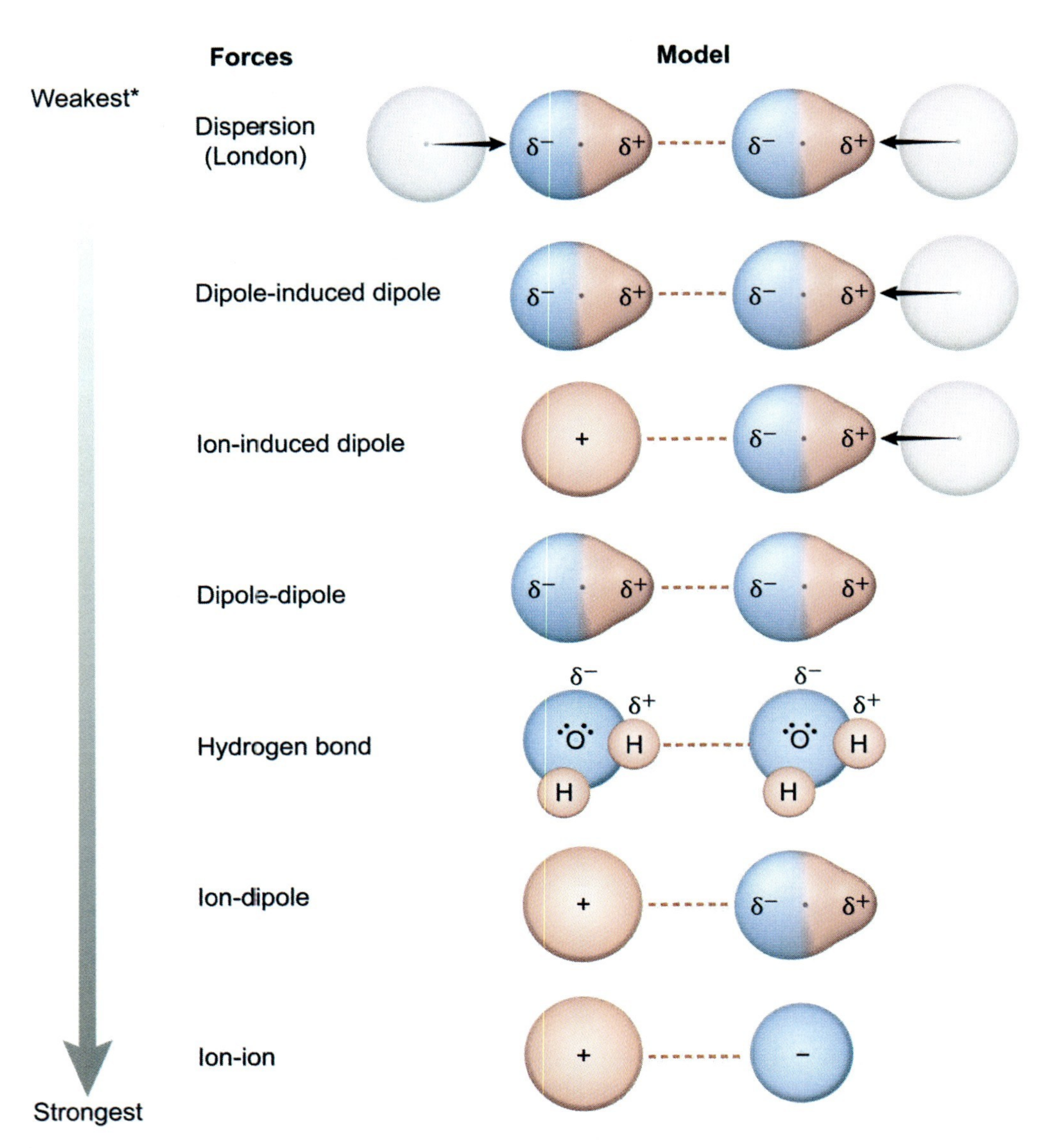

*Note: Strength comparisons are made based on individual interactions. The combined effect of multiple dispersion force interactions may sometimes overpower other stronger forces.

Figure 2.8 Relative strength of intermolecular forces.

Attractive intermolecular forces are a result of the electrostatic attraction between oppositely charged molecules, and the strength of a given interaction is proportional to the magnitude of charges. Induced dipoles have transient charge separation and generate the weakest intermolecular forces. Dipole-associated forces are stronger due to the *permanent* partial charges of dipoles; hydrogen bonding is a particularly strong type of dipole-dipole interaction. Interactions between fully charged molecules (ions) represent the strongest interactions.

It is important to note that this relative strength hierarchy is based on *single interactions*. However, the relative number of interactions can have a significant impact on the result of these forces.

Molecular dipoles are generated in molecules containing polar bonds, and dipole-associated intermolecular forces are regularly observed in polar organic molecules. Their intermediate relative strength, coupled with their intermediate prevalence, makes dipole-associated forces an important component of the behavior of organic molecules.

Although induced dipole-associated forces are the weakest type of interaction listed, they are also *especially numerous* in occurrence. Nearly all organic molecules demonstrate forces arising from induced dipoles. In some cases, the *sum* of individual induced dipole interactions is a *greater attractive force* than that of a single dipole-dipole interaction. This effect is especially pronounced for molecules with entirely nonpolar character.

For example, Table 2.3 lists physical properties of the diatomic halogens. The physical properties (eg, boiling point, melting point, density, solubility) of a substance are generally related to the intermolecular forces present in samples of that substance (see Lesson 2.4). All halogens have the same valence electron configuration, and their diatomic molecules are all purely nonpolar. However, the diatomic halogens differ in number of total electrons and atomic radius, factors that impact the London dispersion forces. These differences are responsible for the significant increases in boiling point and melting point down the Group 7A column of the periodic table.

Table 2.3 Impact of London dispersion forces on diatomic halogens.

Halogen	Diatomic Lewis structure	Total electrons (diatomic)	Atomic radius (pm)	Melting point (°C)	Boiling point (°C)	Phase at point 25 °C
Fluorine	:F̤̈–F̤̈:	18	71	−220	−188	Gas
Chlorine	:C̤̈l–C̤̈l:	34	99	−101	−35	Gas
Bromine	:B̤̈r–B̤̈r:	70	114	−7	59	Liquid
Iodine	:Ï̤–Ï̤:	106	133	114	184	Solid

Increasing London dispersion forces

Increasing melting and boiling points

Concept Check 2.4

What are the major intermolecular forces present in a pure sample of the following molecule?

$H_3N^+–CH_2–C(=O)–O^-$

Solution

Note: The appendix contains the answer.

Lesson 2.4

2.4 Overview of Physical Properties

Introduction

Organic chemistry is concerned with compounds primarily composed of carbon and their physical and chemical properties. A **physical property** of a material depends only on its current state and does not directly describe its capacity to undergo a chemical transformation. Although there are many types of physical properties, some properties most applicable to organic chemistry include melting point, boiling point, solubility, specific rotation, and density.

In contrast, a **chemical property** of a material describes its capacity to change into other materials. Unit 2 discusses the chemical properties of organic molecules.

This lesson provides an introduction to the physical properties of boiling point, solubility, and density. Unit 2 will discuss these properties as they relate to functional groups.

2.4.01 Boiling Point

The **boiling point (bp)** of a compound is the temperature at which its vapor pressure is equivalent to the ambient pressure of the system. It may also be defined as the temperature at which a substance undergoes a phase transition from liquid to gas.

The transition from liquid to gas requires sufficient energy to disrupt the intermolecular forces present in the liquid state. Therefore, the magnitude of a compound's boiling point is generally proportional to the types and magnitude of intermolecular forces present; compounds with greater or stronger intermolecular forces require more energy (ie, a higher temperature) to undergo the phase transition to gas.

In general, a compound's boiling point is proportional to its molecular weight. Compounds with a higher molecular weight have more total electrons and larger London dispersion forces. For many types of organic compounds, the magnitude of London dispersion forces is a primary determinant of boiling point.

If two compounds have the *same* molecular weight, the compound with a larger *surface area* will have a higher boiling point. The presence of branching in a compound's structure typically *decreases* its surface area and causes a lower relative boiling point (Figure 2.9).

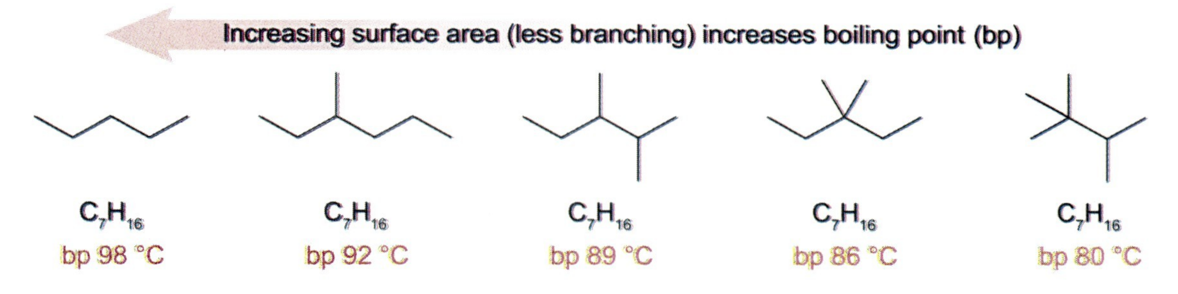

Figure 2.9 Impact of surface area (branching) on boiling point.

Because dipole-associated intermolecular forces are individually stronger than London dispersion forces, the boiling point of a compound is also generally proportional to its molecular polarity. Compounds with more polar structures often have higher boiling points.

In particular, the capacity to hydrogen bond can have a profound impact on a compound's boiling point. The data in Figure 2.10 emphasize several important trends. The compounds shown have nearly identical molecular weight, minimizing the role of London dispersion forces in accounting for differences in boiling point. Structures with *both* hydrogen bond donors and acceptors are capable of hydrogen bonding and have higher boiling points (37–97 °C) than those with *only* hydrogen bond acceptors (3.5–7.4 °C).

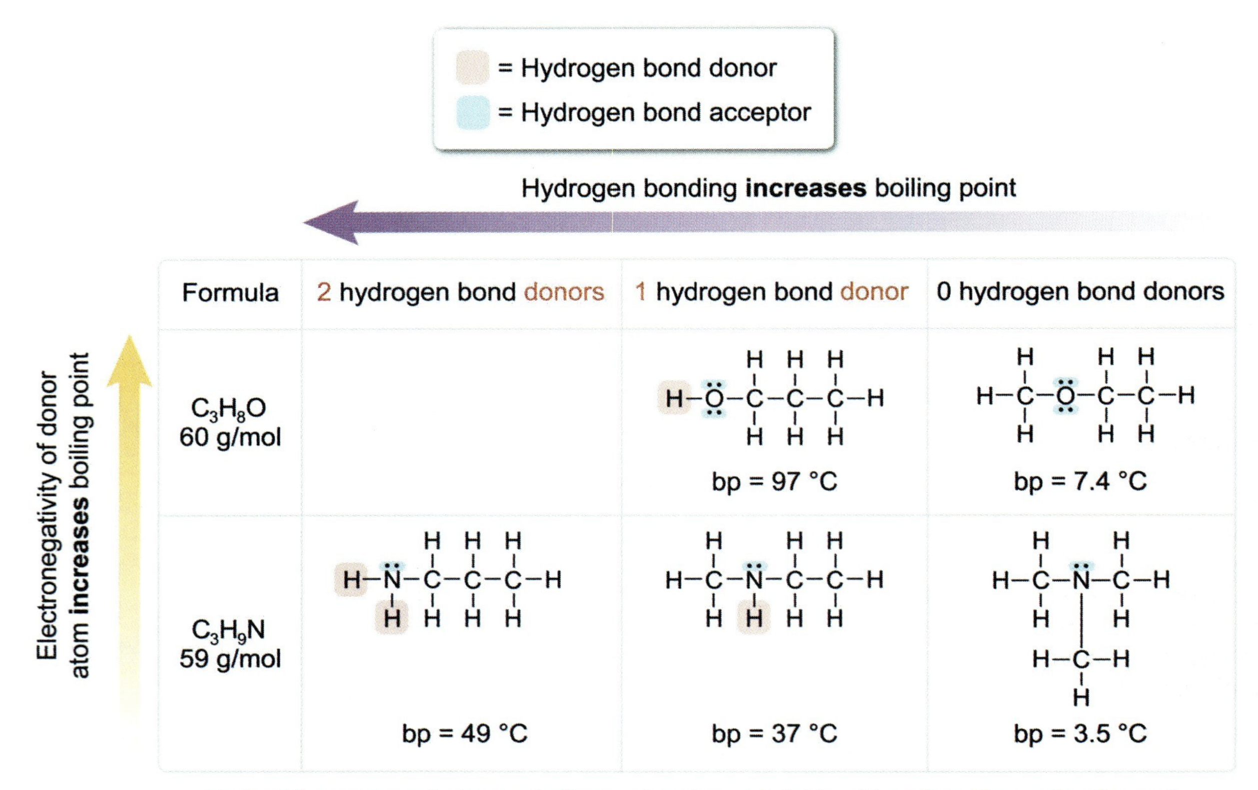

Figure 2.10 Impact of hydrogen bonding on boiling point.

The boiling point of a compound with two hydrogen bond donors (49 °C) is modestly higher than a similar compound with only one hydrogen bond donor (37 °C). Because oxygen is more electronegative than nitrogen, hydrogen bonds involving oxygen are stronger than those involving nitrogen, leading to an even higher boiling point (97 °C).

2.4.02 Solubility

A compound's **solubility** is a quantitative measure of the maximum mass that can dissolve in a quantity of a liquid. In this context, the dissolved compound is known as the **solute**, while the liquid it is dissolved in is called the **solvent**. Often, solubility is reported in units of grams of solute per 100 mL solvent, where larger values indicate a more soluble compound.

In organic chemistry, polarity is the dominant factor in determining the solubility of organic molecules. This relationship is often expressed via the phrase "like dissolves like" (Figure 2.11), which summarizes the following observations:

- Nonpolar solutes tend to be more soluble in nonpolar solvents.
- Polar solutes tend to be more soluble in polar solvents.
- Solutes with both polar and nonpolar character (amphipathic) have partial solubility in both polar and nonpolar solvents.

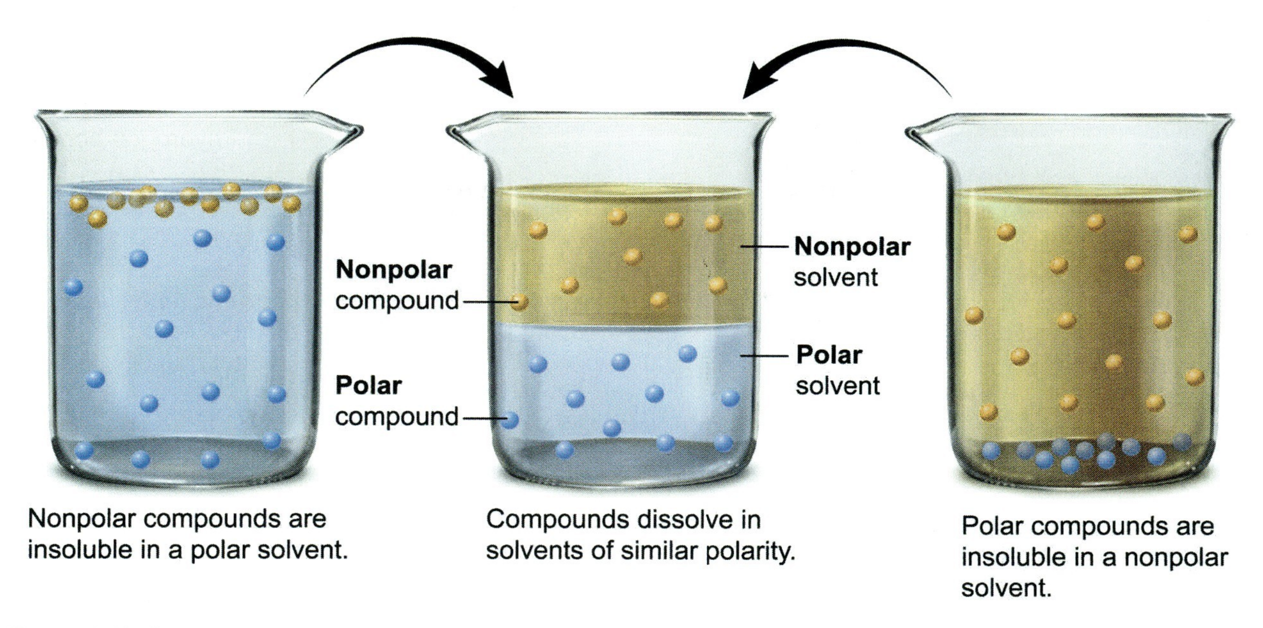

Figure 2.11 Conceptual illustration of "like dissolves like."

At a molecular level, the principle of "like dissolves like" is derived from the favorable presence of similar intermolecular forces between solute and solvent (Figure 2.12). Polar solutes generate polar intermolecular forces (ie, ion-associated, dipole-associated) with polar solvent molecules, while nonpolar solutes generate nonpolar intermolecular forces (ie, induced dipole) with nonpolar solvent molecules. When solute and solvent are unable to generate attractive intermolecular forces, solute molecules prefer to interact with other solute molecules and solvent molecules prefer to interact with other solvent molecules (ie, **insolubility**).

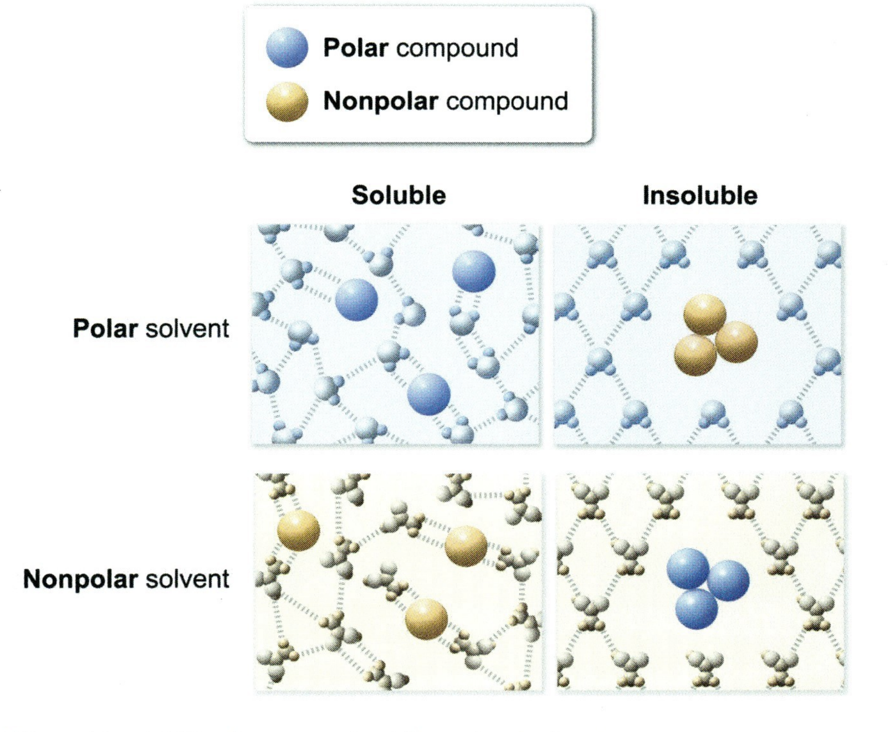

Figure 2.12 Solubility and insolubility of compounds at the molecular level.

While most solutes have a solubility limit in a solvent, some liquid solutes have infinite solubility in a solvent. When one liquid is capable of dissolving (or mixing) with another liquid in *any* ratio, the pair is said to be **miscible**. For example, the common organic solvents methanol (MeOH), ethanol (EtOH), and tetrahydrofuran (THF) are all miscible in water.

The principles associated with solubility are integral to several aspects of organic chemistry. Chemical reactions that require two or more reactants typically take place in the solution phase, using a solvent with an appropriate solubility profile. In reactions that occur in a system with two or more immiscible phases, **phase-transfer catalysts** are often employed to bring reactants across phase boundaries and into contact with other reactants. The most common phase transfer catalysts are **amphiphilic** molecules with regions bearing ionic formal charges (eg, tetrabutylammonium iodide). Solubility also provides the underlying basis for many separation and purification techniques (eg, liquid-liquid extraction), discussed further in Unit 3.

2.4.03 Density

Density (*d*) is a physical property that correlates the volume a substance occupies and its mass. Densities are normally provided in grams per mL (or cubic centimeters, cm^3). The volume of a material changes with temperature; therefore, density is also a temperature-sensitive quantity that typically *decreases* with increased temperature. Often, densities are reported at a temperature of either 20 °C or 25 °C (room temperature).

As a point of reference, the density of liquid water is approximately 1 g/mL. Although there are exceptions, most organic compounds have a density less than 1 g/mL. Consequently, in an immiscible mixture of an organic compound and water, the less-dense organic layer will usually be above the more-dense water layer (Figure 2.13). Some exceptions to this trend are discussed in Lesson 7.1.

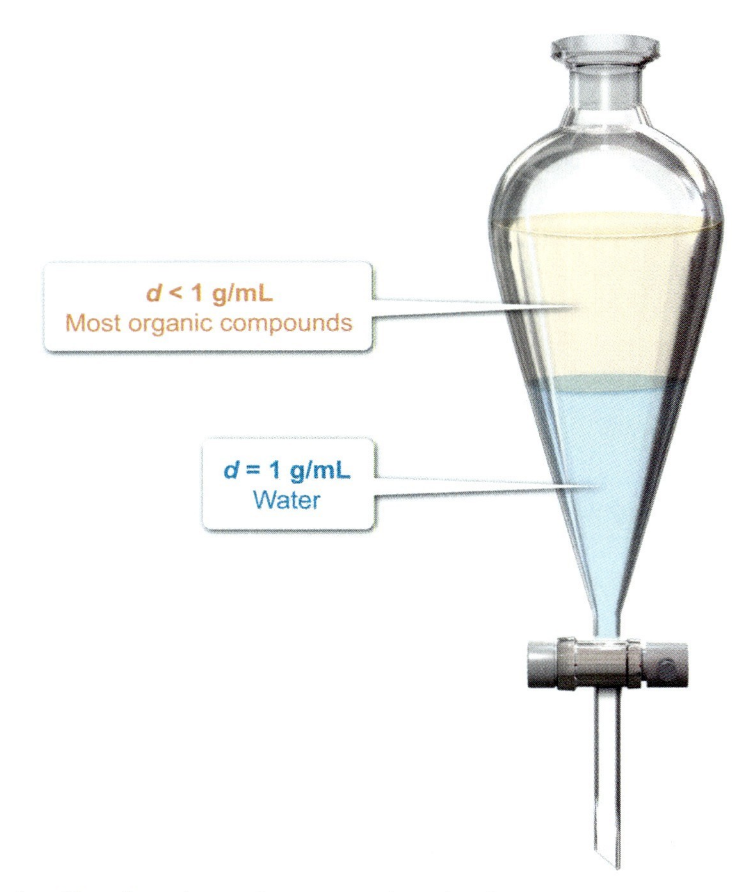

Figure 2.13 Relative densities of most organic compounds and water.

Lesson 2.5

Drawing Organic Molecules

Introduction

The **molecular formula** of a molecule denotes the type and quantity of atoms present in it. However, aside from very simple molecules (eg, H_2, H_2O, NH_4), a molecular formula does *not* typically provide a complete picture of the chemical bonds in a molecule (ie, its structure).

Because structure is an essential aspect of organic chemistry, additional depictions are used that more explicitly portray it. Figure 2.14 illustrates how the sole use of molecular formulas leads to the loss of significant information and context. In contrast, more structurally descriptive methods (eg, line-angle formulas, see Concept 2.5.02) provide much more information.

Structural depiction	Toluene		Acetyl chloride	Catalyst	Products		
Molecular formula	C_7H_8	+	C_2H_3ClO	$\xrightarrow{AlCl_3}$	$C_9H_{10}O$	+	HCl
Line-angle formula	CH$_3$	+	O, H$_3$C, Cl	$\xrightarrow{AlCl_3}$	CH$_3$, H$_3$C, O	+	HCl

Figure 2.14 Limitations of using molecular formulas to depict organic structure.

This lesson describes common ways of representing the structure of organic molecules. As any of these methods may be used to depict a molecule, it is important to understand the strengths, limitations, underlying assumptions, and potential applications of each approach.

2.5.01 Condensed Structural Formulas

Condensed structural formulas emphasize the sigma bond skeleton (carbon-carbon and carbon-heteroatom covalent bonds). As a reminder, a **heteroatom** is any atom that is *not* carbon or hydrogen (eg, oxygen, nitrogen, sulfur). Because carbon-hydrogen single bonds are among the least reactive bonds, carbon-hydrogen bonds are *de-emphasized*. Condensed structural formulas provide a way to depict structures within a line of text without the need for an illustration; the effectiveness of condensed structural formulas is *reduced* for *larger molecules* or those with *cyclic* components. The general guidelines for condensed structural formulas include:

- A molecule is viewed as a sequence of central (ie, nonhydrogen) **parent chain** atoms.
- Atoms directly bonded to a parent chain atom are listed immediately after it.
- **Subscript** numbers denote the presence of multiple, identical groups.
- **Parentheses** are used to either signify a group of atoms, a repeating unit, or atoms that branch from the parent chain.
- Multiple bonds can be either explicitly shown or implied based on the connectivity of atoms and the octet rule.

- Lone pairs are usually *not* shown and must be deduced from the octet rule and common bonding patterns.

Figure 2.15 depicts two common methods for converting a Lewis structure (Concept 1.1.03) to a condensed structural formula. In the **atom accounting method**, parent chain atoms are numbered and used as reference points during the conversion. In the **grouping method**, each atom in the parent chain (and those bound to it) are condensed in sequence.

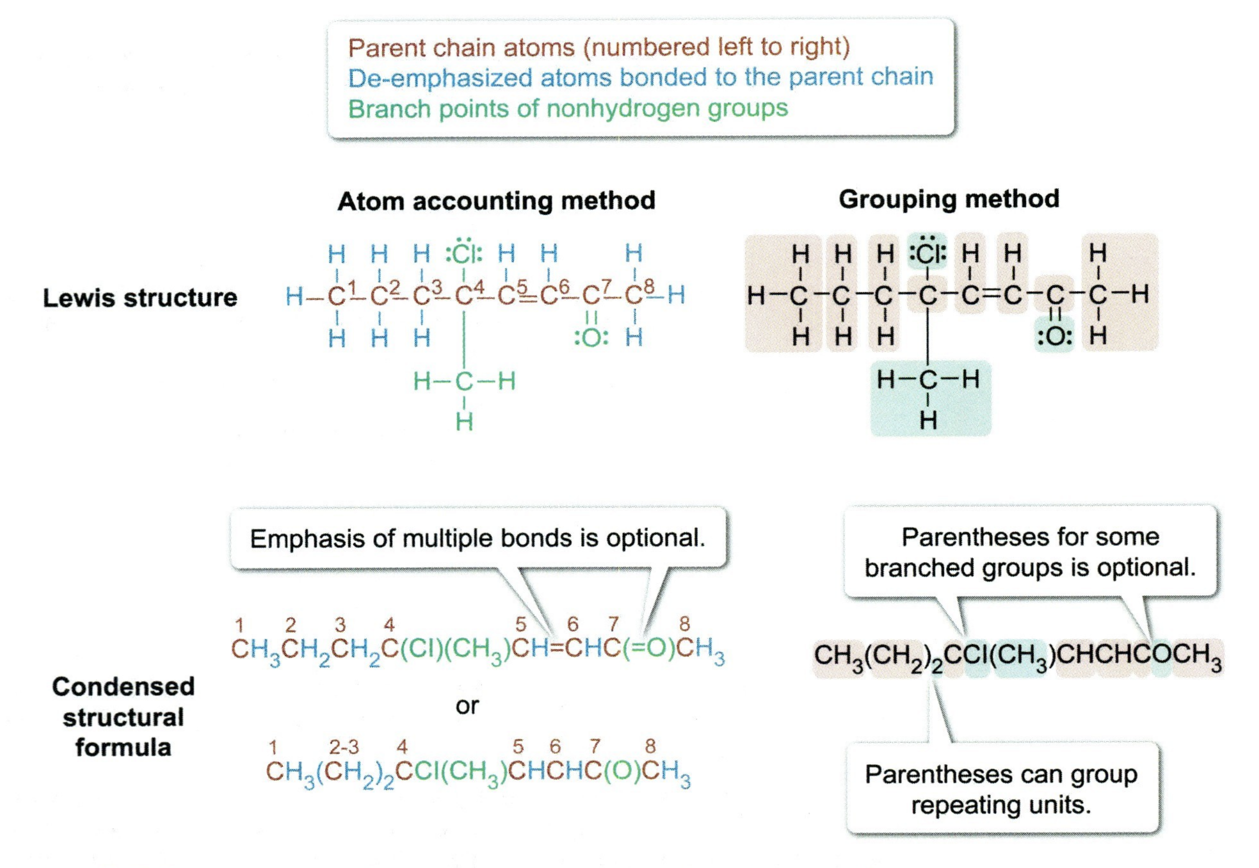

Figure 2.15 Methods of converting a Lewis structure to a condensed structural formula.

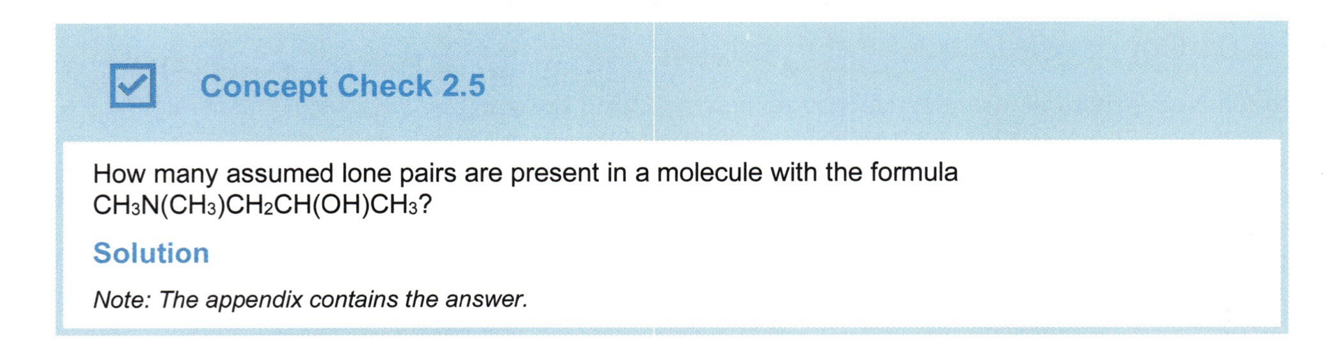

Concept Check 2.5

How many assumed lone pairs are present in a molecule with the formula $CH_3N(CH_3)CH_2CH(OH)CH_3$?

Solution

Note: The appendix contains the answer.

2.5.02 Line-Angle Formulas

A **line-angle formula** is the most common way of depicting the structure of an organic molecule. Line-angle formulas further simplify the depiction of structural information with the goal of emphasizing the carbon and heteroatom skeleton of a molecule. Some advantages of line-angle formulas include the ease of depicting *cyclic* structures, *larger molecules*, and molecules represented in *three-dimensional space* (see Concept 2.5.03). Line-angle formulas follow these general guidelines:

- Chemical bonds are represented by lines (one line = single bond, two lines = double bond, three lines = triple bond).
- Elemental symbols are rarely used for carbon (C) and hydrogen (H) but are *always* used for heteroatoms.
- A carbon atom is implied at every location where two lines meet (a vertex) or a line ends (a terminal carbon).
- Sufficient hydrogen atoms to fulfill an octet are implied on every carbon.
- Hydrogen atoms bonded to heteroatoms are usually depicted explicitly.
- Lone pairs can be either explicitly depicted or implied (similar to condensed structural formulas).
- Formal charges are always depicted.
- Often, common bonding arrangements (eg, **functional groups**, **alkyl groups**) can be shortened to a label. These designations are discussed in Lesson 2.8 and Lesson 4.1, respectively.

The most important skill in working with (and interpreting) line-angle formulas is the ability to **count carbons** (Figure 2.16). To reduce the risk of losing track of the total number of atoms in a structure, carbon atoms can be labeled with arbitrary numbers.

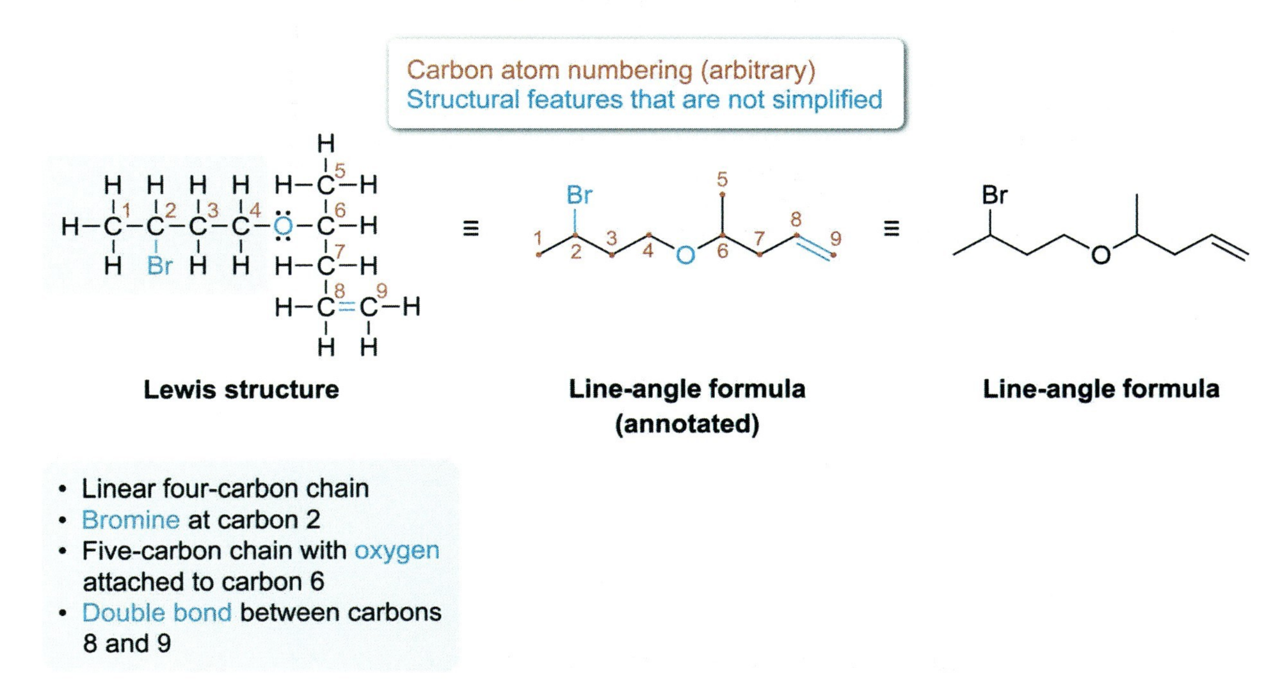

Figure 2.16 Converting a Lewis structure to a line-angle formula.

Table 2.4 provides additional examples of molecules represented in three different formats: the full Lewis structure, the condensed structural formula, and the line-angle formula.

Table 2.4 Ways of depicting organic molecular structures.

Full (Lewis) structure (all bonds and electrons shown)	Condensed structural formula (de-emphasized C–H bonds)	Line-angle formula (C and H atoms implied)
	OH, CH, H_2C, CH, H_2C, CH_2, C, F	OH, F
	$(CH_3CH_2)_2NCH_2CH(CH_3)NH_2$	N, NH_2
	$CHCCHCHCO_2CH_3$	O, O

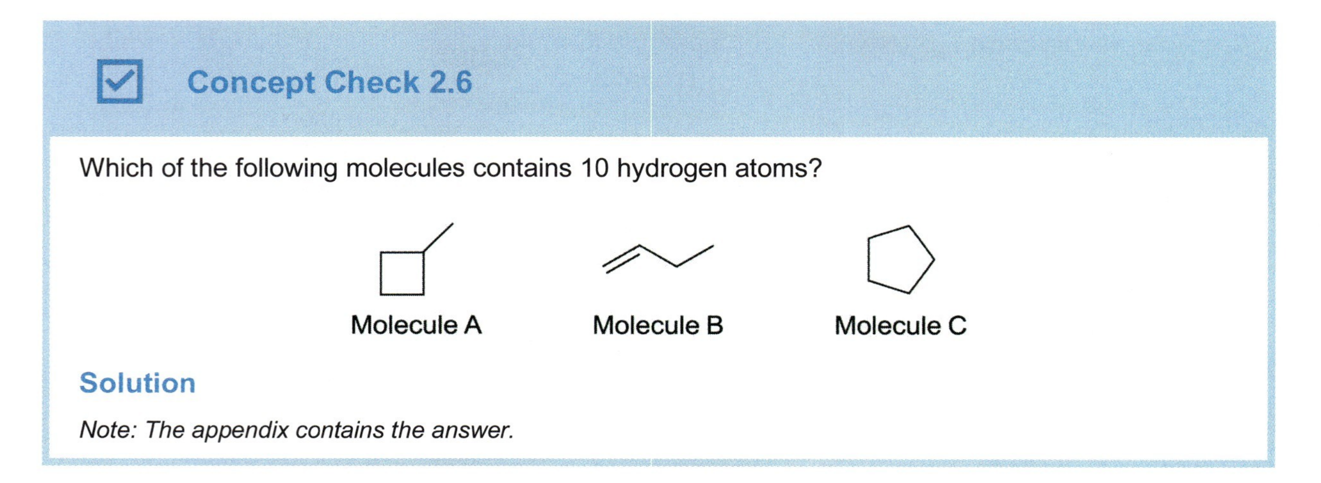

Concept Check 2.6
Which of the following molecules contains 10 hydrogen atoms?
Molecule A
Molecule B
Molecule C
Solution
Note: The appendix contains the answer.

2.5.03 Depicting Three-Dimensional Molecules

Lesson 2.1 introduced the notion that organic molecules have a three-dimensional shape. Chapter 3 explores this further with the topics of conformational isomers and stereoisomers. Given that most chemical structures are depicted on a two-dimensional surface (such as a piece of paper or a computer screen), chemists have developed conventions to allow a three-dimensional molecule to be represented on a flat surface.

Until this point, a bond has been represented by a simple line. The style of a bond can be changed if a chemist desires to represent the bond with a particular *orientation* in three-dimensional space (Figure 2.17).

- A **plain bond** is used to represent a bond with either a nonspecific orientation or one that is within the plane of the drawing.
- A **wedge bond** is used to represent a bond that is oriented *toward* the viewer (coming *out of the plane* of the drawing).
- A **dash bond** is used to represent a bond that is oriented *away* from the viewer (going *into the plane* of the drawing).
- A **wavy bond** is used to represent a bond that has an equal chance of being either a wedge or a dash within the population of molecules in a sample. Wavy bonds are frequently used to represent 1:1 mixtures of configuration at a particular atom (see Lesson 3.3).

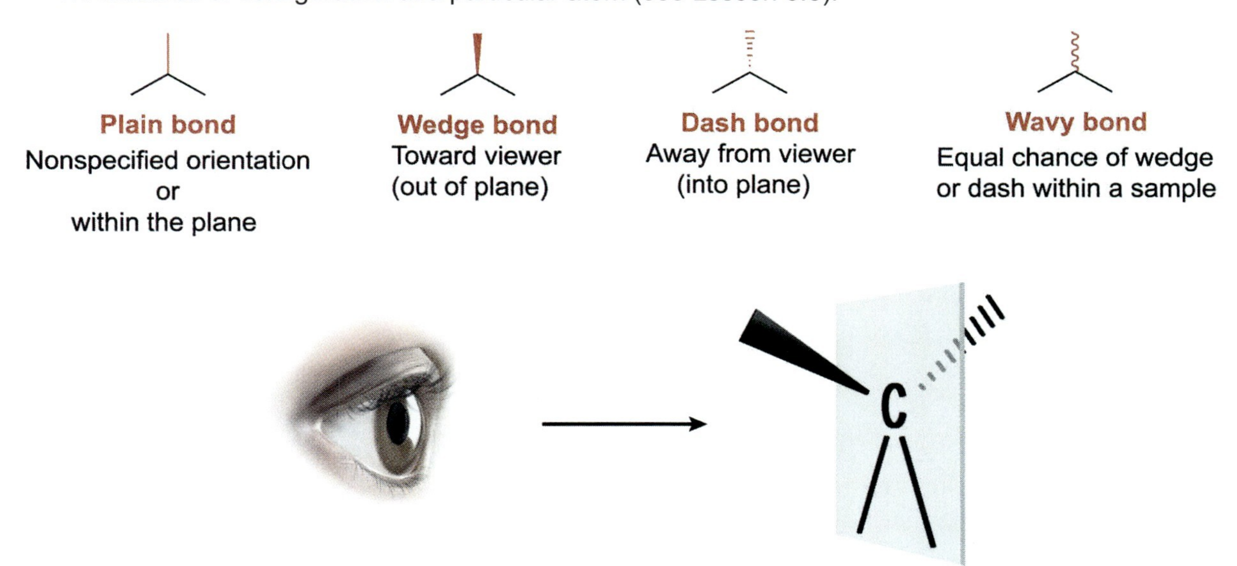

Figure 2.17 Bond types which represent three-dimensional orientation.

The most frequently encountered three-dimensional shape of an atom in organic chemistry is the tetrahedron, which is present for all sp^3 hybridized atoms. If a tetrahedral atom is oriented such that two of its bonds are in the two-dimensional plane, one bond will be oriented out of the plane (a wedge) and one bond will be oriented into the plane (a dash). This premise forms the basis for the way line-angle formulas are represented.

Consequently, if a tetrahedral atom has more than two bonds represented within a line-angle formula, the orientation of the other bonds should correctly mirror the shape of the tetrahedron (Figure 2.18). In general, when two bonds are in the plane, a wedge or dash should always be oriented *away* from them. In general, when *only* a wedge or a dash is shown on an atom, an *implied hydrogen* atom has the *opposite* orientation.

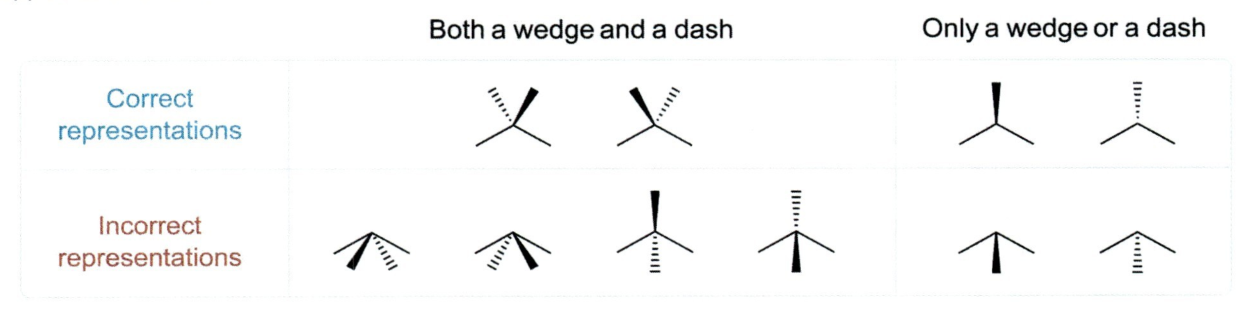

Figure 2.18 Correct and incorrect usage of wedge and dash notation.

Lesson 2.6

Degrees of Unsaturation

Introduction

Although Lesson 2.5 discusses why a molecular formula often provides *incomplete* information about an organic molecule's structure, the molecular formula is still extremely useful. For example, the molecular formula can be used to calculate a compound's molecular weight or molar mass. Knowledge of molecular formulas is useful for certain aspects of spectroscopy (see Chapter 14).

This lesson describes how the molecular formula is used to calculate a molecule's **degrees of unsaturation** and how this information provides partial structural information about an organic compound.

2.6.01 Guidelines for Hydrocarbons

Noncyclic molecules (whether linear or branched) with only carbon-hydrogen single bonds (ie, **alkanes**, see Lesson 2.8) always have a molecular formula of $C_nH_{(2n+2)}$, in which n is a counting number. Table 2.5 shows how these molecules consist of a chain of $–CH_2–$ units capped by two additional hydrogen atoms. Such molecules are called **saturated**, as they contain the *maximum* number of hydrogen atoms per carbon atom and are *unable* to react with additional hydrogen gas (H_2).

Table 2.5 Linear and branched saturated alkanes with the formula $C_nH_{(2n+2)}$.

Name	Lewis structure	Molecular formula	n in $C_nH_{(2n+2)}$
Methane	CH_4	CH_4	1
Ethane	CH_3-CH_3	C_2H_6	2
Propane	$CH_3-CH_2-CH_3$	C_3H_8	3
Butane	$CH_3-CH_2-CH_2-CH_3$	C_4H_{10}	4
Isobutane	$CH_3-CH(CH_3)-CH_3$	C_4H_{10}	4
Pentane	$CH_3-CH_2-CH_2-CH_2-CH_3$	C_5H_{12}	5
Isopentane	$CH_3-CH(CH_3)-CH_2-CH_3$	C_5H_{12}	5
Neopentane	$CH_3-C(CH_3)_2-CH_3$	C_5H_{12}	5

Similarly, an **unsaturated** molecule is one that has the capacity to add a molecule of H_2. While the terms saturated and unsaturated describe double- and triple-bond containing molecules (see Lessons 6.3 and 6.4), a **degree of unsaturation** is more broadly defined as *any structural feature* that decreases the maximum number of hydrogen atoms in a molecular formula by two (relative to the number of carbon atoms).

One degree of unsaturation is present for every π bond *or* cyclic (ring) structure. When a molecule becomes cyclized, two hydrogen atoms are removed and a new carbon-carbon bond is formed (Figure 2.19). Formation of a π bond requires a similar loss of two hydrogen atoms.

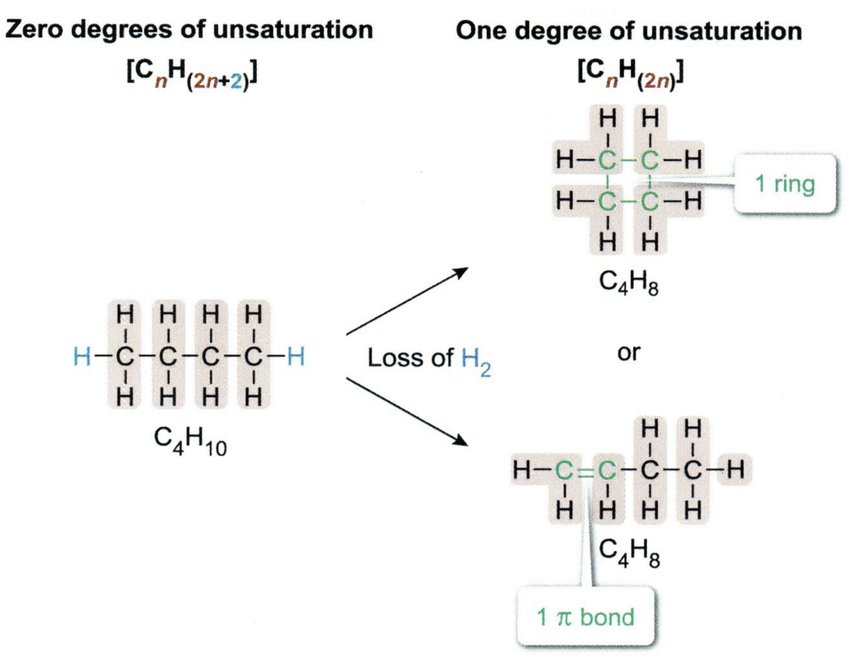

Figure 2.19 Ways to generate a degree of unsaturation.

Because the carbon-to-hydrogen ratio of a saturated molecule has a molecular formula of $C_nH_{(2n+2)}$, a molecule with:

- One degree of unsaturation has a molecular formula of $C_nH_{(2n)}$
- Two degrees of unsaturation has a molecular formula of $C_nH_{(2n-2)}$
- Three degrees of unsaturation has a molecular formula of $C_nH_{(2n-4)}$

Consequently, knowing the molecular formula for a compound allows its degrees of unsaturation to be calculated, providing partial structural information about the compound. For example, C_6H_{10} follows the $C_nH_{(2n-2)}$ formula and has *two* degrees of unsaturation. Therefore, it must have one of the following: two π bonds, one π bond and one ring, or two rings.

To calculate the degrees of unsaturation for a molecular formula:

1. Add up the total number of carbon atoms (or carbon atom equivalents, see Concept 2.6.02).
2. Multiply this number by 2, then add 2. This value is the saturated maximum number.
3. Subtract the number of hydrogen atoms in the molecular formula (or hydrogen atom equivalents, see Concept 2.6.02) from the saturated maximum number, then divide by 2 (2 hydrogen atoms = 1 degree of unsaturation).

Concept Check 2.7

What is the molecular formula of a hydrocarbon if each molecule has seven carbon atoms and contains two π bonds and one ring?

Solution

Note: The appendix contains the answer.

2.6.02 Guidelines for Heteroatoms

In hydrocarbons, degrees of unsaturation are derived from the carbon-to-hydrogen ratio; however, degrees of unsaturation can also be calculated for molecules that contain heteroatoms (Figure 2.20).

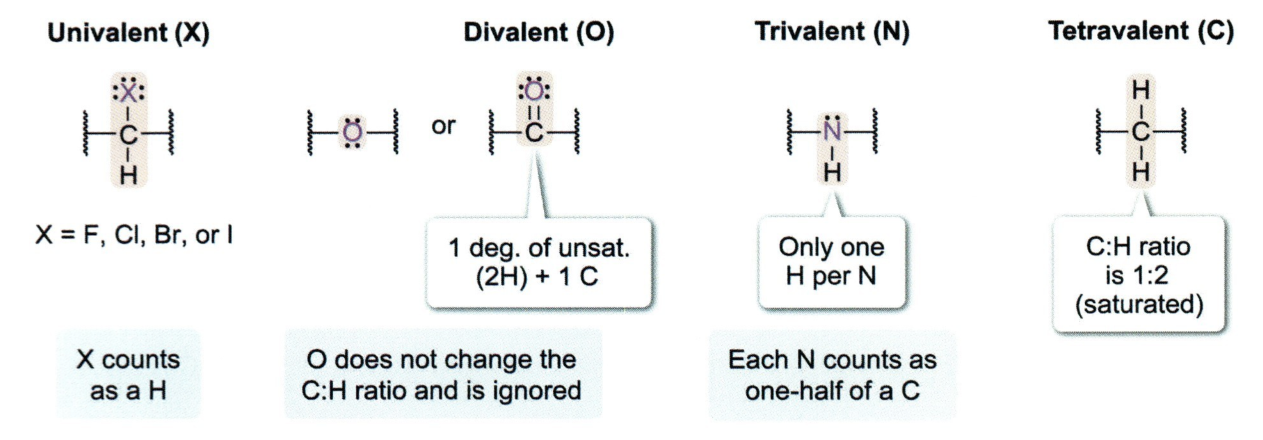

Figure 2.20 Impacts of heteroatom valency on degrees of unsaturation.

Similar to hydrogen atoms, the halogens (F, Cl, Br, I) prefer to form *one* covalent bond with another atom (**univalent**). Therefore, each halogen atom is *equivalent* to one hydrogen atom in a degrees of unsaturation calculation. For example, C_4H_9Br has 4 carbon atoms and *10* hydrogen atom equivalents (9 H + 1 Br) and has zero degrees of unsaturation.

Oxygen typically forms *two* covalent bonds with neighboring atoms (**divalent**). Oxygen atoms can either be incorporated as part of the linear chain or as a double-bonded branch point (a **carbonyl**, see Lesson 2.8). In either scenario, an oxygen atom does not change the carbon-to-hydrogen ratio. Therefore, oxygen atoms can be *ignored* when performing degrees of unsaturation calculations.

Nitrogen typically forms three covalent bonds with other atoms (**trivalent**). Within a chain of atoms (accounting for two covalent bonds), only one covalent bond remains available for additional atoms. Since a standard carbon atom is **tetravalent** (forms four total bonds) and can have up to *two* covalent bonds to hydrogen (within the context of a chain), each nitrogen atom is counted as *one-half* of a carbon atom equivalent. For example, $C_4H_{11}N$ has a total of 4.5 carbon equivalents (4 carbon atoms + 0.5 carbon equivalent from the nitrogen). This molecule has a maximum of 11 hydrogen atoms and therefore is saturated (0 degrees of unsaturation).

Figure 2.21 provides additional examples of degrees of unsaturation calculations involving heteroatoms. The usefulness of this calculation is decreased for third (or higher) row elements (eg, phosphorus, sulfur), due to complexities arising from the potential of an expanded octet.

A saturated molecule has a molecular formula of $C_nH_{(2n+2)}$

Example 1 C_5H_9Cl

Carbon equivalents = 5C

Saturated hydrogen equivalents = (5C × 2) + 2 = 12H

Hydrogen equivalents = 9H + 1H(Cl) = 10H

Degrees of unsaturation = $12H - 10H = 2H \times \left(\frac{\text{deg. of unsat.}}{2H}\right)$ = 1 deg. of unsat.

Example 2 $C_7H_{12}O_2$

Carbon equivalents = 7C

Saturated hydrogen equivalents = (7C × 2) + 2 = 16H

Hydrogen equivalents = 12H (ignore oxygen)

Degree of unsaturation = $16H - 12H = 4H \times \left(\frac{\text{deg. of unsat.}}{2H}\right)$ = 2 deg. of unsat.

Example 3 $C_6H_{17}N_3$

Carbon equivalents = $6C + \left(\frac{1}{2}\right)(3C) = 7.5C$

Saturated hydrogen equivalents = (7.5C × 2) + 2 = 17H

Hydrogen equivalents = 17H

Degrees of unsaturation = $17H - 17H = 0H \times \left(\frac{\text{deg. of unsat.}}{2H}\right)$ = 0 deg. of unsat.

Figure 2.21 Example analyses of degrees of unsaturation involving heteroatoms.

Concept Check 2.8

The molecular formula for a compound was determined to be $C_5H_{12}N_2O_2$. If an experiment provides evidence that the compound contains a π bond between carbon atoms, what conclusions can be drawn about the structure of the compound?

Solution

Note: The appendix contains the answer.

Lesson 2.7

Resonance

Introduction

This lesson reviews the use of Lewis to represent the electronic structure of molecules. While many molecules have a single valid Lewis structure, some have *multiple* valid Lewis structures. For these molecules, a single Lewis structure *cannot* fully encompass the molecule's electronic features. Instead, the *array* of Lewis structures is most representative of the molecule's true structure, properties, and reactivity. For these molecules, the concept of **resonance** is used to provide a more complete description.

This lesson is focused on the foundational principle of resonance and introduces its *many* applications within organic chemistry.

2.7.01 Definition of Resonance

Resonance is used to rationalize the electronic structure of molecules for which multiple Lewis structures are valid. Resonance builds on the idea that a molecule's electrons are distributed *throughout* atoms in a molecule (and are not dedicated to particular atoms). In general, the electron **delocalization** of resonance is a *stabilizing* feature that can greatly impact a molecule's properties (see Concept 2.7.04).

The Lewis structure of a carbonate ion (CO_3^{2-}), for example, involves a central carbon connected to three oxygen atoms. Two of the oxygen atoms have a −1 formal charge and are connected to the carbon by a single bond; the third oxygen (which lacks a formal charge) is connected by a double bond.

However, *which* oxygen atom is bonded through a double bond? Figure 2.22 shows three separate, but equally valid, Lewis structures that differ in the identity of the double-bonded oxygen. Note that these multiple valid Lewis structures (**resonance structures**) are *not* due simply to rotation (ie, *the relative position of atoms does not change*). Rather, only the positions of the *electrons* change between resonance structures, which can lead to differences in the locations of pi bonds and formal charges.

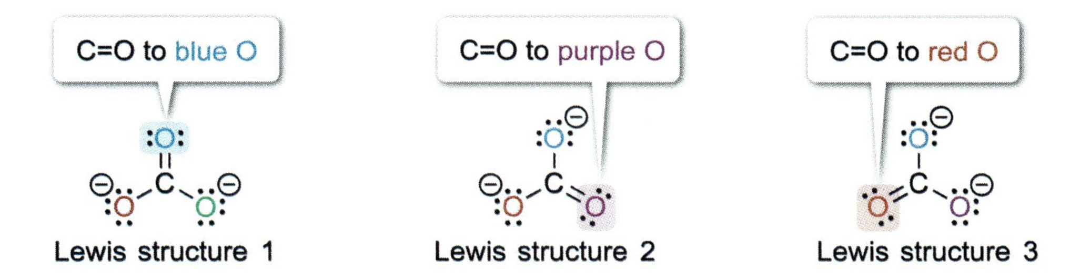

Figure 2.22 Lewis structures of the carbonate ion.

The three different resonance structures of the carbonate ion are different electronic representations of the *same molecule*. Importantly, molecules do *not* oscillate (or equilibrate) between individual resonance structures. To signify the difference between resonance and equilibrium, resonance structures are drawn with a **double-headed arrow** between individual structures (Figure 2.23), and all of the structures are enclosed in **brackets**. The principles of *curved* arrow electron flow are addressed in Concept 2.7.02.

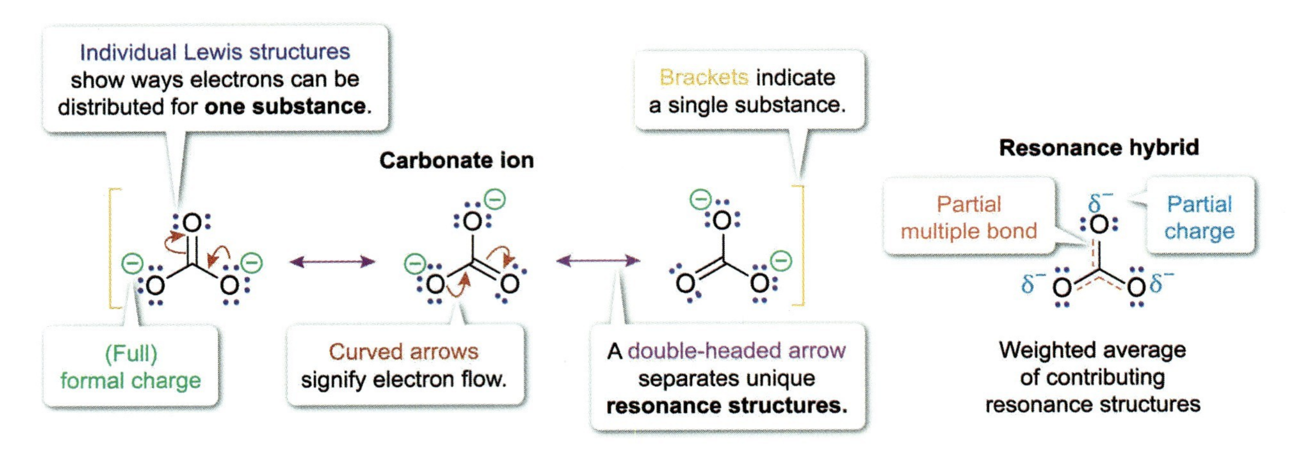

Figure 2.23 Anatomy of the resonance structures and resonance hybrid of the carbonate ion.

The weighted average of individual resonance structures contributes to the overall character of the **resonance hybrid**, the representation that *best* describes the electronic nature of a molecule. In the example of the carbonate ion, all three oxygen atoms have individual resonance structures that form *either* a double bond *or* a single bond and have either a neutral *or* a negative formal charge. To reconcile these differences, resonance hybrid structures frequently depict **partial electronic character**.

A **partial multiple bond** is denoted by a dashed second or third line and indicates that the true bond order lies between two integer values. A **partial charge (δ^+ or δ^-)** indicates that an atom carries a non-integer amount of charge. Furthermore, the sum of the partial charges in a resonance hybrid structure must equal the sum of the formal charges for the molecule or ion (as represented by any one resonance structure).

Importantly, resonance hybrids are not simply a means to consolidate various representations, but they also have important implications in a molecule's physical and chemical properties (eg, geometry, reactivity, see Concept 2.7.04).

2.7.02 How to Identify and Draw Resonance Structures

When generating the full array of resonance structures, it is important to start with a single, valid initial structure that includes nonbonding electrons and formal charges. As mentioned in Concept 2.7.01, the positions of the atoms *remain the same* between resonance structures and only the positions of the *electrons differ*.

Curved arrow formalism is used to depict electron movement. The curved arrow proceeds *from* the original position of the electron(s) *to* their new location, and the number of barbs on the arrowhead indicates the number of electrons that are moving (Figure 2.24). An arrow with a one-barb head signifies the movement of one electron, while an arrow with a two-barb head signifies the movement of an electron pair (two electrons).

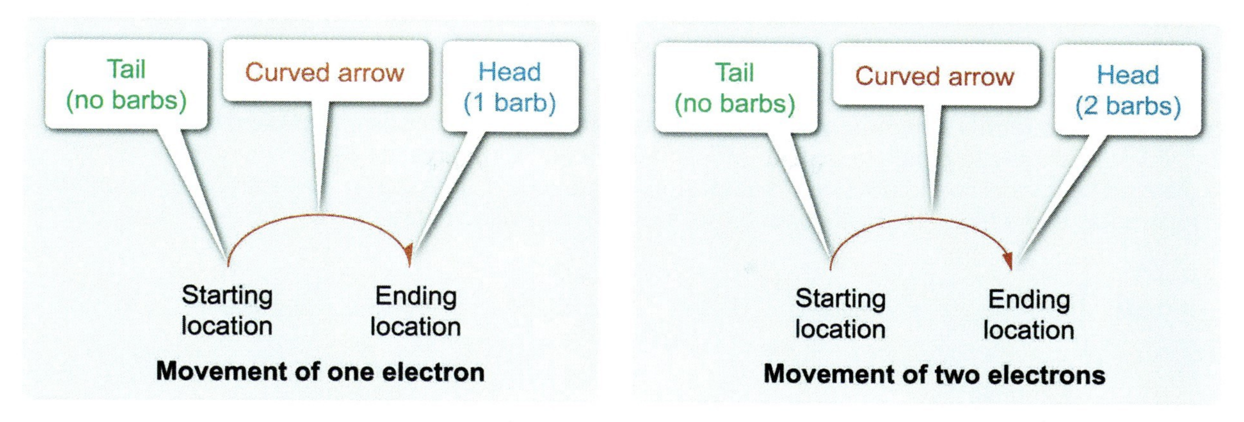

Figure 2.24 Curved arrow formalism.

Resonance structures *must* have the *same* sigma bond network and atom composition. Stated another way, resonance structures *cannot* gain or lose atoms and *cannot* change the position of any sigma bonds, as this would indicate a **chemical transformation** and formation of a *new* substance (Figure 2.25). Resonance structures are different representations of the *same* substance.

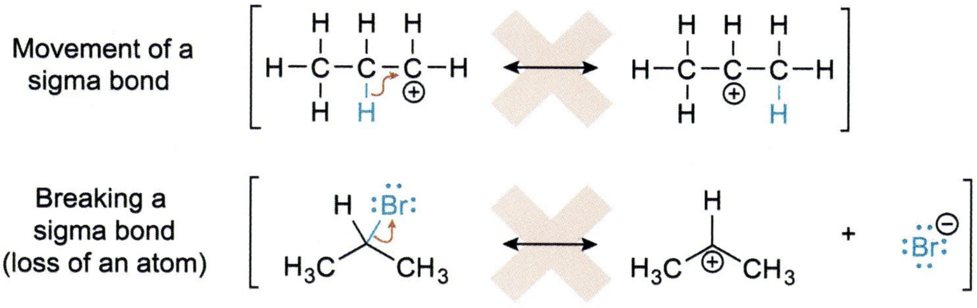

Movement of a sigma bond

Breaking a sigma bond (loss of an atom)

Figure 2.25 Chemical transformations do not qualify as resonance.

Valid electron valence must also be maintained between resonance structures. Second-row elements (eg, carbon, nitrogen, oxygen) are limited to only eight valence electrons. Recall that each chemical bond requires two electrons, as does a lone pair (see Lesson 2.7).

Consequently, it is *not possible* for second-row elements to participate in a structure in which the element has more than four bonds or lone pairs, as this violates the octet rule (Figure 2.26). However, resonance structures *may* result in one or more atoms having an *incomplete* octet (eg, a carbon atom with only three bonds and a +1 formal charge).

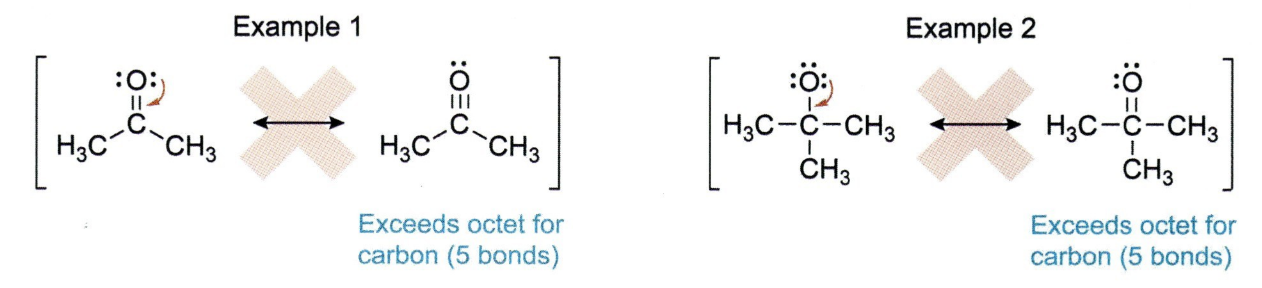

Figure 2.26 Second-row elements cannot violate the octet rule.

Given these restrictions, the only electrons that can be changed or moved between resonance structures of second-row elements are *nonbonding electrons (lone pairs)* and *electrons in π bonds*.

These two categories of electrons interconvert between resonance structures. For example, a lone pair may be moved to form a new π bond to an adjacent atom, or the electrons in a π bond may become localized on a single atom as a lone pair. Most often, these two processes happen simultaneously during the electron flow transition to another resonance structure (see Figure 2.27). This can be remembered by the mnemonic expression "make a bond, break a bond" (and vice versa).

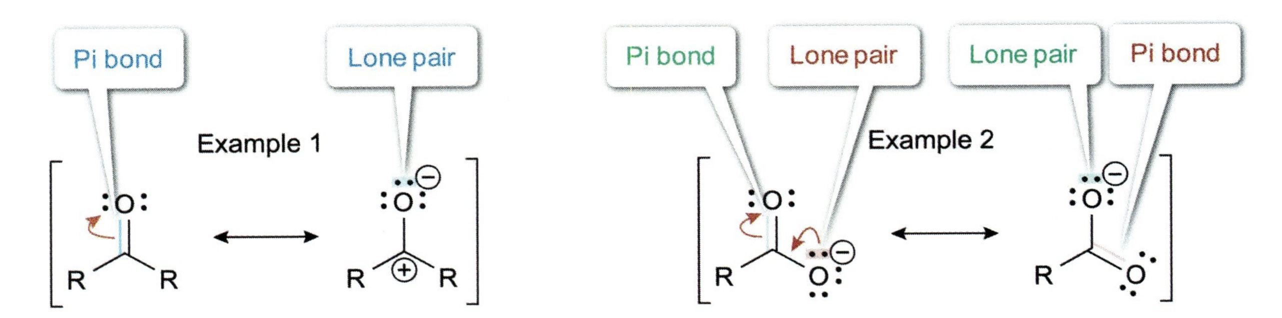

Figure 2.27 Interconversion of lone pairs and pi bonds through resonance.

Figure 2.28 shows curved arrow formalism with the delocalization of formal charges. Because a *positive* formal charge represents an atom that *lacks* electron density relative to its valence electrons, curved arrow flow must originate at a source of electrons (eg, lone pair or pi bond) and flow *toward* the electron-deficient cation. After electron flow, the cation then appears at a new location—usually the position that was *vacated* by the moved electrons. In contrast, a negative formal charge is representative of an atom that has an *abundance* of electron density and can therefore be a *source* for curved arrow flow.

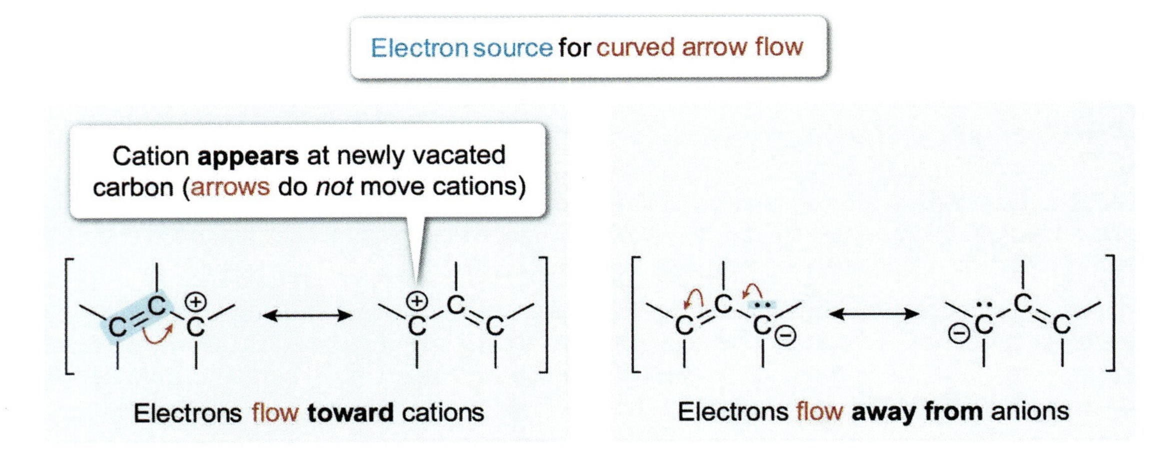

Figure 2.28 Resonance delocalization of cations versus anions.

Even when a molecule contains lone pairs or π bonds, the ability to draw other resonance structures is dependent on the proximity of eligible, mobile electrons to one another. Mobile electrons require adjacency and cannot be separated by more than one sigma bond in the next resonance structure. When working with line-angle structures, it is especially important to account for assumed hydrogen atoms or lone pairs that do not appear in an initial structure.

The most common patterns for resonance delocalization include:

- A lone pair one atom removed from either a π bond or a carbon bearing a +1 formal charge (a **carbocation**, see Lesson 5.2)
- A polar multiple bond between atoms

- Rings with an alternating pattern of single and double bonds (aromatic compounds, see Lesson 6.5)

Figure 2.29 depicts frequently encountered resonance patterns in several neutral, anionic, and cationic groups.

	Group	Contributing resonance structures	Resonance hybrid
Neutral molecules	Ketone		
	Amide		
	Aromatic (Benzene)		
Anionic molecules	Carboxylate ion		
	Enolate ion		
	Phenoxide ion		
Cationic molecules	Allylic cation		
	Guanidinium ion		

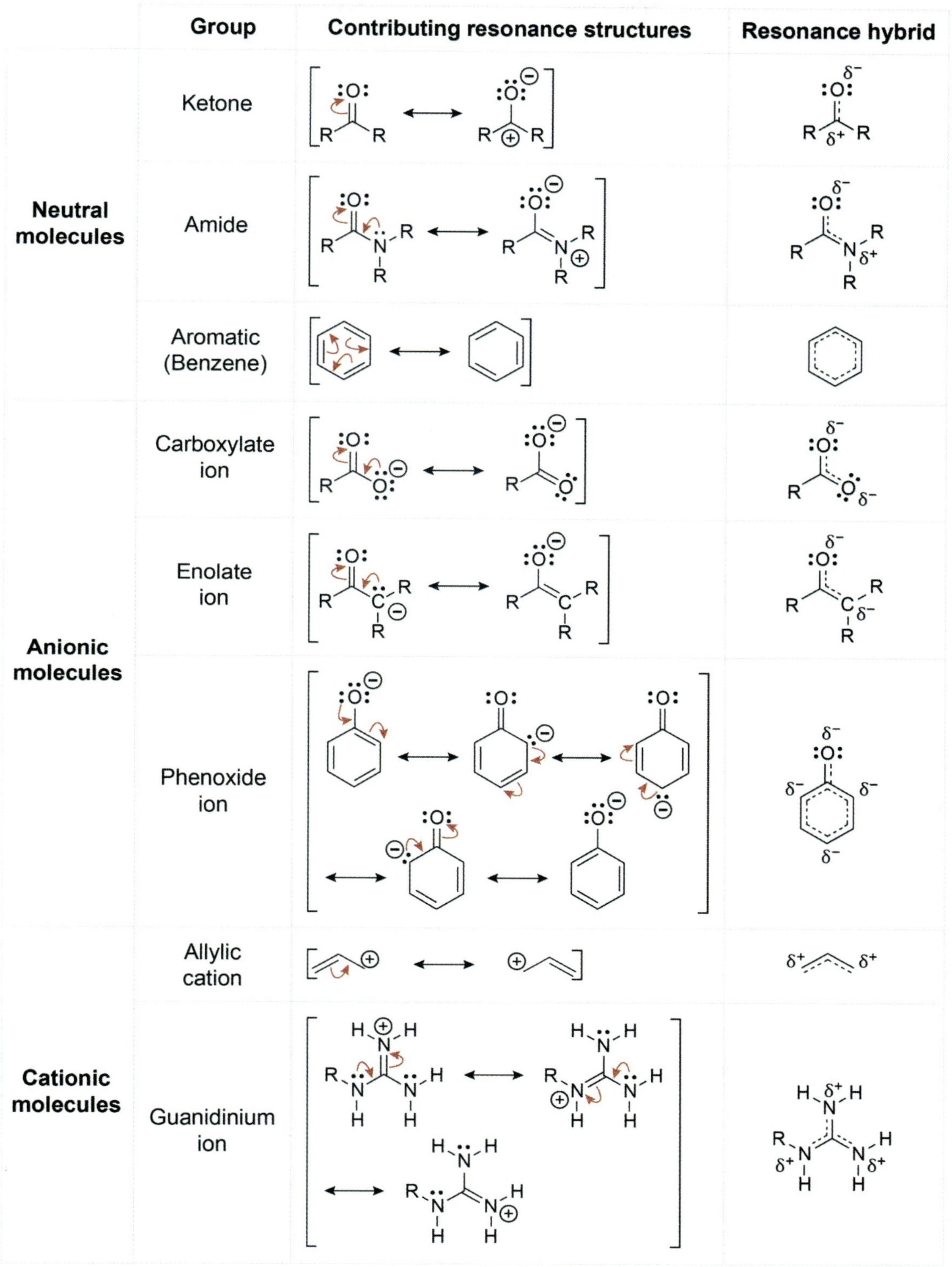

Figure 2.29 Resonance structures for common neutral, anionic, and cationic groups.

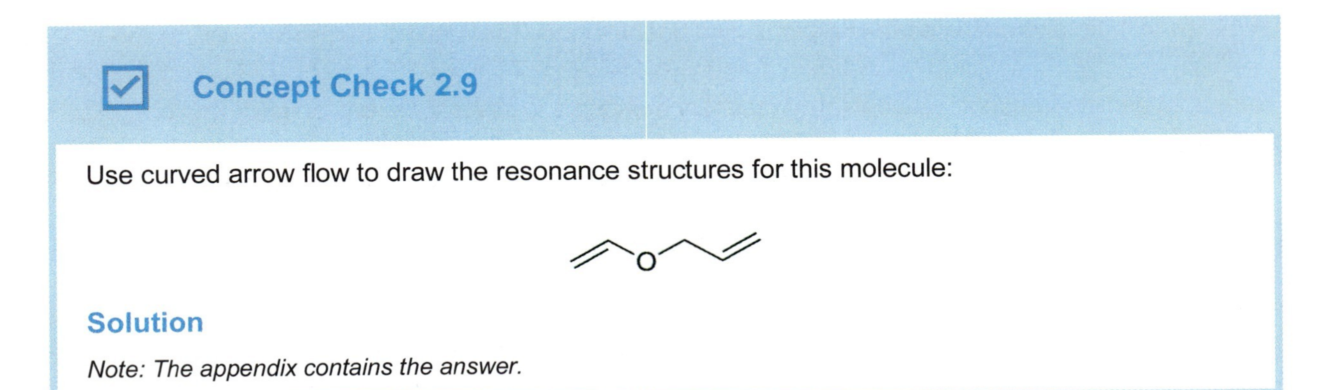

Concept Check 2.9

Use curved arrow flow to draw the resonance structures for this molecule:

Solution

Note: The appendix contains the answer.

2.7.03 Major and Minor Resonance Contributors

Not all resonance structures contribute equally to the electronic structure for a substance. The best resonance structures typically have the lowest energy (and therefore, the greatest stability). The relative terms **major contributor** and **minor contributor** are used to describe a valid resonance structure's contribution to the hybrid resonance structure. This concept covers three guidelines that can be used to distinguish major contributors from minor contributors, discussed in their relative order of importance.

The best resonance structures maximize the number of atoms that have a *complete octet* of electrons. Because sharing electrons through covalent bonds is energetically favorable (Lesson 1.1), the best resonance structures maximize the number of covalent bonds and minimize the number of nonbonding electrons. Figure 2.30 shows examples of this principle through the resonance structures of a carbonyl group. The left resonance structure provides all atoms with a complete octet and is the major contributor; the right resonance structure contains a carbon with only six valence electrons and is a minor contributor.

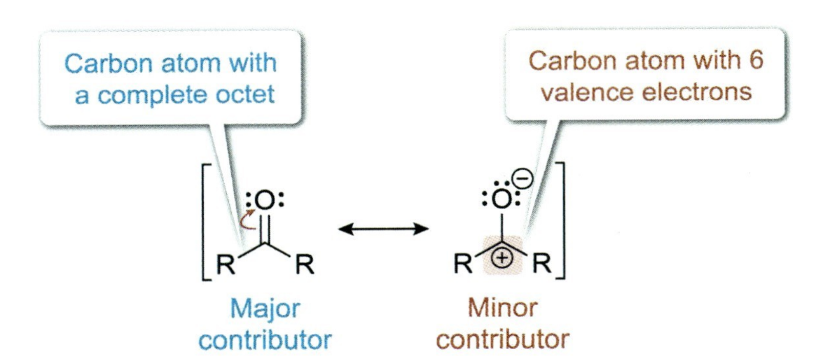

Figure 2.30 Resonance structures of a carbonyl group.

The best resonance structures also *minimize the presence and separation of formal charge*. In general, a resonance structure without any atoms that bear a formal charge will be the most important resonance contributor (Figure 2.31). However, in cases in which formal charges cannot be fully avoided, it is most favorable to have the smallest number of formal charges. Having formal charges of the same sign in close proximity leads to electrostatic repulsion and further reduces the contribution of the resonance structure.

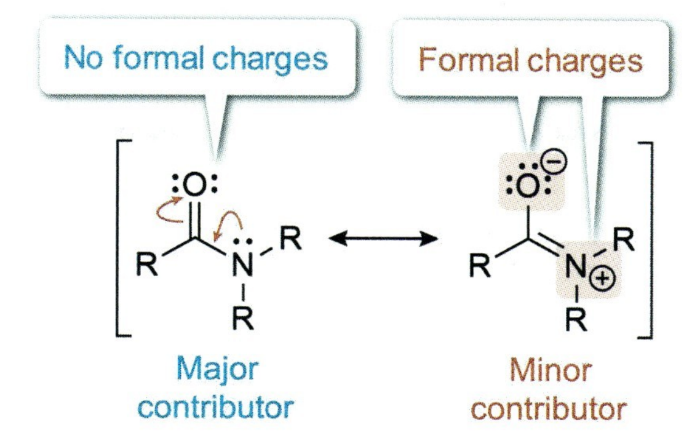

Figure 2.31 Resonance structures of an amide.

If there is ever a choice between atoms that could bear a formal charge, the best resonance structure places *negative* formal charges on the *most* electronegative atom (and vice versa) (Figure 2.32).

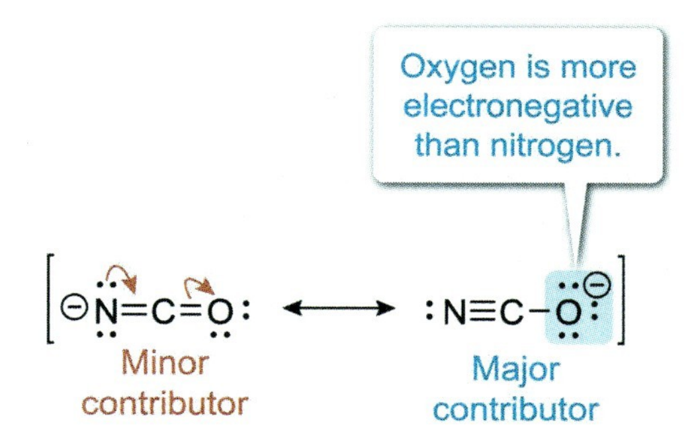

Figure 2.32 Resonance structures of the isocyanate ion.

Partial charges **(**δ^+ or δ^-**)** in a hybrid resonance structure (see Concept 2.7.01) are often qualitative and can represent a range of values. However, analysis of major and minor resonance structures allows additional information to be inferred regarding the relative *magnitude* of a given partial charge.

For example, the phenoxide ion has five valid resonance structures (Figure 2.33) that demonstrate that its negative formal charge is delocalized across its oxygen atom and three carbon atoms. Since oxygen is more electronegative than carbon, the resonance structures that place the negative formal charge on the oxygen atom are the *major contributors*. Because a hybrid resonance structure is a weighted average of contributing resonance structures, it can be inferred that the *hybrid* structure has a more negative partial charge on the oxygen atom than on any of the three carbon atoms.

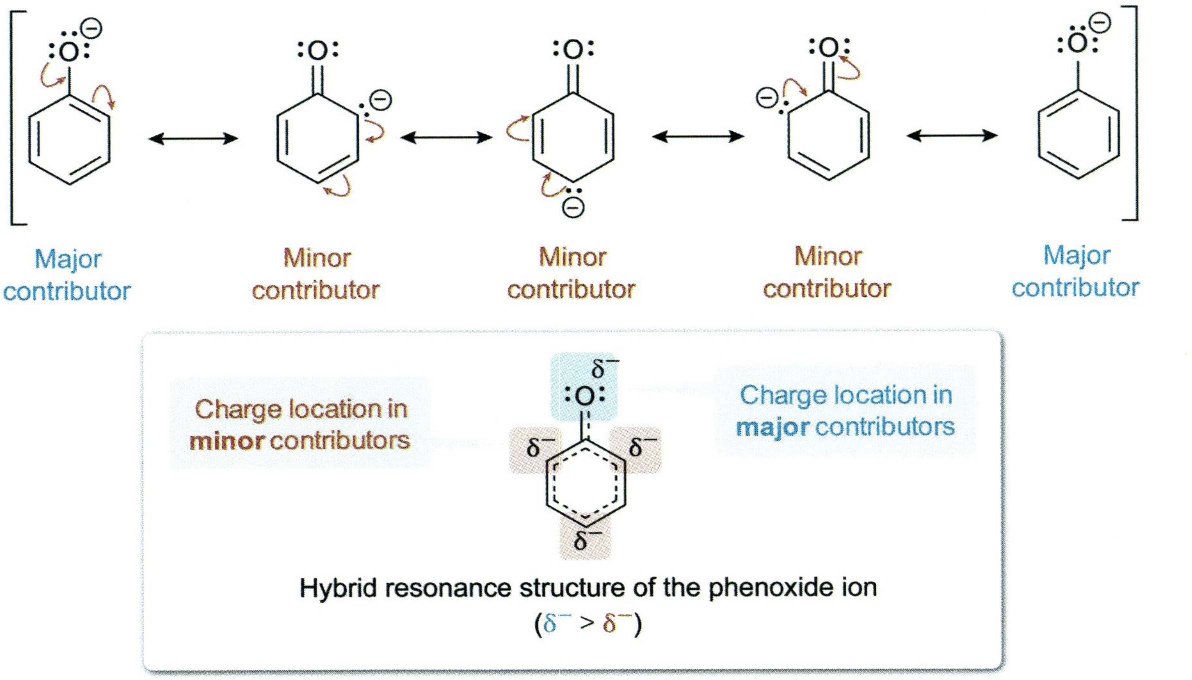

Major contributor

Minor contributor

Minor contributor

Minor contributor

Major contributor

Charge location in **minor** contributors

Charge location in **major** contributors

Hybrid resonance structure of the phenoxide ion

($\delta^- > \delta^-$)

Figure 2.33 Resonance structures and hybrid resonance structure of the phenoxide ion.

Importantly, even though the *partial charge* described by a minor contributor is weighted less than that of major contributors, the *geometry* required by a minor contributor (ie, electron domain geometry, see Lesson 2.1)—and of *all* valid contributors—*must be achievable* for every structure. This has important implications in, for example, the peptide bond in biochemistry. Concept 2.7.04 further discusses applications of resonance in organic chemistry.

2.7.04 Applications of Resonance in Organic Chemistry

Resonance delocalization of electrons *requires* the participating electron orbitals to overlap in a side-to-side fashion. This means that electrons of second-row elements must be in a pi bond or in an unhybridized *p* orbital. Figure 2.34 demonstrates how the lone pair on the nitrogen atom in pyridine is located in an sp^2 orbital that is *perpendicular* to the *p* orbitals in the ring.

Although it might initially appear that the lone pair could delocalize into the ring and provide the indicated resonance structure, the lack of orbital alignment prevents resonance delocalization of the nitrogen lone pair—the lone pair *cannot* participate in side-to-side overlap with the adjacent pi bond. As a general guideline, it is important to evaluate each possible scenario for resonance.

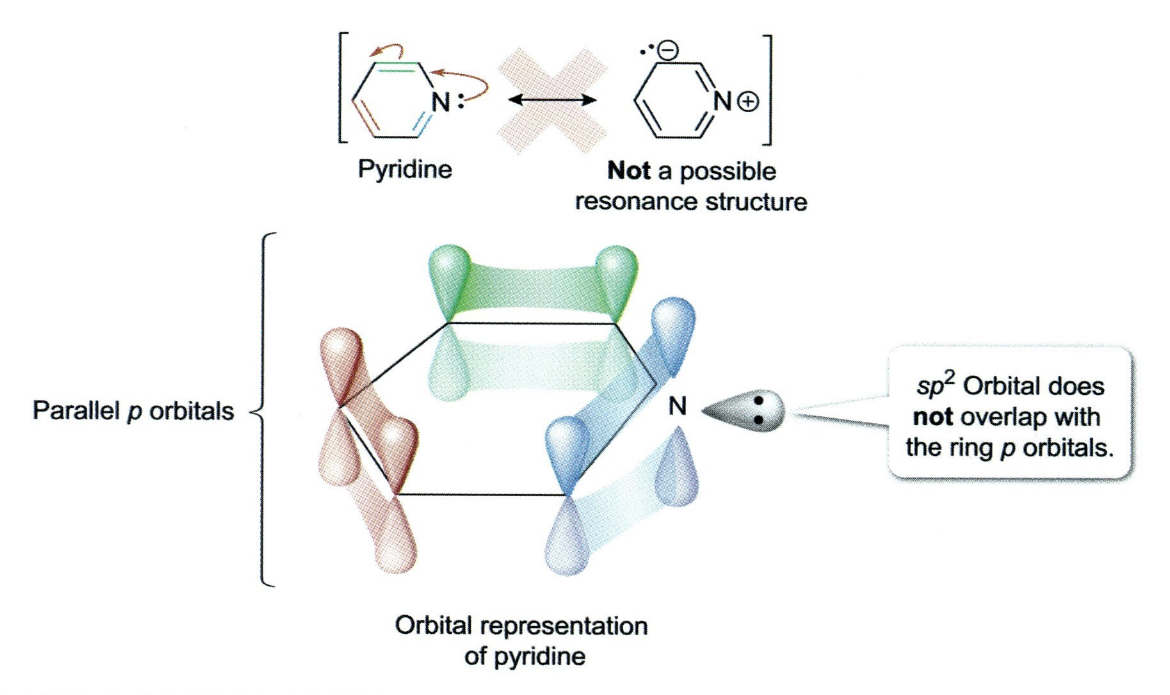

Figure 2.34 Orbital depiction of pyridine.

Because of the requirement that lone pairs be in unhybridized orbitals (to allow for side-to-side overlap with adjacent pi bonds), the actual hybridization of some atoms bearing lone pairs may be different than what might be *initially* concluded from a line-angle structure.

For example, the line-angle structure of the major resonance contributor of an amide shows a nitrogen atom with *apparently* four electron domains, leading to an *initial* conclusion that the nitrogen atom has sp^3 hybridization. However, a valid minor resonance contributor shows that the nitrogen atom *must* have only *three* electron domains, which corresponds to sp^2 hybridization. For the hybrid resonance structure to exist, the carbon, oxygen, and nitrogen must *all* have an unhybridized *p* orbital. Therefore, because of the stabilizing qualities of resonance delocalization of the lone pair, the nitrogen atom adopts an sp^2 hybridization (Figure 2.35).

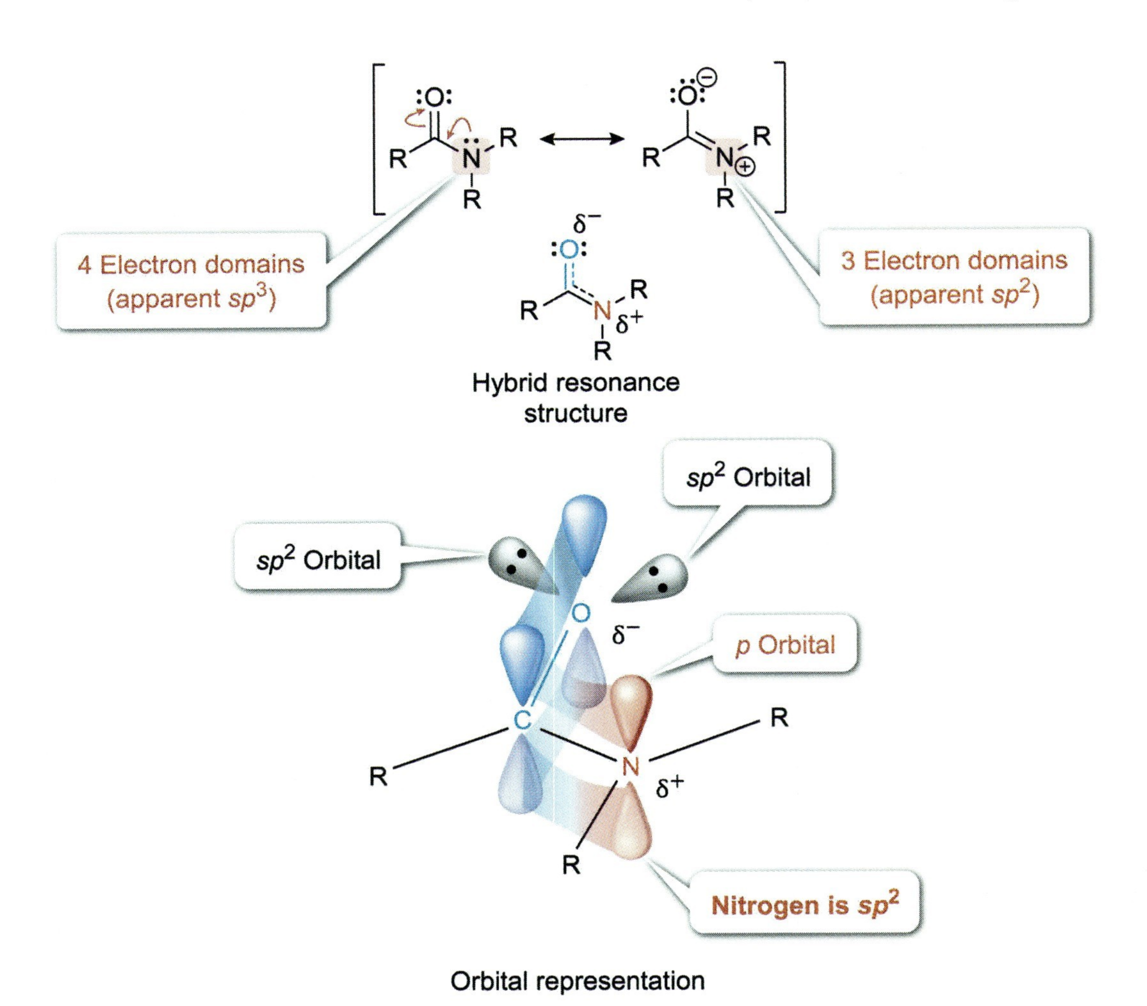

Figure 2.35 Orbital depiction of an amide.

The principle of resonance delocalization also serves as the theoretical basis for both conjugation and aromaticity. A **conjugated** system typically refers to a structure with an alternating pattern of single and multiple (double, triple) bonds (Figure 2.36). An **aromatic** compound has a similar pattern, arranged in a ring. Lesson 6.5 provides an overview of the properties of conjugated and aromatic molecules.

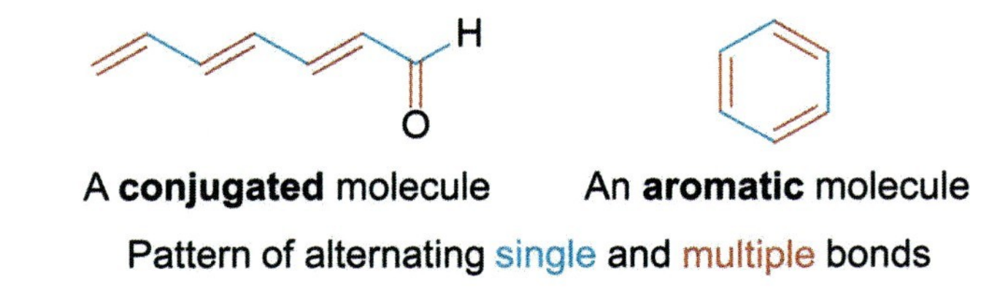

Figure 2.36 Conjugated and aromatic molecules.

Resonance delocalization of electrons is almost always considered a stabilizing quality (ie, resonance decreases a substance's energy). Consequently, resonance has profound impacts on the equilibrium and favorability of a reaction. As a general guideline, reactions favor the lower-energy substance (Figure 2.37). Stated another way, if a reaction involves a resonance-stabilized *product*, then product formation (ie, the forward reaction) is favored; conversely, if a reaction involves a resonance-stabilized *reactant*, then reactant formation (ie, the *reverse reaction* is favored).

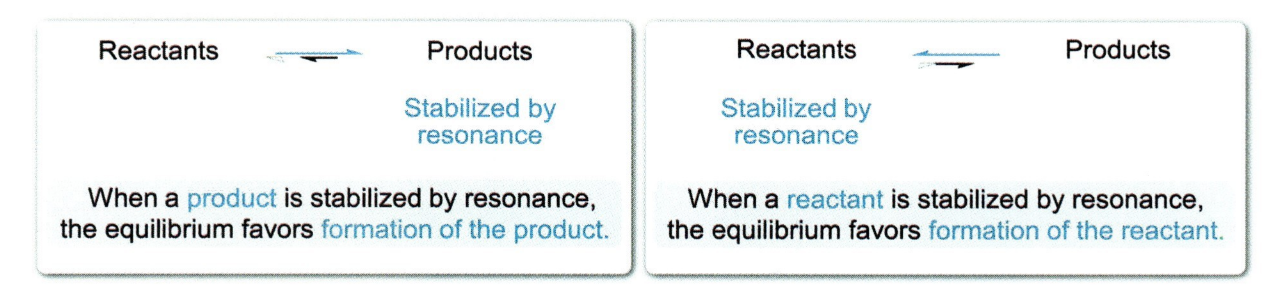

Figure 2.37 Relationship between resonance stabilization and equilibrium.

Applications of this principle are discussed throughout this book, including acid-base reactivity (Lesson 5.1), carbocation stability trends (Lesson 5.2), chemical mechanisms (Lesson 5.4), and the formation of enolates (Lesson 9.4).

As discussed in General Chemistry Lesson 8.1, a Brønsted-Lowry **acid** is a proton (H^+) donor, while a **base** is a proton acceptor. After an acid donates a proton, it becomes a conjugate base; after a base accepts a proton, it becomes a conjugate acid.

Many organic acids and bases (or their conjugates) are stabilized by resonance delocalization of electrons, which therefore affects acid-base equilibrium. Figure 2.38 depicts the reactions of two organic acids (a carboxylic acid and a guanidinium ion) with a generic base. In the top scenario, the conjugate base reaction *product* (the carboxylate ion) is stabilized by resonance, so equilibrium favors *formation of the products*. Consequently, a carboxylic acid is a relatively stronger acid, whereas a carboxylate ion is a relatively weaker base.

In the bottom scenario, the acid *reactant* (guanidinium ion) is stabilized by resonance, so equilibrium favors formation of the *reactants*. Consequently, a guanidinium ion is a relatively weaker acid and a guanidine is a relatively stronger base. See Lesson 5.1 for more on acids and bases from an organic chemistry perspective.

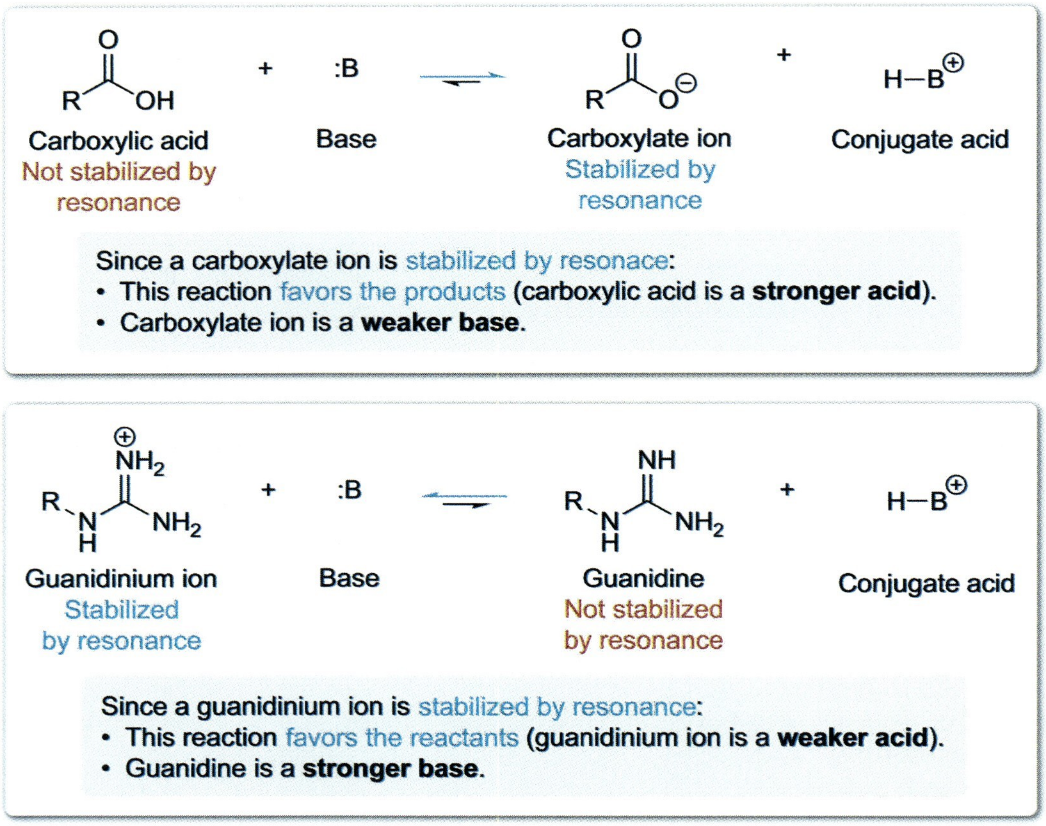

Carboxylic acid + :B ⇌ Carboxylate ion + H–B⊕

Carboxylic acid
Not stabilized by resonance

Base

Carboxylate ion
Stabilized by resonance

Conjugate acid

Since a carboxylate ion is stabilized by resonance:
- This reaction favors the products (carboxylic acid is a **stronger acid**).
- Carboxylate ion is a **weaker base**.

Guanidinium ion + :B ⇌ Guanidine + H–B⊕

Guanidinium ion
Stabilized by resonance

Base

Guanidine
Not stabilized by resonance

Conjugate acid

Since a guanidinium ion is stabilized by resonance:
- This reaction favors the reactants (guanidinium ion is a **weaker acid**).
- Guanidine is a **stronger base**.

Figure 2.38 Impact of resonance on acid-base strength.

Lesson 2.8

Overview of Functional Groups

Introduction

While organic molecules are composed primarily of carbon and hydrogen atoms, they may contain other heteroatoms, including oxygen, nitrogen, and the halogens. Molecules are often classified by **functional groups**, recurring bonding patterns between atoms that are frequently the reactive portions of a molecule. The term functional group can be used to represent either a certain group of atoms within a molecule or the entire molecule that bears the certain group of atoms.

Figure 2.39 shows the major functional group classifications. This lesson covers the general structure of the major functional groups. A comprehensive explanation of the nomenclature rules and guidelines for each of the functional groups is presented in Chapter 4, and the properties and reactivity of each functional group will be discussed throughout Unit 2.

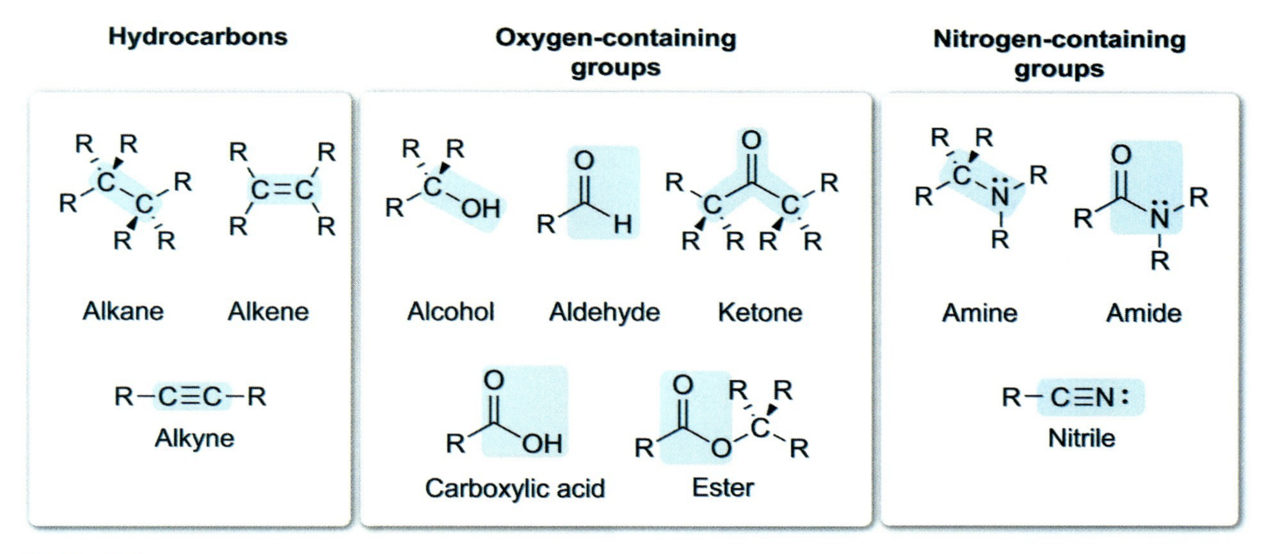

Figure 2.39 Major functional group classifications and their general structures.

2.8.01 Hydrocarbons

Hydrocarbons are organic molecules composed of solely carbon and hydrogen atoms, which makes these molecules both nonpolar and hydrophobic (see Lesson 6.1). Figure 2.40 shows the four major types of hydrocarbons.

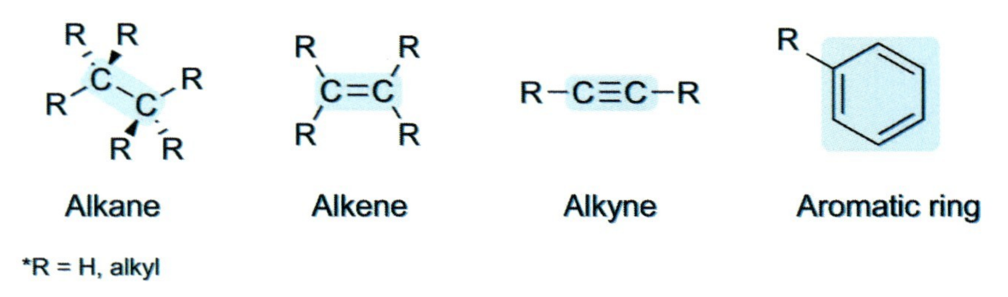

Figure 2.40 General structure of hydrocarbon functional groups.

The different hydrocarbons differ in the types of carbon-carbon bonds that are present. **Alkanes** contain *only* carbon-carbon single bonds (ie, only σ bonds and no π bonds). In contrast, **alkenes** contain at least one carbon-carbon double bond. **Alkynes** contain at least one carbon-carbon triple bond. Therefore, alkenes and alkynes contain both σ and π bonds. Double and triple bonds can be located at any position within a chain (ie, either at the end of the chain or at an internal position).

Aromatic hydrocarbons contain an aromatic structure, such as a phenyl group (ie, a six-membered ring with alternating single and double bonds between carbon atoms). Phenyl rings are derived from a common aromatic hydrocarbon called benzene. Concepts 6.5.02 and 6.5.03 explore the criteria of aromaticity and other examples of aromatic molecules.

Alkanes can contain a linear chain or a branched chain of carbon atoms (see Lesson 3.1). Alkenes can adopt different configurations (*cis*/*trans* or *E*/*Z*) depending on the relative orientation of double bond substituents (Concept 3.3.11). Alkanes, alkenes, and alkynes also have cyclic variants known as **cycloalkanes**, **cycloalkenes**, and **cycloalkynes**, respectively (Figure 2.41). Cycloalkynes are only possible for rings with eight or more carbon atoms; the triple bond forces the near-linear relationship of four carbons, causing angle strain too great for smaller rings to accommodate.

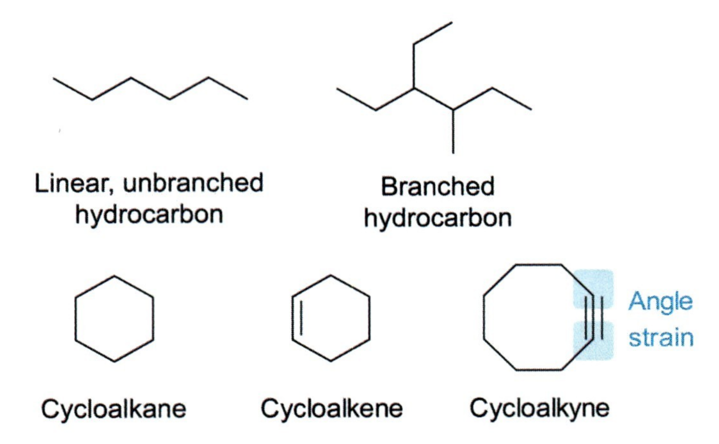

Figure 2.41 Linear, branched, and cyclic hydrocarbons.

2.8.02 Oxygen-Containing Groups

The major classes of organic functional groups that contain oxygen atoms include alcohols, ethers, aldehydes, ketones, carboxylic acids, and carboxylic acid derivatives. The oxygen-containing functional groups are particularly prevalent in biological molecules.

Alcohols have the general formula R–OH, in which R is an **alkyl group** (ie, an alkane missing one hydrogen atom) (Figure 2.42). Because a **hydroxy group (–OH)** can form hydrogen bonds, alcohols are polar, hydrophilic molecules. Alcohols are discussed further in Lesson 8.1.

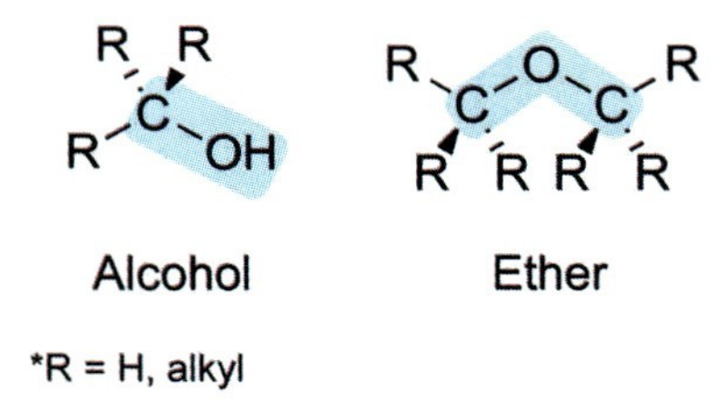

Figure 2.42 General structure of an alcohol and an ether.

Ethers have the general formula R–O–R′ (Figure 2.42). Although ethers contain a hydrogen bond acceptor, ethers are significantly *less* polar than alcohols because they lack a hydrogen bond donor (see Lesson 7.2).

Aldehydes and **ketones** are two functional groups that contain a **carbonyl group**, a carbon-oxygen double bond (Figure 2.43). Aldehydes have the general formula R–CH(O), in which the carbonyl carbon is bonded to an alkyl group (R) and a hydrogen atom. Ketones have the general formula R–C(O)–R′, in which the carbonyl carbon is bonded to two alkyl groups (R and R′).

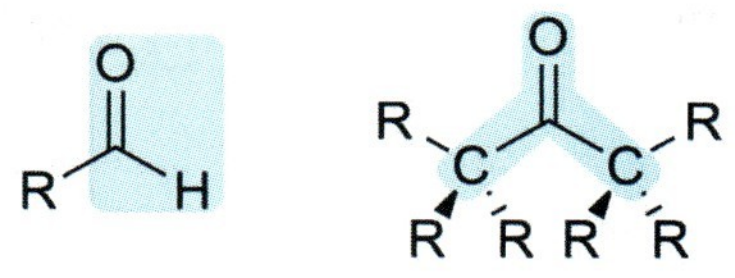

Aldehyde

Ketone

*R = H, alkyl

Figure 2.43 General structure of an aldehyde and a ketone.

Because the carbonyl oxygen atom is a hydrogen bond acceptor, aldehydes and ketones can form hydrogen bonds with molecules containing a hydrogen bond donor. Alone, aldehydes and ketones form dipole-dipole interactions and have an intermediate polarity among the organic functional groups (less polar than alcohols but more polar than hydrocarbons and ethers). Aldehydes and ketones are covered in greater detail in Chapter 9.

Carboxylic acids contain a **carboxyl functional group** and have the general formula R–C(O)OH (Figure 2.44). Even though a carboxyl group contains both a carbonyl group *and* a hydroxy group, it is inappropriate to refer to the carbonyl of a carboxylic acid as a ketone, or the hydroxyl group of a carboxylic acid as an alcohol. Instead, the carboxyl group should be referred to as *one unit.*

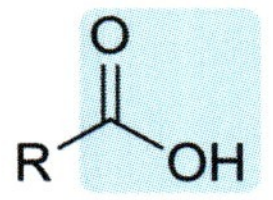

Carboxylic acid

*R = H, alkyl

Figure 2.44 General structure of a carboxylic acid.

Carboxylic acids are highly polar molecules (even more polar than alcohols) due to the intermolecular forces the carboxyl group experiences. A carboxylic acid can form strong dipole-dipole interactions, including hydrogen bonding (see Chapter 10). The acidity of a carboxylic acid is attributed to the resonance stabilization of its conjugate base, the **carboxylate ion ($-COO^-$)** (see Concept 5.1.07).

Carboxylic acids can be converted into several other functional groups known as **carboxylic acid derivatives** (Lesson 10.3 discusses reactions of carboxylic acids). A carboxylic acid derivative contains a carbonyl carbon bonded to an alkyl group and either oxygen, nitrogen, or halogen. The common carboxylic acid derivatives (Figure 2.45) include:

- **Acid halides** have the general formula R–C(O)–X, in which X is a halogen.
- **Carboxylic acid anhydrides** have the general formula R–C(O)–O–C(O)–R′.
- **Esters** have the general formula R–C(O)–OR′.
- **Amides** have the general formula R–C(O)–N(R′)$_2$, in which the groups on the nitrogen can be any combination of alkyl group (R) or hydrogen atom (H).

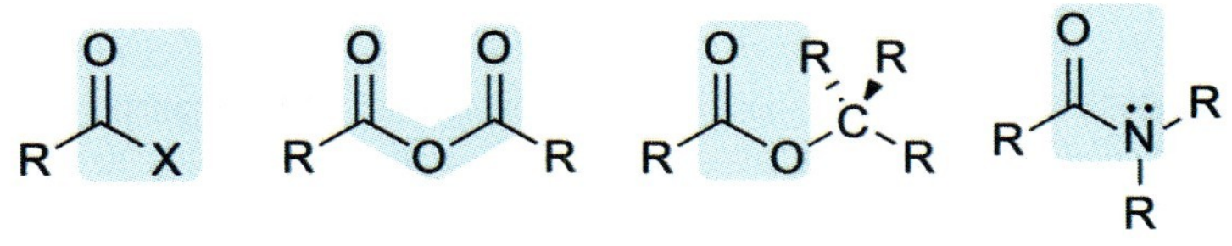

Acid halide Anhydride Ester Amide

*R = H, alkyl
X = F, Cl, Br, I

Figure 2.45 General structure of the carboxylic acid derivatives.

Because the carboxylic acid derivatives each contain a carbonyl, these compounds experience intermolecular forces similar to those experienced by other carbonyl-containing compounds. Carboxylic acid derivatives all demonstrate dipole-dipole interactions and have the capacity to act as hydrogen bond acceptors. Chapter 11 covers the physical properties, reactivity, and reactions of carboxylic acid derivatives.

2.8.03 Nitrogen-Containing Groups

The nitrogen-containing functional groups include the amines, amides, and nitriles (Figure 2.46) and are relevant to several important classes of biological molecules (eg, amino acids and nucleic acids).

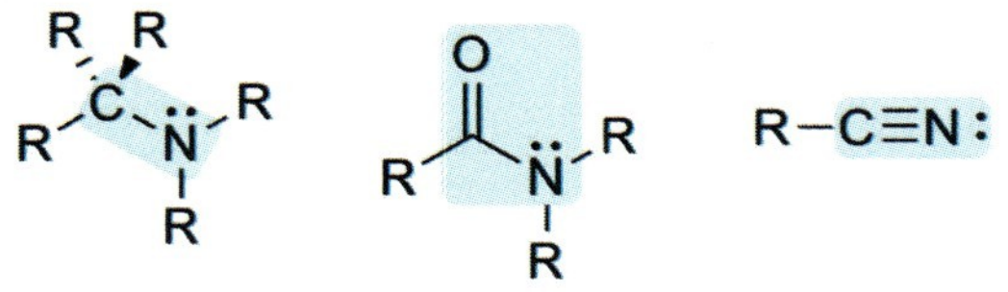

R–C≡N:

Amine Amide Nitrile

*R = H, alkyl

Figure 2.46 General structure of the nitrogen-containing functional groups.

Amines contain a nitrogen atom with a lone pair of electrons and three σ bonds to either alkyl groups or hydrogen atoms (see Concept 12.1.02). The lone pair on an amine nitrogen atom is a hydrogen bond acceptor; if the nitrogen is bonded to at least one hydrogen atom, an amine is also a hydrogen bond donor. The lone pair of an amine may also be capable of acting as a **base** (see Lesson 5.1).

As discussed in the Concept 2.8.02, **amides** are carboxylic acid derivatives and have the general formula R–C(O)–N(R′)$_2$.

Nitriles contain a **cyano group** (a carbon-nitrogen triple bond) at the end of an alkyl chain (R–C≡N). Like the other oxygen-containing and nitrogen-containing functional groups, nitriles are hydrogen bond acceptors and among the most polar classes of organic compounds.

2.8.04 Miscellaneous Groups

Several other important functional groups (Figure 2.47) are relevant to organic chemistry and biochemistry.

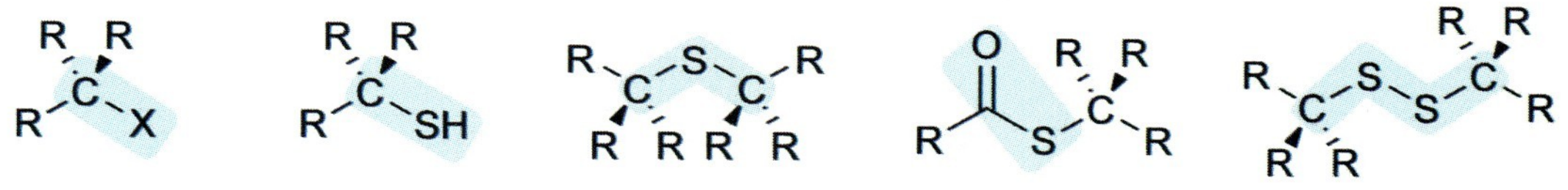

Alkyl halide | Thiol | Thioether | Thioester | Disulfide

*R = H, alkyl
X = F, Cl, Br, I

Figure 2.47 General structure of an alkyl halide and the sulfur-containing functional groups.

Alkyl halides are alkanes in which at least one hydrogen atom has been replaced with a halogen atom.

Given that oxygen and sulfur are both Group 6A elements, there are sulfur variations to a few key oxygen-containing functional groups. **Thiols** are analogous to alcohols and have the general formula R–SH. **Thioethers** are similar in structure to ethers and contain a sulfur atom bonded to two alkyl groups (R–S–R′). These two sulfur-containing functional groups are present in amino acid side chains (Biochemistry Chapter 1). **Thioesters** are comparable to esters but contain a sulfur atom bonded to the carbonyl carbon rather than an oxygen atom (R–C(O)–SR′). Finally, **disulfides** are formed from two thiols and have the general formula R–S–S–R′. Disulfide linkages are frequently present in the structure of proteins (see Biochemistry Chapter 2).

Concept Check 2.10

Consider the following condensed structural formulas and line-angle formulas. Identify which formula(s) contain a ketone functional group and which formula(s) contain an amine functional group.

1)

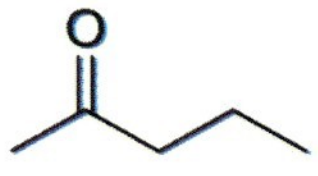

2)

3)

4) $CH_3(CH_2)_3CN$

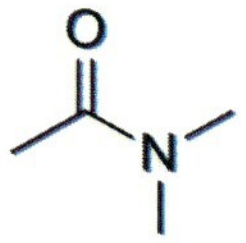

5) $CH_3CH_2O(CH_2)_2COOCH_3$

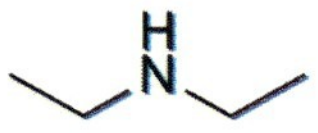

6) $CH_3NH(CH_2)_2COCH_2CH_3$

Solution

Note: The appendix contains the answer.

Lesson 3.1

Constitutional Isomers

Introduction

Isomers are molecules with the *same* molecular formula but *different* structures. Isomers are classified based on the connectivity and spatial orientation of the atoms in the molecular structure. The three main classes of isomers are **constitutional isomers** (Lesson 3.1), **conformational isomers** (Lesson 3.2), and **stereoisomers** (Lesson 3.3). Figure 3.1 outlines the main classes of isomers and how they relate to one another.

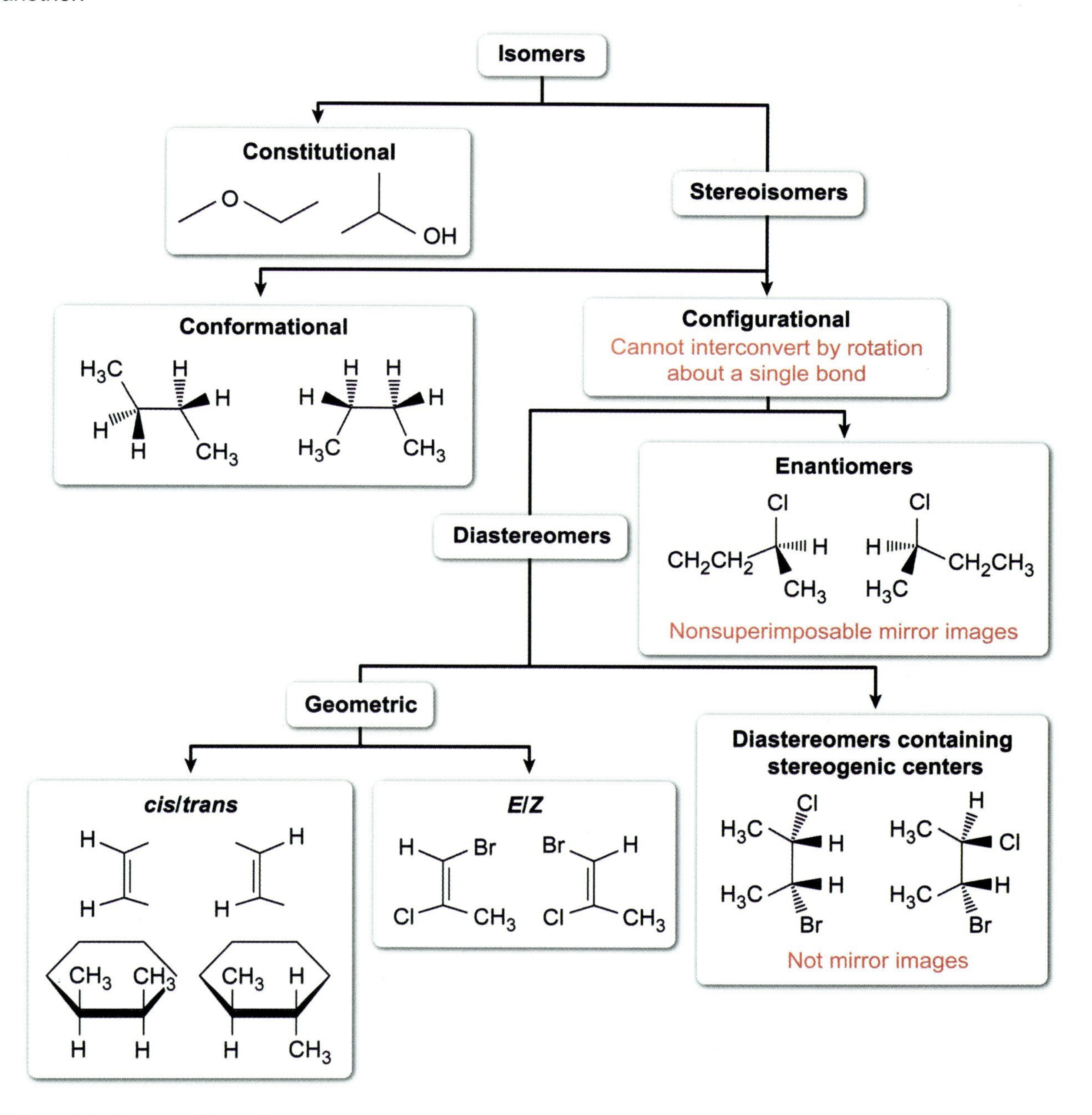

Figure 3.1 Classes of isomers.

3.1.01 Classes of Constitutional Isomers

Constitutional isomers, or **structural isomers**, are molecules that have the *same* molecular formula but *different* connectivity between their atoms. The main types of constitutional isomers are skeletal isomers, positional isomers, and functional group isomers.

Skeletal isomers have different arrangements of their carbon skeletons (eg, a linear hydrocarbon versus a branched hydrocarbon). Figure 3.2 illustrates that a hydrocarbon with the molecular formula C_5H_{12} has three possible carbon skeletons. The number of possible carbon skeletons increases exponentially as the number of carbon atoms increases. Changes to the carbon skeleton also impact the name of the molecule (ie, its nomenclature, see Chapter 4).

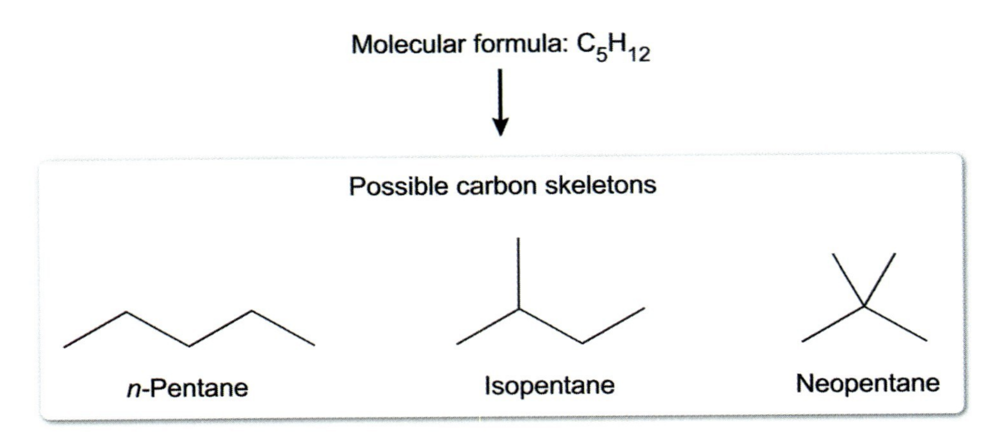

Figure 3.2 Skeletal isomers of C_5H_{12}.

Positional isomers have functional groups in different locations (or positions) within a molecule. Figure 3.3 shows two examples of positional isomerism: one in which the carbons that different functional groups (a hydroxy group and a carbonyl) are attached to change and one in which the position of a pi bond changes. Because functional groups location is indicated in the name of a molecule, positional isomers usually differ in the numbers present in their chemical names (see Chapter 4).

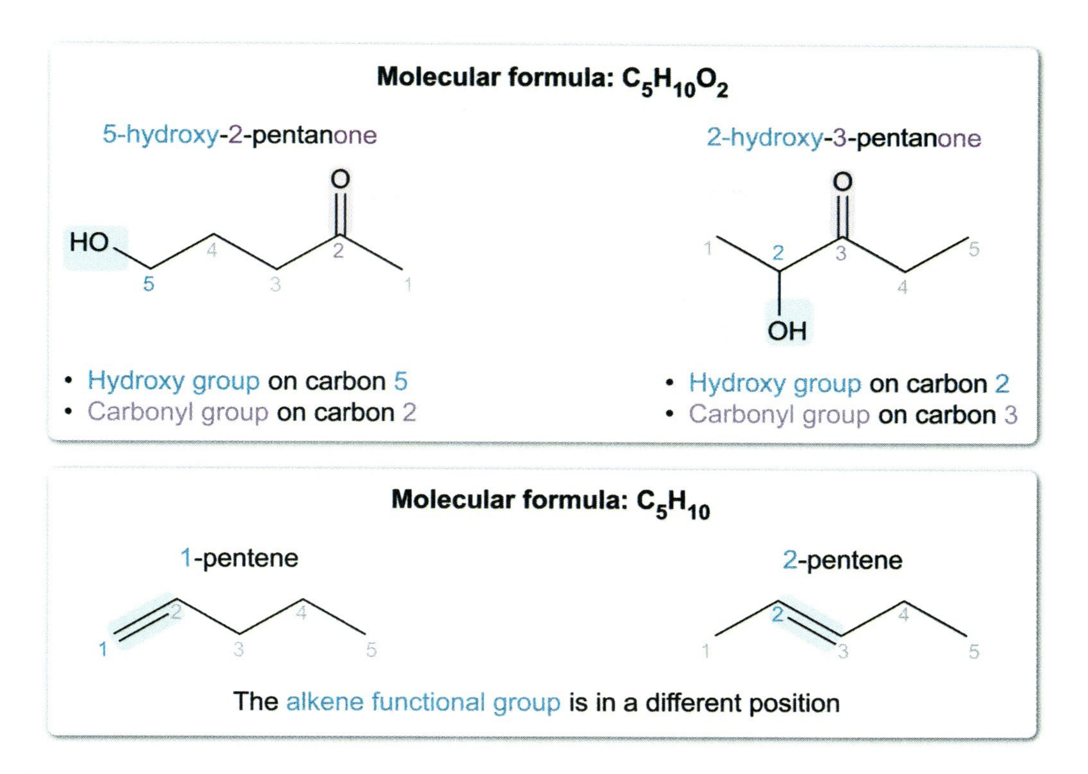

Figure 3.3 Examples of positional isomers.

Because functional group position affects carbon numbering (Concept 4.1.03), nomenclature guidelines should be used to number the carbon chain *before* determining whether two structures are positional isomers. Even if the functional group appears to be in a different location in a given structural depiction, the structures are the *same* molecule as long as the molecules have the same systematic name.

For example, Figure 3.4 shows four-carbon chains with the bromine atom on either end of the carbon chain. After numbering, it is evident that the two structures are the *same* molecule: 1-bromobutane. In contrast, the structure with the bromine atom in the middle of the carbon chain (2-bromobutane) *is* a positional isomer of 1-bromobutane, as evidenced by the difference in the bromine's numerical position.

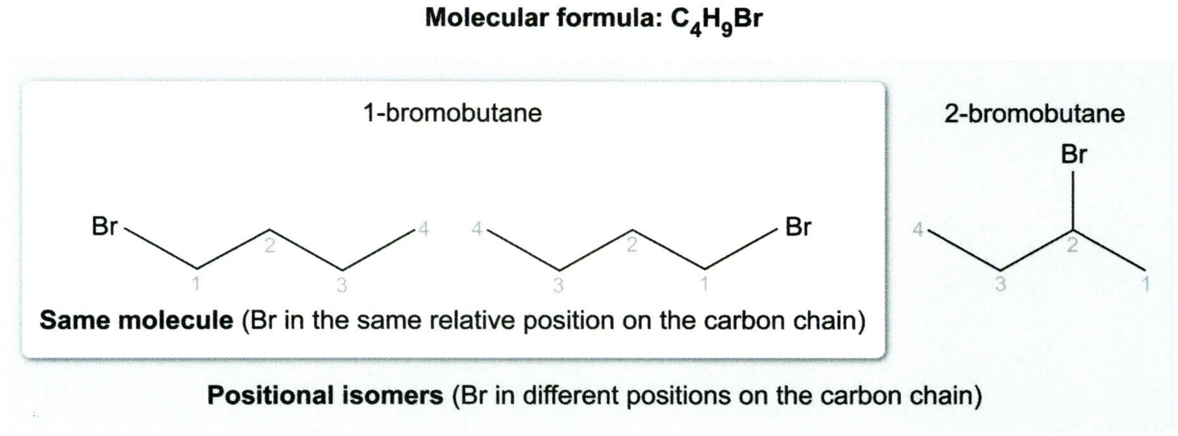

Molecular formula: C_4H_9Br

1-bromobutane

2-bromobutane

Br Br Br

Same molecule (Br in the same relative position on the carbon chain)

Positional isomers (Br in different positions on the carbon chain)

Figure 3.4 Identifying that two structures are the same molecule and not positional isomers.

Functional group isomers are molecules with the same molecular formula and local bonding environments but *different* functional groups. One way to identify functional group isomers is to consider a molecule as being composed of molecular blocks. Although the identity of the blocks does not change, changing the *order* of the blocks may lead to the presence of *different functional groups*. Some common examples of functional group isomers include a ketone and an aldehyde, or an ether and an alcohol (Figure 3.5).

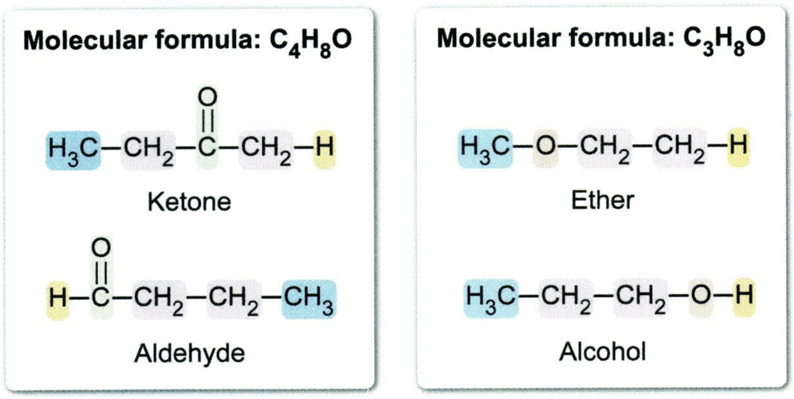

Molecular formula: C_4H_8O

$H_3C{-}CH_2{-}C({=}O){-}CH_2{-}H$

Ketone

$H{-}C({=}O){-}CH_2{-}CH_2{-}CH_3$

Aldehyde

Molecular formula: C_3H_8O

$H_3C{-}O{-}CH_2{-}CH_2{-}H$

Ether

$H_3C{-}CH_2{-}CH_2{-}O{-}H$

Alcohol

Figure 3.5 Examples of functional group isomer pairings.

Because constitutional isomers are *different* molecules, they have ***different*** **physical properties** (eg, melting point, boiling point, density, solubility) (see Lesson 2.4) based on their functional groups. The physical properties of individual functional groups are covered throughout Unit 2.

Concept Check 3.1

Identify whether each of the following pairs of molecules are skeletal isomers, positional isomers, functional group isomers, the same molecule, or non-isomers.

1) 

2) 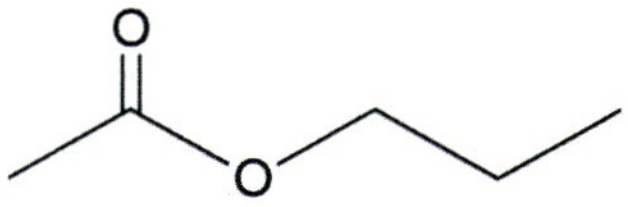 O O; O OH

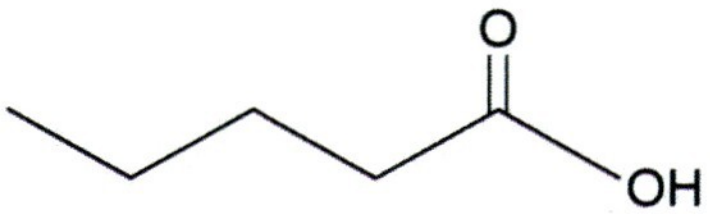

3)

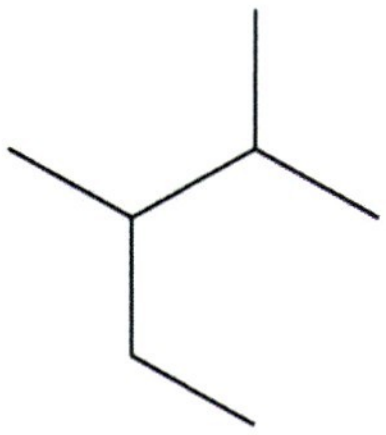

4)  H_2N; NH_2

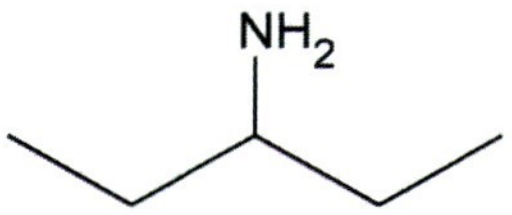

5) 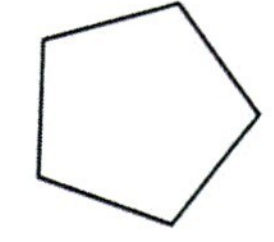 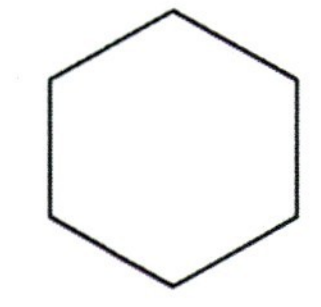

Solution

Note: The appendix contains the answer.

Lesson 3.2

Conformational Isomers

Introduction

Isomers that have an *identical connectivity* of atoms but *different* three-dimensional *orientations* of atoms are broadly classified as **stereoisomers**. **Conformational isomers** are stereoisomers formed through the *rotation* of atoms relative to the axis of a single bond, while **configurational isomers** are stereoisomers that differ in *rigid* orientations of atoms in three-dimensional space. This lesson focuses on topics related to *conformational isomers*; configurational isomers are discussed in greater detail in Lesson 3.3.

According to molecular orbital theory, all single bonds result from the head-to-head overlap of orbitals on each atom (ie, a sigma [σ] bond). In most cases, the atoms joined by *only* a σ bond are *not* rigidly fixed; instead, the atoms are capable of rotating with respect to one another along the σ bond (Figure 3.6). This produces an array of **conformations**, and any one conformation is called a **conformer**.

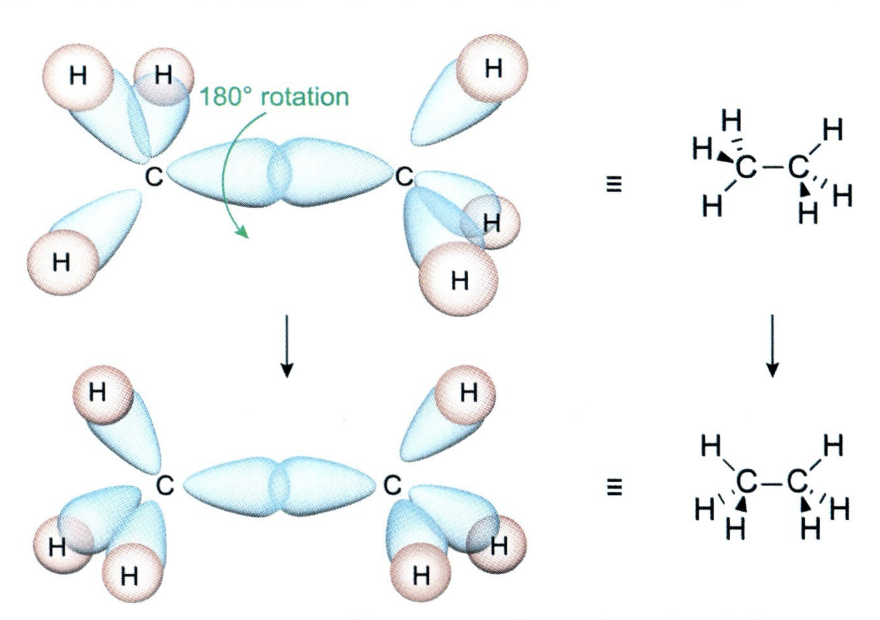

Figure 3.6 Conformations.

This lesson discusses how to draw molecular conformations, the types of relative relationships that exist among conformers, conformational analysis, and the conformations of cycloalkanes.

3.2.01 Newman Projections

Although molecular conformations of simple molecules can be depicted using wedges and dashes in a line-angle format, such illustrations quickly become challenging with larger molecules. The most common tool for representing chemical conformations is a **Newman projection**, which provides the ability to draw and analyze three-dimensional conformations using two dimensions.

The first step in drawing a Newman projection is to identify the bond to be analyzed and a viewpoint from which to look down the identified bond (Figure 3.7). A Newman projection represents the **front atom** (ie, the atom *closer* to the viewer) with a point. The **back atom**, by contrast, is represented with a large circle.

Newman projections depict the spatial relationship of R groups during conformational rotation about a sigma (single) bond. The front atom of the analyzed bond is represented by a point, and the back atom is represented by a large circle.

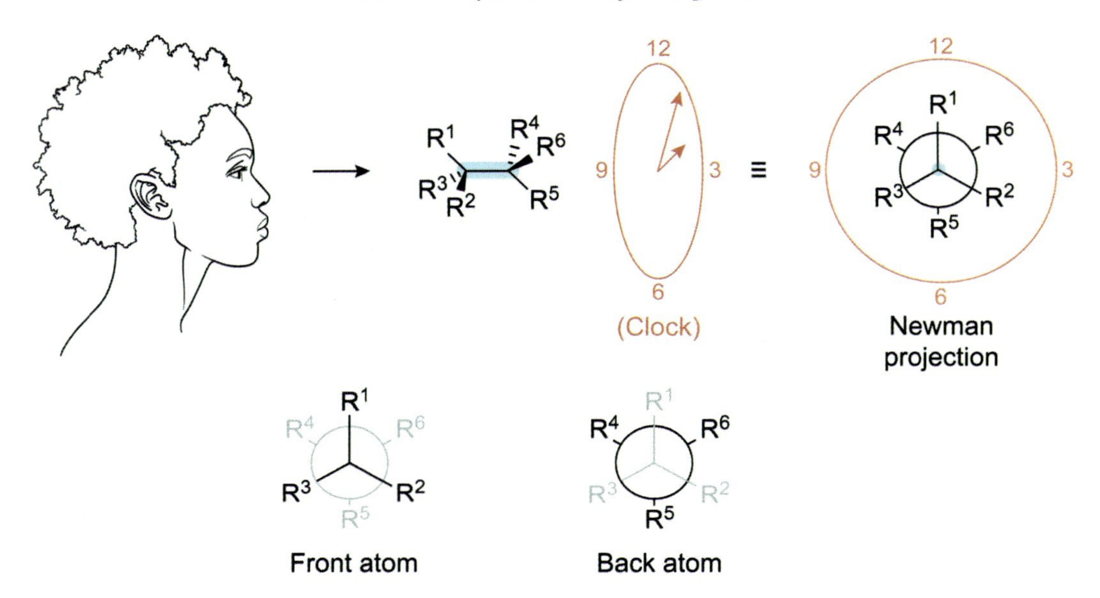

Figure 3.7 Newman projections.

The groups on the front atom connect to the center point, while the groups on the back atom end on the large circle. In both cases, the lines on a given atom are spaced evenly around the Newman projection. If either the front atom or the back atom contains four different groups (including the other group along the bond), it may be classified as a stereocenter, and the relative placement of groups around that atom is critical (see Lesson 3.3). In these cases, it can be helpful to envision the viewpoint as looking toward a clockface, as in Figure 3.7.

A given Newman projection represents one conformation of many. From this initial conformation, the groups on *either* the front or back atom can be rotated to generate a panel of conformations. Conformational analysis is discussed in greater detail in Concept 3.2.03.

Concept Check 3.2

Draw the line-angle structure of the hydrocarbon represented by the following Newman projection.

CH_3
$(H_3C)_2HC$ H
H CH_3
H

Solution

Note: The appendix contains the answer.

3.2.02 Conformational Relationships

An advantage of Newman projections is their ability to depict the spatial relationships between groups on the front atom and groups on the back atom. A **dihedral angle (θ)** describes the *rotational angle* between the bond for a chosen group on the front atom relative to the bond for a chosen group on the back atom (Figure 3.8). More generally, a dihedral angle describes the geometric relationship between *any* two groups that are separated by exactly three bonds.

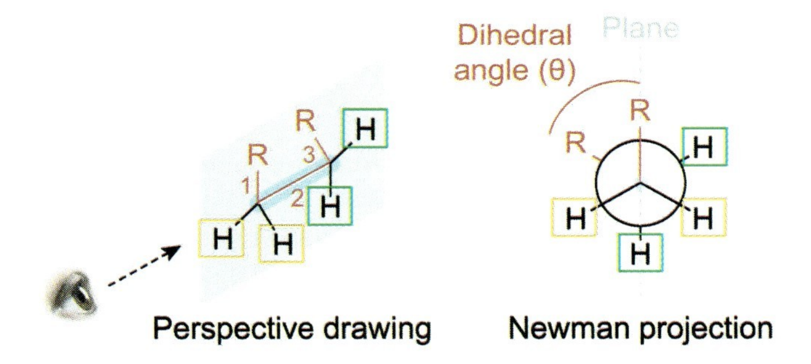

Figure 3.8 Dihedral angle.

Figure 3.9 shows some particularly important conformational relationships. In an **eclipsed conformation** the bonds are perfectly aligned between the front and back atom, and in a **staggered conformation** bonds on the front and back atom are evenly spaced (θ = 60° between all groups). As discussed in Concept 3.2.03, a molecule's eclipsed conformations always have higher energy than its staggered conformations. A **skew conformation** refers to any conformation that is neither eclipsed nor staggered.

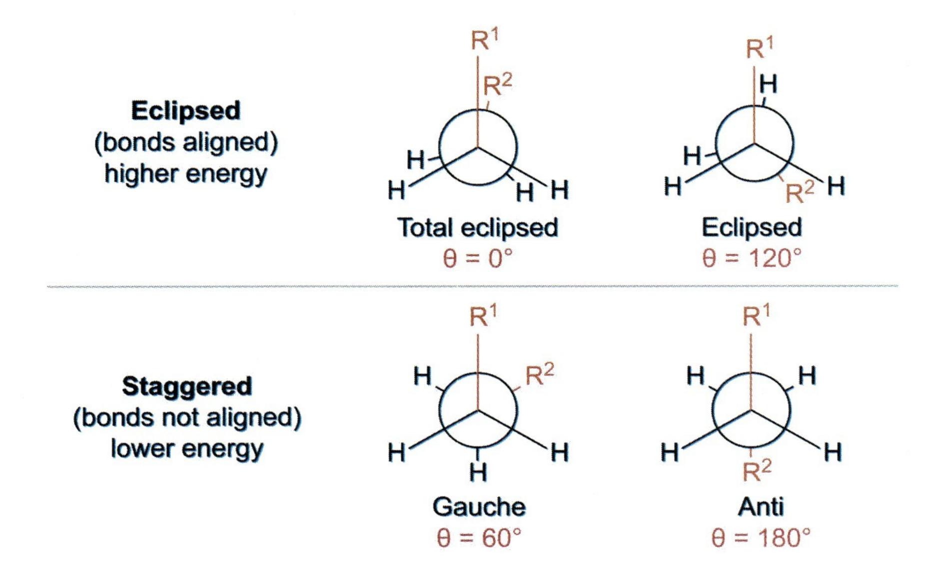

Figure 3.9 Conformational relationships in Newman projections.

A **total eclipsed conformation** is a specific eclipsed conformation in which two chosen groups (eg, R^1 and R^2) have a dihedral angle of 0°. Although the two groups can be any two groups chosen by the analyzer, the term most commonly refers to the *largest* groups on the front and back atoms, respectively. Among the staggered conformations, a **gauche conformation** has the two groups with a dihedral angle of ±60°, and an **anti conformation** has the two groups on opposite sides (θ = 180°).

3.2.03 Conformational Analysis

Conformational analysis refers to processes that compare a compound's possible conformations in order to draw conclusions about its properties or reactivity. The most commonly analyzed feature is the energies of various conformations. Because conformations are generated through the rotation (or *torsion*) of a sigma bond, conformational energy is known as **torsional energy** (or **torsional strain**). For convenience, conformational analysis frequently focuses on six conformers with dihedral angles that differ in 60° increments (Figure 3.10).

Figure 3.10 illustrates the torsional energy for the rotation of the carbon-carbon bond in ethane (CH_3CH_3). The torsional energy plot demonstrates a sinusoidal pattern in which the torsional energy is highest for eclipsed conformations and lowest for staggered conformations, a generalization that applies to nearly all conformational relationships. Consequently, it can be concluded that the relative torsional energy of a conformation is proportional to the number and type of eclipsed bonds in its Newman projection.

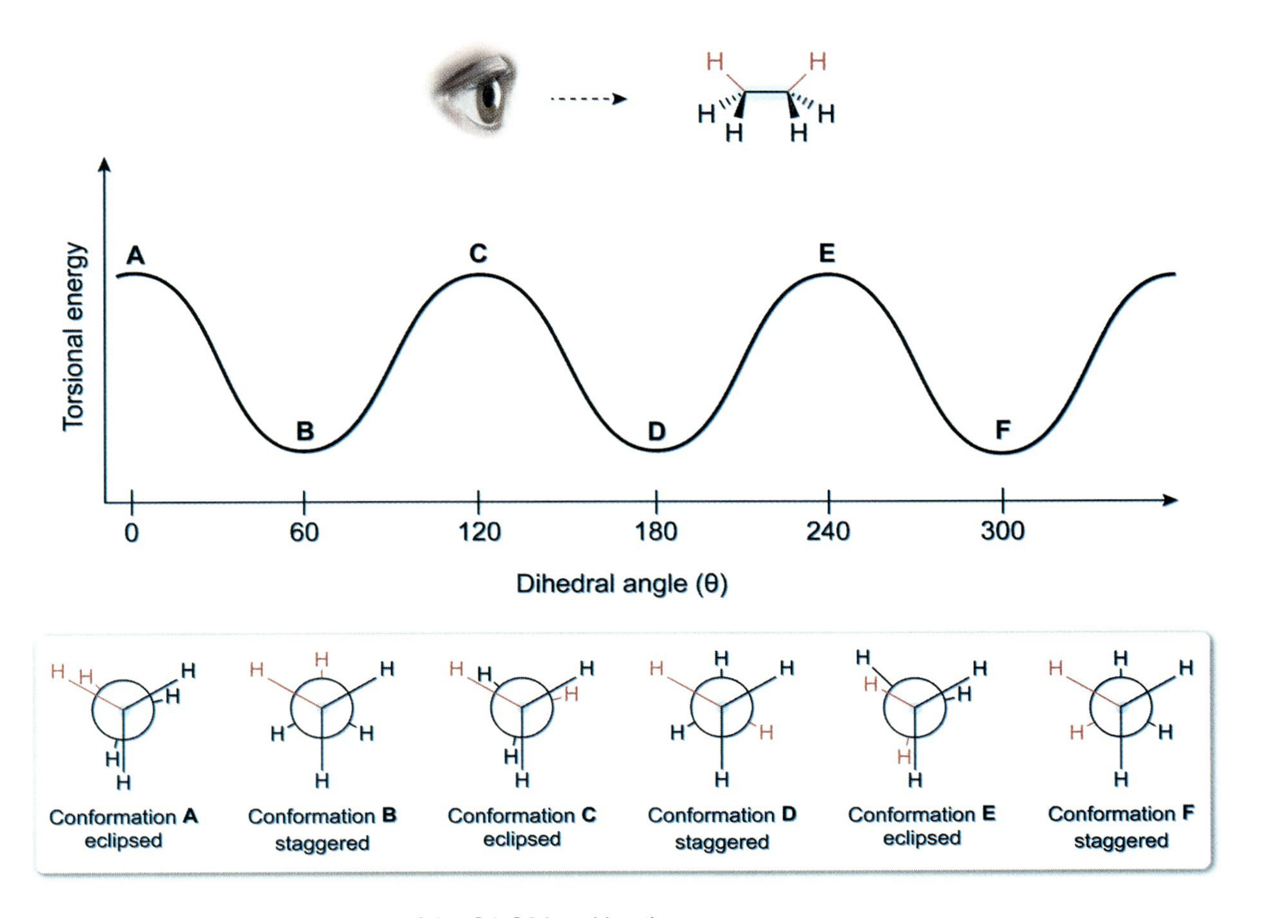

Figure 3.10 Conformational analysis of the C1-C2 bond in ethane.

At the atomic level, torsional strain is a manifestation of **steric strain (steric hindrance)** that takes place when the electron orbitals of nearby atoms are brought *too close* together in space. In general, the amount of steric strain is proportional to the *sizes* of the nearby groups (ie, strain is affected by the number and types of atoms in the group as well as its electron orbital surface area). For example, steric strain increases with the number of carbon atoms in an alkyl group (eg, methyl [$-CH_3$] < ethyl [$-CH_2CH_3$] < propyl [$-CH_2CH_2CH_3$] < butyl [$-CH_2CH_2CH_2CH_3$]).

A more complicated scenario is observed when analyzing the conformations of butane ($CH_3CH_2CH_2CH_3$), viewed down the middle carbon-carbon bond (Figure 3.11). While the torsional energy plot again demonstrates that eclipsed conformations have the highest energy (peaks) and staggered conformations have the lowest energy (valleys), the peaks and valleys do *not* have uniform magnitudes. Conformation A has the highest torsional energy due to steric strain that arises from the methyl groups with total eclipsed conformation. By comparison, eclipsed Conformations C and E have identical and *lower* torsional energies relative to Conformation A, since each methyl group is eclipsed with a smaller hydrogen atom.

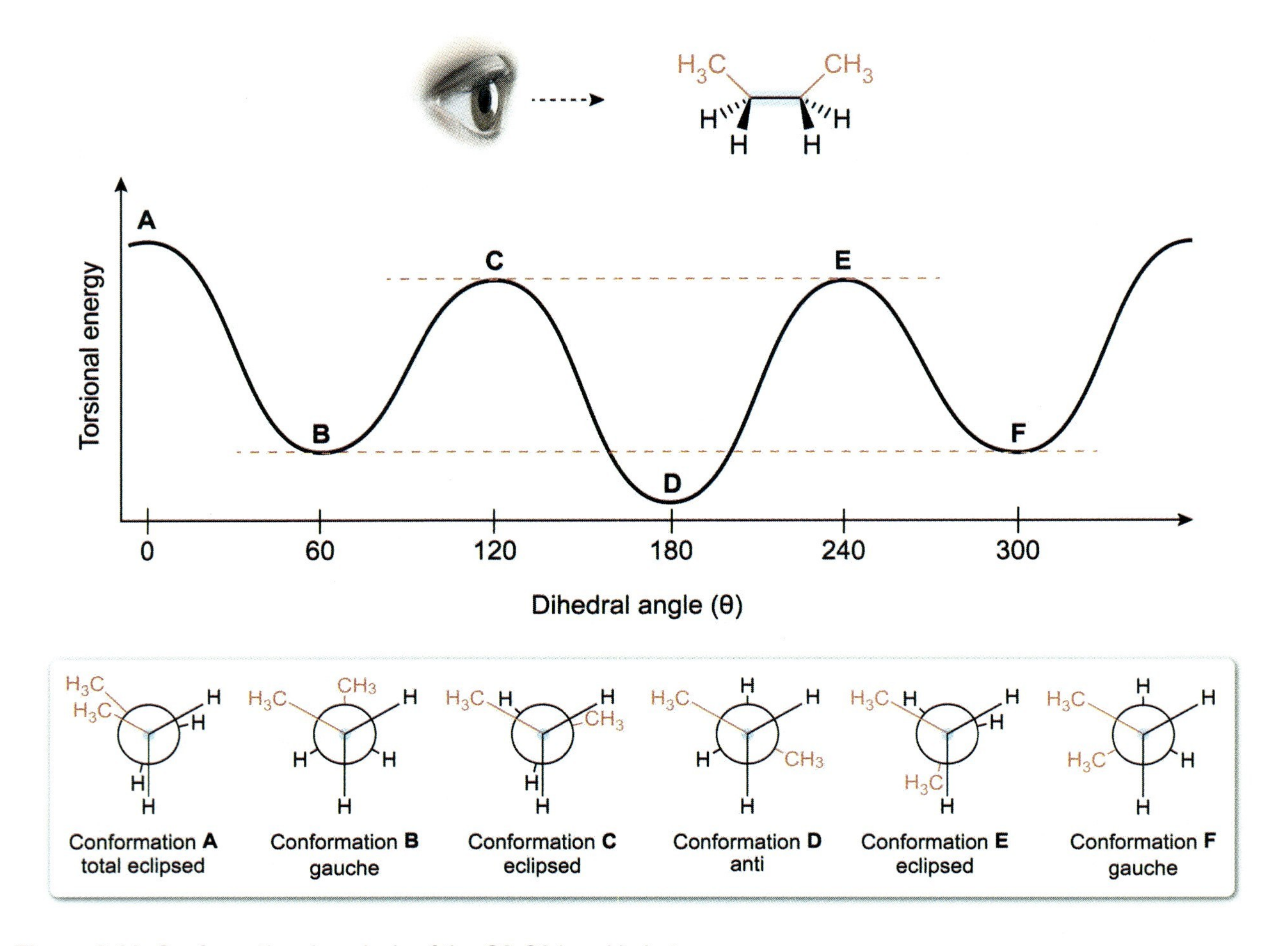

Figure 3.11 Conformational analysis of the C2-C3 bond in butane.

Of the staggered conformations (B, D, and F), Conformation D has the lowest torsional energy due to the methyl groups being positioned as far apart as possible (anti conformation). By comparison, the gauche Conformations B and F have identical torsional energies that are slightly larger than the torsional energy of Conformation D. This is because the electron clouds of the methyl groups are still close enough together to generate a small amount of torsional strain. Consequently, anti conformations are generally the lowest-energy conformation for a given bond.

Concept Check 3.3

If conformational analysis is performed for the indicated bond at 60° intervals, what percentage of the conformations have at least one gauche interaction that contributes to the conformation's total torsional strain?

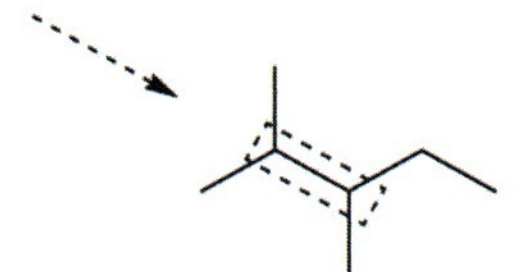

Solution

Note: The appendix contains the answer.

3.2.04 Conformations of Cycloalkanes

Although alkanes contain only carbon-carbon and carbon-hydrogen single bonds, which are typically poorly reactive (see Chapter 6), cycloalkanes containing between 3 and 6 carbon atoms have *widely* different stabilities and therefore reactivities (Figure 3.12). In this chain-length range, cycloalkane reactivity is inversely related to the number of carbon atoms: the three-carbon cycloalkane (**cyclopropane**) is *exceptionally* reactive for a hydrocarbon, while a six-carbon cycloalkane (**cyclohexane**) is especially stable. For these compounds, the reason for the difference in chemical reactivity is due to molecular geometry (bond angles) and conformation (torsional strain).

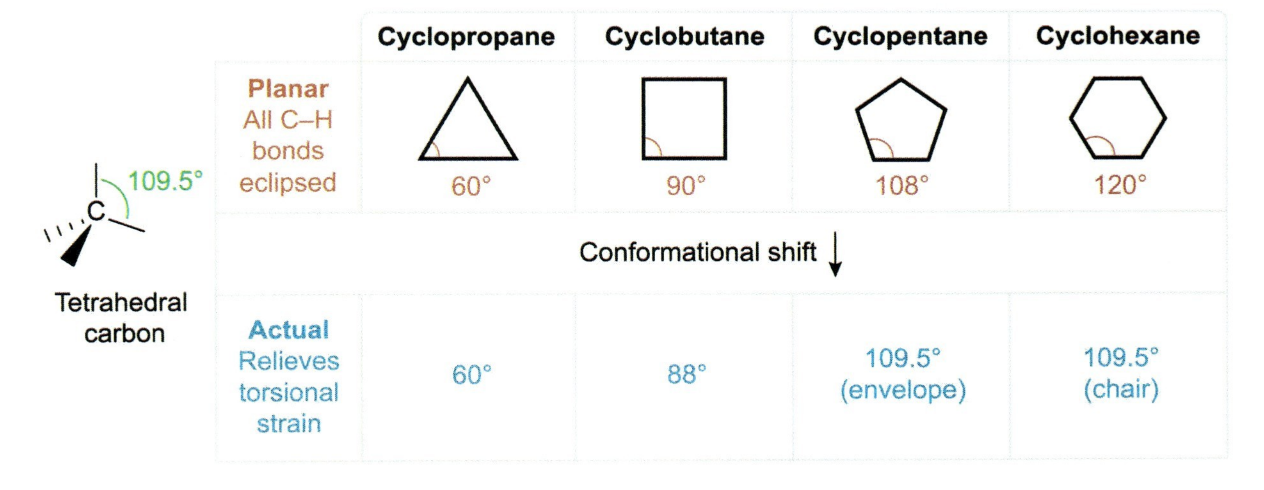

Figure 3.12 Bond angles in cycloalkanes.

Alkane carbon atoms each have sp^3 hybridization (ie, tetrahedral shape and a preferred bond angle of 109.5°). If the cycloalkane carbon atoms were all coplanar (ie, present in one plane), their geometric bond angles would vary from 60° to 120° depending on the number of carbon atoms (Figure 3.12). Particularly for smaller ring sizes, the deviation from an ideal 109.5° bond angle (**angle strain**) is especially unfavorable and is a significant factor in increased reactivity (which breaks open the ring to relieve angle strain). The bond angle compression in cyclopropane is so severe that the carbon-carbon sigma bond overlap occurs *outside* the line connecting the nuclei. These bonds are often called **banana bonds** and are much weaker than a standard carbon-carbon sigma bond (Figure 3.13).

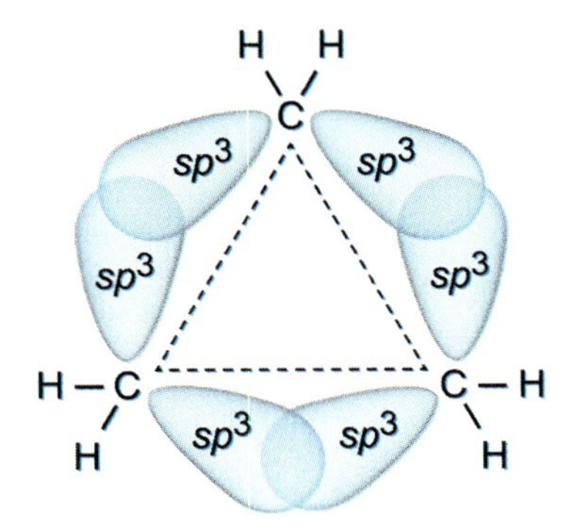

The sigma bond orbital overlap in cyclopropane deviates from a head-to-head relationship, leading to weaker banana bonds.

Figure 3.13 Banana bonds in cyclopropane.

A second factor that influences cycloalkane ring reactivity is molecular conformation. In a planar scenario, every C–H bond is eclipsed with C–H bonds on adjacent carbon atoms, resulting in *significant* torsional strain. Although rotation could reduce the amount of torsional strain (Concept 3.2.03), cyclopropane has only three ring carbon atoms and is *unable* to adopt other conformations. Therefore, cyclopropane is rigidly locked in an eclipsed conformation (Figure 3.14). In contrast, a planar cyclohexane molecule *can* rotate its sigma bonds, bringing its groups into a lower-energy, staggered relationship (Concept 3.2.05).

H2 H4
H1 H3
Cyclopropane
≡
H_2C
H3
H2 H4 H1
Newman projection
C–H are permanently eclipsed

H1 H5
H3
H7
H2
H6
H4 H8
Cyclohexane
≡
H1 H3
H5 H7
H2 H4
H6 H8
Newman projection
C–H bonds are staggered

Figure 3.14 Newman projections of cyclopropane and cyclohexane.

The combination of angle strain and torsional strain is called **ring strain**, which is the principal factor in the difference in cycloalkane chemical reactivity. Relative to planar cycloalkanes, each molecule independently adjusts its angle strain and torsional strain to minimize total ring strain.

As discussed, cyclopropane is unable to adopt alternate conformations, leaving its carbons rigidly held in a planar shape (see Figure 3.14). Cyclobutane adopts a slight pucker, causing a small *increase* in angle strain (from 90° to 88°) that is offset by a *greater reduction* in torsional strain (C–H bonds move from eclipsed to skew). Cyclopentane adopts a more pronounced pucker (an **envelope conformation**), achieving *nearly* staggered C–H bond relationships with ideal bond angles (109.5°). Cyclohexane also adopts a puckered conformation (a **chair conformation**) with ideal bond angles *and* staggered C–H bond relationships (see Concept 3.2.05). Unsurprisingly, five- and six-membered cycloalkanes are especially stable and the most common ring sizes found in natural products and biochemical systems.

3.2.05 Chair Conformations of Cyclohexanes

The six-membered cyclohexane ring is generally considered the most stable cycloalkane variant primarily due to its ability to adopt a chair conformation (see Concept 3.2.04). In a **chair conformation**, one of the ring atoms is puckered in one direction and its opposing ring atom is puckered in the opposite direction (see Figure 3.15). The puckered atoms are often referenced as either the "headrest" or the "footrest," relative to the square "seat" of the central four atoms: a **half-chair conformation** has only one puckered atom alongside five coplanar carbon atoms, and a **boat conformation** has both puckered atoms oriented in the same direction.

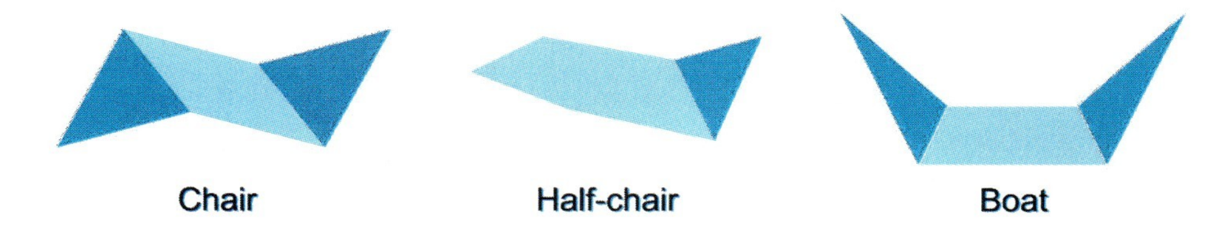

Figure 3.15 Common conformations of a cyclohexane ring.

In a chair conformation, two bonds from each carbon combine to form the core ring. The remaining two bonds are oriented either axially or equatorially. **Axial positions** are oriented *above and below* the plane of the cyclohexane ring, while **equatorial positions** are *radially* oriented *away* from the ring (Figure 3.16). Identifying whether a position is axial or equatorial is an essential skill when working with chair conformations.

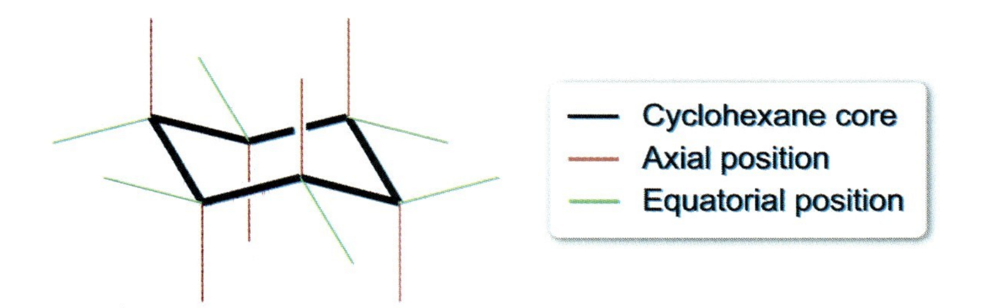

Figure 3.16 Axial and equatorial positions on the chair conformation of a cyclohexane ring.

To draw a chair conformation, draw the seat of the chair, followed by the headrest and footrest atoms (Figure 3.17). The six axial positions (if explicitly depicted) are drawn as vertical lines relative to the core chair structure and alternate between an upward and downward orientation around the ring—the axial position should point up on the headrest and down on the footrest. The six equatorial positions (if explicitly depicted) also alternate upward and downward, but with more angled slopes that match the slope of a non-adjacent ring bond.

1. Draw two slanted parallel lines offset halfway to each other. These form the "seat" of the chair.

2. Draw two lines from each side to form the "headrest" and "footrest" of the chair.

then

Note: The ring bonds have one of three slopes.

Slope **1** Slope **2** Slope **3**

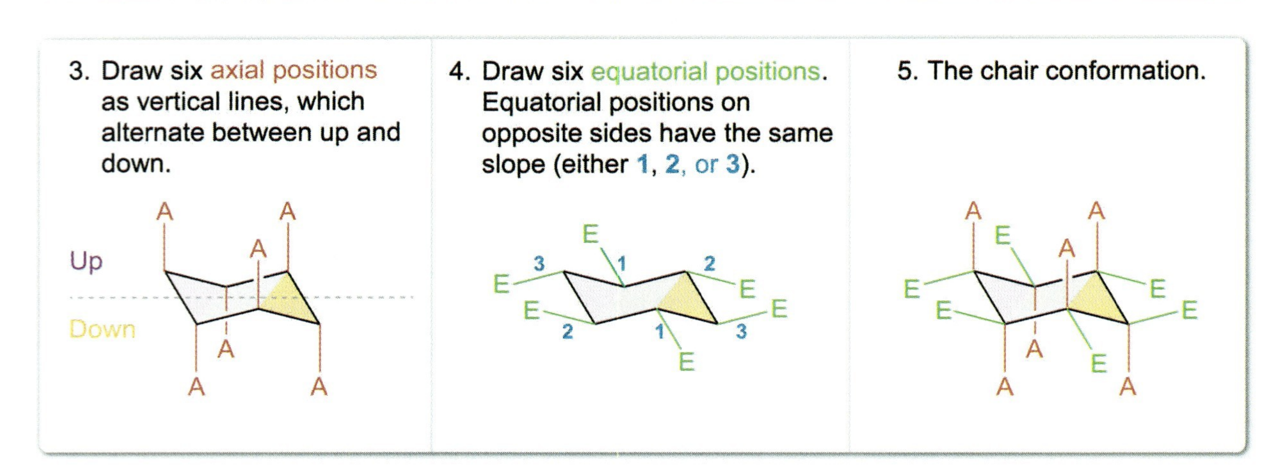

Figure 3.17 How to draw a cyclohexane ring in a chair conformation.

Conformational analysis of a cyclohexane ring is best accomplished with a double Newman projection looking down *two* parallel bonds of interest (Figure 3.18). In between the two projections are the headrest and footrest. The Newman projection of a chair conformation shows a staggered relationship between groups on the front and back atoms, demonstrating lower torsional energy relative to the eclipsed conformation of a planar cyclohexane.

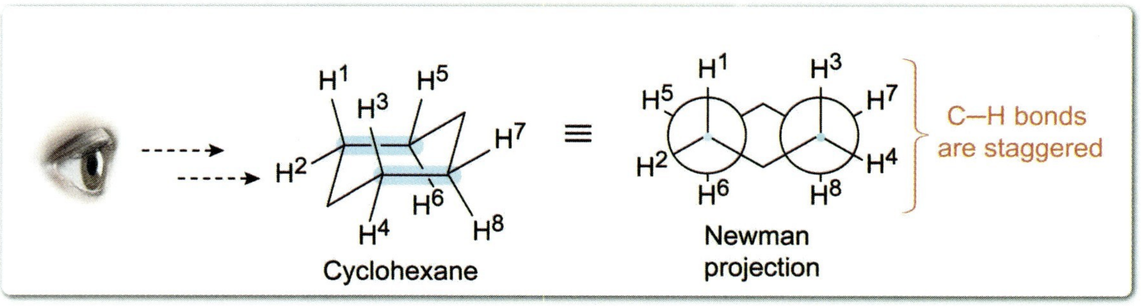

Cyclohexane ≡ Newman projection

C–H bonds are staggered

Figure 3.18 Newman projection of a chair conformation.

Newman projections are also used to evaluate substituted cyclohexane ring conformations with one or more alkyl groups (Figure 3.19). When an alkyl group is in an axial position, a dihedral angle of 60° is formed between the group and the headrest carbon atom (a *gauche* relationship), causing some amount of torsional strain. Furthermore, this also leads to a **1,3-diaxial interaction** (ie, steric strain generated by adjacent axial groups in the 1 and 3 positions).

In contrast, when an alkyl group is in an equatorial position, there is a lower-energy *anti* relationship with the headrest carbon atom and the alkyl group no longer experiences a 1,3-diaxial interaction. Consequently, cyclohexane rings are lower energy (and more stable) when its alkyl groups are in equatorial positions. The difference in torsional energy is proportional to alkyl group size; larger alkyl groups prefer even more strongly to be positioned equatorially.

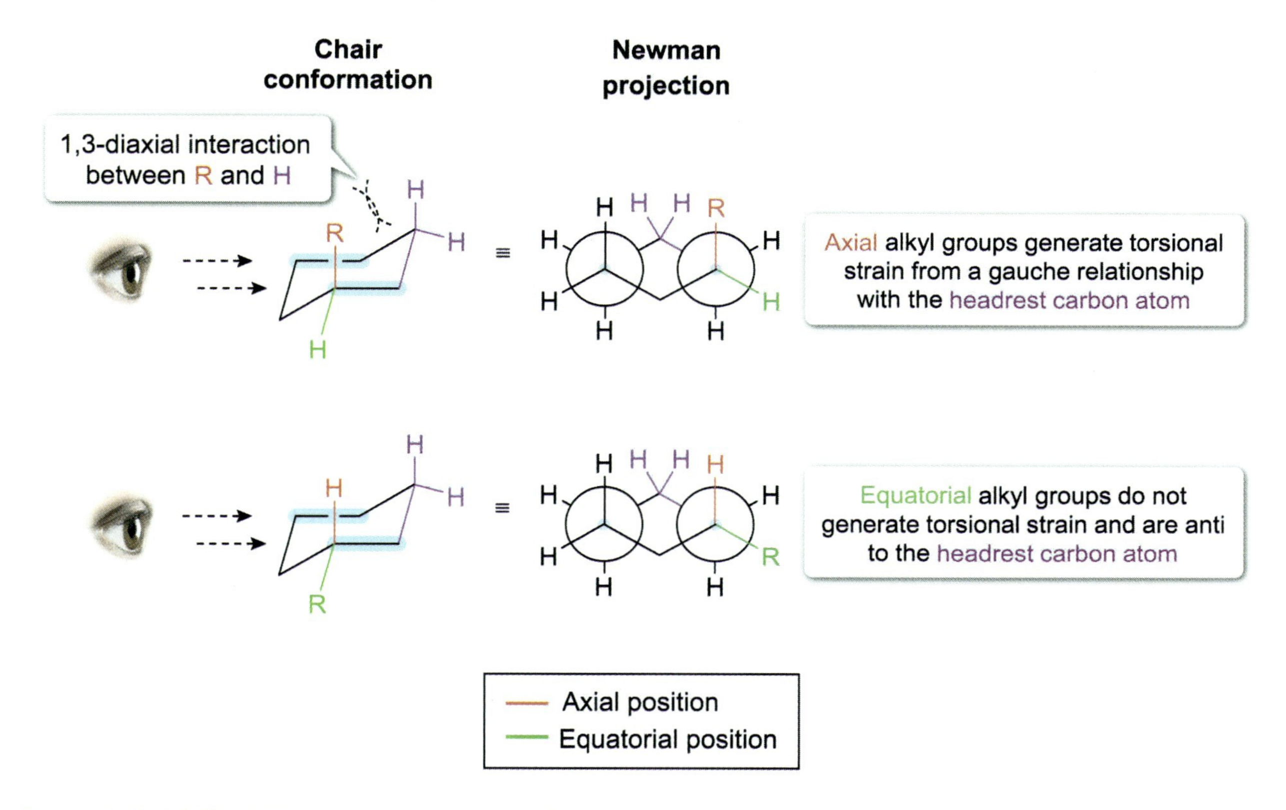

Figure 3.19 1,3-Diaxial interactions generate torsional strain.

Cyclohexane chair conformations can undergo the equilibrium process of a **chair flip**, to interconvert between two related chair conformations. Effectively, a chair flip causes the headrest carbon to become the footrest carbon, and vice versa. The chair flip process proceeds through several key intermediate and transition state conformations (Figure 3.20).

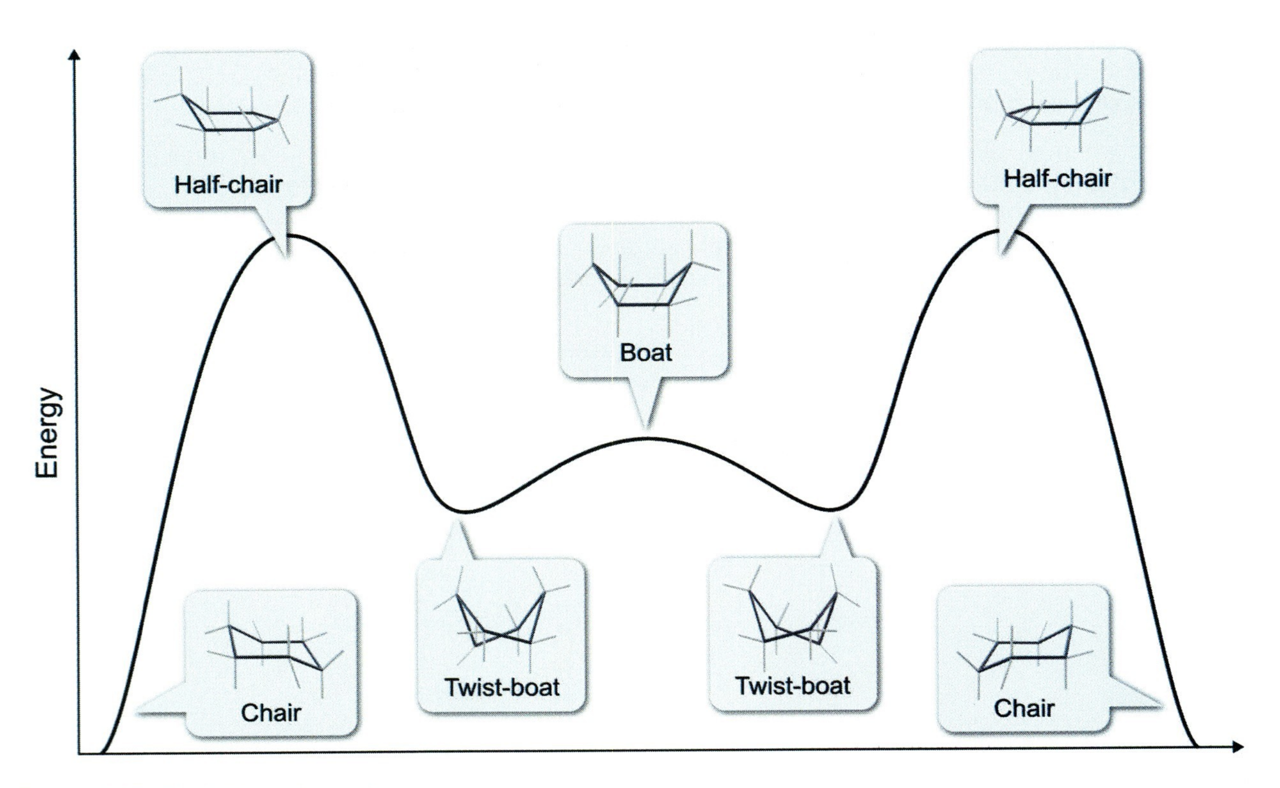

Figure 3.20 Chair flip conformations.

Chair flips *do not change* the relative orientation of groups on ring carbon atoms (ie, a group that is positioned on the top face of the ring remains on the top face). However, chair flips *do* interconvert the axial and equatorial designations for *all* groups (Figure 3.21).

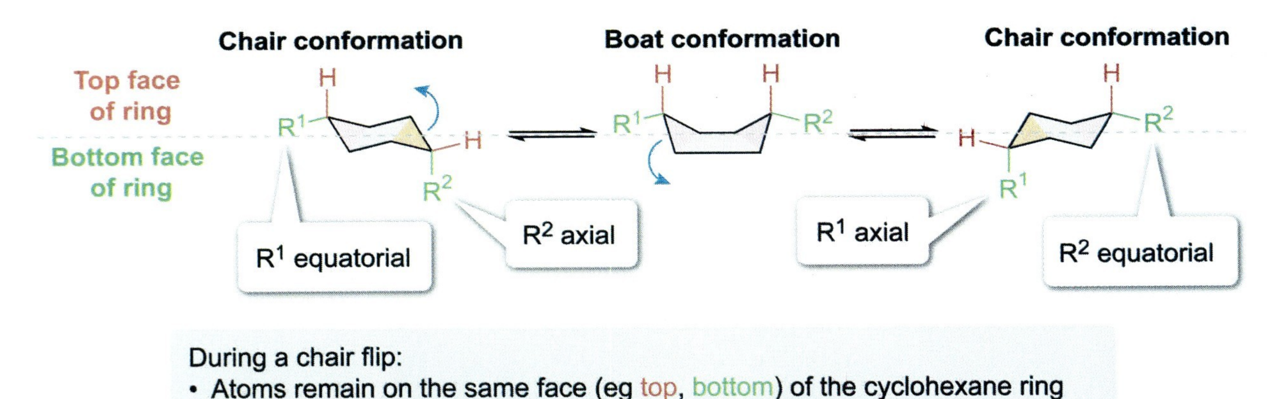

Figure 3.21 The relationship between chair flip and axial/equatorial designations.

Because the chair flip process is a dynamic equilibrium, substituents can interconvert between axial and equatorial positions. However, cyclohexane rings *preferentially* adopt the chair conformation that places both:

- the *greatest number* of alkyl groups in an equatorial position, and
- the *largest* alkyl groups in an equatorial position

Monosubstituted cyclohexane rings *always* prefer placing the alkyl group in an equatorial position. With two (or more) alkyl groups, both chair flip conformers must be evaluated to determine which has the lower torsional energy. In some cases, both conformers may have identical torsional energy, representing a chair flip equilibrium that favors neither one.

When working with chair conformations, interconverting chair conformations with perspective drawings using wedges and dashes is also important. For any ring carbon atom, one of its two groups is oriented above the "plane" of the ring, and one is oriented below (Figure 3.22). Axial and equatorial positions may be *either* above or below the plane and in fact alternate above and below around the ring; consequently, knowing wedge/dash orientation will *not* lead to knowing axial/equatorial positioning without further information (eg, knowing axial/equatorial position of another substituent or inferring the lower-energy conformation).

The interconversion between the two representations relies heavily on skills involving perspective and three-dimensional orientation, which is also relevant to Lesson 3.3.

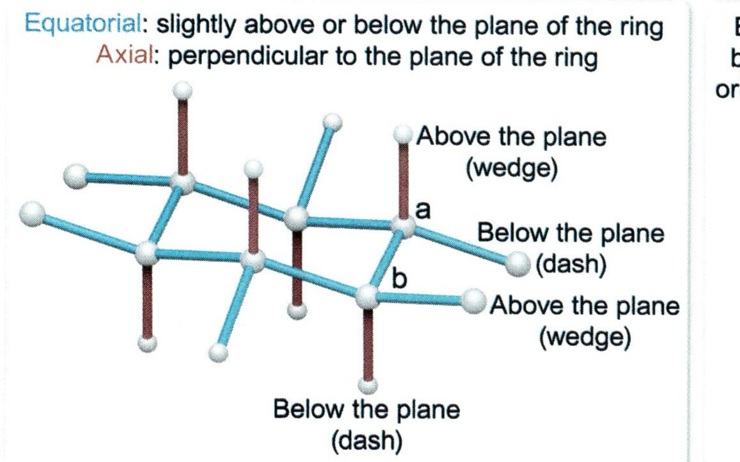

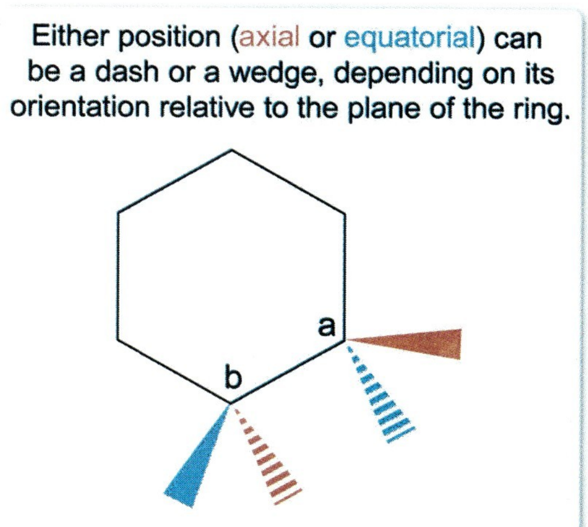

Figure 3.22 Interconversion of chair conformations with perspective drawings.

Concept Check 3.4

Draw the lowest energy chair conformation for the molecule *trans*-1,3-dimethylcyclohexane.

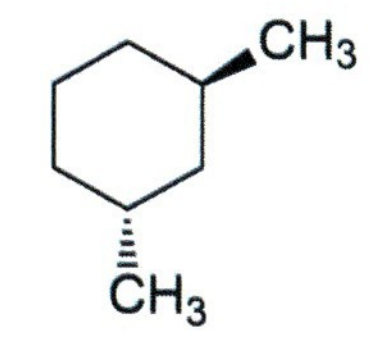

trans-1,3-Dimethylcyclohexane

Solution

Note: The appendix contains the answer.

Lesson 3.3

Stereoisomers

Introduction

As discussed in Lesson 3.2, **stereoisomers** are isomers with the same connectivity between atoms but a *different orientation* of those atoms in three-dimensional space. Lesson 3.2 discussed **conformational isomers** (formed by the rotation of atoms along the axis of a bond). This lesson focuses on **configurational isomers**, covering topics including chirality, stereocenters, stereochemical relationships, and Fischer projections. Identification of a **stereocenter** (Figure 3.23), any atom for which changing the position of two substituents forms a different stereoisomer, is an important skill not only for this lesson but for all organic chemistry.

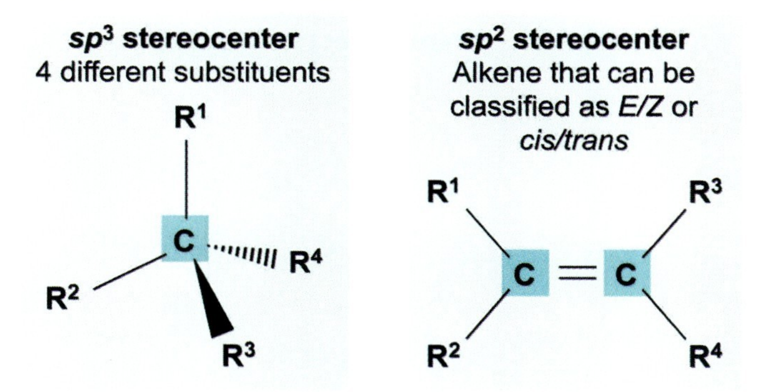

Figure 3.23 Stereocenters.

3.3.01 Chirality and Handedness

Three-dimensional orientation is an important topic in organic chemistry and plays an essential role in the function of medically relevant molecules. Some medications have configurational isomers that display vastly different properties (eg, one isomer is a useful drug while another has harmful side effects), demonstrating why understanding spatial orientation is so important.

All objects have a mirror image (or reflection). Some objects can be superimposed on their mirror image while others cannot. Objects with a **plane of symmetry** (eg, a spoon, a chair) are *superimposable*, meaning they have the *same* three-dimensional spatial orientation as their mirror images, and are called **achiral**. In contrast, the mirror image of a right hand is a left hand. While the two hands may have the same components (ie, four fingers and a thumb), they are *nonsuperimposable* because they do *not* have a plane of symmetry. Such nonsuperimposable objects, which have "handedness," are described as **chiral** (Figure 3.24).

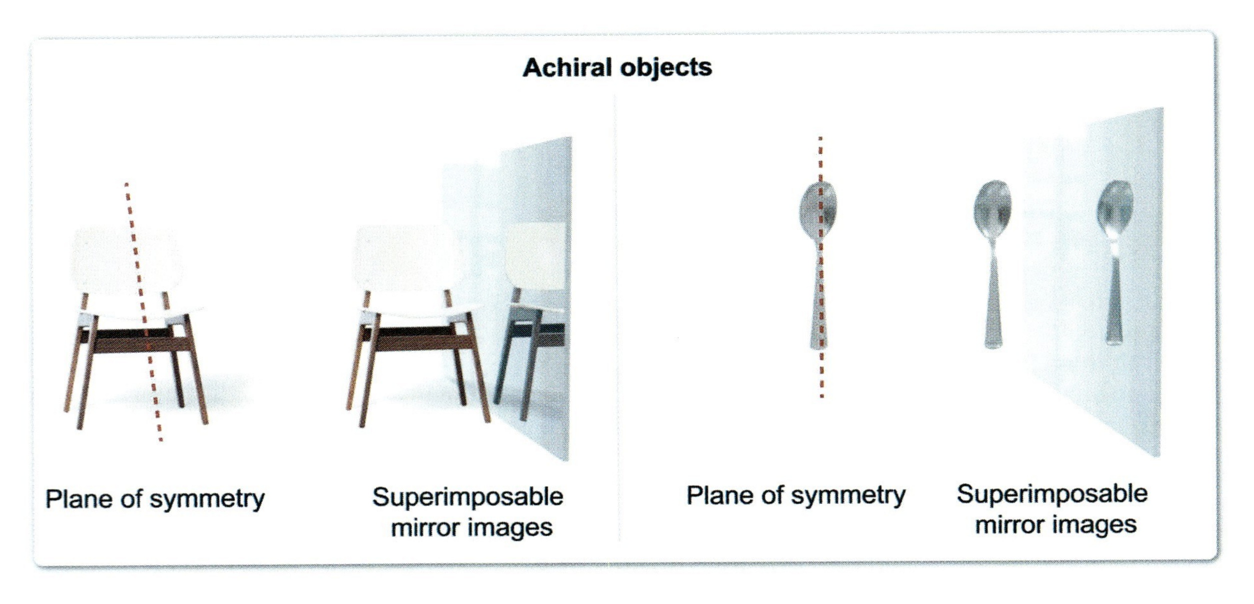

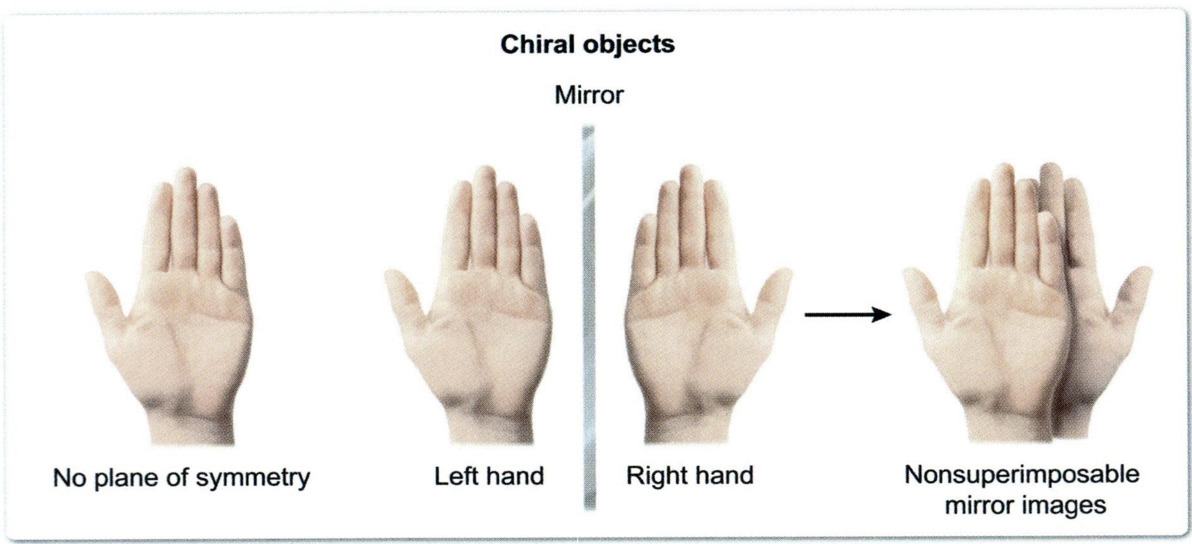

Figure 3.24 Chiral and achiral objects.

Chemical molecules can also be classified as either chiral or achiral. Achiral molecules have a plane of symmetry and have superimposable mirror images, meaning the original molecule and its mirror image are the *same molecule*. Conversely, chiral molecules do *not* have a plane of symmetry and have nonsuperimposable mirror images—the original molecule and its mirror image are *different molecules* (Figure 3.25).

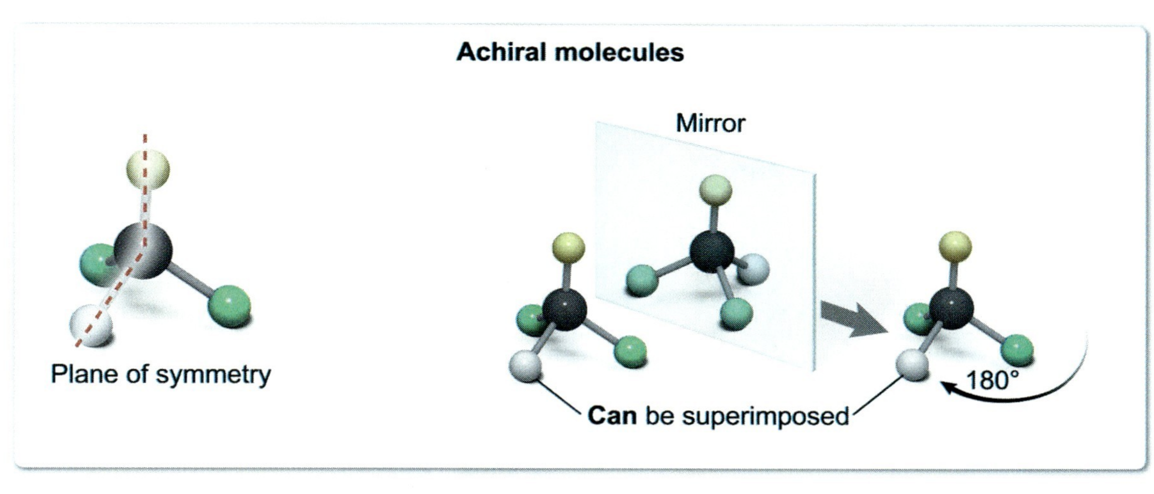

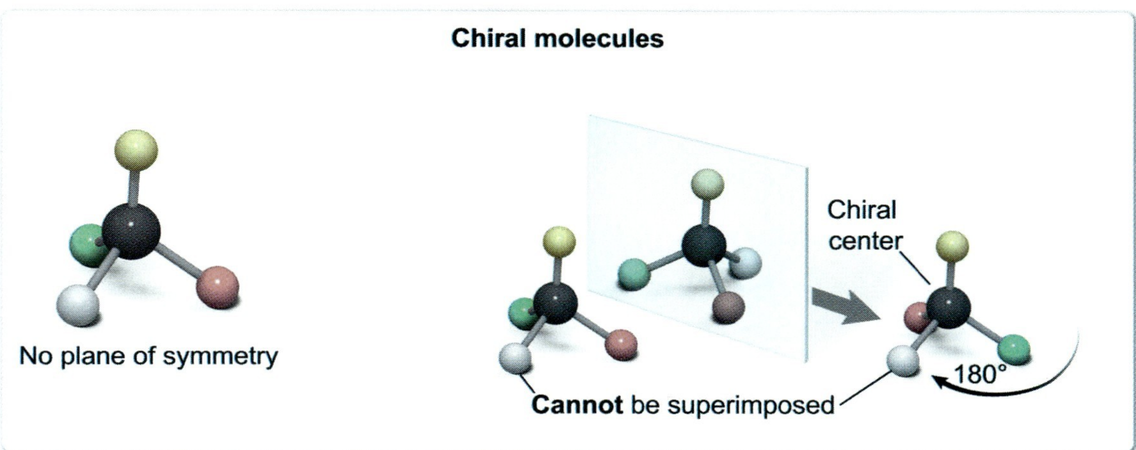

Figure 3.25 Chiral and achiral molecules.

3.3.02 Chiral Carbon Atoms

A common example of chirality is a molecule that contains at least one sp^3 hybridized atom (typically carbon) with *four* different substituents, which is known as a **chiral center** or an **asymmetric carbon** (Figure 3.26). Note that the *entire* substituent must be considered when determining chirality—a methyl ($-CH_3$) and an ethyl ($-CH_2CH_3$) group count as *different* substituents, even if they both connect to a chiral center through an identical atom (ie, carbon).

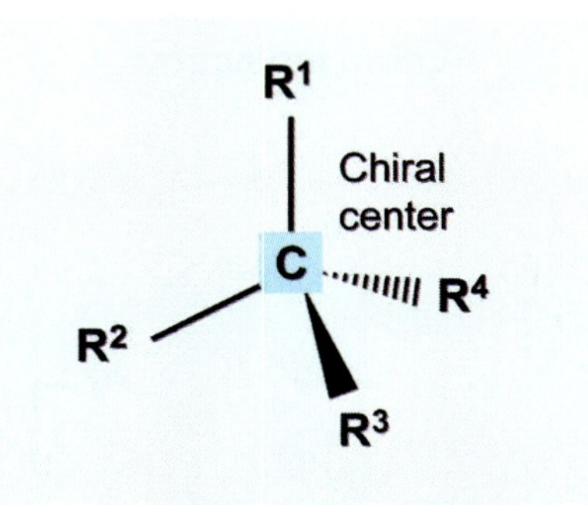

A chiral center is an atom bonded to four different constituents (R^1–R^4) and not superimposable on its mirror image

Figure 3.26 Chiral carbon.

The **number of stereoisomers** of a molecule is related to the number of stereocenters in the compound. (Recall that a stereocenter is an atom for which the switching of two of its substituents will form a different stereoisomer.) The *maximum* number of stereoisomers possible for a compound is determined by the expression 2^n, where n is the number of stereocenters. The actual number of stereoisomers may be lower than the maximum value if certain conditions are met (eg, when a compound contains an *internal* plane of symmetry; see Concept 3.3.04).

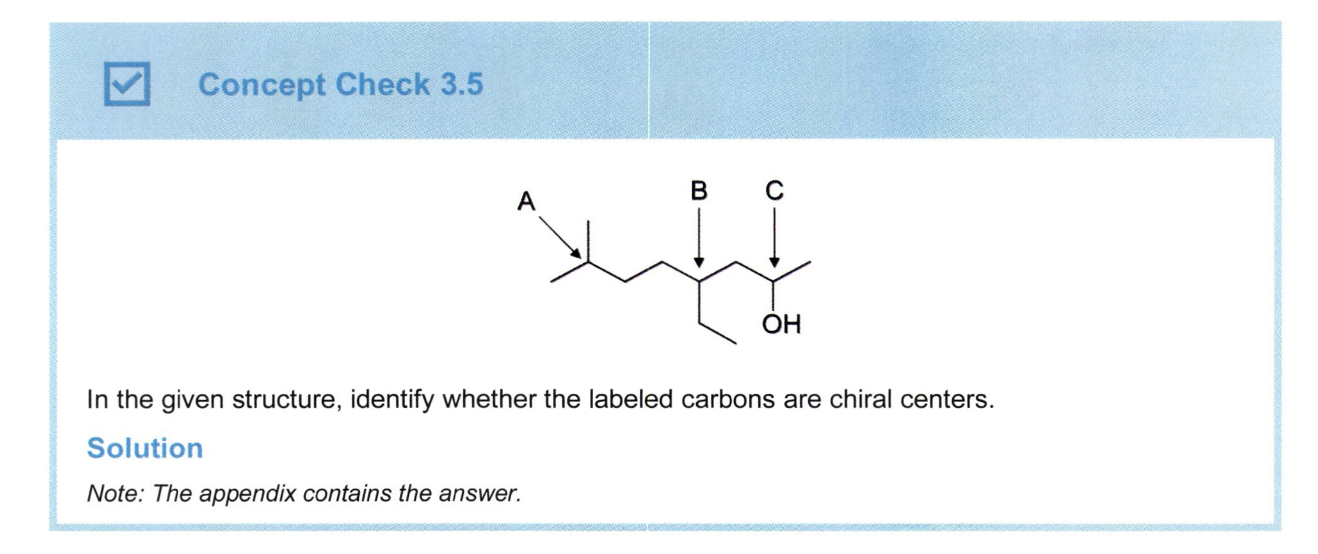

Concept Check 3.5

In the given structure, identify whether the labeled carbons are chiral centers.

Solution

Note: The appendix contains the answer.

3.3.03 Absolute Configuration (*R*/*S*)

Every asymmetric carbon atom has one of two spatial arrangements, or **configurations**: the original and its mirror image. The **Cahn-Ingold-Prelog guidelines** are a systematic process used to *assign* an **absolute configuration** of substituents around a chiral carbon. Groups attached to the chiral center are given ranked priorities and, based on the arrangement of the substituent priorities, each chiral carbon atom is assigned either an ***R*** or ***S*** absolute configuration.

To assign the priority (1–4) of each substituent based on Cahn-Ingold-Prelog guidelines:

1. Atoms bonded to the stereocenter are prioritized based on *atomic number* (higher atomic number = higher priority).
2. If two or more atoms bonded to the stereocenter have the same atomic number, a tiebreaker must be performed by comparing the atoms *attached* to the tied atoms. Arrange the attached atoms in order of decreasing atomic number. If a multiple bond is encountered, *each bond* in the multiple bond is counted as a separate bond to that atom.
3. For each group, compare the attached atoms with the highest atomic number until a difference is evident.
4. If no difference is observed for the attached atoms, repeat steps 1–3 for the *next atom in the chain* until all four ranked group priorities are determined.

Note: If two groups attached to a potential chiral carbon atom are identical *at all levels* of tiebreakers, then the groups are identical, and the carbon is achiral.

Figure 3.27 shows an application of the Cahn-Ingold-Prelog guidelines to assign the priority of each substituent in a chiral molecule.

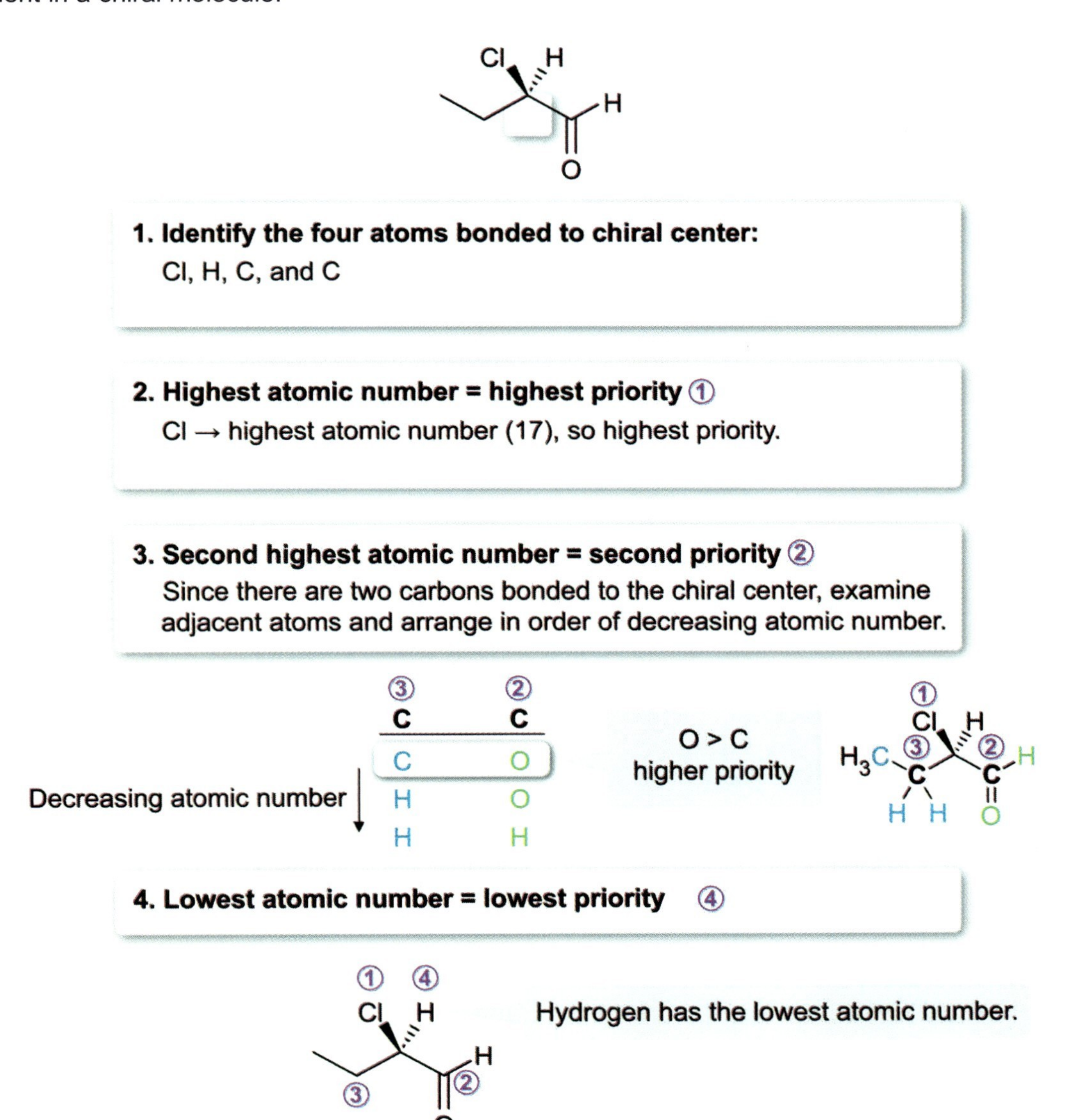

Figure 3.27 Application of Cahn-Ingold-Prelog priority ranking for *R*/*S* configuration determination.

The priority and circular arrangement of the substituents must both be considered to determine *R*/*S* configuration (Figure 3.28). To determine whether a chiral center has an *R* or *S* configuration:

1. Draw a circle starting from the highest priority (1) to priority 2, and then priority 3.
2. If the lowest-priority group (ie, priority 4) is pointed *behind* the plane (dashed), a clockwise arrangement of groups 1, 2, and 3 is defined as an *R* configuration, whereas a counterclockwise arrangement is defined as an *S* configuration.
3. If the lowest-priority group is pointed in *front* of the plane (wedged), a clockwise arrangement of groups 1, 2, and 3 is defined as an *S* configuration, whereas a counterclockwise arrangement is defined as an *R* configuration.

Note: If the lowest-priority group is in the *same* plane as the chiral center (neither wedged or dashed), the absolute configuration *cannot* be determined without rotating the molecule so that the priority 4 is *either behind* or in *front* of the plane.

Draw an arrow from the highest priority (1), to the second (2), then third (3) priority substituent.

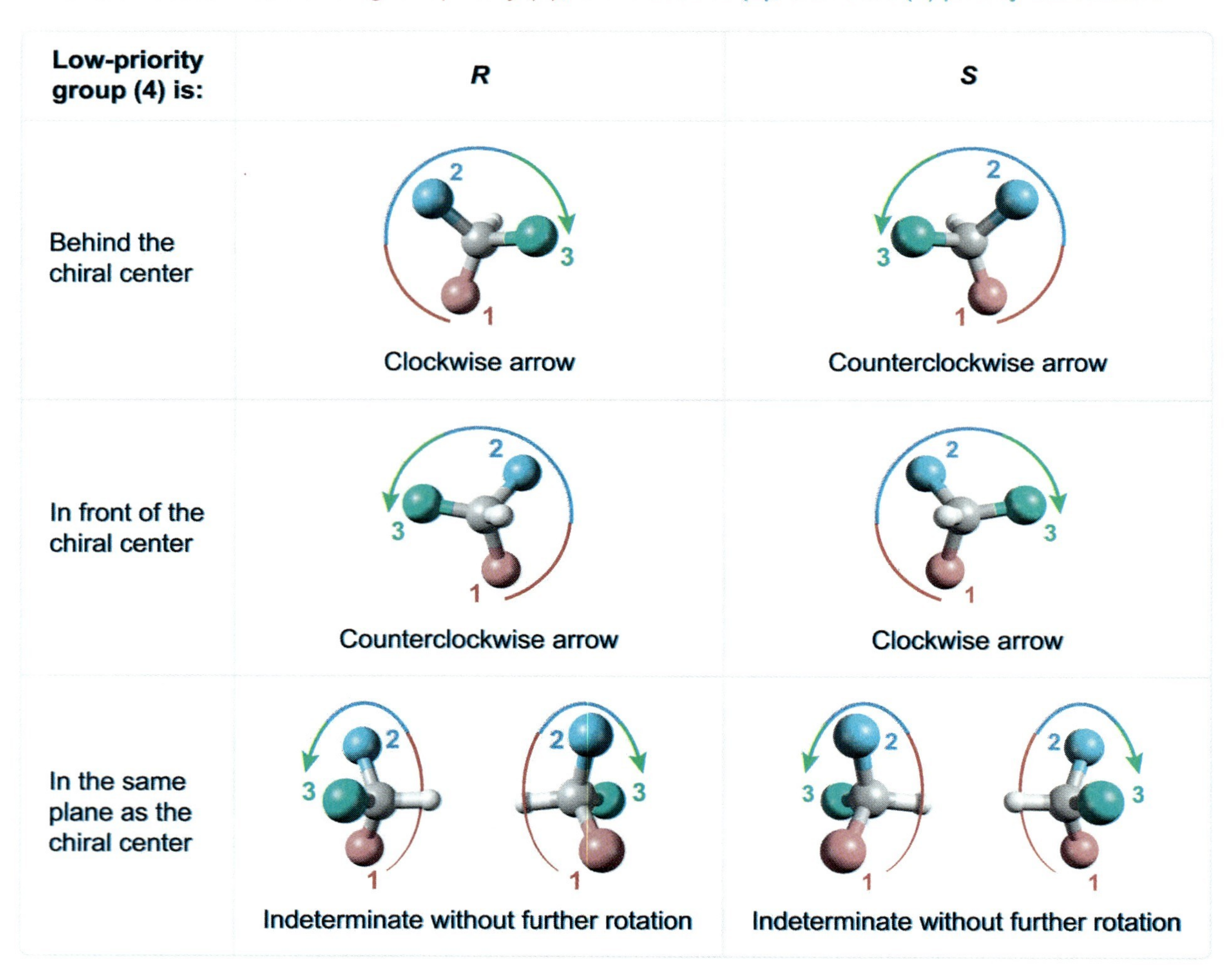

Figure 3.28 *R* and *S* absolute configurations.

Concept Check 3.6

Determine the absolute configuration (*R* or *S*) of the chiral center (*) in the following molecule.

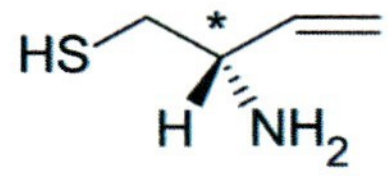

HS, *, H, NH_2

Solution

Note: The appendix contains the answer.

3.3.04 Stereochemical Relationships

Stereoisomers can be classified as enantiomers, diastereomers, epimers, anomers, or meso. To determine the relationship between two stereoisomers, the absolute configuration of each chiral center in the molecule must be assigned.

Stereoisomers that are nonsuperimposable mirror images are called **enantiomers**. Figure 3.29 shows how enantiomers have the opposite absolute configuration at *every* chiral center (ie, if a chiral center in a molecule has an *R* configuration, its enantiomer has an *S* configuration). Therefore, a chiral molecule can have only *one* enantiomer. Enantiomers have *identical* chemical and physical properties, with two exceptions: the direction they rotate plane-polarized light (see Lesson 14.2) and their interactions with other chiral substances.

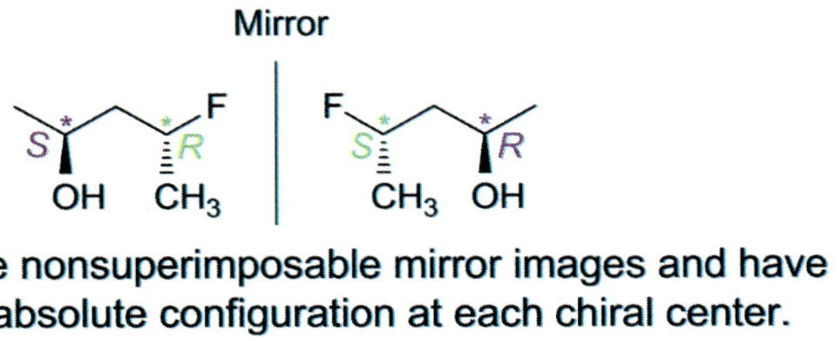

Mirror

S, R, F, OH, CH_3 | F, S, R, CH_3, OH

Enantiomers are nonsuperimposable mirror images and have the opposite absolute configuration at each chiral center.

Figure 3.29 Enantiomers.

Diastereomers are defined as stereoisomers that are *not* mirror images of each other. One type of diastereomers involves molecules with *at least two* chiral centers, where some (but *not all*) of their chiral centers have opposite absolute configurations. Therefore, it is possible for a chiral molecule to have multiple diastereomers (Figure 3.30). Diastereomers have *different* chemical and physical properties, which can be used to separate a mixture of diastereomers.

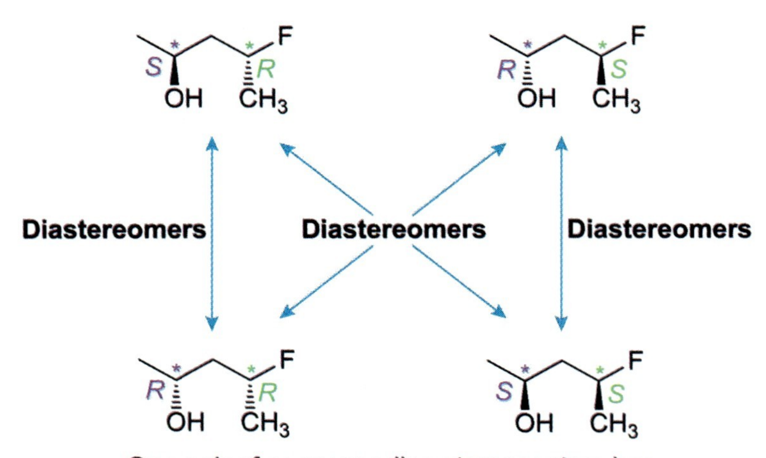

Figure 3.30 Diastereomers.

Epimers are a subset of diastereomers that differ in spatial orientation at only *one* chiral center and are most commonly encountered in the context of carbohydrates. For example, D-glucose and D-mannose are epimers that differ in configuration at only the C-2 position (Figure 3.31). If two sugars are diastereomers and differ in absolute configuration only at the anomeric carbon, such as α-D-fructose and β-D-fructose (Figure 3.31), they are classified as **anomers**.

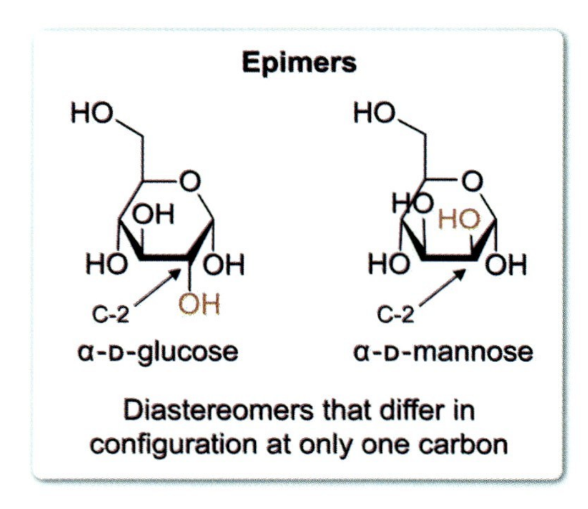

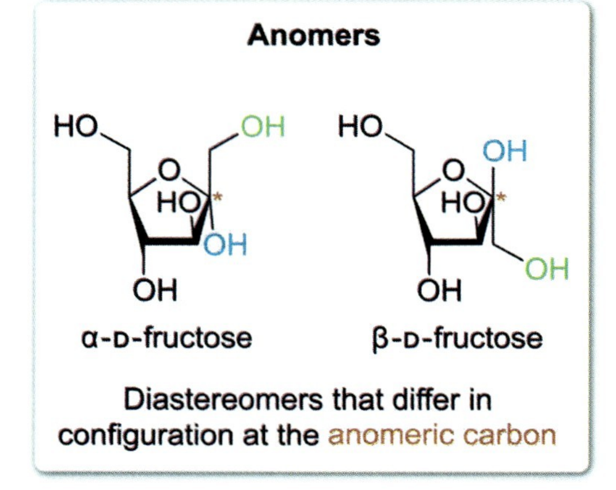

Figure 3.31 Epimers and anomers.

Meso compounds are molecules that contain *both* an internal plane of symmetry and chiral centers. A unique feature of a meso compound is that its mirror image *is* superimposable on itself (due to the internal plane of symmetry). Since the mirror image of a meso compound is *itself* (the same molecule), a meso compound does not have an enantiomer and is achiral (Concept 3.3.01) (Figure 3.32). The presence of meso compounds is one of the ways a structure might have fewer total stereoisomers than predicted by 2^n, in which n is the number of stereocenters (Concept 3.3.02).

- Achiral molecules containing chiral centers
- Superimposable mirror images

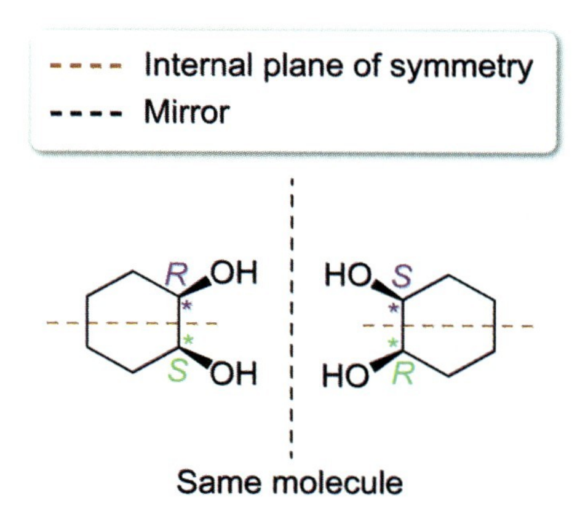

Figure 3.32 Meso compounds.

Concept Check 3.7

Label the following pairs of molecules using the most specific relationship term: enantiomers, diastereomers, epimers, or anomers.

1)

2)

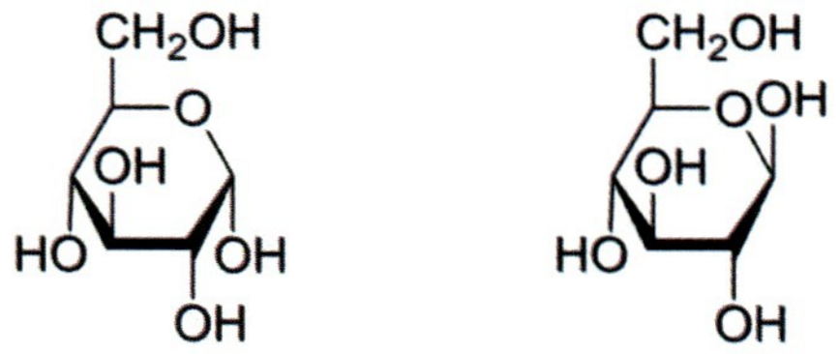

3)

4)

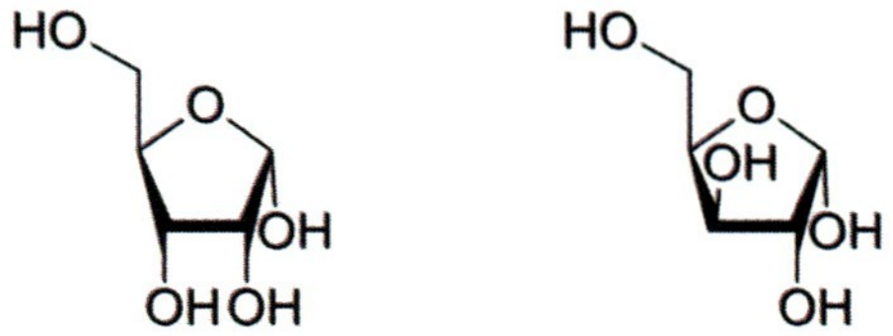

Solution

Note: The appendix contains the answer.

3.3.05 Racemic Mixtures

A 50:50 mixture of enantiomers is known as a **racemic mixture** (or **racemate**). Many types of chemical reactions that result in the formation of a new chiral carbon atom produce a 1:1 mixture of absolute configurations at the new stereocenter (see Unit 2). Figure 3.33 illustrates an example of a reaction producing a racemic mixture.

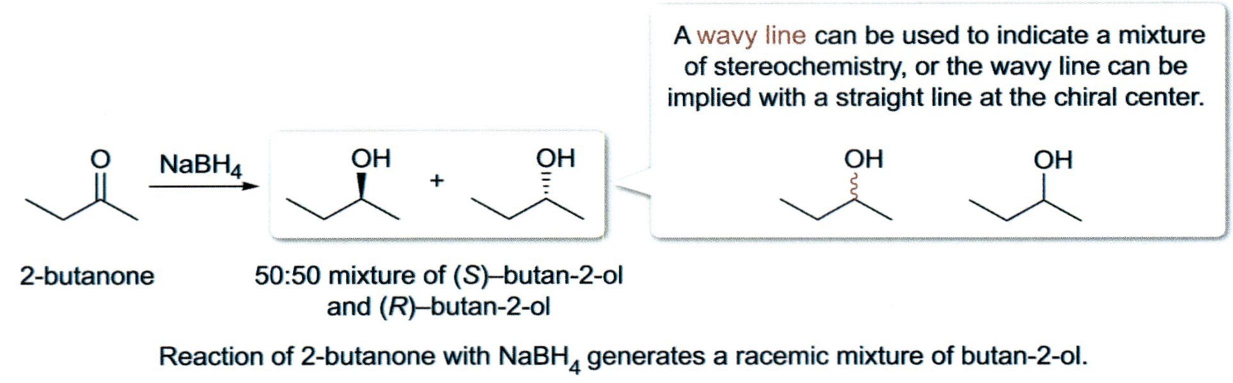

Figure 3.33 Formation of a racemic mixture.

Because enantiomers have identical chemical and physical properties, racemic mixtures cannot be easily separated without methods that change the physical properties of the molecules (eg, reversible chemical reactions to change their chemical identity).

3.3.06 Fischer Projections

Fischer Projections

Like Newman projections (see Concept 3.2.01), **Fischer projections** are a way to depict the three-dimensional orientation of a molecule as a flat drawing. Fischer projections are commonly used to visualize the stereochemistry of sugars (which contain multiple stereocenters per molecule).

Fischer projections are made up of vertical lines, which are oriented either *away* from the viewer or in the plane of the drawing, and horizontal lines, which are oriented *toward* the viewer. Figure 3.34 shows the proper angle from which to view a molecule in a tetrahedral wedge/dash structure to produce its corresponding Fischer projection.

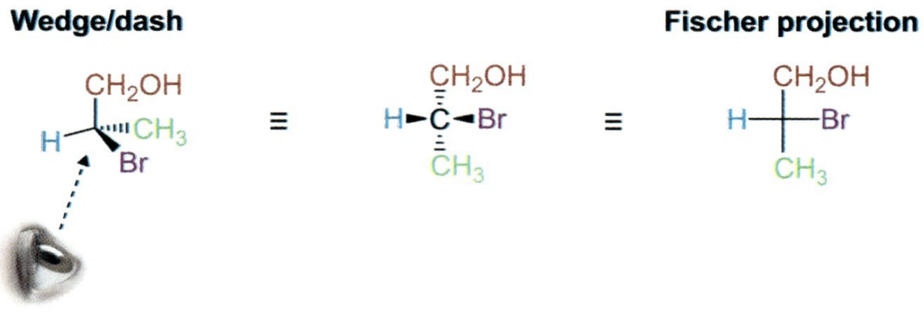

Figure 3.34 Wedge/dash perspective and Fischer projections.

In a Fischer projection, the carbon chain is positioned along the vertical line, and the substituents are the groups placed on the horizontal lines. Each intersection in a Fischer projection represents an sp^3 stereocenter. The highest oxidized carbon atom, often an aldehyde or a carboxylic acid, is at the top of the Fischer projection and labeled C1. Alternatively, the top position is the carbon atom with position 1 within the molecule's name (see Concept 4.1.01).

A 180° rotation of a Fischer projection maintains the relative positions and configuration around each stereocenter. Therefore, a Fischer projection that has been rotated 180° is the *same molecule* (Figure 3.35). In contrast, rotating a Fischer projection 90° *changes* the implied orientation (wedge/dash) of all the groups, meaning that a Fischer projection that has been rotated 90° is *no longer* the same molecule.

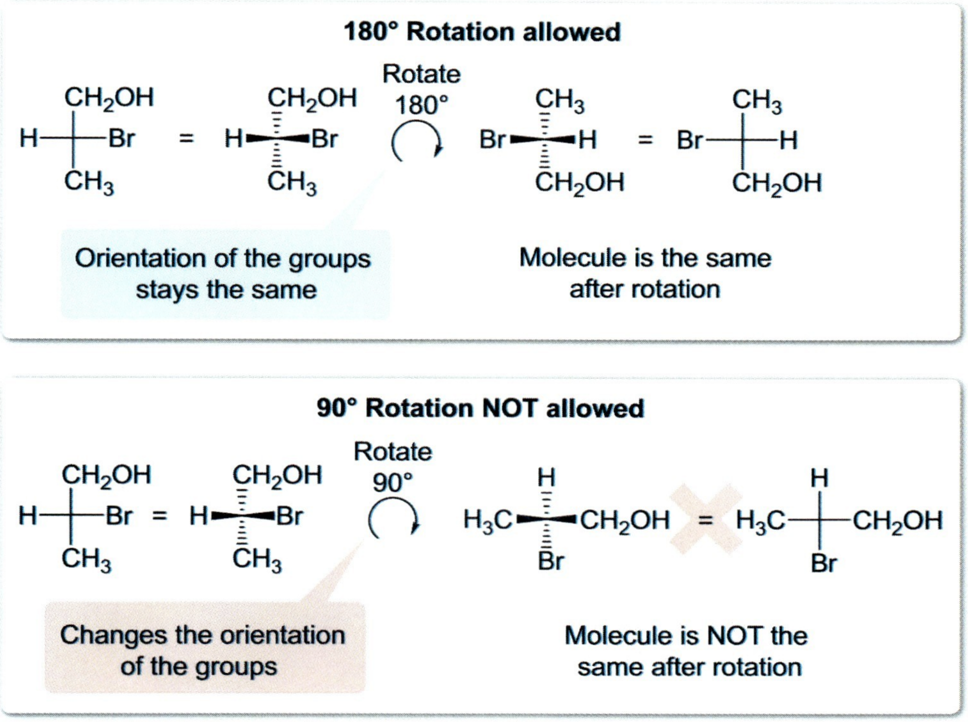

Figure 3.35 Rotating Fischer projections.

The assignment of absolute configuration of chiral centers in a Fischer projection follows the same guidelines as in a line-angle structure (Figure 3.36). Note that the fourth priority group on a chiral center in a Fischer projection is typically on one of the horizontal lines, which point *toward* the viewer. When this is the case, a *clockwise* arrangement of priorities 1, 2, and 3 has an *S* configuration, and a *counterclockwise* arrangement of priorities 1, 2, and 3 has an *R* configuration.

Figure 3.36 Absolute configurations in Fischer projections.

There are several ways to convert a line-angle structure to a Fischer projection. One method involves three-dimensional perspective, like the skill used for the drawing of Newman projections (see Concept 3.2.01) (Figure 3.37). The key to this approach is to ensure that the viewpoint is oriented toward the chiral carbon atom so that the groups at the top (T) and bottom (B) are oriented *away*. Then, each chiral center must be viewed from the appropriate perspective to populate the horizontal arms of that carbon's Fischer projection.

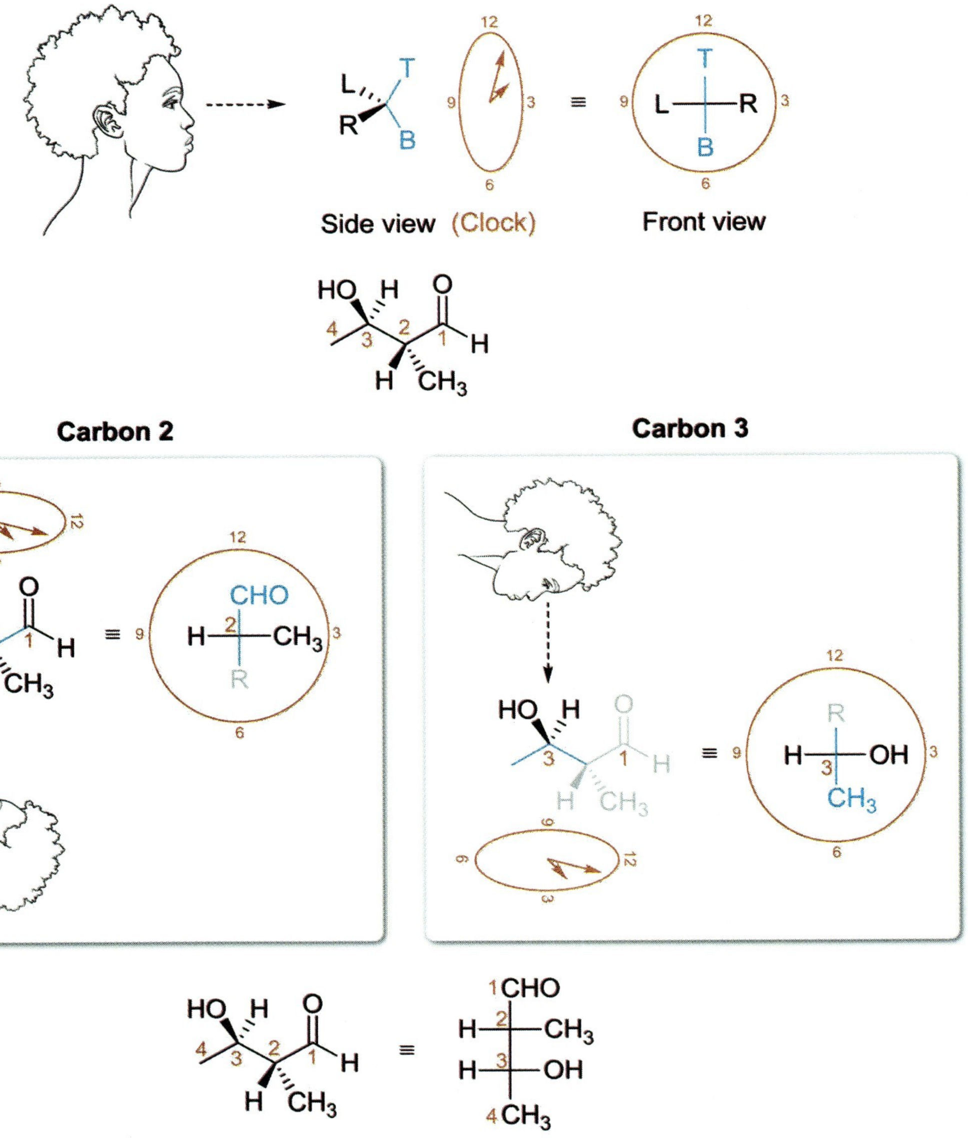

Figure 3.37 Using three-dimensional perspective to convert a line-angle structure to a Fischer projection.

Alternatively, Fischer projections can be drawn using absolute configuration assignments (Figure 3.38). In this method, the first step is to number the carbons in the chain and draw a Fischer projection skeleton containing the correct number of carbons. The number of horizontal bars should equal the number of stereocenters along the carbon chain. Next, add the substituents at the ends of the horizontal lines (it does not matter at this point whether they should go on the right or left—the last step will take care of this). Finally, assign *R* or *S* to each stereocenter and compare it to the corresponding configuration in the line-angle structure. If the configurations do not match, swap the two substituents at that chiral center.

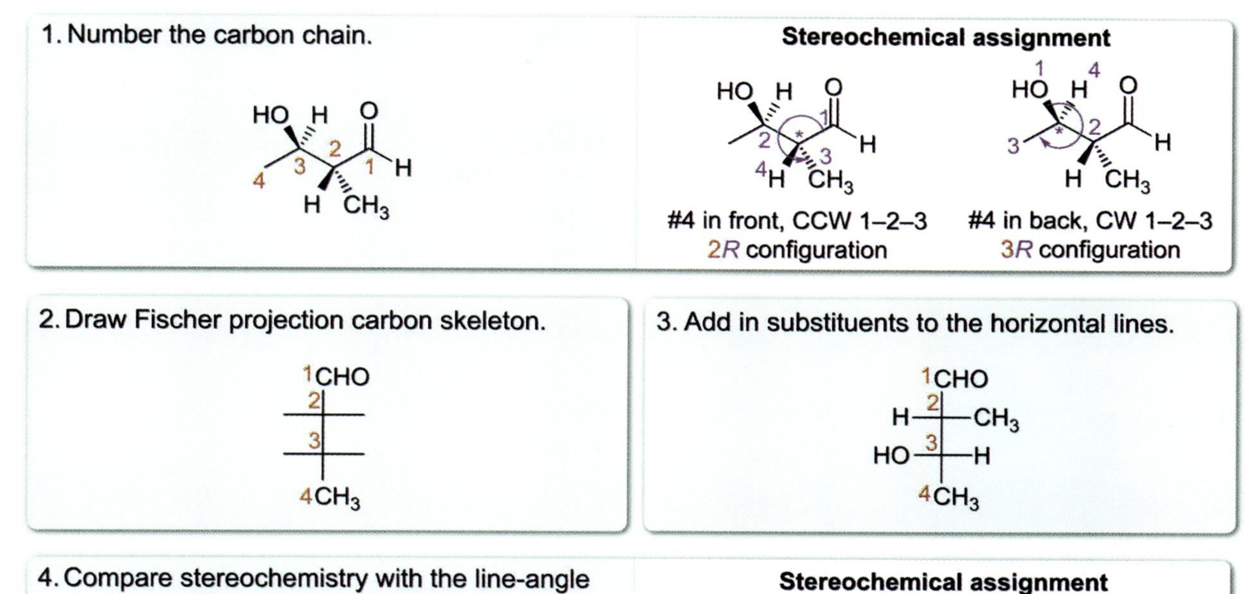

4. Compare stereochemistry with the line-angle structure – configuration at C3 does not match.

Stereochemical assignment

#4 in front, CCW 1–2–3
R configuration

#4 in front, CW 1–2–3
S configuration

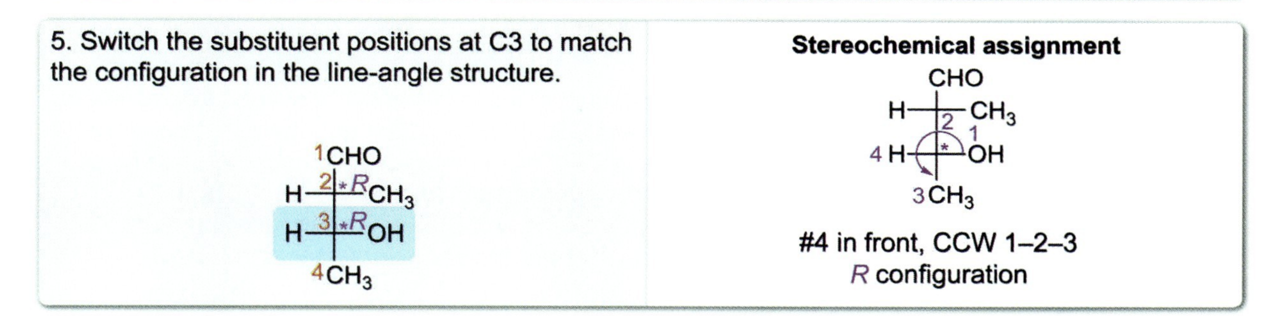

Figure 3.38 Using absolute stereochemistry to convert a line-angle structure to a Fischer projection.

Figure 3.39 shows another way to convert a line-angle structure to a Fischer projection. Rotate the line-angle structure so that the top and bottom carbons are pointing away from the viewer and the wedged and dashed groups are pointing toward the viewer. The vertical carbon chain becomes the vertical portion of the Fischer projection, and the wedge/dash groups are on the ends of the horizontal Fischer projection lines.

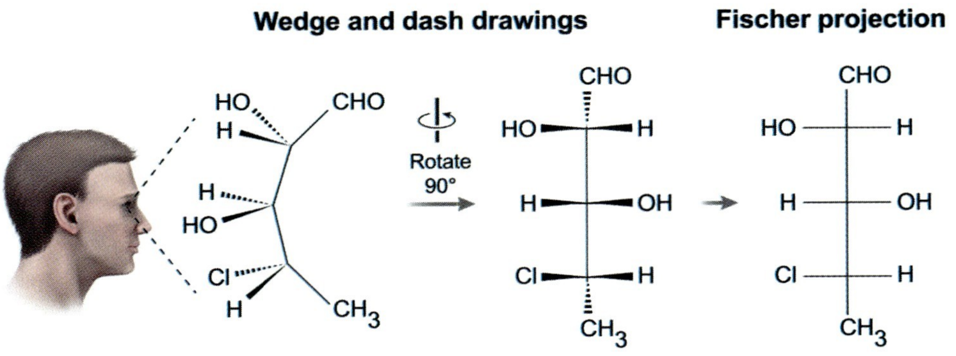

Figure 3.39 Converting line-angle structures to Fischer projections.

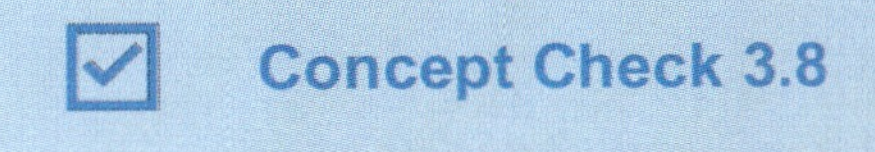

Concept Check 3.8

Convert the following Fischer projection to a line-angle structure.

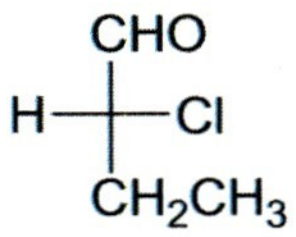

Solution

Note: The appendix contains the answer.

Fischer projections can be used to easily compare and identify the relationship between stereoisomers. Figure 3.40 shows Fischer projections of different tartaric acid stereoisomers and their stereochemical relationships. If the stereoisomers are nonsuperimposable mirror images (all stereocenters are flipped), they are **enantiomers**. The stereoisomers that are not mirror images and have one, not both, stereocenters flipped are **diastereomers**. (2*R*,3*S*)-tartaric acid and (2*S*,3*R*)-tartaric acid appear to be enantiomers, but due to an internal plane of symmetry, they are the same molecule and **meso**.

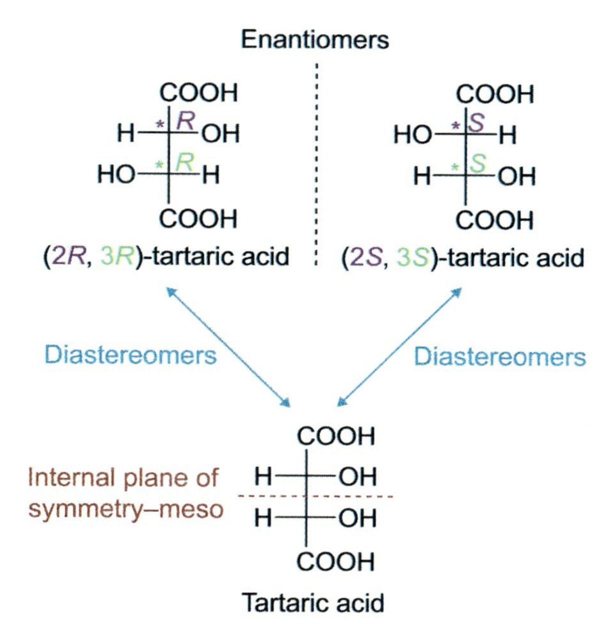

Figure 3.40 Using Fischer projections to identify stereochemical relationships of different tartaric acid stereoisomers.

D/L (Relative) Configurations

Before 1951, when x-ray crystallography was first used to determine the absolute configuration of molecules, the **D/L system** was devised to compare the chiral configuration of an amino acid or sugar to the configuration of a standard compound (a type of **relative configuration**). Because the chiral standards were D-glyceraldehyde and L-glyceraldehyde, all molecules could be assigned a D/L configuration by comparing the *final* chiral center in a Fischer projection to the chiral center in D- or L-glyceraldehyde. Figure 3.41 shows that when the nonhydrogen group of the final chiral carbon is on the right side of a Fischer projection, the molecule has a D configuration, and when the nonhydrogen groups are on the left side, the molecule has an L configuration.

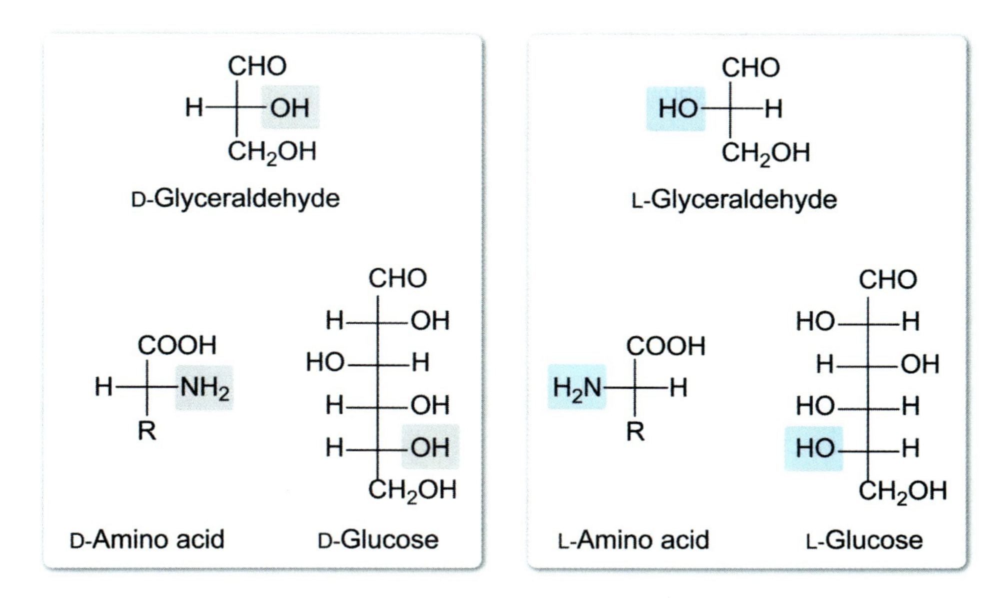

Figure 3.41 Relative configurations using the D/L system.

Note that the D/L system has *no* correlation to either the *R* and *S* absolute configuration (see Concept 3.3.03) or experimental + or − designations (see Lesson 14.2).

3.3.07 Geometric Isomers

Geometric isomers (also called *cis/trans* isomers) are a type of diastereomer that differ in the relative position of two nonhydrogen substituents on different carbon atoms in either a cycloalkane or an alkene. Because the bonds in a cycloalkane and the double bond in an alkene cannot freely rotate, the substituents cannot change orientation.

Cycloalkanes can be drawn in different ways to represent the stereochemistry of the ring substituents. The ring can be drawn in a Haworth projection, in which the bold bonds are closest to the viewer and the substituents point either to the top or bottom face of the ring. Alternatively, cycloalkanes can also be drawn using wedges and dashes to indicate spatial orientation of the substituents. Regardless, when two substituents point toward the same face of the ring, they are ***cis*** to each other. When two substituents point toward opposite faces of a cycloalkane ring, the substituents are ***trans*** to one another (Figure 3.42).

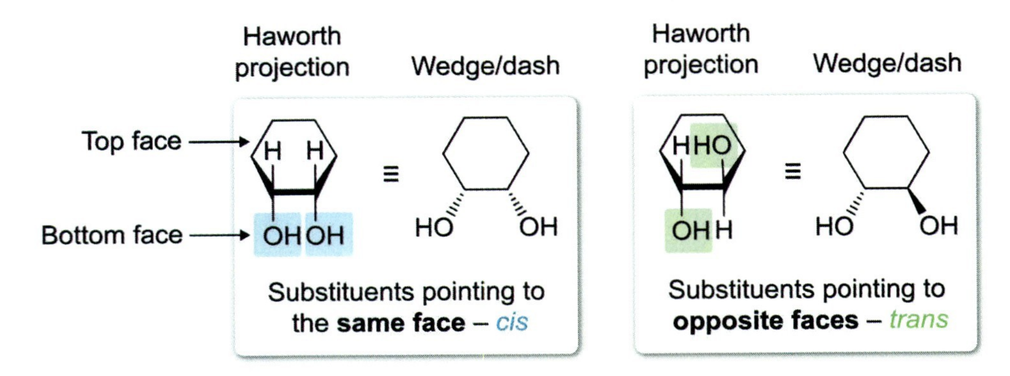

Figure 3.42 An example of *cis* and *trans* designations in cycloalkanes.

While the *cis/trans* designation is useful, it gives only the *relative* configuration of the substituents on the cycloalkanes and does not indicate the *absolute* configuration (ie, *R* or *S* configuration) of the chiral center.

Cis and *trans* can be used to describe the spatial relationship between alkene substituents *only* if each alkene carbon contains *one* nonhydrogen substituent. Like cycloalkanes, when both alkene substituents are on the *same* side of the double bond, the substituents are *cis.* When the alkene substituents are on *opposite* sides of the double bond, the substituents are *trans* (Figure 3.43).

H_3C CH_3 H H

Substituents on **same side** – *cis*

H_3C H H CH_3

Substituents on **opposite sides** – *trans*

Figure 3.43 An example of *cis* and *trans* designations in alkenes.

3.3.08 *E*/*Z* Determination

Because *cis* and *trans* designations are common nomenclature (Concept 4.7.02) limited to only alkenes containing one nonhydrogen substituent on each carbon atom, a way of indicating the stereochemistry of more complex alkenes (ie, alkenes containing tri- and tetra-substituted double bonds) is needed. Similar to assigning *R* or *S* in Concept 3.3.03, Cahn-Ingold-Prelog priorities can be assigned to unambiguously assign the absolute configuration of an alkene stereocenter as either ***E* or *Z***.

To assign Cahn-Ingold-Prelog priorities and determine *E*/*Z* configuration:

1. Examine the two atoms bonded to each alkene carbon atom. For each alkene carbon, the substituent containing the atom with the higher atomic number is the higher priority substituent (Figure 3.44).

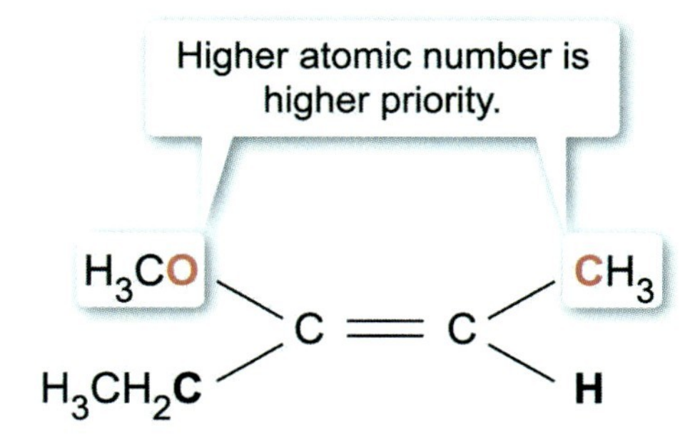

Figure 3.44 Higher priority substituent determination in alkenes.

2. If both atoms are the same, each atom's substituents are compared. If those are also the same, the process extends to the next atom in the chain until a point of difference is reached. The substituent containing the atom with the higher atomic number at the point of difference is the higher priority substituent (Figure 3.45).

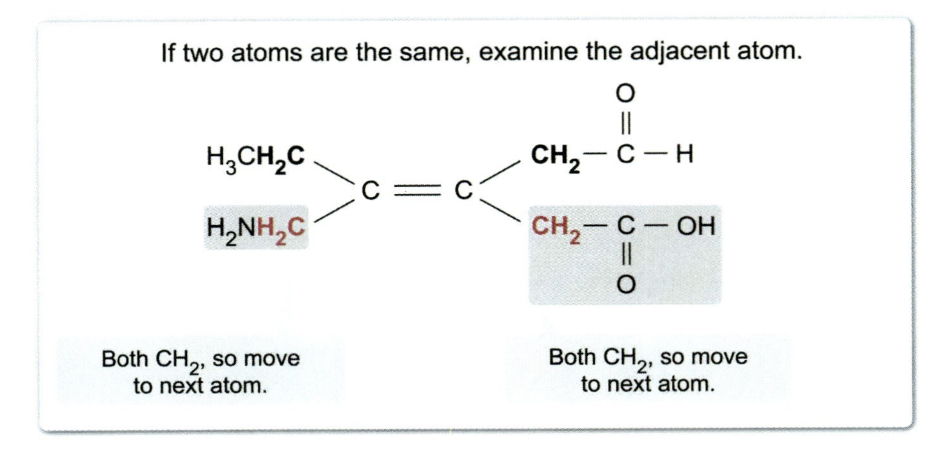

Note: if C is double bonded to X, count C as bonded to X twice (X= any atom that can take part in double bond).

CH_3 versus NH_2 [H_3CH_2C / H_2NH_2C] C = C [$CH_2-C(=O)-H$ / $CH_2-C(=O)-OH$] Carbon bonded to O, O, H versus O, O, O

Higher atomic number, so higher priority.

More bonds to oxygen, so higher priority.

Repeat the process until the higher priority substituent is determined.

Figure 3.45 Application of Cahn-Ingold-Prelog priority ranking for *E*/*Z* configuration determination.

3. If the higher priority groups on each alkene carbon are on the *same* side of the double bond, the alkene is given an absolute configuration of *Z*, whereas if the higher priority groups on each alkene carbon are on *opposite* sides of the double bond, the alkene is given an absolute configuration of *E* (Figure 3.46).

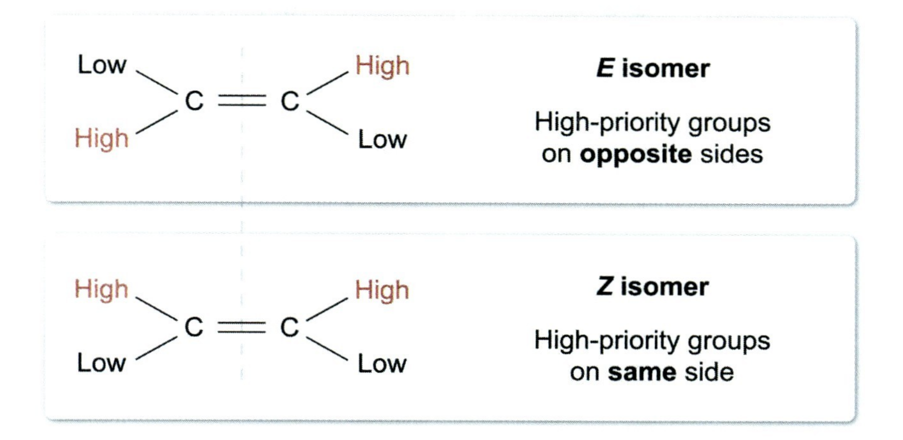

Figure 3.46 Assigning *E* and *Z* configurations.

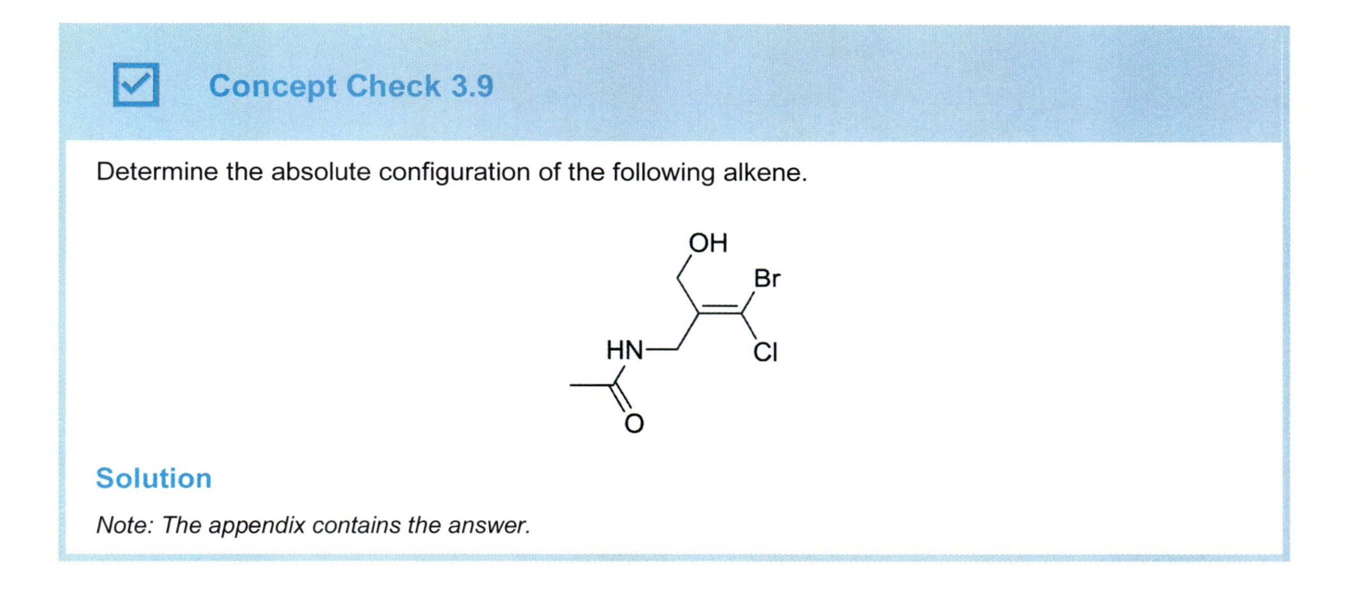

Concept Check 3.9

Determine the absolute configuration of the following alkene.

Solution

Note: The appendix contains the answer.

Lesson 4.1

Introduction to IUPAC Nomenclature (The Alkanes)

Introduction

Nomenclature is the use of chemical names to describe the details of molecular structure. Before 1900, different scientific bodies often used their own names for chemicals, with chemists and biologists sometimes using a different term for the same chemical. Some of these diverse naming conventions persist today and are generally described as **common nomenclature**.

In 1919, the **International Union of Pure and Applied Chemistry (IUPAC)** was created to develop a worldwide set of guidelines for many facets of chemistry, including nomenclature. Under the **IUPAC nomenclature system**, all unique chemical structures are assigned a unique name. Consequently, if two structures have the same systematic name, they *must* be the same substance—a feature that can be particularly helpful when working with isomers (Chapter 3).

This lesson discusses the foundational basics of IUPAC nomenclature for the simplest functional group: the alkanes. Subsequent lessons (Lessons 4.2 to 4.5) expand the IUPAC guidelines to other functional group classes. The IUPAC nomenclature of multifunctional molecules is discussed in Lesson 4.6. An overview of common nomenclature is presented in Lesson 4.7.

4.1.01 The Parent Chain

The alkanes are the most basic class of functional groups, containing solely carbon and hydrogen atoms connected by single bonds. The first step toward generating a name for an alkane is to identify its parent chain. For the alkanes, the **parent chain** is defined as the *longest continuous chain* of carbon atoms.

If the longest continuous carbon chain can be drawn two or more ways, IUPAC guidelines identify the parent chain as the one with the *greatest* number of branch points (ie, substituents, see Concept 4.1.02). Figure 4.1 depicts an alkane with two possible ways of identifying the longest continuous carbon chain (seven carbon atoms). Option A indicates a parent chain with *two* branch points, which is less than Option B with *three* branch points.

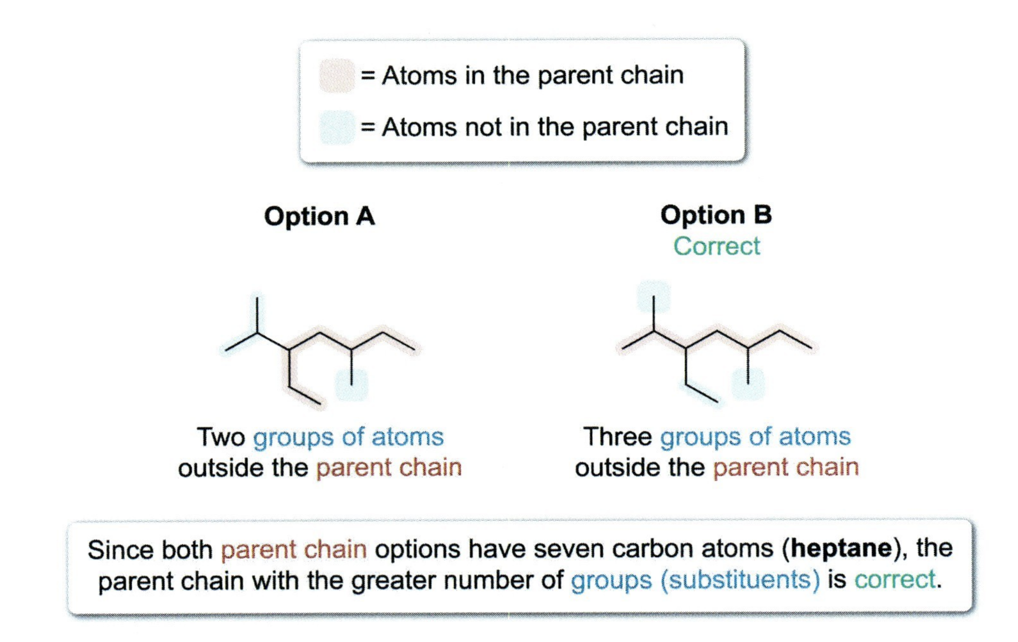

Figure 4.1 Selecting the correct parent chain based on the number of substituents.

For an alkane with no other functional groups, the name of the parent chain consists of a prefix that denotes the number of carbon atoms followed by the suffix *–ane*. For example, a parent chain of four carbon atoms is named *butane*, and a parent chain of eight carbon atoms is named *octane*. Table 4.1 lists the IUPAC prefixes for chain lengths from one to ten carbon atoms.

Table 4.1 IUPAC prefixes for the parent chain.

Number of carbons	Prefix	Number of carbons	Prefix
1	meth–	6	hex–
2	eth–	7	hept–
3	prop–	8	oct–
4	but–	9	non–
5	pent–	10	dec–

The next step is to provide a numerical location for each carbon atom in the parent chain. The atoms in the parent chain are numbered sequentially, starting from one end. Typically, a linear parent chain can be numbered two ways, starting from either end (Figure 4.2). IUPAC guidelines state that the correct parent chain numbering provides the *lowest possible values* for any appended groups. Often, the easiest way to determine the correct parent chain numbering is to start from the end closer to a substituent.

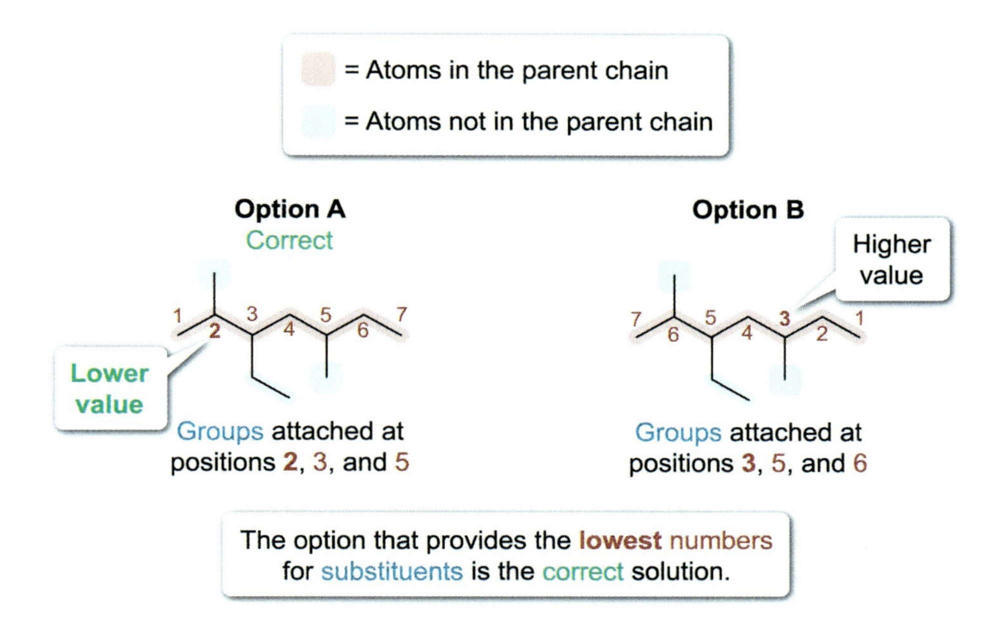

Figure 4.2 Selecting the best numbering system for a parent chain.

If both numbering schemes provide the same values for all substituents, *alphabetical order* for the substituent names becomes the tiebreaker (see Concept 4.1.02).

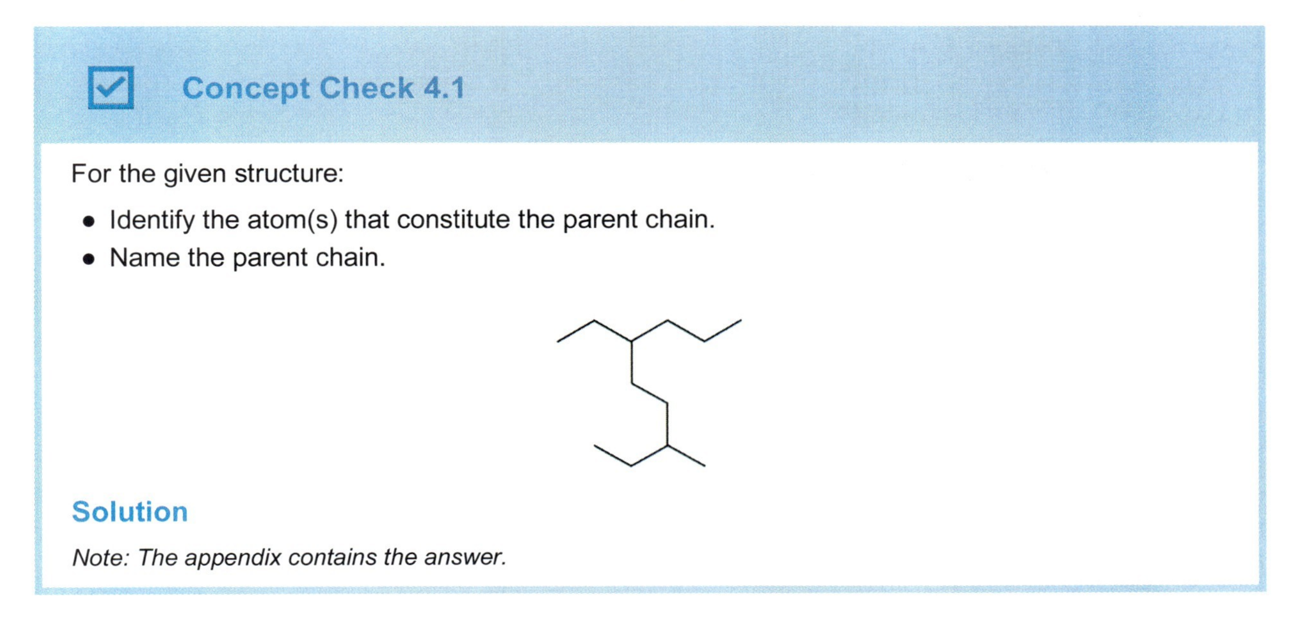

Concept Check 4.1

For the given structure:

- Identify the atom(s) that constitute the parent chain.
- Name the parent chain.

Solution

Note: The appendix contains the answer.

4.1.02 Substituents

The groups of atoms appended to the parent chain are known as **substituents**. When a substituent is composed of an alkane functional group, it is called an **alkyl group**. Alkyl groups are named using a similar set of guidelines as the parent chain (Table 4.2):

- A prefix denotes the longest continuous chain of carbon atoms *from the point of attachment to the parent chain.*
- A *–yl* suffix denotes that the group is a substituent of something else.
- Numbering starts *from the point of attachment to the parent chain.*

Table 4.2 Alkanes versus alkyl groups.

Alkane (whole molecule)			Alkyl group (substituent)		
Name	Condensed structural formula	Line-angle formula	Name	Condensed structural formula	Line-angle formula
Methane	$H_3\overset{1}{C}–H$	N/A	Methyl	$H_3\overset{1}{C}–$	
Ethane	$H_3\overset{2}{C}H_2\overset{1}{C}–H$	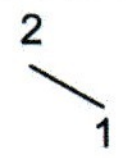	Ethyl	$H_3\overset{2}{C}H_2\overset{1}{C}–$	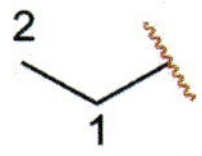
Propane	$H_3\overset{3}{C}H_2\overset{2}{C}H_2\overset{1}{C}–H$	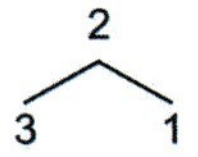	Propyl	$H_3\overset{3}{C}H_2\overset{2}{C}H_2\overset{1}{C}–$	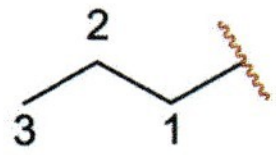
Butane	$H_3\overset{4}{C}H_2\overset{3}{C}H_2\overset{2}{C}H_2\overset{1}{C}–H$	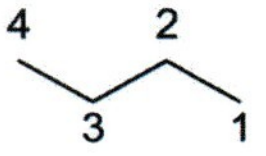	Butyl	$H_3\overset{4}{C}H_2\overset{3}{C}H_2\overset{2}{C}H_2\overset{1}{C}–$	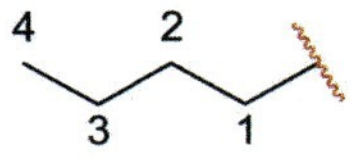

R– = Point of attachment to another group

An **alkyl group** is an alkane substituent of a larger molecule, signified by a –yl suffix.

Nomenclature requires identification of each alkyl group's point of attachment to the parent chain. The appropriate parent chain number is listed before the substituent, separated by a hyphen. If two or more of the same alkyl group (eg, methyl, ethyl) are present, each type of group is mentioned only once, and a Greek prefix is added to denote the number of groups present (Table 4.3). A numerical location is added in ascending order for *each* instance of the alkyl group, separated by commas.

For alkyl groups that are attached to the *same* carbon on the parent chain, each instance of the alkyl group is provided its own number. The list of number values must always match the value of the Greek prefix. For example, a molecule that contains four methyl groups located at positions 2, 4, 4, and 7 would have 2,4,4,7-tetramethyl in its name.

Table 4.3 Greek prefixes.

Number of identical groups	Prefix	Number of identical groups	Prefix
1	—	6	hexa–
2	di–	7	hepta–
3	tri–	8	octa–
4	tetra–	9	nona–
5	penta–	10	deca–

The full compound name is assembled by listing the alkyl group substituents (along with their corresponding location numbers) in *alphabetical order*, followed by the name of the parent chain (Figure 4.3). Importantly, the Greek prefixes (eg, *di–*, *tri–*) are *not* considered for alphabetical ordering; for example, the group *2,3-dimethyl* is alphabetized starting with the "m" of methyl. Substituents are separated by a hyphen with one exception: the final substituent continues into the name of the parent chain *without* a hyphen.

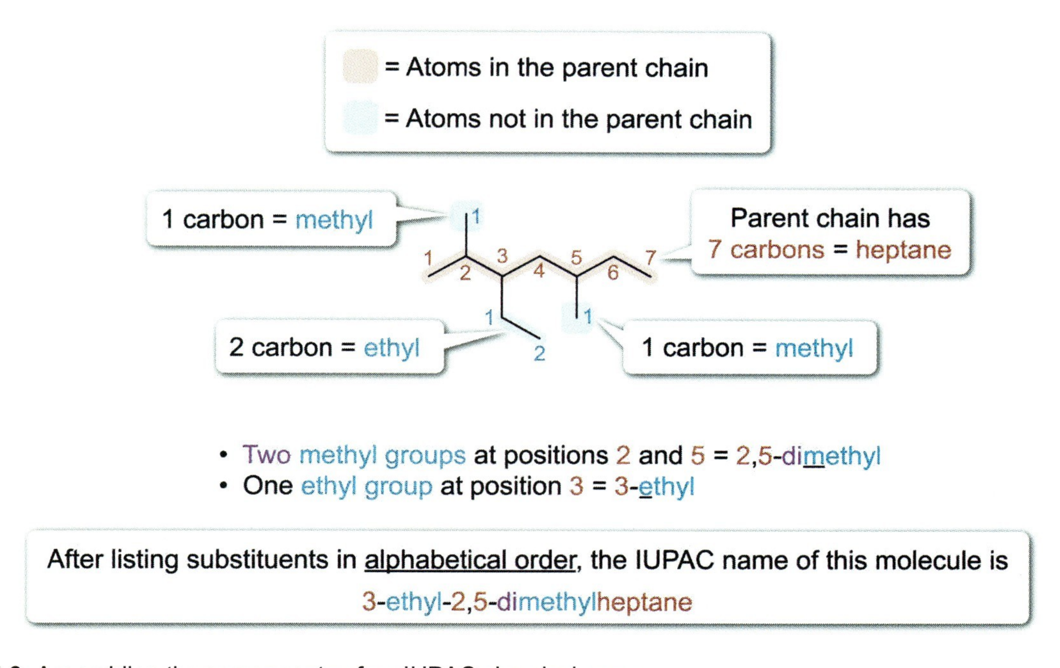

Figure 4.3 Assembling the components of an IUPAC chemical name.

Although simple linear alkyl groups can be named using the criteria in Table 4.2, alkyl groups that contain branch points must be named as **complex alkyl substituents**. In a complex alkyl substituent, the **parent substituent chain** is the longest, continuous chain of carbon atoms *from the point of attachment to the parent chain*.

Naming a complex alkyl substituent has the following modified criteria:

- The parent substituent chain is numbered *from the point of attachment to the parent chain* (like other alkyl groups) and has a *–yl* suffix because it is itself a substituent.
- Groups that branch off the parent substituent chain are named like other alkyl groups and are provided numerical locations based on the number of their *branch point.*
- Complex alkyl substituents are enclosed within parentheses. The relevant numerical locator *to the parent chain* is located *outside* the parentheses.
- These criteria can be extended to any number of additional levels, denoted by additional sets of parentheses.

Figure 4.4 provides several examples of complex alkyl substituent nomenclature.

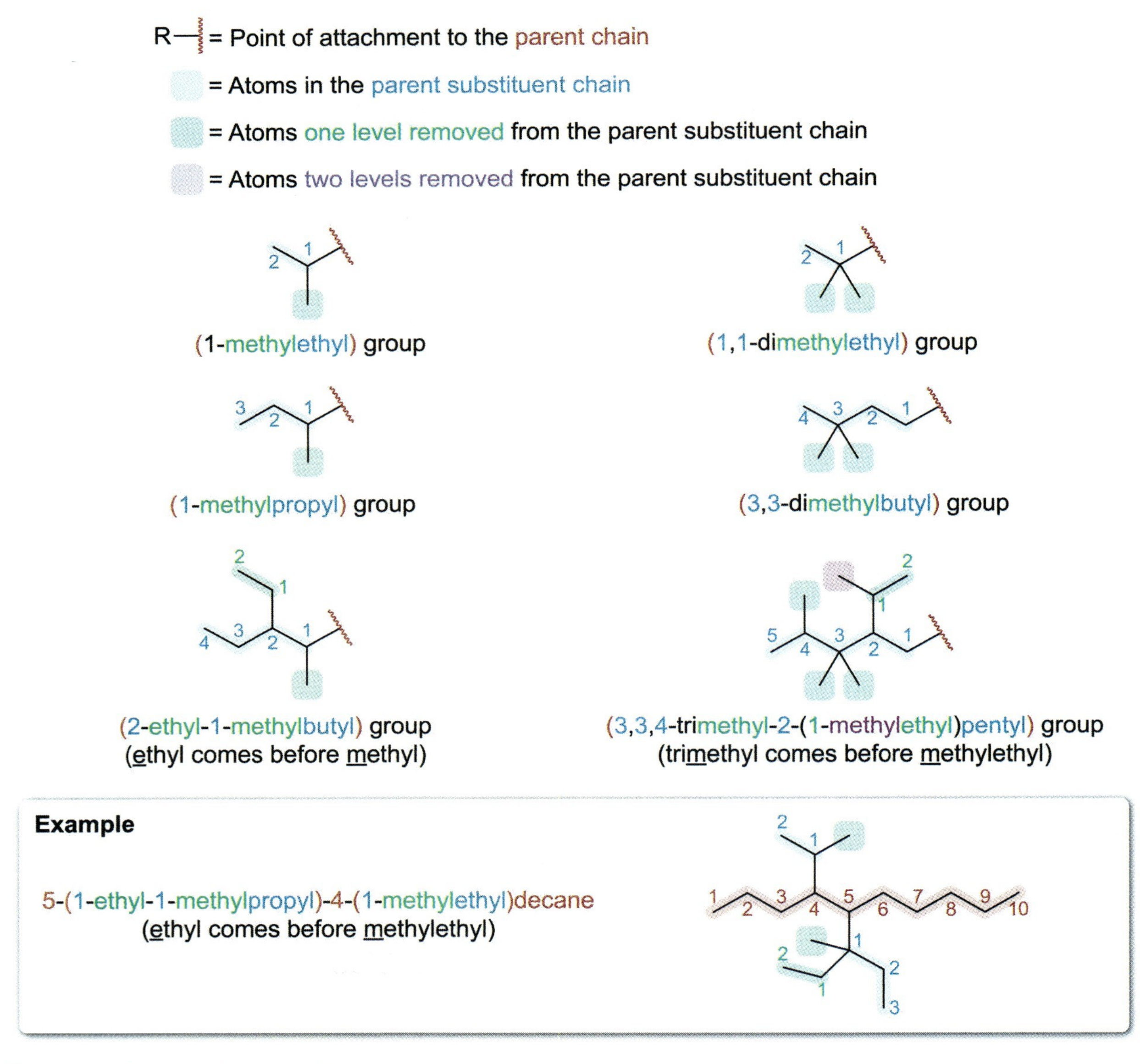

Figure 4.4 Complex alkyl substituents.

Concept Check 4.2

Draw a line-angle structure for the molecule 3-ethyl-2,3-dimethylhexane.

Solution

Note: The appendix contains the answer.

4.1.03 Absolute Configuration

Alkanes that contain one or more stereocenters have IUPAC names that include stereochemical configurations (see Lesson 3.3). If an atom can be assigned either an *R* or *S* absolute configuration, those designations are listed at the *beginning* of the IUPAC name, enclosed in parentheses (Figure 4.5). Each absolute configuration is provided with a number corresponding to the position of the atom with the absolute configuration.

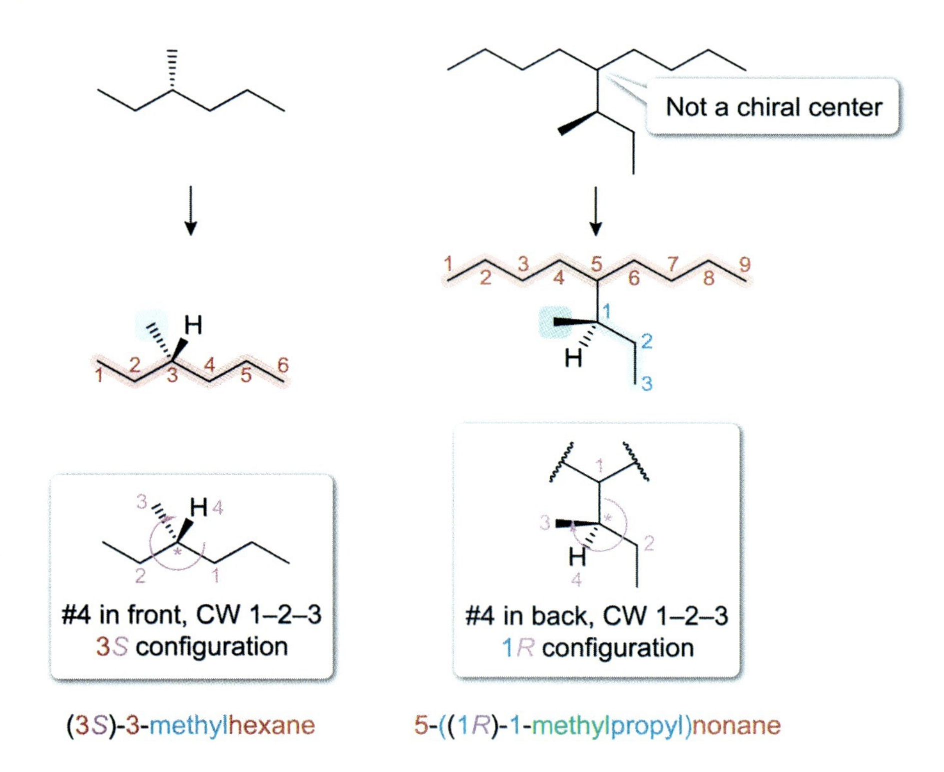

Figure 4.5 Representing *R* and *S* absolute configuration in IUPAC names.

Stereocenters within a *complex* alkyl group substituent have the *R* or *S* designation listed at the beginning of the name for the complex alkyl group (inside its parentheses) with a numerical location that corresponds to the position of the chiral atom *within the substituent*. If more than one absolute configuration must be included for an IUPAC name, they are listed in *ascending* order, separated by commas, within the same set of parentheses at the start of the name.

IUPAC names with *E/Z* configurations (alkenes; see Concept 4.2.02) are treated in the same manner as *R/S* configurations, where the *E/Z* configuration is listed at the start of the name, in ascending order, enclosed within parentheses (Lesson 3.3). A *single* numerical locator is provided for each *E* or *Z* configuration—the *lower* value of the two positions that participate in the double bond (Figure 4.6). If a molecule contains a combination of both *R*/*S* and *E*/*Z* configurations, they are *all* enclosed within a single set of parentheses and listed in ascending order.

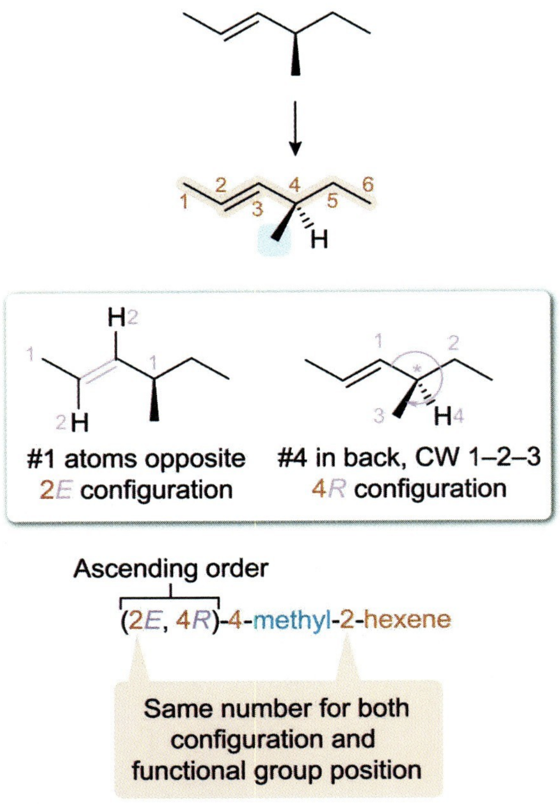

H2
1
1
2H
#1 atoms opposite
2*E* configuration

1 2
3 H4
#4 in back, CW 1–2–3
4*R* configuration

Ascending order
(2*E*, 4*R*)-4-methyl-2-hexene

Same number for both configuration and functional group position

Figure 4.6 Representing *E* and *Z* absolute configuration in IUPAC names.

Lesson 4.2

IUPAC Nomenclature of Hydrocarbons and Related Compounds

Introduction

Lesson 4.1 discussed the systematic (IUPAC) nomenclature guidelines for the alkanes and introduced the concepts of the parent chain, substituents, and the inclusion of absolute configuration information.

This lesson covers the IUPAC nomenclature guidelines for the other hydrocarbon functional groups (alkenes, alkynes, and aromatic compounds), cyclic hydrocarbons, and the alkyl halides.

4.2.01 Alkyl Halides

An alkyl halide is an alkane with one or more of its hydrogen atoms substituted for a **halogen atom** (eg, fluorine, chlorine, bromine, iodine). Halogen atoms have the same priority as an alkyl group and are *always* named as *substituents*. Each halogen atom is referenced using a unique substituent term (Table 4.4) alongside the appropriate numerical locator for its point of attachment.

As an example, an IUPAC name that includes *3-bromo* indicates that a bromine atom is attached at position 3 (of whatever parent chain being referenced).

Table 4.4 IUPAC group terms for the halogens.

Halogen	Atomic symbol	IUPAC group term
Fluorine	F	fluoro
Chlorine	Cl	chloro
Bromine	Br	bromo
Iodine	I	iodo

Figure 4.7 includes examples of IUPAC names for alkyl halides.

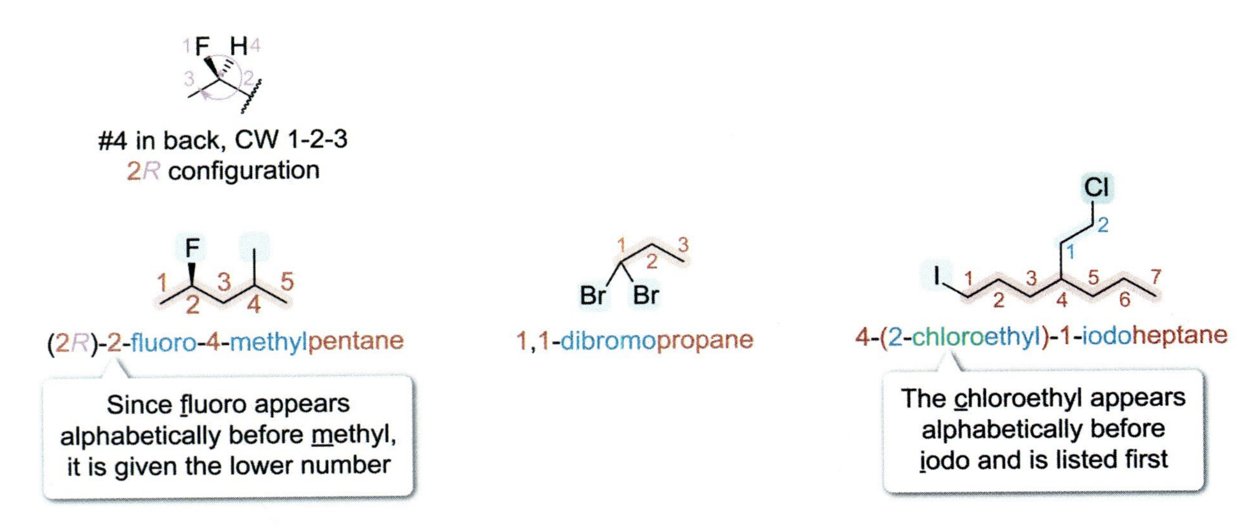

Figure 4.7 Examples and IUPAC names for alkyl halides.

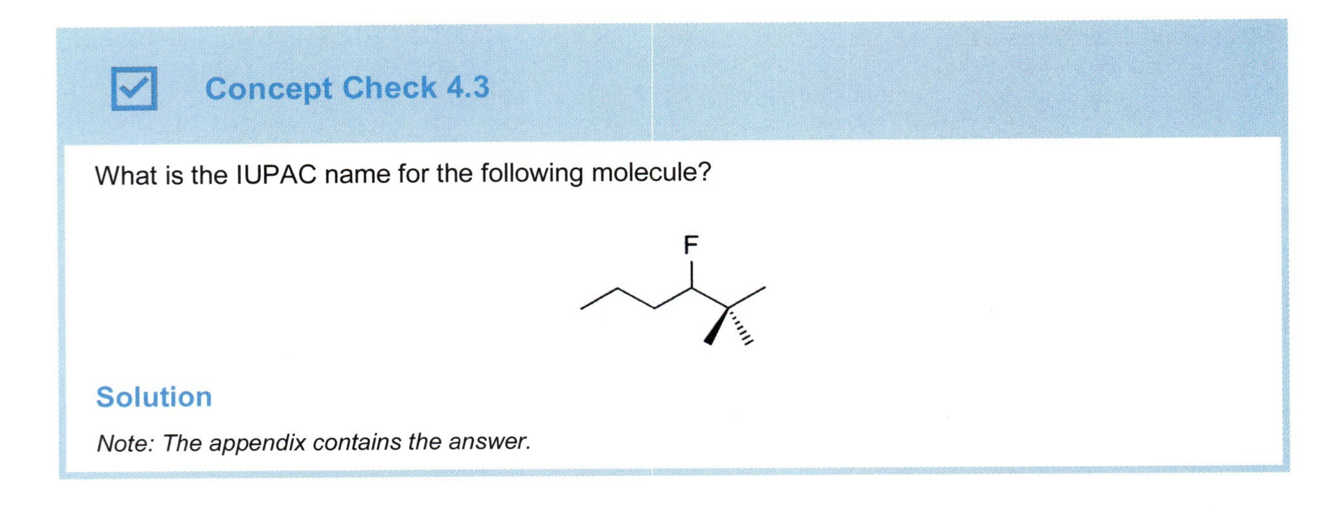

4.2.02 Alkenes

Alkenes are the first example of a functional group that modifies the name of the *parent chain*. Generating the systematic IUPAC name for an alkene follows the same general steps as used for alkanes, with a few caveats (Figure 4.8):

- The parent chain of an alkene is the longest chain of carbon atoms *that contains both carbon atoms of the alkene functional group*.
- The parent chain of an alkene is numbered to give the *alkene carbon atoms* the lowest possible values.
- The parent chain is named using the prefix for the number of carbon atoms with the suffix *–ene*.
- The location of the double bond chain is provided by a single number: the *lower* value of the two carbon atoms that participate in the double bond.

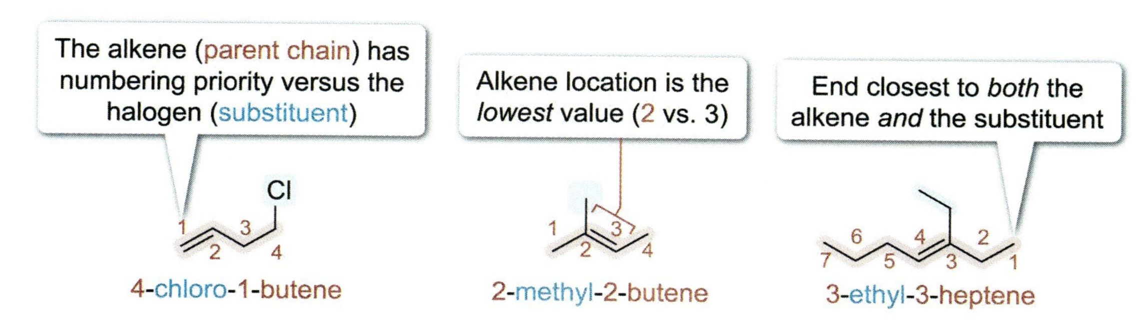

Figure 4.8 Examples and IUPAC names for simple alkenes.

If a molecule contains multiple alkene functional groups, the parent chain is the longest carbon chain that contains the *maximum number of alkene functional groups*. Only alkene functional groups with *both* their carbon atoms *within the parent chain* modify the name of the parent chain.

The parent chain includes the appropriate prefix for the number of carbon atoms; the letter *a*; the appropriate Greek prefix for the number of double bonds (in the parent chain); and the suffix *–ene*. As usual, each double bond has a single numerical location based on the alkene carbon atom with the lower value. Figure 4.9 depicts examples of how to name molecules with multiple alkene functional groups.

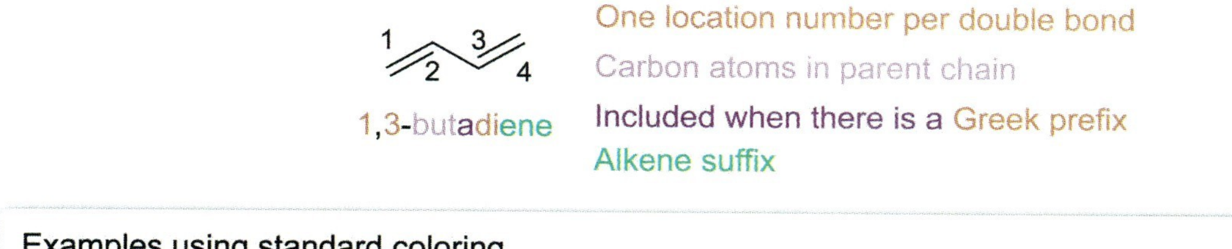

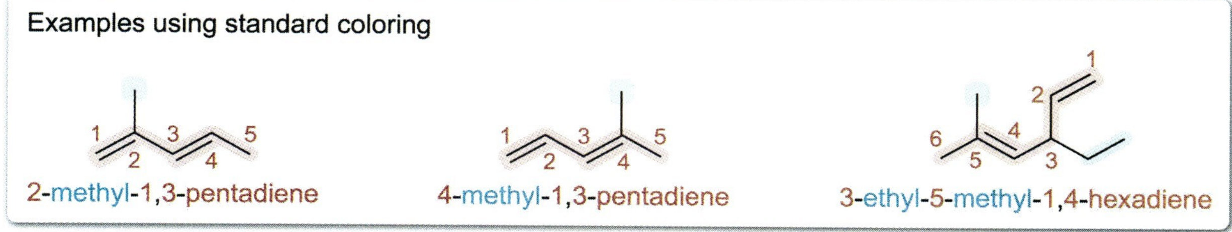

Figure 4.9 Names of molecules containing multiple alkene functional groups.

Older IUPAC guidelines placed the numerical location of functional groups that modified the parent chain *before* the *entire* parent chain (eg, 1-butene). This becomes more challenging when molecules contain multiple different functional groups that impact the name of the parent chain (see Lesson 4.6). In 1993, IUPAC adopted revisions to streamline the nomenclature process. By the revised guidelines, the location number is now listed *immediately before* the nomenclature component it describes (eg, but-1-ene). Although these revisions have now been in place for decades, both styles are still commonly seen.

Figure 4.10 provides examples of IUPAC names using the original and the revised format.

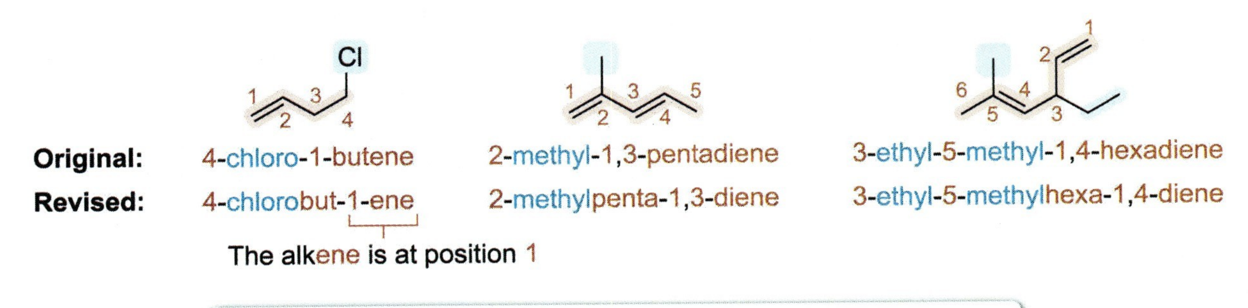

Figure 4.10 Comparison between original and revised IUPAC names of alkenes.

Alkene functional groups that are *not* part of the parent chain are named as **alkenyl groups**. Alkenyl groups are named like alkyl groups:

- The parent substituent chain is the longest chain of carbon atoms *containing the alkene functional group* starting from the point of attachment to the parent chain.
- The parent substituent chain is numbered from the point of attachment and has a *–yl* suffix.
- The position of each alkene functional group within the parent substituent chain is denoted by a single numerical locator based on the lower-value alkene carbon atom.

A one-carbon alkenyl group substituent is called a **methylidene group**. Methylidene groups occur when one carbon of an alkene functional group is within a parent chain and the other carbon is a substituent. Other examples of alkenyl groups are depicted in Figure 4.11.

To maintain an emphasis on nomenclature of substituent groups, many examples throughout the rest of this chapter utilize a carboxylic acid parent chain as the highest-priority functional group (see Lesson 4.6 for comprehensive coverage of functional group priority in nomenclature).

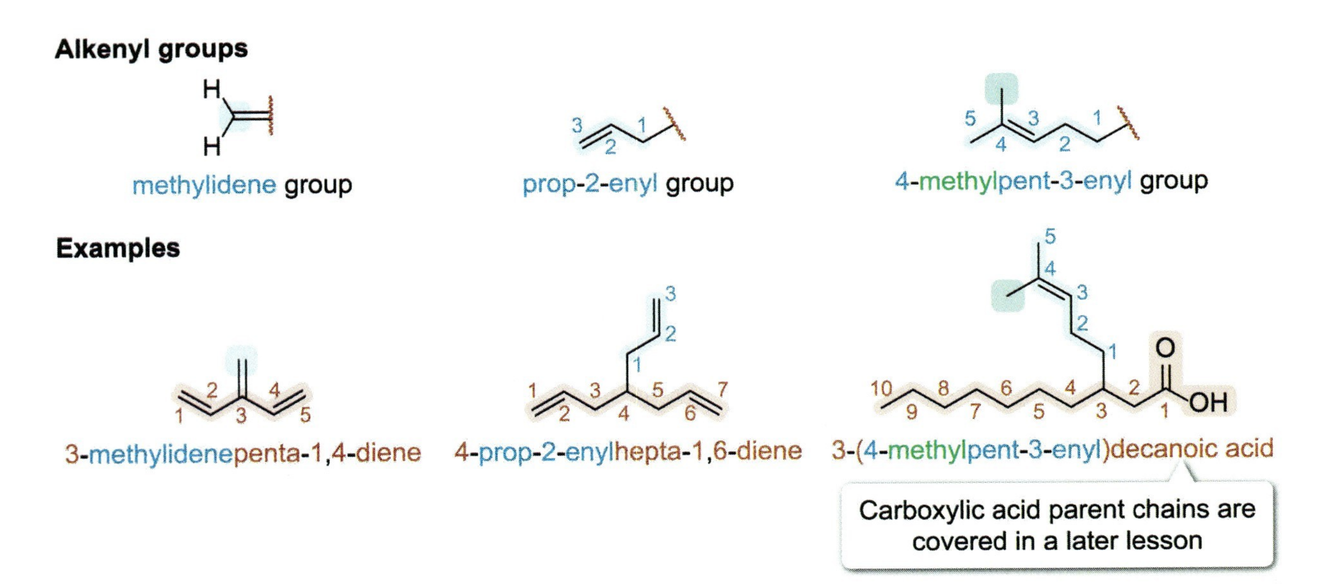

Figure 4.11 Alkenyl groups and examples.

Systematic IUPAC names must also include the absolute stereochemistry of any double bonds (see Concept 4.1.03).

Concept Check 4.4

While working in the lab, you observe that your laboratory partner has incorrectly named the following molecule (3*E*)-4-ethyl-2-methylidenehex-3-ene. Provide a brief explanation to help the laboratory partner arrive at the correct systematic name.

Solution

Note: The appendix contains the answer.

4.2.03 Alkynes

IUPAC nomenclature guidelines for alkynes are similar to the guidelines for alkenes (Concept 4.2.02):

- The parent chain of an alkyne is the longest chain of carbon atoms *that contains both carbon atoms of the alkyne functional group*.
- The parent chain of an alkyne is numbered to give the *alkyne carbon atoms* the lowest values.
- The parent chain is named using the prefix for the number of carbon atoms with a *–yne* ending.
- The location of the triple bond within the parent chain is provided by a single number: the *lower* value of the two carbon atoms that participate in the triple bond.

Parent chains that contain two or more alkyne functional groups include the appropriate prefix for the number of carbon atoms, the letter *a*, the appropriate Greek prefix for the number of triple bonds (in the parent chain), and the ending *–yne*. Figure 4.12 contains several examples of IUPAC names for alkynes.

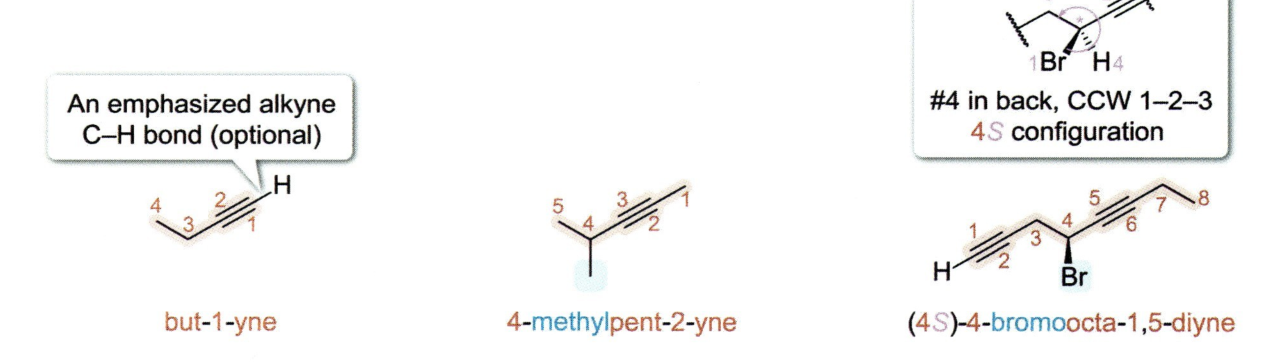

Figure 4.12 Examples and IUPAC names for alkynes.

Alkyne functional groups *not* part of the parent chain are named as **alkynyl groups** (Figure 4.13):

- The parent substituent chain is the longest chain of carbon atoms *containing the alkyne functional group*, starting from the point of attachment to the parent chain.
- The parent substituent chain is numbered from the point of attachment to the parent chain and has a *–yl* suffix.
- The position of the alkyne functional group within the parent substituent chain is denoted by a single numerical locator based on the lower-value alkene carbon atom.

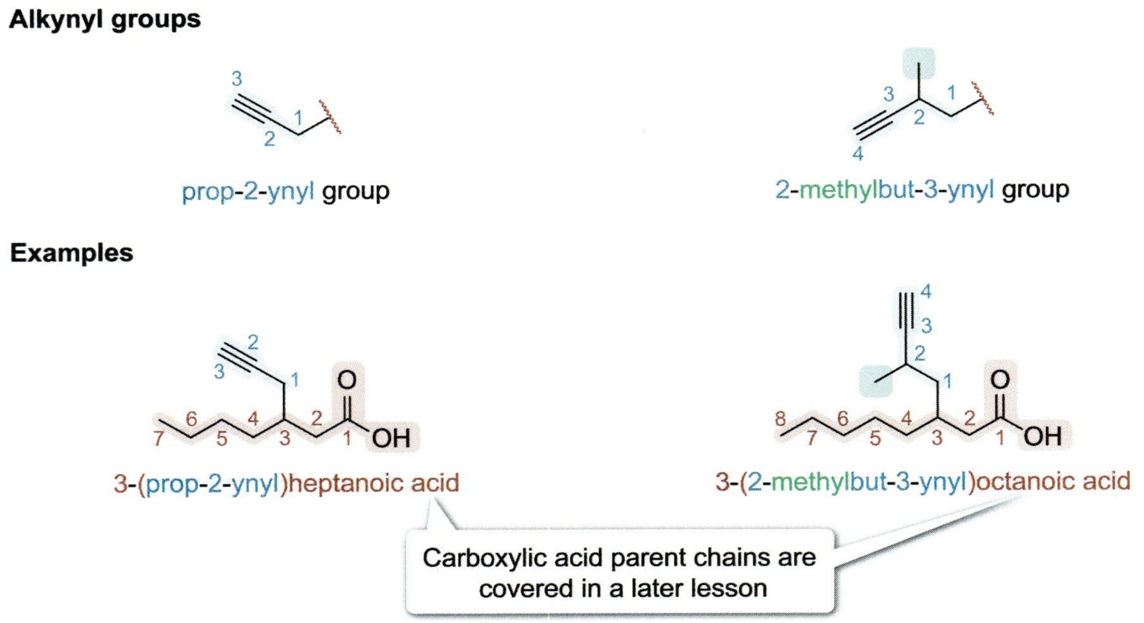

Alkynyl groups

prop-2-ynyl group

2-methylbut-3-ynyl group

Examples

3-(prop-2-ynyl)heptanoic acid

3-(2-methylbut-3-ynyl)octanoic acid

Carboxylic acid parent chains are covered in a later lesson

Figure 4.13 Alkynyl groups and examples.

4.2.04 Cycloalkanes, Cycloalkenes, and Cycloalkynes

Cyclic hydrocarbons contain one or more rings of carbon atoms. A **cycloalkane** contains only single bonds, a **cycloalkene** contains at least one double bond, and a **cycloalkyne** contains at least one triple bond. Aromatic rings represent a special type of cyclic molecule (see Concept 4.2.05).

A key difference between linear and cyclic hydrocarbons is the use of the *cyclo–* prefix, which appears directly before the prefix used to denote the number of carbon atoms in the parent chain. A comparison of linear alkanes versus their cycloalkane counterparts is depicted in Table 4.5.

Table 4.5 Alkanes versus cycloalkanes.

Number of carbon atoms	Linear (alkane)	Cyclic (cycloalkane)
1	1 CH_4 methane	—
2	2 1 $H_3C–CH_3$ ethane	—
3	3 2 1 propane	3 1 2 cyclopropane
4	4 3 2 1 butane	4 1 3 2 cyclobutane
5	5 4 3 2 1 pentane	4 5 1 3 2 cyclopentane
6	6 5 4 3 2 1 hexane	6 5 1 4 2 3 cyclohexane

IUPAC guidelines for cycloalkanes generally follow the guidelines for alkanes (Concept 4.1.01), with a few exceptions:

- The parent chain is the longest continuous chain of carbon atoms (*either* linear *or* cyclic).
- Parent chain numbering occurs sequentially for the atoms around the ring (*either* clockwise *or* counterclockwise) to achieve the lowest numbering for substituents.
- Cyclic hydrocarbons with only one substituent automatically have the substituent at position 1, a value that is typically not included in the final IUPAC name (as it is assumed).
- If two numbering methods are equivalent, the alphabetical order of the substituents determines the relative priority.

Table 4.6 Incorrect and correct examples of how to number the parent chain of a cycloalkane.

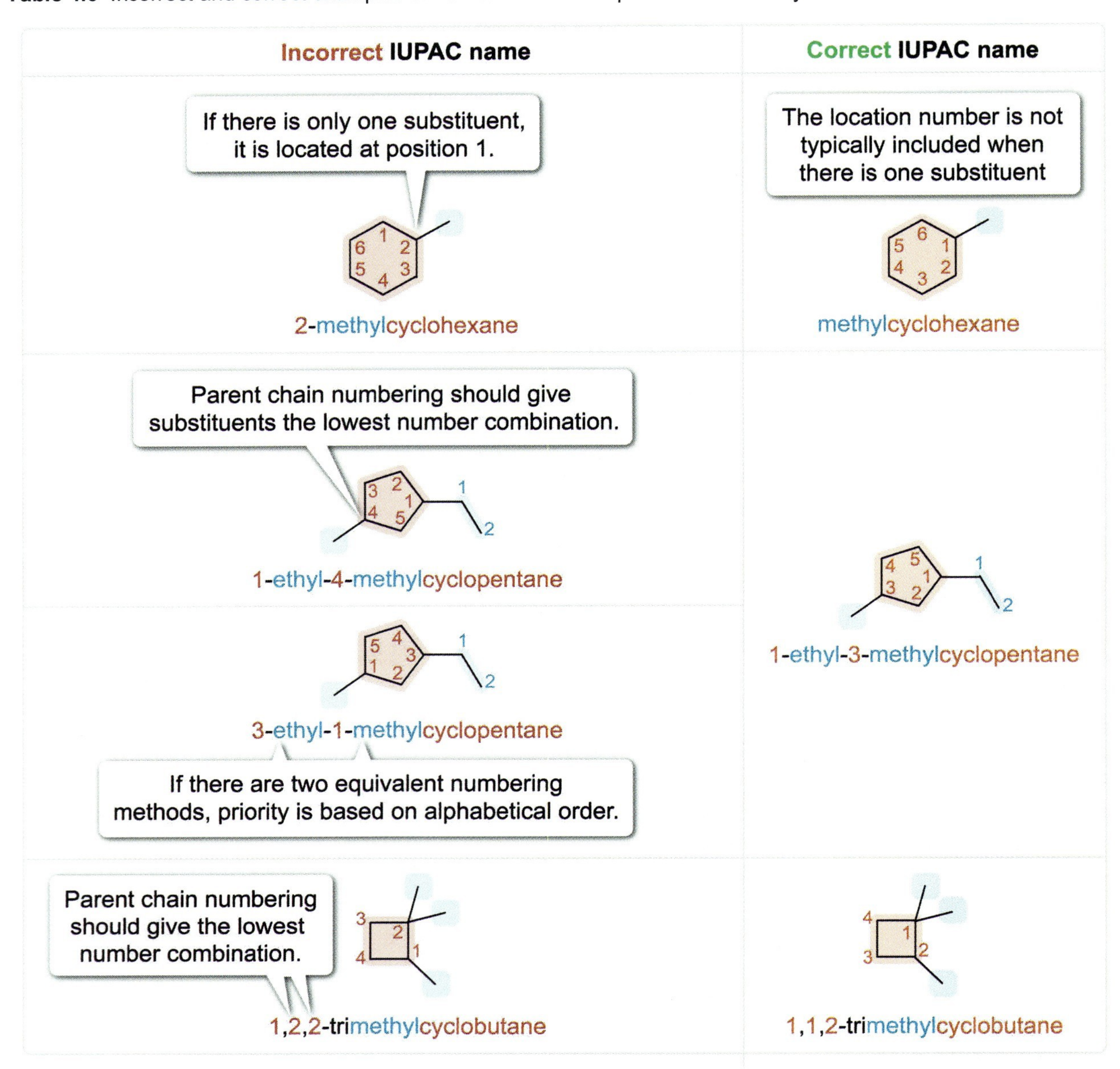

Because cycloalkenes and cycloalkynes contain double or triple bonds (higher priority functional groups relative to an alkane), the numbering of a cycloalkene or cycloalkyne chain *always* places position 1 at one carbon of the multiple bond. Ring numbering then proceeds to the other carbon of the multiple bond (position 2) and around the rest of the ring.

The parent rings of cycloalkenes and cycloalkynes are numbered with two tiers of priority. The greatest priority is given to ensuring that the carbon atoms of any *double or triple bonds* receive the lowest position numbering. If a tiebreaker is needed, the next highest priority is to have *substituents* receive the lowest possible numbering. Alphabetical order of substituents is used as the final tiebreaker between equivalent numbering methods. Figure 4.14 provides several examples of IUPAC names for cycloalkenes and cycloalkynes.

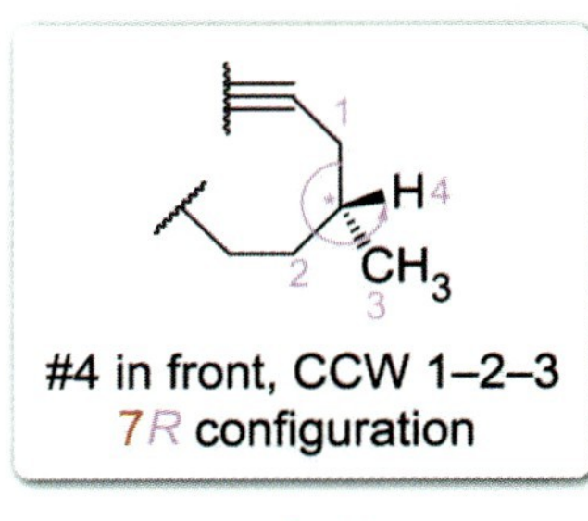

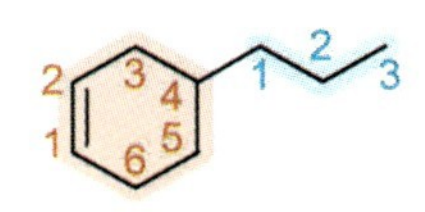

4-propylcyclohexene

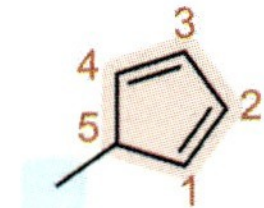

5-methylcyclopenta-1,3-diene

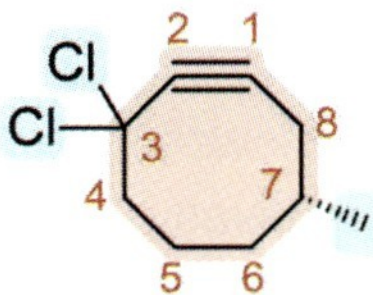

(7*R*)-3,3-dichloro-7-methylcyclooctyne

Cycloalkene and cycloalkyne rings have the following numbering priorities

1. Parent chain groups (double or triple bonds) with the lowest values
2. Substituents with the lowest values

Figure 4.14 Examples and IUPAC names for cycloalkenes and cycloalkynes.

When a cyclic hydrocarbon falls within a substituent group, it is named as either a **cycloalkyl group**, a **cycloalkenyl group**, or a **cycloalkynyl group**. The IUPAC names of these groups follow the guidelines for alkyl groups (Concept 4.1.02), alkenyl groups (Concept 4.2.02), or alkynyl groups (Concept 4.2.03), with the appropriate *cyclo–* prefix.

It *may not* always be necessary to include an *E* or *Z* absolute configuration within the name for a cycloalkene. Due to bond angle limitations, a *trans* double bond is not possible within cycloalkene rings smaller than seven carbon atoms. Cycloalkene Cmolecules with six or fewer ring carbons frequently do not include an *E* or *Z* absolute configuration for double bonds *within* the ring.

Concept Check 4.5

If the following molecular fragment is a substituent group, what would be the systematic name for this group?

Br
Br

R— = Point of attachment to another group

Solution

Note: The appendix contains the answer.

4.2.05 Aromatic Compounds

Benzene is a six-membered ring with alternating single and double bonds. While the terms "benzene" and "aromatic" can sometimes be used interchangeably, the modern definition of an aromatic compound is broader and not limited to benzene (see Lesson 6.5).

The name *benzene* has been adopted by IUPAC and can be used as the parent chain for substituted benzene derivatives (Figure 4.15). The IUPAC naming guidelines for substituted benzene derivatives closely mirror the conventions for cycloalkanes:

- If a high-priority functional group is present (Lesson 4.6), the benzene ring carbon that bears the high-priority functional group becomes position 1, and numbering proceeds around the ring to provide the lowest numbering for substituents.
- Benzene derivatives with only one substituent do not typically include location 1 in an IUPAC name.
- Benzene derivatives with two or more substituents number the ring carbons to provide substituents with the lowest possible numerical locations. Alphabetical order of the substituents determines the relative priority of equivalent numbering methods.

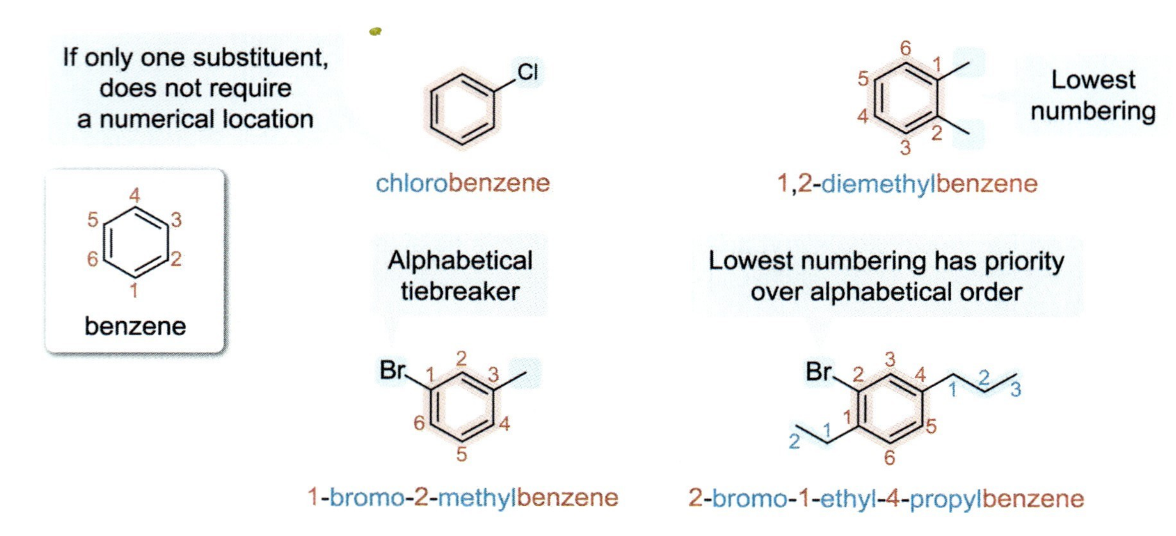

Figure 4.15 IUPAC names for substituted benzene derivatives.

Aromatic benzene rings can also be present as substituents (Figure 4.16). Two aromatic substituents that are most commonly encountered are the *phenyl* and *benzyl* groups. A **phenyl group** refers to a benzene ring substituent (C_6H_5–) and is often abbreviated "Ph," whereas a **benzyl group** refers to a benzene ring offset by one methylene group (C_6H_5–CH_2–) and is abbreviated "Bn."

In both cases, these terms can be used to describe the parent substituent chain. Both aromatic rings are numbered following the same IUPAC convention, where the ring carbon at the point of attachment is position 1. Similar to how "R" is used to indicate a generic alkyl group, sometimes the term **aryl group** (abbreviated "Ar") is used to indicate a generic aromatic-containing structural region.

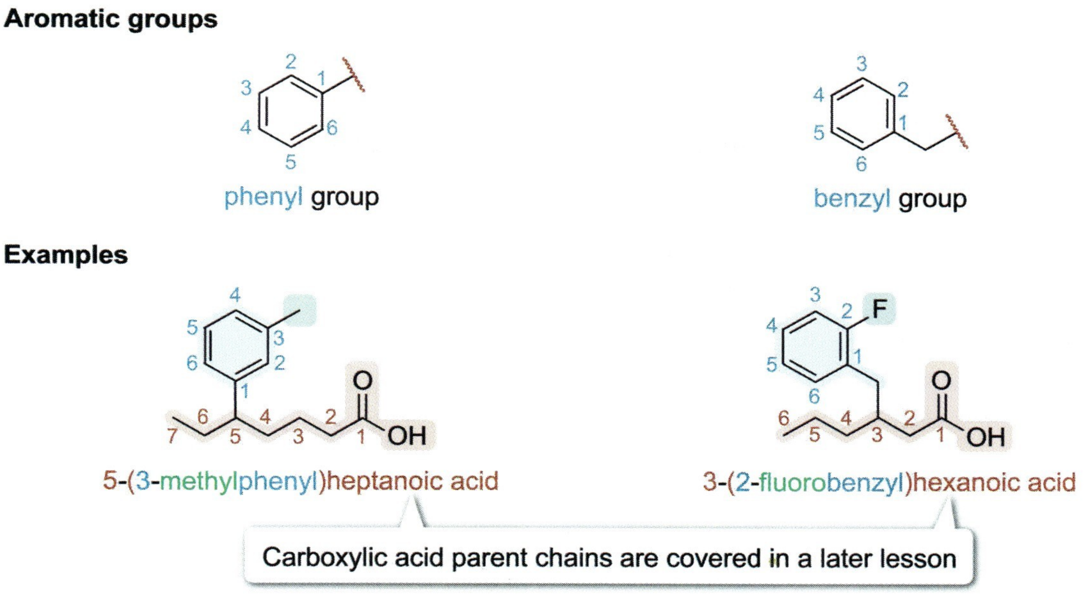

Aromatic groups

phenyl group

benzyl group

Examples

5-(3-methylphenyl)heptanoic acid

3-(2-fluorobenzyl)hexanoic acid

Carboxylic acid parent chains are covered in a later lesson

Figure 4.16 Aromatic substituent groups and examples.

Just as the term **heteroatom** refers to any atom that is *not* carbon or hydrogen (eg, oxygen, nitrogen, halogens), the term **heterocycle** describes a cyclic structure containing one or more heteroatoms *within the ring*. **Aromatic heterocycles** are a subclass of molecules that can also meet the criteria of aromaticity (see Concept 6.5.02).

Figure 4.17 depicts the names, structures, and IUPAC parent chain numbering for several aromatic heterocycles:

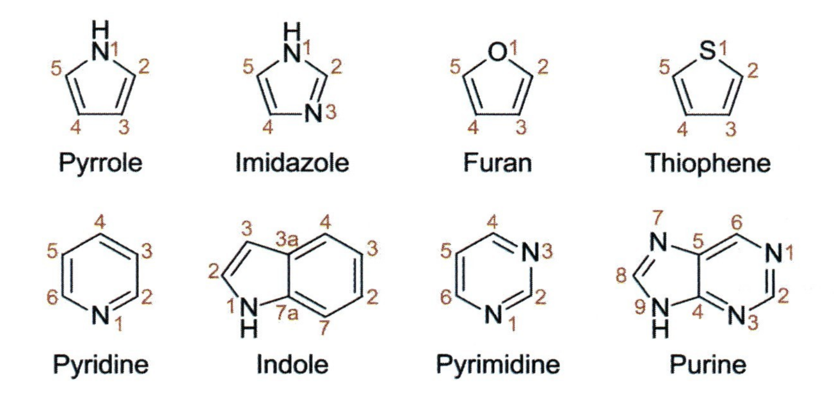

Figure 4.17 Structures, IUPAC names, and parent chain numbering for aromatic heterocycles.

Lesson 4.3

IUPAC Nomenclature of Oxygen-Containing Groups

Introduction

Lesson 4.1 introduced how systematic (IUPAC) chemical names are generated for the alkanes, concepts that were expanded to the other classes of hydrocarbons in Lesson 4.2. This lesson explores IUPAC nomenclature guidelines for the major organic functional groups that contain oxygen atoms.

4.3.01 Alcohols

An alcohol is a molecule that contains one or more **hydroxyl groups (–OH)** attached to a carbon atom within an alkyl or aryl group.

Whereas the IUPAC names for alkenes and alkynes modify the *middle portion* of a parent chain with either *–en–* or *–yn–*, alcohols are the first example of a functional group that modifies the *suffix* of the parent chain. These are the steps for generating a systematic IUPAC name for an alcohol:

- The parent chain is the longest continuous chain of carbon atoms *that contains the carbon attached to the hydroxyl functional group*.
- The parent chain is numbered to give the *carbon attached to the hydroxyl group* the lowest value.
- The parent chain is named using the prefix for the number of carbon atoms, a middle modifier of *–an–*, and an *–ol* suffix (eg, *but* + *an* + *ol* = butanol).
- The numerical location of the hydroxyl groups is located either before the parent chain (eg, 1-butanol, original IUPAC guidelines) or before the *–ol* suffix (eg, butan-1-ol, revised IUPAC guidelines; see Concept 4.2.02).

If a molecule contains multiple alcohol functional groups, the parent chain includes the appropriate prefix for the number of carbon atoms; a middle modifier of *–ane–*; the appropriate Greek prefix for the number of hydroxyl groups (in the parent chain); and the suffix *–ol*. Each hydroxyl group is given a numerical location. Figure 4.18 depicts examples of IUPAC names for alcohols.

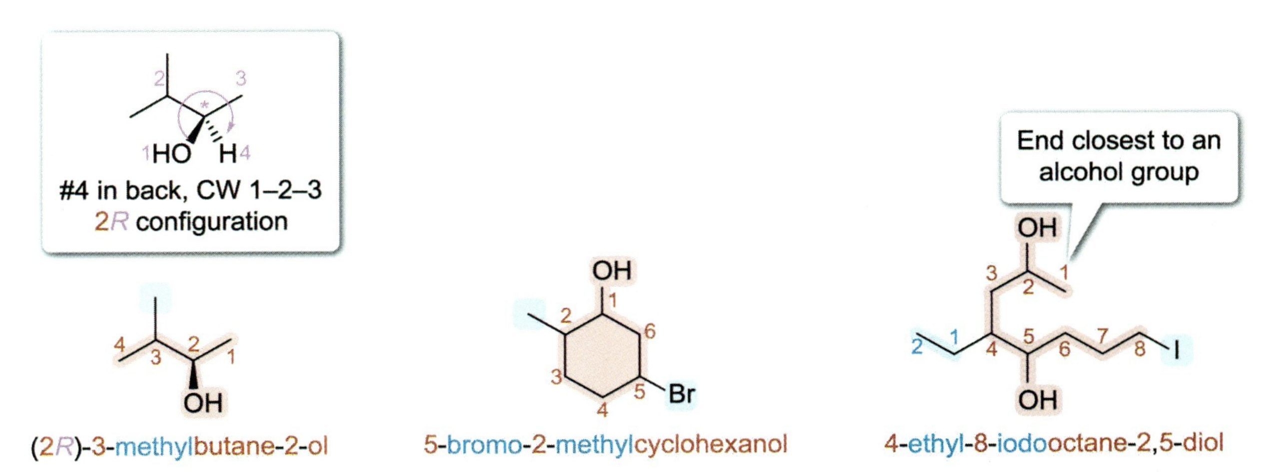

Figure 4.18 Examples and IUPAC names for alcohols.

When a hydroxyl group is attached to an aromatic benzene ring, it is called a **phenol** (Figure 4.19). The name *phenol* has been adopted by IUPAC and can be used as the parent chain for substituted phenol derivatives.

The IUPAC naming guidelines for substituted phenol derivatives closely mirror the conventions for aromatic rings: The aromatic ring carbon bearing the hydroxyl group is position 1, and numbering proceeds around the ring to provide the lowest numbering for substituents. If two numbering methods are equivalent, the alphabetical order of the substituents determines the relative priority.

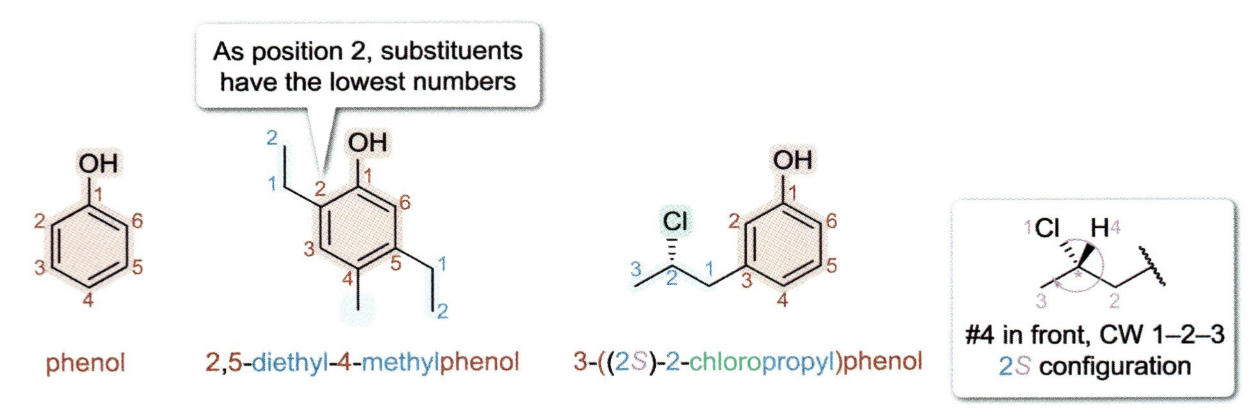

Figure 4.19 Examples and IUPAC names for phenols.

An alcohol that either is *not* the highest-priority functional group (Lesson 4.6) or is present within a substituent chain is named as a substituent group. In these cases, the term *hydroxy* is used in combination with an appropriate numerical location (Figure 4.20).

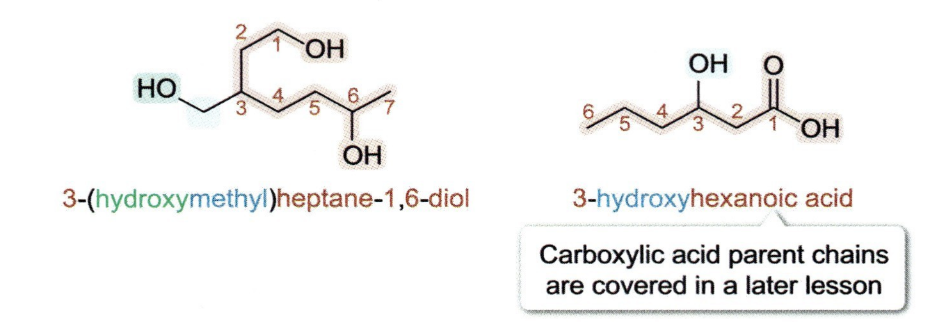

Figure 4.20 Examples and IUPAC names of molecules containing alcohol substituents.

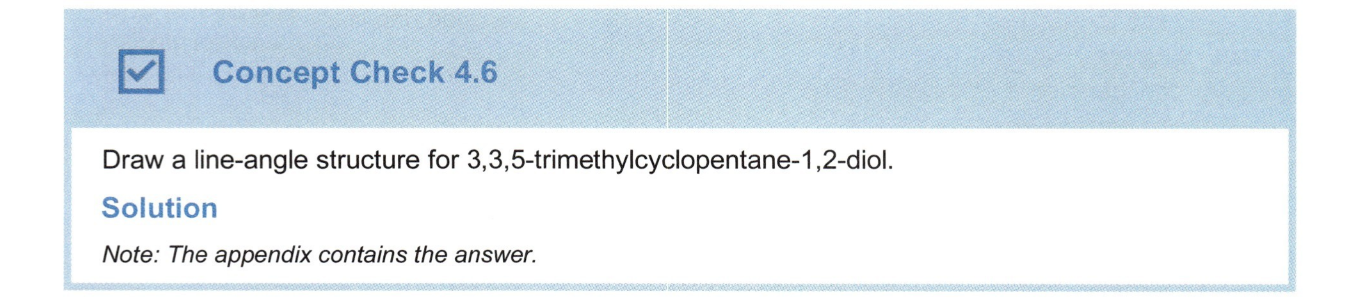

Concept Check 4.6

Draw a line-angle structure for 3,3,5-trimethylcyclopentane-1,2-diol.

Solution

Note: The appendix contains the answer.

4.3.02 Ethers

In an ether, the alkyl or aryl "R" groups can have either the same structure (**symmetrical ether**) or different structures (**unsymmetrical ether**). Only one of the attached carbon groups can be taken as either the parent chain or the substituent parent chain. The rest of the atoms in an ether (ie, oxygen atom + one alkyl group) are named as an **alkoxy group**, where the *alk–* is replaced by the appropriate prefix that indicates the number of carbon atoms in the R group. For example, a one-carbon $–OCH_3$ group is named *methoxy*, and a two-carbon $–OCH_2CH_3$ group is named *ethoxy*.

Alkoxy groups include the appropriate numerical locator for its point of attachment to the parent chain. Figure 4.21 provides several examples of ether-containing molecules and their corresponding IUPAC names.

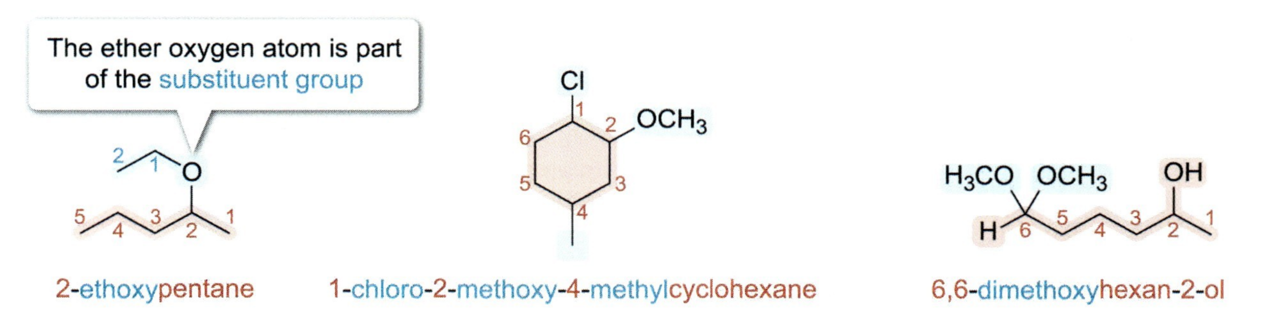

Figure 4.21 Examples and IUPAC names for ethers.

Figure 4.22 depicts the names, structures, and IUPAC parent chain numbering for the **heterocyclic ethers**.

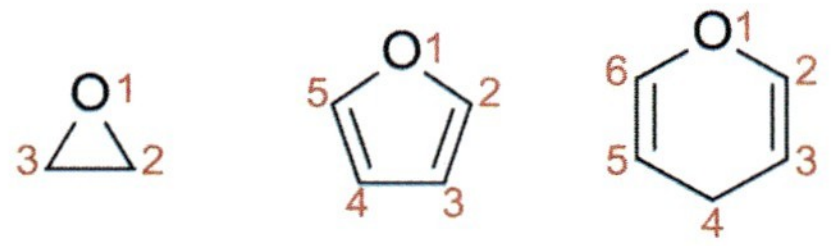

Epoxide (Oxirane) Furan Pyran

Figure 4.22 Structures, IUPAC names, and parent chain numbering for select heterocyclic ethers.

An **epoxide** is a heterocyclic ether that forms a three-membered ring. When an epoxide ring is named as a substituent, the term *epoxy* is used alongside *two* numerical locators, one number for each point of attachment of the epoxide oxygen atom to the parent chain (Figure 4.23). As an example, an IUPAC name that includes *4,5-epoxy* indicates that the oxygen atom of an epoxide ring is attached to positions 4 and 5 (of whatever parent chain being referenced).

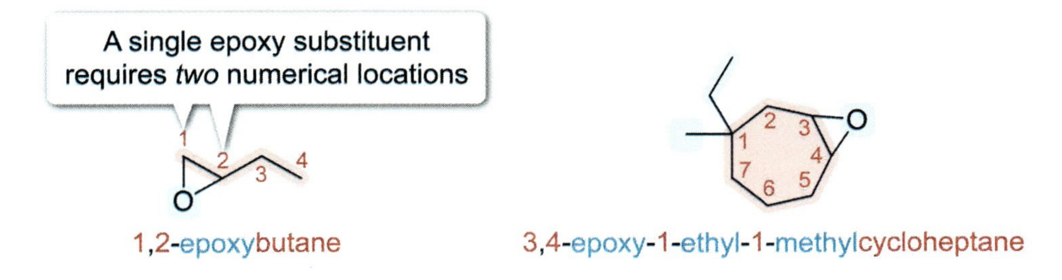

Figure 4.23 Examples and IUPAC names for epoxide-containing molecules.

4.3.03 Aldehydes

An aldehyde functional group is *always* at the end of a carbon chain and may be represented by the notation –CHO. Like alcohols (Concept 4.3.01), aldehydes are a functional group that modifies the *suffix* of the parent chain. These are the steps for generating a systematic IUPAC name for an aldehyde:

- The parent chain is the longest continuous chain of carbon atoms *that contains the carbon of the aldehyde functional group*.
- The parent chain is numbered to give the *carbon of the aldehyde functional group* the lowest value. For most aldehydes, the aldehyde carbon atom is at position 1.
- The parent chain is named using the prefix for the number of carbon atoms, a middle modifier of –*an*–, and an –*al* suffix (eg, *but* + *an* + *al* = butanal).
- If the parent chain contains only one or two aldehydes, numerical locations are not necessary (the aldehyde functional groups are assumed to be at the first and last carbon atom of the linear parent chain).

If a parent chain contains two aldehyde functional groups, its name includes the appropriate prefix for the number of carbon atoms; a middle modifier of –*ane*–; the Greek prefix *di*–; and the suffix –*al* (eg, pentanedial). When a –CHO group is directly attached to an atom in a ring, its IUPAC name is the name of the ring followed by the suffix –*carbaldehyde* (eg, cyclopentanecarbaldehyde). Figure 4.24 depicts examples of IUPAC names for aldehydes.

2-methylpropanal

3-fluoro-2-(methylethyl)pentanedial

2-ethylcyclobutanecarbaldehyde

Aldehyde functional groups rarely require a numerical locator

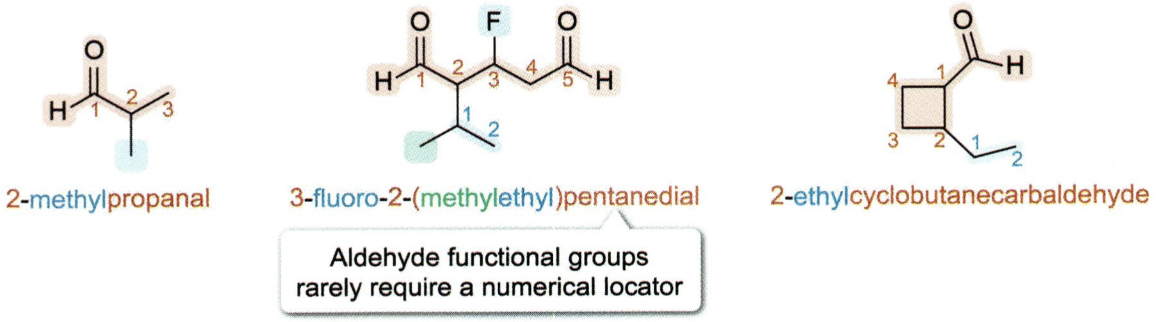

Figure 4.24 Examples and IUPAC names for aldehydes.

An aldehyde may be named as a substituent if it either is *not* the highest-priority functional group (Lesson 4.6) or is present within a substituent chain. In these cases, the term *formyl* is used to refer to the –CHO substituent. Unlike most other aldehyde groups, a formyl group requires an appropriate numerical point of attachment (Figure 4.25).

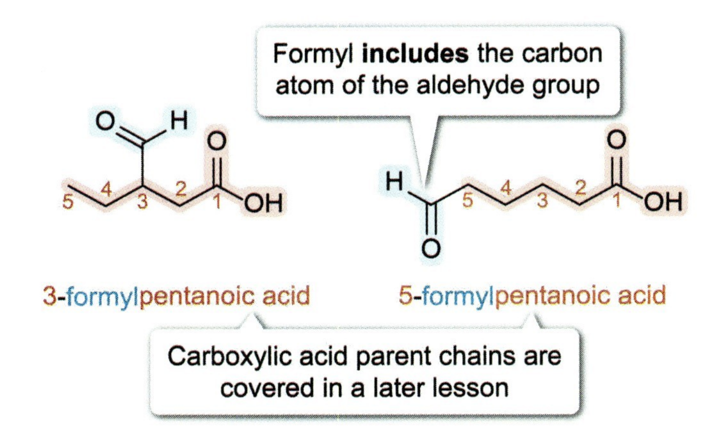

Figure 4.25 Examples and IUPAC names of molecules containing aldehyde substituents.

4.3.04 Ketones

While an aldehyde functional group is *always* at the end of a carbon chain, a ketone can *never* be at the end of a carbon chain. The notations –C(O)– or –CO– are commonly used to represent a ketone group. These are the steps for generating a systematic IUPAC name for a ketone:

- The parent chain is the longest continuous chain of carbon atoms *that contains the carbon of the ketone functional group.*
- The parent chain is numbered to give the *carbon of the ketone functional group* the lowest value.
- The parent chain is named using the prefix for the number of carbon atoms, a middle modifier of –*an*–, and an –*one* suffix (eg, *but* + *an* + *one* = butanone).
- The numerical location of the carbonyl group is located either before the parent chain (eg, 2-butanone, original IUPAC guidelines) or before the –*one* suffix (eg, butan-2-one, revised IUPAC guidelines; see Concept 4.2.02).

If a parent chain contains two (or more) ketone functional groups, its name includes the appropriate prefix for the number of carbon atoms; a middle modifier of –*ane*–; the appropriate Greek prefix for the number of ketones (in the parent chain); and the suffix –*one*. As usual, each ketone group is given a numerical location. Figure 4.26 depicts examples of IUPAC names for ketones.

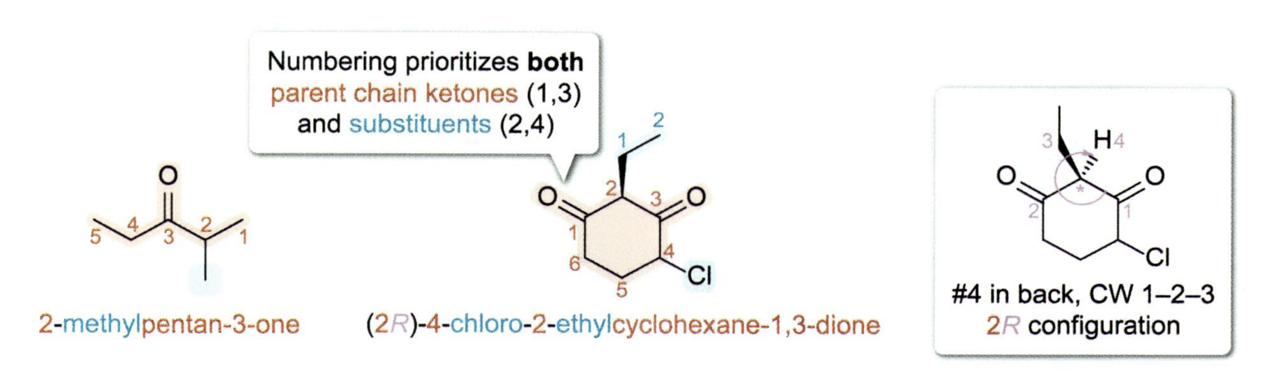

Figure 4.26 Examples and IUPAC names for ketones.

A ketone may be named as a substituent group if it either is *not* the highest-priority functional group (Lesson 4.6) or is present within a substituent chain. In these cases, the term *oxo* is used in combination with an appropriate numerical location.

An **acyl group** is a general term for the structure R–CO–, where the carbonyl carbon is the point of attachment of the acyl group to the rest of the molecule. By IUPAC nomenclature guidelines, acyl group substituents also classified as ketones can be named in one of two ways:

- The position of the carbonyl can be indicated using an *oxo*– prefix (eg, 1-oxopropyl group).
- The acyl group can be named as an **alkanoyl group**, where the *alk* is replaced by the appropriate prefix for the number of carbon atoms in the substituent (eg, propanoyl group).

Figure 4.27 provides examples and names of molecules containing ketone substituents.

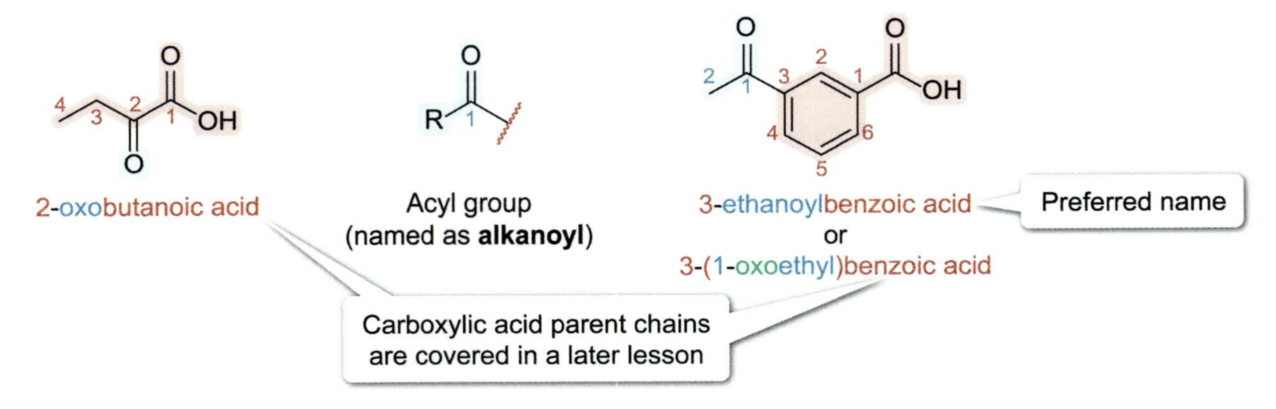

Figure 4.27 Examples and IUPAC names of molecules containing ketone substituents.

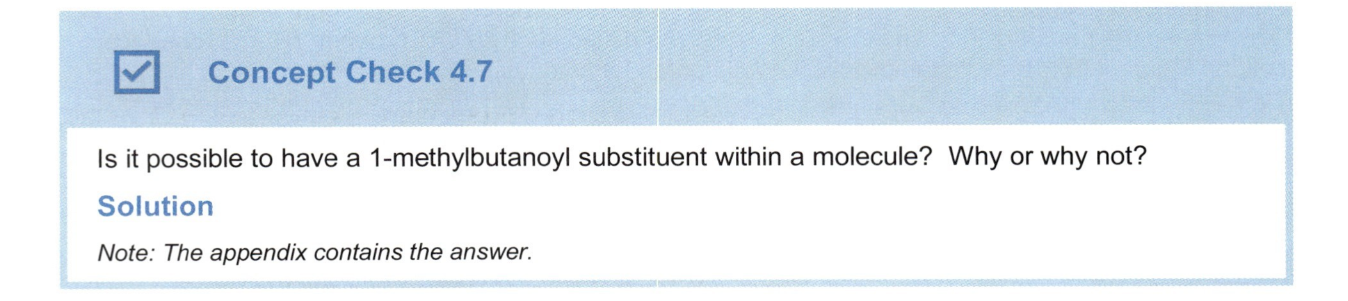

Concept Check 4.7

Is it possible to have a 1-methylbutanoyl substituent within a molecule? Why or why not?

Solution

Note: The appendix contains the answer.

4.3.05 Carboxylic Acids

Like an aldehyde, a carboxylic acid functional group is *always* at the end of a carbon chain. The notations –CO_2H or –COOH are commonly used to represent a carboxylic acid group. These are the steps for generating a systematic IUPAC name for a carboxylic acid:

- The parent chain is the longest continuous chain of carbon atoms *that contains the carbon of the carboxylic acid functional group*.
- The parent chain of a carboxylic acid is numbered to give the *carbon of the carboxylic acid functional group* the lowest value. For most carboxylic acids, the carboxylic acid carbon atom is at position 1 in the parent chain.
- The parent chain is named using the prefix for the number of carbon atoms, a middle modifier of –*an*–, and an –*oic acid* suffix (eg, *but* + *an* + *oic acid* = butanoic acid).
- If the parent chain contains only one or two carboxylic acids, numerical locations are not necessary (the carboxylic acid functional groups are assumed to be at the first and last carbon atoms of the linear parent chain).

If a parent chain contains two carboxylic acid functional groups, its name includes the appropriate prefix for the number of carbon atoms, a middle modifier of –*ane*–, the Greek prefix *di*–, and the suffix –*oic acid* (eg, hexanedioic acid). Similar to aldehydes, when a –CO_2H group is directly attached to an atom in a ring, the IUPAC name of the molecule is the name of the ring followed by the suffix –*carboxylic acid* (eg, cyclopentanecarboxylic acid). Figure 4.28 depicts examples of IUPAC names for carboxylic acids.

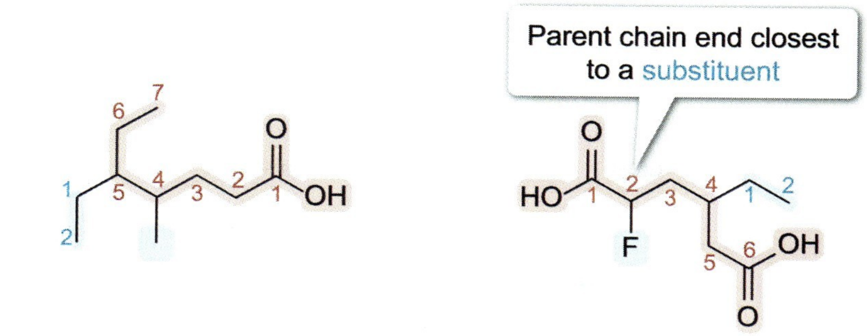

Parent chain end closest to a substituent

5-ethyl-4-methylheptanoic acid

4-ethyl-2-fluorohexanedioic acid

Figure 4.28 Examples and IUPAC names for carboxylic acids.

When a carboxylic acid group is attached to an aromatic benzene ring, it is called **benzoic acid** (Figure 4.29). Like phenol (Concept 4.3.01), benzoic acid has been adopted by IUPAC and can be used as the parent chain for substituted benzoic acid derivatives.

By the IUPAC naming guidelines for substituted benzoic acid derivatives, the aromatic ring carbon bearing the carboxylic acid group is position 1, and numbering proceeds around the ring to provide the lowest numbering for substituents. If two numbering methods are equivalent, the alphabetical order of the substituents determines the relative priority.

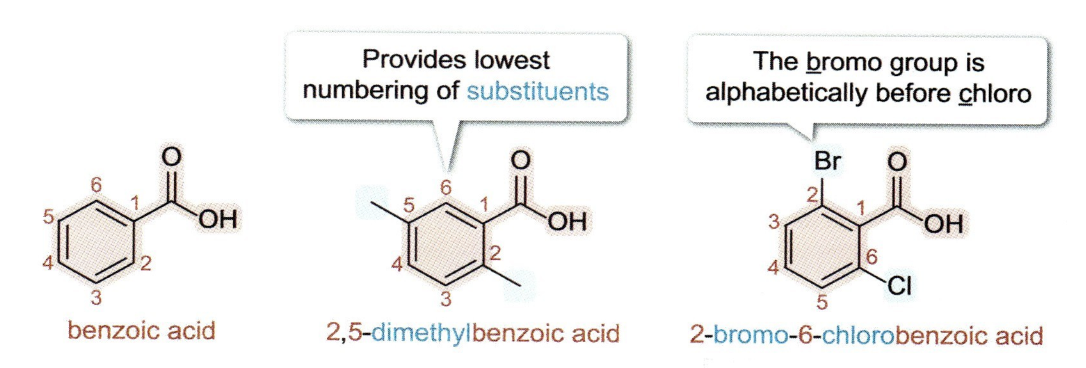

Figure 4.29 Examples and IUPAC names for benzoic acids.

If a carboxylic acid in a molecule is present within a substituent chain, the term *carboxy* is used to refer to the $-CO_2H$ substituent. Unlike most other carboxylic acid groups, a carboxy group requires an appropriate numerical point of attachment (Figure 4.30).

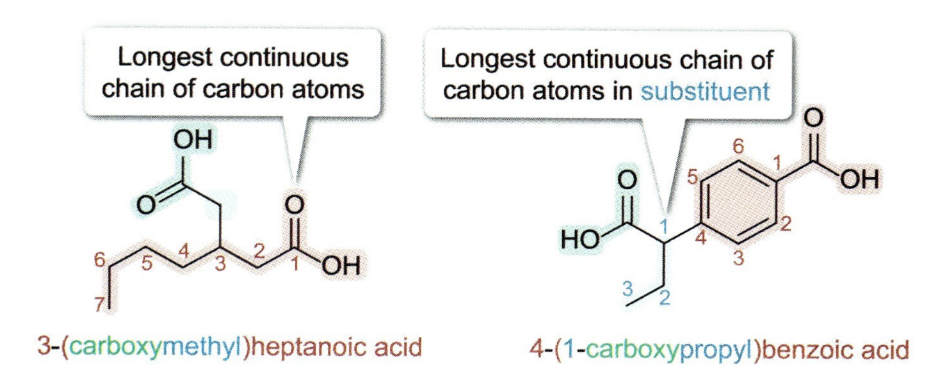

Figure 4.30 Examples and IUPAC names of molecules containing carboxylic acid substituents.

A carboxylic acid (–CO_2H) that has lost its acidic proton (H^+) is called a **carboxylate ion** (–CO_2^-). The IUPAC nomenclature guidelines for carboxylate ions are like the guidelines for carboxylic acids:

- When a carboxylate ion is the primary functional group in a parent chain, the suffix changes from –*oic acid* to –*oate*.
- Molecules that would normally be named using the suffix –*carboxylic acid* are instead named using the suffix –*carboxylate*.
- When a proton is lost, the parent chain *benzoic acid* becomes *benzoate*.
- A carboxylate ion named as a substituent uses the term *carboxylato*.

Table 4.7 Comparison of IUPAC nomenclature guidelines for carboxylic acids versus carboxylate ions.

	Carboxylic acid		Carboxylate	
Structure type	**IUPAC suffix or term**	**Example**	**IUPAC suffix or term**	**Example**
Parent chain suffix	–oic acid	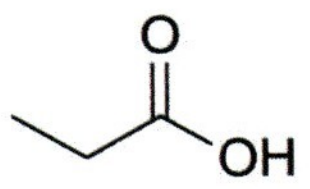 propanoic acid	–oate	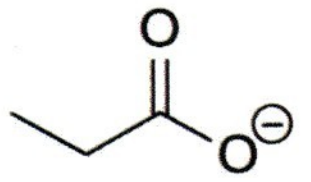 propanoate
Cycloalkane suffix	–carboxylic acid	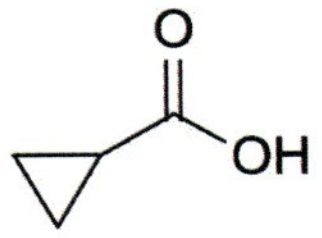 cyclopropanecarboxylic acid	–carboxylate	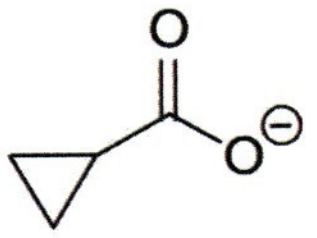 cyclopropanecarboxylate
Aromatic parent chain	–benzoic acid	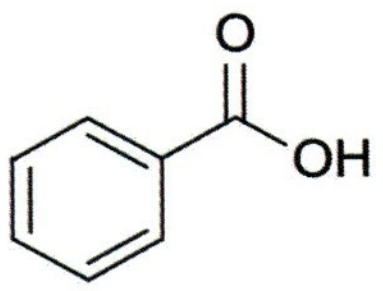 benzoic acid	benzoate	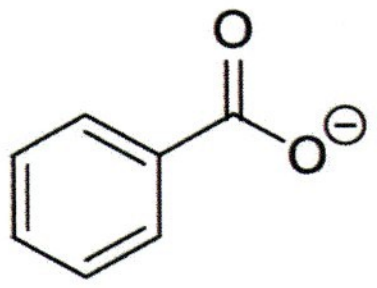 benzoate
Substituent	carboxy	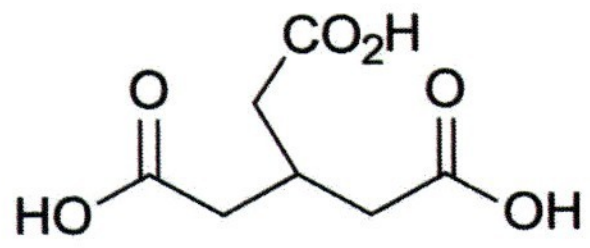 3-(carboxymethyl)pentanedioic acid	carboxylate	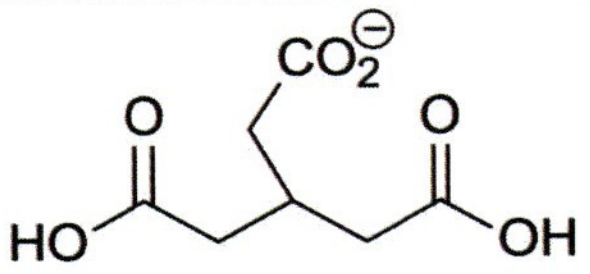 3-(carboxymethyl)pentanedioic acid

4.3.06 Carboxylic Acid Derivatives

Carboxylic acid derivatives are molecules that have different heteroatom groups (–Z) substituted for the –OH group in a carboxylic acid. The relevant carboxylic acid derivatives include:

- **Esters**, Z = –OR
- **Acid halides**, Z = halogen
- **Anhydrides**, Z = –OC(O)R
- **Amides**, Z = N(R_1)(R_2)

Esters

These are the steps for generating an IUPAC name for an ester:

- The name contains *two* components and is named as an *alkyl carboxylate*. Each component has its own system of numbering the carbon atoms.
- The *alkyl* portion of the ester is named and numbered the same as other alkyl groups, where the point of attachment in the alkyl group is given position 1 (Concept 4.1.02).
- The *carboxylate* portion of the ester is named and numbered the same as a carboxylate ion, which uses the *–oate* suffix and the carboxylate carbon atom is given position 1 (Concept 4.3.05).
- Substituents attached to *either* component of an ester are included within the names for their respective portions.

If a parent chain contains two ester functional groups, its name includes the appropriate prefix for the number of carbon atoms; a middle modifier of *–ane–*; the Greek prefix *di–*; and the suffix *–oate*. Ester alkyl groups that are identical are named as a *dialkyl* group (eg, dimethyl). Ester alkyl groups that are different are listed in alphabetical order. Figure 4.31 depicts examples of IUPAC names for esters.

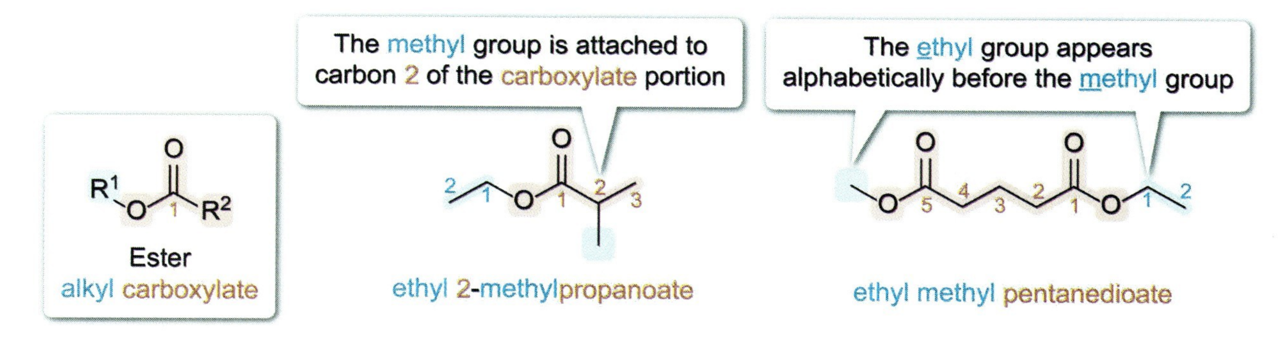

Figure 4.31 Examples and IUPAC names for esters.

A lactone is a cyclic variant of an ester where the alkyl chain and the carboxylate chain are connected to one another. The IUPAC names for lactones includes the appropriate prefix for the number of carbon atoms in the ring, the middle modifier of *–an–*, and the suffix *–olactone*. The carbon atoms in the lactone chain are numbered starting from the carbonyl carbon. The numerical point of attachment of the lactone oxygen atom is inserted between the "o" and *lactone*, which is also one less than the number of atoms in the lactone ring. Figure 4.32 depicts examples of IUPAC names for lactones.

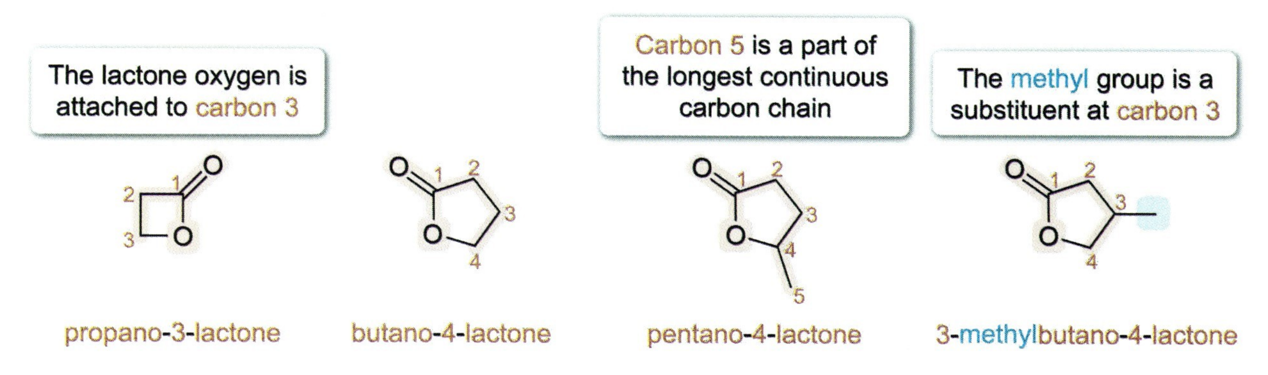

Figure 4.32 Examples and IUPAC names for lactones.

If an ester is present within a substituent chain:

- The term *alkoxycarbonyl* refers to a –C(O)OR substituent when the substituent is connected through the ester carbonyl carbon. The *alk* is replaced by the appropriate prefix for the number of carbon atoms in the substituent (eg, methoxycarbonyl).
- The term *alkanoyloxy* refers to a –OC(O)R substituent when the substituent is connected through the ester oxygen atom (eg, ethanoyloxy).

Alkoxycarbonyl and alkanoyloxy groups both require an appropriate numerical point of attachment (Figure 4.33).

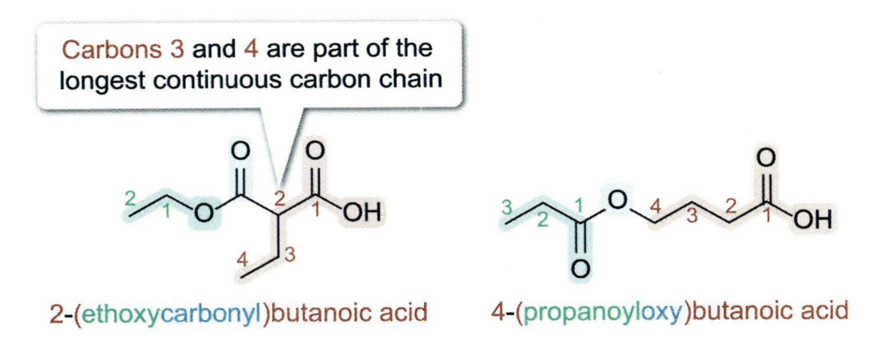

Figure 4.33 Examples and IUPAC names of molecules containing ester substituents.

Acid Halides

These are the steps for generating an IUPAC name for an acid halide:

- The name of an acid halide contains *two* components and is named as an *alkanoyl halide*.
- The *alkanoyl* portion is named and numbered the same as other alkanoyl groups, where the point of attachment is given position 1 (Concept 4.3.04).
- The *halide* portion of the acid halide uses the names for the halogen anions: *fluoride*, *chloride*, *bromide*, and *iodide*.
- Substituents attached to the alkanoyl component of an acid halide are included within the name for that portion.

If a parent chain contains two acid halide functional groups, the parent chain includes the appropriate prefix for the number of carbon atoms; a middle modifier of *–ane–*; the Greek prefix *di–*; and the suffix –*oyl* (eg, pentanedioyl). Normally, the halide groups of the acid halide are identical and named as a *dihalide* group (eg, dichloride). Figure 4.34 depicts examples of IUPAC names for acid halides.

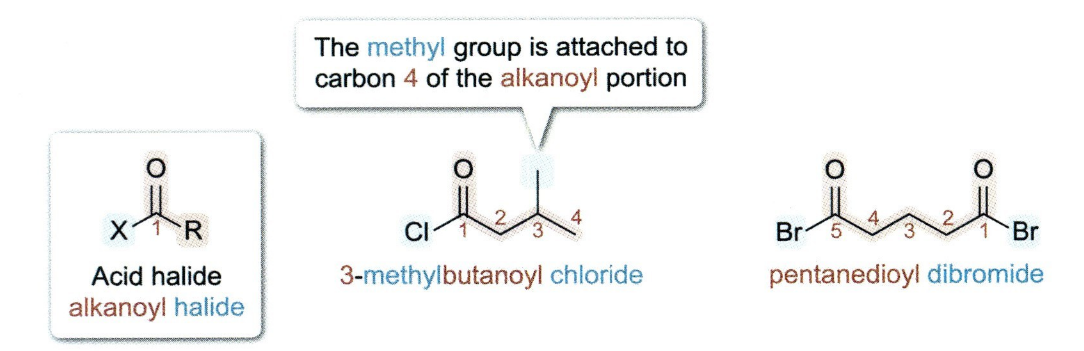

Figure 4.34 Examples and IUPAC names for acid halides.

For an acid halide within a substituent chain, the term *halocarbonyl* refers to the –C(O)X substituent, where the *halo* is replaced by the appropriate halogen term. A halocarbonyl group requires an appropriate numerical point of attachment (Figure 4.35).

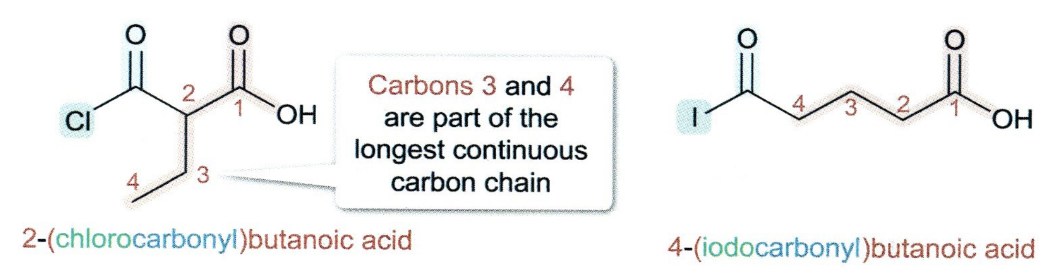

Figure 4.35 Examples and IUPAC names of molecules containing acid halide substituents.

Anhydrides

These are steps for generating an IUPAC name for an anhydride:

- The name contains up to *three* components and is named as an *acid (acid) anhydride*. Each of the *acid* components has its own system of numbering the carbon atoms.
- The *acid* portions of the anhydride are named and numbered the same as a carboxylic acid, which uses the *–oic* suffix and the carbonyl carbon atom is given position 1 (Concept 4.3.05).
- If the anhydride is **symmetrical**, only *one* acid is named (*without* the Greek prefix *di–*) and followed with *anhydride*. If the anhydride is **mixed (unsymmetrical)**, *both* acids are named (in alphabetical order) and followed by *anhydride*.
- Substituents attached to *either* acid component of an anhydride are included within the names for their respective portions.

Figure 4.36 depicts examples of IUPAC names for anhydrides.

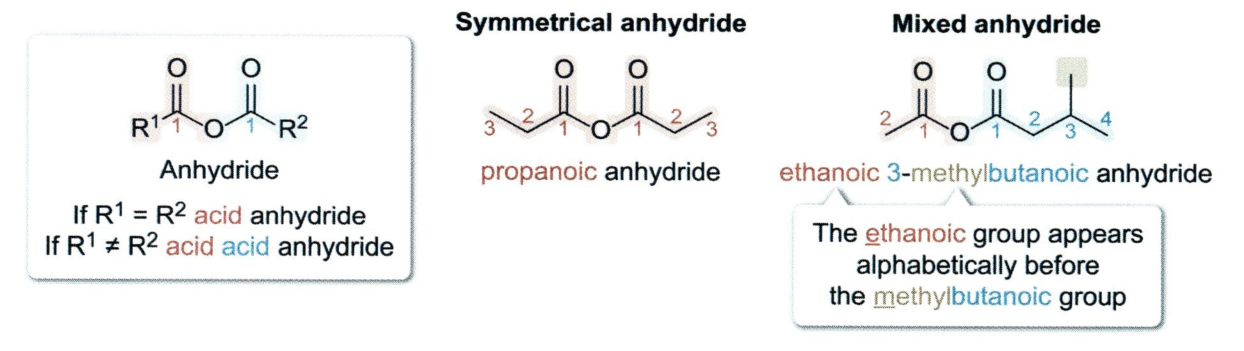

Figure 4.36 Examples and IUPAC names for anhydrides.

Amides

- Amide nomenclature is comprehensively discussed in Concept 4.4.02.

Concept Check 4.8

Identify the carboxylic acid derivative found in the following molecule, then provide its systematic (IUPAC) name.

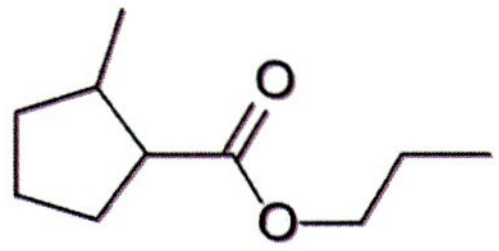

Solution

Note: The appendix contains the answer.

Lesson 4.4

IUPAC Nomenclature of Nitrogen-Containing Groups

Introduction

This lesson continues the exploration of IUPAC nomenclature guidelines, discussing the major organic functional groups that contain nitrogen atoms.

4.4.01 Amines

The IUPAC nomenclature guidelines for amines are similar to the guidelines for alcohols (Concept 4.3.01), with a few key differences. These are the steps for generating a systematic IUPAC name for an amine:

- The parent chain of an amine is the longest chain of carbon atoms *that contains the carbon attached to the amine nitrogen atom.*
- The parent chain of an amine is numbered to give the *carbon attached to the nitrogen atom* the lowest value.
- The parent chain is named using the prefix for the number of carbon atoms, a middle modifier of –*an*–, and an –*amine* suffix (eg, *but* + *an* + *amine* = butanamine).
- The numerical location of the nitrogen atom is placed either before the parent chain (eg, 1-butanamine, original IUPAC guidelines) or before the –*amine* suffix (eg, butan-1-amine, revised IUPAC guidelines; see Concept 4.2.02).

If a molecule contains two (or more) amine functional groups, the parent chain is the longest carbon chain that contains the *maximum number carbons bound to amine nitrogen atoms*. In these cases, the parent chain includes the appropriate prefix for the number of carbon atoms, a middle modifier of –*ane*–, the appropriate Greek prefix for the number of amine groups (in the parent chain), and the suffix –*amine* (eg, butanediamine). The attachment point of each amine group is given a numerical location. Figure 4.37 depicts examples of IUPAC names for amines.

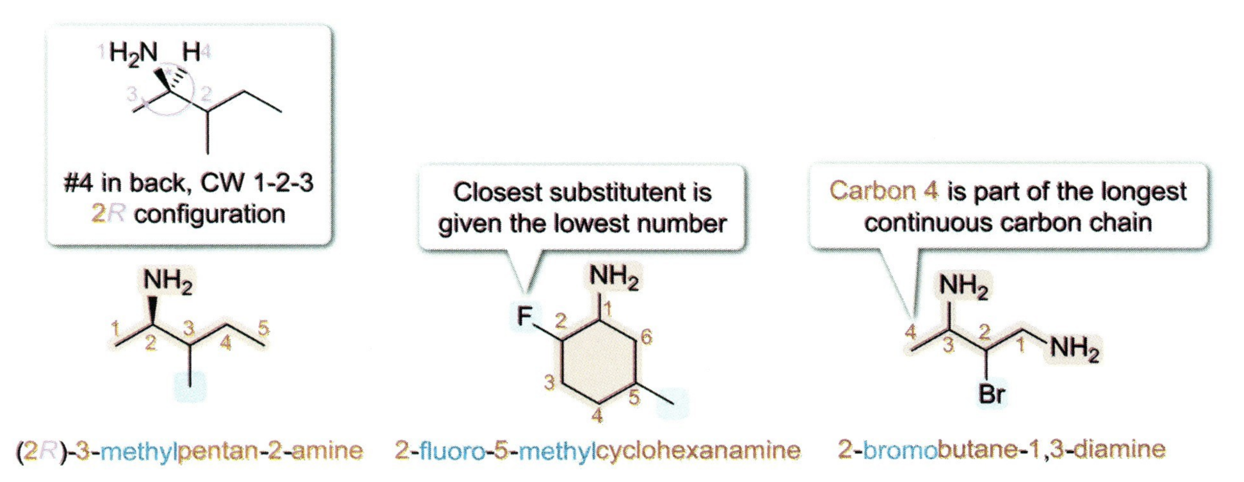

Figure 4.37 Examples and IUPAC names for amines.

If an amine contains alkyl or aryl groups attached to the nitrogen atom *beyond* the one classified as the parent chain for the molecule, the additional groups are named as an alkyl group and given a "numerical" location of *N*.

N designations appear at the appropriate place within an IUPAC name instead of a traditional number (Figure 4.38). If numerical and *N* locations are *both* present for a substituent grouping, the *N* comes after the ascending numerical order. For example, a (3,*N*-dimethyl) substituent indicates that two single-carbon substituents are present—one attached to position 3 and one attached to the nitrogen atom.

The parent chain is the longest continuous carbon chain

N-methylpropan-1-amine

N-cyclobutyl-*N*-ethylcyclohexanamine

2-ethyl-3,*N*-dimethylbutan-1-amine

N- positions are listed *after* numerical positions

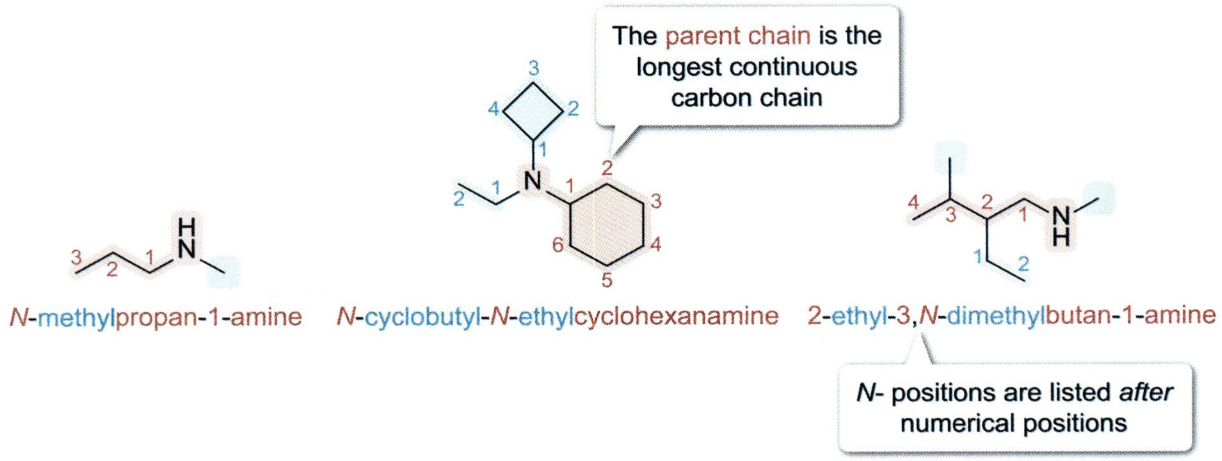

Figure 4.38 Use of *N*- designations in amine nomenclature.

When an amine nitrogen atom is attached to an aromatic benzene ring, it is called an **aniline** (Figure 4.39). The name *aniline* has been adopted by IUPAC and can be used as the parent chain for substituted aniline derivatives. The IUPAC naming guidelines for substituted aniline derivatives mirrors the conventions for aromatic rings: the aromatic ring carbon bearing the aniline nitrogen atom is position 1, and numbering proceeds around the ring to provide the lowest numbering for substituents. If two numbering methods are equivalent, the alphabetical order of the substituents determines the relative priority.

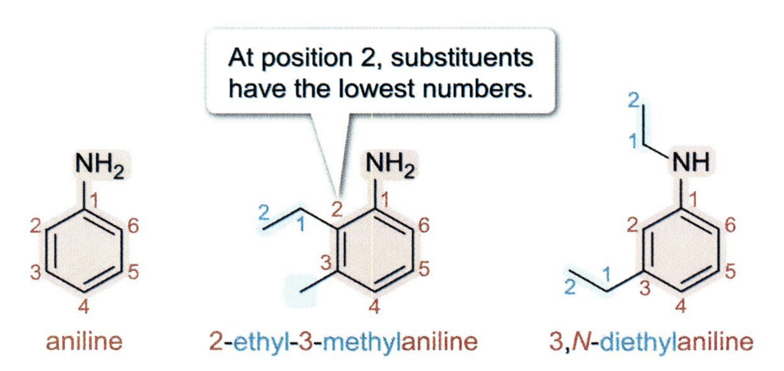

Figure 4.39 Examples and IUPAC names for anilines.

An amine group may be named as a substituent if it is either *not* the highest-priority functional group (Lesson 4.6) or if it is present within a substituent chain. In these cases, the term *amino* is used in combination with an appropriate numerical or *N* locations within an IUPAC name (Figure 4.40).

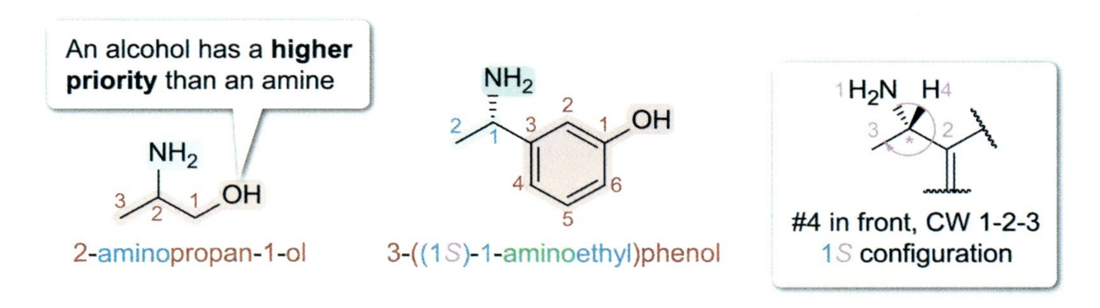

Figure 4.40 Examples and IUPAC names of molecules containing amine substituents.

An amine (R_3N) that has gained an acidic proton (H^+) is called an **ammonium ion** (R_3HN^+). An ammonium ion can also be generated if an amine acts as a **nucleophile** and uses its lone pair to attack an **electrophilic** alkyl group, generating a nitrogen atom with *four sigma bonds to alkyl groups* (R_4N^+; see Lesson 5.3).

Similar to the relationship between a carboxylic acid and a carboxylate ion (Concept 4.3.05), the IUPAC nomenclature guidelines for ammonium ions are like the guidelines for amines. When an ammonium ion (from a protonated amine) is the primary functional group in a parent chain, the suffix changes from –*amine* to –*aminium*. However, for ammonium ions with four R groups, the suffix changes from –*amine* to –*ammonium*. The ammonium ion form of *aniline* is called an *anilinium ion*.

Table 4.8 Comparison of IUPAC nomenclature guidelines for amines versus ammonium ions.

	Amine		Ammonium	
Structure type	**IUPAC suffix or term**	**Example**	**IUPAC suffix or term**	**Example**
Parent chain suffix	–amine	NH_2 propan-1-amine	–aminium	$\overset{+}{N}H_3$ propan-1-aminium
			–ammonium	$\overset{+}{N}(CH_3)_3$ *N*,*N*,*N*-trimethylpropan-1-ammonium
Aromatic parent chain	aniline	NH_2 aniline	anilinium	$\overset{+}{N}H_3$ anilinium
Substituent	amino	H_2N OH 2-aminoethanol	–	–

Figure 4.41 depicts the names, structures, and IUPAC parent chain numbering for **heterocyclic amines**.

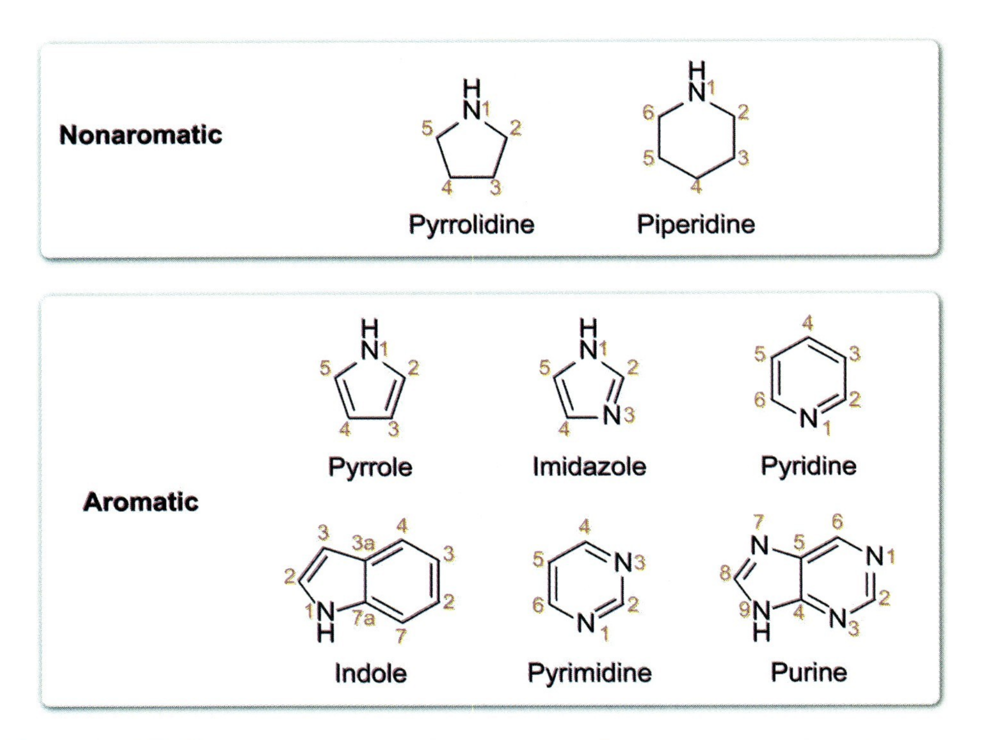

Figure 4.41 Structures, IUPAC names, and parent chain numbering for select heterocyclic amines.

Concept Check 4.9

What is the minimum total number of carbon atoms that must be present in an organic amine for the molecule's IUPAC name to contain an *N*-propyl group?

Solution

Note: The appendix contains the answer.

4.4.02 Amides

The IUPAC nomenclature guidelines for amides contain similarities to both ester nomenclature (Concept 4.3.06) and amine nomenclature (Concept 4.4.01). These are the steps for generating a systematic IUPAC name for an amide:

- The parent chain is the longest continuous carbon atom chain *that contains the carbonyl carbon of the amide*.
- The parent chain is numbered to give the *carbonyl carbon of the amide* position 1.
- The parent chain is named using the prefix for the number of carbon atoms, a middle modifier of *–an–*, and an *–amide* suffix (eg, *but* + *an* + *amide* = butanamide).
- Any *additional* alkyl groups attached to the nitrogen atom of the amide are named as alkyl groups with an attachment location of *N* (see Concept 4.4.01).

If a parent chain contains two amide functional groups, the parent chain includes the appropriate prefix for the number of carbon atoms; a middle modifier of *–ane–*; the Greek prefix *di–*; and the suffix *–amide* (eg, butanediamide). When the carbonyl carbon of an amide is directly attached to an atom in a ring, the

IUPAC name of the molecule is the name of the ring followed by the suffix *–carboxamide* (eg, cyclopentanecarboxamide). Figure 4.42 depicts examples of IUPAC names for amides.

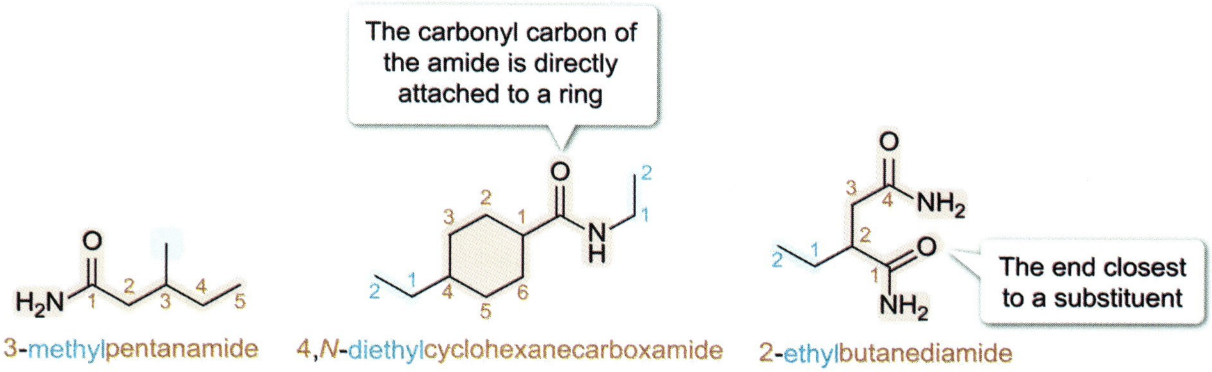

The carbonyl carbon of the amide is directly attached to a ring

The end closest to a substituent

3-methylpentanamide 4,*N*-diethylcyclohexanecarboxamide 2-ethylbutanediamide

Figure 4.42 Examples and IUPAC names for amides.

A lactam is a cyclic variant of an amide where the amide nitrogen is connected to the parent chain. The IUPAC names for lactams include the appropriate prefix for the *longest continuous carbon chain* that *includes the carbon* that cyclizes with the amide nitrogen (for lactams, this chain can extend beyond the ring), the middle modifier of *–an–*, and the suffix *–olactam*. The carbon atoms in the lactam chain are numbered starting from the carbonyl carbon. The numerical position of the carbon that cyclizes with the lactam nitrogen atom is inserted before the suffix *–lactam*; this number is *one fewer than* the number of atoms in the lactam ring. Figure 4.43 depicts examples of IUPAC names for lactams.

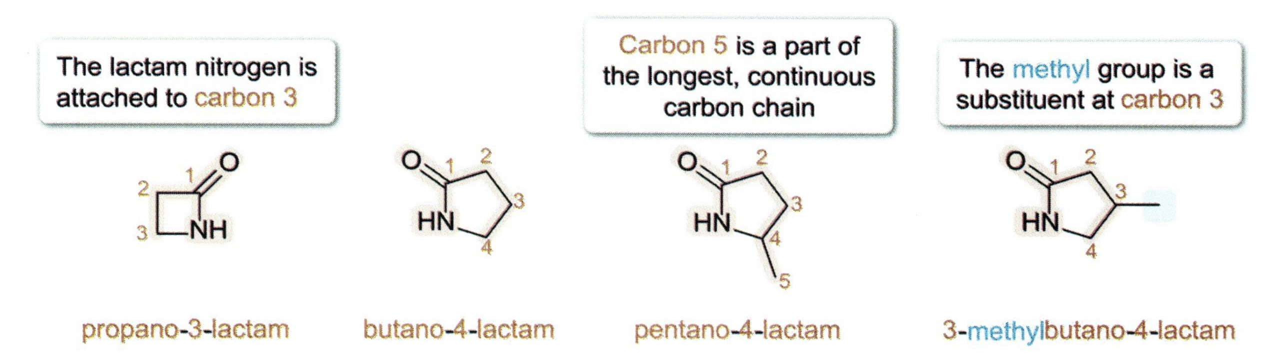

Figure 4.43 Examples and IUPAC names for lactams.

If an amide is present within a substituent:

- The term *carbamoyl* refers to the –C(O)N(R^1)(R^2) substituent (ie, the attachment to the parent chain is through the *carbonyl carbon*), where R^1 and R^2 groups are named as alkyl groups, if present (eg, methylcarbamoyl).
- The term *amido* refers to the –N(R^1)C(O)R^2 substituent (ie, the attachment to the parent chain is through the amide nitrogen), where a nonhydrogen R^1 group (if present) is named as an *N*-alkyl group and the amide carbonyl + R^2 group is part of the substituent parent chain (eg, butanamido).

Carbamoyl and amido groups both require an appropriate numerical point of attachment (Figure 4.44).

2-carbamoylpropanoic acid   3-butanamidopropanoic acid

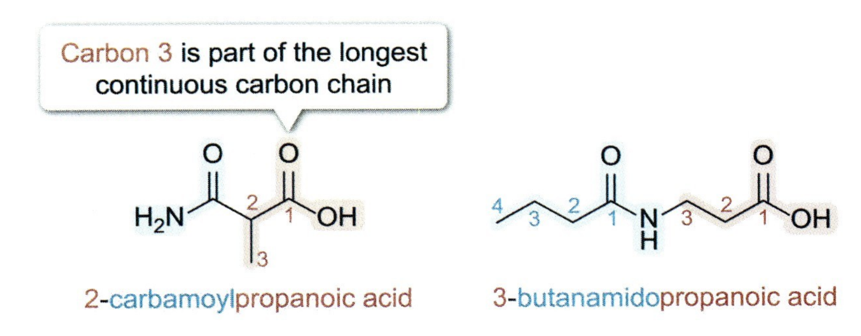

Figure 4.44 Examples and IUPAC names of molecules containing amide substituents.

4.4.03 Nitriles

The notation –CN is commonly used to represent a nitrile group. These are the steps for generating a systematic IUPAC name for a nitrile:

- The parent chain is the longest continuous carbon atom chain *that contains the carbon of the nitrile group*.
- The parent chain is numbered to give the *carbon of the nitrile group* the lowest value.
- The parent chain is named using the prefix for the number of carbon atoms, a middle modifier of –*ane*–, and a –*nitrile* suffix (eg, *but* + *ane* + *nitrile* = butanenitrile).
- If the parent chain contains either one or two nitrile groups, numerical locations are not necessary (ie, the nitrile functional groups are assumed to be at the first and last carbon atom of the linear parent chain).

If a molecule contains two nitrile functional groups, the parent chain includes the appropriate prefix for the number of carbon atoms, a middle modifier of –*ane*–, the Greek prefix *di*–, and the suffix –*nitrile*. When a –CN group is directly attached to an atom in a ring, the IUPAC name of the molecule is the name of the ring (eg, cyclopentane) followed by the suffix –*carbonitrile* (eg, cyclopentanecarbonitrile). Figure 4.45 depicts examples of IUPAC names for nitriles.

3-methylpentanenitrile   3-(1-methylethyl)cycloheptanecarbonitrile   hexanedinitrile

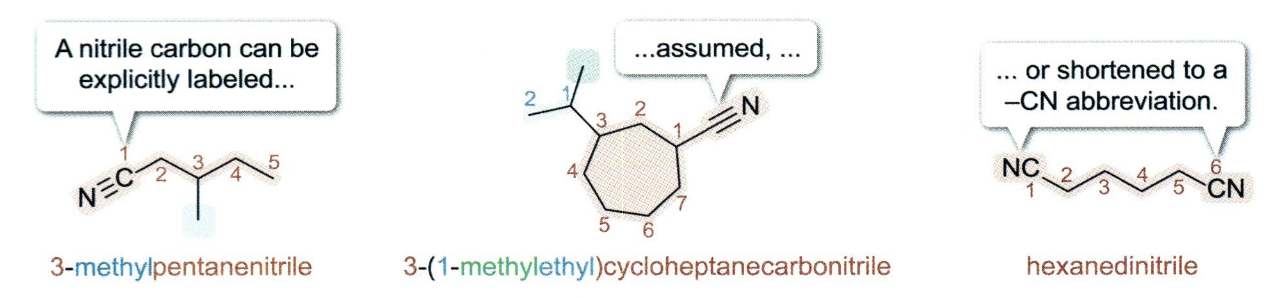

Figure 4.45 Examples and IUPAC names for nitriles.

If a nitrile group is present within a substituent chain, the term *cyano* is used to refer to the –CN substituent. A cyano group requires an appropriate numerical point of attachment (Figure 4.46).

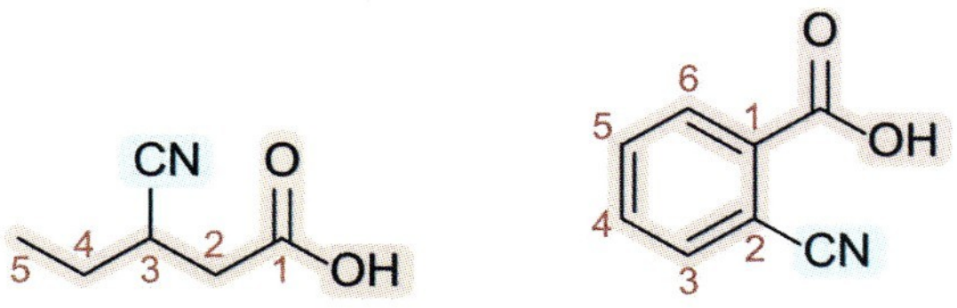

3-cyanopentanoic acid   2-cyanobenzoic acid

Figure 4.46 Examples and IUPAC names of molecules containing nitrile substituents.

Concept Check 4.10

What is the systematic (IUPAC) name for the following molecule?

NC–CH₂–(cyclopentane)–CN

Solution

Note: The appendix contains the answer.

Lesson 4.5

IUPAC Nomenclature of Sulfur-Containing Groups

Introduction

This lesson continues the exploration of nomenclature guidelines, discussing the major organic functional groups that contain sulfur atoms, with emphasis on how these guidelines are derived from their respective oxygen-containing analogs.

4.5.01 Thiols

A thiol is an organic molecule that contains one or more **sulfhydryl groups (–SH)** attached to an alkyl or aryl group carbon atom. Given that thiols are structurally similar to alcohols, the IUPAC nomenclature guidelines for thiols and alcohols (Concept 4.3.01) share many common themes:

- The parent chain is the longest continuous chain of carbon atoms *that contains the carbon attached to the sulfhydryl functional group*.
- The parent chain is numbered to give the *carbon attached to the sulfhydryl group* the lowest value.
- The parent chain is named using the prefix for the number of carbon atoms, a middle modifier of –*ane*–, and a –*thiol* suffix (eg, *but* + *ane* + *thiol* = butanethiol).
- The numerical location of the sulfhydryl group is located either before the parent chain (eg, 1-butanethiol, original IUPAC guidelines) or before the –*thiol* suffix (eg, butane-1-thiol, revised IUPAC guidelines; see Concept 4.2.02).

If a molecule contains multiple thiol functional groups, the parent chain is the longest continuous carbon chain attached to the *maximum number of sulfhydryl groups*. In these cases, the parent chain includes the appropriate prefix for the number of carbon atoms, a middle modifier of –*ane*–, the appropriate Greek prefix for the number of sulfhydryl groups (on the parent chain), and the suffix –*thiol*. Each sulfhydryl group is given a numerical location. Figure 4.47 depicts examples of IUPAC names for thiols.

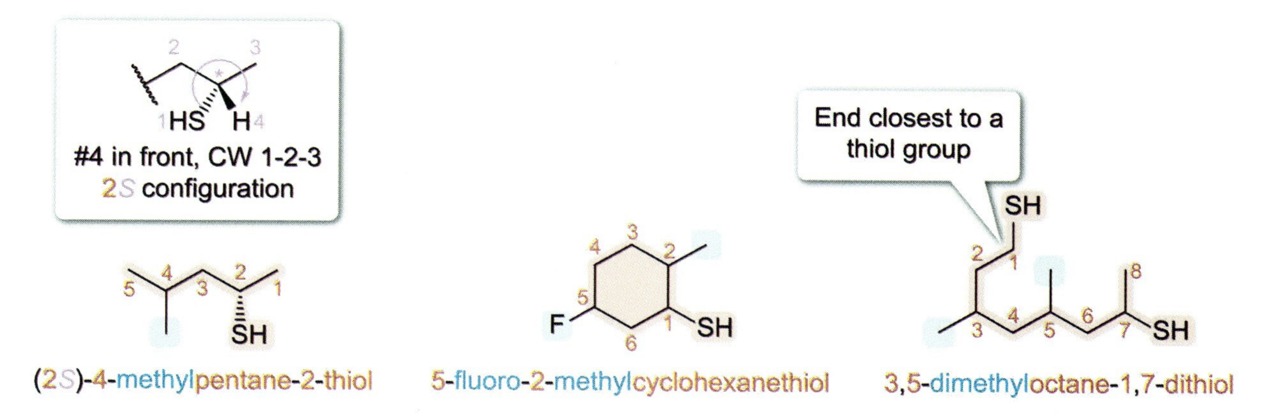

Figure 4.47 Examples and IUPAC names for thiols.

A thiol may be named as a substituent if it either is *not* the highest-priority functional group (Lesson 4.6) or is present within a substituent chain. In these cases, *sulfanyl* is the current IUPAC-recommended term used in combination with an appropriate numerical location (Figure 4.48). *Mercapto* is an outdated term that has not been recommended for years but is still seen in some historical or common names (eg, β-mercaptoethanol).

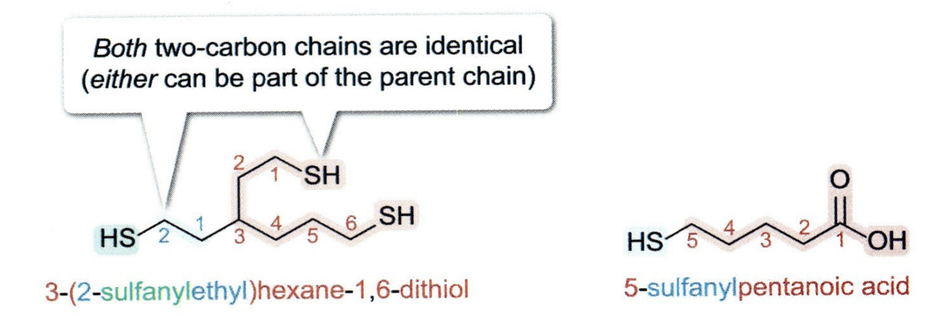

Figure 4.48 Examples and IUPAC names of molecules containing thiol substituents.

Concept Check 4.11

The chemical 3-methylbutane-1-thiol is one of the molecules responsible for the odor of skunk spray. Draw a line-angle structure of 3-methylbutane-1-thiol.

Solution

Note: The appendix contains the answer.

4.5.02 Thioethers

A thioether contains an sp^3-hybridized sulfur atom bound to two alkyl or aryl groups (R groups). The R groups can have either the same structure (**symmetrical thioether**) or different structures (**unsymmetrical thioether**).

Like an ether (Concept 4.3.02), the IUPAC name for a thioether *always* has the thioether functional group named as a substituent. Only one of the two carbon atoms in a thioether functional group is included in either the parent chain or substituent parent chain.

The rest of the atoms in a thioether functional group (sulfur atom + one alkyl group) are named as an *alkylsulfanyl* group, where the *alk* is replaced by the appropriate prefix that indicates the number of carbon atoms in the R group. For example, a one-carbon $-SCH_3$ group is named *methylsulfanyl*, and a two-carbon $-SCH_2CH_3$ group is named *ethylsulfanyl*. Alternatively, thioether substituents may be named as an *alkylthio* group (eg, $-SCH_3$ = *methylthio*, $-SCH_2CH_3$ = *ethylthio*).

Alkylsulfanyl or alkylthio groups must include the appropriate numerical locator for its point of attachment to the parent chain. Figure 4.49 provides several examples of thioether-containing molecules and their corresponding IUPAC names.

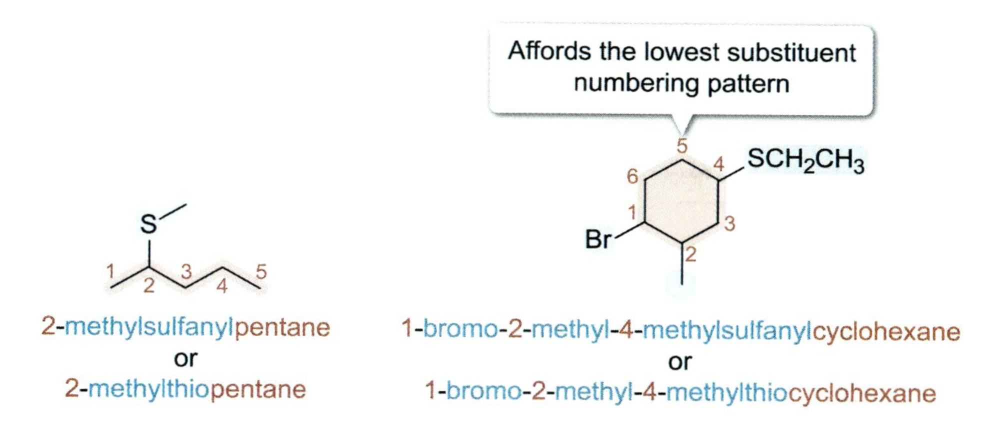

Figure 4.49 Examples of IUPAC names for thioethers.

4.5.03 Thioesters

A thioester is a sulfur-containing variant of an ester that is also classified as a carboxylic acid derivative. Thioesters have a general formula of R–C(O)–SR, where an acyl group is attached to an alkylsulfanyl (–SR) group (Concept 4.5.02).

Thioester IUPAC nomenclature is very similar to the IUPAC nomenclature for regular esters (Concept 4.3.06). These are the steps for generating a systematic IUPAC name for a thioester:

- The name contains *two* components and is named as an *S-alkyl alkanethioate.* Each component has its own system of numbering the carbon atoms.
- The *S-alkyl* portion is named and numbered the same as other alkyl groups, where the point of attachment within the alkyl group is given position 1 (Concept 4.1.02).
- The *alkanethioate* portion is named and numbered similarly to a carboxylate ion (Concept 4.3.05), only using a *–thioate* suffix (eg, hexanethioate).
- Substituents attached to *either* component are included within the names for their respective portions.

When a thioester (–C(O)SR) group is directly attached to an atom in a ring, the IUPAC name of the alkanethioate portion of the thioester is the name of the ring followed by the suffix *–carbothioate* (eg, cyclopentanecarbothioate). Figure 4.50 depicts examples of IUPAC names for thioesters.

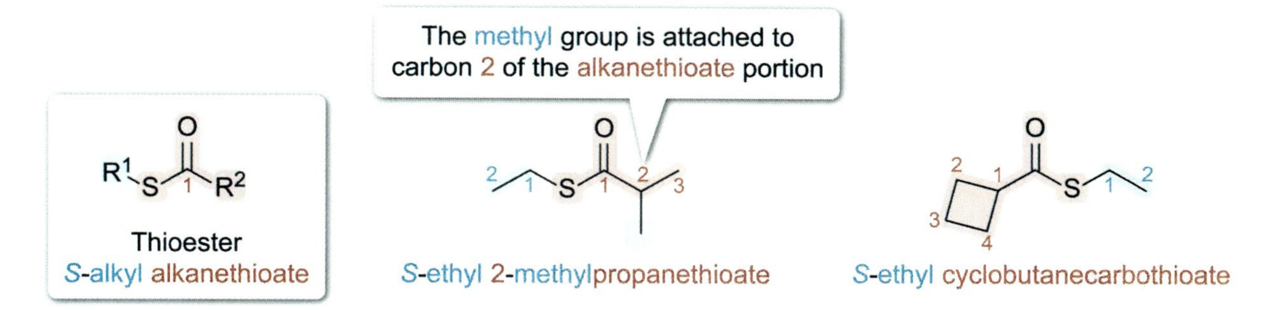

Figure 4.50 Examples and IUPAC names for thioesters.

Lesson 4.6

IUPAC Nomenclature of Multi-Functional Compounds

Introduction

This lesson describes the general steps toward generating an IUPAC name for *any* molecule, including those with two or more major functional groups (ie, **multifunctional compounds**). A secondary goal of this lesson is to collate and summarize many of the key points raised throughout the rest of the chapter.

4.6.01 Functional Group Priority

The first step toward generating a name for an organic structure is to identify its parent chain (Lessons 4.1 through 4.5). For multifunctional compounds, the parent chain is the longest continuous chain of carbon atoms *that contains the highest-priority functional group.*

IUPAC has developed a systematic, relative ranking (priority) of functional groups (Table 4.9). The first three tiers aggregate functional groups by the number of carbon-heteroatom bonds, where tiers with more heteroatom bonds have a higher priority. The highest-priority functional group is the carboxylic acid, followed by the related carboxylic acid derivatives. Next are the carbonyl-containing functional groups (ketones and aldehydes), followed by groups containing only one carbon-heteroatom bond (eg, alcohols, thiols, amines).

The pi bond–containing hydrocarbons (alkenes and alkynes) have the *same* priority level; however, the alkene (*–en–*) comes before alkyne (*–yn–*) due to alphabetical order. As noted in Concept 4.1.01, the lowest priority functional group that can be named as the parent chain are the alkanes; all lower-priority groups (ie, ethers, thioethers, alkyl halides) are *always* named as substituents.

Table 4.9 Priority of functional groups in IUPAC nomenclature.

Functional group	Structure	
Carboxylic acid	RC(=O)OH	3 Bonds to heteroatom
Anhydride	RC(=O)OC(=O)R	
Ester	RC(=O)OR	
Acid halide	RC(=O)X	
Amide	RC(=O)N̈R₂	
Nitrile	R–C≡N:	
Aldehyde	RC(=O)H	2 Bonds to heteroatom
Ketone	RC(=O)R	

Functional group	Structure	
Alcohol	R–OH	1 Bond to heteroatom
Thiol	R–SH	
Amine	R₂N̈–R	
Alkene	R₂C=CR₂	Hydrocarbons with pi bonds
Alkyne	R–C≡C–R	
Alkane	R₃C–CR₃	Lowest priority parent chain
Ether	R–O–R	Eternal substituents
Thioether	R–S–R	
Alkyl halide	R₃C–X	

Note: Structures listed in decreasing order of priority

After identification of the highest-priority functional group, the next step is to identify the longest continuous chain of carbon atoms *that includes the maximal number of highest-priority groups* (Figure 4.51). This parent chain is then numbered to give the *highest-priority group* the *lowest* possible value(s). Finally, the full name of the parent chain is assembled as described in Lessons 4.1 through 4.5.

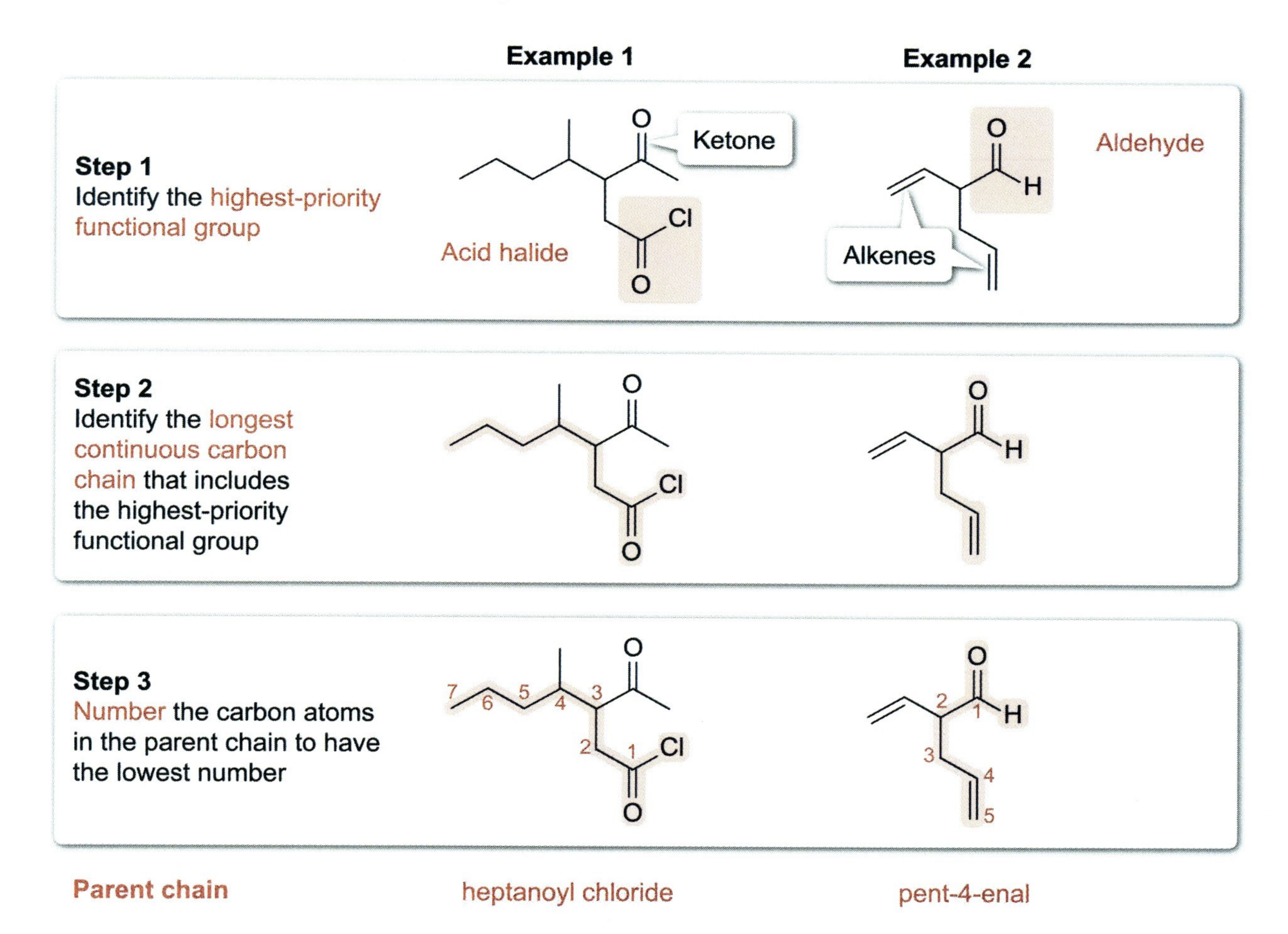

Figure 4.51 Identifying the parent chain of a multifunctional compound.

Any remaining atoms are named as substituent groups, each of which requires a numerical point of attachment to the parent chain. Finally, the complete IUPAC name is assembled from substituent groups (listed in alphabetical order) and the parent chain (Figure 4.52).

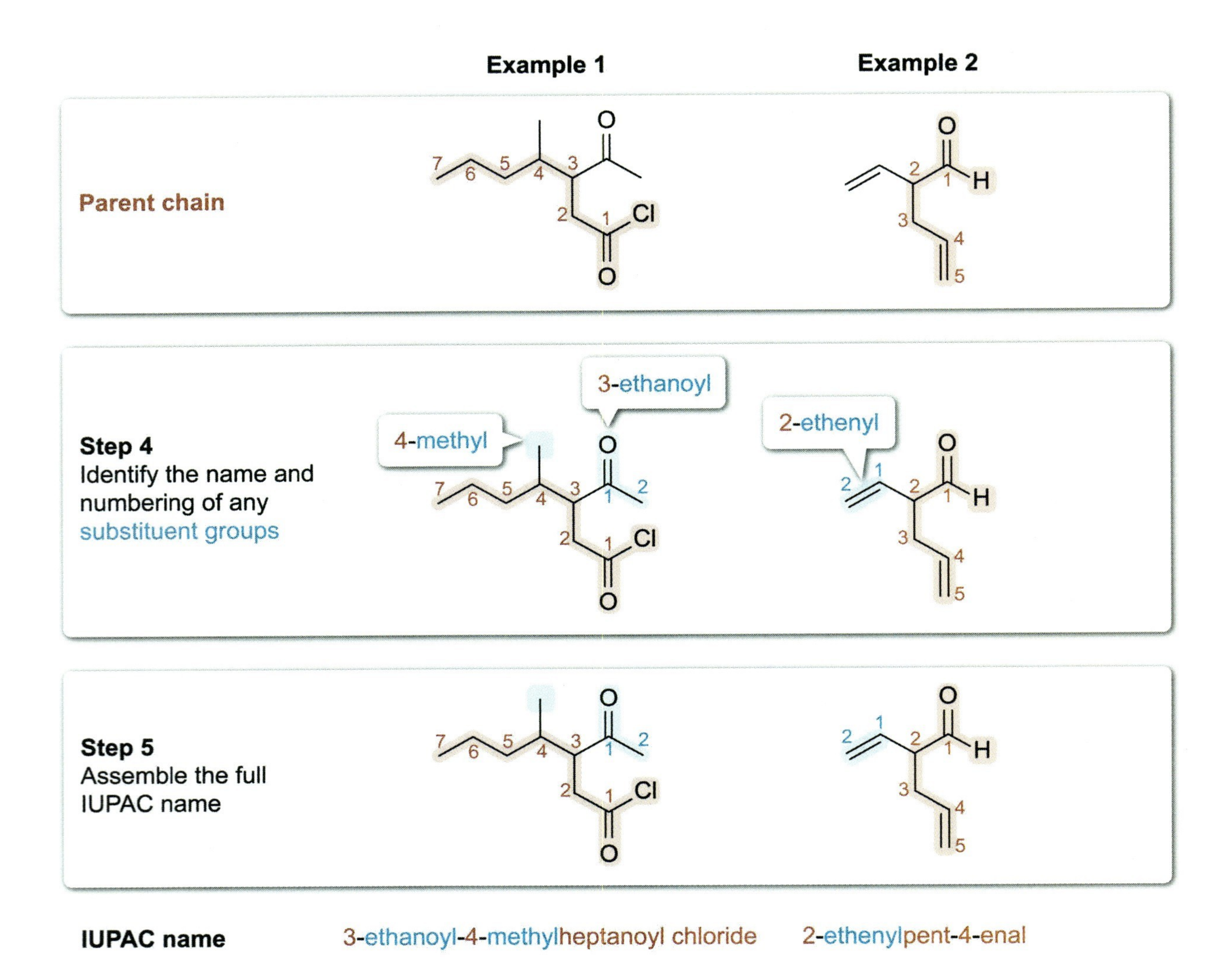

Figure 4.52 Substituents and IUPAC names of multifunctional compounds.

Table 4.10 provides a summary of functional groups (in priority order), their parent chain suffixes, and the terms used to denote substituent groups.

Table 4.10 Summary of IUPAC functional group suffixes and substituent terms.

	Parent chain (if highest priority)		Substituent (if not highest priority)	
Functional group	**Suffix (linear)**	**Suffix (cyclic)**	**Group**	**Substituent term**
Carboxylic acid	–oic acid	–carboxylic acid	$-CO_2H$	carboxy
Anhydride	–oic anhydride	—	—	—
Ester	–oate	–carboxylate	–C(O)OR	alkoxycarbonyl
			–OC(O)R	alkanoyloxy
Acid halide	–oyl halide	–carbonyl halide	–C(O)X	halocarbonyl
Amide	–amide	–carboxamide	$-C(O)N(R_2)$	carbamoyl
			–N(R)C(O)R	amido
Nitrile	–onitrile	–carbonitrile	–CN	cyano
Aldehyde	–al	–carbaldehyde	–CHO	formyl
Ketone	–one	—	=O	oxo
			–C(O)R	alkanoyl
Alcohol	–ol	—	–OH	hydroxy
Thiol	–(e)thiol	—	–SH	sulfanyl (mercapto)
Amine	–amine	—	$-N(R_2)$	amino
Alkene	–ene	—	—	alkenyl
Alkyne	–yne	—	—	alkynyl
Alkane	–ane	—	–R	alkyl
Ether	—	—	–OR	alkoxy
Thioether	—	—	–SR	alkylsulfanyl
Alkyl halide	—	—	–X	fluoro chloro bromo iodo

Substituent terms containing "alk" are replaced by the appropriate prefix for the number of carbon atoms.

Concept Check 4.12

What is the systematic (IUPAC) name for mevalonate, an intermediate that is biosynthesized from acetyl CoA?

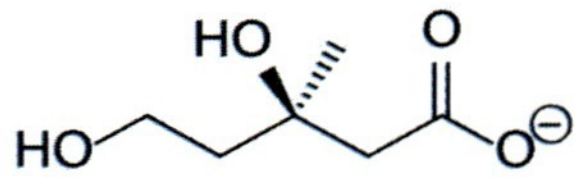

Solution

Note: The appendix contains the answer.

Lesson 4.7

Common Nomenclature

Introduction

Although IUPAC was created in 1919 to develop worldwide guidelines for chemical nomenclature, many chemical names were already in use, some of which are still used today. Broadly, nonsystematic (ie, non-IUPAC) names are described as **common nomenclature**.

Lessons 4.1 through 4.6 provided an overview of IUPAC nomenclature for the major classes of functional groups. This lesson introduces aspects of common nomenclature that are relevant to exam preparation.

4.7.01 Historical Terminology for Alkyl Groups

Concept 4.1.01 introduced the IUPAC prefixes to describe the number of carbon atoms in carbon chains. Table 4.11 shows the common short-chain prefixes used *prior* to IUPAC standardization. Even today, some molecules are almost always named using these common prefixes (eg, formic acid and acetic acid as one- and two-carbon carboxylic acids, respectively).

Table 4.11 Comparison of common and IUPAC prefixes.

Number of carbons	Common prefix	IUPAC prefix
1	form–	meth–
2	acet–	eth–
3	propion–	prop–
4	butyr–	but–

Several common terms are also used to describe the branching patterns of short-chain alkyl groups. A linear, unbranched alkyl chain can be given a designation of *n*- for "normal" (eg, *n*-propyl) for added clarity. Within an alkyl group:

- The designation *sec*- (or *s*-, short for "secondary") is used if the attached carbon atom is bonded to *two* carbons within the alkyl group.
- The designation *tert*- (or *t*-, short for "tertiary") is used if the attached carbon atom is bonded to *three* carbons within the alkyl group.
- The prefix iso– is used if a one-carbon branch is located one carbon from the *end* of the carbon chain (eg, isobutyl, isopentyl).

The common names of important short-chain alkyl groups are shown in Figure 4.53. Although systematic names of complex alkyl groups are typically preferred, isopropyl, isobutyl, *sec*-butyl, and *tert*-butyl have been grandfathered by IUPAC and *may* appear in chemical names. Of these, only the prefix *iso*– impacts alphabetical order.

Branch prefixes

1 R 2 3

normal group (linear, *n*–, n = 2+)

iso group (n = 0+)

secondary group (*sec*– or *s*–)

tertiary group (*tert*– or *t*–)

Alkyl groups

n-propyl

isopropyl

isobutyl

sec-butyl

tert-butyl

The underlined letter represents the first letter used in determining alphabetical order.

Figure 4.53 Common names of select alkyl groups.

4.7.02 Bonding Relationships

Several aspects of common nomenclature describe the bonding relationships between groups.

Cis/*Trans* Isomers

The terms *cis* and *trans* describe the relative relationship of two groups in reference to a side (or face) of a molecule's structure (see Concept 3.3.07). A ***cis*** **relationship** occurs when the groups are on the same side, and a ***trans*** **relationship** occurs when they are on opposite sides. In common nomenclature, *cis*/*trans* terminology is most often observed in two different contexts: cycloalkane rings and (disubstituted) alkenes (Figure 4.54).

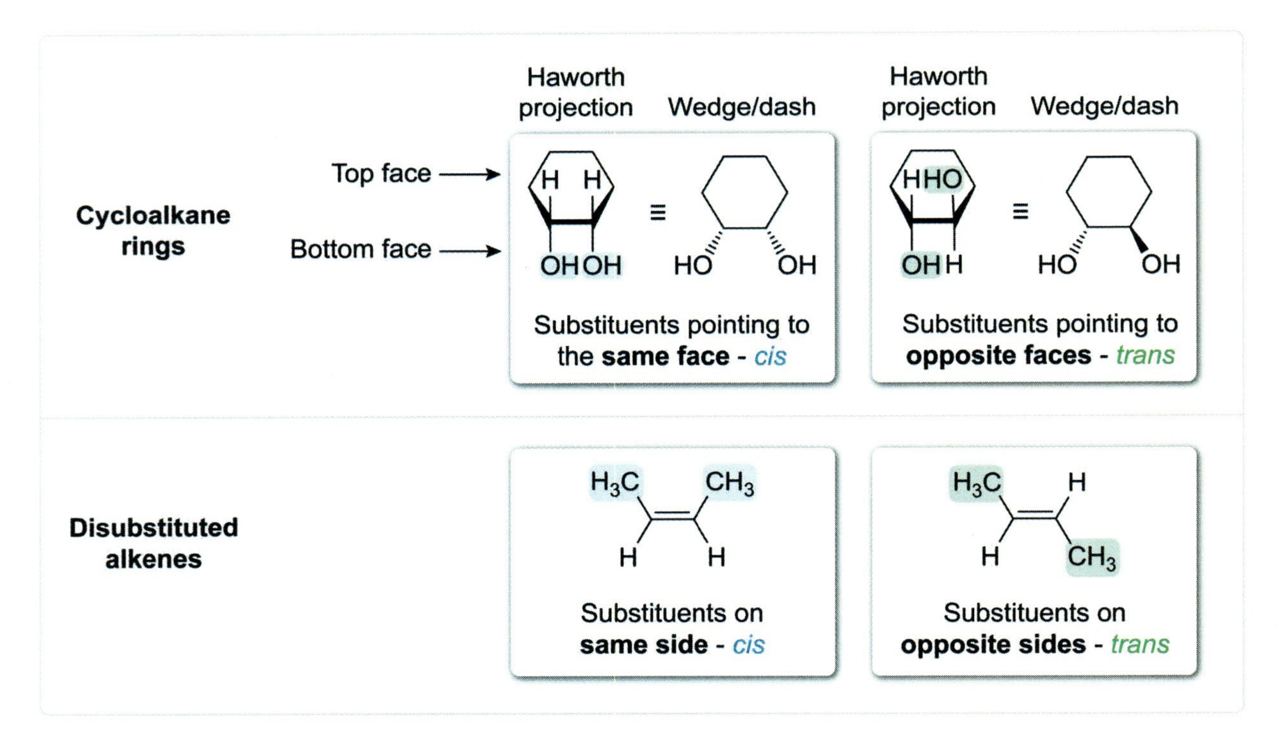

Figure 4.54 Uses of *cis*/*trans* nomenclature to describe relative orientation.

Cis/*trans* terminology does *not* provide information about the *absolute* configuration of any stereocenters. If absolute configurations are required, cycloalkane rings are better described using *R*/*S* configurations (see Concept 3.3.03), whereas alkenes use *E*/*Z* configurations (see Concept 3.3.08).

Ortho, *Meta*, and *Para* Relationships

Historically, the constitutional isomers of disubstituted benzene rings were named using the common terms *ortho* (*o*-), *meta* (*m*-), and *para* (*p*-) (Figure 4.55). In an ***ortho* relationship**, two groups are on *adjacent* ring atoms in a 1,2-relationship (numbering here is independent of the actual ring numbering). A ***meta* relationship** has two groups in a 1,3-relationship, and a ***para* relationship** has two groups on opposite sides of the benzene ring (a 1,4-relationship). *Ortho*, *meta*, and *para* can also be generalized to describe the relative relationship of *any* two groups on a benzene ring, even when there are more than two substituents.

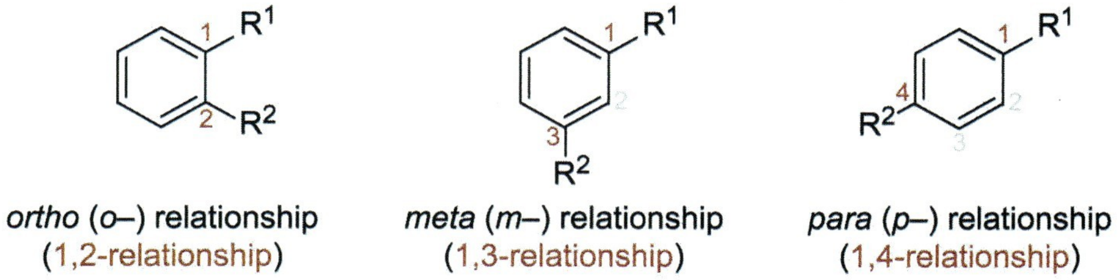

R1, R2

ortho (*o*–) relationship (1,2-relationship) | *meta* (*m*–) relationship (1,3-relationship) | *para* (*p*–) relationship (1,4-relationship)

Figure 4.55 *Ortho*, *meta*, and *para* relationships on a benzene ring.

Common Nomenclature for Diols

A **diol** is a molecule that contains two hydroxyl groups (Figure 4.56). When the hydroxyl groups in a diol are attached to *neighboring* carbon atoms in a chain, the molecule is called a **vicinal diol** or a **glycol**. If the hydroxyl groups in a diol are attached to the *same carbon atom*, the molecule is called a **geminal diol**.

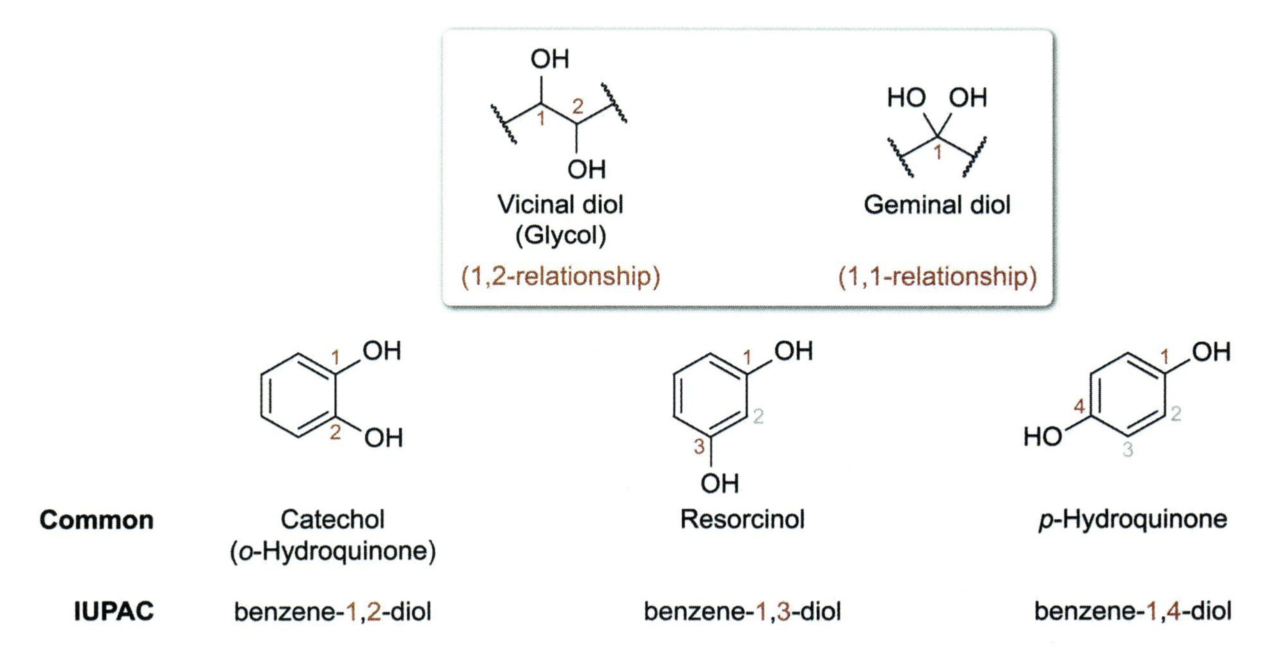

Figure 4.56 Common and IUPAC nomenclature for diols.

Benzene rings substituted with two hydroxyl groups also have important common names. The *ortho* isomer of benzenediol is **catechol** (or ***ortho*-hydroquinone**), the *meta* isomer is **resorcinol**, and the *para* isomer is ***para*-hydroquinone**.

Spatial Relationships Using Greek Letters

One historical nomenclature carryover that remains prevalent within organic chemistry and biochemistry involves the use of Greek letters to designate positions relative to a reference point. This is particularly prominent when describing the carbon positions relative to a carbonyl. With the exception of ketones, most other carbonyl carbon atoms are at the *end* of a continuous chain and would be given position 1 using IUPAC nomenclature.

Greek letters refer to each position extending *away* from the carbonyl carbon (which itself is *not* assigned a Greek letter) (Figure 4.57). The important Greek letters to know are **α (alpha)**, **β (beta)**, **γ (gamma)**, **δ (delta)**, and **ε (epsilon)**. The *final* carbon atom in the chain (of whatever length) is referenced using the Greek letter **ω (omega)**. The ω designation may also be used in combination with a number to refer to a defined number of positions *closer* to the carbonyl from the ω position (ie, the ω−1 position is one position closer to the carbonyl from the ω position).

IUPAC numbering* | Greek letter position

*Assumes that the carbonyl carbon is given position 1

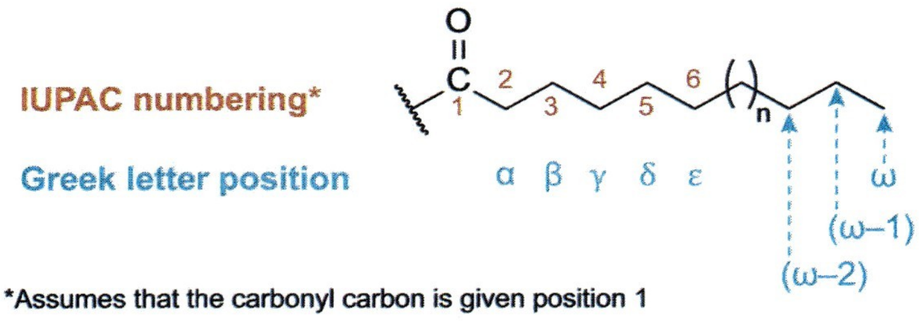

Figure 4.57 Greek letter system of relative location reference.

For carboxylic acids, carboxylic acid derivatives, and aldehydes, Greek letters extend only *one* direction away from the carbonyl carbon, as these groups are inherently at the end of a continuous chain. A ketone has Greek letters extending in *both* directions away from the carbonyl carbon.

Concept Check 4.13

Pyruvate is one of several molecules in biochemistry classified as an α-ketoacid. Draw the structure of a general α-ketoacid.

Solution

Note: The appendix contains the answer.

4.7.03 Functional Groups

Although IUPAC names are generally the preferred way to name molecules, most functional groups retain some amount of common nomenclature usage, particularly for molecules with simple structures.

For many functional groups, common nomenclature places the *functional group* as the group listed *last* within a name (similar to a parent chain in IUPAC nomenclature). For alcohols, ethers, ketones, amines, and thioethers, the rest of the molecule is named using one or more alkyl groups (Table 4.12).

Table 4.12 Examples of common nomenclature for various functional groups.

Functional group	Parent term	Examples	
Alcohol	alcohol	OH *t*-butyl alcohol	H_3C-OH methyl alcohol
Ether	ether	O ethyl methyl ether	O diethyl ether (ethyl ether)
Ketone	ketone	O ethyl methyl ketone	
Amine	amine	N triethylamine	N diisopropylethylamine
Thioether	sulfide	S dimethyl sulfide	

Common names for aldehydes, carboxylic acids, and the carboxylic acid derivatives use the common nomenclature numerical prefixes alongside functional group–specific suffixes (Table 4.13).

Table 4.13 Examples of common nomenclature for aldehyde, carboxylic acid, ester, acid halide, anhydride, amide, and nitrile functional groups.

Functional group	Common nomenclature suffix	Examples	
Aldehyde	–aldehyde	acetaldehyde	butyraldehyde
Carboxylic acid	–ic acid	acetic acid	butyric acid
Ester	–ate	ethyl acetate	
Acid Halide	–yl halide	acetyl chloride	
Anhydride	–ic anhydride	acetic anhydride	
Amide	–amide	formamide	acetamide
Nitrile	–onitrile	$H_3C{-}CN$ acetonitrile	

For symmetrical ethers and ketones, the prefix *di–* may or may not be used to specify that the surrounding alkyl groups are identical. For example, diethyl ether and ethyl ether are synonymous common names, as the *di–* can be assumed without a second named alkyl group. If *different* alkyl groups are on the parent functional group, they are listed in alphabetical order. Uniquely, common names for amines list alkyl groups in alphabetical order *without spaces.*

Common nomenclature for lactones and lactams uses the Greek letter system to denote the size of the ring, where the Greek letter corresponds to the point of attachment of the ring heteroatom to the parent chain (Figure 4.58).

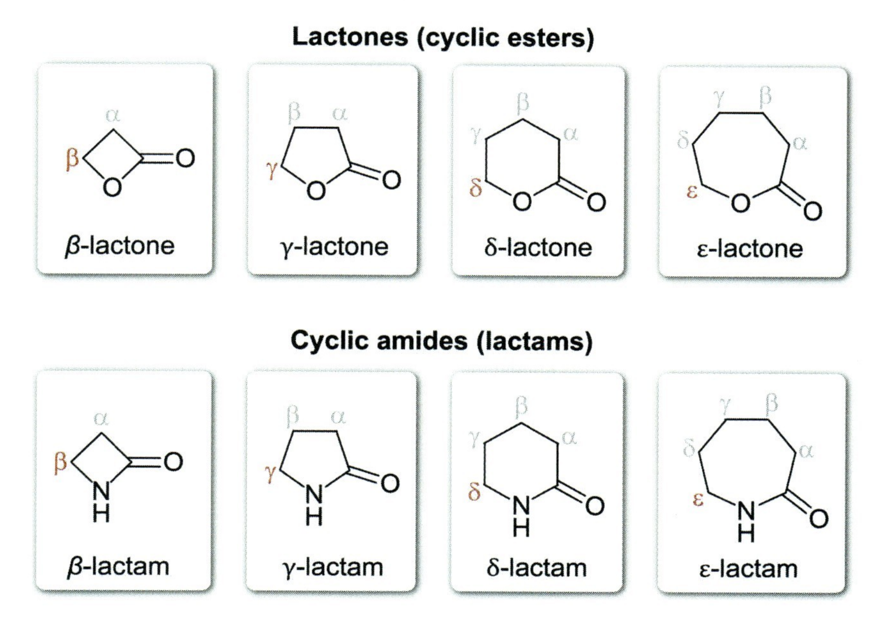

Figure 4.58 Common nomenclature for lactones and lactams.

The common names of short-chain, alkyl halides list the alkyl chain followed by the appropriate halide (similar to the other functional groups in Table 4.12). For example, CH_3CH_2Br is named ethyl bromide. One-carbon alkyl halides have unique historical names that are still used regularly (Table 4.14).

Table 4.14 Common versus IUPAC nomenclature of one-carbon alkyl halides.

Name	CH_3X	CH_2X_2	CHX_3	CX_4
Common	methyl halide	methylene halide	haloform	carbon tetrahalide
IUPAC	halomethane	dihalomethane	trihalomethane	tetrahalomethane

Note: *halide = flouride, chloride, bromide, or iodide; halo = fluoro, chloro, bromo, or iodo.*

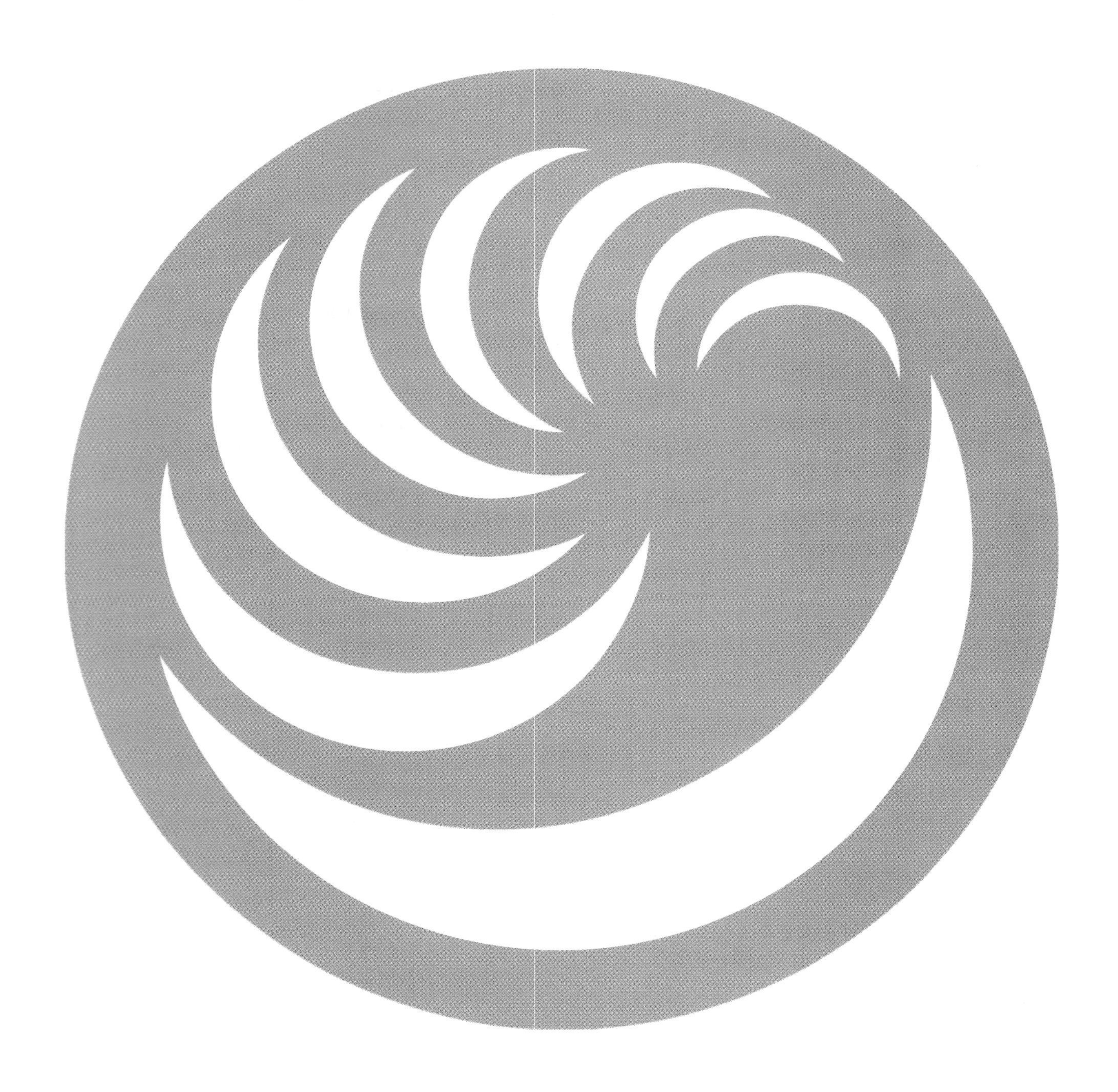

Lesson 5.1

Acid-Base Reactions

Introduction

One important aspect of chemistry involves the study of chemical reactions. A **chemical reaction** is a process that involves the creation or breaking of covalent bonds (see Lesson 1.1). This chapter provides an overview of the major classifications of chemical reactions:

- Acid-base (neutralization) reactions
- Substitution reactions
- Addition reactions
- Elimination reactions
- Oxidation-reduction (redox) reactions

The current lesson focuses on **neutralization reactions** (ie, reactions between acids and bases).

5.1.01 Definitions of Acids and Bases

There are three different definitions of **acids** and **bases**: Arrhenius, Brønsted-Lowry, and Lewis.

Arrhenius Acids and Bases

The Arrhenius definition is the narrowest of the three and pertains to situations where *water is the solvent*. An **Arrhenius acid** is a molecule that dissociates in water to release a proton, which then reacts with water to form **hydronium (H_3O^+) ions**. An **Arrhenius base** is a molecule that dissociates in water to form **hydroxide (HO^-) ions** (Figure 5.1).

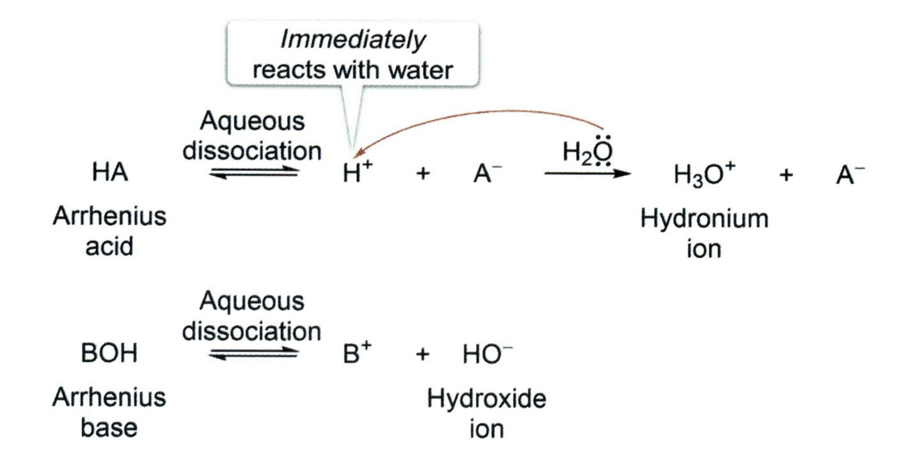

Figure 5.1 Arrhenius acids and bases.

Brønsted-Lowry Acids and Bases

The Brønsted-Lowry definition of acids and bases uses the **proton (H^+)** as the currency. This broader definition does *not* require reaction in aqueous solvent, making it more useful in organic chemistry. A **Brønsted-Lowry acid** is a molecule that can *donate* H^+ to another molecule, and a **Brønsted-Lowry base** is a molecule that can *accept* H^+ from another molecule (Figure 5.2). Brønsted-Lowry bases are either neutral or have a negative formal charge.

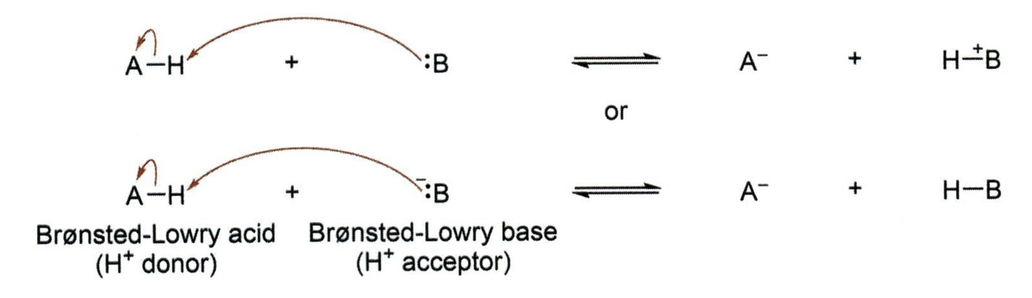

Figure 5.2 Brønsted-Lowry acids and bases.

Lewis Acids and Bases

The broadest definition of acids and bases is the Lewis definition, which describes acids and bases in relation to electron density. A **Lewis acid** is an electron-deficient molecule and can *accept* a lone pair of electrons from another molecule (Figure 5.3); a **Lewis base** is an electron-rich molecule and can *donate* a lone pair of electrons to another molecule.

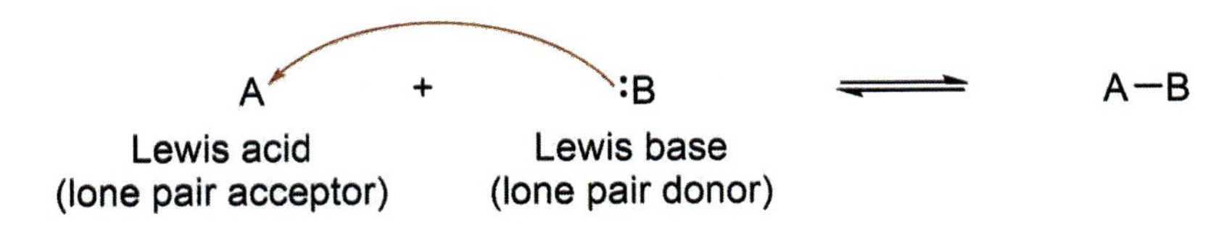

Figure 5.3 Lewis acids and bases.

Lewis acids and bases are closely related to the concepts of **electrophiles** and **nucleophiles**, respectively (see Lesson 5.3).

The Lewis definition encompasses *all* acid-base reactions, whereas the Brønsted-Lowry definition is restricted to reactions involving proton transfers. The Arrhenius definition is further restricted to the generation of hydronium and hydroxide ions in water.

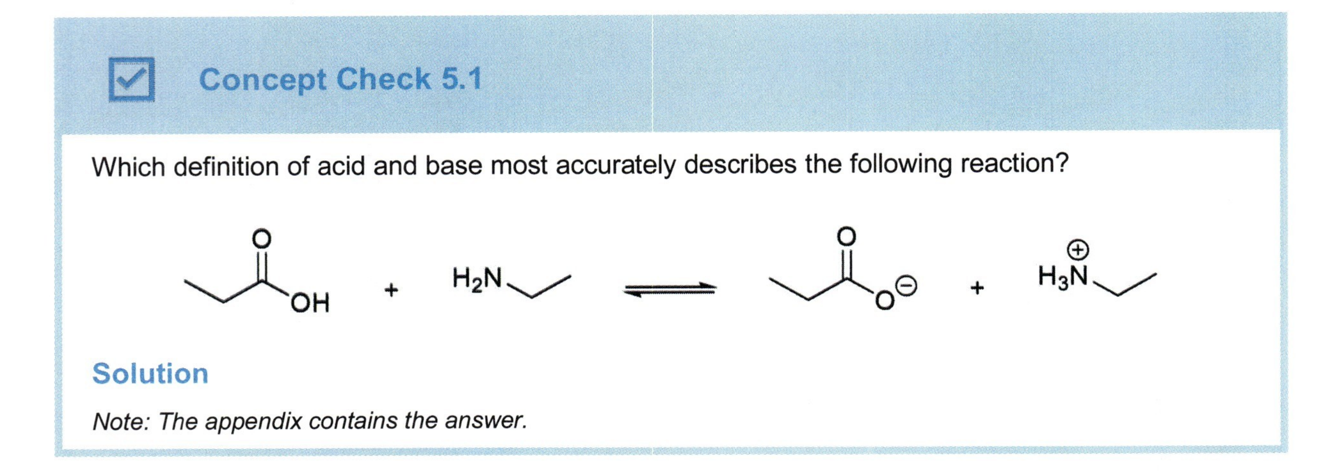

Concept Check 5.1

Which definition of acid and base most accurately describes the following reaction?

Solution

Note: The appendix contains the answer.

5.1.02 Conjugate Acids and Bases

An important outcome of the Brønsted-Lowry definition is that *any* Brønsted-Lowry acid-base reaction results in the formation of *another* Brønsted-Lowry acid-base pair. This equilibrium relationship is described in terms of **conjugate acid-base pairs** (Figure 5.4).

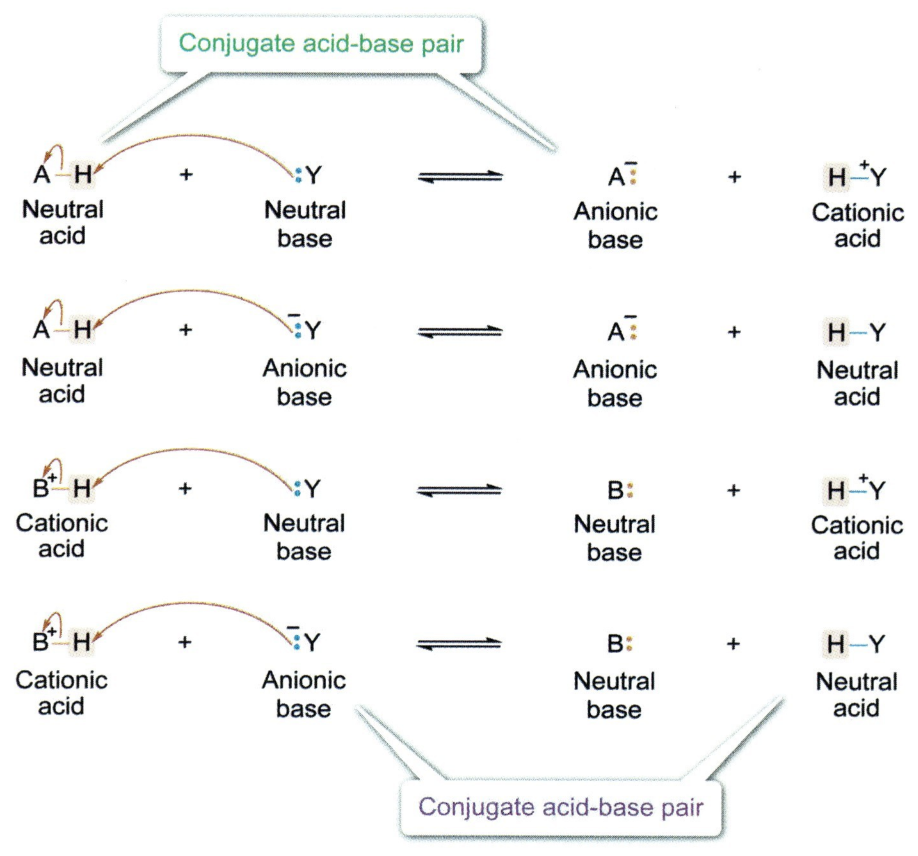

Figure 5.4 Conjugate acid-base pairs.

When a molecule acting as a Brønsted-Lowry acid *donates* its acidic proton to another molecule, the left-behind electrons become a new lone pair of the resultant product molecule (its **conjugate base**), which could itself later *accept* a proton. Conversely, when a molecule acting as a Brønsted-Lowry base *accepts* an acidic proton from another molecule, the resultant product (its **conjugate acid**) now possesses an acidic proton that could, in turn, be *donated* to another molecule.

Conjugate acid-base pairs take on one of two general forms, depending on the formal charges:

- A neutral acid (HA), with an acidic proton. When HA loses a proton (H^+), its conjugate base (A^-) gains the lone pair of electrons from the H–A bond ($HA - H^+ = A^-$).
- A cationic acid (BH^+), with an acidic proton. When BH^+ loses a proton (H^+), its conjugate base (B) gains the lone pair of electrons from the $H–B^+$ bond ($BH^+ - H^+ = B$).

Figure 5.5 depicts examples of conjugate acid-base pairs that will be encountered in later lessons.

Functional group	Conjugate acid	Conjugate base
Alkane	H–CH$_3$	$^{\ominus}$:CH$_3$
Amine	H–N̈H$_2$	$^{\ominus}$:N̈H$_2$
Alkyne	H–≡–R	$^{\ominus}$:≡–R
Water	H–ÖH	$^{\ominus}$:ÖH
Protonated amine	H–N$^{\oplus}$H$_3$	:NH$_3$
Carboxylic acid	HO–C(=O)–R	$^{\ominus}$:O–C(=O)–R
Hydrochloric acid	H–Cl:	$^{\ominus}$:Cl:

Figure 5.5 Examples of conjugate acid-base pairs in organic chemistry.

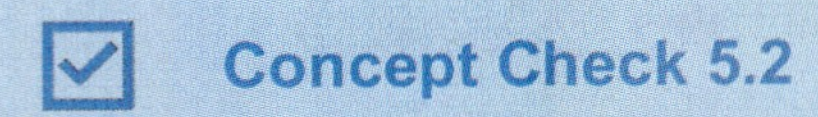

Concept Check 5.2

The indicated hydrogen atom is acidic. Write a neutralization reaction (with curved arrow flow) between this molecule and the hydroxide ion (^{-}OH).

O–H

Solution

Note: The appendix contains the answer.

5.1.03 pH, p*K*a, p*K*b, and the Henderson-Hasselbalch Equation

Due to the autoionization of water, neutral water has a hydronium ion concentration ([H_3O^+]) and a hydroxide ion concentration ([OH^-]) both equal to 10^{-7} M. Since H_3O^+ provides a source of protons (H^+), H_3O^+ and H^+ can be used interchangeably within most contexts.

The pH Scale

The **pH scale** is used to measure the acidity (or basicity) of an aqueous solution. A solution's pH is the negative logarithm of its [H^+] (Figure 5.6). Neutral water has a pH of exactly 7 ($-\log(10^{-7}$ M$) = 7$). **Acidic** solutions have an *increased* [H^+] and pH values *less than* 7, whereas **basic (alkaline)** solutions have a *decreased* [H^+] and pH values *greater than* 7.

As a logarithmic quantity, *each unit* of the pH scale reflects a *ten-fold difference* in acidity or basicity (eg, a solution at pH 6 is ten times more acidic than a solution at pH 7).

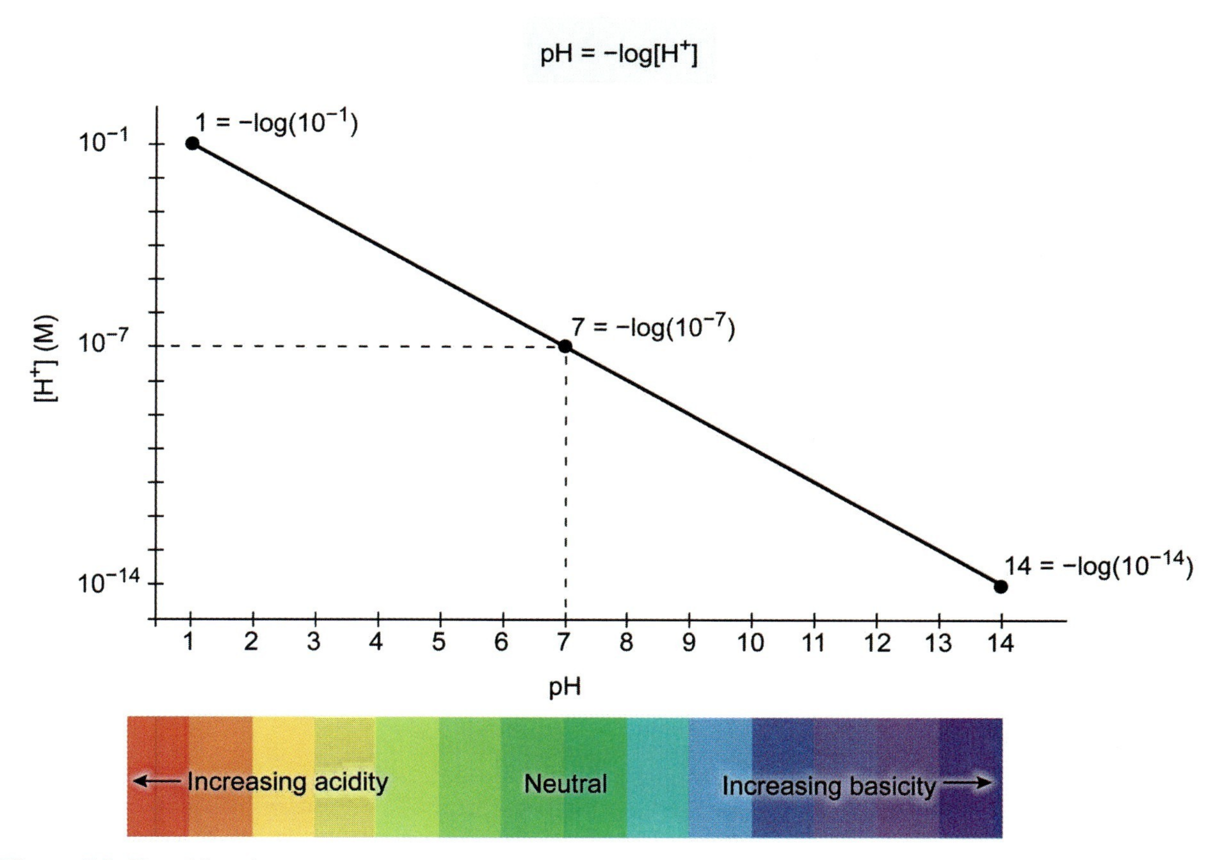

Figure 5.6 The pH scale.

Weak-Acid Dissociation Constant, K_a

A **strong acid** is one that ionizes almost completely in solution; however, relatively few strong acids exist. Most organic acids are weak acids and only *partially* ionize to release H^+, so a dynamic equilibrium model is more appropriate to describe their characteristics.

The equilibrium governing the dissociation of a weak acid in water is described by the **weak acid dissociation constant K_a**, which is defined in terms of the concentrations of the conjugate acid [HA], the conjugate base [A^-], and [H^+] (Figure 5.7).

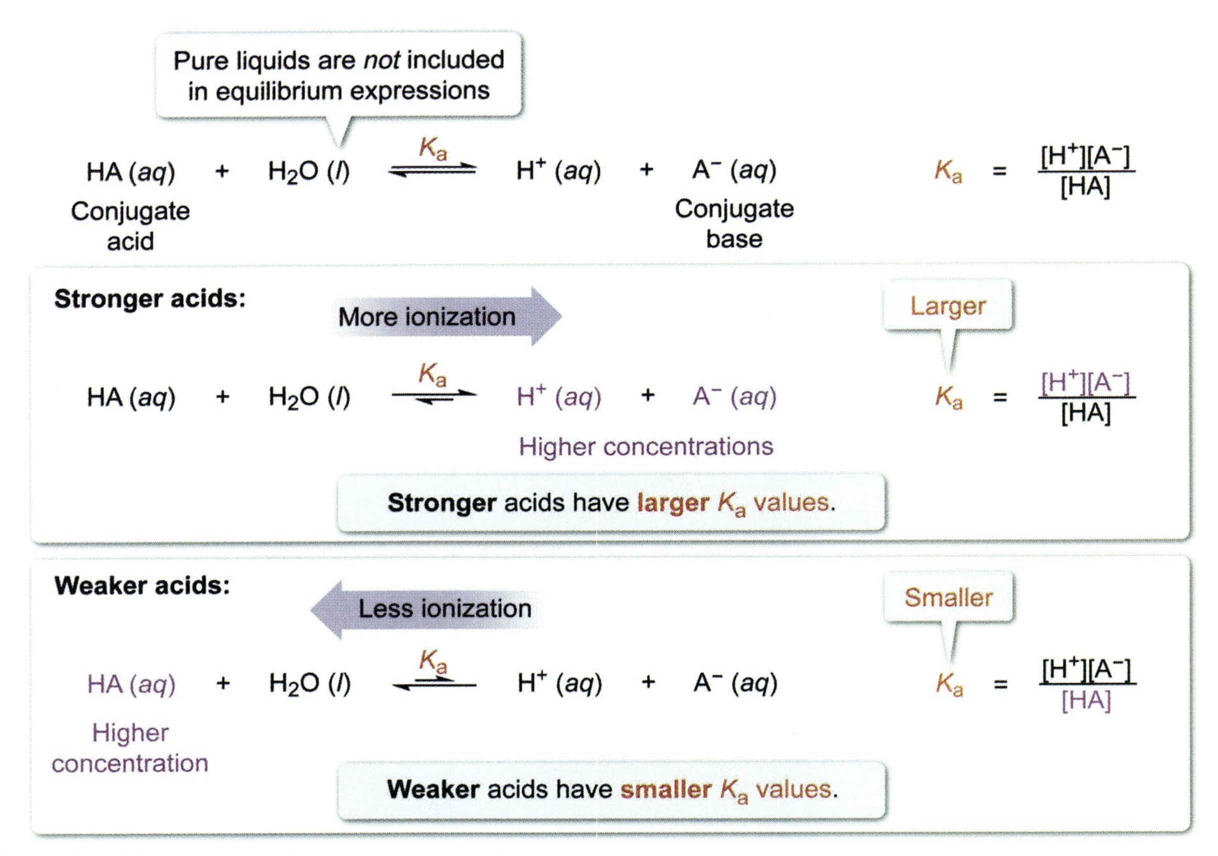

Figure 5.7 The weak acid dissociation constant, K_a.

The magnitude of K_a reflects the amount of ionization for a weak acid: The stronger the acid, the more equilibrium favors greater amounts of the products (ie, H^+ and A^-), and therefore *larger* K_a values. In contrast, weaker acids do *not* dissociate very much, so the equilibrium favors greater amounts of the *reactant* (HA), which have *smaller* K_a values.

Weak-Base Dissociation Constant, K_b

Most **strong bases** are hydroxide salts of Group IA or Group IIA ions. In contrast, most organic bases are *weak* bases that only *partially* ionize by accepting an H^+.

The equilibrium governing the dissociation of a weak base in water is described by the **weak base dissociation constant K_b**, which is defined in terms of the concentrations of the conjugate acid ([HA]), the conjugate base ([A^-]), and hydroxide ([OH^-]) (Figure 5.8). For the purposes of later calculations, [A^-] generally refers to the conjugate base of an acid-base pair, even if the conjugate base is not necessarily an anion.

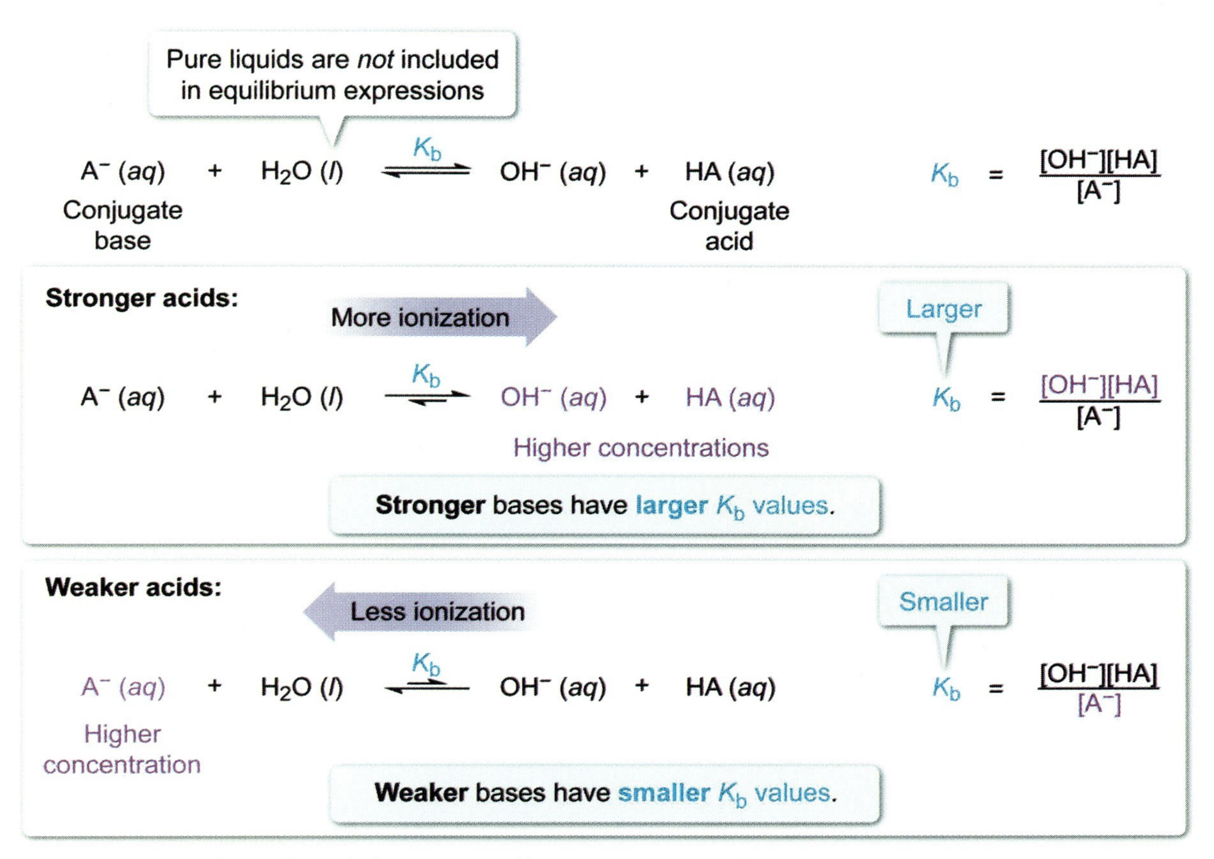

Figure 5.8 The weak base dissociation constant, K_b.

The magnitude of K_b reflects the amount of ionization for a base: The stronger the base, the more equilibrium favors formation of the products (ie, OH^- and HA), and therefore *larger* K_b values. In contrast, weaker bases do not greatly ionize water, so equilibrium favors the reactant (A^-) and leads to *smaller* K_b values.

The Relationship Between pK_a and pK_b for a Conjugate Acid-Base Pair

Chemists frequently report K_a and K_b on a logarithmic scale. The **pK_a** value is the negative logarithm of K_a, and the **pK_b** value is the negative logarithm of K_b (Figure 5.9). Unlike with K_a and K_b, acids and bases get *stronger* with *smaller* pK_a and pK_b values, respectively. Each unit on the pK_a or pK_b scale refers to a ten-fold difference in chemical property (eg, a compound with pK_a 6 is ten times more acidic than a compound with pK_a 7).

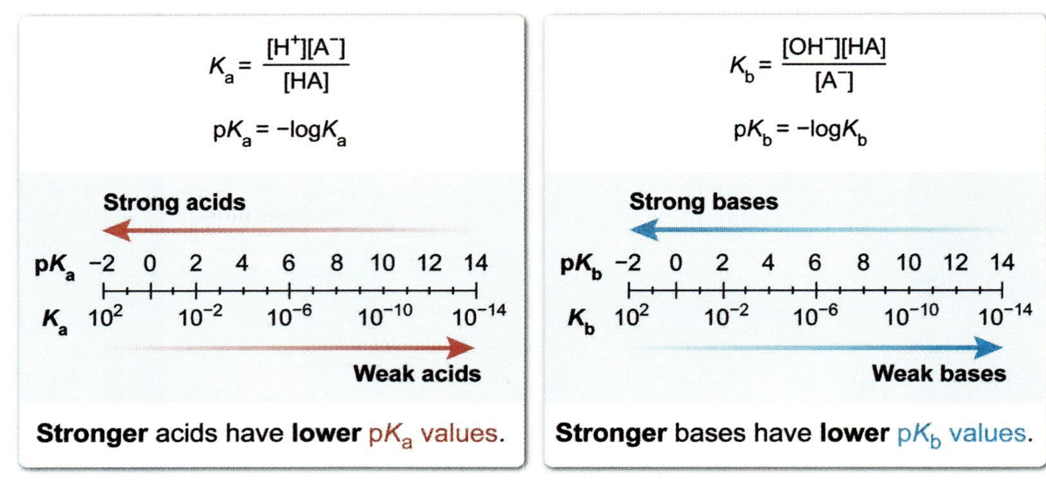

Figure 5.9 Acid strength in relation to K_a and pK_a, and base strength in relation to K_b and pK_b.

For a conjugate acid-base pair (ie, HA versus A^-), an *inverse* relationship exists between the strength of the conjugate acid (K_a or pK_a) and the strength of the conjugate base (K_b or pK_b) (Figure 5.10). Multiplication of K_a and K_b leads to an expression that simplifies to $[H^+][OH^-]$, which is the **equilibrium dissociation constant for water K_w** (10^{-14}). A similar mathematical expression on a logarithmic scale leads to pK_a + pK_b = pK_w = 14.

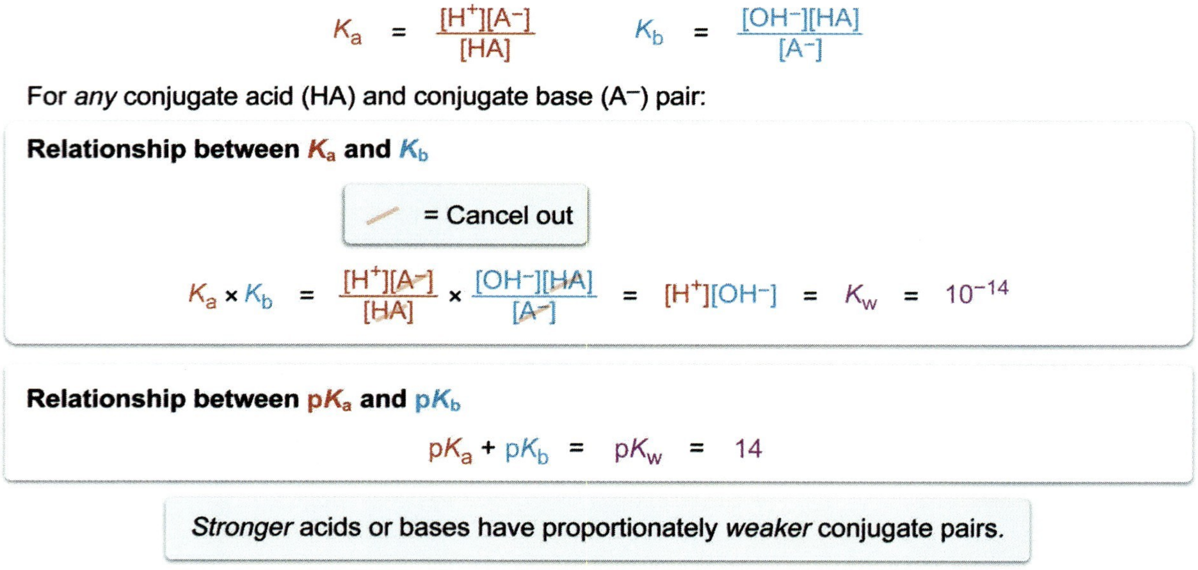

$$K_a = \frac{[H^+][A^-]}{[HA]} \qquad K_b = \frac{[OH^-][HA]}{[A^-]}$$

For *any* conjugate acid (HA) and conjugate base (A^-) pair:

Relationship between K_a and K_b

/ = Cancel out

$$K_a \times K_b = \frac{[H^+][A^-]}{[HA]} \times \frac{[OH^-][HA]}{[A^-]} = [H^+][OH^-] = K_w = 10^{-14}$$

Relationship between pK_a and pK_b

$$pK_a + pK_b = pK_w = 14$$

Stronger acids or bases have proportionately *weaker* conjugate pairs.

Figure 5.10 Mathematical relationship between K_a and K_b, and between pK_a and pK_b for a conjugate acid-base pair.

Therefore, knowing any one constant among K_a, K_b, pK_a, and pK_b for a conjugate acid-base pair allows *all* the other constants to be calculated. For example, if acetic acid (CH_3CO_2H) has a pK_a of 4.76, the pK_b of its conjugate base (acetate ion, $CH_3CO_2^-$) can be calculated as pK_b = 14 − pK_a = 14 − 4.76 = 9.24. Between acetic acid and the acetate ion, acetic acid is a stronger relative acid than the acetate ion is as a base (pK_a 4.76 < pK_b 9.24).

Predicting Reaction Outcomes Based on pK_a Values

Combined with thermodynamic considerations, wherein *the position of an equilibrium tends to prefer molecules that are lower in energy* (ie, more stable, or less reactive), an important use of the acid and base dissociation constants is the prediction of a proposed neutralization reaction's outcome.

The first step toward predicting the outcome is the identification of the acids and bases on both sides of the reaction equilibrium. By the Brønsted-Lowry definition, a neutralization reaction can be generalized to an acid and a base reacting to form a base and an acid (Figure 5.11). Consequently, each side of the equilibrium should contain exactly one acid and one base.

Equilibrium prefers reaction *by* the stronger acid (ie, the acid with the lower pK_a) and *formation of* the *weaker* acid (ie, the acid with the higher pK_a). Therefore, if pK_a values are known for the reactant and product acids, the favored reaction direction (forward or reverse) can be determined. Due to the reciprocal nature of conjugate acid-base strengths, the side of the equilibrium with the weaker acid *always* also has the weaker base (and vice versa); therefore, *either* the acids *or* the bases can be compared with identical results.

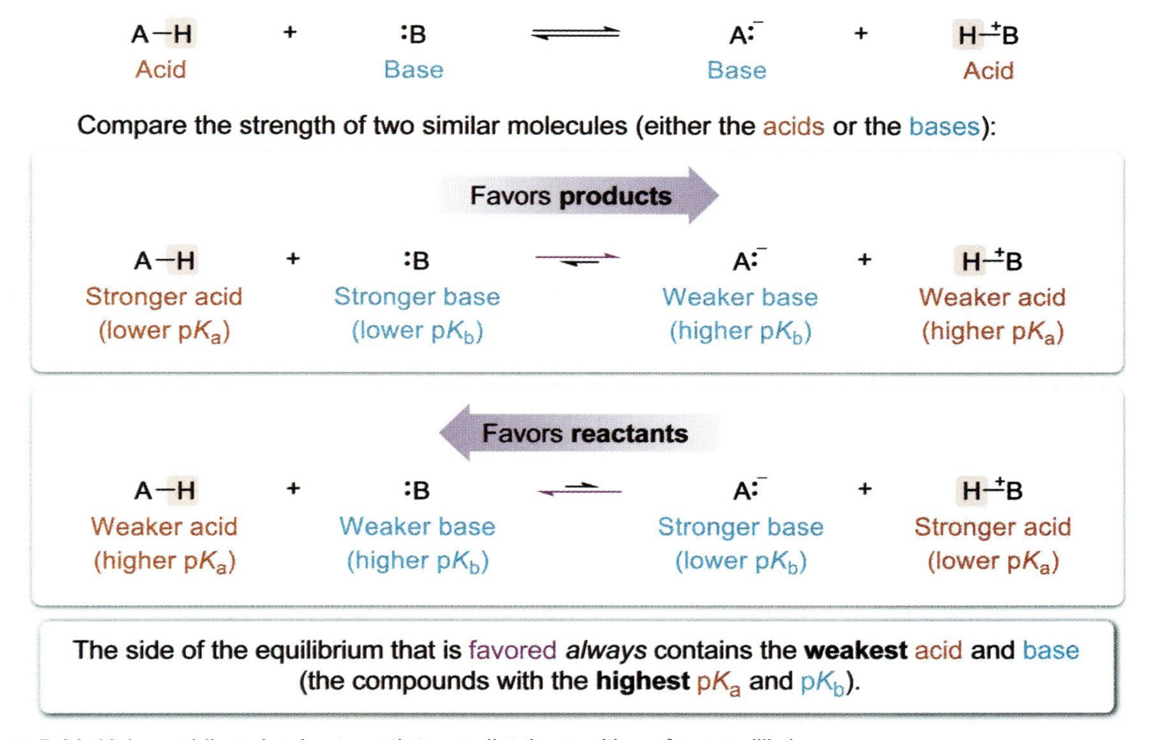

Figure 5.11 Using acidic or basic strength to predict the position of an equilibrium.

If pK_a values are not available, qualitative assessment of acidic strength can often be inferred based on either molecular structures or an understanding of underlying chemical principles. Concepts 5.1.04 through 5.1.07 provide an overview of some chemical phenomena that can impact molecular acidity.

Henderson-Hasselbalch Equation

The equation describing K_a can be mathematically manipulated into an alternative form that is especially useful for organic chemistry and biochemistry. The **Henderson-Hasselbalch equation**, derived in Figure 5.12, describes the equilibrium ratio of a conjugate acid-base pair in terms of two variables: the pH of the solution and the pK_a of the weak acid.

Dissociation of a weak acid HA:

$$HA\,(aq) + H_2O\,(l) \xrightleftharpoons{K_a} H^+\,(aq) + A^-\,(aq)$$

$$K_a = \frac{[H^+][A^-]}{[HA]}$$

Step 1: Solve for $[H^+]$

$$[H^+] = K_a\frac{[HA]}{[A^-]}$$

Step 2: Take logarithms of both sides

$$\log[H^+] = \log\left(K_a\frac{[HA]}{[A^-]}\right)$$

$$\log[H^+] = \log K_a + \log\left(\frac{[HA]}{[A^-]}\right)$$

Recall that $\log(ab) = \log(a) + \log(b)$

Step 3: Multiply both sides by –1

$$\underbrace{-\log[H^+]}_{pH} = \underbrace{-\log K_a}_{pK_a} - \log\left(\frac{[HA]}{[A^-]}\right)$$

Step 4: Rewrite in terms of pH and pK_a

$$pH = pK_a - \log\left(\frac{[HA]}{[A^-]}\right)$$

Recall that $\log\left(\frac{a}{b}\right) = -\log\left(\frac{b}{a}\right)$

or

$$pH = pK_a + \log\left(\frac{[A^-]}{[HA]}\right)$$

The Henderson-Hasselbalch equation

Figure 5.12 Deriving the Henderson-Hasselbalch equation for a weak acid.

Implications of the Henderson-Hasselbalch Equation

The primary benefit of the Henderson-Hasselbalch equation within organic chemistry is an understanding of the relationship between pH, pK_a, and the equilibrium ratio of conjugate base to conjugate acid. Since most situations requiring the use of the Henderson-Hasselbalch equation take place in the same solution, the *volume* components of [HA] and $[A^-]$ usually have the *same value*. In these situations, the number of *moles* of HA and A^- can be used instead of their *concentrations*.

When the pH value is less than pK_a, the high $[H^+]$ results in protonation of the conjugate base, meaning the conjugate *acid* form becomes predominant (Figure 5.13). In these situations, the *difference* between pH and pK_a determines the ratio of $[A^-]$ to [HA]. Since pH and pK_a are both logarithmic values, the ratio of acid to base increases by a factor of ten for each unit of difference between pH and pK_a. The conjugate acid form does not become quantitatively formed (ie, >99.9%) until the pH is adjusted to *3 units* below the pK_a.

$$pH = pK_a + \log\left(\frac{[A^-]}{[HA]}\right) \longrightarrow 10^{(pH - pK_a)} = \frac{[A^-]}{[HA]}$$

Two forms of the Henderson-Hasselbalch equation

pH	pK_a – 3	pK_a – 2	pK_a – 1	pK_a	pK_a + 1	pK_a + 2	pK_a + 3
$\frac{[A^-]}{[HA]}$	$\frac{1}{1000}$	$\frac{1}{100}$	$\frac{1}{10}$	$\frac{1}{1}$	$\frac{10}{1}$	$\frac{100}{1}$	$\frac{1,000}{1}$
% Major component	99.9%	99.0%	90.9%	50%	90.9%	99.0%	99.9%

Conjugate acid predominates ← → Conjugate base predominates

Figure 5.13 Relationship between pH, pK_a, and the ratio of conjugate base to conjugate acid.

A similar situation occurs when the pH value is *greater than* the pK_a value, where each unit of difference between pH and pK_a leads to a ten-fold increase in the amount of conjugate base relative to the conjugate acid.

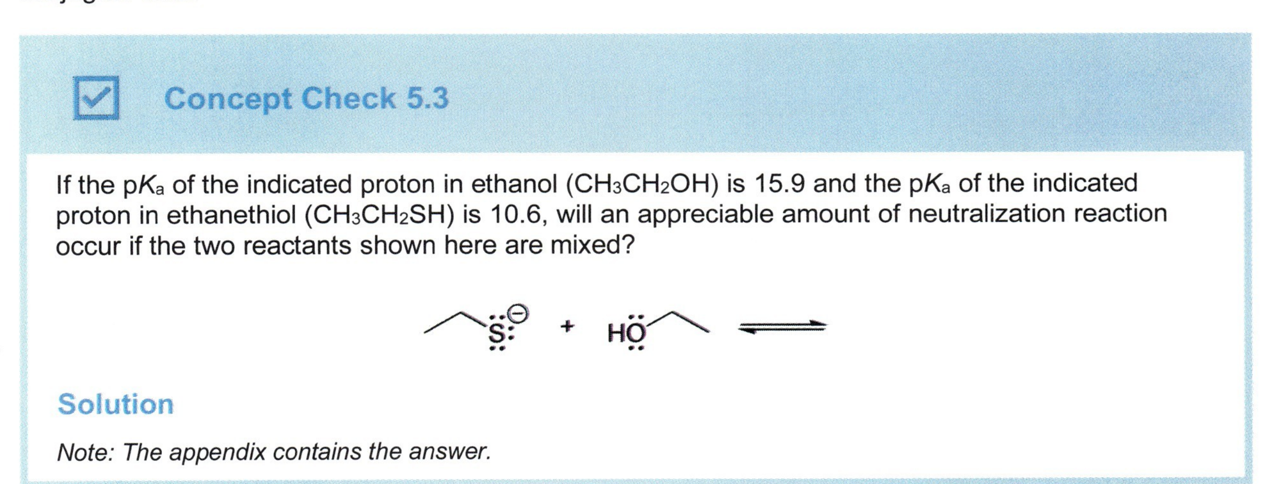

Concept Check 5.3

If the pK_a of the indicated proton in ethanol (CH_3CH_2OH) is 15.9 and the pK_a of the indicated proton in ethanethiol (CH_3CH_2SH) is 10.6, will an appreciable amount of neutralization reaction occur if the two reactants shown here are mixed?

Solution

Note: The appendix contains the answer.

5.1.04 Impact of Electronegativity on Acidity

The presence or absence of electronegative atoms in a molecule can have a profound effect on acid strength. The electronegativity of an atom (X) *directly connected* to a proton impacts both the polarity of the X–H bond *and* the acidity of the attached proton (Figure 5.14). The bonds between protons and strongly electronegative atoms are very polar and relatively easier to break (ie, the proton is more easily donated, and therefore more acidic).

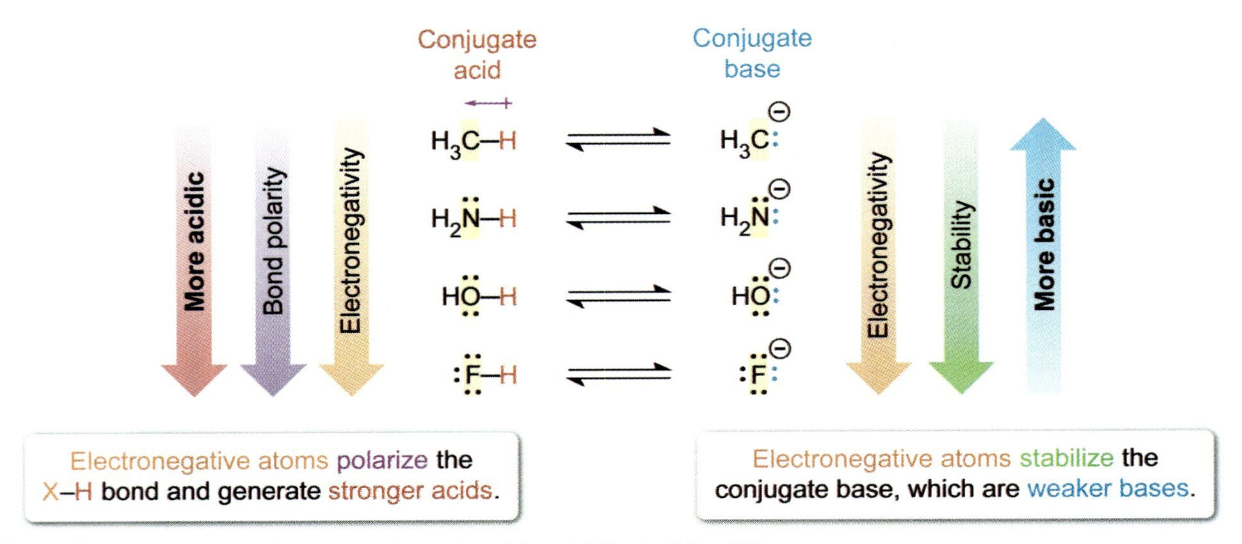

Figure 5.14 Relationship between electronegativity, acidity, and basicity.

Strongly electronegative atoms also more strongly stabilize the lone pair of electrons left behind from breaking the X–H bond, thereby stabilizing the conjugate base. Since conjugate acid-base properties are inversely related, stabilization of the conjugate base (*decreased* reactivity) promotes *increased* reactivity of the corresponding conjugate acid (acidic strength).

Electronegative atoms also impact acid-base properties even when they are located elsewhere on the molecule. The **inductive effect** describes how electron density can be added (donated) or removed (withdrawn) *through* sigma bonds (Figure 5.15). The trends for the presence of electronegative atoms in an acid are identical to those described in Figure 5.15; they simply act through the inductive effect across a greater distance.

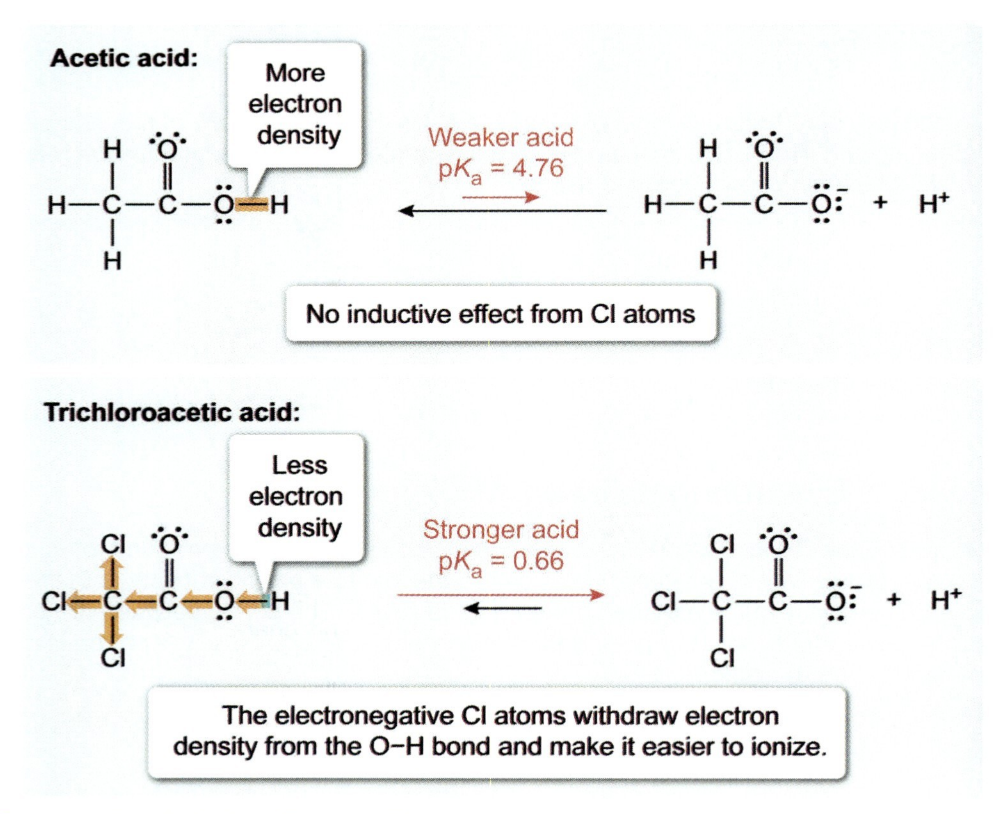

Figure 5.15 Impact of the inductive effect on acidity.

The magnitude of the observed inductive effect for an electronegative atom is dependent on:

- Its distance from the acidic proton (closer = more impact).
- Its electronegativity (greater value = more impact).
- The number of electronegative atoms (more = more impact).

Because an **electron-donating group** (eg, alkyl group) *contributes* electron density to the molecule through the inductive effect, the X–H bond becomes *more* stable and the conjugate base is *less* able to accommodate the additional lone pair. Consequently, electron-donating groups tend to *decrease* the acidity of a molecule. For example, acetic acid (CH_3CO_2H, pK_a 4.76) is *less* acidic than formic acid (HCO_2H, pK_a 3.74) due to its additional methyl group.

5.1.05 Impact of Atomic Size on Acidity

Atomic or ionic size can also have a significant effect on the relative strength of an acid. Using hydrohalic acids as an example, the H–X bond length increases down a group on the period table (Figure 5.16). The increase in bond length results in a *decrease* in bond *strength*, as the halogen's orbital overlap with the hydrogen's 1*s* orbital decreases. Consequently, weaker H–X bond strengths lead to an *increase* in acidity.

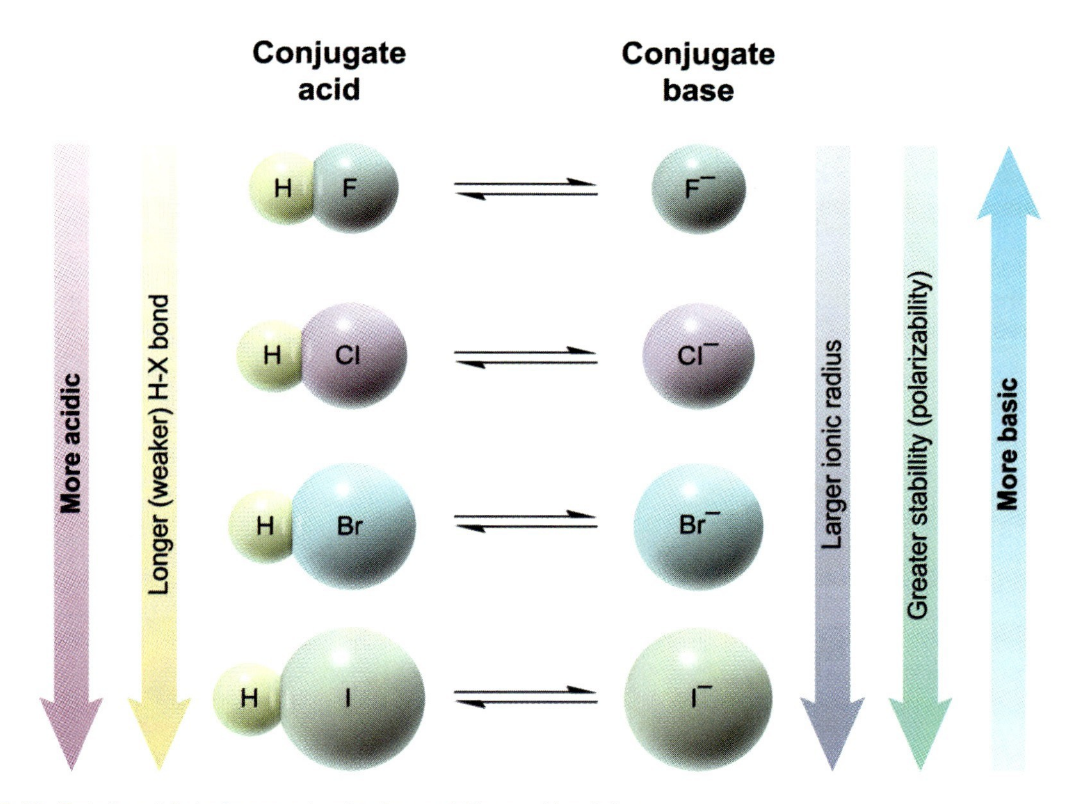

Figure 5.16 Relationship between atomic size, acidity, and basicity.

Similarly, the increasing ionic radius of the halide anions (ie, the conjugate bases of hydrohalic acids) down a group results in *increasing stability* of the conjugate base. This characteristic is due to the concept of **polarizability**; larger anions have a larger surface over which to distribute the increased electron density, leading to a lower charge density. In contrast, smaller anions have their excess electron density distributed over a much smaller surface area (ie, a more concentrated charge distribution).

5.1.06 Impact of Hybridization on Acidity

Hybridization can have a substantial effect on the relative strength of an acid (Figure 5.17). Once the acidic proton has been removed from an alkane, alkene, or alkyne, their respective conjugate bases have the resultant lone pair of electrons in *different* hybrid orbitals. Since the *s* orbital is one of the four atomic orbitals that combine to form sp^3 hybrid orbitals ($s + p + p + p$), the lone pair in the conjugate base of an alkane has 25% ***s* character**, whereas alkenes and alkynes have 33% and 50% *s* character, respectively.

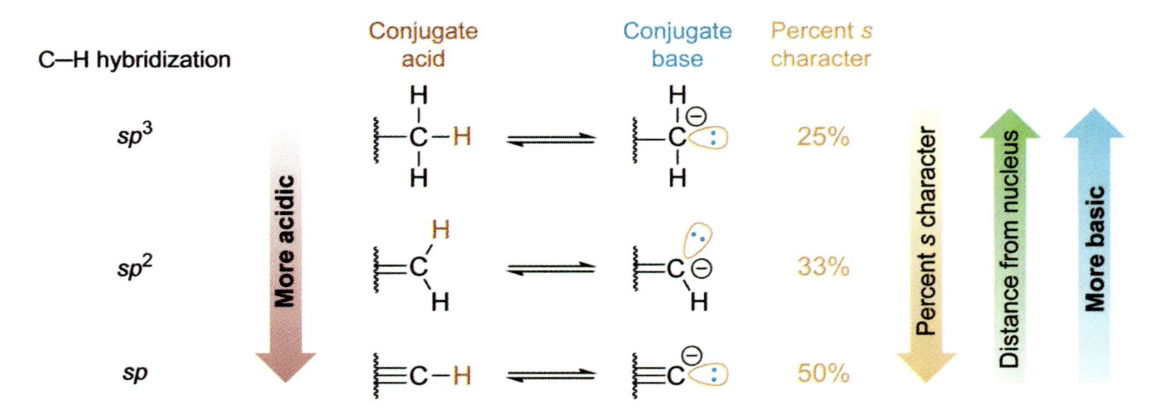

Figure 5.17 Relationship between hybridization, acidity, and basicity.

Because an *s* atomic orbital holds its electrons *closer to the nucleus* than a *p* orbital, the *greater* the percent *s* character, the *closer* the hybrid orbital holds its electrons to the nucleus. Electrons held closer to the nucleus have *greater* stability due to the closer proximity of the positively charged protons in the nucleus. Therefore, the conjugate base of a terminal alkyne (with its lone pair in an *sp* orbital) is *more stable* than the conjugate bases of alkenes or alkanes (which would have their lone pairs in an sp^2 and sp^3 orbital, respectively).

Since conjugate acid-base properties are inversely related, conjugate base stabilization (decreased reactivity) promotes increased acidic reactivity (acidic strength). Therefore, the acidity of a C–H bond *increases* as its hybridization changes from $sp^3 \rightarrow sp^2 \rightarrow sp$.

5.1.07 Impact of Resonance on Acidity

As a general guideline, resonance delocalization tends to be a *stabilizing* feature that makes a substance *less* reactive—*particularly* if the reactivity would disrupt the resonance. This attribute can have a meaningful effect on the acid or base strength (Figure 5.18). Acids or bases that demonstrate resonance delocalization are generally more stable and less reactive, and therefore *weaker* acids or bases.

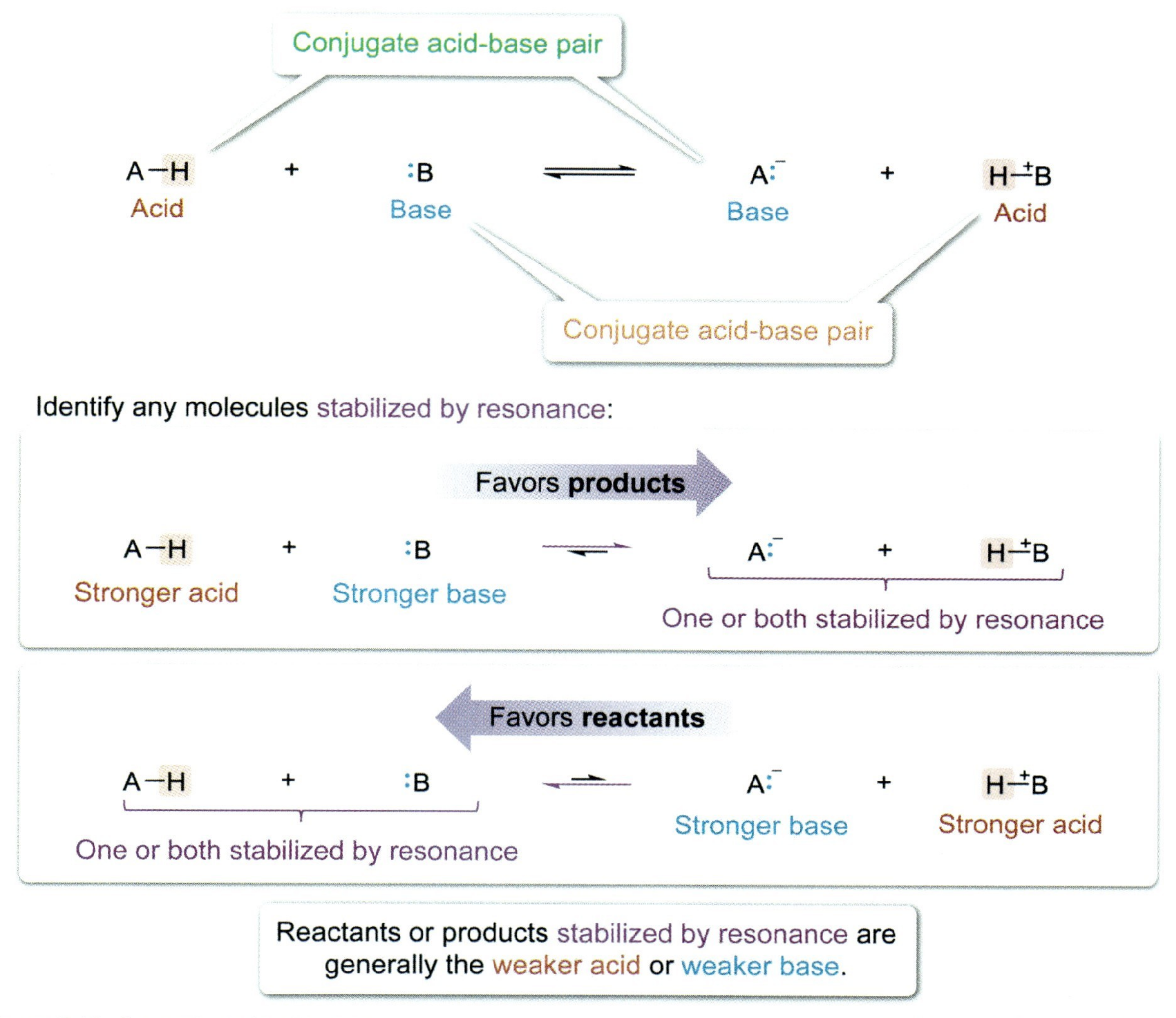

Figure 5.18 General relationship between molecules demonstrating resonance delocalization and the strength of acid or base.

The specific impact that resonance-stabilized molecules have on equilibrium is dependent on whether they are present in a *reactant* or a *product*. Products that are resonance-stabilized promote an equilibrium that favors their formation (ie, the formation of products). Reactants that are resonance-stabilized promote an equilibrium that favors their formation (ie, the formation of reactants). See Concept 2.7.04 for more on resonance and acid-base equilibria.

Lesson 5.2

Reactive Intermediates

Introduction

Many chemical reactions in organic chemistry involve the formation of **reactive intermediates**—transient (ie, short-lived) molecular fragments with an unusual number of bonds that play crucial roles in the reaction mechanism. Because reactive intermediates are consumed quickly after formation, they are rarely found in significant concentrations within the reaction mixture.

This lesson focuses on three reactive intermediates encountered in traditional Organic Chemistry: carbocations, carbanions, and free radicals. Each intermediate contains a carbon atom with three bonds and a different number of nonbonding electrons.

5.2.01 Carbocations

A **carbocation** is a carbon atom with three sigma bonds, a vacant *p* orbital, and a positive formal charge. A carbocation does *not* have any nonbonding electrons. Since only three regions of electron density (electron domains) are present, a carbocation is sp^2 hybridized, exhibits a trigonal planar geometry, and has 120° bond angles (Figure 5.19). A carbocation's vacant *p* orbital is unhybridized and oriented perpendicular to the plane of the three sp^2 orbitals.

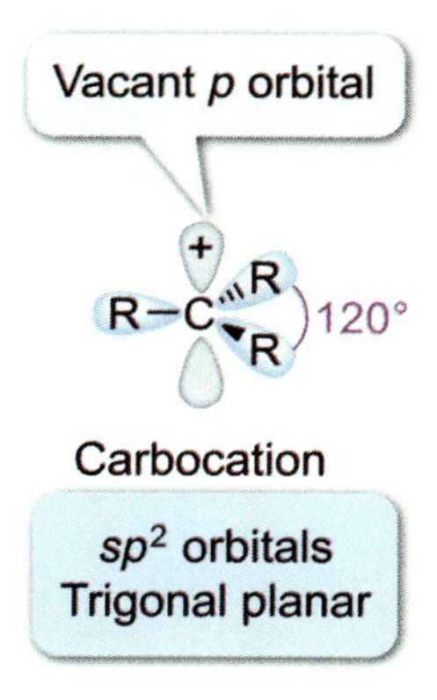

Figure 5.19 The orbital structure of a carbocation.

Since a carbocation has only six valence electrons (ie, less than a complete octet), *all* carbocations are electron-deficient and electrophilic (Concept 5.3.02). Therefore, carbocations readily react with nucleophiles in reactions such as the S_N1 reaction (see Concept 5.4.02).

Carbocations are classified based on the number of attached alkyl substituents (Figure 5.20). A **methyl carbocation** has zero alkyl groups, a **primary (1°) carbocation** has one alkyl group, a **secondary (2°) carbocation** has two alkyl groups, and a **tertiary (3°) carbocation** has three alkyl groups.

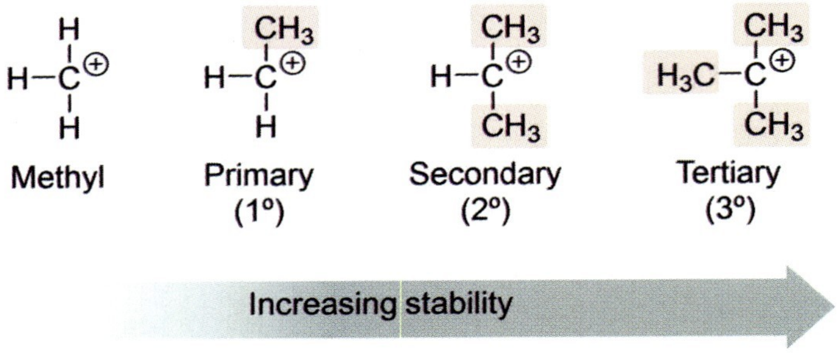

Methyl | Primary (1°) | Secondary (2°) | Tertiary (3°)

Increasing stability

Figure 5.20 Carbocation substitution and stability.

Carbocation stability is directly related to the number of attached alkyl groups. Alkyl groups stabilize the positive charge of a carbocation via electron donation by the **inductive effect** (Figure 5.21). Because carbon is more electron-rich than hydrogen, the electron cloud of a C–C sigma bond is more polarizable than a C–H sigma bond. Therefore, the electron-deficient carbocation carbon causes the C–C sigma bonds of adjacent alkyl groups to polarize and donate electron density toward the positively charged carbon.

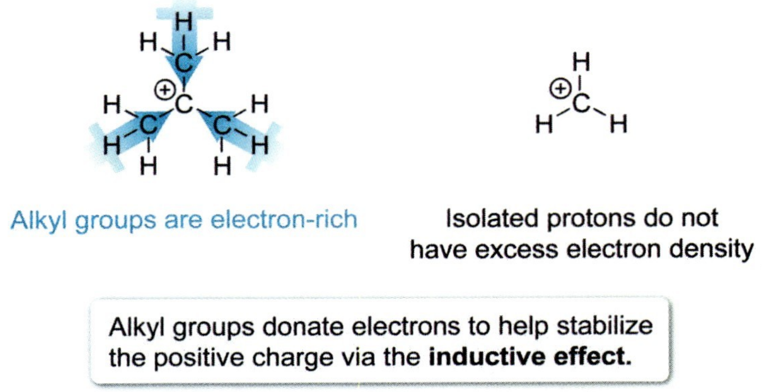

Alkyl groups are electron-rich

Isolated protons do not have excess electron density

Alkyl groups donate electrons to help stabilize the positive charge via the **inductive effect.**

Figure 5.21 The inductive effect and carbocation stabilization.

Alkyl group stabilization of carbocations is additive; a methyl carbocation has the *least* stability, and a tertiary carbocation has the *greatest* stability. Additional alkyl groups (ie, a more substituted carbocation) provide a greater number of atoms over which to distribute the positive charge, leading to carbocations with *increased* stability.

Alkyl groups also stabilize the vacant *p* orbital of a carbocation through hyperconjugation (Figure 5.22). **Hyperconjugation** occurs when the sp^3 orbitals of an alkyl group provide secondary (weak) overlap with the unhybridized *p* orbital, stabilizing the molecule. Conformational rotation of the sigma bond provides each C–H or C–R bond in an alkyl group the possibility of hyperconjugation with the vacant *p* orbital.

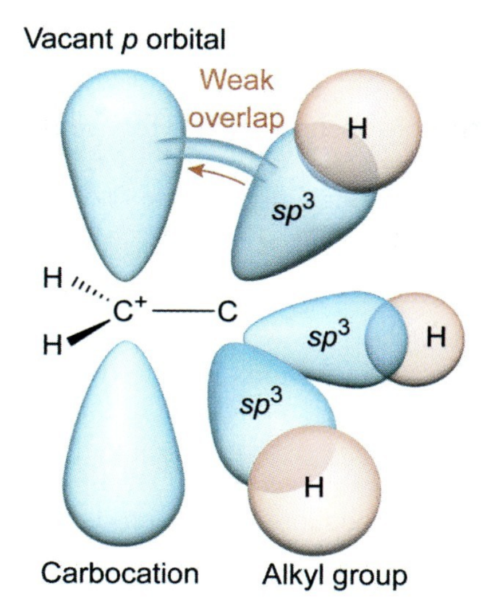

The sp^3 orbital of an alkyl group overlaps with a vacant *p* orbital via **hyperconjugation** to stabilize the carbocation.

Figure 5.22 Hyperconjugation.

Other factors that influence the stability of a carbocation include resonance and conjugation. A carbocation adjacent to a double bond (eg, an allylic carbocation) can delocalize its positive charge over multiple atoms through resonance. Figure 5.23 shows how the positive charge of an allylic carbocation is distributed over two carbon atoms, where each carbon carries a partial positive charge δ^+. Charge delocalization increases the stability of a carbocation by lowering the charge density.

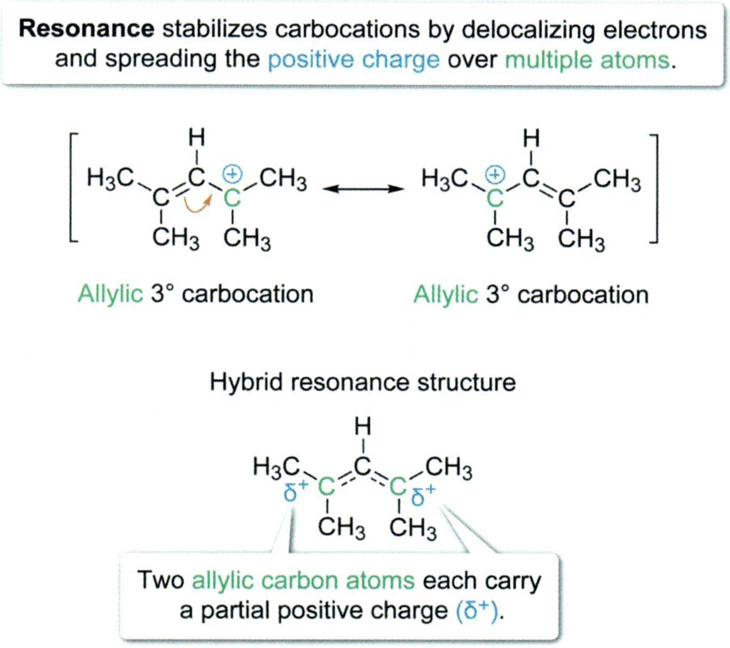

Resonance stabilizes carbocations by delocalizing electrons and spreading the positive charge over multiple atoms.

Allylic 3° carbocation ⟷ Allylic 3° carbocation

Hybrid resonance structure

Two allylic carbon atoms each carry a partial positive charge (δ^+).

Figure 5.23 Resonance delocalization of an allylic cation.

Carbocation Rearrangements

All carbocations have the possibility of undergoing a type of chemical reaction known as a **rearrangement**, a chemical transformation that *changes* the sigma bond skeleton of the molecule. Carbocation rearrangements typically *only* occur when the rearrangement leads to a *more stable carbocation*. For example, the rearrangement of a 1° carbocation to either a 2° or 3° carbocation may be favorable, whereas the rearrangement of a 3° carbocation to either a 2° or 1° carbocation is unlikely to be favorable.

Two types of carbocation rearrangements are common (Figure 5.24):

- A **hydride shift** occurs when a hydrogen atom and its two electrons (ie, a hydride ion, H:⁻) *moves* to the carbocation carbon from an *adjacent* carbon. A hydride shift is an example of a **1,2-shift** that ultimately changes the location of the carbocation to a more substituted (and more stable) position.
- A **methyl shift** is similar to a hydride shift except that a *methyl* group and its pair of electrons ($H_3C:^-$) undergo a 1,2-shift to the carbocation carbon, producing a more stable carbocation.

1,2-hydride shift: 1° carbocation → 3° carbocation

1,2-methyl shift: 1° carbocation → 2° carbocation

Carbocation rearrangements are a type of chemical reaction because the sigma bond scaffold is altered.

Figure 5.24 Rearrangements of a carbocation through a hydride shift and a methyl shift.

Hydride and methyl shifts are *chemical reactions* because sigma bonds are broken and formed to create a *new molecule*. Any time a reaction mechanism results in the formation of a carbocation, it is appropriate to consider possible 1,2-rearrangements that could occur, as these may influence the reaction outcome.

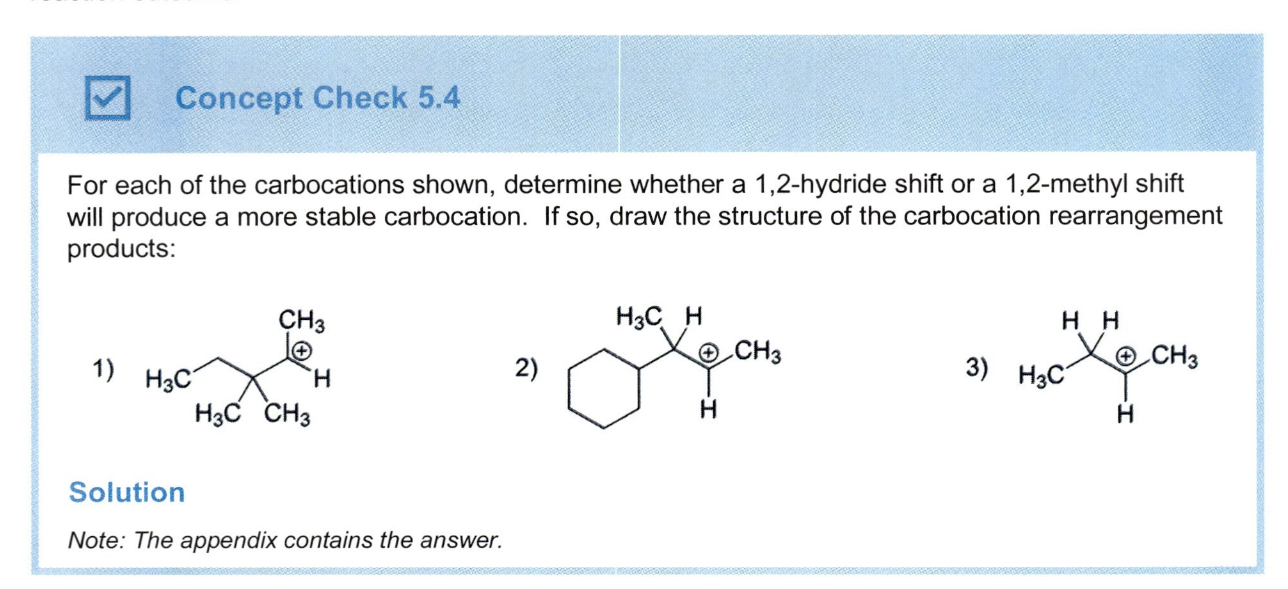

Concept Check 5.4

For each of the carbocations shown, determine whether a 1,2-hydride shift or a 1,2-methyl shift will produce a more stable carbocation. If so, draw the structure of the carbocation rearrangement products:

Solution

Note: The appendix contains the answer.

5.2.02 Carbanions

A **carbanion** is a carbon atom with three sigma bonds, one lone pair of electrons, and a *negative* formal charge (Figure 5.25). A carbanion generally has four electron domains, is sp^3 hybridized, has tetrahedral electron domain geometry, and has approximate bond angles of 109.5°.

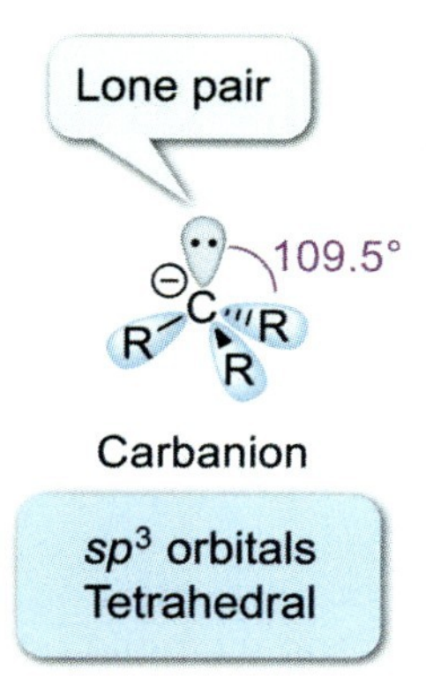

Figure 5.25 The orbital structure of a carbanion.

Unlike carbocations, carbanions are electron-rich and have an abundance of electron density; consequently, carbanions are very strong Lewis bases (Concept 5.1.01) and nucleophiles (Concept 5.3.01). For example, methane (CH_4) has a pK_a of approximately 50, and the pK_b of its carbanion conjugate base ($H_3C:^-$) is calculated to be −36; therefore, a methyl carbanion is a *very* strong base.

Although the hydroxide ion (HO^-) is generally considered a strong base, methyl carbanion is a *much* stronger base. Figure 5.26 shows the acid-base reaction that occurs between a methyl carbanion and water. Because equilibrium favors the formation of the weaker acid and base, a methyl carbanion is a strong enough base to form the hydroxide ion.

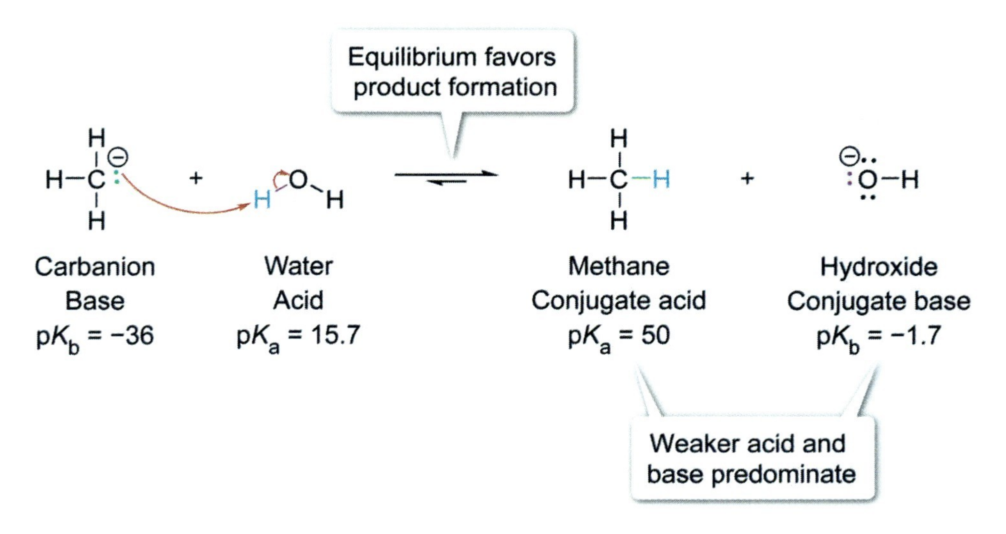

Figure 5.26 Acid-base reaction between a carbanion and water.

Like carbocations, carbanions are classified based on the number of attached alkyl substituents (Figure 5.27). A **methyl carbanion** has zero alkyl groups, a **primary (1º) carbanion** has one alkyl group, a **secondary (2º) carbanion** has two alkyl groups, and a **tertiary (3º) carbanion** has three alkyl groups.

Tertiary (3°) Secondary (2°) Primary (1°) Methyl

Increasing stability

Figure 5.27 Carbanion substitution and stability.

As described in Concept 5.2.01, an alkyl group is electron-donating through the inductive effect. Because carbanions are already electron-rich, alkyl groups *destabilize* carbanions—an effect that increases with each additional alkyl group. Therefore, carbanions have the *opposite* stability trend as carbocations (Concept 5.2.01).

A carbanion is *stabilized* by neighboring electron-withdrawing groups (eg, halogens), which pull electron density away from the negatively charged carbon atom through the inductive effect. Figure 5.28 shows that as the number of electron-withdrawing groups on a carbanion *increases*, *more* electron density is withdrawn and carbanion stability *increases*.

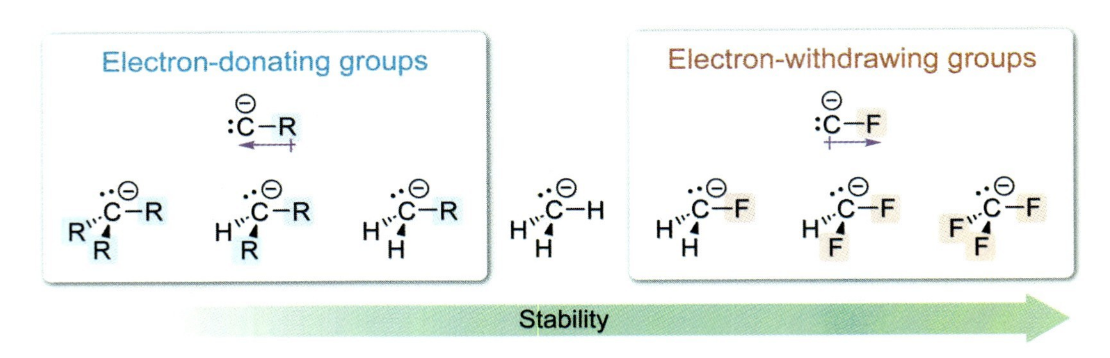

Addition of electron-withdrawing groups (eg, halogens) stabilizes carbanions via the **inductive effect**.

Figure 5.28 The inductive effect and carbanion stability.

Carbanions adjacent to a carbonyl or an alkene (eg, an allylic carbanion) are stabilized by resonance delocalization, which distributes the negative charge over multiple atoms. Because resonance delocalization *only* occurs when pi orbitals have side-to-side overlap with the lone pair of electrons on the carbanion, resonance stabilization of a carbanion *requires* the carbanion lone pair to be in an unhybridized *p* orbital (Figure 5.29).

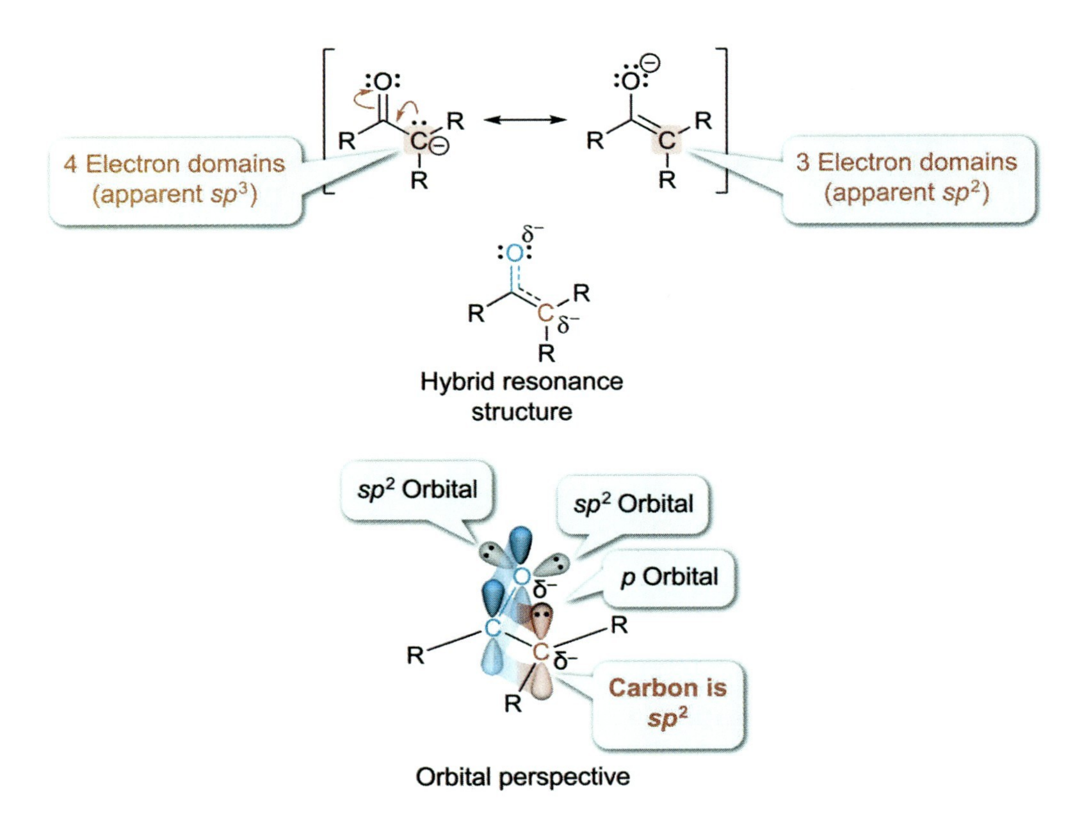

Figure 5.29 Resonance, hybridization, and carbanion stability.

Although a resonance-stabilized carbanion carbon may initially appear to be sp^3 hybridized (four electron domains), its resonance structure contains a C–C double bond (three electron domains) and thus is sp^2 hybridized. Consequently, resonance-stabilized carbanions are sp^2 hybridized. Resonance-stabilized carbanions are *more stable* than even a methyl carbanion.

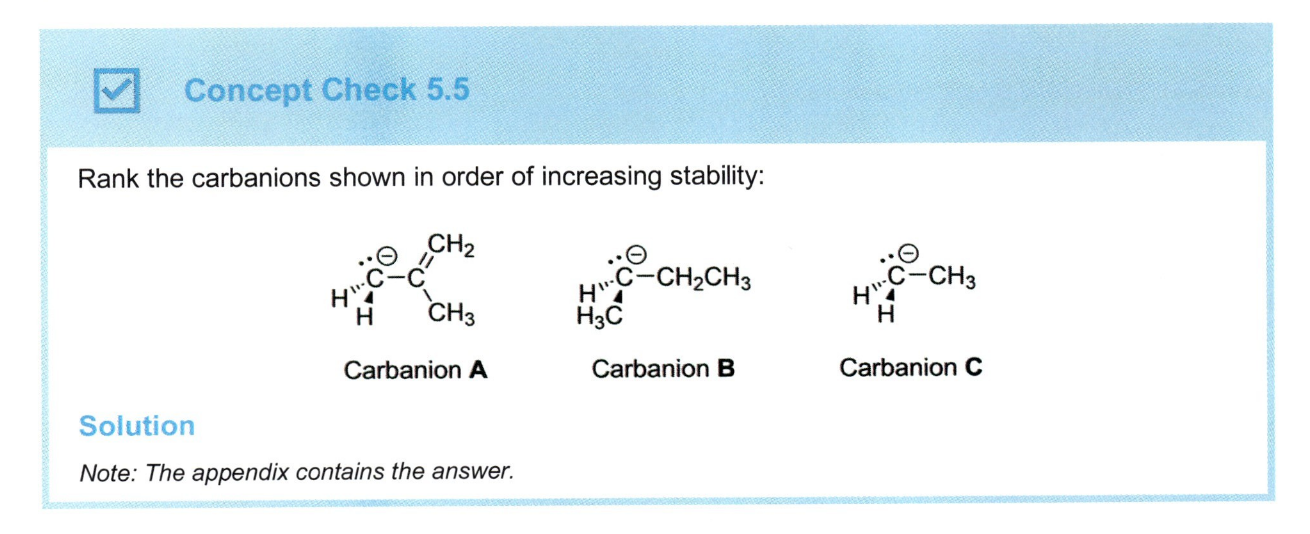

Concept Check 5.5

Rank the carbanions shown in order of increasing stability:

Solution

Note: The appendix contains the answer.

5.2.03 Free Radicals

A **free radical** is a reactive intermediate containing an *unpaired* nonbonding electron; a **carbon free radical** contains three sigma bonds and the unpaired electron.

Free radicals are structurally similar to a carbocation (see Concept 5.2.01) in terms of hybridization (sp^2) and geometry (trigonal planar). However, the unhybridized *p* orbital of a free radical contains the unpaired electron (Figure 5.30). Since a free radical has only seven valence electrons (ie, less than a full octet), it is also **electron-deficient**.

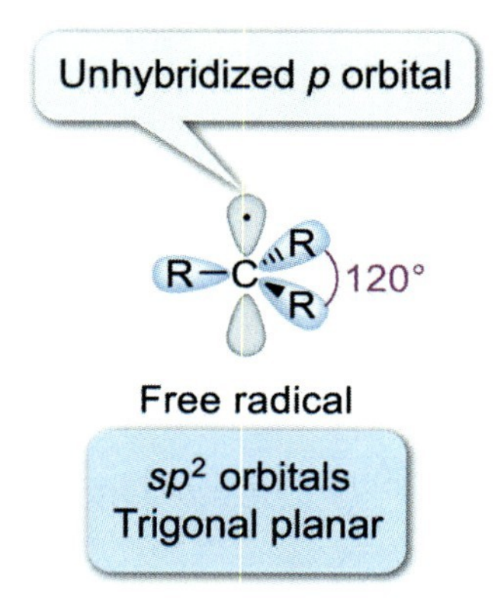

Figure 5.30 The orbital structure of a carbon free radical.

Free radicals are classified based on the number of attached alkyl substituents. A **methyl radical** has zero alkyl groups, a **primary (1°) radical** has one alkyl group, a **secondary (2°) radical** has two alkyl groups, and a **tertiary (3°) radical** has three alkyl groups (Figure 5.31).

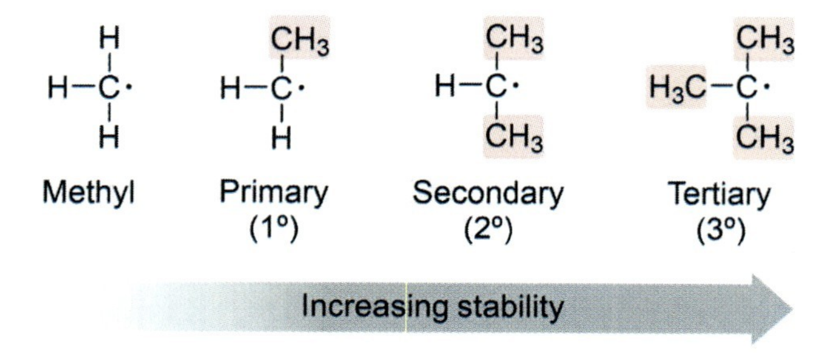

Figure 5.31 Free radical substitution and stability.

Free radicals follow the *same* stability trends as carbocations due to the inductive effect and hyperconjugation. Alkyl groups donate electron density toward the *electron-deficient* free radical, stabilizing the unpaired electron. The stability of a free radical increases with the number of alkyl groups; a methyl radical has the least stability, and a tertiary free radical has the greatest stability.

Allylic and benzylic free radicals are stabilized by **resonance**. The *p* orbitals of the pi bond overlap with the free radical's unhybridized *p* orbital (Figure 5.32). Note that the curved arrows use an arrowhead with only *one* barb to indicate the movement of only *one* electron.

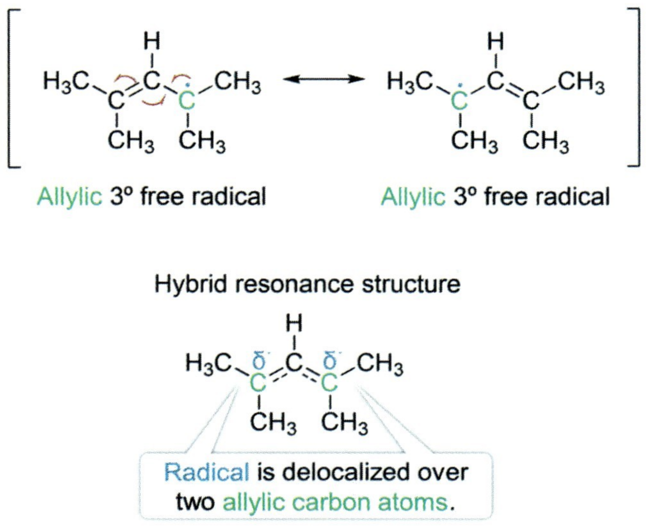

Allylic 3° free radical

Allylic 3° free radical

Hybrid resonance structure

Radical is delocalized over two allylic carbon atoms.

Figure 5.32 Resonance and free radical stability.

Lesson 5.3

Nucleophiles, Electrophiles, and Leaving Groups

Introduction

Lesson 5.1 defined chemical reactions as transformations that involve the creation or breaking of covalent bonds. Since a covalent bond is formed from the constructive overlap of electron orbitals (Lesson 1.1), chemical reactions can alternatively be viewed as processes that utilize *electrons* as their primary currency. This designation parallels Lewis acid and base definitions, which *also* focus on the role of electrons in defining acidic or basic character.

Many classifications of chemical reactions occur between molecules with high electron density and molecules with low electron density. This lesson introduces three essential components widely used to define roles in organic chemical reactions:

- Nucleophiles
- Electrophiles
- Leaving groups

5.3.01 Nucleophiles

A **nucleophile** ("lover of nuclei") is a substance with one or more lone pairs of electrons that can be donated or shared with an electron-deficient, nonhydrogen atom (Figure 5.33). Since the *lone pair* is the defining feature of a nucleophile, the abbreviations Nu: or Nuc: refer to a general, undefined nucleophile.

The specific atom that bears the lone pair within the nucleophile's structure is the **nucleophilic atom**. A nucleophilic atom can be either neutral or have a negative formal charge.

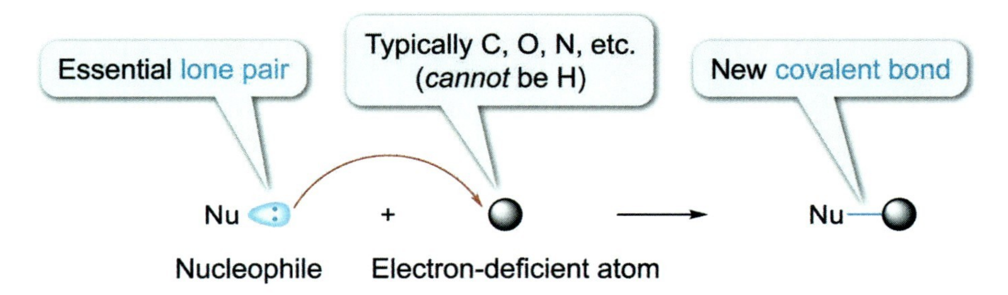

Figure 5.33 General structure and reactivity of a nucleophile.

Different nucleophiles have differing capacities to donate their lone pair to a given electron-deficient nucleus. The term **nucleophilicity** is used to describe the relative capacity of a substance to act as a nucleophile. Nucleophilicity is often described in nonquantitative terms (eg, strong, weak).

Differentiating Nucleophilicity and Basicity

Because a nucleophile and a (Lewis) base both donate a lone pair of electrons to another substance, chemists frequently invoke the Brønsted-Lowry definition to differentiate the processes of nucleophilicity and **basicity**. Bases donate their lone pairs to electron-deficient *hydrogen* atoms, whereas nucleophiles donate their lone pairs to electron-deficient nonhydrogen atoms (Figure 5.34). In either case, a new covalent bond is formed between the electron donor and the electron-deficient atom.

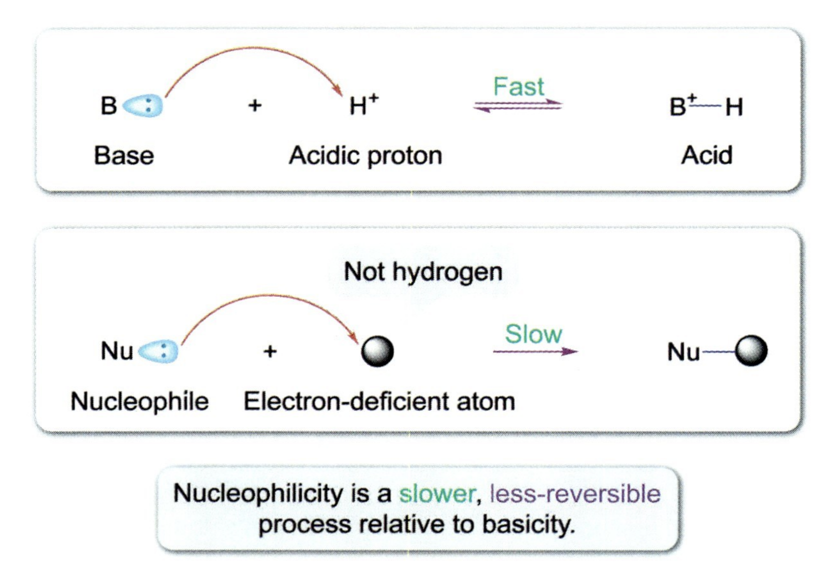

Figure 5.34 Differences between basicity and nucleophilicity.

The process of a base abstracting an acidic proton is a dynamic equilibrium, defined by expressions like K_b. In contrast, nucleophilic attack is generally viewed to be *less* reversible and is often represented by a standard reaction arrow. An acid-base reaction typically occurs orders of magnitude faster than the reaction between a nucleophile and an electron-deficient atom.

Periodic Trends for Nucleophilicity

The nucleophilicity of a substance is impacted by two distinct periodic trends: trends across a period (row) and trends down a group (column).

Across a row, nucleophilicity is *inversely* related to the electronegativity of the nucleophilic atom (Figure 5.35). Atoms with a high electronegativity are more inclined to *hold lone pairs tightly* (and *less* inclined to donate them away); therefore, electronegative atoms are *less* nucleophilic, and nucleophilicity *decreases* from left to right.

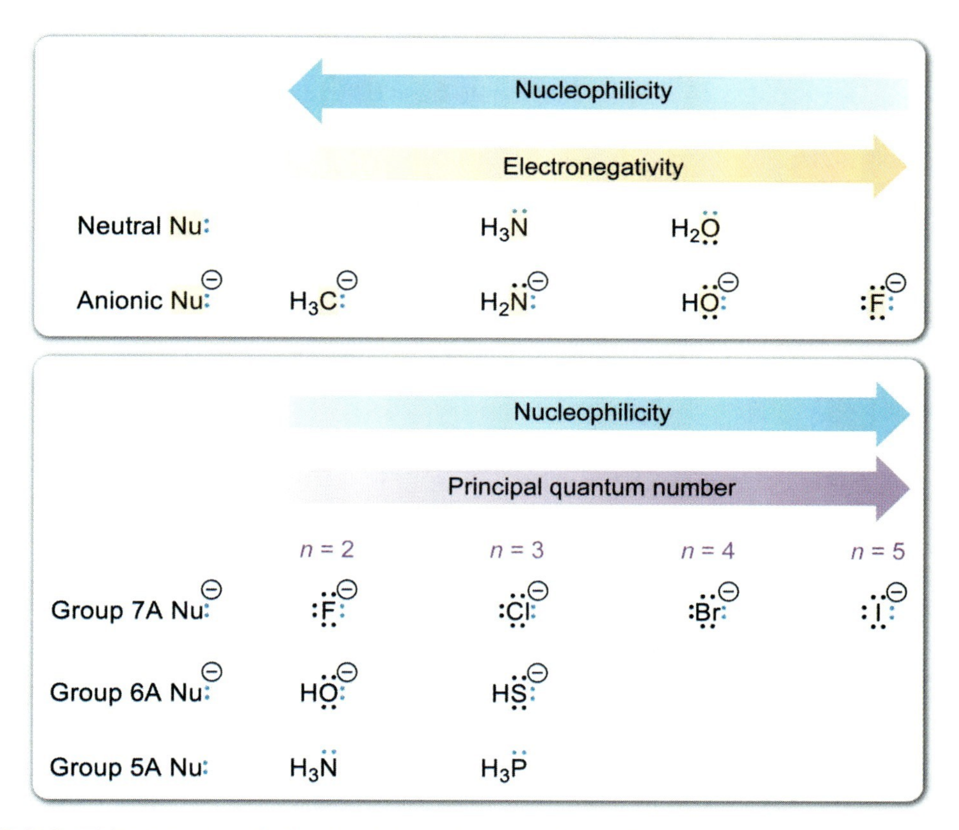

Figure 5.35 Relationships between electronegativity, valence electron principal quantum number, and nucleophilicity.

Down a column, nucleophilicity is also related to the principal quantum number (n) of the atom's valence electrons. Atoms with a *larger* radius for their valence electron cloud have a *greater* capacity to overlap the lone pair orbitals with another species while also reducing the risk of repulsion between the approaching nuclei. Larger electron clouds are also more **polarizable** during a nucleophilic attack (Figure 5.36). Consequently, nucleophilicity *increases* from top to bottom.

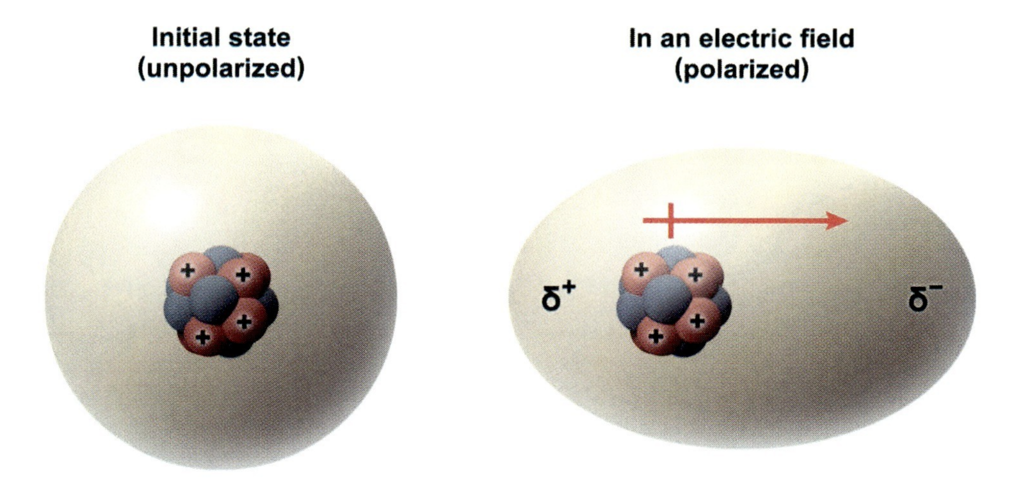

Figure 5.36 Polarizability of an electron cloud.

Conjugate Acid-Base Pairs and Nucleophilicity

A general relationship exists between conjugate acid-base pairs and their relative nucleophilicity (Figure 5.37). A Brønsted-Lowry conjugate acid and conjugate base differ by a single H^+, and the conjugate base possesses an additional lone pair of electrons from the broken X–H sigma bond. Since nucleophilicity *requires* a donatable source of electron density, a conjugate base is *always* more nucleophilic than its conjugate acid.

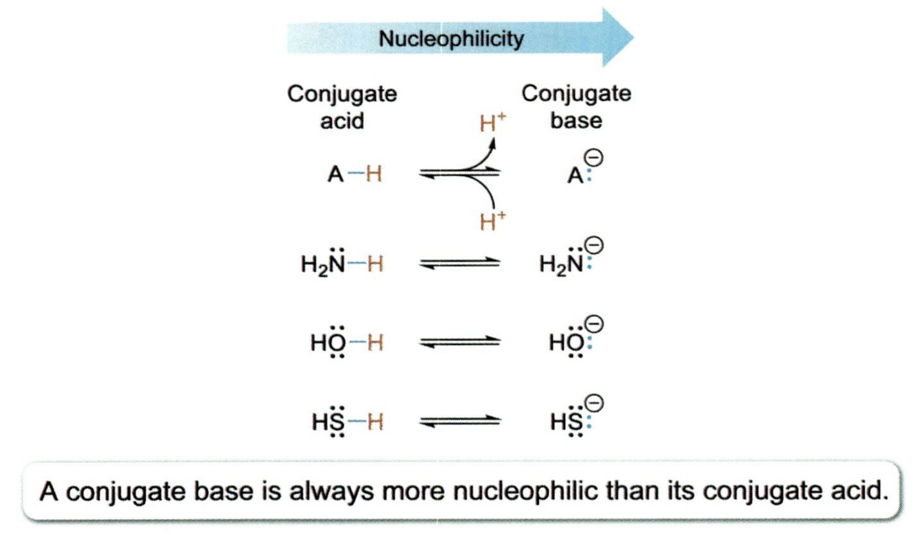

Figure 5.37 Relationship between nucleophilicity and the molecules in a conjugate acid-base pair.

Impact of Steric Hindrance on Nucleophilicity

In organic chemistry, atoms can be classified based on the number of attached alkyl (R) groups. In general, the terms **primary**, **secondary**, **tertiary**, and **quaternary** refer to local bonding patterns with between one and four R groups, respectively (Figure 5.38). For carbon, these classifications also have associated names: A methyl group (1°) and a methylene group (2°) should be familiar from IUPAC and common nomenclature (see Chapter 4); a tertiary carbon group (3°) is called a **methine group**; and a quaternary carbon group (4°) does not have an associated name.

Primary (1°) carbon	Secondary (2°) carbon	Tertiary (3°) carbon	Quaternary (4°) carbon
R–CH₃ (H, H, H on C)	R–CH₂–R	R₃C–H	R₄C
Methyl	Methylene	Methine	

Figure 5.38 Primary, secondary, tertiary, and quaternary carbon atoms.

For a molecule to act as a nucleophile, the nucleophilic atom must first get close enough to the other molecule for their electron clouds to overlap. If the nucleophilic atom (or other nearby atoms) has many R groups, the nucleophile has a *reduced* ability to achieve a close approach. Consequently, **bulky nucleophiles** have a *slower rate* of nucleophilic reaction and therefore have *reduced* nucleophilicity.

The phrase **steric hindrance** describes situations where bulky functional groups limit or preclude a reaction that would otherwise occur. The relative nucleophilicity of the **alkoxide ions** (conjugate bases of alcohols) are depicted in Figure 5.39. Although all these molecules are strong organic bases, they demonstrate a range of nucleophilic character.

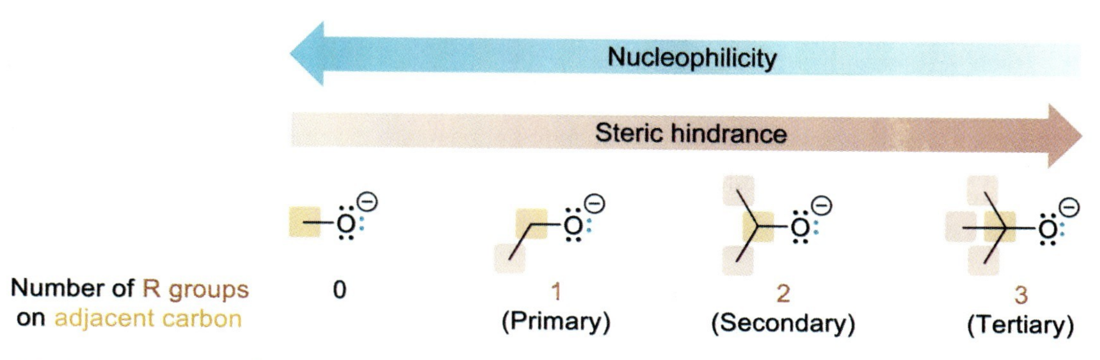

Figure 5.39 Steric hindrance of the alkoxide nucleophiles.

The molecule where the adjacent carbon atom has zero methyl groups on it (**methoxide**) has the *least* steric hindrance and is the *strongest nucleophile*. The molecule with one R group (**ethoxide**) is the next most nucleophilic, followed by the molecule with two R groups (**isopropoxide**). The alkoxide with three R groups (***t*-butoxide**) has such high steric hindrance and poor nucleophilicity that it is *rarely used as a nucleophile* in experiments.

Impact of Resonance on Nucleophilicity

Since a nucleophile is defined by its ability to donate a lone pair of electrons to something else, lone pairs delocalized by resonance (Lesson 2.7) affect nucleophilicity (Table 5.1). Resonance delocalization can *decrease* the electron density of the nucleophilic atom, therefore *reducing* its nucleophilicity. Although an amine, amide, and aniline initially appear to have a similar nucleophilic functional group (R–NH_2), a *significant* difference exists in their nucleophilicities.

Table 5.1 Relationship between nucleophilicity and resonance delocalization.

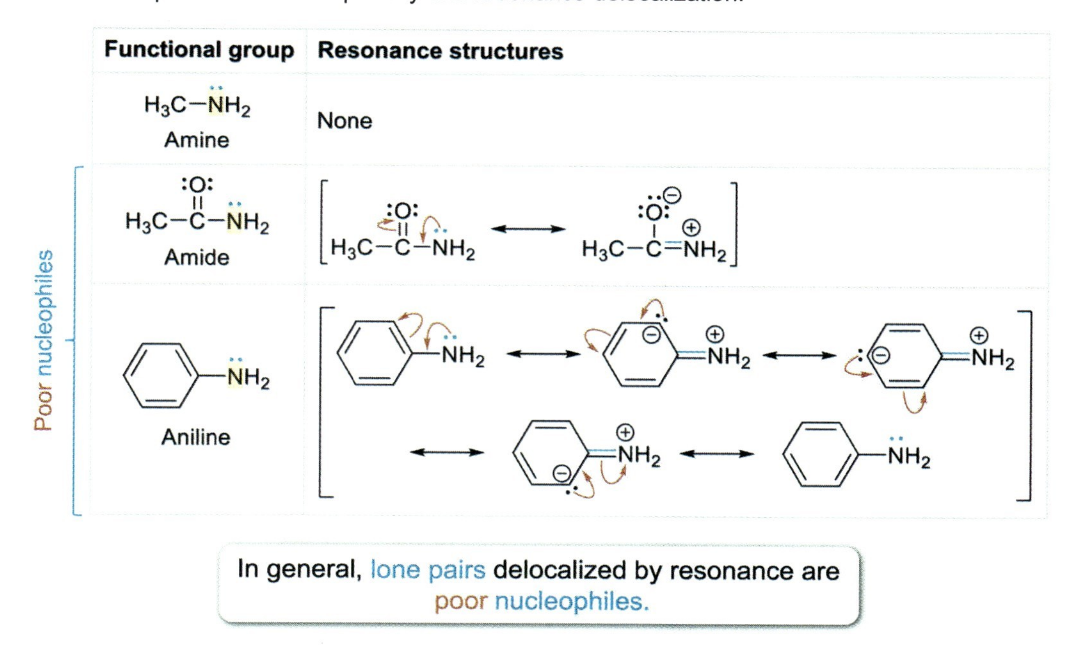

Functional group	Resonance structures
$H_3C–NH_2$ Amine	None
$H_3C–C(=O)–NH_2$ Amide	[$H_3C–C(=O)–NH_2$ ⟷ $H_3C–C(–O^-)=NH_2^+$]
Aniline ($C_6H_5–NH_2$)	[$C_6H_5–NH_2$ ⟷ $=NH_2^+$ ⟷ $=NH_2^+$ ⟷ $=NH_2^+$ ⟷ $C_6H_5–NH_2$]

In general, lone pairs delocalized by resonance are poor nucleophiles.

Impact of the Reaction Solvent on Nucleophilicity

Most reactions occur in a liquid **solvent**, which promotes the mixing of reactants, and the choice of solvent can impact reactant nucleophilicity. Most reactions involving nucleophiles proceed through polar intermediates or transition states, which require the use of a *polar organic solvent*. Polar solvents are further divided into **polar protic solvents** (which *contain* acidic N–H or O–H bonds) and **polar aprotic solvents** (which *lack* these acidic bonds).

Since nucleophiles share many similarities with Lewis bases, dissolving a nucleophile in a polar protic solvent causes the Lewis acidic N–H or O–H bonds to encapsulate the nucleophile, *limiting* its capacity for reaction (Figure 5.40). Because of this effect, polar protic solvents *decrease* nucleophilicity. Examples of polar protic solvents include water, alcohols, and amines containing at least one N–H bond.

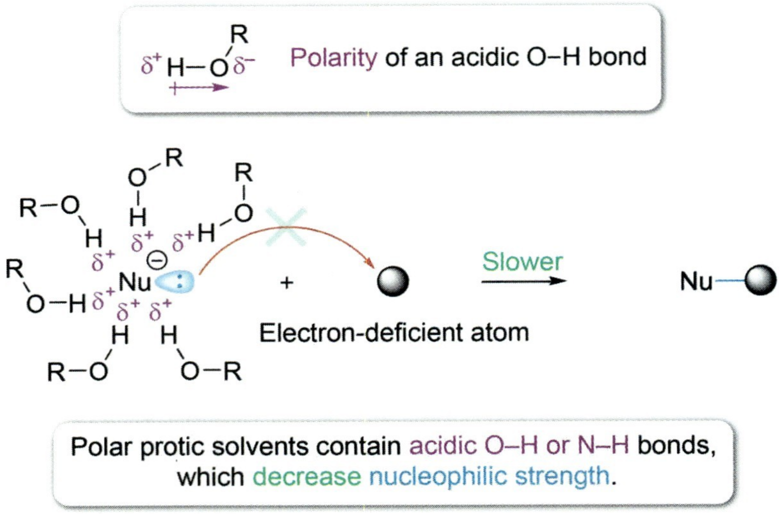

Polarity of an acidic O–H bond

Slower

Electron-deficient atom

Polar protic solvents contain acidic O–H or N–H bonds, which decrease nucleophilic strength.

Figure 5.40 Interaction between a nucleophile and a polar protic solvent.

In contrast, dissolving a nucleophile in a polar aprotic solvent provides the necessary polar environment *without* decreasing the nucleophilicity of any reactants. Common polar aprotic solvents are depicted in Figure 5.41. Although polar aprotic solvents can be used in many types of chemical reactions, seeing one used within a reaction can sometimes be a clue that the reaction involves the use of a strong nucleophile.

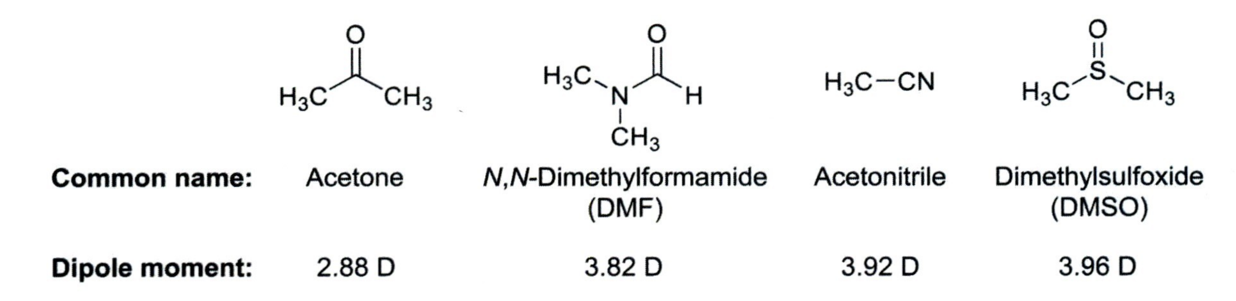

Figure 5.41 Common polar aprotic solvents.

Summary of Relative Nucleophilic Strength

The many considerations that influence nucleophilicity allow functional groups to be classified as weak, moderate, or strong nucleophiles based on their relative rates of nucleophilic reaction (Table 5.2).

Table 5.2 Relative nucleophilicity of common functional groups.

Nucleophilicity →

Weak		Moderate		Strong	
Water	H–ÖH	Chloride	$:\ddot{Cl}:^{\ominus}$	Alkoxides	$R–\ddot{O}:^{\ominus}$
Alcohols	R–ÖH	Carboxylates	$R–C(=O)–\ddot{O}:^{\ominus}$	3° Amines	$R–\ddot{N}(R)–R$
Fluoride	$:\ddot{F}:^{\ominus}$	Thiols	R–S̈H	Cyanide	$N{\equiv}C:^{\ominus}$
Carboxylic acids	R–C(=O)–ÖH	1° Amines	$R–\ddot{N}(H)–H$	Iodide	$:\ddot{I}:^{\ominus}$
		2° Amines	$R–\ddot{N}(R)–H$	Thiolates	$R–\ddot{S}:^{\ominus}$
		Bromide	$:\ddot{Br}:^{\ominus}$	Phosphines	$R–\ddot{P}(R)–R$

Nucleophilicity ↓

Weak nucleophiles are the slowest to react and often require the addition of heat to promote a reasonable rate of nucleophilic attack. **Moderate nucleophiles** have a greater likelihood of reaction at room temperature and are regularly used within organic synthesis. **Strong nucleophiles** are among the most powerful nucleophiles, with reactions often conducted at cold temperatures to *lower* the reaction rate to a point where it can be better controlled.

Concept Check 5.6

For each of the pairs of molecules shown, identify the nucleophilic atoms and determine which molecule is the stronger nucleophile.

1) CH_3OH vs. CH_3SH

2) NH_2 (propylamine) vs. NH_2 (neopentylamine)

3) $Br^{\ominus}$ in OH (ethanol) vs. $Br^{\ominus}$ in N,N-dimethylformamide (N, O, H)

Solution

Note: The appendix contains the answer.

5.3.02 Electrophiles

An **electrophile** ("lover of electrons") is a substance that has a *deficiency* of electron density and can *accept* a lone pair of electrons from another group (the nucleophile).

The specific atom that is electron-deficient is called the **electrophilic atom**. An electrophilic atom can be either neutral or have a positive formal charge; however, neutral electrophilic atoms typically have at least a *partial* positive charge. Therefore, a general electrophile is often abbreviated as either E or E^+.

Electrophiles in organic chemistry can be further classified into one of three general categories (Figure 5.42); for the purposes of this book, these are later referenced as Type 1, Type 2, and Type 3 electrophiles.

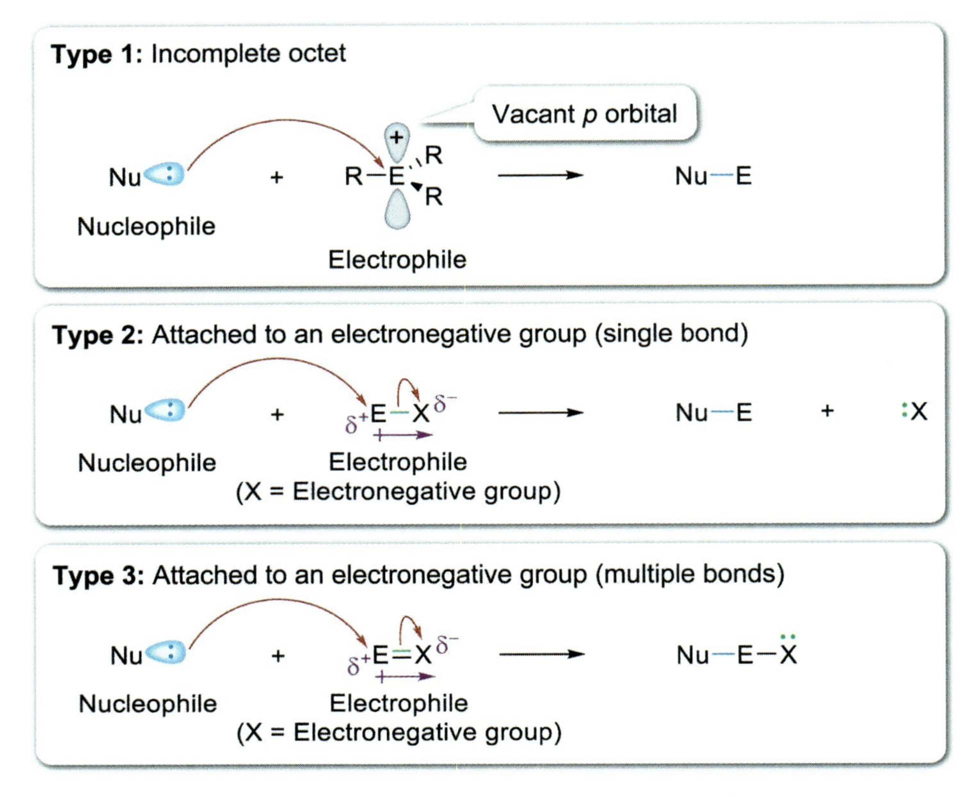

Figure 5.42 General structure and reactivity of an electrophile (Types 1, 2, and 3).

Type 1 electrophiles are electron-deficient due to an incomplete octet of valence electrons. In general, this type of electrophile has a vacant (unhybridized) *p* orbital and reacts directly with a nucleophile to form a covalent bond (Nu–E). Carbocations are examples of Type 1 electrophiles (Concept 5.2.01).

Type 2 electrophiles have the electrophilic atom attached to a highly electronegative group through a polar *single* bond, giving the electronegative atom a partial positive charge. When a nucleophile attacks the electrophilic atom, the single (sigma) bond breaks and becomes a *lone pair* on the electronegative group. Alkyl halides are examples of Type 2 electrophiles (Concept 7.1.02).

Type 3 electrophiles have the electrophilic atom attached to a group with high electronegativity through either a *double* bond or *triple* bond. Although the electrophilic atom has a partial positive charge, a nucleophilic attack breaks a *pi* bond and converts its electrons to a lone pair on the electronegative group. Aldehydes and ketones are examples of Type 3 electrophiles (Lesson 9.3).

The term **electrophilicity** describes the relative capacity of a substance to act as an electrophile.

Relationship between Electrophilicity and Acidity

The terms *electrophile* and *Lewis acid* are synonymous with one another in the context of introductory organic chemistry, as both are defined as something that can accept a lone pair of electrons. Even under the more restrictive Brønsted-Lowry definition, the acidic proton in a Brønsted-Lowry acid can be classified as a Type 2 electrophilic atom (see Figure 5.42).

Strong Brønsted-Lowry acids also have an ability to *enhance* the electrophilicity of another substance through **acidic catalysis**. Recall that the base in a conjugate acid-base pair is always more nucleophilic than its acid counterpart (Concept 5.3.01); conversely, an *acid* is always more *electrophilic* than its conjugate base (Figure 5.43).

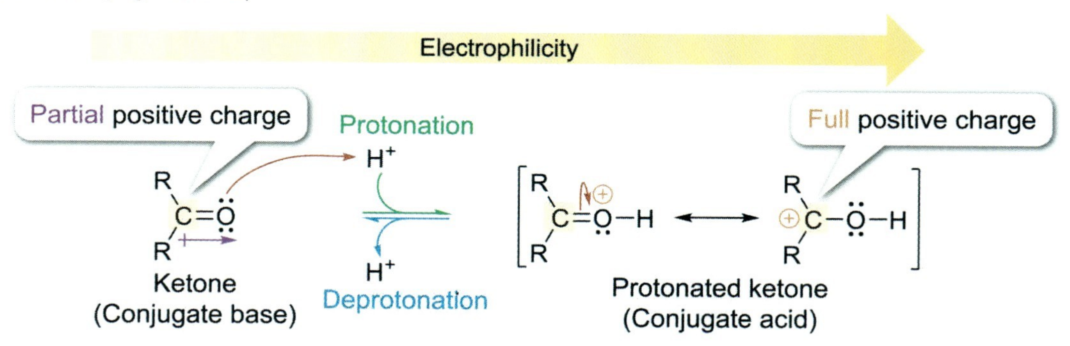

Figure 5.43 Acidic activation of a Type 3 electrophile.

Acidic catalysis is commonly observed for Type 3 electrophiles—protonation of the electronegative group generates a conjugate acid that has a resonance form with a formal positive charge on the electrophilic carbon atom. Since the conjugate acid form *increases* the amount of positive charge (electron-deficiency) on the carbon, the carbon becomes a stronger electrophile.

Impact of Steric Hindrance on Electrophilicity

For an electrophile to react with a nucleophile, the electrophilic atom must get close enough to the nucleophilic atom for their electron clouds to overlap. If the electrophilic atom (or any other nearby atom) has many R groups, the electrophile has a *reduced* ability to achieve a close approach with a nucleophile (Figure 5.44).

Consequently, **bulky electrophiles** have a slower rate of reaction with nucleophiles due to **steric hindrance**. This trend is most relevant to Type 2 and Type 3 electrophiles. A Type 1 electrophile (ie, a planar carbocation) is *so* reactive and short-lived that the rate of nucleophilic attack is not substantially affected by steric hindrance.

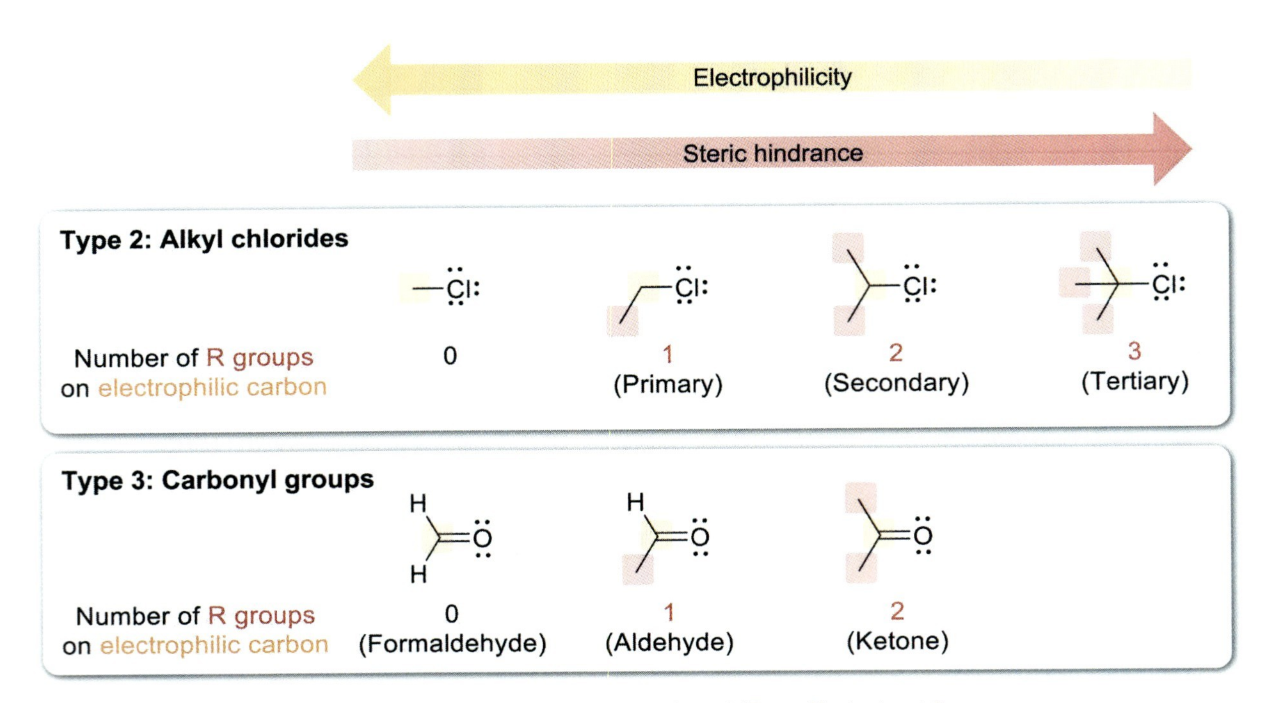

Figure 5.44 Steric hindrance of alkyl chloride (Type 2) and carbonyl (Type 3) electrophiles.

The formation of a carbocation (Step 1, Figure 5.45) is a separate and *earlier* mechanistic step than the reaction with a nucleophile (Step 2). Because the **activation energy E_a** for the carbocation formation (E_{a1}) is greater than the activation energy for its reaction with a nucleophile (E_{a2}), *carbocation formation* is the **rate-limiting step** (Concept 5.4.02).

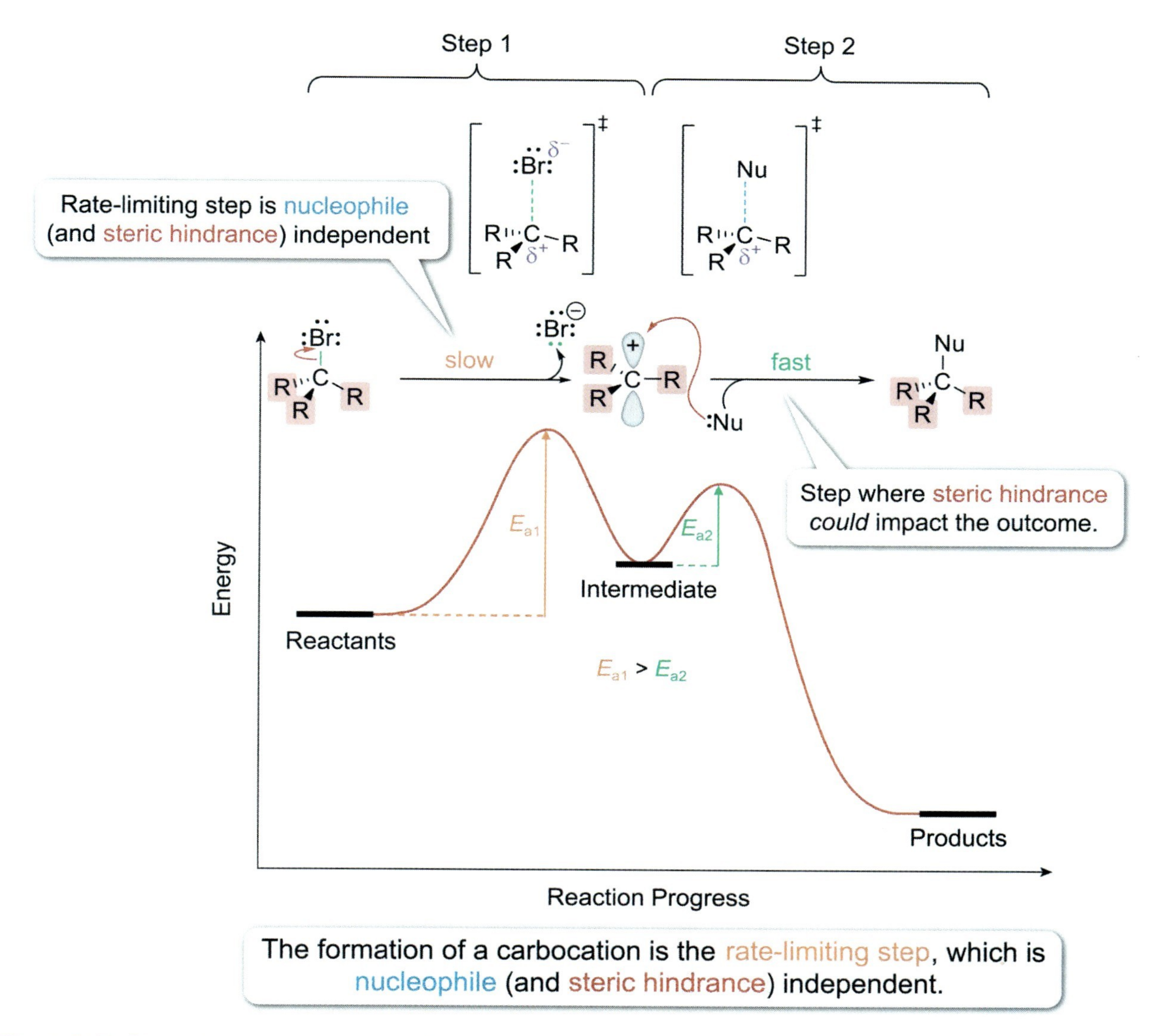

Figure 5.45 Steric hindrance and carbocations.

Because the *rate-limiting step* controls the overall rate of a multistep process and, in the case of carbocations, occurs *independently* of a steric-influenced collision with the nucleophile, *steric hindrance* does *not* impact the overall electrophilicity of carbocations.

However, alkyl (R) groups surrounding the carbocation carbon atom *stabilize* the vacant *p* orbital through a combination of the inductive effect and hyperconjugation (Concept 5.2.01). Since the formation of a carbocation is the rate-limiting step, structural features (like alkyl groups) that stabilize a carbocation *decrease* the activation energy barrier for the rate-limiting step and *enhance* the rate of its formation. Consequently, instead of providing steric hindrance, alkyl groups on a carbocation serve the opposite function and *increase* its rate of reaction.

Impact of Electronegativity on Electrophilicity

The impact of nearby electronegative atoms on electrophilicity varies depending on the class of electrophile (Figure 5.46).

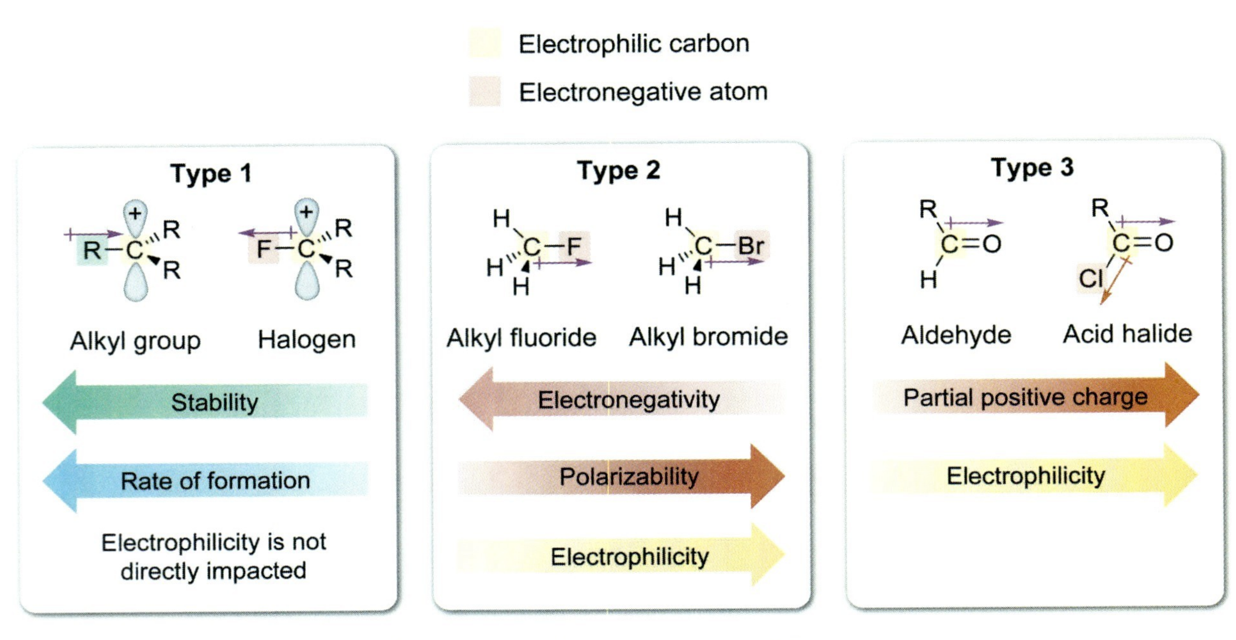

Figure 5.46 Relationship between electronegative atoms and electrophilicity.

Type 1 electrophiles (eg, carbocations) are *further destabilized* by nearby electronegative atoms, making them more challenging to form and leading to a *decrease* in reaction rate. The electrophilicity of a Type 1 electrophile is not *directly* affected by nearby electronegative atoms.

Although a greater C–X electronegativity difference *increases* the amount of partial positive charge on a Type 2 electrophile, carbon electrophilicity is *more significantly* impacted by the ability of the C–X bond to be broken and for X to *leave* with the pair of electrons. As such, the electrophilicity of Type 2 electrophiles tends to be proportional to the polarizability of X, which is itself *inversely related* to the electronegativity of X (see Concept 5.3.01).

Electronegative atoms *increase* the electrophilicity of a Type 3 electrophile through the inductive effect.

Impact of Resonance on Electrophilicity

The delocalization of electron density through resonance, conjugation (Concept 6.5.01), or aromaticity (Concept 6.5.02) can influence the electrophilicity of a substance. The three categories of electrophiles are affected by delocalization differently (Figure 5.47):

- Type 1 electrophiles (resonance-delocalized carbocations) have *increased* stability and a *higher* rate of formation (Concept 5.2.01). Once formed, the electrophilicity (rate of reaction) of a carbocation is not generally affected by electron-delocalization.
- Type 2 electrophiles that have the electrophilic atom at a conjugated position (ie, allylic, benzylic) are *stronger* electrophiles, as the adjacent pi bonds *stabilize* the transition state following nucleophilic attack (see Concept 5.4.03 and Concept 6.5.01).
- Type 3 electrophiles that have *greater* electron delocalization demonstrate a *decrease* in electrophilicity (see Concept 11.2.02).

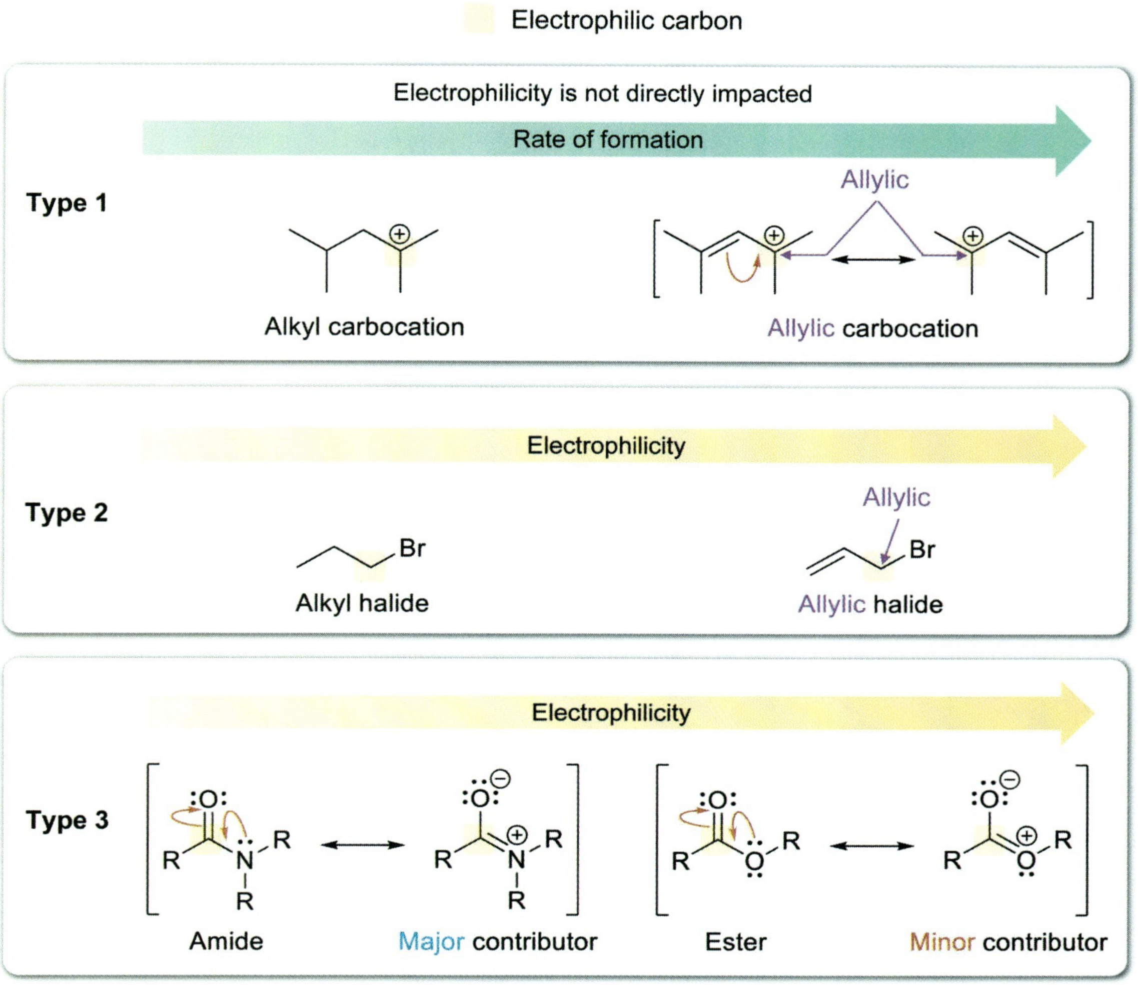

Figure 5.47 Relationship between resonance-delocalization and electrophilicity.

Impact of the Leaving Group on Electrophilicity

One of the most important considerations that impacts the electrophilicity of Type 2 and Type 3 electrophiles is the identity of the electronegative group (X) attached to the electrophilic atom. Electronegative groups that can both *polarize* the C–X bond and *accept* the lone pair of electrons *enhance* a molecule's electrophilicity.

For Type 2 electrophiles, the electronegative group is known as a **leaving group** (see Concept 5.3.03).

Concept Check 5.7

For each pair of electrophiles shown, identify:

- the electrophilic atom.
- the type of electrophile.
- the primary reason why the indicated molecule in each pair has greater electrophilicity.

1) Br I

Stronger electrophile

2) O NH

Stronger electrophile

Solution

Note: The appendix contains the answer.

5.3.03 Leaving Groups

In Concept 5.3.02, a Type 2 electrophile was defined as having an electronegative group (X) attached to the electrophilic atom through a single bond. Given that the electronegative group *leaves* with the lone pair of electrons from the broken bond, this group is generally called a **leaving group** (Figure 5.48). A general leaving group may also be abbreviated as LG.

Leaving groups have two main roles:

- A leaving group is electronegative and makes any attached atoms electron-deficient (electrophilic) through the inductive effect.
- A leaving group accepts and *stabilizes* a lone pair of electrons from the broken bond.

Type 2: Electrophile is attached to a leaving group through a single bond.

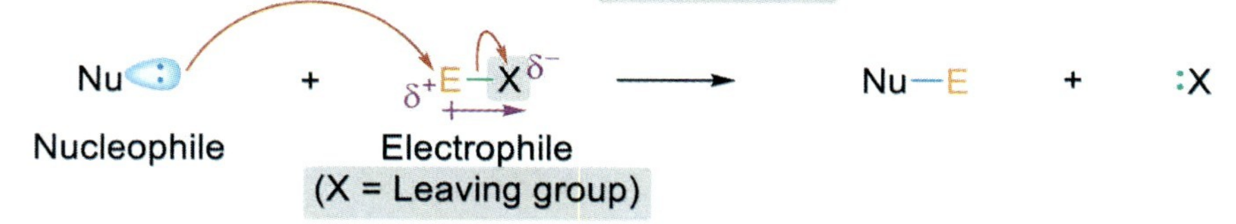

Figure 5.48 General structure and reactivity of a leaving group.

Impact of Acid-Base Chemistry on Leaving Groups

Acid-base chemistry impacts leaving group function in two ways. The Lewis and Brønsted-Lowry definitions require a base to possess a lone pair of electrons (Concept 5.1.01). Given that a leaving group ultimately accepts a lone pair of electrons from the broken bond to the electrophile, a leaving group (once it has left) is *also* a base.

Typically, only *weak bases* are considered to be good leaving groups. A weak base generally *stabilizes* its lone pair of electrons in one or more ways. Some weak bases delocalize lone pairs through resonance (eg, carboxylate ions), whereas others contain polarizable atoms that can distribute the electrons over a larger surface area (eg, iodide ion).

Due to the inverse relationship between conjugate acid and conjugate base strength, the *best* leaving groups are usually *weak* conjugate bases of strong acids (Table 5.3). For example, the bromide ion (Br^-), the weak conjugate base of hydrobromic acid (HBr), is an excellent leaving group. In contrast, *strong* conjugate bases of weak acids are typically *poor* leaving groups. For example, the hydroxide ion (HO^-) is a strong conjugate base of water (H_2O) and does *not* typically act as a leaving group.

Table 5.3 Relationship between basicity and leaving group strength.

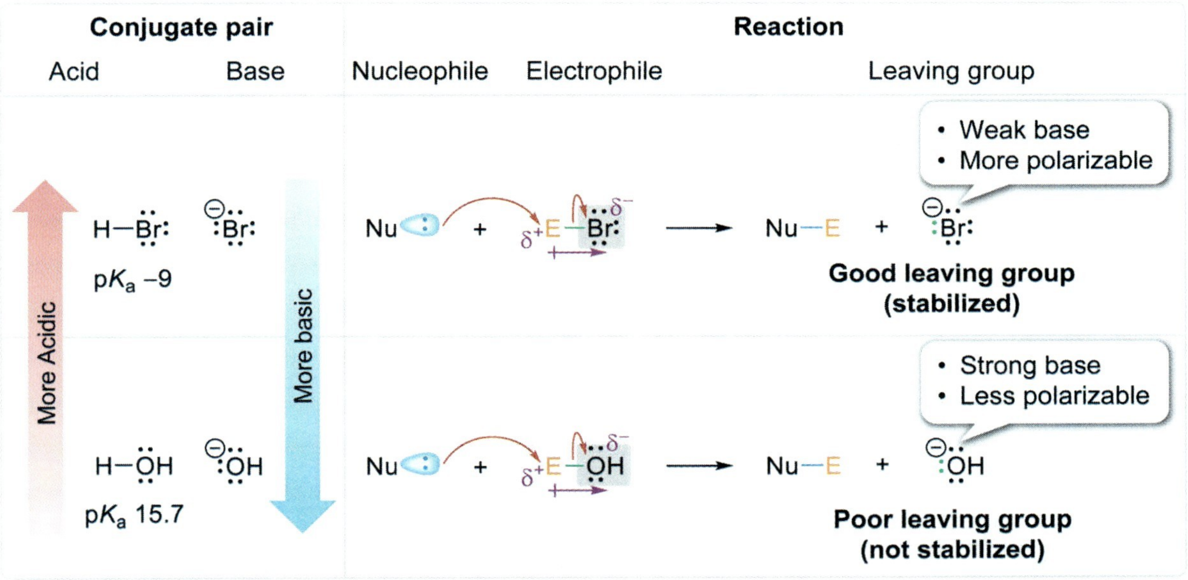

Conjugate pair		Reaction			
Acid	Base	Nucleophile	Electrophile		Leaving group
H–Br: (pK_a −9)	$^{\ominus}$:Br:	Nu: +	δ^+E–Br:δ^-	⟶ Nu–E +	$^{\ominus}$:Br: • Weak base • More polarizable **Good leaving group (stabilized)**
H–OH (pK_a 15.7)	$^{\ominus}$:OH	Nu: +	δ^+E–OHδ^-	⟶ Nu–E +	$^{\ominus}$:OH • Strong base • Less polarizable **Poor leaving group (not stabilized)**

More Acidic (↑) | More basic (↓)

Treatment with a strong acid can sometimes convert a poor leaving group into an excellent leaving group. For example, treatment with a strong acid converts an alcohol into an **oxonium ion**, which can leave as a molecule of neutral water (a weak base).

The curved arrow formalism for a regular leaving group (eg, iodide ion, I^-) and a protonated alcohol (eg, oxonium ion, $-^+OR_2$) are *identical* (once the acidic proton has been added) (Figure 5.49). In both cases, the formal charge on the leaving group atom directly attached to the electrophilic carbon *decreases* by 1 when it gains possession of the lone pair.

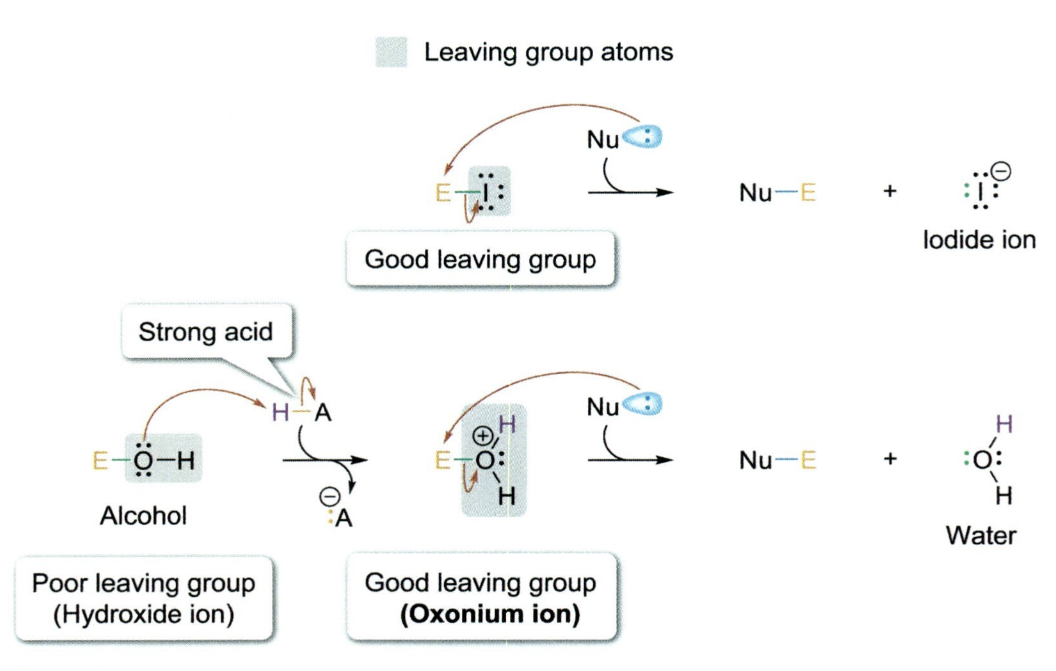

Figure 5.49 Use of acid to convert an alcohol into a good leaving group.

One limitation of using a strong acid to convert an alcohol into a good leaving group is that not all nucleophiles are amenable to strongly acidic experimental conditions. The strong acid (typically used in stoichiometric excess) *also* protonates the nucleophile's lone pair, ultimately decreasing its nucleophilicity.

Summary of Leaving Group Strength

A good leaving group meets the following criteria:

- It is either *electronegative* or *electron-deficient*, making it capable of polarizing the bond to the electrophilic atom.
- It is a *weak base*.
- It can *stabilize* the accepted pair of electrons (eg, polarizability, resonance).

Table 5.4 provides a ranked overview of leaving groups common to organic chemistry. For clarity, entries are named by both the initial functional group (containing the leaving group) and the leaving group atoms *after* they have accepted the lone pair. In most situations, the entries labeled as poor leaving groups are *unlikely* to act as a leaving group during a chemical reaction.

Table 5.4 Leaving groups in organic chemistry.

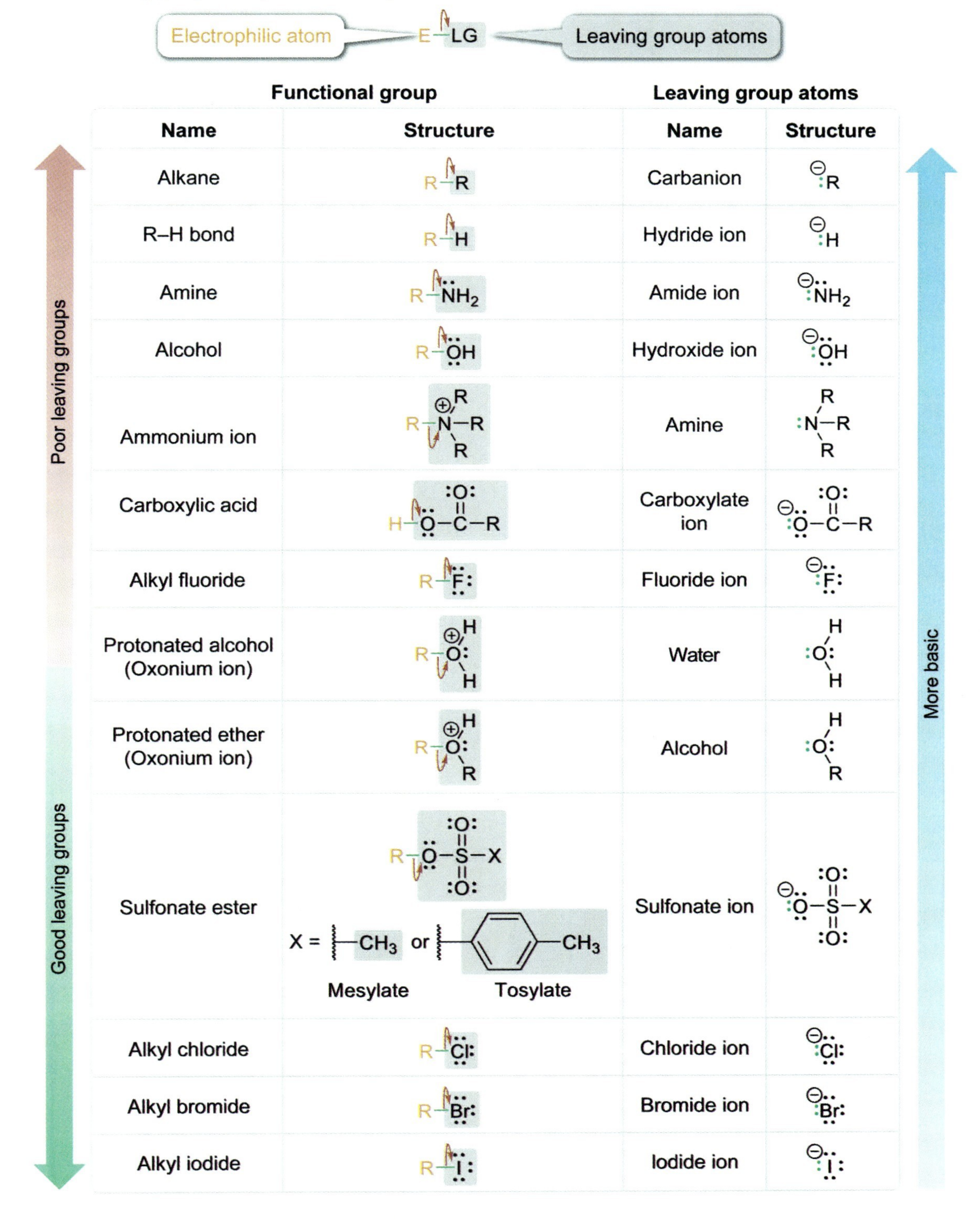

Functional group: Name	Functional group: Structure	Leaving group atoms: Name	Leaving group atoms: Structure
Alkane	R–R	Carbanion	$^{\ominus}$R
R–H bond	R–H	Hydride ion	$^{\ominus}$H
Amine	R–NH_2	Amide ion	$^{\ominus}NH_2$
Alcohol	R–OH	Hydroxide ion	$^{\ominus}$OH
Ammonium ion	R–$N^{\oplus}$R(R)	Amine	:N(R)(R)–R
Carboxylic acid	H–O–C(=O)–R	Carboxylate ion	$^{\ominus}$O–C(=O)–R
Alkyl fluoride	R–F:	Fluoride ion	$^{\ominus}$F:
Protonated alcohol (Oxonium ion)	R–$O^{\oplus}$(H)(H)	Water	H–O:–H
Protonated ether (Oxonium ion)	R–$O^{\oplus}$(H)(R)	Alcohol	H–O:–R
Sulfonate ester	R–O–S(=O)(=O)–X; X = CH₃ (Mesylate) or p-C₆H₄–CH₃ (Tosylate)	Sulfonate ion	$^{\ominus}$O–S(=O)(=O)–X
Alkyl chloride	R–Cl:	Chloride ion	$^{\ominus}$Cl:
Alkyl bromide	R–Br:	Bromide ion	$^{\ominus}$Br:
Alkyl iodide	R–I:	Iodide ion	$^{\ominus}$I:

Molecules containing good leaving groups include:

- **Oxonium ions**, either as protonated alcohols (Concept 8.3.05 and Concept 8.3.07) or protonated ethers (Concept 7.3.04).
- **Mesylate esters** and **tosylate esters**, which can be prepared from alcohols (Concept 8.3.04).
- **Alkyl halides**, excluding alkyl fluorides (Lesson 5.4 and Concept 7.1.02).

Lesson 5.4

Substitution Reactions

Introduction

The current lesson explores the roles of nucleophiles, electrophiles, leaving groups, and reactive intermediates in substitution reactions. The following types of substitution reactions are discussed:

- Unimolecular nucleophilic substitution (S_N1)
- Bimolecular nucleophilic substitution (S_N2)
- Nucleophilic acyl substitution
- Aromatic substitution

5.4.01 Definition of a Substitution Reaction

A **substitution reaction** is a chemical reaction in which one group is replaced by another group (Figure 5.50). In a **nucleophilic substitution (S_N) reaction**, a nucleophile becomes the newly added group (Y) after formation of a covalent bond with the electrophilic substrate (R), which breaks the bond to the leaving group (X).

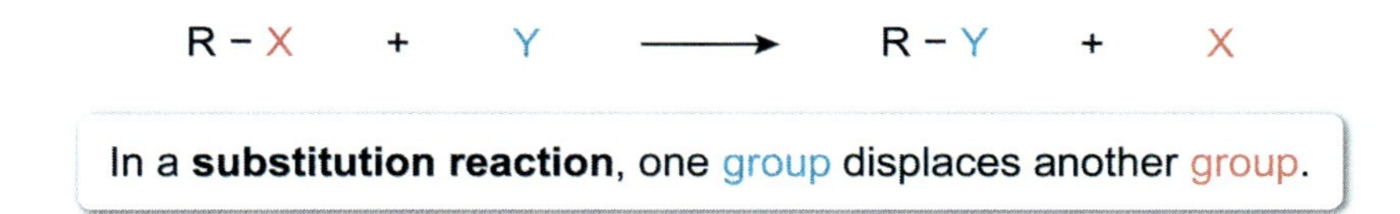

Figure 5.50 A general substitution reaction.

A **reaction coordinate diagram** depicts the energy of the components of a reaction (ie, reactants, intermediates, transition states, products) versus the progress of a reaction (Figure 5.51). Energy minima (ie, low points) correspond to reactants, intermediates, and products. An intermediate is a substance formed and later consumed in a chemical reaction; therefore, intermediates do not appear in the *net* reaction but only at energy minima *between* reactants and products. Each energy change between *adjacent minima* represents a discrete **step** in the reaction (ie, a single set of changes that occurs simultaneously).

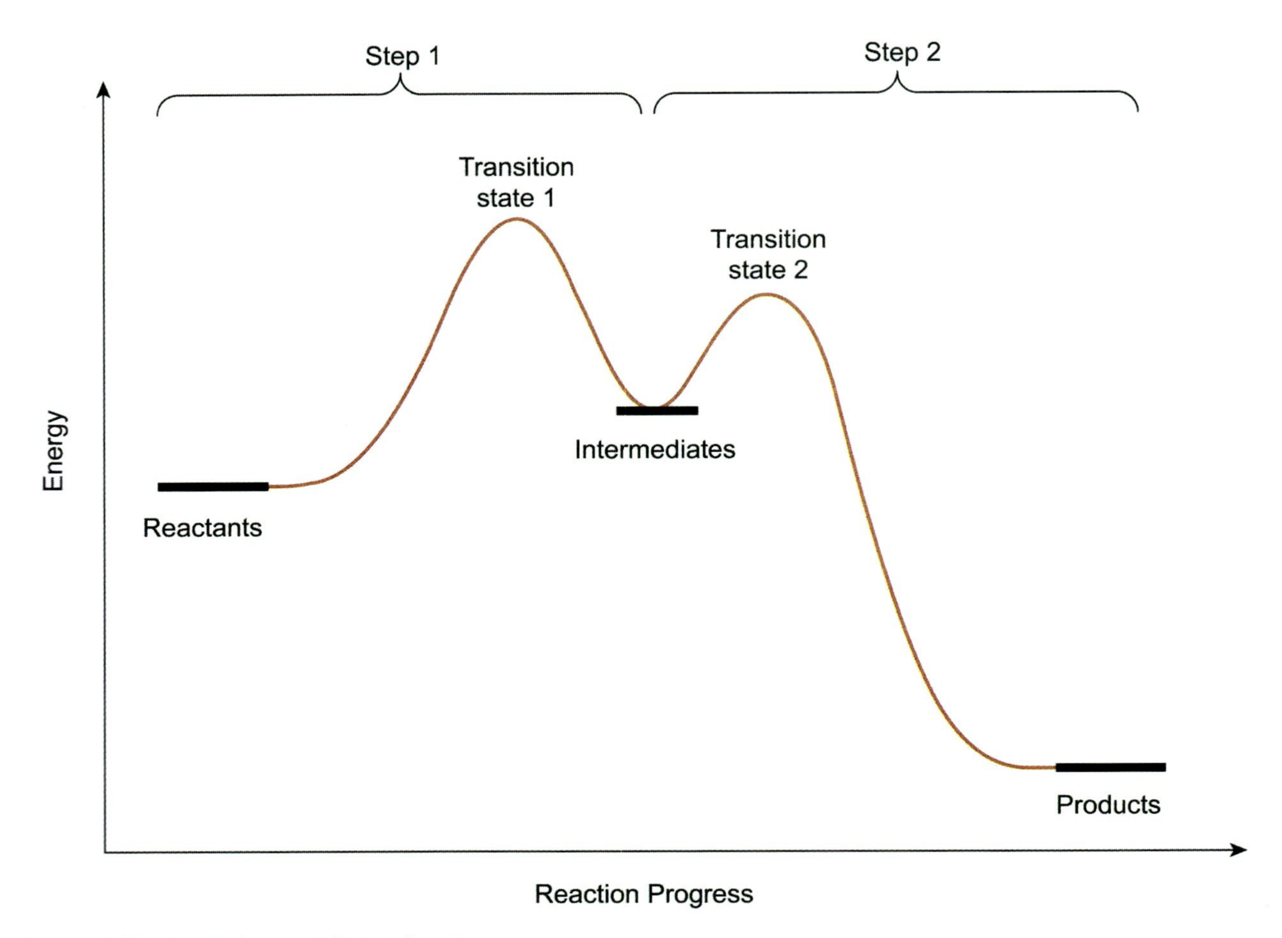

Figure 5.51 A reaction coordinate diagram.

A **transition state** is a high-energy substance formed during the transition *between* reactants, intermediates, and products. Transition states *cannot* be isolated, as they would quickly decompose to a lower-energy substance.

A reaction coordinate diagram also provides information about the kinetics (rate) of a reaction. For a reaction with two or more separate steps, the **rate-limiting step** is the step with the greatest activation energy E_a. The overall reaction rate is constrained by the rate-limiting step and is mathematically described by the rate law.

The overall **reaction order** often serves as an indication of the number of molecules present in the rate-limiting step; it is equal to the sum of the exponents in the rate law. In a **first-order (unimolecular) reaction**, the rate depends on the concentration of only *one* molecule, and in a **second-order (bimolecular) reaction**, the rate depends on the concentration of *two* molecules.

Three types of nucleophilic substitution reactions are examined in this lesson: the S_N1 reaction, the S_N2 reaction, and the nucleophilic acyl substitution reaction. Although these are all classified as nucleophilic substitutions, they vary in the number of steps, the reactive intermediate formed, and the overall reaction order (Table 5.5).

Table 5.5 Comparison of nucleophilic substitution reactions.

Reaction type	Number of steps	Reactive intermediate	Overall reaction order
S_N1	2	Carbocation	First
S_N2	1	None	Second
Nucleophilic acyl substitution	2	Tetrahedral intermediate	Second

5.4.02 S_N1 Reaction

For a **unimolecular nucleophilic substitution (S_N1) reaction**, the number (1) refers to the number of molecules present in the rate-limiting step. First, an electrophilic substrate (C–X) loses its leaving group (X), and then a nucleophile attacks the substrate, forming a new sigma bond.

All S_N1 reactions share the following essential features:

- At least two mechanistic steps
- The formation of a carbocation intermediate
- A reaction rate *independent* of the nucleophile (ie, only dependent on the substrate)
- Results that include a mixture of stereoisomers (if the reaction results in a chiral carbon).

S_N1 Reaction Mechanism

The mechanism of any S_N1 reaction consists of at least two steps (Figure 5.52):

- The polar bond between the leaving group and the electrophilic carbon breaks, forming a carbocation intermediate. The leaving group gains the pair of electrons.
- A nucleophile attacks the carbocation to form a new sigma bond.
- Note: If the nucleophile is *neutral*, the product is deprotonated in a *third* mechanistic step.

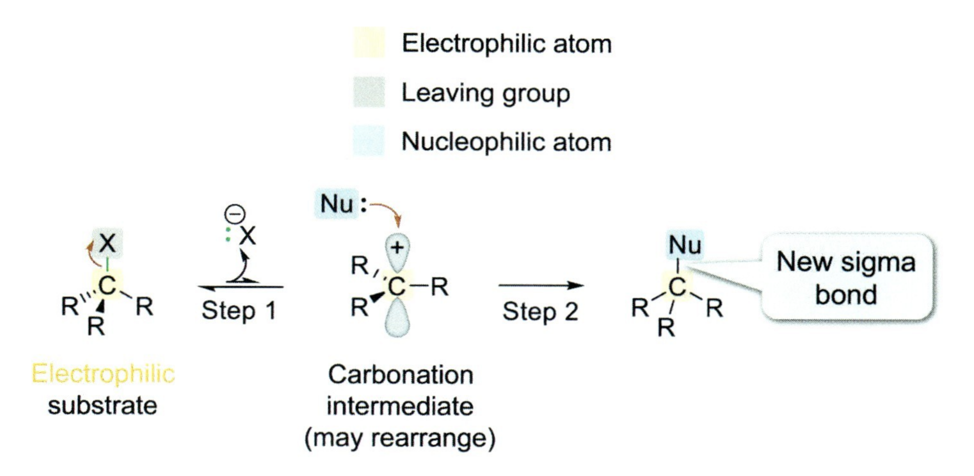

Figure 5.52 The S_N1 reaction mechanism.

Since *all* S_N1 reactions include a carbocation intermediate, an experimenter must *always* analyze the carbocation formed during Step 1 to determine if rearrangement through a hydride shift or a methyl shift is possible. Carbocation rearrangement *changes* the carbon atom attacked by the nucleophile and necessitates performing this analysis before proceeding.

In a **hydride shift**, the carbon skeleton does not change, whereas a **methyl shift** does change the carbon skeleton (Figure 5.53).

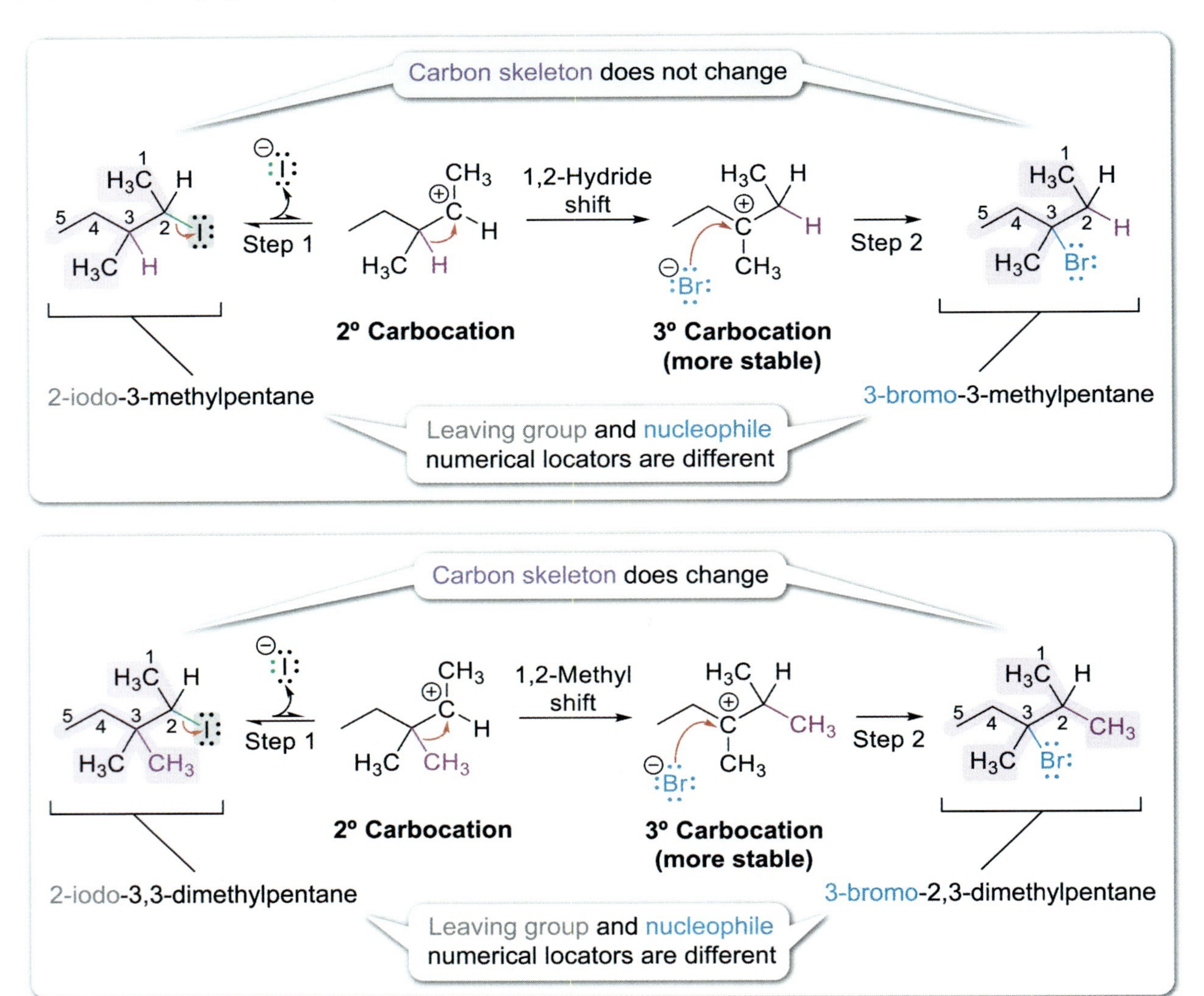

Figure 5.53 A hydride shift and a methyl shift in an S_N1 mechanism.

S_N1 Reaction Order

The reaction coordinate diagram of an S_N1 reaction has at least two peaks representing the transition state occurring in each reaction step. Carbocation formation (Step 1) in an S_N1 reaction is slow and has a larger activation energy (E_{a1}) than the nucleophilic attack's activation energy (Step 2, E_{a2}). Therefore, carbocation formation is the rate-limiting step for an S_N1 reaction. Since carbocation formation depends *only* on the electrophilic substrate, the overall rate of an S_N1 reaction depends *only* on the concentration of the electrophilic substrate (Figure 5.54).

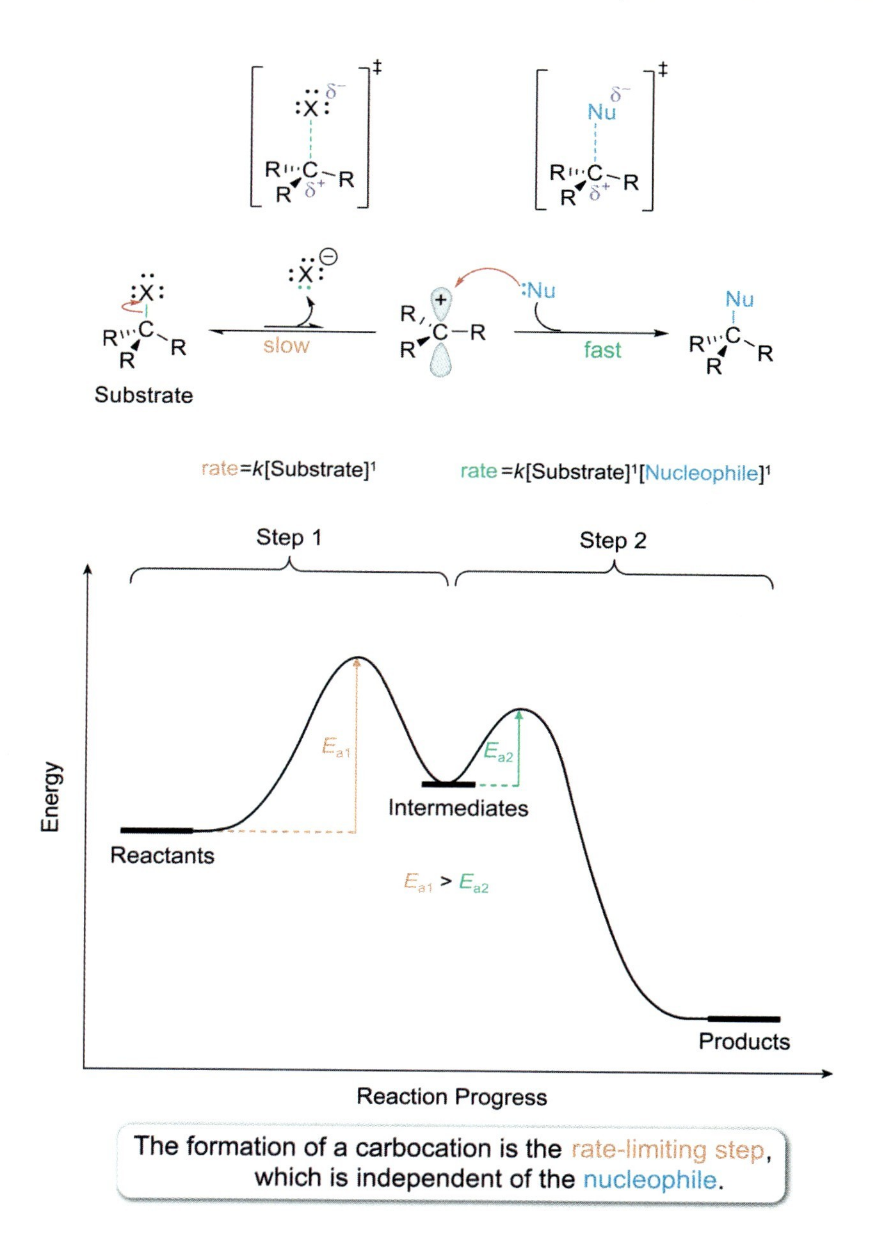

Figure 5.54 Reaction coordinate diagram for the S_N1 mechanism.

The rate law for an S_N1 reaction is expressed by the equation:

$$\text{rate} = k[\text{Substrate}]^1[\text{Nucleophile}]^0$$

$$= k[\text{Substrate}]^1$$

which is first-order with respect to the substrate and zero-order with respect to the nucleophile. The sum of the exponents in the rate law indicates the reaction is first-order *overall*.

Variables in an S_N1 Reaction

Because a carbocation is a strong Lewis acid and slow to form, it is *not possible* to generate a carbocation under strongly (Lewis) basic (eg, nucleophilic) reaction conditions. Therefore, a weak nucleophile in a polar protic solvent is typically required for an S_N1 reaction (Figure 5.55). Sometimes, the polar protic solvent (eg, alcohol, water) functions as *both* the nucleophile and the solvent; such reactions are called a **solvolysis**.

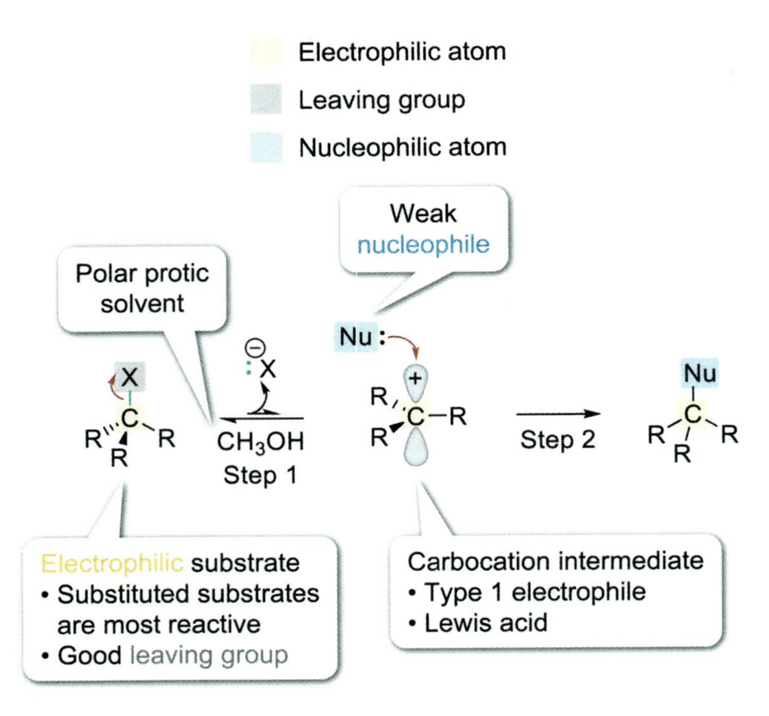

Figure 5.55 Summary of variables that promote S_N1 reactions.

The carbocation carbon atom is the electrophilic atom in an S_N1 reaction, as it has an incomplete octet. Recall that steric hindrance does *not* impact the electrophilicity of a planar carbocation. Instead, the number of alkyl groups on a carbocation influence the *stability* of the carbocation and the *rate* of its formation.

The best substrates for an S_N1 reaction are highly substituted (eg, 3° carbon atoms) due to the inductive effect and hyperconjugation (Figure 5.56). A 2° substrate is also capable of undergoing an S_N1 reaction. A 1° substrate *only* undergoes an S_N1 reaction if its carbocation can be resonance-stabilized (eg, an allylic substrate). Methyl substrates rarely react via an S_N1 reaction.

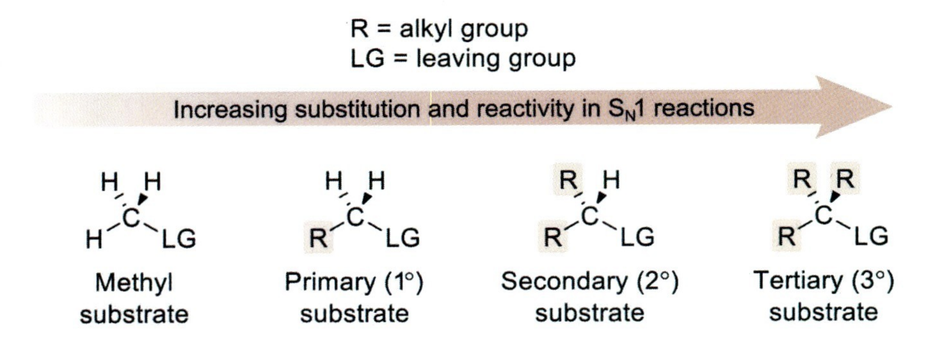

Figure 5.56 Substrate substitution and reactivity trend for S_N1 reactions.

In Step 1 of an S_N1 reaction, the bond between the leaving group and the electrophilic atom breaks, and the leaving group accepts the pair of electrons from the broken bond. Some of the most common leaving groups for an S_N1 reaction are the halides chloride, bromide, and iodide.

Stereochemistry of an S_N1 Reaction

If an S_N1 reaction occurs at an asymmetric carbon atom, the asymmetric carbon loses all stereochemical information when the planar carbocation forms. Therefore, the nucleophile in Step 2 can attack the carbocation from *either* above *or* below the plane, resulting in a mixture of two products (Figure 5.57). If the substrate contains only one asymmetric carbon atom, the product is a racemic mixture.

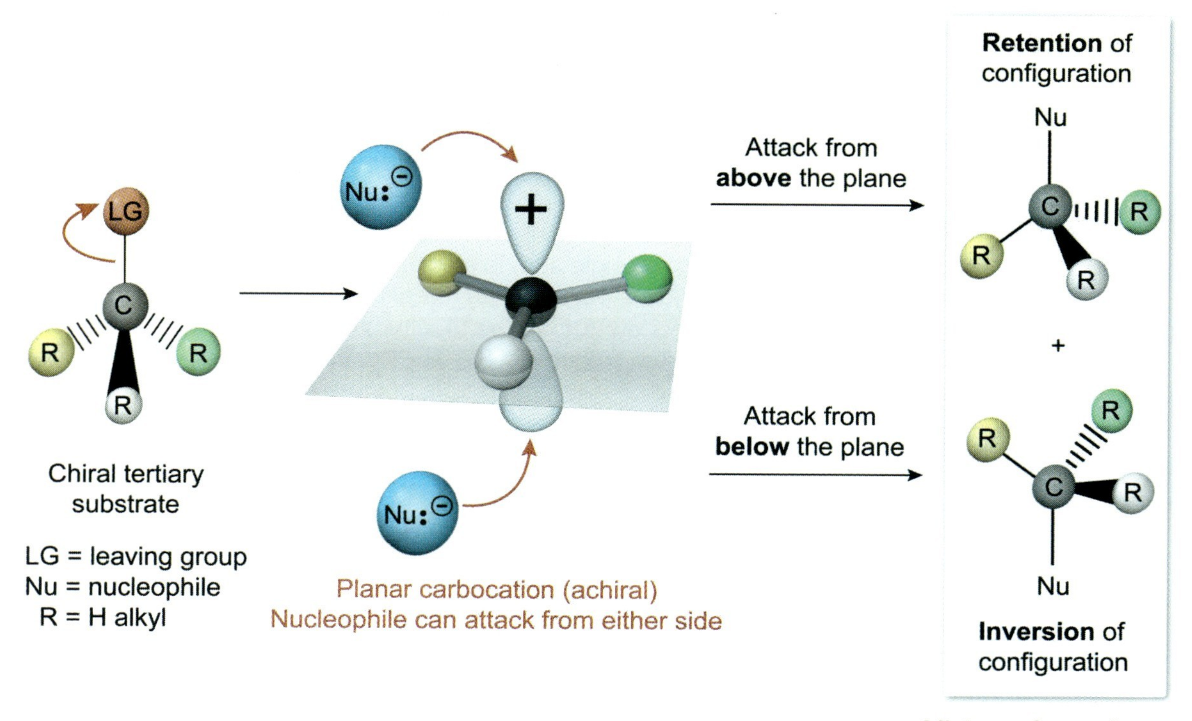

Figure 5.57 Stereochemical outcome of an S_N1 reaction.

If the substrate contains two or more asymmetric carbon atoms, *only* the carbocation carbon atom becomes racemized; in such a case, the products are a mixture of diastereomers. Furthermore, because racemization results from the *planar* carbocation structure, substrates that undergo carbocation rearrangement (where the position of the carbocation migrates) experience racemization of *every* carbon atom that bore a carbocation during the rearrangement process.

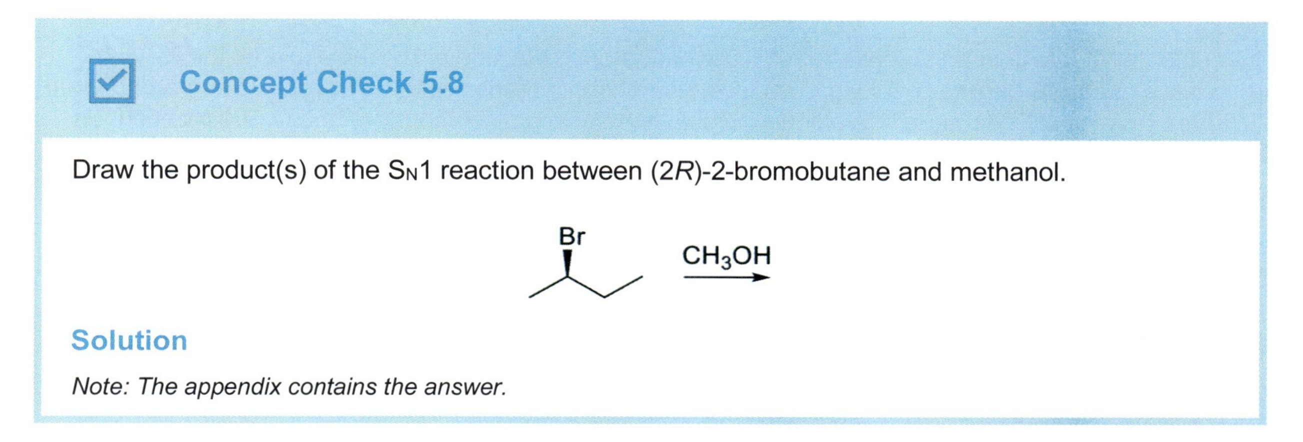

Concept Check 5.8

Draw the product(s) of the S_N1 reaction between (2*R*)-2-bromobutane and methanol.

Solution

Note: The appendix contains the answer.

5.4.03 S_N2 Reaction

For a **bimolecular nucleophilic substitution (S_N2) reaction**, the number (2) refers to the number of molecules present in the rate-limiting step. All S_N2 reactions share the following essential features:

- One mechanistic step (ie, a **concerted** reaction), without any intermediates
- A reaction rate dependent on *both* the nucleophile and the substrate
- Results that include a Walden inversion of stereochemistry (if the reaction occurs at a chiral carbon)

S_N2 Reaction Mechanism

The mechanism of any S_N2 reaction consists of a single step in which the nucleophile attacks the electrophilic carbon atom of the substrate (a Type 2 electrophile), breaking the polar bond between the substrate and the leaving group. The transition state for an S_N2 reaction has partial bonds for both the new sigma bond and the sigma bond being broken (Figure 5.58). The concerted S_N2 mechanism ensures that only *one* product is ever formed (ie, no rearrangements and no mixtures of stereochemistry).

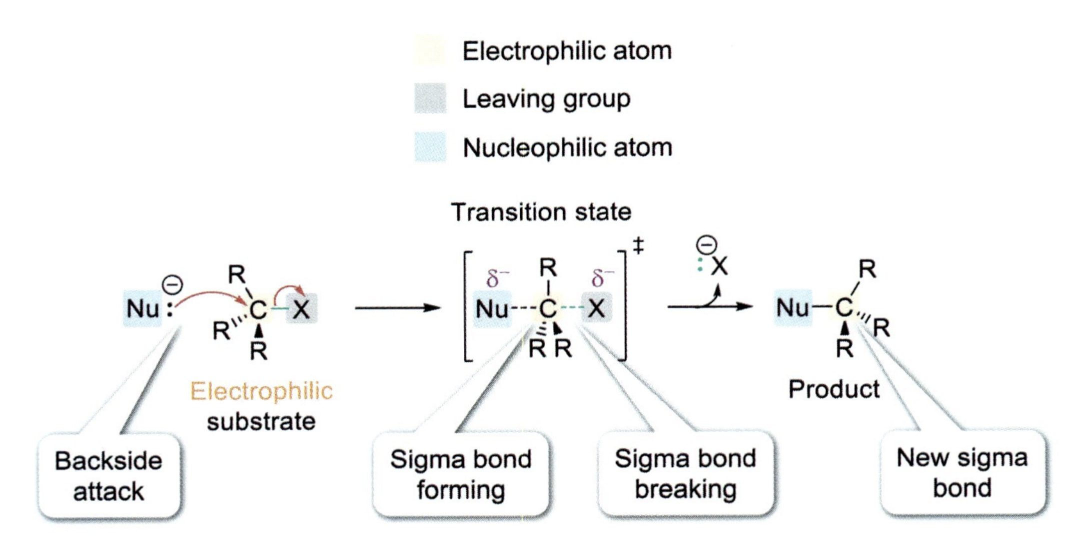

Figure 5.58 The S_N2 reaction mechanism.

The nucleophilic attack in an S_N2 reaction occurs on the *opposite* side of the molecule as the leaving group (ie, a backside attack), resulting in an inversion of configuration at the electrophilic carbon atom (ie, a **Walden inversion**). Because a Walden inversion is an *essential* component of any S_N2 reaction, an S_N2 reaction *cannot* occur if a Walden inversion is not possible (for any reason).

Stereochemistry of an S_N2 Reaction

For achiral carbons, it is not possible to experimentally verify that a Walden inversion has occurred (Figure 5.59). However, if the electrophilic carbon atom is chiral, the inversion of configuration may be experimentally observed through a *change* in stereochemistry. In a Walden inversion, *only* the electrophilic carbon atom changes its stereochemistry.

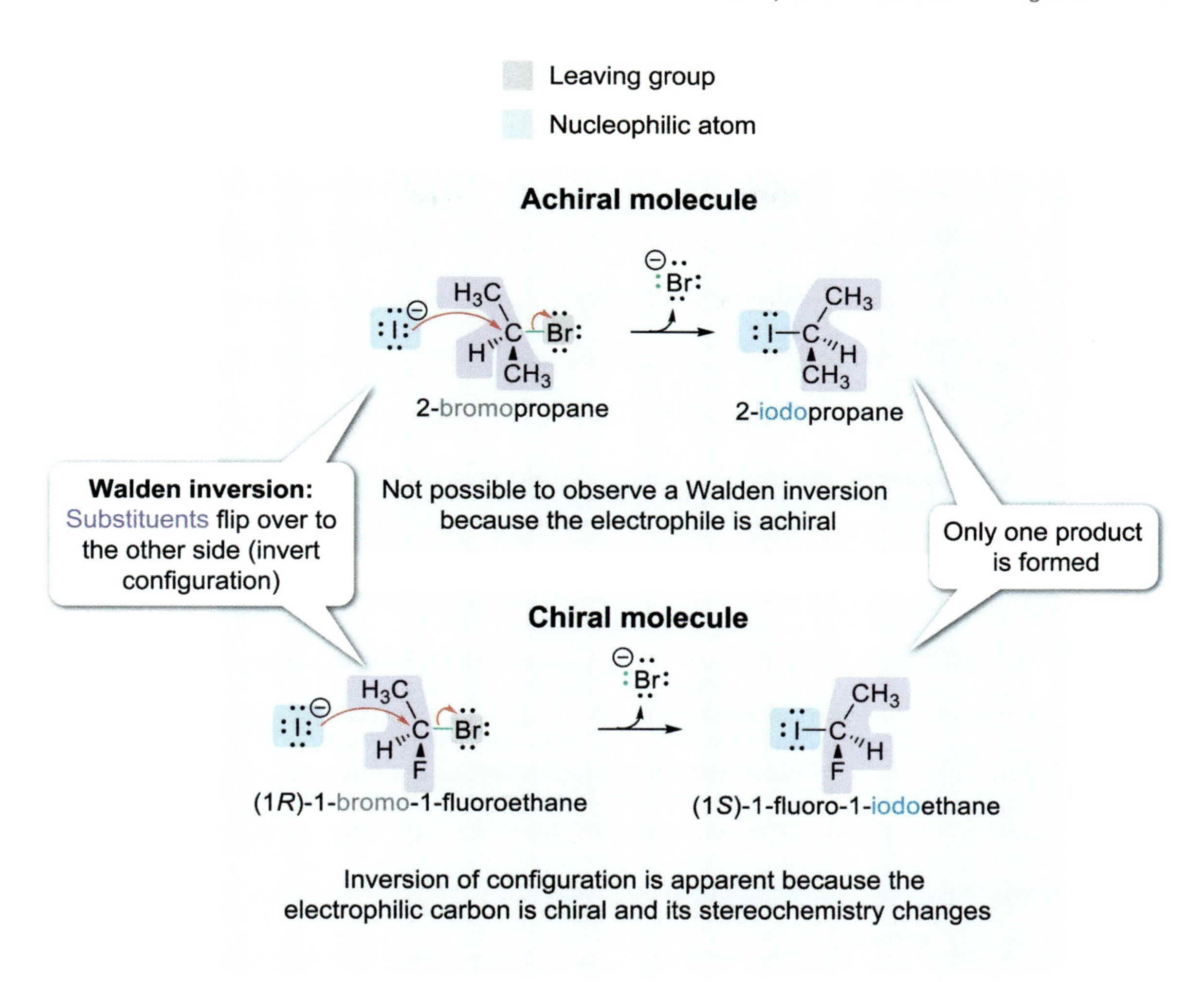

Figure 5.59 Walden inversion in an S_N2 reaction.

S_N2 Reaction Order

The S_N2 reaction is concerted, does not form any intermediates, and has a reaction coordinate diagram with a single peak representing the transition state (Figure 5.60). Transition state formation is the rate-limiting step.

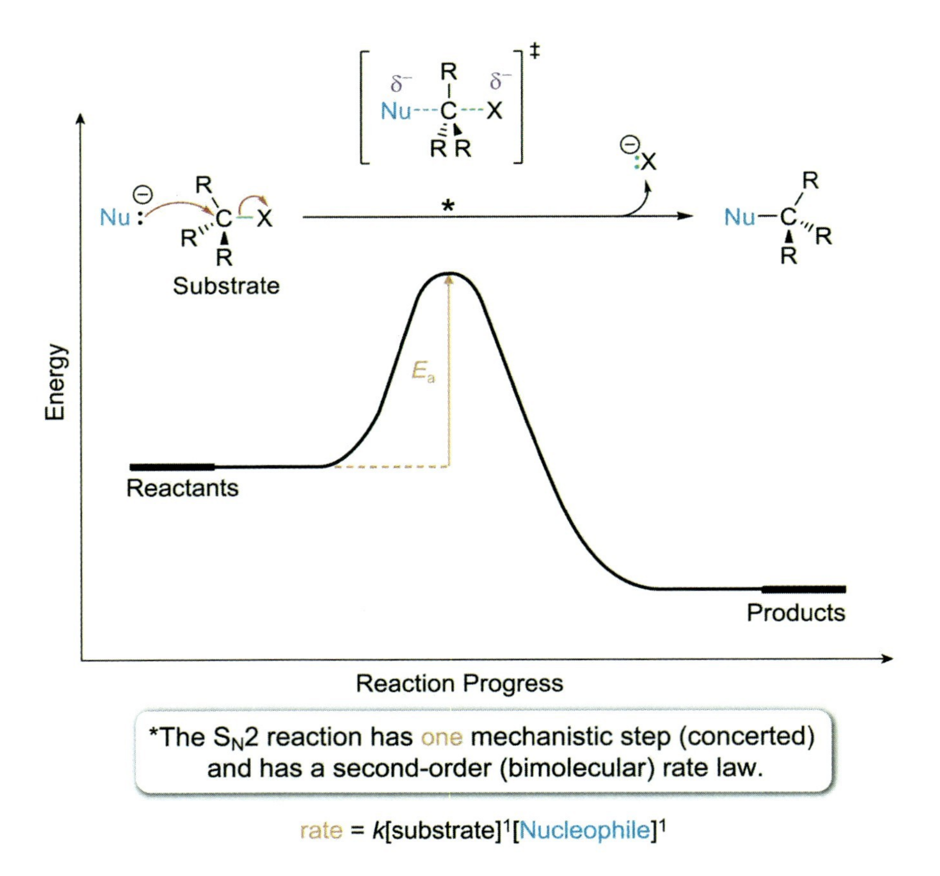

Figure 5.60 Reaction coordinate diagram for the S_N2 mechanism.

An S_N2 reaction is bimolecular and requires a productive collision between the nucleophile and the electrophilic substrate. Because *both* the nucleophile and the substrate are required for a collision, *both* molecules are present in the rate-limiting step, and the concentrations of *both* the nucleophile and the substrate impact the reaction rate.

The rate law of an S_N2 reaction is expressed by the equation:

$$\text{rate} = k[\text{Substrate}]^1[\text{Nucleophile}]^1$$

which is first-order with respect to *both* the electrophilic substrate *and* the nucleophile. The sum of the exponents in the rate law indicates an S_N2 reaction is second-order *overall*.

Variables in an S_N2 Reaction

The following variables influence the nucleophile in an S_N2 reaction (Figure 5.61):

- A strong and sterically unhindered nucleophile is required. Bulky nucleophiles (eg, a tertiary alkoxide) react more slowly, if at all.
- A polar aprotic solvent is ideal, as it lacks acidic N–H or O–H bonds that would encapsulate the nucleophile and *reduce* its reactivity.

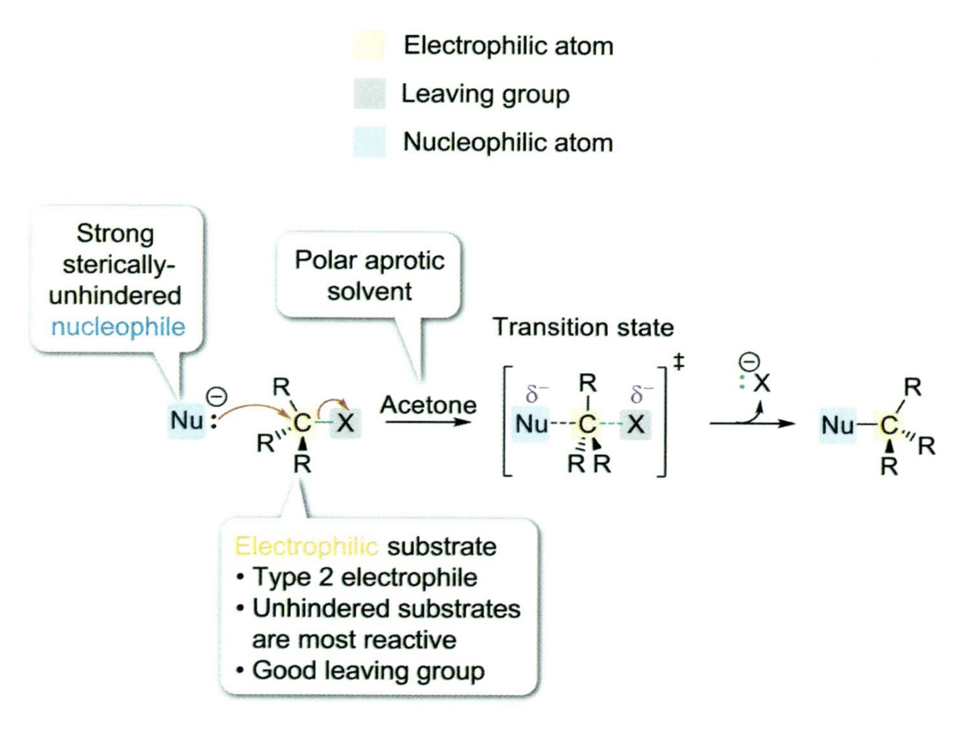

Figure 5.61 Variables that promote S_N2 reactions.

The following variables influence the electrophilic substrate in an S_N2 reaction:

- Sterically unhindered electrophiles are the best substrates. An S_N2 reaction *cannot* occur at a tertiary carbon.
- A good leaving group is required.

Differentiating S_N1 and S_N2 Reactions

Determining whether a given reaction proceeds via an S_N1 or S_N2 mechanism requires an analysis of both the nucleophile and the electrophile. The solvent may also provide clues toward the likely reaction mechanism.

Nucleophile

- A *strong* nucleophile suggests an S_N2 reaction, as it *cannot* participate in an S_N1 reaction.
- A *weak* nucleophile suggests an S_N1 reaction, as they *rarely* undergo an S_N2 reaction.
- Reactions involving an alkoxide ion in its corresponding alcohol are determined by the properties of the *alkoxide ion*. Methyl, 1°, and 2° alkoxide ions suggest an S_N2 reaction, but a 3° alkoxide ion (eg, *t*-butoxide) is too sterically hindered for an S_N2 reaction.

Electrophile

- A methyl or 1° electrophile suggests an S_N2 reaction.
- A 3° electrophile suggests an S_N1 reaction.
- Because a 2° electrophile can undergo *either* an S_N2 or S_N1 reaction, the *nucleophile* is generally the differentiating factor. A *strong* nucleophile suggests an S_N2 reaction, whereas a *weak* nucleophile suggests an S_N1 reaction.

Solvent

- A polar aprotic solvent suggests an S_N2 reaction.
- A polar protic solvent suggests an S_N1 reaction. However, an alcohol (a polar protic solvent) together with its conjugate alkoxide ion (a *strong nucleophile*) suggests an S_N2 reaction.

Table 5.6 summarizes the key components of S_N1 and S_N2 reactions.

Table 5.6 Comparison of S_N1 and S_N2 reaction components.

	S_N1 reactions	**S_N2 reactions**
Nucleophile	Weak (may be the solvent)	Strong
Electrophile	3° > 2°	Methyl > 1° > 2°
Solvent	Polar protic	Polar aprotic

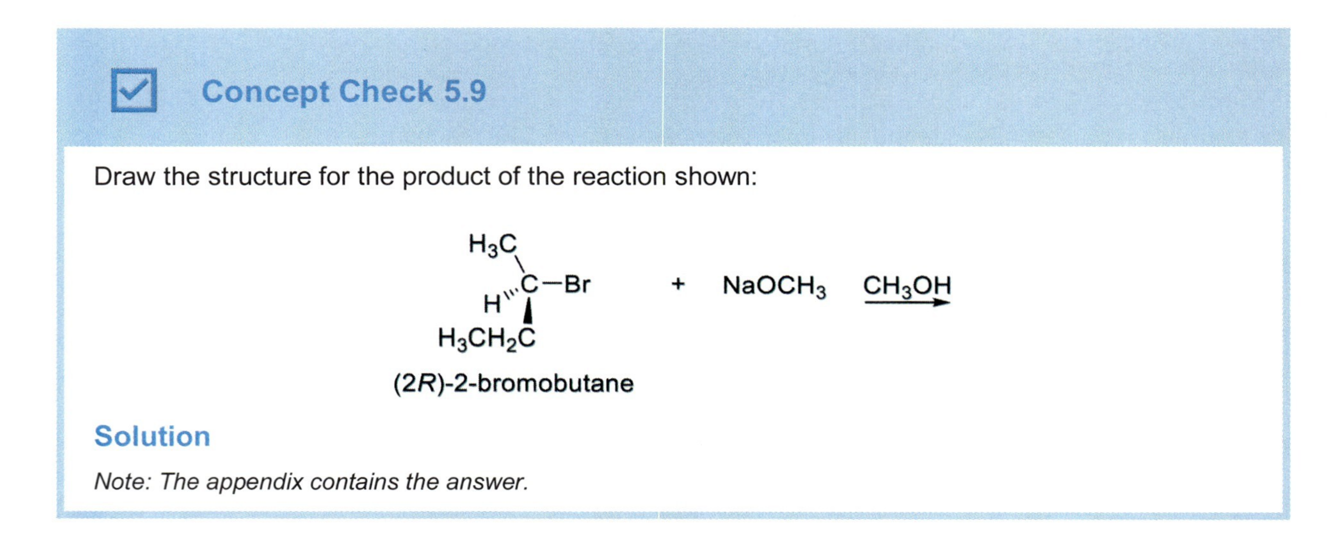

Concept Check 5.9
Draw the structure for the product of the reaction shown:
H3C
C—Br
H
H3CH2C
+
NaOCH3
CH3OH
(2R)-2-bromobutane
Solution
Note: The appendix contains the answer.

5.4.04 Nucleophilic Acyl Substitution

A **nucleophilic acyl substitution** is a substitution reaction where a nucleophile displaces a leaving group attached to the electrophilic carbon atom of an acyl group. A nucleophilic acyl substitution reaction requires *at least* two mechanistic steps and is sometimes referred to as an **addition-elimination reaction**.

Nucleophilic Acyl Substitution Mechanism

The first step in a nucleophilic acyl substitution mechanism is an **addition reaction**, where a nucleophile reacts with a Type 3 electrophile (Figure 5.62). Nucleophilic attack causes the electrons of the pi bond to become a lone pair on the oxygen atom. Because the hybridization of the electrophilic carbon temporarily changes from sp^2 to sp^3, this is called a **tetrahedral intermediate**.

The second mechanistic step is an **elimination reaction**, where the pi bond is reformed and the leaving group is lost.

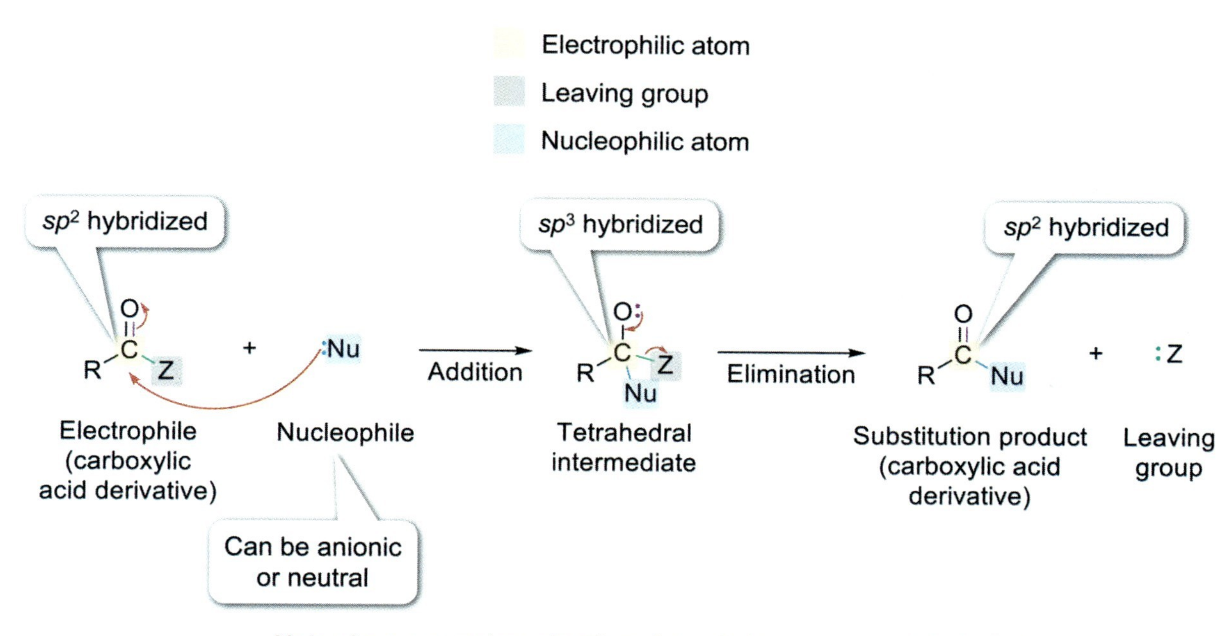

Figure 5.62 Electron flow for a general nucleophilic acyl substitution.

Nucleophilic Acyl Substitution Reaction Order

Because the mechanism of a nucleophilic acyl substitution has at least *two* steps, the corresponding reaction coordinate diagram has at least *two* peaks representing the transition state for each reaction step. The formation of the tetrahedral intermediate has the larger E_a and is the rate-limiting step (Figure 5.63).

Substrate

:Nu slow

:Z fast

rate = k [Substrate]^1[Nucleophile]^1 rate = k [Tetrahedral intermediate]^1

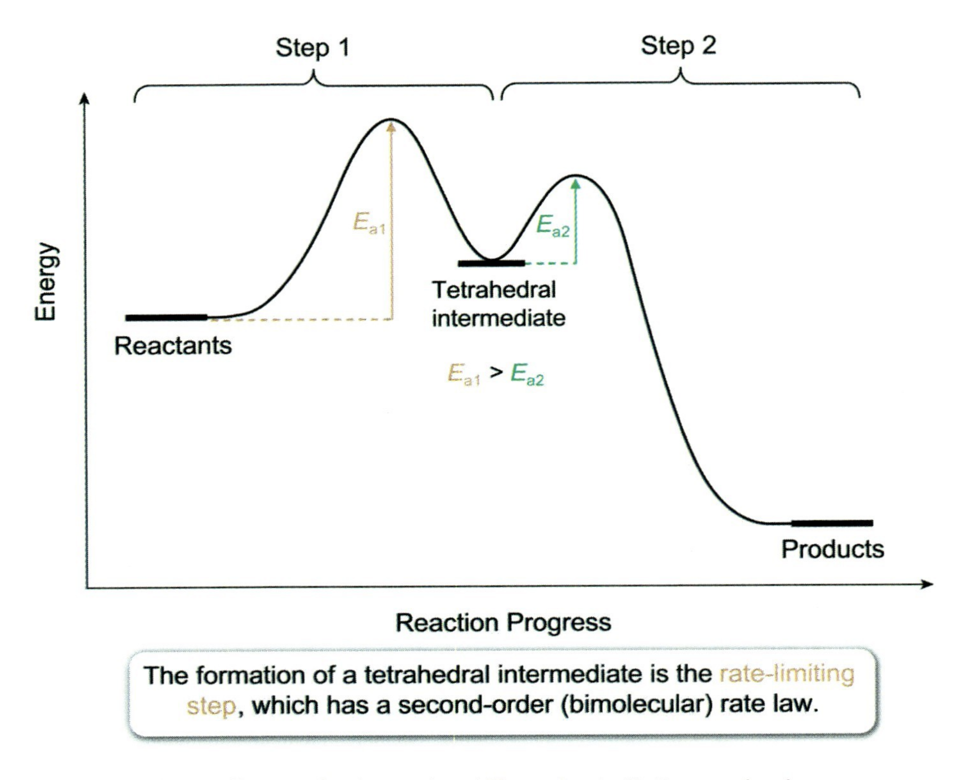

Figure 5.63 Reaction coordinate diagram for the nucleophilic acyl substitution mechanism.

Because both the nucleophile and the substrate are present in the rate-limiting step, the rate law for a nucleophilic acyl substitution is dependent on *both* the nucleophile and the substrate. The rate law is expressed by the equation:

$$\text{rate} = k[\text{Substrate}]^1[\text{Nucleophile}]^1$$

which is first-order with respect to *both* the substrate *and* the nucleophile. The sum of the exponents in the rate law indicates a nucleophilic acyl substitution reaction is second-order *overall.*

Like an S_N2 reaction, a nucleophilic acyl substitution is *bimolecular* and requires a productive collision between *two* molecules to form the tetrahedral intermediate. Although an S_N2 reaction and a nucleophilic acyl substitution are both second-order reactions, the S_N2 reaction has a single mechanistic step and nucleophilic acyl substitution requires at least two mechanistic steps.

5.4.05 Aromatic Substitution

Substitution reactions may also take place on aromatic rings through electrophilic aromatic substitution and nucleophilic aromatic substitution. The key difference between these two reactions is the role the aromatic ring plays in the reaction mechanism (Figure 5.64).

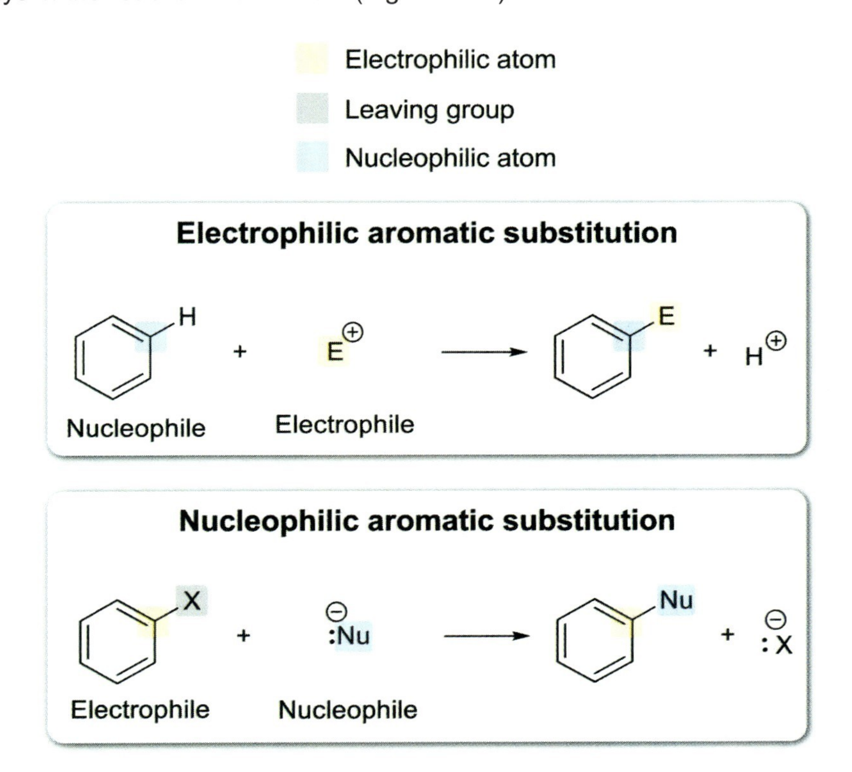

Figure 5.64 Aromatic substitution reactions.

In an **electrophilic aromatic substitution**, an aromatic ring carbon acts as the *nucleophile* and attacks an electrophile. The electrophile ultimately displaces an aromatic hydrogen atom.

In a **nucleophilic aromatic substitution**, a carbon atom on an aromatic ring acts as an *electrophile*. The electrophilic carbon is attached to a leaving group, which is ultimately displaced by a *very* strong nucleophile (eg, H_2N^-).

Lesson 5.5

Addition Reactions

Introduction

The focus of this lesson is addition reactions. After a brief general introduction to addition reactions, two specific types of addition reactions are discussed:

- Nucleophilic addition
- Electrophilic addition

5.5.01 Definition of an Addition Reaction

An **addition reaction** is a reaction where two groups are added across a pi bond, consuming the pi bond and producing two new sigma bonds (Figure 5.65). Although the figure depicts an addition reaction across a carbon-carbon double bond, addition reactions can also occur with triple bonds and with carbon-heteroatom (eg, oxygen, nitrogen) multiple bonds.

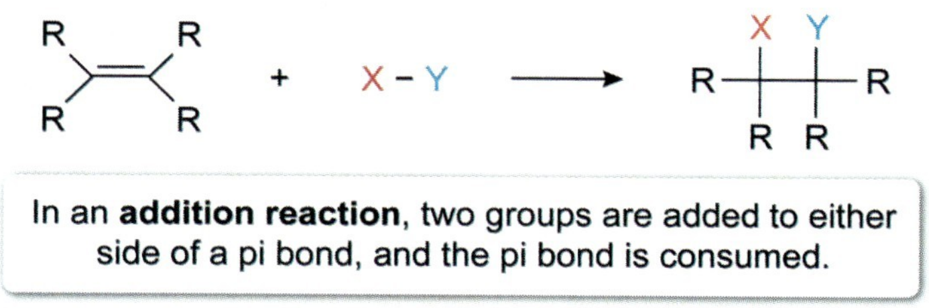

$R_2C{=}CR_2 + X{-}Y \longrightarrow R_2C(X){-}C(Y)R_2$

In an **addition reaction**, two groups are added to either side of a pi bond, and the pi bond is consumed.

Figure 5.65 A general addition reaction.

In traditional Organic Chemistry, two classes of addition reactions are introduced: nucleophilic addition and electrophilic addition (Figure 5.66). In a nucleophilic addition, the pi bond acts as the electrophile and the groups added act as the nucleophile. Conversely, in an electrophilic addition the pi bond acts as the *nucleophile* and the groups added act as the *electrophile*.

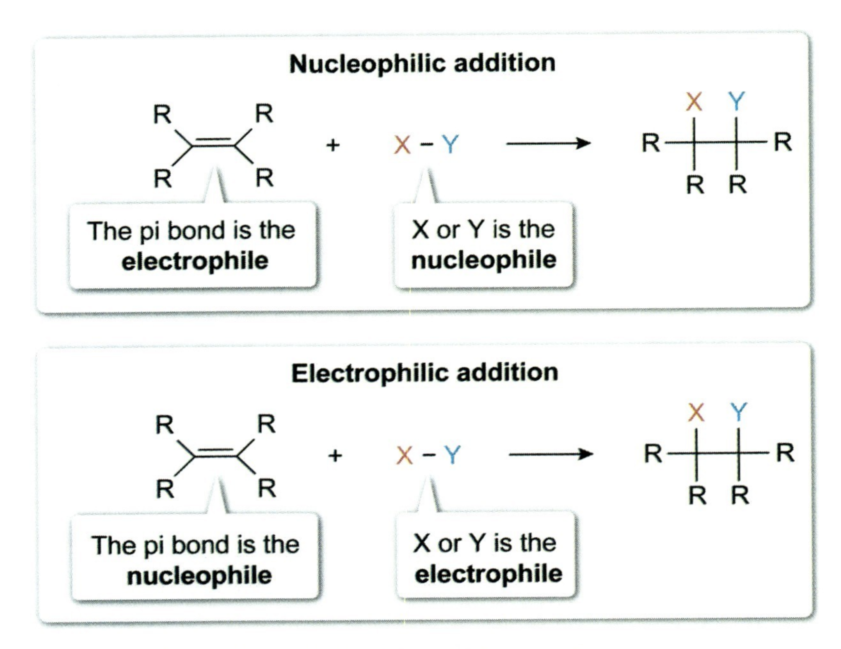

Figure 5.66 Comparing nucleophilic addition electrophilic addition reactions.

5.5.02 Nucleophilic Addition

A **nucleophilic addition reaction** is a multistep reaction where two groups (typically a nucleophile and a hydrogen atom) add across the pi bond of a polar carbon-heteroatom multiple bond C=X. The heteroatom is usually either an electronegative oxygen or a nitrogen atom.

Figure 5.67 shows a nucleophile adding to an electrophilic carbon and a hydrogen atom adding to an electronegative atom to give an addition product.

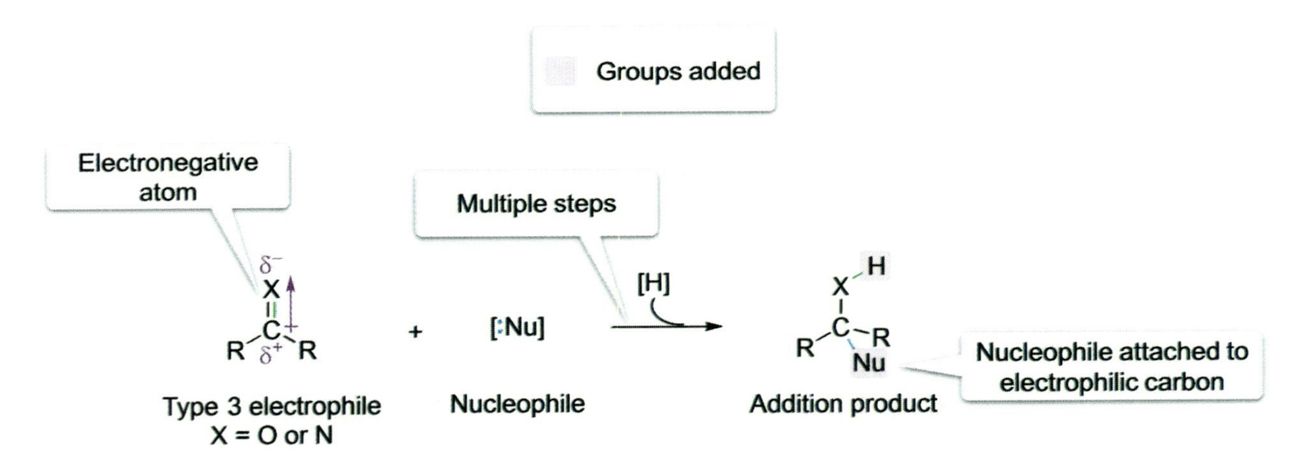

Figure 5.67 A general nucleophilic addition reaction.

A nucleophilic addition reaction can take place under either acidic or basic conditions. The details of these parallel mechanisms are covered in Lesson 9.3.

The Type 3 electrophile (Concept 5.3.02) in a nucleophilic addition reaction contains a polar carbon-heteroatom multiple bond and an electrophilic carbon atom (Figure 5.68). Resonance (see Lesson 2.7) and the inductive effect both contribute to the polarity of the carbon-heteroatom bond, making the carbon atom electron-deficient (δ^+), electrophilic, and susceptible to nucleophilic attack.

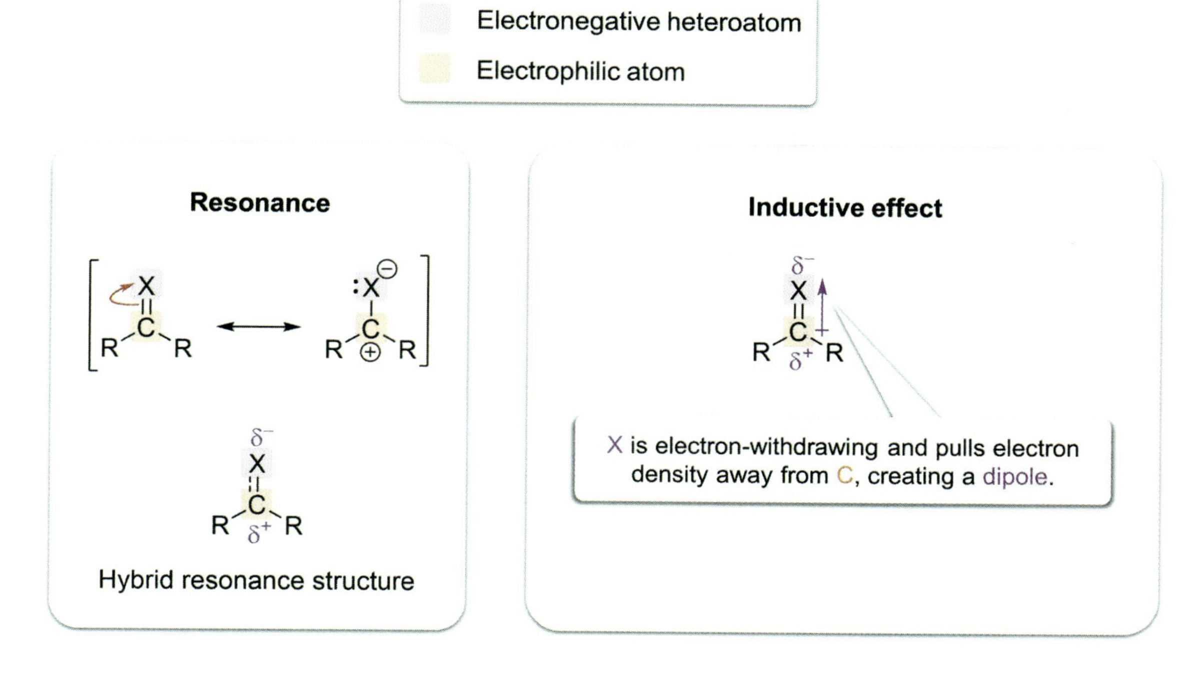

Resonance and the inductive effect create a δ⁺ on carbon, making it electrophilic and susceptible to nucleophilic attack.

Figure 5.68 Impact of resonance and the inductive effect on a Type 3 electrophile.

Comparing a Nucleophilic Addition to other Mechanisms

Although a nucleophilic addition reaction is in some ways similar to a nucleophilic acyl substitution reaction, key differences also exist (Figure 5.69). Both mechanisms begin with an addition step between a nucleophile and the carbon atom of a Type 3 electrophile, but the mechanisms *diverge* due to a difference in the groups attached to the electrophilic carbon atom.

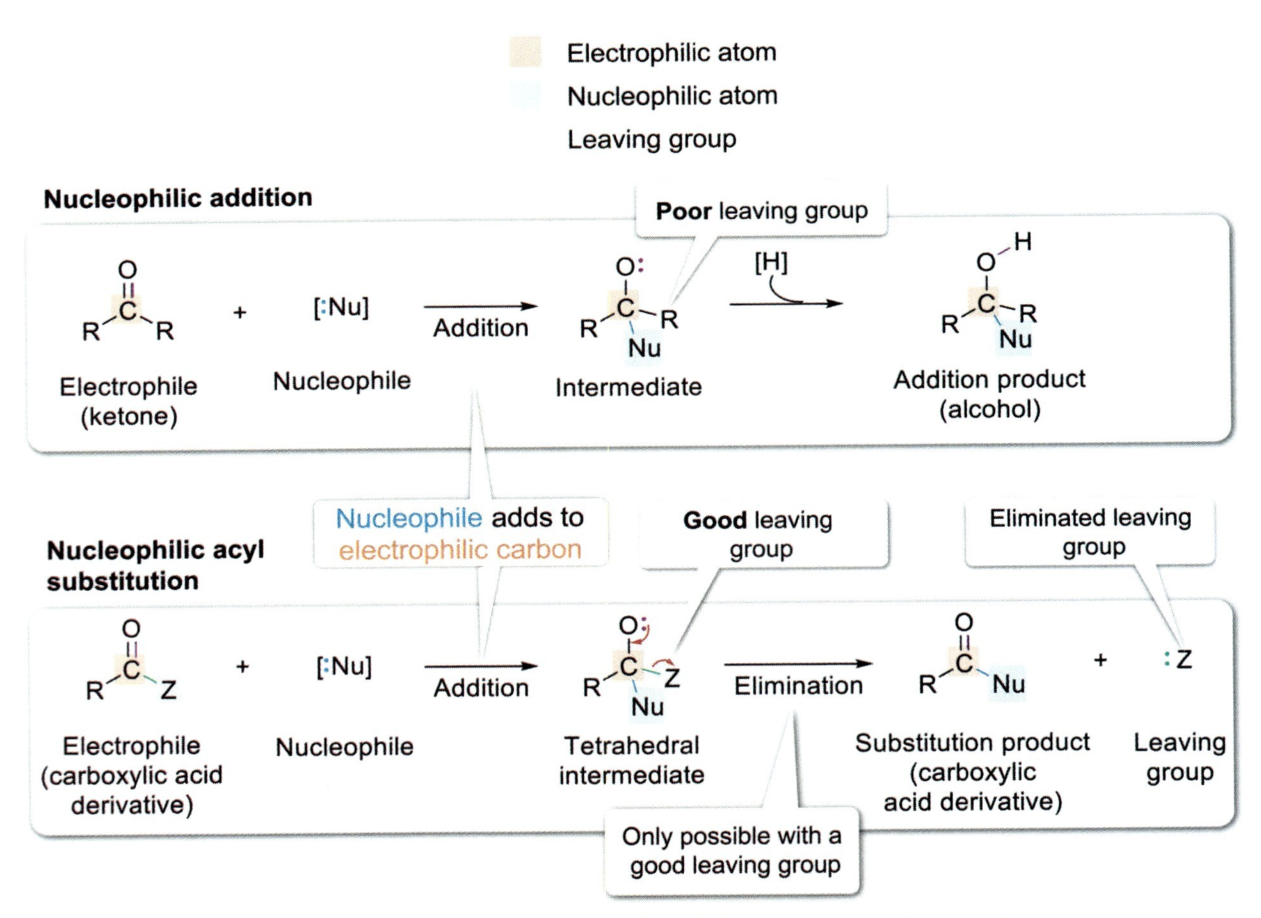

Figure 5.69 Comparing a nucleophilic addition and a nucleophilic acyl substitution for a Type 3 electrophile.

The R group in a ketone is a *poor* leaving group (ie, an alkyl group), whereas the Z group in a carboxylic acid derivative is a *good* leaving group (eg, a heteroatom). Consequently, the intermediate in a nucleophilic acyl substitution *can* undergo an elimination step, while the intermediate in an addition reaction *cannot*.

Stereochemistry in a Nucleophilic Addition

Because the electrophilic carbon in a Type 3 electrophile must contain at least one unhybridized *p* orbital for the required pi bond, it is inherently achiral. Nucleophilic attack can occur from *either side* of the pi bond and leads to the formation of potentially *two* products (Figure 5.70). If a nucleophilic attack results in the carbon atom becoming a chiral center, the two products are stereoisomers.

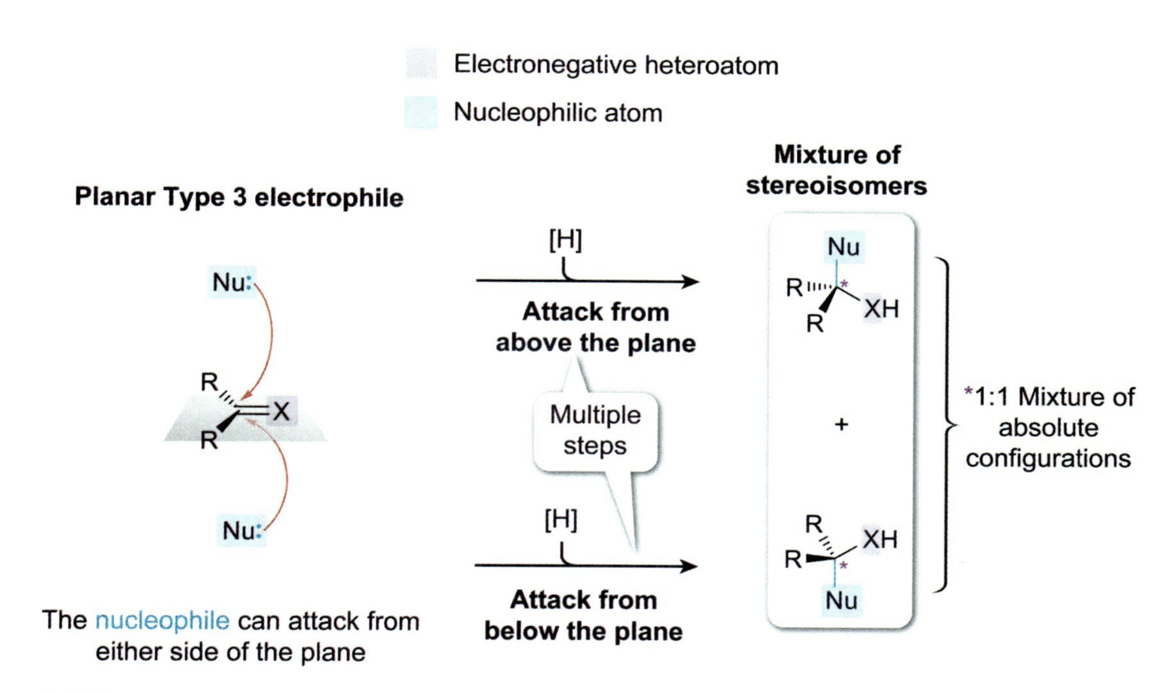

Figure 5.70 Stereochemical outcome of a nucleophilic addition.

If the new chiral center is the only stereocenter among the product molecules, the products are *enantiomers* and the product mixture is racemic. If *other* stereocenters are also present, the products are *diastereomers*; configuration at other stereocenters is not impacted by the nucleophilic addition.

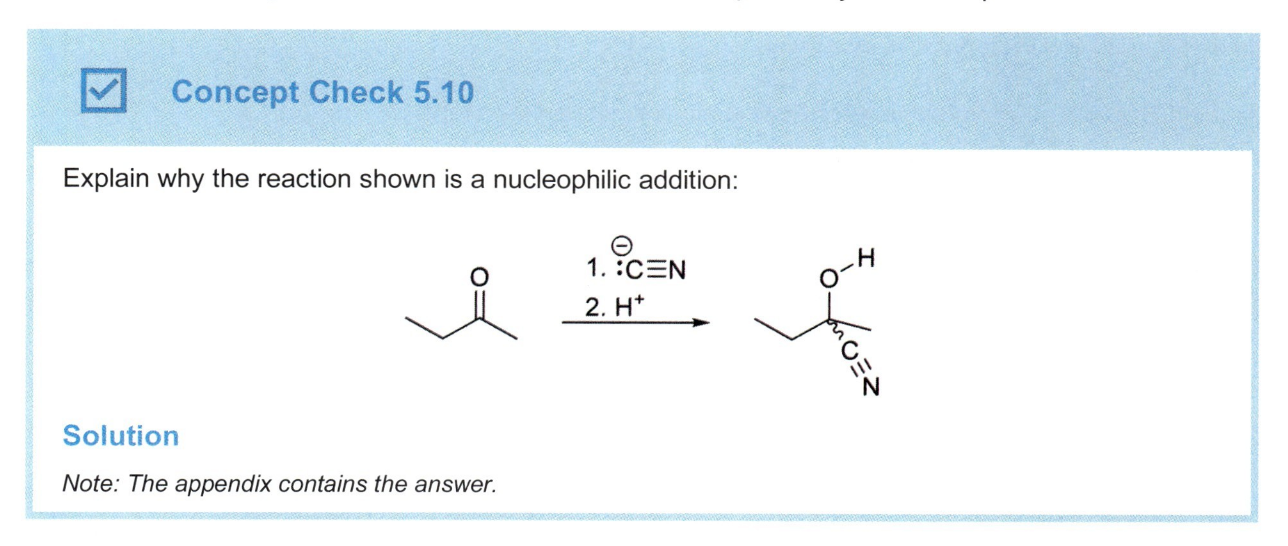

Concept Check 5.10
Explain why the reaction shown is a nucleophilic addition:
O
1. :C≡N
2. H+
O
H
C
N
Solution
Note: The appendix contains the answer.

5.5.03 Electrophilic Addition

An **electrophilic addition reaction** is a multistep reaction where two groups (an electrophile and another group) add across the pi bond of an alkene or an alkyne, consuming the pi bond and forming two new sigma bonds (Figure 5.71).

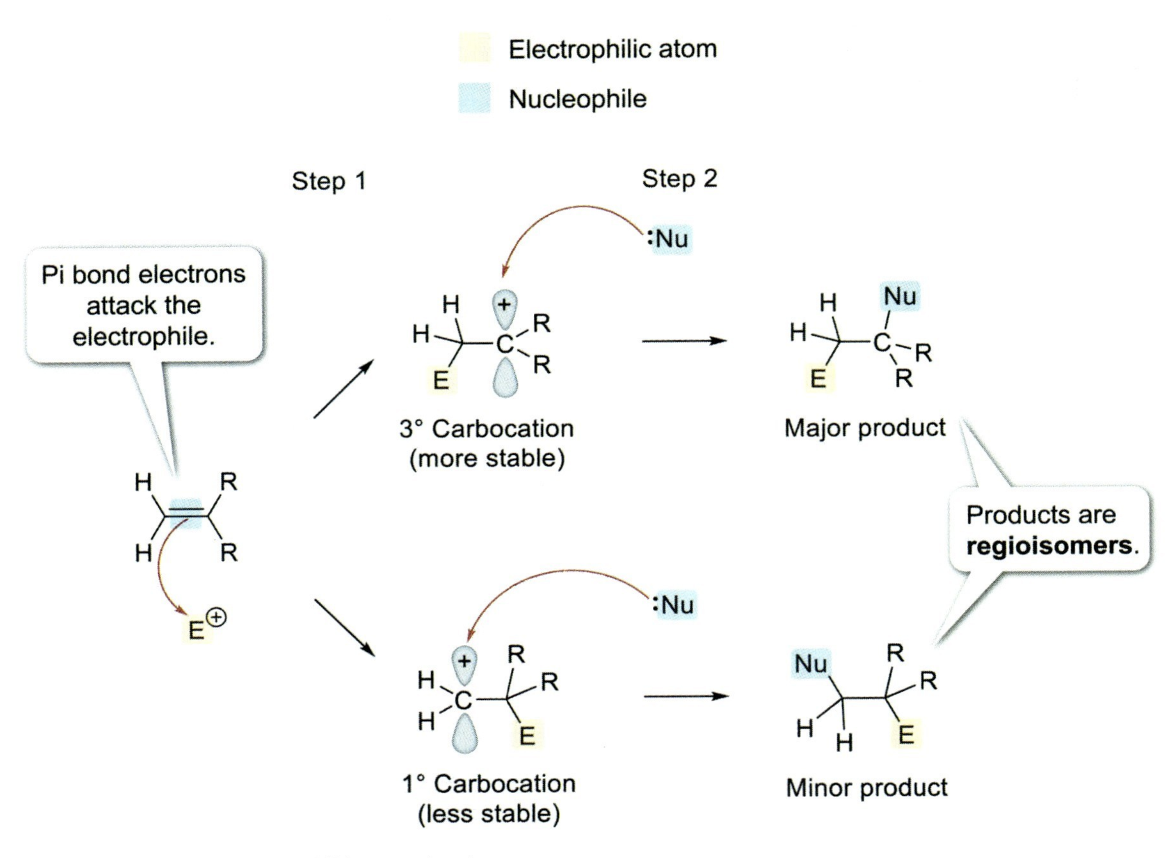

Figure 5.71 The electrophilic addition mechanism.

Most electrophilic addition reactions have at least two mechanistic steps:

- The pi bond acts as a nucleophile and attacks an electrophile (E^+), forming an intermediate. The atom with the new sigma bond to the electrophile retains its octet, whereas the other atom becomes an electron-deficient carbocation. This is typically the rate-limiting step.
- A nucleophile attacks the carbocation and forms the second new sigma bond.

If the electrophilic reactant is *not* symmetrical, an electrophilic addition can produce two possible products that are regioisomers, a type of positional isomer (see Lesson 3.1).

Lesson 5.6

Elimination Reactions

Introduction

This lesson focuses on a third general class of reactions that occur between a Brønsted-Lowry base and an electrophile—the elimination reaction. After providing an overview of the general qualities of an elimination reaction, two specific elimination reactions are discussed:

- The first-order elimination (E1) reaction
- The second-order elimination (E2) reaction

5.6.01 Definition of an Elimination Reaction

An **elimination reaction** is a reaction where two groups are lost (eliminated) from adjacent carbon atoms in a molecule, resulting in the formation of a new pi bond (Figure 5.72). Although the identities of the two eliminated groups (ie, X, Y) can vary significantly, one of the two groups often acts as a leaving group (see Concept 5.3.03).

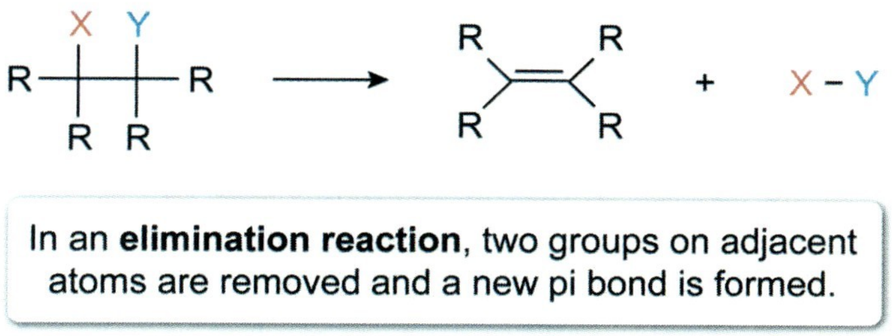

In an **elimination reaction**, two groups on adjacent atoms are removed and a new pi bond is formed.

Figure 5.72 A general elimination reaction.

Because an addition reaction involves two groups adding across a pi bond (see Concept 5.5.01), the terms "addition" and "elimination" refer to *reciprocal* processes (Figure 5.73).

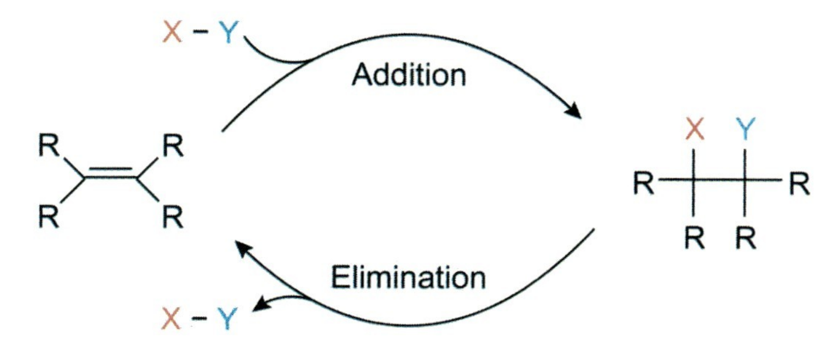

Figure 5.73 Reciprocal nature of addition and elimination reactions.

Some of the more advanced mechanisms in later lessons have addition and elimination processes as *separate* elementary steps within the *same* mechanistic sequence. Such reaction mechanisms can be described as either **addition-elimination reactions**, or **elimination-addition reactions**. Nucleophilic acyl substitution (Concept 5.4.04) is an example of an addition-elimination reaction.

Like S_N1 and S_N2 reactions (Lesson 5.4), the two types of elimination reactions discussed within this lesson are differentiated primarily by their **reaction order**. A first-order elimination mechanism is abbreviated as E1, and a second-order elimination mechanism is abbreviated as E2.

5.6.02 E1 Reaction

The **E1 reaction** (first-order elimination mechanism) is a unimolecular elimination where a substrate (C–X) loses its leaving group (X) and a proton (H) to form a new a pi bond (eg, an alkene) (Figure 5.74).

An E1 reaction requires *at least* two mechanistic steps:

- The polar bond between an electrophilic carbon and its leaving group is ionized to create a carbocation intermediate.
- A base removes a proton from one of the carbon atoms *adjacent* to the carbocation, and the electrons from the broken C–H bond are shared with the carbocation carbon to form a pi bond.

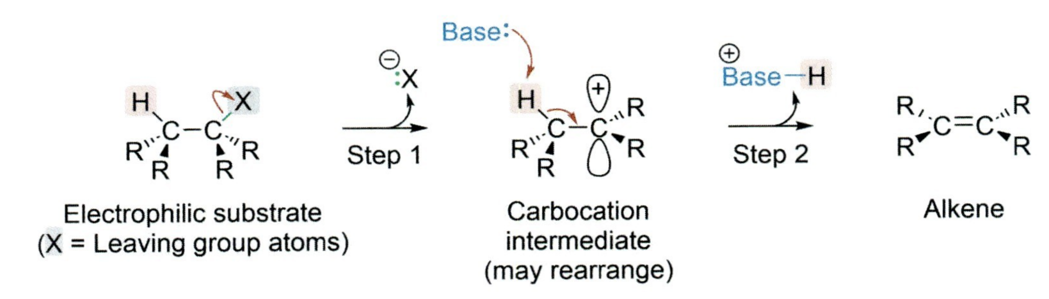

Figure 5.74 The E1 mechanism.

The E1 reaction and the SN1 reaction (Concept 5.4.02) share many of the same traits and features. Because formation of the carbocation (Step 1) is the rate-limiting step, the overall rate of an E1 reaction is dependent *only* on the concentration of the electrophilic substrate. As a result, the rate law for an E1 mechanism is *first-order* with respect to the electrophilic substrate and *zero-order* with respect to the base (Figure 5.75).

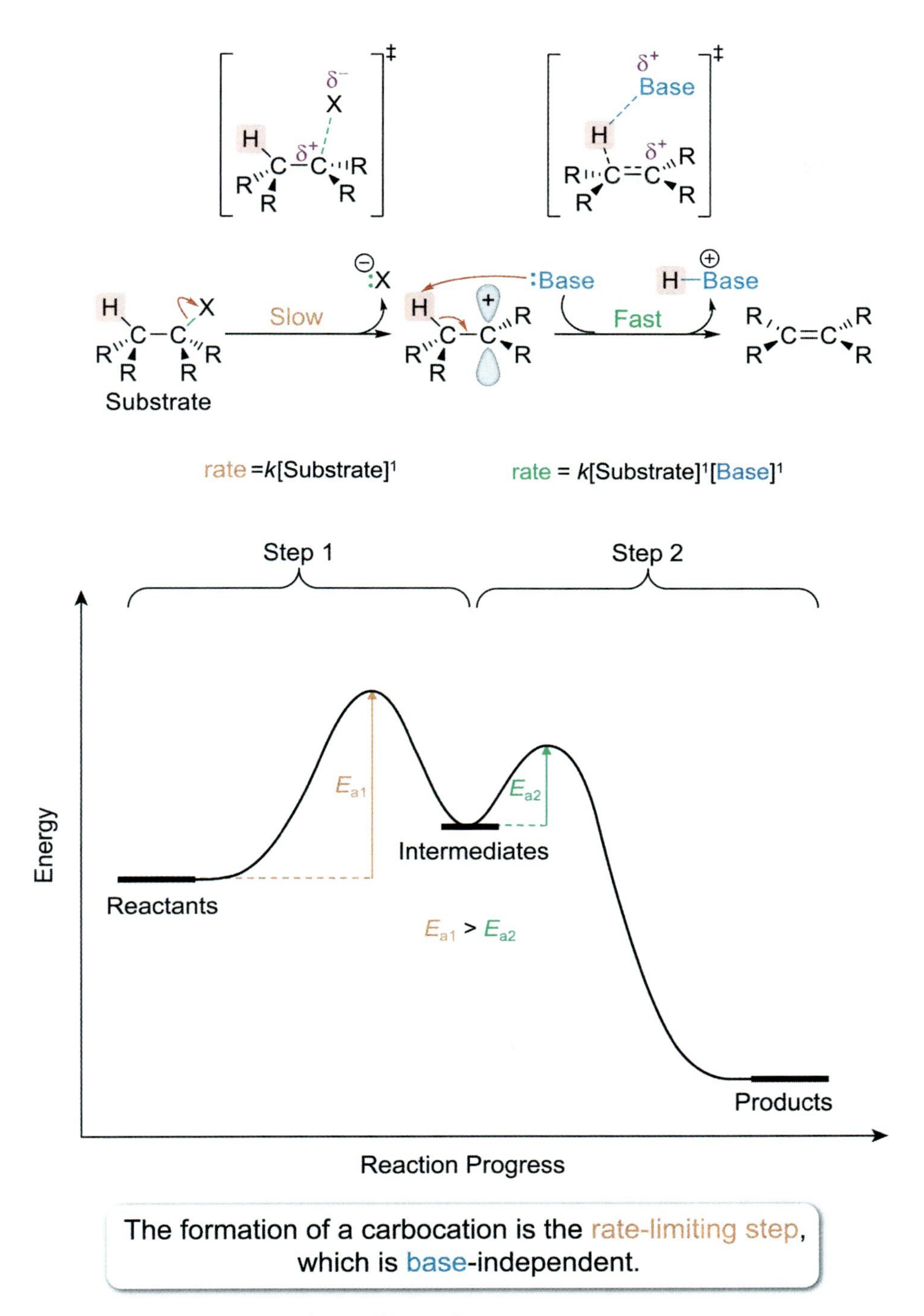

Figure 5.75 Reaction coordinate diagram for an E1 reaction.

The carbocation intermediate in an E1 reaction *may* be subject to rearrangement through either a hydride shift or an alkyl shift, depending on the structure of the substrate (see Concept 5.2.01). As a result, multiple possible positions may exist for the carbocation within a molecule, which may lead to a greater number of possible E1 alkene products.

During the deprotonation step in an E1 mechanism (Step 2), more than one adjacent carbon atom can have a proton. As a result, an E1 reaction may result in the formation of multiple alkene regioisomers, a type of positional isomer where the double bond is located between *different* sets of carbon atoms (see Lesson 3.1). An E1 reaction tends to prefer formation of the most-stable alkene (ie, the alkene with the greatest number of R groups attached to the alkene carbon atoms), a manifestation of thermodynamic control commonly known as **Zaitsev's rule**.

The E1 Dehydration of an Alcohol

Heating an alcohol with a strong acid can promote the loss of water (**dehydration**) and conversion of the alcohol to an alkene (Figure 5.76). First, the strong acid protonates the alcohol functional group to convert the alcohol (a poor leaving group) into an oxonium ion (a good leaving group; see Concept 5.3.03).

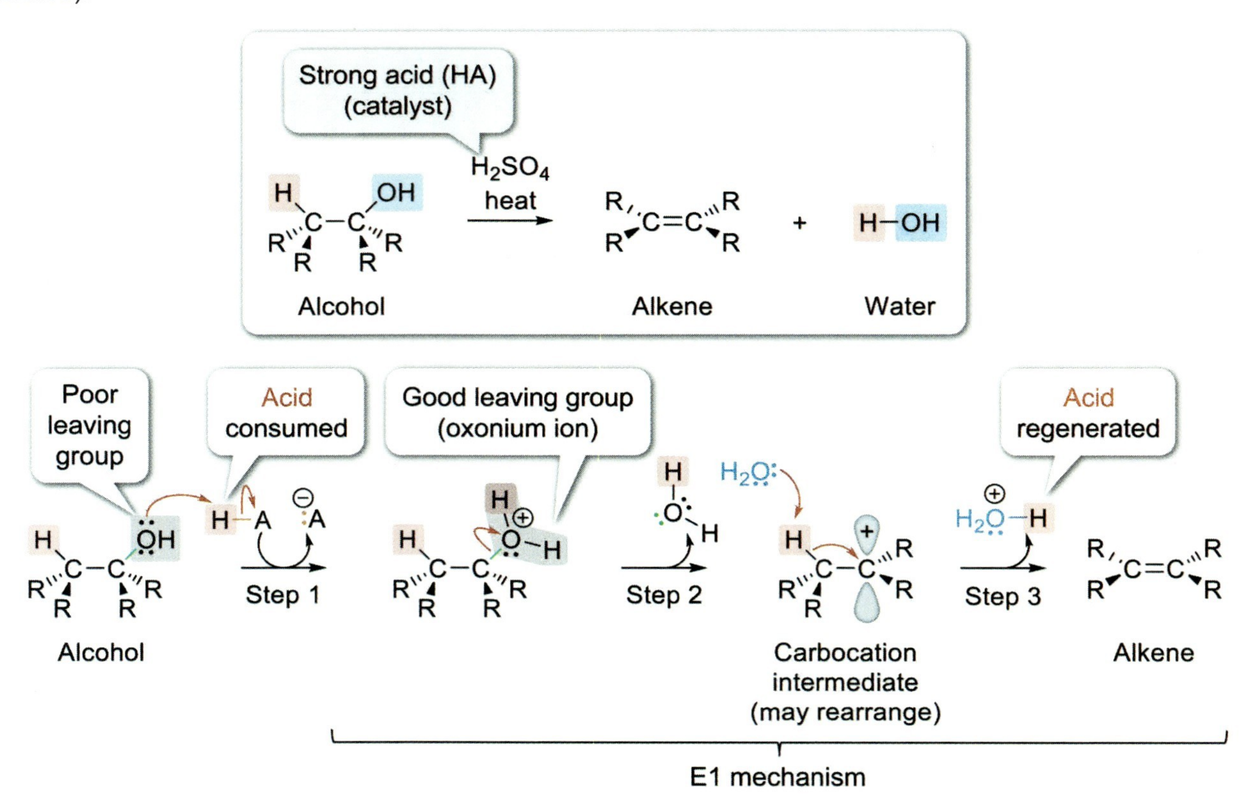

Figure 5.76 Acidic dehydration of an alcohol to an alkene.

The rest of the acidic dehydration of an alcohol proceeds through an E1 mechanism. The oxonium ion is lost as a molecule of neutral water to form a carbocation. Following any carbocation rearrangements, a molecule of water acts as a base to remove a proton on a carbon adjacent to the carbocation, forming the pi bond of the alkene product. Since a molecule of strong acid is *consumed* (Step 1) and later *regenerated* (Step 3), the acidic dehydration is considered to be acid-catalyzed.

5.6.03 E2 Reaction

The **E2 reaction** (second-order elimination mechanism) is a bimolecular elimination where a substrate (C–X) loses its leaving group (X) and a proton (H) to form a new a pi bond (eg, an alkene) (Figure 5.77).

The entirety of an E2 reaction takes place in only *one* mechanistic step. Reactions that take place in one mechanistic step are called **concerted reactions** and do *not* generate intermediates (eg, carbocations).

In an E2 reaction, the following events occur *all at the same time*:

- A base removes a proton from one of the carbon atoms *adjacent* to the electrophilic carbon atom.
- The electrons from the broken C–H bond are shared with the electrophilic carbon to form a pi bond.

- The polar bond between the electrophilic carbon and its leaving group is broken, with the electrons becoming a lone pair on the leaving group.

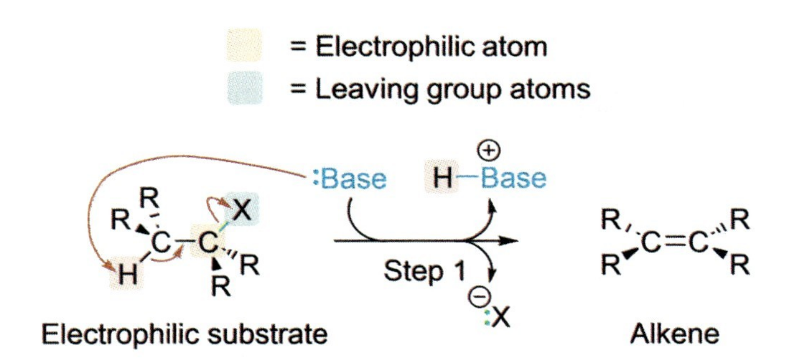

Figure 5.77 The E2 mechanism.

The E2 reaction and the S_N2 reaction (see Concept 5.4.03) share many characteristics. Given that both E2 and S_N2 are concerted reactions, the rate of an E2 reaction is dependent on the concentrations of *both* the electrophilic substrate *and* the base. As a result, the rate law for an E2 mechanism is *first-order* with respect to the substrate, *first-order* with respect to the base, and a combined *second-order* (bimolecular) reaction overall (Figure 5.78).

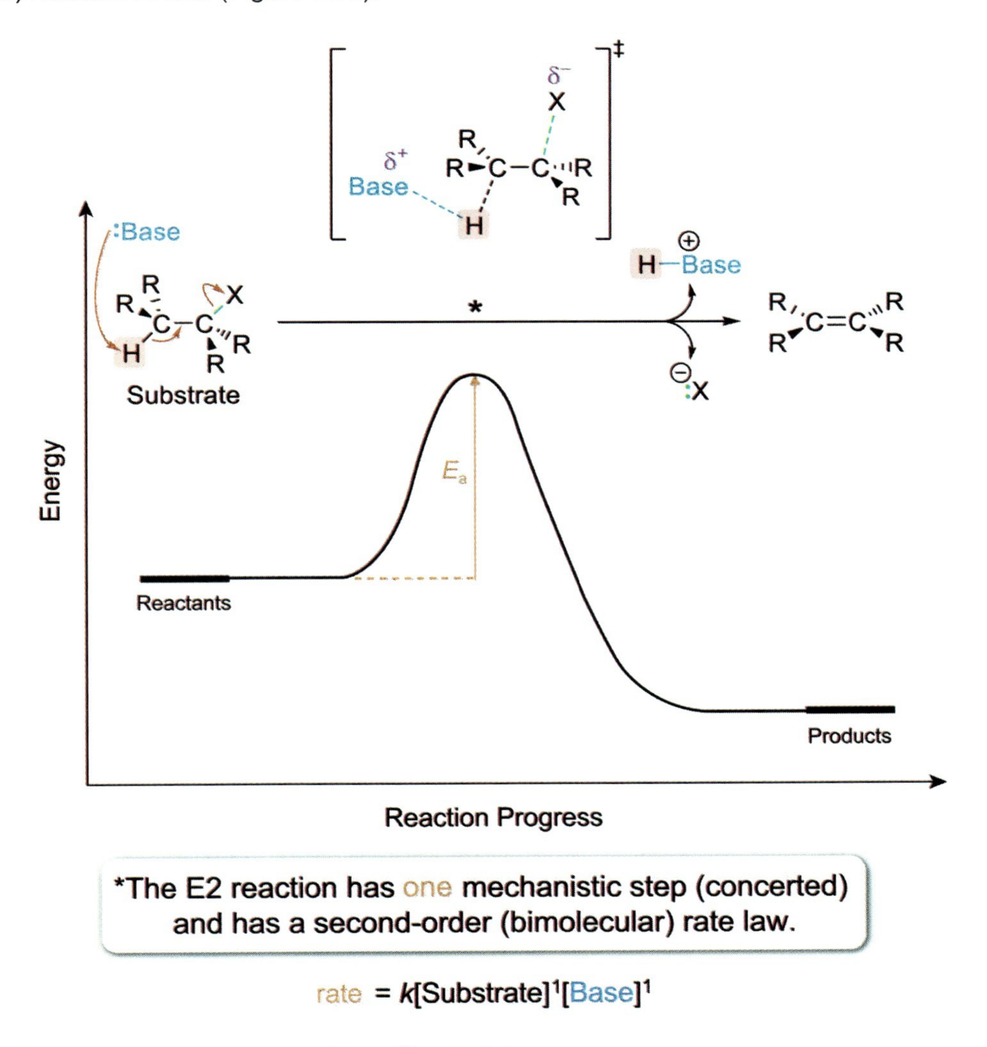

Figure 5.78 Reaction coordinate diagram for an E2 reaction.

Because the E2 mechanism *requires* the use of a strong base, the experimental conditions for an E2 reaction tend to be very strongly basic and can potentially limit the functional groups, which likely remain intact and unchanged. The most common bases that facilitate an E2 reaction are hydroxide (HO^-) and alkoxide (RO^-) salts.

Concerted reactions (eg, S_N2, E2) often have specific stereochemical requirements for the reactants. In an E2 mechanism, the acidic proton and the leaving group must be geometrically aligned to allow the necessary orbital overlap. The preferred dihedral angle between these groups is 180°, an anti (or antiperiplanar) relationship.

In general, an E2 reaction produces the *most-stable* (most highly substituted) alkene as the major product, called the **Zaitsev product** for its adherence to Zaitsev's rule (see Concept 5.6.02). However, if a bulky, sterically hindered base is used (eg, *tert*-butoxide), the major product tends to be the *least-substituted* alkene. The use of a bulky base is an example of kinetic control, which forms the higher-energy (but least stable) product at a faster rate of formation. For an E2 reaction, the kinetic product is called the **Hoffmann product**.

Lesson 5.7

Oxidation-Reduction Reactions

Introduction

In previous reactions discussed in this chapter, electron *pairs* are donated by nucleophiles and accepted by electrophiles. Although the final class of reactions discussed here also involves electron transfers, oxidation-reduction (redox) reactions can involve *any* number of electrons. This lesson discusses oxidation and reduction reactions in the context of organic chemistry.

5.7.01 Organic Definitions of Oxidation and Reduction

Oxidation is a chemical process involving the *loss* of electrons, and **reduction** is a chemical process involving the *gain* of electrons (Figure 5.79). Because these processes are inherently reciprocal in nature, they *cannot* occur in isolation—an oxidation is always associated with a corresponding reduction. For this reason, **oxidation-reduction reactions** are often described using the contraction **redox** (*red* = REDuction, *ox* = OXidation).

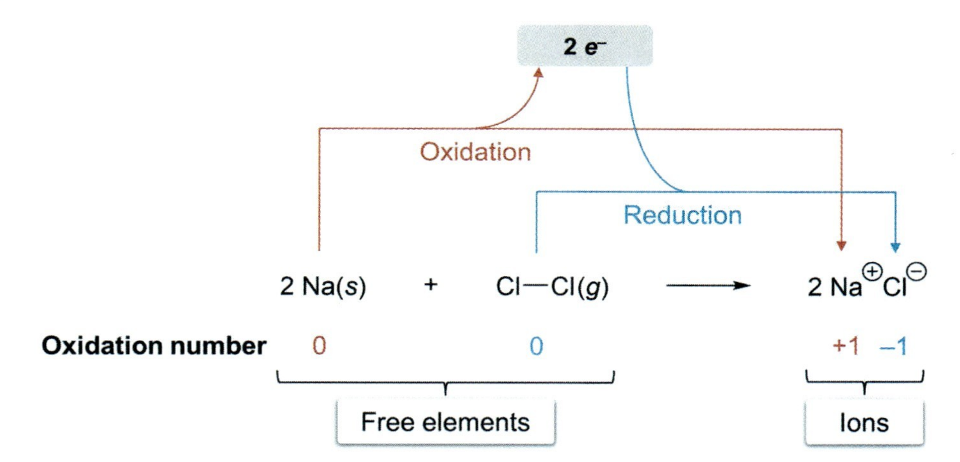

Figure 5.79 An example of a simple oxidation-reduction reaction.

The **oxidation number** (or **oxidation state**) of an atom is its hypothetical formal charge if each of its bonds were *purely ionic*. The oxidation number trends from general chemistry illustrate the general definition of oxidation number (including most of the noted exceptions). Oxidation of an atom leads to an *increase* in oxidation number, whereas reduction of an atom leads to a *decrease* in oxidation number.

Determining the oxidation number for atoms with primarily *covalent bonds* (including organic molecules) requires a different approach. Bonds between atoms of the *same* element do not alter the oxidation number, since there is no difference in electronegativity; *all* other covalent bonds are treated as ionic bonds. For example, carbon is more electronegative than hydrogen, so a C–H bond would *decrease* the oxidation state of the carbon atom by 1 and *increase* the oxidation state of the hydrogen atom by 1. The oxidation state of an atom is calculated from the sum of the impacts from all its covalent bonds (Figure 5.80).

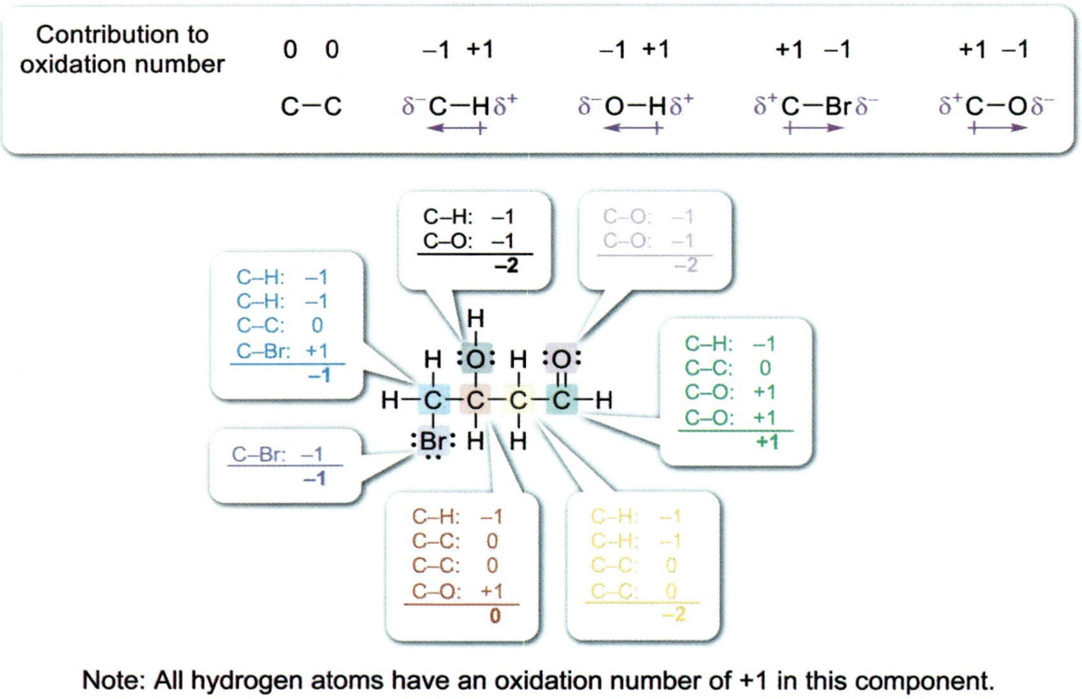

Contribution to oxidation number	0 0	−1 +1	−1 +1	+1 −1	+1 −1
	C−C	δ⁻C−Hδ⁺	δ⁻O−Hδ⁺	δ⁺C−Brδ⁻	δ⁺C−Oδ⁻

C–H: −1
C–H: −1
C–C: 0
C–Br: +1
−1

C–Br: −1
−1

C–H: −1
C–O: −1
−2

C–O: −1
C–O: −1
−2

C–H: −1
C–C: 0
C–O: +1
C–O: +1
+1

C–H: −1
C–C: 0
C–C: 0
C–O: +1
0

C–H: −1
C–H: −1
C–C: 0
C–C: 0
−2

Note: All hydrogen atoms have an oxidation number of +1 in this component.

Figure 5.80 Calculating the oxidation number of atoms with covalent bonds.

In organic chemistry, it is often more convenient to define oxidation and reduction in terms of other criteria more relevant to organic transformations involving carbon (Table 5.7). An **organic oxidation reaction** is a process that results in a *gain* of either atomic oxygen ([O]), molecular oxygen (O_2), or a molecular halogen (X_2), or a *loss* of molecular hydrogen (H_2). In contrast, an **organic reduction reaction** is the reciprocal set of processes: a *gain* of H_2 or hydride ion, or a *loss* of [O], O_2, or X_2.

Table 5.7 Identifying an organic oxidation reaction and organic reduction reaction.

Reference carbon

	Gain of:		Loss of:	
	Atoms	**Example**	**Atoms**	**Example**
Oxidation	[O]	H_3C-H → ([O]) H_3C-O-H	H_2	H_3C-OH → (H_2) $H_2C=O$
	O_2	—		
	X_2	alkene → (X_2) vicinal dihalide (X, X)		
Reduction	H_2	alkyne → (H_2) alkene (H, H)	[O]	$HCOOH$ → ([O]) $H_2C=O$
	$H:^{\ominus}$	$H_2C=O$ → ($H:^{\ominus}$) $H_3C-O^{\ominus}$	O_2	—
			X_2	vicinal dihalide (X, X) → (X_2) alkene

Note that some types of organic transformations are *neither* an oxidation *nor* a reduction. Some examples of reactions that fall into this category include the gain *or* loss of:

- proton (H^{+})
- water (H_2O)
- hydrohalic acid (HX).

Concept Check 5.11

With respect to the indicated carbon atom, is the reaction shown an oxidation, a reduction, or neither an oxidation nor a reduction?

$R-C(=O)-N(CH_3)_2 \longrightarrow R-CH_2-N(CH_3)_2$

Solution

Note: The appendix contains the answer.

5.7.02 Oxidation States of Carbon

Carbon is a Group 4A element with four valence electrons; consequently, a carbon atom could foreseeably either gain *or* lose four valence electrons to be isoelectronic with a noble gas. Therefore, the oxidation number of a carbon atom can range from −4 (CH_4, fully reduced) to +4 (CO_2, fully oxidized) (Figure 5.81).

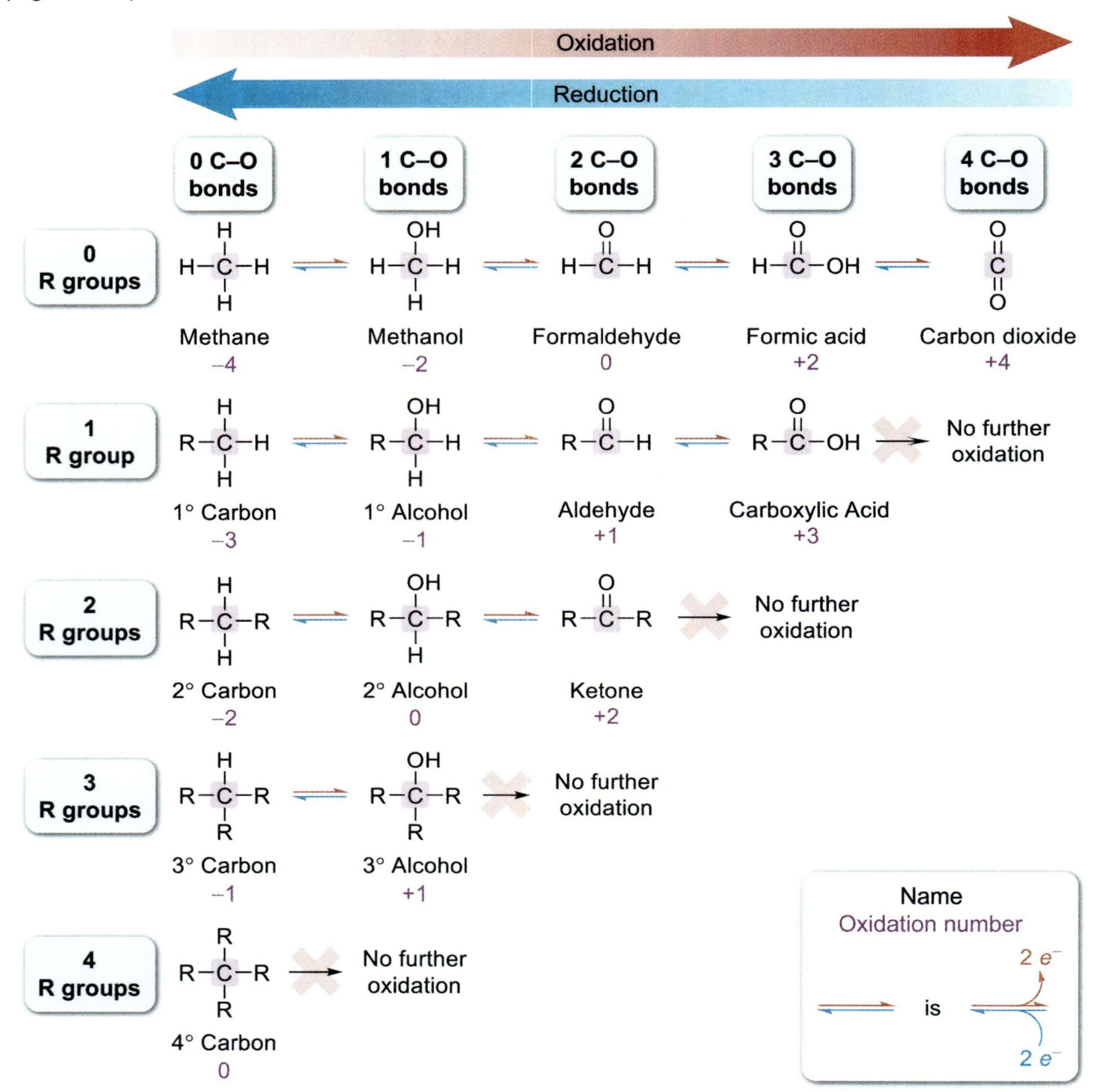

Figure 5.81 The oxidation states of carbon with respect to the number of carbon-carbon and carbon-oxygen bonds.

Redox transformations between the extremes are typically *two-electron* processes, where an oxidation results in a loss of two e^- and a reduction results in a gain of two e^-. Successive two-electron oxidations of an alkane carbon atom generate the following oxidation state groups:

- alcohol (one C–O bond)
- carbonyl (ketone or aldehyde, two C–O bonds)

- carboxylic acid (three C–O bonds)
- carbon dioxide (four C–O bonds)

The number of alkyl (R) groups on the carbon atom impacts the maximum amount of oxidation that can occur. Although many ways exist to exchange a C–H bond for a C–O bond through an oxidation reaction, oxidatively breaking a carbon-carbon bond is *much* more difficult. A 1° carbon atom becomes a carboxylic acid, a 2° carbon becomes a ketone, and a 3° carbon becomes a 3° alcohol. Methane is the only carbon atom that can be fully oxidized to carbon dioxide through organic reactions, and quaternary (4°) carbon atoms cannot be easily oxidized.

The oxidation states of carbon can also be viewed from the perspective of multiple bonds (Figure 5.82). Each of these multiple-bond redox transformations corresponds to a two-electron process. When working with oxidation numbers for multiple-bond transformations, *both* carbon atoms that participate in the multiple bond must be included. The two-electron oxidation of an alkane forms an alkene (one pi bond), and the two-electron oxidation of an alkene forms an alkyne (two pi bonds).

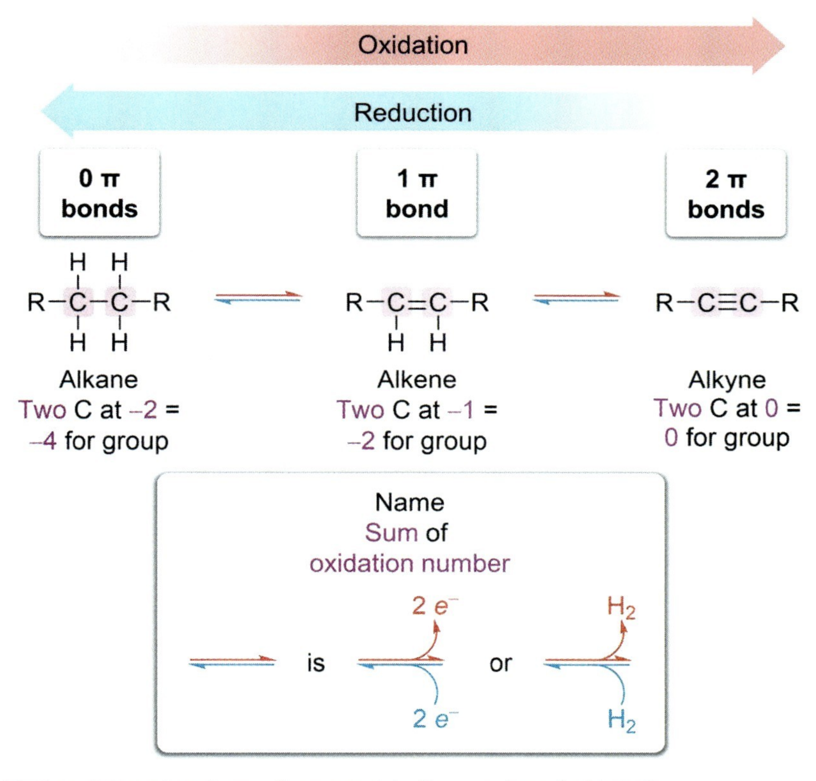

Figure 5.82 The oxidation states of carbon with respect to the number of pi bonds.

Concept Check 5.12

Apply an understanding of oxidation numbers to demonstrate how the addition of water to an alkene is neither an oxidation reaction nor a reduction reaction.

Solution

Note: The appendix contains the answer.

5.7.03 Oxidizing Agents and Reducing Agents

Most synthetic redox transformations utilize a chemical reagent to promote a redox reaction on a carbon-containing reactant; these reagents are called oxidizing and reducing agents (Figure 5.83). An **oxidizing agent** is a chemical entity that typically has a *high oxidation state* and causes *other substances* to become oxidized. Through the process of oxidizing another substance, the oxidizing agent becomes reduced by the electrons provided by that substance.

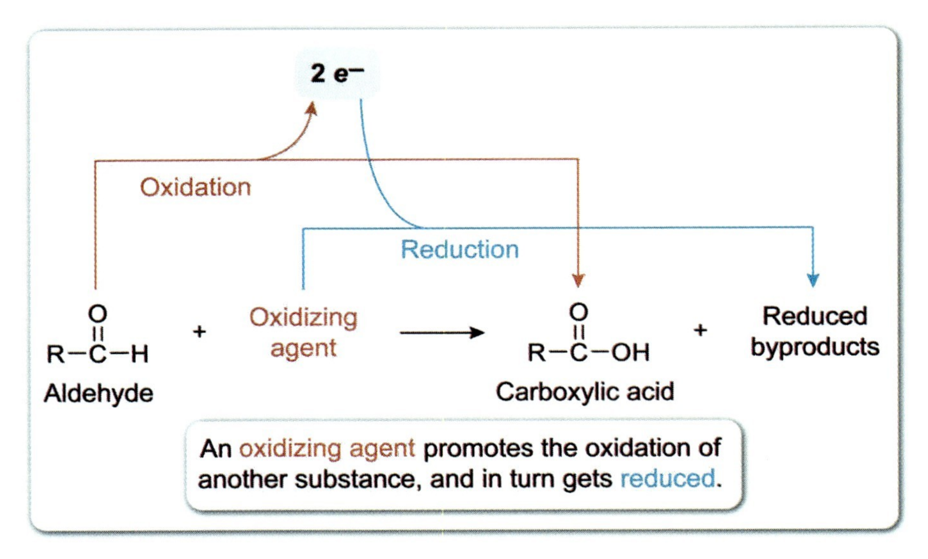

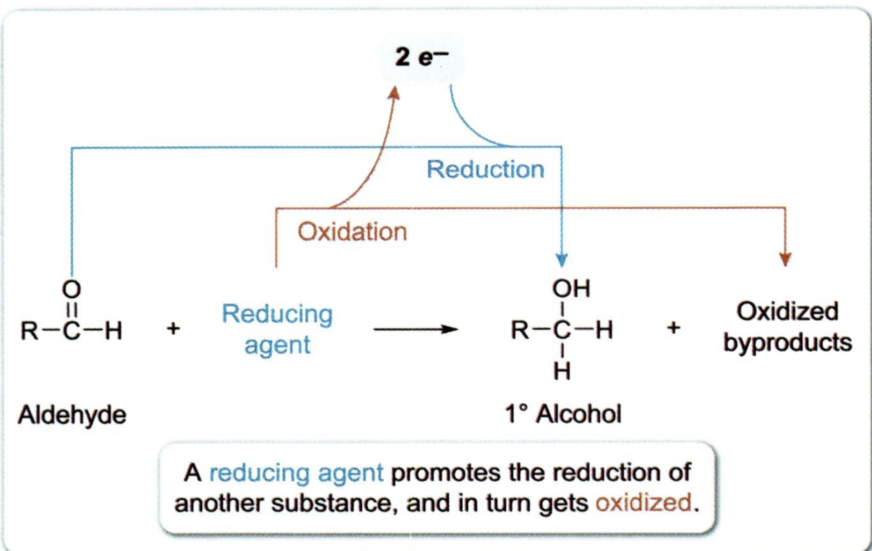

Figure 5.83 The general reaction of an aldehyde with an oxidizing agent and a reducing agent.

In comparison, a reducing agent is a chemical entity that typically has a *low oxidation state* and causes *other substances* to become reduced. Through the process of reducing another substance (ie, giving electrons away), the reducing agent becomes oxidized.

Several oxidizing agents are frequently encountered in organic chemistry (Table 5.8). Chromic acid (H_2CrO_4) is a powerful oxidizing agent containing a Cr^{6+} ion that promotes the *complete* oxidation of carbon atoms bearing at least one C–O bond. The dichromate ion ($Cr_2O_7^{2-}$) and Jones reagent (CrO_3 in H_2SO_4) both generate chromic acid within a reaction and serve similar functional purposes. The permanganate ion (MnO_4^-) performs similar oxidations on molecules containing C–O bonds, in addition to its ability to oxidize the pi bonds in alkenes and alkynes.

Table 5.8 Common oxidizing agents in organic chemistry.

	Oxidizing agent	Formula	Synthetic uses
Strong oxidants	Chromic acid	H_2CrO_4	
	Dichromate ion (Sodium dichromate or potassium dichromate)	$Cr_2O_7^{2-}$ ($Na_2Cr_2O_7$ or $K_2Cr_2O_7$)	1° Alcohol → Carboxylic acid Aldehyde → Carboxylic acid 2° Alcohol → Ketone
	Jones reagent	CrO_3 in H_2SO_4	
	Permanganate ion (Potassium permanganate)	MnO_4^- ($KMnO_4$)	1° Alcohol → Carboxylic acid Aldehyde → Carboxylic acid 2° Alcohol → Ketone
Mild oxidants	Hypochlorous acid	HClO	1° Alcohol → Aldehyde 2° Alcohol → Ketone
	Pyridinium chlorochromate	PCC	
	Silver (I) ion (Tollens's test)	Ag^+	Aldehyde → Carboxylic acid

Hypochlorous acid (HClO), hydrogen peroxide (H_2O_2), and pyridinium chlorochromate (PCC) are relatively mild oxidizing agents for alcohols and are among the *few* methods that chemists use to *selectively* oxidize a 1° alcohol to an aldehyde. Since an aldehyde is generally *easier* to oxidize than an alcohol, most *other* oxidizing agents (eg, chromic acid) *overoxidize* the aldehyde intermediate all the way to a carboxylic acid.

Some common reducing agents in organic chemistry are listed in Table 5.9. Many of these reducing agents are a source of the hydride ion. Lithium aluminum hydride (LAH, $LiAlH_4$) is a powerful reducing agent that fully reduces a carboxylic acid or most carboxylic acid derivatives to a 1° alcohol. The exception to this generalization is an amide, which is reduced to an amine. Ketones and aldehydes are reduced to a 2° alcohol and a 1° alcohol, respectively.

Table 5.9 Common reducing agents in organic chemistry.

	Reducing agent	Formula	Synthetic uses
Strong reductant	Lithium aluminum hydride (LAH)	$LiAlH_4$	Carboxylic acid → 1° Alcohol Acid chloride → 1° Alcohol Anhydride → 1° Alcohol Ester → 1° Alcohol Nitrile → 1° Amine Amide → Amine Ketone → 2° Alcohol Aldehyde → 1° Alcohol
Mild reductants	Borohydride ion (Lithium borohydride or sodium borohydride)	BH_4^- ($LiBH_4$ or $NaBH_4$)	Ketone → 2° Alcohol Aldehyde → 1° Alcohol
	Lithium tri-*tert*-butoxy aluminum hydride	$LiAlH(O\mathit{t}Bu)_3$	Acid chloride → Aldehyde
Specialized reductants	Borane	BH_3	Carboxylic acid → 1° Alcohol
	Hydrogen gas (with transition metal catalyst)	H_2 (with Pd or Pt)	Alkyne → Alkane Alkene → Alkane

By comparison, the borohydride ion (BH_4^-) is a mild source of hydride ions and is only capable of reducing a carbonyl (ie, ketone or aldehyde) to an alcohol; BH_4^- does not generally react with molecules at the carboxylic acid (or derivative) oxidation state. The reagent lithium tri-*tert*-butoxy aluminum hydride ($LiAl(O\mathit{t}Bu)_3$) is a sterically hindered source of the hydride ion that *selectively* reduces an acid halide to an aldehyde.

Borane (BH_3) provides a unique ability to *selectively* reduce a carboxylic acid to a 1° alcohol, even in the presence of *other* carbonyl-containing functional groups (eg, carboxylic acid derivatives, ketones, aldehydes). The use of hydrogen gas (H_2) with one of several transition metal catalysts (eg, platinum, palladium) promotes the complete reduction of the carbon-carbon pi bonds in alkenes and alkynes, providing an alkane product.

END-OF-UNIT MCAT PRACTICE

Congratulations on completing **Unit 1: Introduction to Organic Chemistry**.

Now you are ready to dive into MCAT-level practice tests. At UWorld, we believe students will be fully prepared to ace the MCAT when they practice with high-quality questions in a realistic testing environment.

The UWorld Qbank will test you on questions that are fully representative of the AAMC MCAT syllabus. In addition, our MCAT-like questions are accompanied by in-depth explanations with exceptional visual aids that will help you better retain difficult MCAT concepts.

TO START YOUR MCAT PRACTICE, PROCEED AS FOLLOWS:

1) Sign up to purchase the UWorld MCAT Qbank
   IMPORTANT: You already have access if you purchased a bundled subscription.
2) Log in to your UWorld MCAT account
3) Access the MCAT Qbank section
4) Select this unit in the Qbank
5) Create a custom practice test

Unit 2 Functional Groups and Their Reactions

Chapter 6 Hydrocarbons

6.1 Physical Properties of Hydrocarbons

6.1.01 Boiling Point, Solubility, and Density

6.2 Alkanes

6.2.01 Structural Features of Alkanes

6.3 Alkenes

6.3.01 Orbital Structure of an Alkene

6.4 Alkynes

6.4.01 Orbital Structure of an Alkyne

6.5 Conjugated and Aromatic Compounds

6.5.01 Conjugation and Conjugated Compounds
6.5.02 Aromaticity and Aromatic Compounds
6.5.03 Applications of Conjugation and Aromaticity

Chapter 7 Alkyl Halides, Ethers, and Sulfur-Containing Groups

7.1 Alkyl Halides

7.1.01 Physical Properties of Alkyl Halides
7.1.02 Alkyl Halides as Electrophiles
7.1.03 Formation of Grignard Reagents

7.2 Ethers

7.2.01 Physical Properties of Ethers
7.2.02 Use of Ethers as Solvents
7.2.03 Williamson Ether Synthesis

7.3 Thiols, Thioethers, Thioesters, and Disulfides

7.3.01 Physical Properties of Sulfur-Containing Functional Groups
7.3.02 Acidity and Nucleophilicity of Thiols
7.3.03 Disulfide Bonds

Chapter 8 Alcohols

8.1 Structure and Physical Properties of Alcohols

8.1.01 Structural Characteristics of Alcohols
8.1.02 Classification of Alcohols
8.1.03 Physical Properties of Alcohols

8.2 Synthesis of Alcohols

8.2.01 Alcohols through Substitution Reactions
8.2.02 Grignard Reaction with an Aldehyde or Ketone
8.2.03 Reduction of Carbonyl-Containing Groups

8.3 Reactions of Alcohols

8.3.01 Oxidation of Alcohols
8.3.02 Acidity of Alcohols

Lesson 6.1

Physical Properties of Hydrocarbons

Introduction

Unit 2 builds on the foundational topics from Unit 1 and focuses on the study of chemical reactions, broken down by functional group. This chapter examines the **hydrocarbon** functional groups: alkanes, alkenes, alkynes, and conjugated and aromatic molecules. This lesson examines the physical properties of hydrocarbons.

6.1.01 Boiling Point, Solubility, and Density

Boiling Point

For vaporization to occur, the intermolecular forces holding individual molecules together in the liquid phase must be overcome (Concept 2.4.01). The magnitude of the overall London dispersion forces is the primary factor that determines the boiling point for many classes of organic molecules. This generalization is *especially* relevant for nonpolar hydrocarbons, which do not experience significant dipole-dipole forces.

Two trends particularly applicable to hydrocarbons are (Table 6.1):

- The boiling point *increases* as molecular weight *increases*.
- The boiling point *decreases* with *increased* branching (and *decreased* surface area).

Table 6.1 Boiling points of six-carbon hydrocarbons.

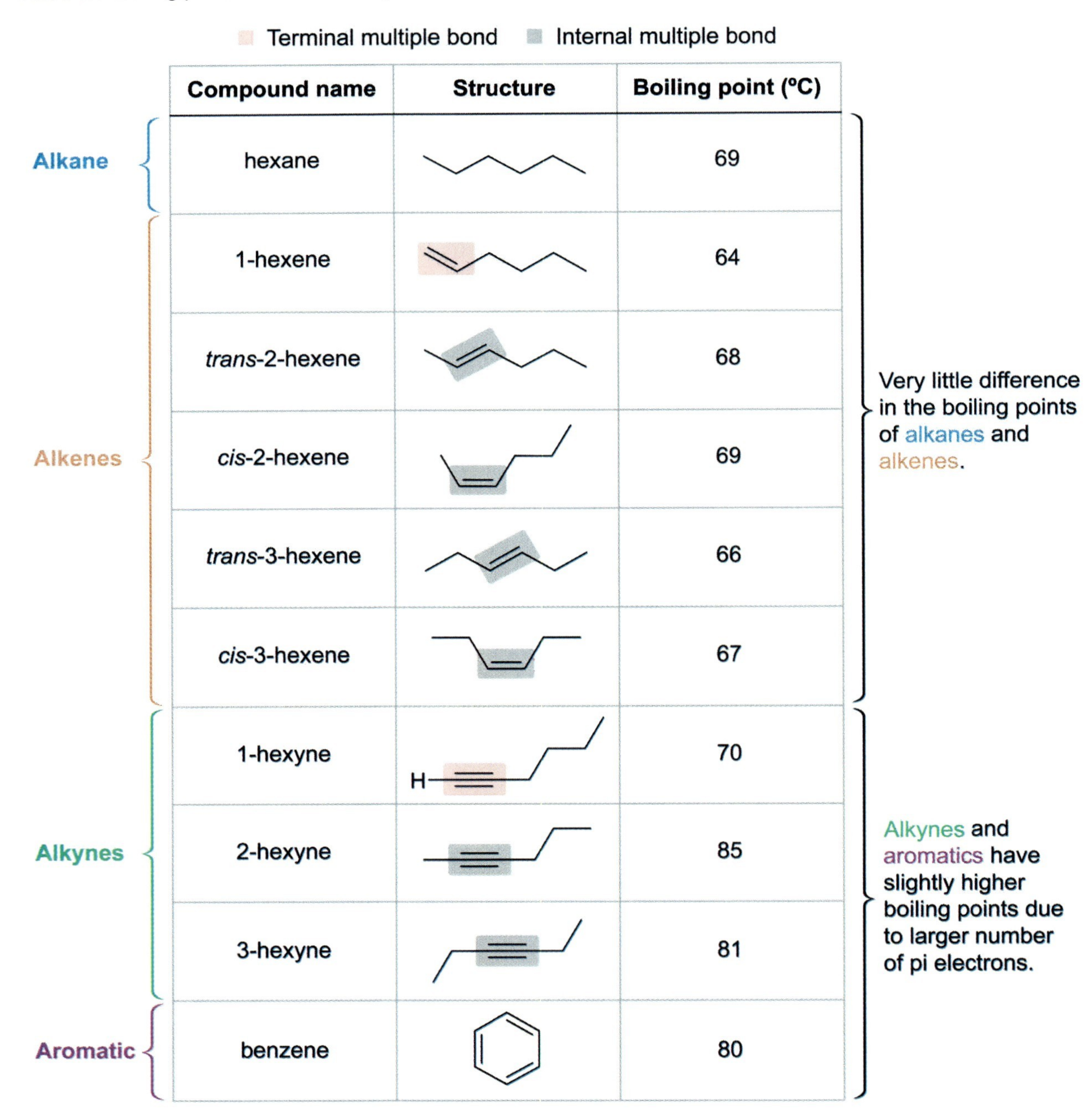

Compound name	Structure	Boiling point (°C)
hexane		69
1-hexene		64
trans-2-hexene		68
cis-2-hexene		69
trans-3-hexene		66
cis-3-hexene		67
1-hexyne		70
2-hexyne		85
3-hexyne		81
benzene		80

Whereas alkane and alkene boiling points are generally similar, the *cis* and *trans* isomers of a disubstituted alkene have slightly different boiling points. A *cis*-alkene has a small net dipole moment and demonstrates dipole-dipole interactions; a *trans*-alkene has no dipole moment (the individual bond moments cancel each other). Consequently, a *cis*-alkene has a slightly *higher* boiling point than its *trans* isomer.

The boiling points of alkynes are usually slightly *higher* because the triple bond contains a greater number of pi electrons and experiences stronger London dispersion forces. Similarly, the boiling points of aromatic compounds also tend to be slightly *higher* due to an increase in attractive forces between the pi electrons of nearby aromatic rings—a nonpolar interaction known as pi stacking.

Disubstituted aromatic compounds have boiling points that depend on the isomeric relationship of the two groups (ie, *ortho*, *meta*, *para*). The *ortho* regioisomer has the largest dipole moment, is the most polar, and has the highest boiling point (Table 6.2).

Table 6.2 Boiling points of disubstituted benzene derivatives.

Compound	Structure	Boiling point (°C)
o-xylene	CH_3, CH_3	144
m-xylene	CH_3, CH_3	139
p-xylene	CH_3, CH_3	138

The *ortho* isomer is the most polar and has the highest boiling point.

Solubility

Because hydrocarbons are *nonpolar* compounds, they typically have the greatest solubility in *nonpolar* organic solvents (eg, hexane). Aromatic hydrocarbons typically have slightly higher solubility in aromatic solvents (eg, benzene, toluene) in which pi stacking with the solvent is possible.

Density

Alkanes, alkenes, and alkynes have densities between 0.6 g/mL and 0.7 g/mL, and aromatic hydrocarbons have densities between 0.8 g/mL and 0.9 g/mL. In nearly all cases, hydrocarbons are less dense than water (1.0 g/mL).

Lesson 6.2

Alkanes

Introduction

An alkane has only carbon-carbon and carbon-hydrogen single bonds (ie, only sigma bonds). This lesson discusses the classification of alkane carbon atoms with a focus on linear alkanes, branched alkanes, and cycloalkanes.

6.2.01 Structural Features of Alkanes

An alkane can be classified by the arrangement of the carbon atoms in the molecule (Figure 6.1). A **linear, straight-chain alkane** has carbon atoms connected to one another in a straight line and contains *only* 1° and 2° carbon atoms. A **branched alkane** has one or more alkyl groups on the parent chain and can contain a combination of 1°, 2°, 3°, and 4° carbon atoms. Most sigma bonds in an alkane can rotate to generate different conformations, which have different torsional energy (see Concept 3.2.02).

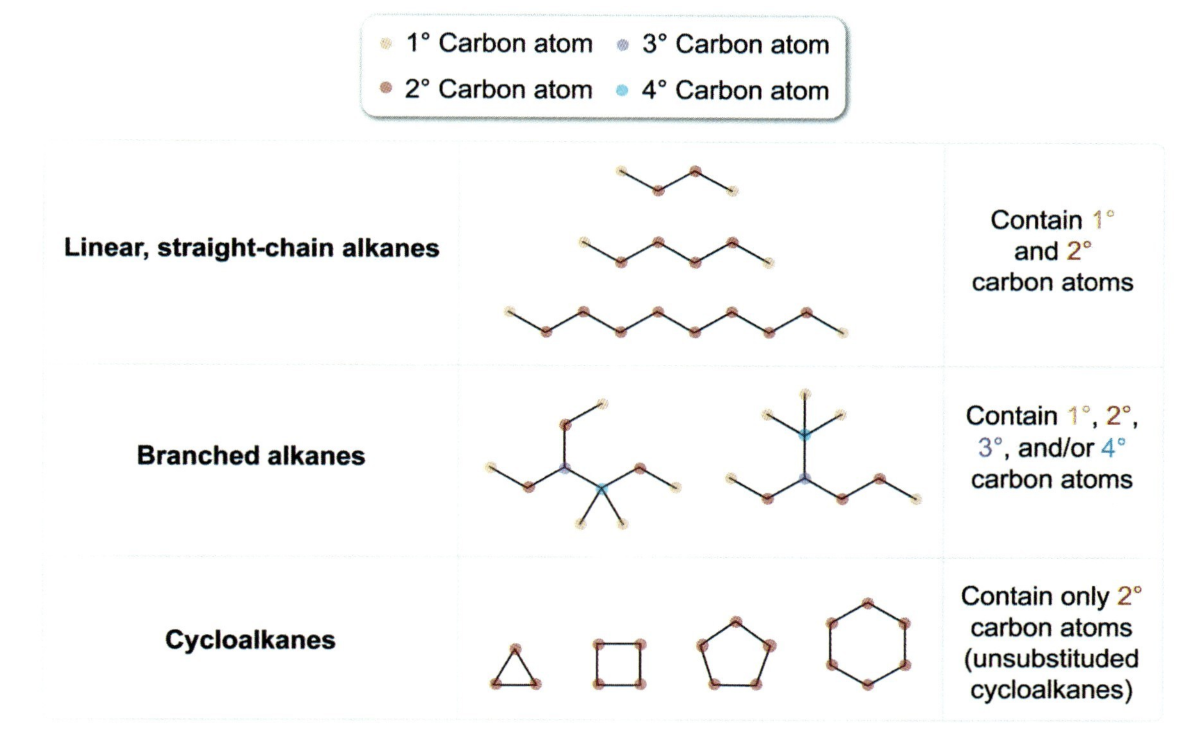

Figure 6.1 Classification of alkanes.

A **cycloalkane** is a cyclic variant of an alkane that contains at least one ring of carbon atoms; unsubstituted cycloalkanes contain only 2° carbon atoms. The different ring sizes have varying amounts of ring strain, where the most stable cycloalkane rings have either five or six carbon atoms (see Concept 3.2.04).

Five- and six-carbon cycloalkane rings adopt envelope and chair conformations, respectively, minimizing torsional strain and providing near-ideal (109.5°) bond angles. More information on cycloalkane conformations is available in Lesson 3.2.

Concept Check 6.1

Identify the 3° carbon atoms in the substituted cyclohexane shown.

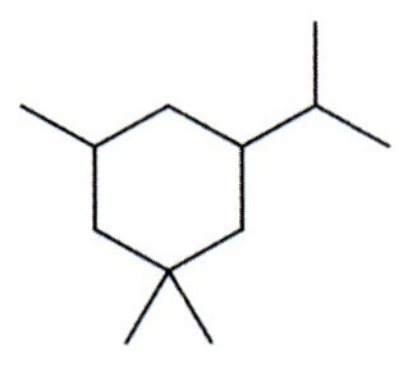

Solution

Note: The appendix contains the answer.

Lesson 6.3

Alkenes

Introduction

Alkenes are hydrocarbons with at least one double bond. This lesson examines the orbital structure of alkenes.

6.3.01 Orbital Structure of an Alkene

Because the double-bonded carbon atoms of an **alkene** are each attached to *three* atoms and have zero nonbonding electrons, these atoms have three electron domains. An atom with three electron domains is sp^2 hybridized and has one unhybridized *p* orbital (see Concept 1.2.02 and Concept 2.1.02). All sp^2 hybridized atoms have a trigonal planar electron domain geometry and 120° bond angles.

A double bond is composed of **one sigma (σ) bond** and **one pi (π) bond** (Concept 1.4.04). In an alkene double bond, the σ bond is formed by head-to-head overlap of an sp^2 hybrid orbital from each carbon atom, and the π bond is formed by side-to-side overlap of the *parallel* unhybridized *p* orbitals (Figure 6.2).

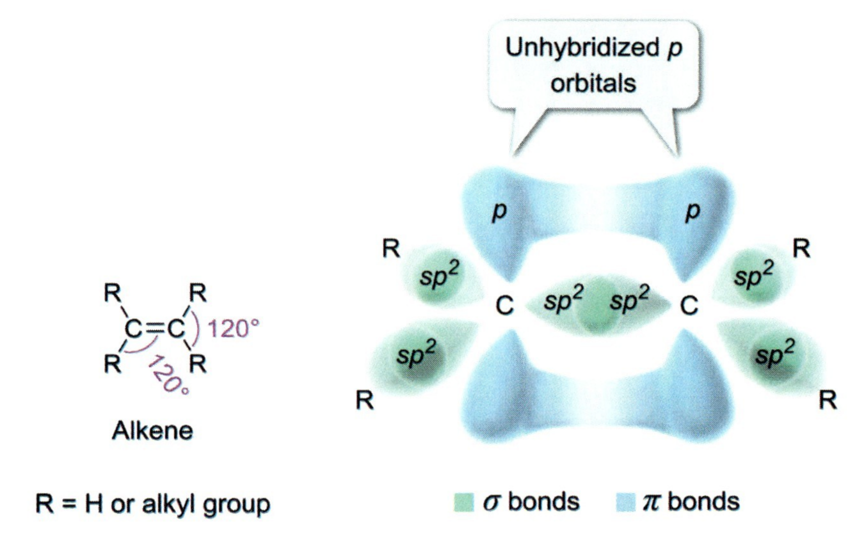

Figure 6.2 Orbital structure of an alkene.

Unlike the σ bonds in an alkane, the sigma bond of an alkene double bond cannot freely rotate because the overlapping *p* orbitals must remain parallel to keep the π bond intact. As a result, the six atoms of an alkene (*two* carbon atoms plus *four* atoms directly attached to the alkene carbons) are all **coplanar** (ie, located within the same plane).

Lesson 6.4

Alkynes

Introduction

Alkynes are molecules with at least one carbon-carbon triple bond. This lesson examines the structure of an alkyne from an atomic orbital perspective.

6.4.01 Orbital Structure of an Alkyne

Each triple-bonded carbon atom in an **alkyne** is bonded to *two* atoms total and has zero nonbonding electrons. Therefore, triple-bonded atoms have two electron domains, are *sp* hybridized, and have *two* unhybridized *p* orbitals. The *sp* hybridized carbon atoms in an alkyne have a linear electron domain geometry, with a 180° bond angle between the *sp* hybrid orbitals.

A triple bond is composed of **one sigma (σ) bond** and **two pi (π) bonds** (see Concept 1.4.04). The σ bond in a triple bond is formed by head-to-head overlap of an *sp* hybrid orbital from each carbon atom, and each π bond is formed by side-to-side overlap of *parallel* unhybridized *p* orbitals. The two π bonds are perpendicular to each other and form a cylinder of π electron density around the axis of the *sp*–*sp* σ bond (Figure 6.3).

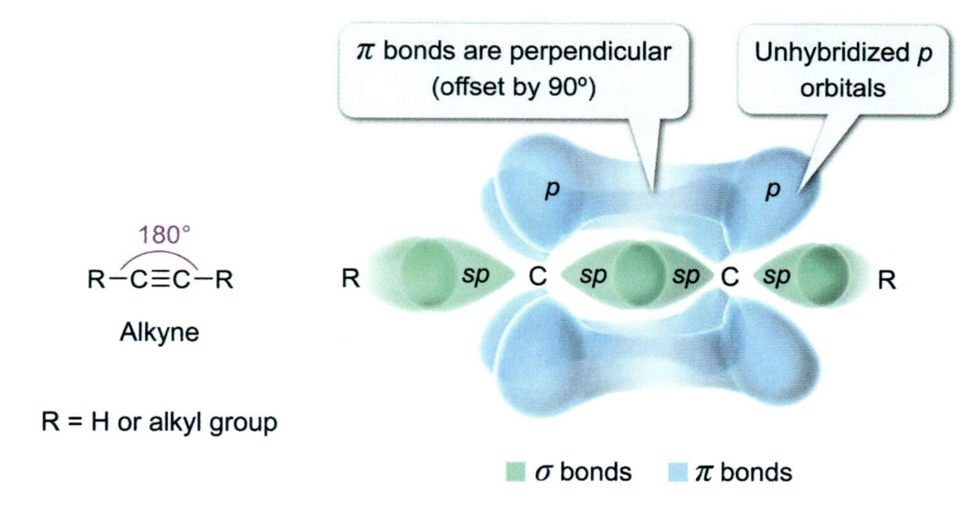

Figure 6.3 Orbital structure of an alkyne.

Like an alkene, the σ bond within the triple bond of an alkyne is unable to freely rotate because the overlapping *p* orbitals must remain parallel to keep the π bond intact. As a result, the four atoms of an alkyne (*two* carbon atoms plus *two* atoms directly attached to the alkyne carbons) are **colinear** (ie, in a linear relationship).

Lesson 6.5

Conjugated and Aromatic Compounds

Introduction

An aromatic compound is a molecule containing an aromatic structure, such as a phenyl group. Although alkenes also contain pi bonds between carbon atoms, the pi bonds in these structures have *vastly* different chemical and physical properties.

This lesson introduces the topics of conjugation and aromaticity and explores applications of conjugation and aromaticity in organic chemistry.

6.5.01 Conjugation and Conjugated Compounds

A carbon-carbon double bond can be further classified into one of three categories (Figure 6.4):

- **Isolated alkenes** are *unable* to delocalize their pi electrons through resonance because both alkene carbon atoms are *at least* two sigma bonds away from the nearest pi bond or lone pair.
- **Conjugated alkenes** *are able* to delocalize their pi electrons through resonance because at least one alkene carbon atom is *exactly* one sigma bond away from the nearest pi bond or lone pair.
- **Aromatic molecules** are a subset of conjugated molecules that contain a cyclic structure (among other criteria described in Concept 6.5.02).

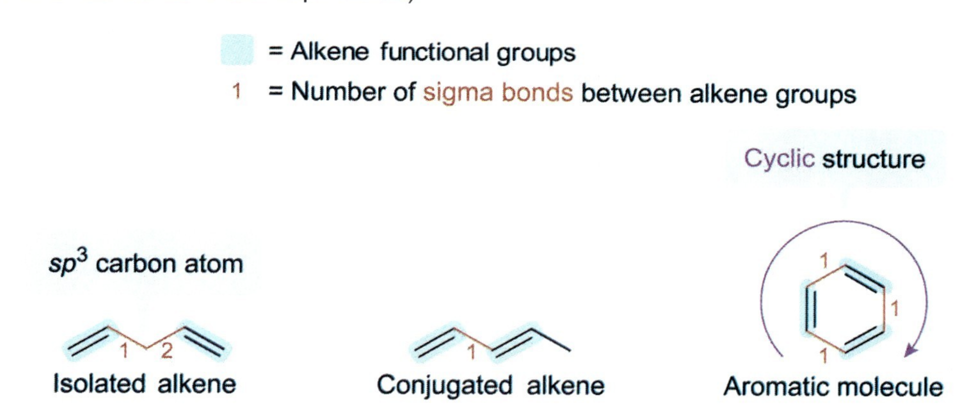

Figure 6.4 Isolated, conjugated, and aromatic molecules.

Resonance delocalization is generally a stabilizing feature that lowers a molecule's energy relative to its localized variant (Concept 2.7.01). The conjugated molecule penta-1,3-diene is approximately 27 kJ/mol lower in energy than the isolated alkene penta-1,4-diene (Table 6.3). The difference in energy (and therefore stability) between isolated and conjugated alkenes is called the **resonance energy**.

Table 6.3 Resonance energy of conjugated alkenes.

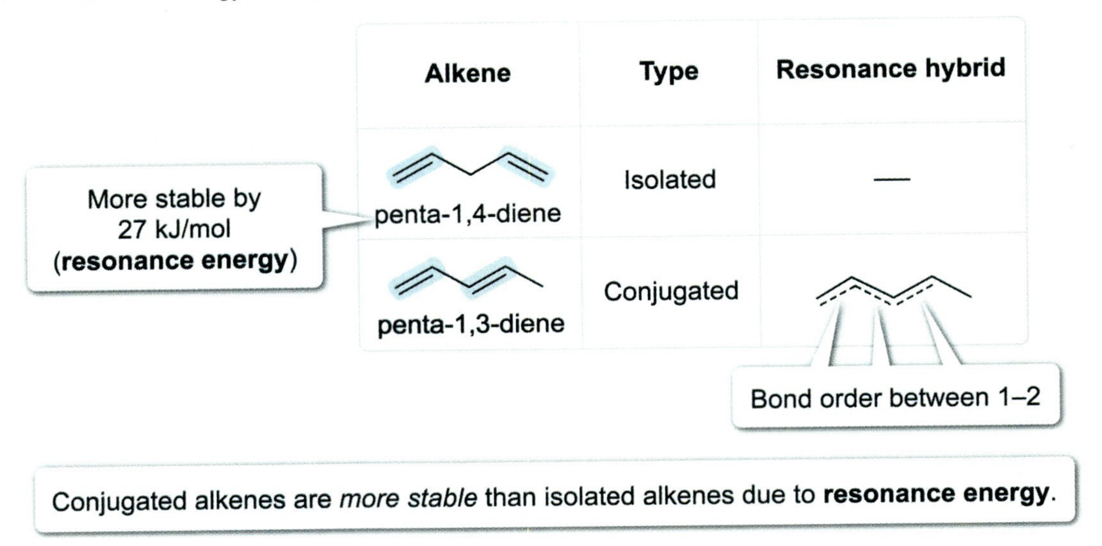

Alkene	Type	Resonance hybrid
penta-1,4-diene	Isolated	—
penta-1,3-diene	Conjugated	

Conjugation also affects the **bond order** between atoms in a conjugated system. Although the line-angle structure of penta-1,3-diene *appears* to indicate two full carbon-carbon double bonds (C1-C2 and C3-C4) separated by a single bond (C2-C3), the resonance hybrid shows that a *partial* double bond exists between *all* these atom pairs (designated by dashed lines).

As introduced in Lesson 1.4, **π molecular orbitals** distribute electron density *across* all adjacent nuclei capable of side-to-side overlap unhybridized *p* orbitals, not just the two atoms shown participating in a multiple bond in the line-angle structure. The number of unhybridized *p* orbitals is equal to the number of atoms in the conjugated system (ie, one unhybridized *p* orbital per atom).

The simplest isolated alkene (ie, ethene) has two carbon atoms, each with one unhybridized *p* orbital. The linear combination of *two p* orbitals generates *two* π molecular orbitals (Figure 6.5). Combination of *p* orbitals with aligned phases forms a **bonding π molecular orbital**, whereas combination of *p* orbitals with nonaligned phases forms an **antibonding π molecular orbital** (which has a node between the nuclei).

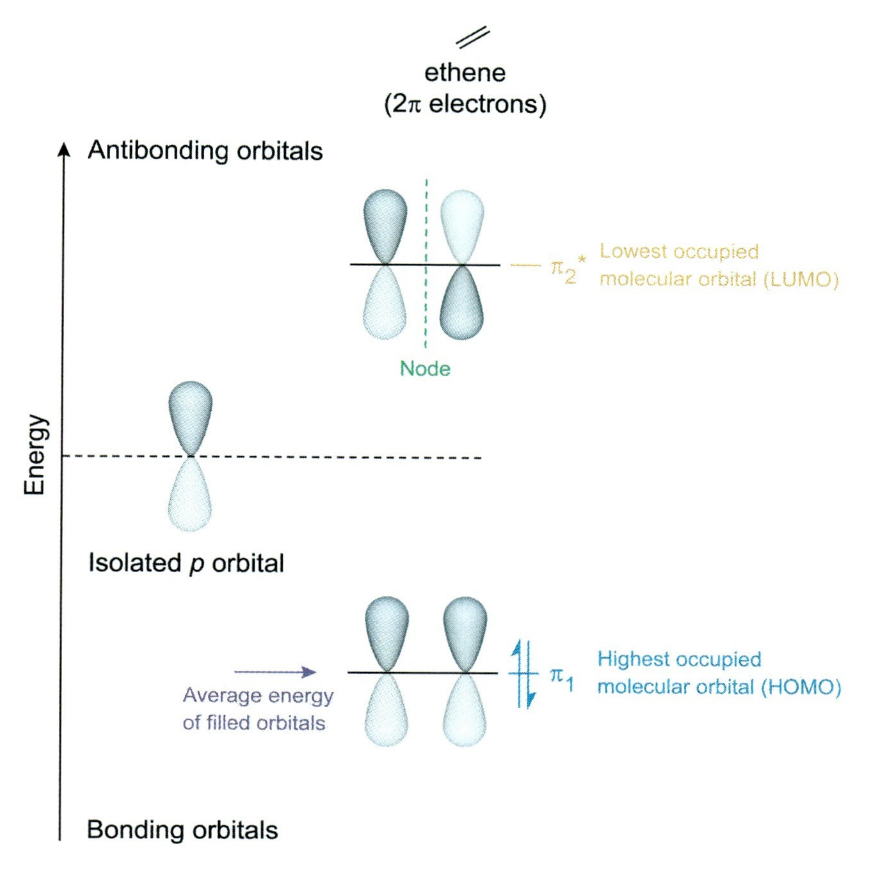

Figure 6.5 Pi molecular orbital diagram for ethene.

Pi molecular orbitals are numbered based on their energy (1 = lowest energy). Antibonding π molecular orbitals are denoted with an asterisk (*). In the molecular orbital diagram for ethene, the bonding molecular orbital is designated π_1 and the antibonding molecular orbital is designated π_2*. In keeping with the **Aufbau principle**, π molecular orbitals are populated starting with the lowest energy orbital; therefore, both π electrons in ethene are located in the π_1 orbital while the π_2* orbital is vacant.

The π molecular orbital diagrams for conjugated molecules have a greater number of molecular orbitals (due to the larger number of atoms in conjugation) (Figure 6.6). The simplest conjugated alkene (buta-1,3-diene) has four carbon atoms in conjugation and four π molecular orbitals (π_1 to π_4*), whereas hexa-1,3,5-triene has *six* carbon atoms in conjugation and *six* π molecular orbitals (π_1 to π_6*).

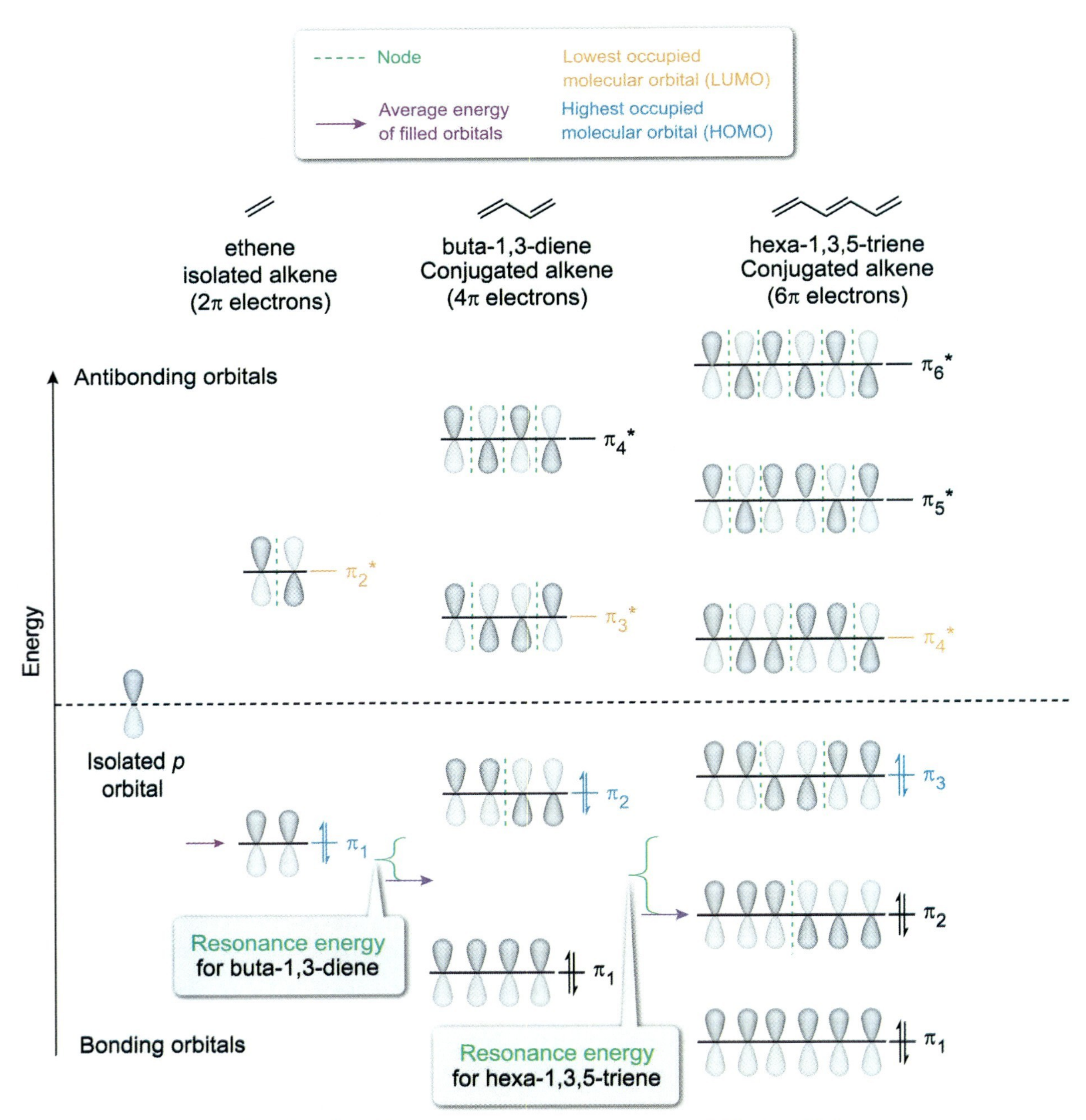

Figure 6.6 Comparison of pi molecular orbital diagrams for isolated and conjugated alkenes.

Conjugated alkenes are more stable than isolated alkenes due to a *decrease* in the average energy of filled orbitals. The average energy of the isolated alkene ethene is the energy of its π_1 orbital, whereas the average energy of the conjugated buta-1,3-diene is halfway between its filled π_1 and π_2 orbitals. The difference in the average energy is the resonance energy, which corresponds to the added stability resulting from the conjugation.

The stabilizing effect *increases* with the number of conjugated atoms. The average energy of hexa-1,3,5-triene is equal to the energy of its π_2 orbital, and the resonance energy for hexa-1,3,5-triene (six conjugated atoms) is *larger* than the resonance energy for buta-1,3-diene (four conjugated atoms).

Concept Check 6.2

If the following two molecules can both be produced from the same chemical reaction, which product is the most likely to be formed if the reaction leads to formation of the lowest-energy product?

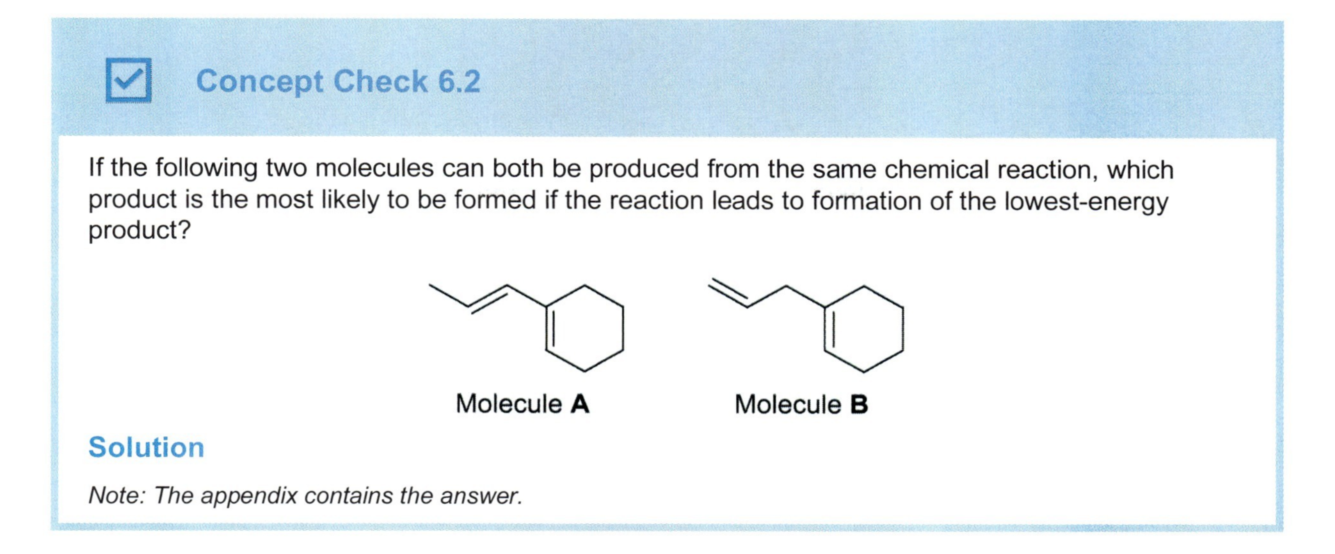

Solution

Note: The appendix contains the answer.

6.5.02 Aromaticity and Aromatic Compounds

The Structure of Benzene

Some conjugated molecules that have a conjugation throughout a ring structure are described as aromatic.

The simplest aromatic molecule is **benzene** (Figure 6.7). Benzene has two *equivalent* resonance structures that differ in the placement of the double bonds. The hybrid resonance structure of benzene has fully delocalized double bonds, and each carbon-carbon bond has a bond order of 1.5. Sometimes a benzene ring is drawn as a six-membered ring containing a circle.

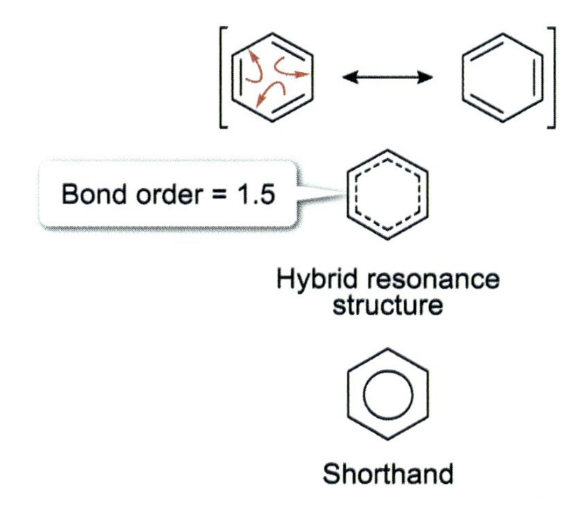

Figure 6.7 Resonance structures of benzene.

Each of the carbon atoms in benzene is sp^2 hybridized and contains one unhybridized *p* orbital. The *p* orbitals in a benzene ring are *all* parallel, and all atoms are **coplanar** (Figure 6.8). The delocalized electrons result in a region of π electron density above and below the plane of the ring.

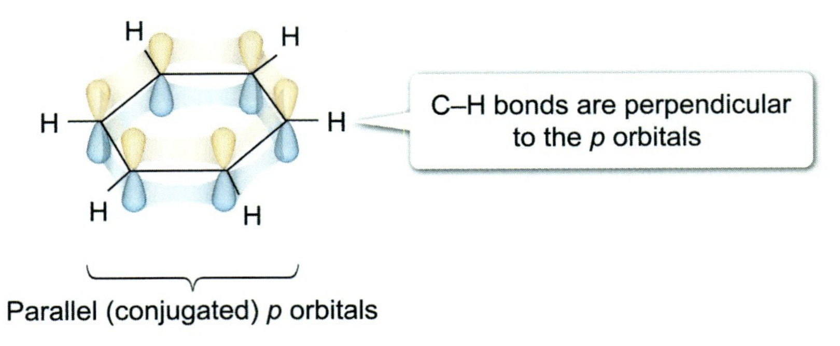

Figure 6.8 Orbital depiction of benzene.

Benzene π Molecular Orbitals

Benzene (a six-atom, *cyclic* conjugated molecule) has a much larger resonance energy (see Concept 6.5.01) than hexa-1,3,5-triene (a six-atom, *linear* conjugated molecule). The difference in resonance energy is best illustrated by examining the π molecular orbitals of benzene.

Although many of the considerations for drawing a molecular orbital (MO) diagram are the same as previously described (see Concept 1.4.01 and Concept 6.5.01), π MO diagrams for cyclic molecules have a few unique features. According to the **polygon rule**, the π MO diagram of a cyclic conjugated molecule has the same shape as the ring, with a MO at each vertex and a single vertex pointed down (ie, lowest energy) (Figure 6.9).

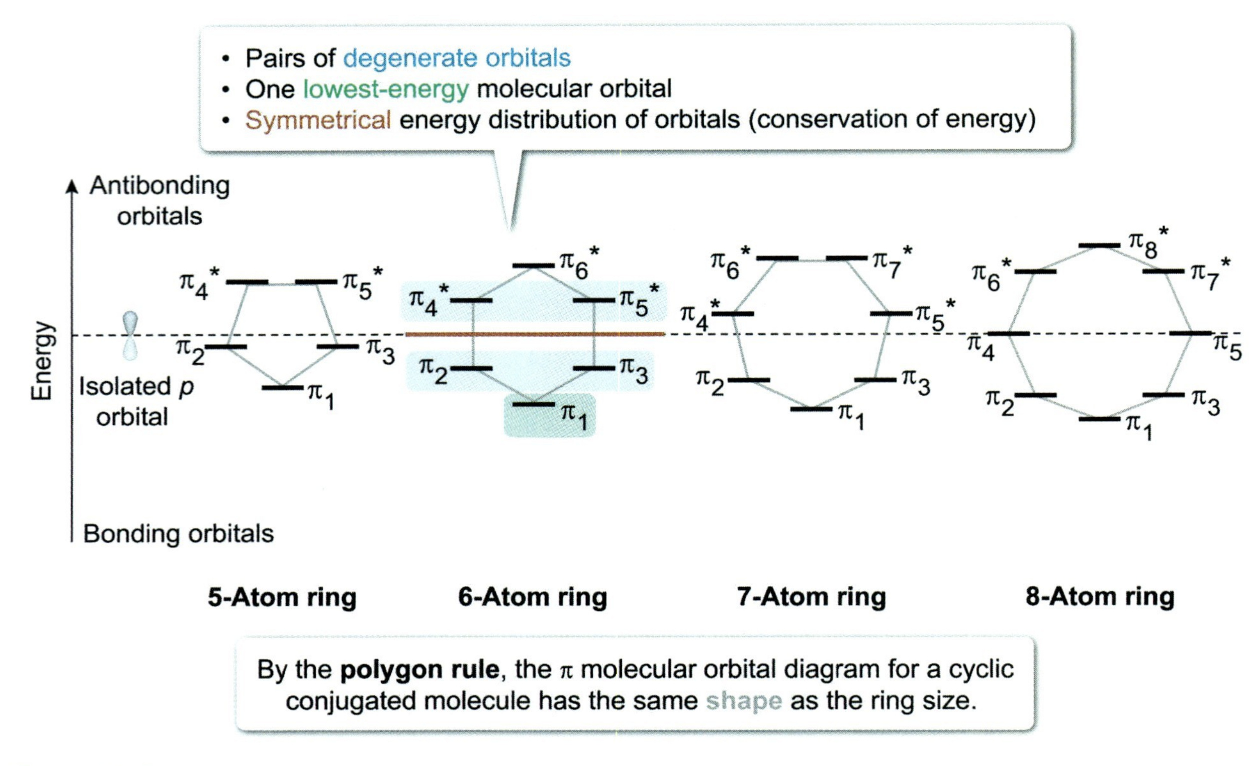

Figure 6.9 The polygon rule.

Because benzene is a *six*-atom ring, it has *six* π MOs symmetrically distributed along the energy axis. Beyond the first, lowest-energy π MO (π_1), conjugated rings have pairs of **degenerate** (equivalent energy) π MOs. Rings with an *even* number of atoms have *one* highest-energy π MO, and rings with an *odd* number of atoms have *two* (degenerate) highest-energy π MOs.

Applying the Aufbau principle, π MOs of the six-membered benzene ring are populated with its six π electrons, starting from the lowest-energy orbital. After the first two π electrons are placed in π_1, the remaining four π electrons fully populate the π_2 and π_3 MOs (Figure 6.10).

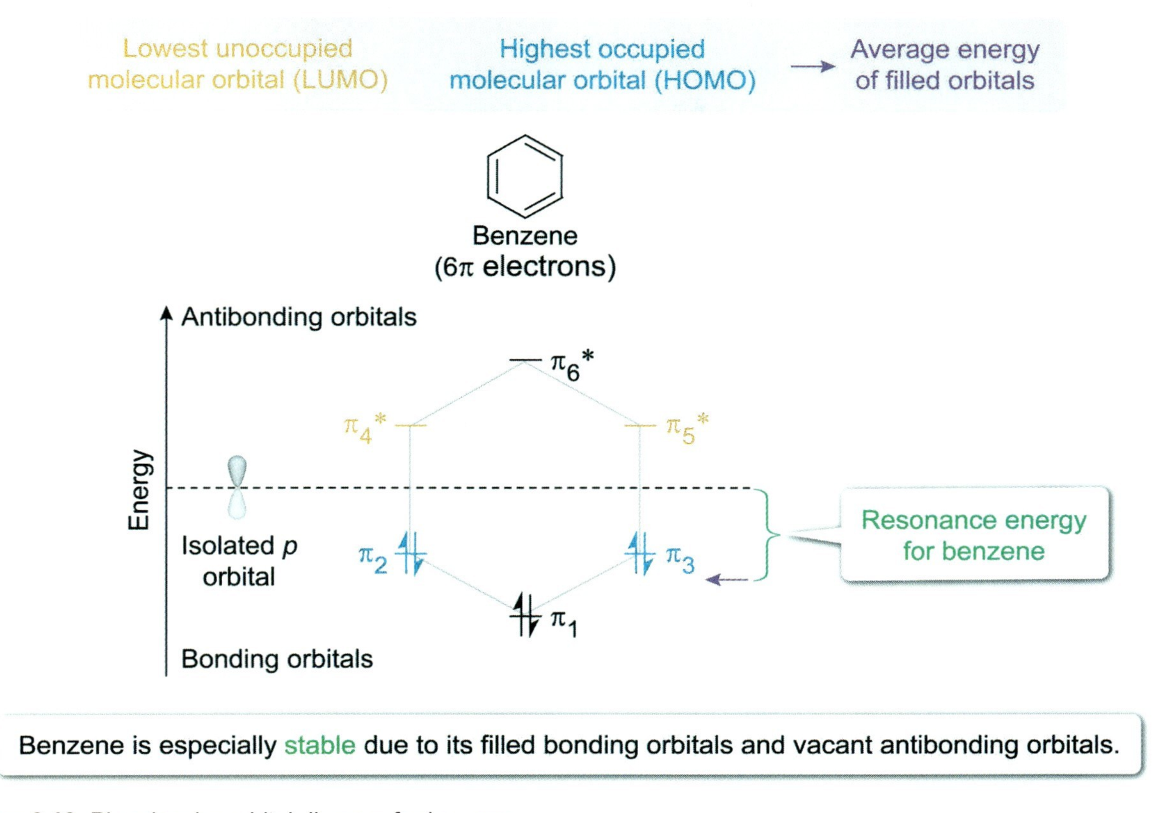

Figure 6.10 Pi molecular orbital diagram for benzene.

The energy of benzene is equal to the weighted average of filled π MOs. For benzene, this occurs between the energy levels of π_2/π_3 and π_1. The difference in energy between this value and the energy of an isolated *p* orbital is the resonance energy for benzene—a much larger value than the resonance energy of hexa-1,3,5-triene.

Nonaromatic, Antiaromatic, and Aromatic Compounds

For a compound to be classified as **aromatic**, it must meet four criteria (Figure 6.11):

1. It must be a cyclic structure containing conjugated π bonds and *p* orbitals.
2. Each atom in the ring must have at least one unhybridized *p* orbital.
3. The ring atoms must be coplanar such that all the unhybridized *p* orbitals are parallel.
4. Delocalization of the π electrons throughout the ring must lead to an *increase* in stability (*decrease* in average energy).

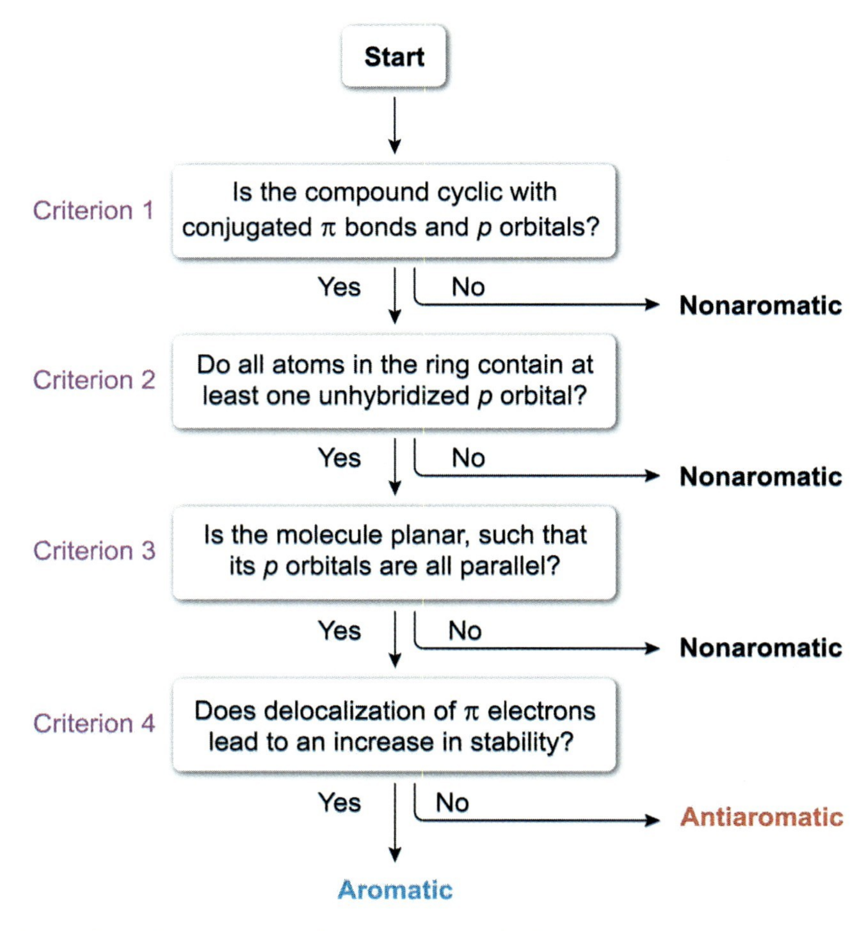

Figure 6.11 Flowchart differentiating nonaromatic, aromatic, and antiaromatic compounds.

A compound that fails to meet any of the first three criteria is **nonaromatic**, a broad designation that encompasses *most* organic molecules and functional groups. A compound that meets the first three criteria but *not* the fourth (ie, π delocalization leads to a *decrease* in stability) is **antiaromatic**. An aromatic compound is especially *stable*, whereas an antiaromatic compound is especially *unstable*.

Cyclobutadiene, a four-membered ring with alternating single and double bonds, is antiaromatic (Figure 6.12). Following the same guidelines used to construct the π MO diagram for benzene and populating them by Hund's rule, the four π electrons in cyclobutadiene are placed in its π_1 (2 e^-), π_2 (1 e^-), and π_3 (1 e^-) orbitals. The electrons in the π_2 and π_3 MOs of cyclobutadiene are *unpaired*, resulting in an *unstable* electron configuration that behaves like a free radical reactive intermediate (see Concept 5.2.03).

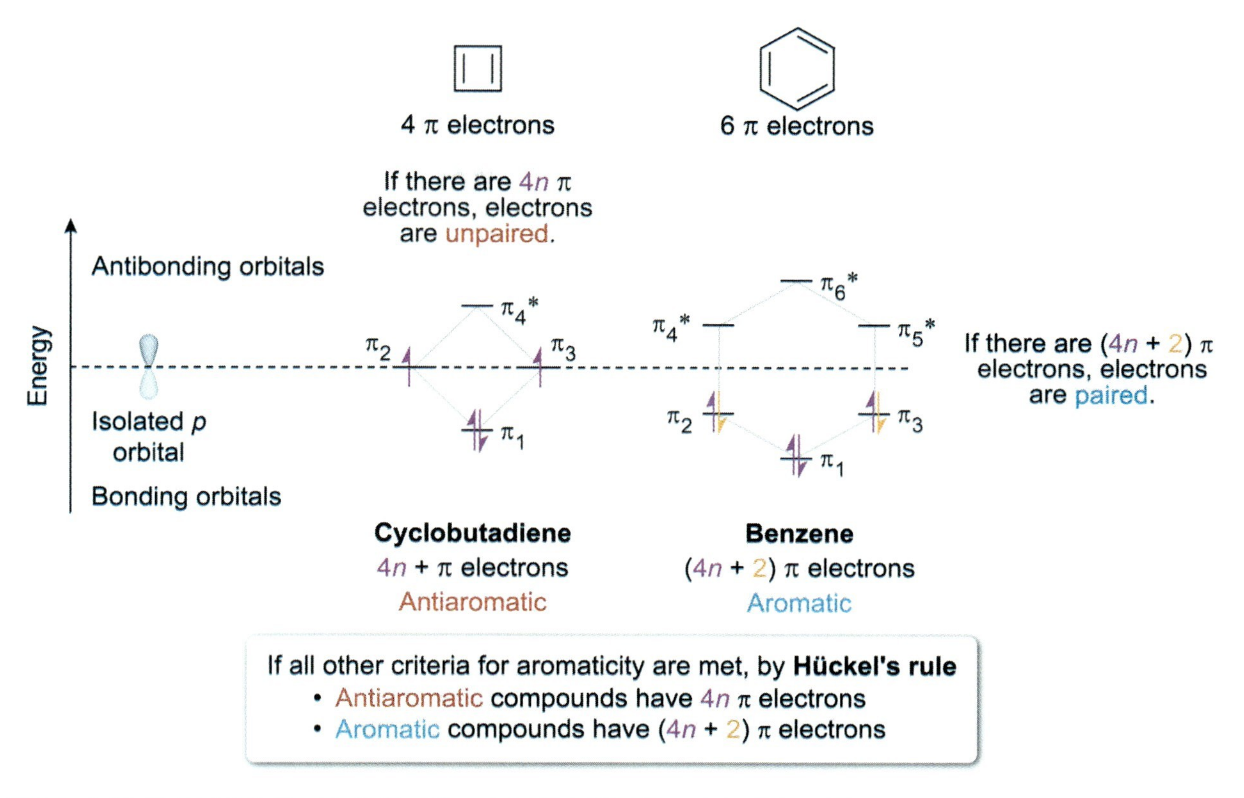

Figure 6.12 Using Hückel's rule to identify aromatic and antiaromatic molecules.

If the first three criteria for aromaticity are met, **Hückel's rule** states that:

- A compound with $4n$ π electrons is antiaromatic (with n = a positive integer).
- A compound with $(4n + 2)$ π electrons is aromatic.

For example, a compound that meets the first three criteria of aromaticity and has 10 π electrons ($4n + 2 = [4 \times 2] + 2 = 10$) is *aromatic*, whereas a compound with 12 π electrons ($4n = [4 \times 3] = 12$) is *antiaromatic*. However, due to the extreme instability associated with antiaromaticity, many compounds that would otherwise be antiaromatic modify their structures or conformations in such a way as to fail one of the first three criteria for aromaticity (eg, coplanarity). In this way, many potentially antiaromatic compounds behave like nonaromatic compounds.

Overview of Aromatic Compounds in Organic Chemistry and Biochemistry

Figure 6.13 depicts some of the aromatic heterocycles that are regularly encountered in organic chemistry. Each of these compounds adheres to Hückel's rule and has $(4n + 2)$ π electrons (present as a combination of π bonds and lone pairs).

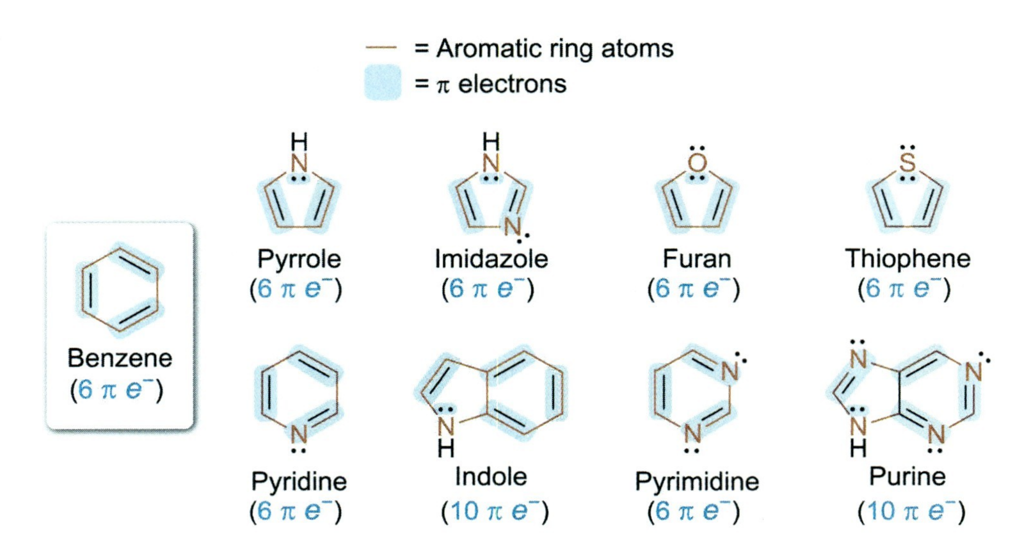

Figure 6.13 Common aromatic compounds.

Concept Check 6.3

Classify each of the following molecules as either nonaromatic, aromatic, or antiaromatic. For the purposes of this question, assume that the ring atoms in each molecule are coplanar.

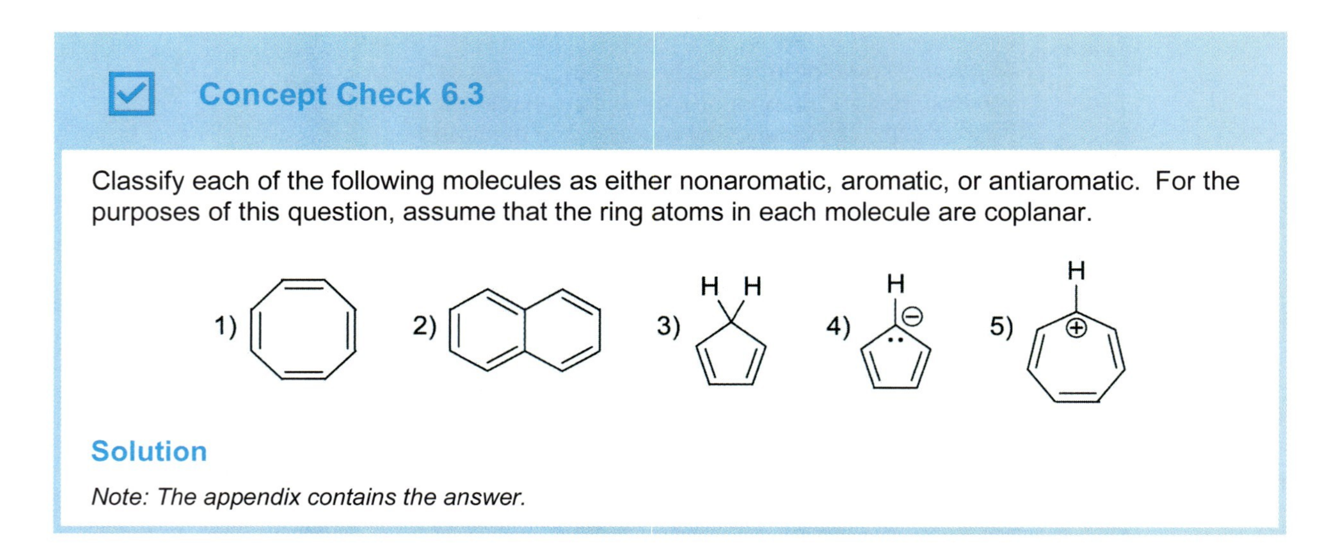

Solution

Note: The appendix contains the answer.

6.5.03 Applications of Conjugation and Aromaticity

This concept provides an overview of select applications of conjugation and aromaticity in organic chemistry.

Electron-Donating Groups and Electron-Withdrawing Groups

Conjugated or aromatic regions of a molecule are typically rich in π electron density. Many types of functional groups can influence π electron density when located near a conjugated or aromatic region. An **electron-donating group (EDG)** *contributes* (ie, *increases*) electron density to a conjugated or aromatic region, and an **electron-withdrawing group (EWG)** *removes* (ie, *decreases*) electron density from the region. For this concept, both conjugation and aromaticity are jointly described as **delocalized π systems**.

Sigma electron-donating groups and **sigma electron-withdrawing groups** contribute or remove electron density, respectively, through the inductive effect (Figure 6.14). Like other inductive donors or withdrawers, the inductive effect is strongest when the groups are closer in proximity. Alkyl groups are examples of electron-rich sigma EDGs. Examples of sigma EWGs include ammonium ions and electronegative halogen atoms.

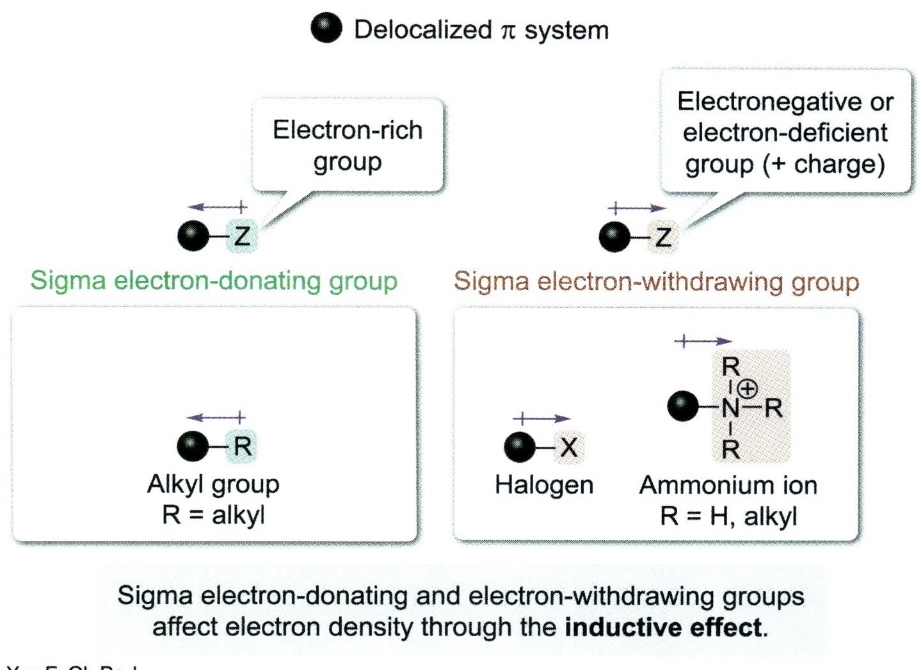

Figure 6.14 Sigma donors and withdrawers of electron density to a delocalized π system.

In contrast, **pi electron-donating groups** and **pi electron-withdrawing groups** add or remove electron density through resonance delocalization (Figure 6.15). A pi EDG typically has one or more lone pairs on the atom *directly* attached to the delocalized π system (ie, the lone pair is conjugated with the delocalized π system).

Examples of pi EDGs include alcohols, ethers, and amines. Carboxylic acid derivatives, such as esters and amides, are pi EDGs *if* they are attached to the delocalized π system through the oxygen or nitrogen atom, respectively.

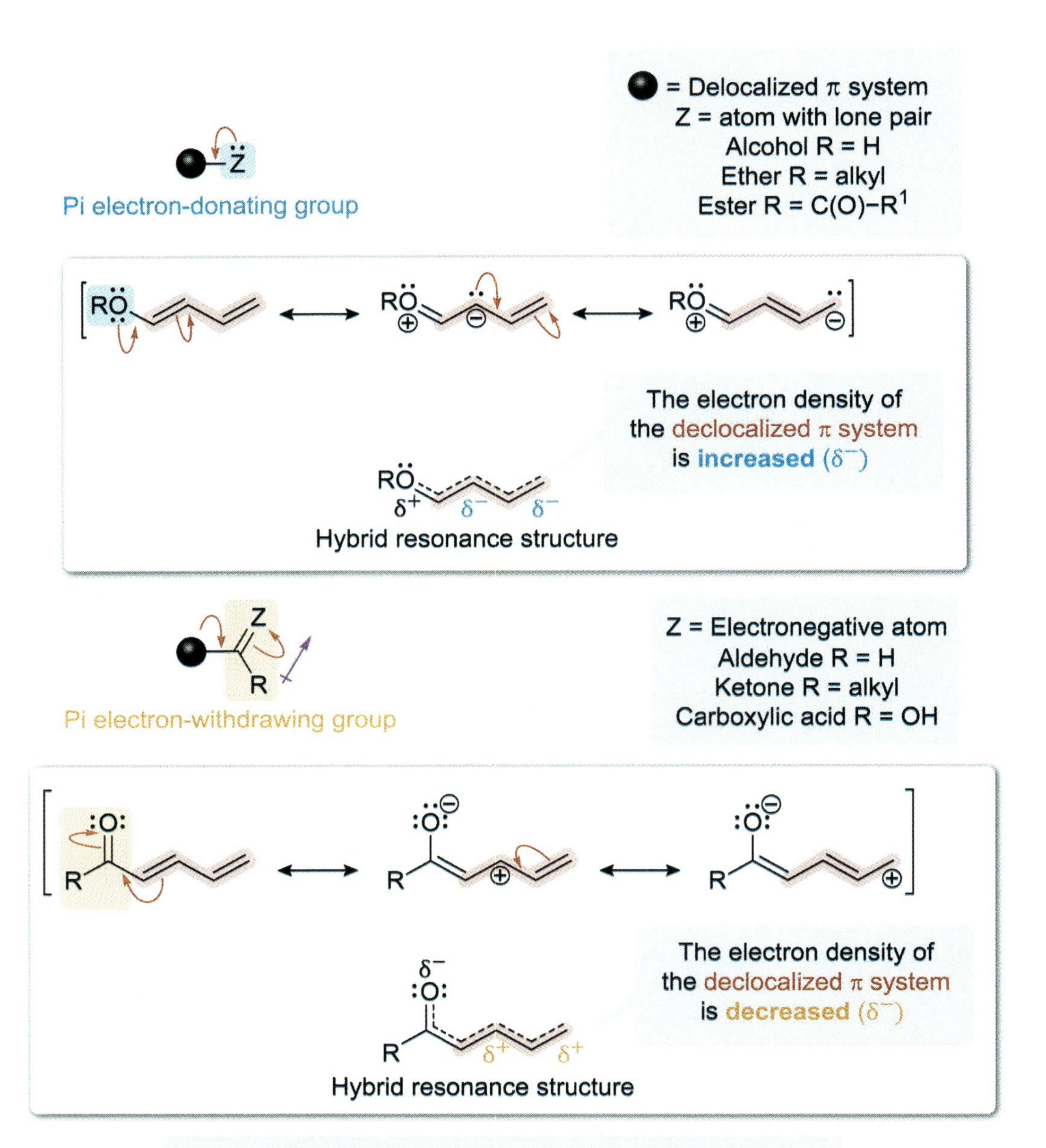

Figure 6.15 Pi donors and withdrawers of electron density to a delocalized π system.

In a pi EWG, the atom *directly* attached to the delocalized π system typically has a pi bond to an electronegative atom. Because the delocalized π system is conjugated with a polar pi bond, the electron density is pulled *toward* the electronegative atom and *away* from the rest of the system. Examples of pi EWG include nitro groups, nitriles, and carbonyl-containing groups (eg, carboxylic acids, carboxylic acid derivatives, ketones, aldehydes) *if* attached through the *carbonyl carbon*.

Table 6.4 provides a summary of the EDGs and EWGs commonly encountered in organic chemistry.

Table 6.4 Common electron-donating and electron-withdrawing groups.

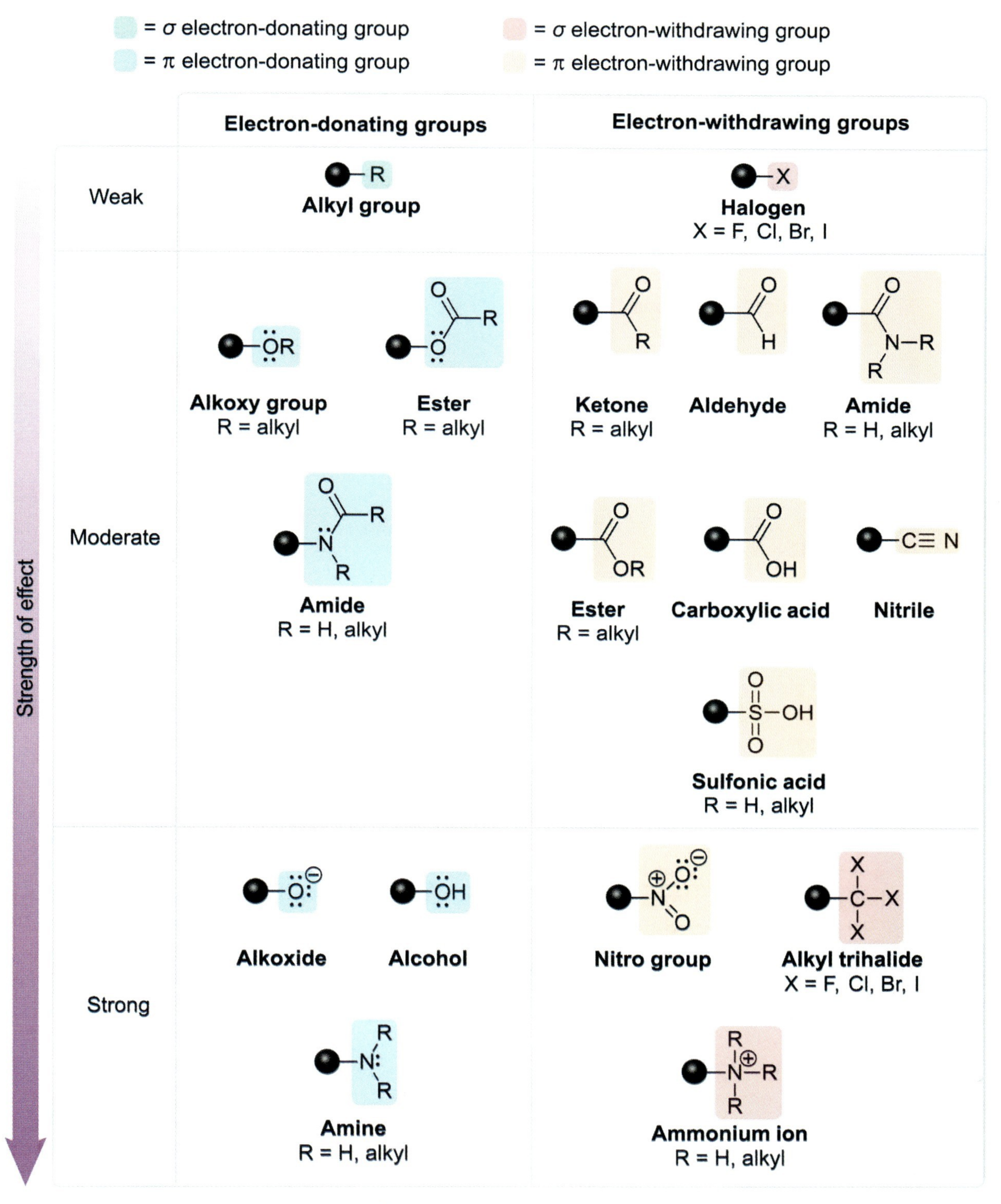

Basic and Nonbasic Heterocyclic Nitrogen Atoms

Although the lone pair on a neutral, trivalent nitrogen atom *often* has the capacity to act as a base or a nucleophile, *some* nitrogen lone pairs *lack* these abilities. The difference in **basic nitrogen atoms** and **nonbasic nitrogen atom** arises in the accessibility of the nitrogen *lone pair*.

For example, nitrogen lone pairs directly adjacent to a pi bond in a *conjugated* or *aromatic* molecule are capable of resonance delocalization. Therefore, *decreased* lone pair electron density exists on the nitrogen atom; in most cases, the nitrogen atoms are considered *nonbasic*.

Figure 6.16 shows that the nitrogen lone pairs in amide and aniline functional groups are conjugated with one or more pi bonds. As a result, amides and anilines are *much* weaker bases than a standard amine like benzylamine, which lacks conjugation of its lone pair.

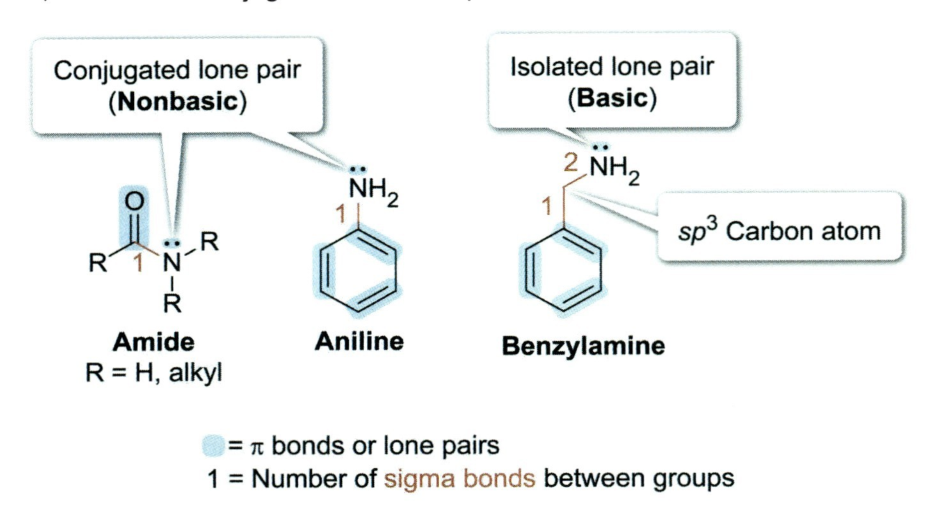

Figure 6.16 Nitrogen basicity and lone pair conjugation.

In addition, many aromatic molecules contain a **heterocyclic nitrogen atom** (ie, a nitrogen within the ring itself). Because aromaticity affords additional stability, a couple of basicity guidelines can be followed:

- A lone pair on a heterocyclic nitrogen atom that is *required* to maintain aromaticity is *nonbasic*.
- A lone pair on a heterocyclic nitrogen atom that is *not required* to maintain aromaticity is *basic*.

The orbital structure of a nitrogen atom in aromatic heterocycles can be compared to one of two local bonding patterns (Table 6.5). Nitrogen atoms with one double bond and one single bond to other atoms in the line-angle structure *must* use their unhybridized *p* orbital for the shown *pi bond*. Consequently, the lone pair *must* be located in an sp^2 orbital *perpendicular* to the pi bond (and *not* conjugated with the system). Since the lone pair can be donated *without* disrupting ring aromaticity, this type of nitrogen is a *basic nitrogen atom*.

Table 6.5 Analysis of nitrogen lone pairs in aromatic heterocycles.

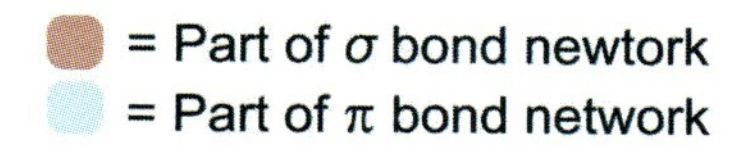

	Basic nitrogen lone pair	**Nonbasic nitrogen lone pair**
Line-angle structure	N; 1 Double bond, 1 single bond	N; H; 3 Single bonds
Lone pair orbital emphasized	N; sp^2 Orbital	Unhybridized p orbital; N; H
3D orbital perspective	Unhybridized p orbital in π bond; N; **Basic** lone pair is perpendicular to π system	**Nonbasic** lone pair is parallel with π system; N; H
Examples	Lone pair outside π system; N; Nitrogen has a π bond; Pyridine	Lone pair within π system; H; N; Nitrogen has 3 single bonds; Pyrrole

Nitrogen atoms that have three single bonds (usually two bonds in the aromatic ring and one bond to a hydrogen atom or alkyl group) *must* have their lone pair in an unhybridized p orbital perpendicular to the plane of the sp^2 orbitals. For the ring to be aromatic, the lone pair *must* be part of the delocalized π system. Donating this type of lone pair to another molecule would require the ring aromaticity to be disrupted. This outcome is *unfavorable*; therefore, this type of nitrogen is a *nonbasic nitrogen atom*.

Lesson 7.1

Alkyl Halides

Introduction

Concept 2.8.04 defines an **alkyl halide** as an alkane with at least one of its hydrogen atoms replaced with a halogen atom. An unspecified halogen atom is represented by the letter X (Figure 7.1). The structures of alkyl halides are differentiated by:

- The identity of the halogen atom.
- The number of alkyl groups on the carbon bearing the halogen atom.

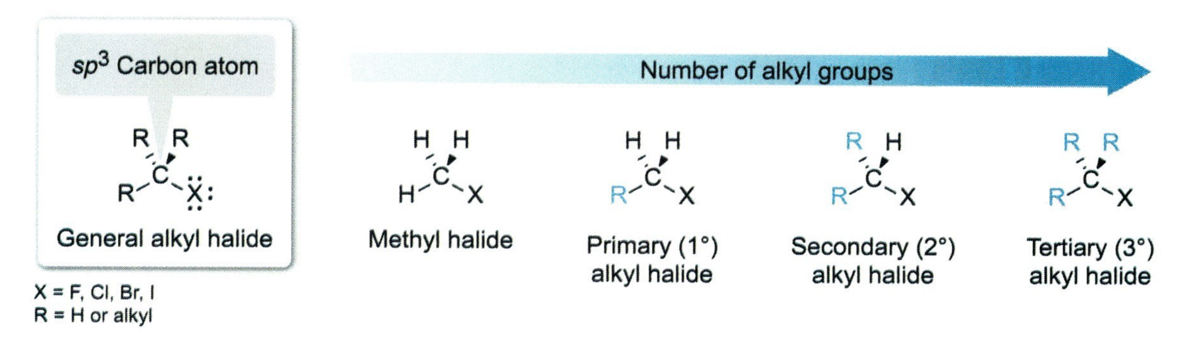

Figure 7.1 Structure and classification of alkyl halides.

This lesson describes the properties and the reactivity of alkyl halides.

7.1.01 Physical Properties of Alkyl Halides

Boiling Point

Because an alkyl halide contains a polar carbon-halogen bond, which generates dipole-dipole intermolecular forces, the boiling point of an alkyl halide is higher than that of its corresponding alkane (Table 7.1). Smaller alkyl halides are more impacted by a single polar bond, leading to larger *differences* in boiling points.

Table 7.1 Boiling points of alkanes and alkyl chlorides.

Alkane	Boiling point (°C)	Alkyl chloride	Boiling point (°C)	Difference in boiling point (°C)
CH_4	−164	CH_3Cl	−24	+140
CH_3CH_3	−89	CH_3CH_2Cl	12	+101
$CH_3CH_2CH_3$	−42	$CH_3CH_2CH_2Cl$	47	+89
$CH_3CH_2CH_2CH_3$	−1	$CH_3CH_2CH_2CH_2Cl$	79	+80
	Increasing boiling point		Increasing boiling point	Impact of halogen atom

Given that alkyl halides share many structural features with alkanes, boiling points of alkyl halides (Table 7.2) follow the same trends (see Concept 6.1.01 for more on alkane boiling point trends):

- Alkyl halide boiling points *increase* with molecular weight.
- Branching *decreases* the surface area, causing a *decrease* in the boiling point.

Table 7.2 Boiling points of alkyl halides.

Alkyl halide	Carbon atoms	Boiling point (°C) X = F 19 g/mol	X = Cl 35.4 g/mol	X = Br 80 g/mol	X = I 127 g/mol
$H_3C–X$	1	−78	−24	4	42
ethyl–X	2	−38	12	38	72
propyl–X	3	3	47	71	102
isopropyl–X	3	−10	36	59	89
butyl–X	4	33	79	102	131
isobutyl–X	4	—	69	92	120
sec-butyl–X	4	—	70	91	118
tert-butyl–X	4	—	52	73	100

Boiling point increases with **molecular weight**

Similar impact of branching

Boiling point decreases with **branching**

Boiling point increases with the **number of carbon atoms**

Solubility

The expression "like dissolves like" (Concept 2.4.02) describes differences in solute and solvent polarity as the dominant factor in the solubility of organic molecules. Despite the polar carbon-halogen bond, most alkyl halides are relatively nonpolar and are preferentially soluble in *organic solvents* (eg, hexane, ethyl acetate, methanol); alkyl halides are typically *poorly soluble* in water.

However, alkyl halides containing a large number of halogen atoms can have *very* different properties, including very low solubility in *both* organic solvents and water. For example, the structure of polytetrafluoroethylene (PTFE), a common nonstick coating for cookware, contains a repeating, two-carbon unit where fluorine has been substituted for *every* hydrogen atom (Figure 7.2) and has low solubility in both organic and aqueous solutions.

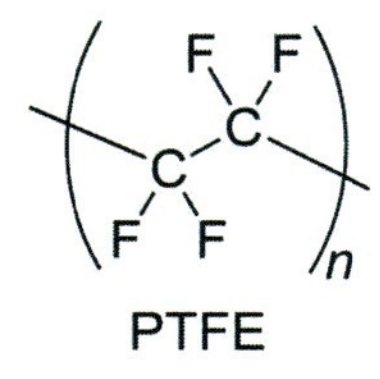

Figure 7.2 The structure of polytetrafluoroethylene (PTFE).

Density

The density (Concept 2.4.03) of most organic compounds is less than water (1 g/mL) with the exception of certain alkyl halides (Figure 7.3). Knowing the density of a molecule relative to water is important for liquid-liquid extraction (Concept 13.1.03).

In general:

- Organic compounds containing one or more fluorine atoms or exactly one chlorine atom have densities *less* than water and are the *top* layer in an organic-aqueous mixture.
- Organic compounds containing two or more chlorine atoms, one or more bromine atoms, or one or more iodine atoms have densities *greater* than water and are the *bottom* layer in an organic-aqueous mixture.

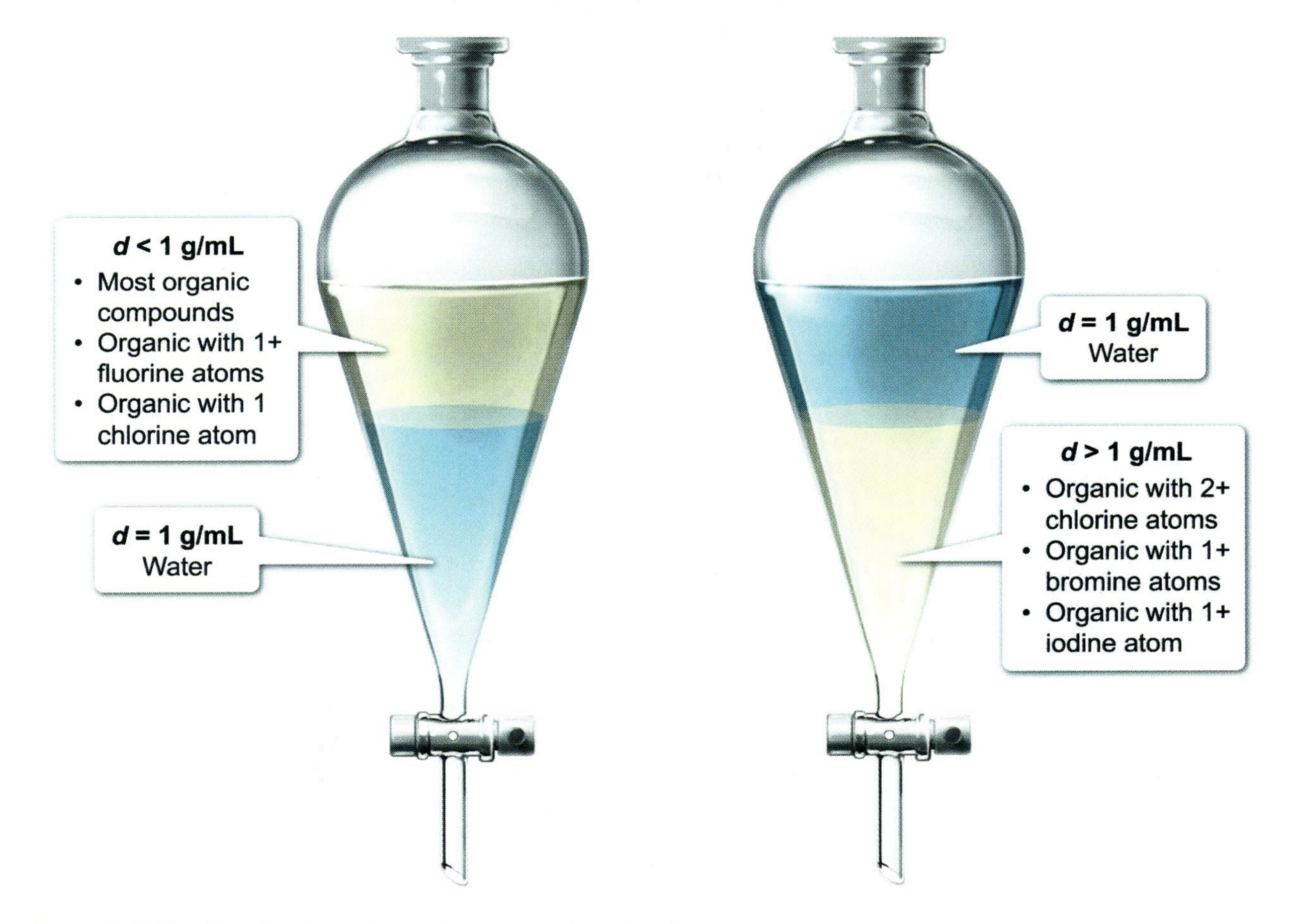

Figure 7.3 Relative densities of organic compounds and water.

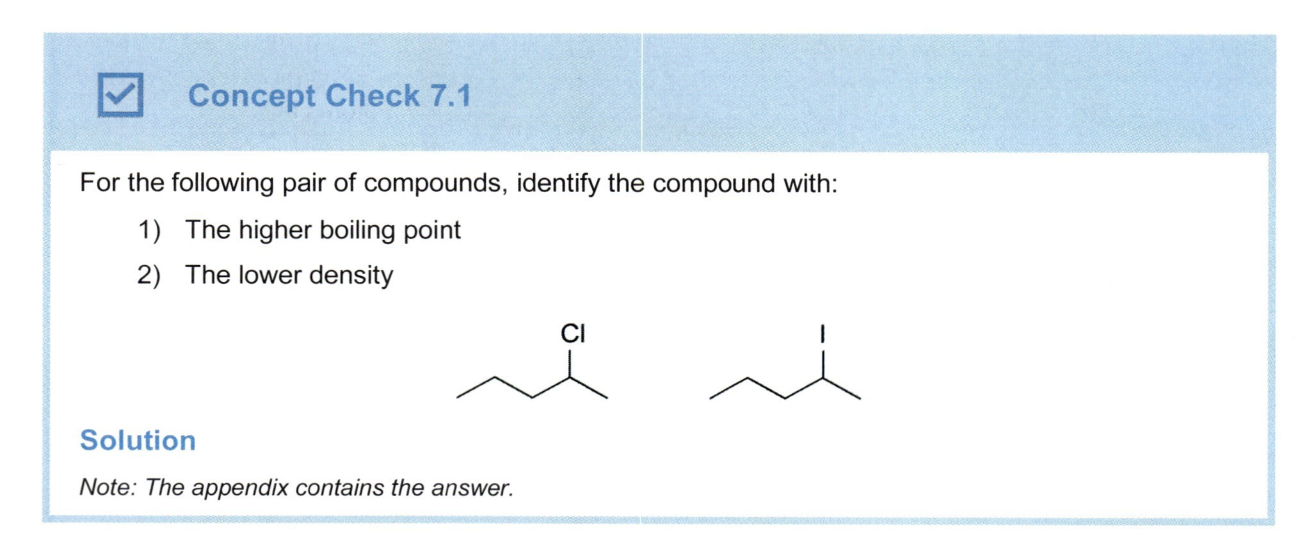

Concept Check 7.1

For the following pair of compounds, identify the compound with:

1) The higher boiling point
2) The lower density

Solution

Note: The appendix contains the answer.

7.1.02 Alkyl Halides as Electrophiles

Alkyl halides are Type 2 electrophiles (Concept 5.3.02) because they contain an electrophilic atom attached by a *single bond* to an electronegative group (halogen). Except for the alkyl fluorides (which have a poor leaving group), *alkyl halides react primarily as electrophiles.*

The relative reactivity of alkyl halides depends on:

- The type of reaction performed (eg, S_N1, S_N2, E1, E2)
- The identity of the halogen atom (Cl < Br << I)
- The amount of steric hindrance surrounding the electrophilic carbon atom.

The S_N1 Reaction

As discussed in Concept 5.4.02, heating an alkyl halide with a *weak* nucleophile (eg, water, alcohols) can promote an S_N1 reaction (Figure 7.4).

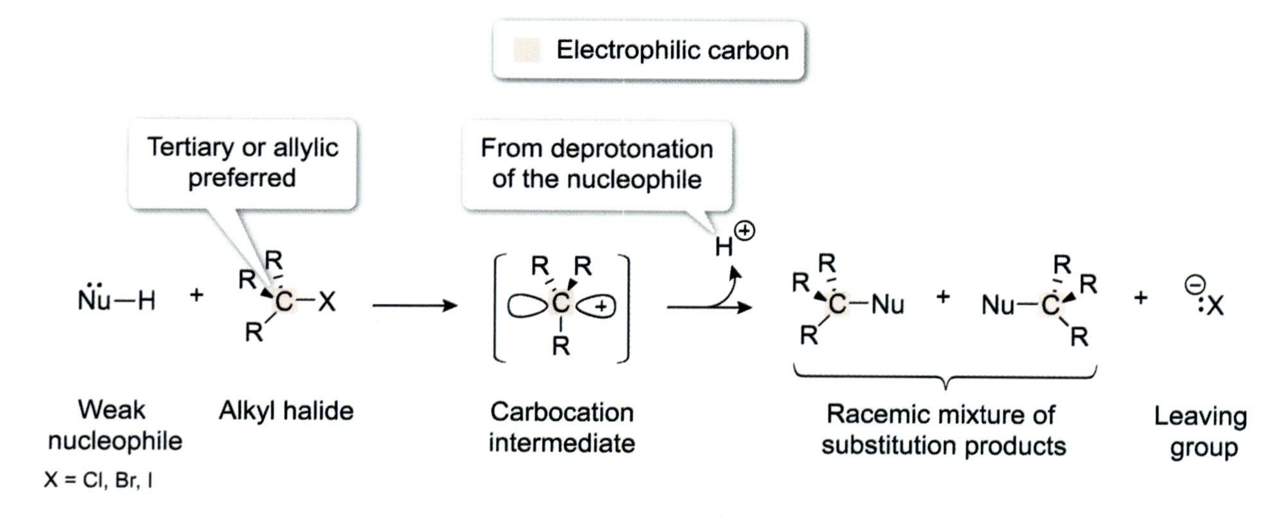

Figure 7.4 S_N1 reaction of an alkyl halide.

Because the effectiveness of an S_N1 reaction *increases* if the electrophilic atom has a *greater* number of alkyl groups, *tertiary* alkyl halides are among the best substrates for an S_N1 reaction. Allylic or benzylic alkyl halides are also very good substrates for S_N1 reactions, given their ability to form a resonance-stabilized carbocation intermediate.

Like other reactions that involve one or more carbocation intermediates, an S_N1 reaction using an alkyl halide may be prone to carbocation rearrangement (eg, a 1,2-hydride shift or a 1,2-methyl shift). S_N1 products contain a racemic mixture of absolute configurations at any carbon atoms that bore a carbocation.

The S_N2 Reaction

As discussed in Concept 5.4.03, treating some alkyl halides with a *strong* nucleophile can promote an S_N2 reaction (Figure 7.5).

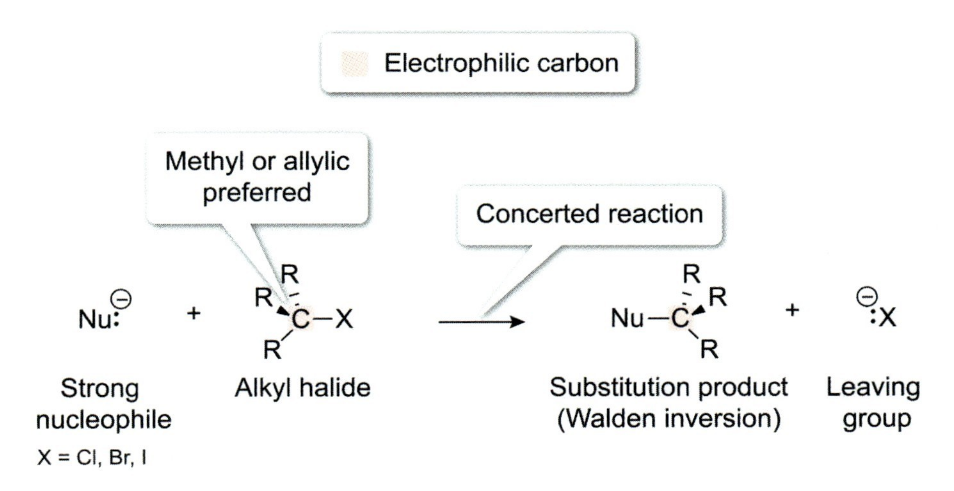

Figure 7.5 S_N2 reaction of an alkyl halide.

The most effective alkyl halides for S_N2 reactions have a good leaving group and minimal steric hindrance around the electrophilic carbon atom. Allylic or benzylic alkyl halides are also very good substrates due to stabilization by the adjacent π system.

Because an S_N2 reaction proceeds through a concerted mechanism, the nucleophile approaches the alkyl halide from the side opposite the polar C–X bond. The resulting Walden inversion can lead to an inversion of configuration if the electrophilic carbon atom is chiral.

Elimination Reactions

As discussed in Concept 5.6.02 and Concept 5.6.03, treating an alkyl halide with a base can promote an elimination reaction (Figure 7.6). Weak bases (eg, water, alcohols) tend to promote an E1 mechanism, whereas strong bases (eg, hydroxide ion, alkoxide ions) tend to promote an E2 mechanism.

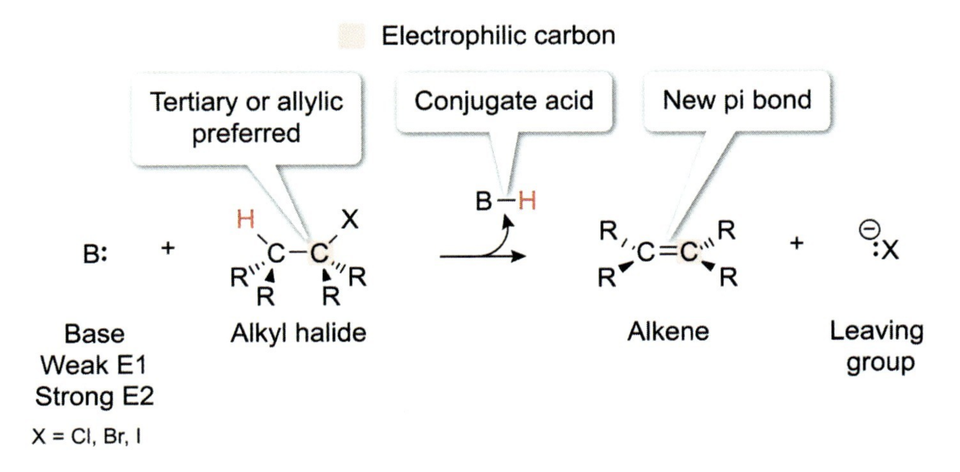

Figure 7.6 General elimination reaction of an alkyl halide.

7.1.03 Formation of Grignard Reagents

Like many other organic reactions, **carbon-carbon bond forming reactions** can be described as reactions between a nucleophile and an electrophile. Carbon-carbon bond forming reactions are one of the principal ways to synthesize larger molecular structures from smaller building blocks (Figure 7.7).

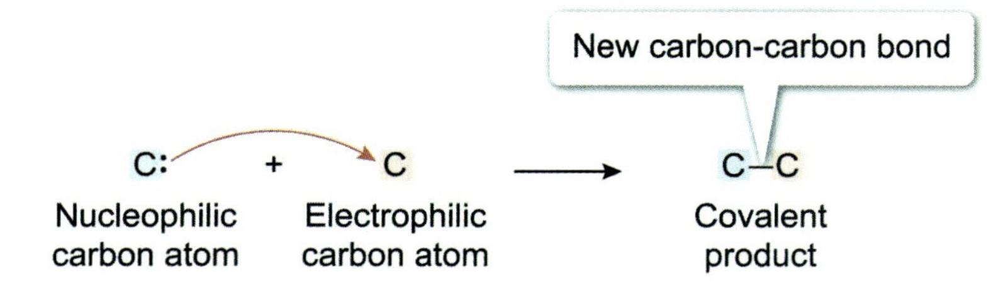

Figure 7.7 General reaction between a carbon nucleophile and a carbon electrophile.

Of the elements most relevant to organic chemistry, carbon is among the *least* electronegative. Most carbon-heteroatom (ie, not carbon or hydrogen) bonds are polar, with a dipole moment pulling electron density *away* from the carbon (Figure 7.8). Therefore, *many* ways exist for a carbon atom to be electron-*deficient* (electrophilic). Alkyl halides are one functional group that contains an electrophilic carbon atom.

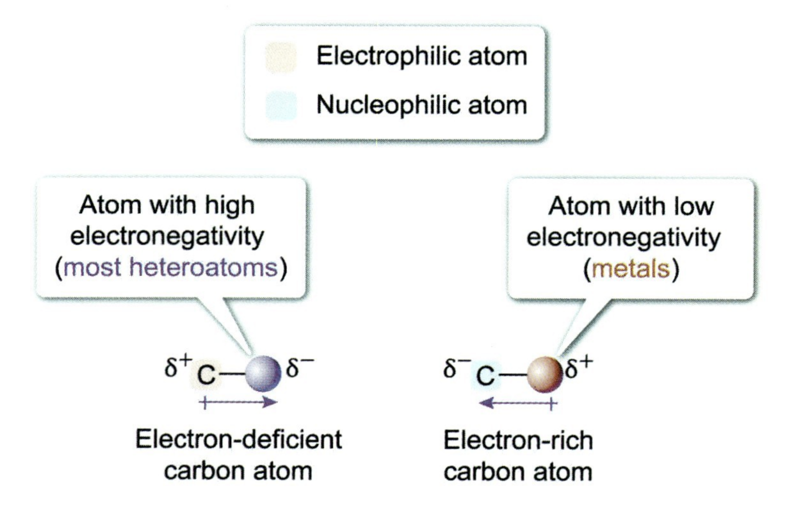

Figure 7.8 Bond dipole moments of electron-deficient and electron-rich carbon atoms.

One way for a carbon atom to become electron-*rich* (nucleophilic) is for it to form a bond with an atom *less* electronegative than carbon. Most elements that meet this criterion are *metals*, which tend to form bonds with a higher percent ionic character due to their lower ionization energies and tendency to form cations.

Organometallic compounds are compounds that contain a carbon-metal covalent bond. Although the bond is *covalent*, it can be helpful to view the carbon atom as behaving like a carbanion (ie, a *very* strong nucleophile).

A **Grignard reagent** is a type of organometallic compound defined as an organomagnesium halide (Figure 7.9), with a general structure of R–MgX. *Any* alkyl halide can readily be converted into a Grignard reagent by treating it with magnesium metal in an ether solvent. The magnesium metal *inserts* itself between the carbon-halogen bond while *retaining* the absolute configuration of the carbon atom.

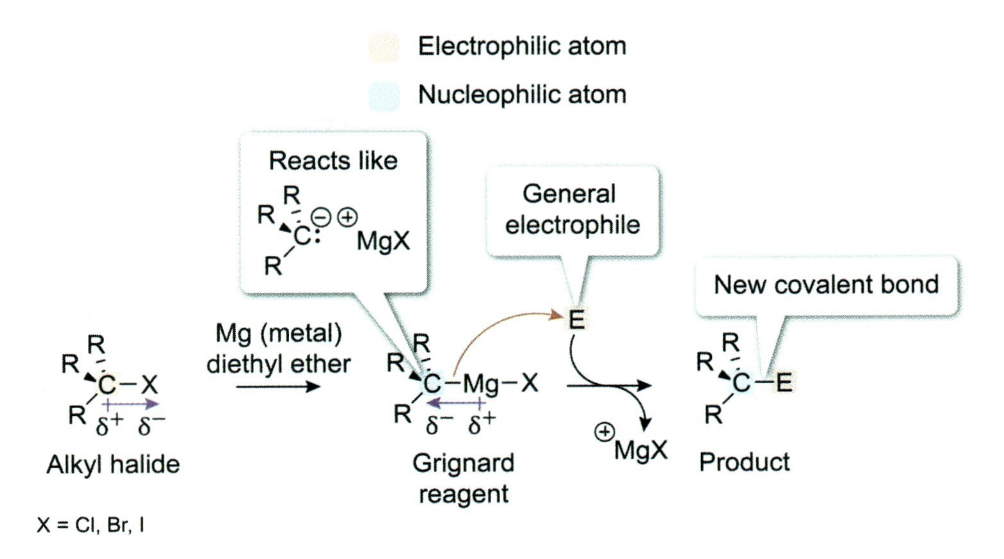

Figure 7.9 Formation and general reaction of a Grignard reagent.

Grignard reagents provide a flexible means of converting an electrophilic carbon atom into a *nucleophilic* carbon atom. Synthetic applications of Grignard reagents are discussed in Concept 8.2.02.

Lesson 7.2

Ethers

Introduction

Ethers are a functional group consisting of an *sp*^{3} hybridized oxygen atom between two carbon groups (R), with a general structure of R–O–R′ (Concept 2.8.02). Ethers with *identical* carbon groups (ie, R = R′) are **symmetrical ethers**, whereas ethers with *different* carbon groups (ie, R ≠ R′) are **unsymmetrical ethers**.

Ethers in which R and R′ are *not connected* through a covalent bond are **linear ethers**, and ethers in which R and R′ are *connected* by a covalent bond are **cyclic ethers** (Figure 7.10).

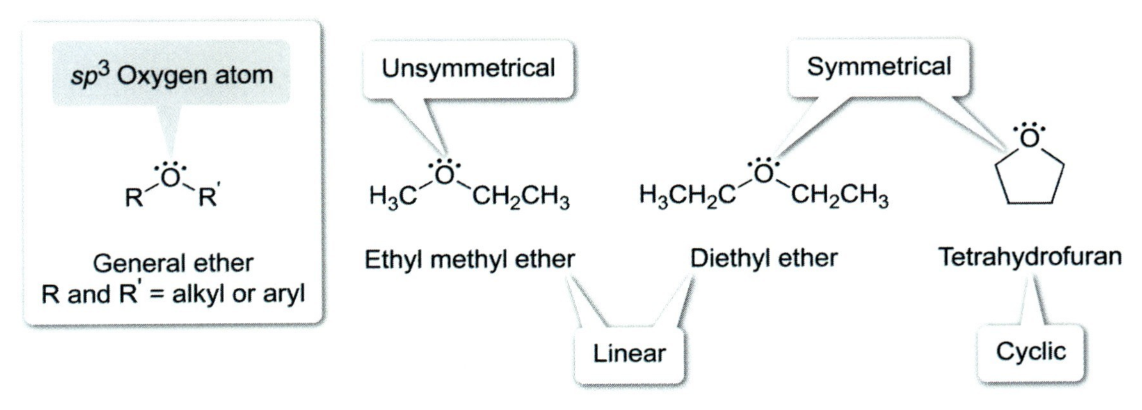

Figure 7.10 Structure and classification of ethers.

This lesson focuses on concepts regarding the chemical properties and reactivity of ethers.

7.2.01 Physical Properties of Ethers

Although an ether contains an electronegative oxygen atom, the bond dipole moments to the *two* R groups partially cancel each other. Consequently, an ether has only a *modest* molecular dipole moment and is considered relatively *nonpolar*. The two lone pairs of electrons on the ether oxygen atom are primarily responsible for the physical properties of ethers.

Boiling Point

Despite having a modest dipole moment, the boiling point of an ether is very similar to the boiling point of an alkane of approximately the same mass (Table 7.3).

Table 7.3 Boiling points of alkanes and ethers.

Alkane	Molecular weight (g/mol)	Boiling point (°C)	Ether	Molecular weight (g/mol)	Boiling point (°C)
$CH_3CH_2CH_3$	44	−42	CH_3OCH_3	46	−24
$CH_3(CH_2)_2CH_3$	58	−1	$CH_3OCH_2CH_3$	60	7
$CH_3(CH_2)_3CH_3$	72	36	$CH_3CH_2OCH_2CH_3$	74	35
			$CH_3OCH(CH_3)_2$	74	31
		Increasing boiling point		Boiling point decreases with **branching**	Increasing boiling point

An alkane and an ether of similar molecular weight have similar boiling points.

Ethers follow the same trends as most functional groups: The boiling point of an ether *increases* with molecular weight (due to an increase in London dispersion forces) and *decreases* for branched compounds.

Solubility

Although ethers are considered nonpolar, the lone pairs on the ether oxygen atom can act as hydrogen bond acceptors. A sample of pure ether contains many hydrogen bond acceptors but it lacks a corresponding hydrogen bond donor (Figure 7.11).

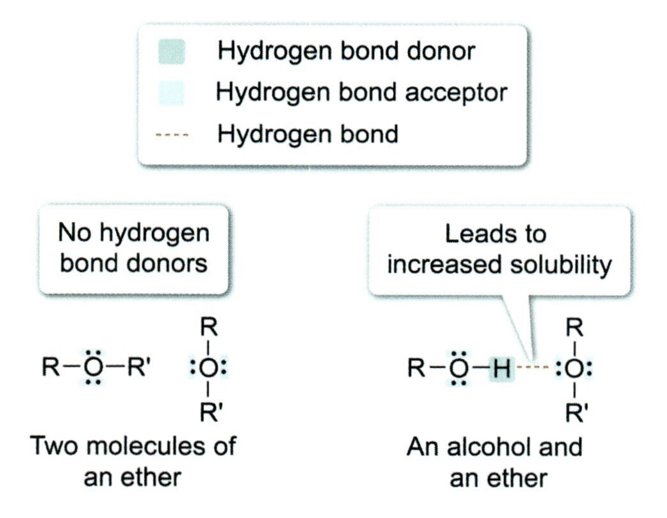

Figure 7.11 Hydrogen bonding in an ether and an ether-alcohol mixture.

However, *mixtures* of an ether and a molecule containing a hydrogen bond donor (ie, water, alcohols) demonstrate hydrogen bonding between molecules (Concept 2.3.03). Consequently, the solubility of an ether in water or an alcohol is generally *much* higher than the solubility of an alkane in these liquids. This effect is particularly evident for low-molecular-weight ethers that have four or fewer total carbon atoms.

7.2.02 Use of Ethers as Solvents

Ethers are among the *least reactive* functional groups (Figure 7.12). Despite the electronegative oxygen atom, the attached carbon atoms are *poorly* electrophilic because alkoxides are poor leaving groups. Therefore, ethers are unlikely to undergo a cleavage reaction under basic (nucleophilic) conditions.

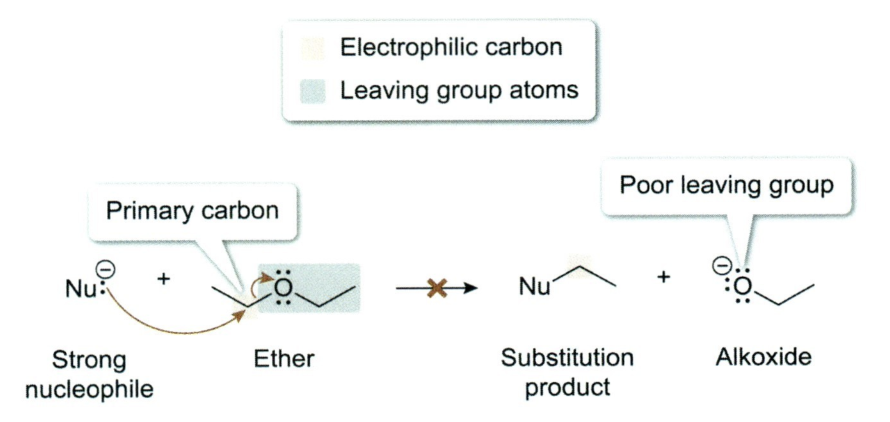

Figure 7.12 Stability of an ether to a strong nucleophile.

However, under heated acidic conditions (ie, treatment with a hydrohalic acid, HX), ethers *can* undergo a cleavage reaction. The strong acid protonates the ether oxygen atom and converts it to an oxonium ion, which is a *good* leaving group (Concept 5.3.03). Figure 7.13 shows that the oxonium ion intermediate can undergo a cleavage reaction via either an S_N2 or S_N1 mechanism, depending on the number of alkyl groups on the electrophilic carbon atoms.

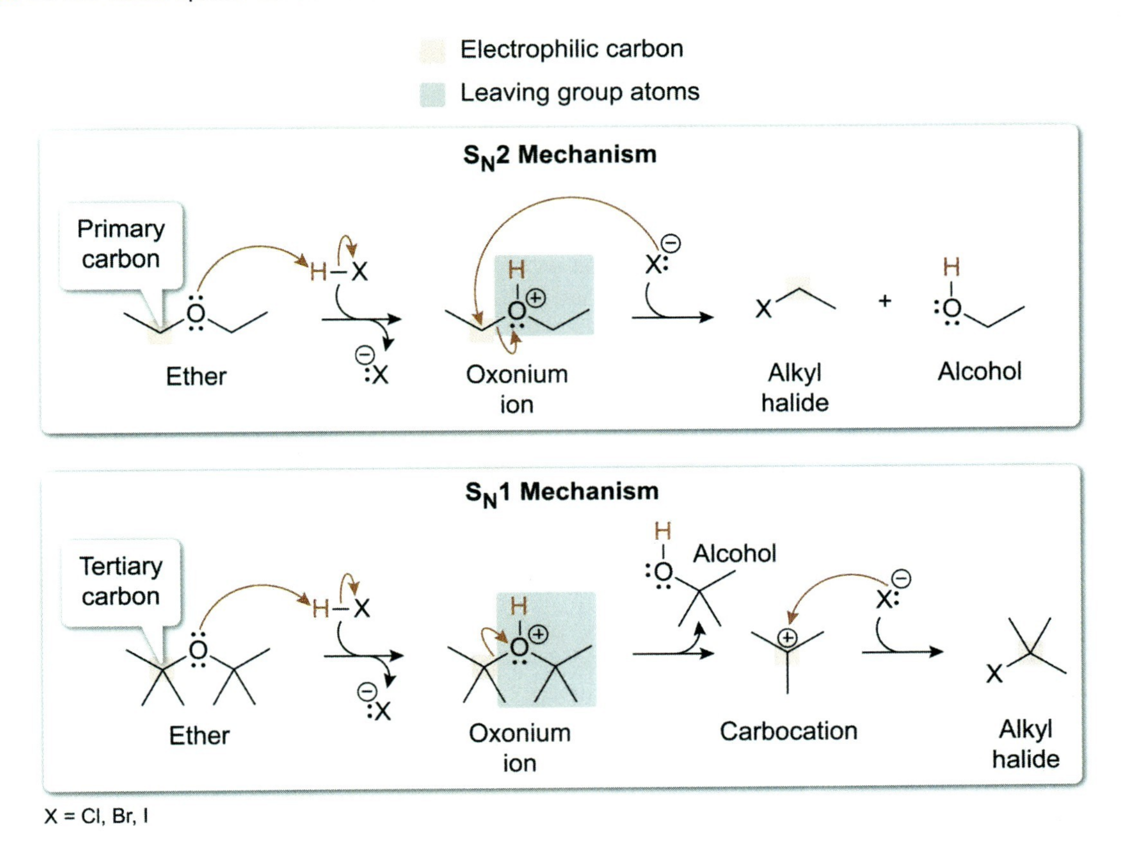

Figure 7.13 Cleavage of an ether by hydrohalic acid (HX).

Ethers can also solvate positively charged cations (Figure 7.14). The electron-rich lone pairs on the ether oxygen atom are hydrogen bond acceptors that can orient toward a cation, allowing ethers to surround a cation with a stable shell of molecules.

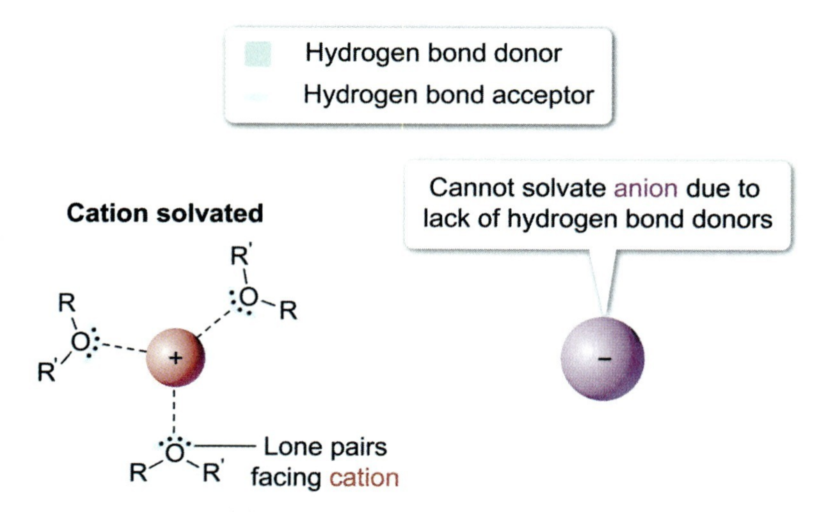

Figure 7.14 Selective solvation of cations by ethers.

However, because an ether lacks a hydrogen bond *donor*, ether molecules are *unable* to solvate *anions*. If a cation-anion pair (eg, an ionic compound) is added to an ether, the anion is poorly solvated by the ether, potentially *enhancing* its reactivity such that the anion demonstrates *increased* nucleophilicity.

Their reactivity profiles and physical properties make ethers useful as reaction solvents that involve basic or nucleophilic conditions, such as the preparation of Grignard reagents (Concept 7.1.03). The ether solvent helps stabilize the Grignard reagent by solvating the magnesium halide cation ($^{+}$MgX) while simultaneously enhancing the nucleophilicity of the anionic nucleophile.

7.2.03 Williamson Ether Synthesis

The Williamson ether synthesis is among the most prevalent and useful approaches to synthesize ethers. The **Williamson ether synthesis** is broadly defined as the S_N2 reaction of an alkoxide ion with an alkyl halide; the alkoxide and alkyl halide each provide one of the alkyl groups (R or R') for the ether product (Figure 7.15).

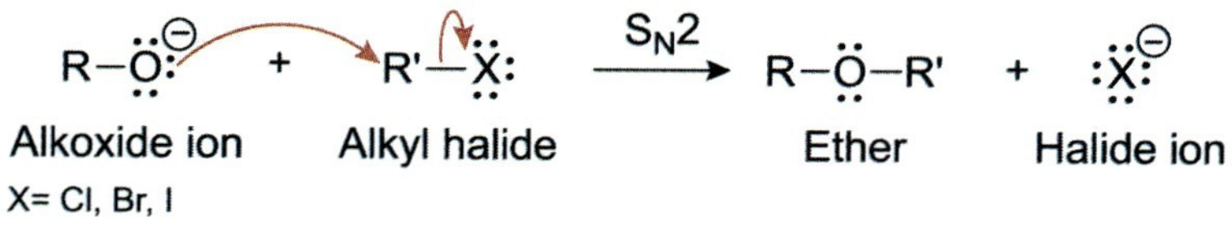

$$\underset{\text{Alkoxide ion}}{R{-}O^{\ominus}} + \underset{\text{Alkyl halide}}{R'{-}X} \xrightarrow{S_N2} \underset{\text{Ether}}{R{-}O{-}R'} + \underset{\text{Halide ion}}{X^{\ominus}}$$

X= Cl, Br, I

Figure 7.15 The Williamson ether synthesis.

Alkoxide ions can be prepared in high yield from corresponding alcohols by treating an alcohol with either an alkali metal or sodium hydride (NaH). For simple alkoxides (eg, methoxide, *t*-butoxide), using an alkali metal and an anhydrous (eg, water-free) alcohol is more common (Figure 7.16). Sodium (Na) is typically used for primary (1°) alcohols, whereas the slightly stronger potassium (K) is typically used for secondary (2°) or tertiary (3°) alcohols. Although an alcohol is a polar protic solvent and is *not generally* preferred for S_N2 reactions (Concept 5.3.01), alkoxides are such strong nucleophiles that the Williamson ether synthesis proceeds without issue.

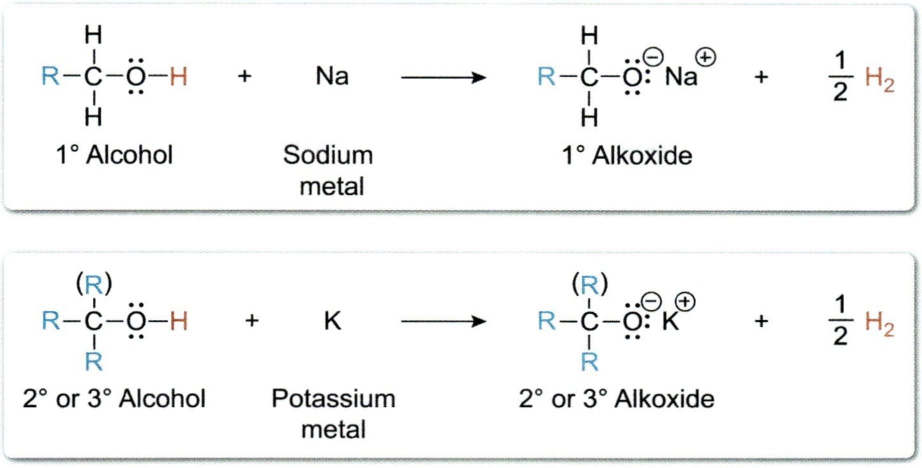

$$R{-}CH_2{-}\ddot{O}{-}H + Na \longrightarrow R{-}CH_2{-}\ddot{O}{:}^{\ominus}\,Na^{\oplus} + \tfrac{1}{2}H_2$$

1° Alcohol + Sodium metal → 1° Alkoxide + ½ H_2

$$R{-}C(R)(R){-}\ddot{O}{-}H + K \longrightarrow R{-}C(R)(R){-}\ddot{O}{:}^{\ominus}\,K^{\oplus} + \tfrac{1}{2}H_2$$

2° or 3° Alcohol + Potassium metal → 2° or 3° Alkoxide + ½ H_2

Figure 7.16 Generation of sodium alkoxides and potassium alkoxides.

Alternatively, sodium hydride (NaH) can be used to fully convert an alcohol to its alkoxide ion (Figure 7.17). This reaction is usually performed in the ether solvent tetrahydrofuran (THF). The hydride ion is a *very* strong base ($pK_b = -21$) that rapidly reacts with the alcohol's acidic proton to form hydrogen gas (a *very* weak acid) and an alkoxide ion.

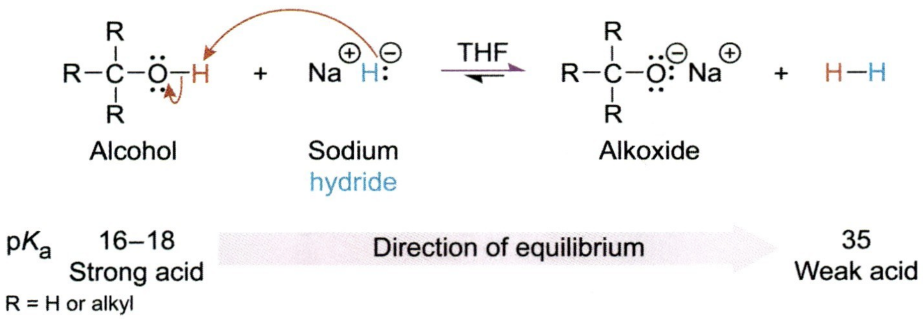

$$R_3C{-}\ddot{O}{-}H + Na^{\oplus}\,H{:}^{\ominus} \xrightleftharpoons{THF} R_3C{-}\ddot{O}{:}^{\ominus}\,Na^{\oplus} + H{-}H$$

Alcohol + Sodium hydride → Alkoxide + H–H

pK_a 16–18 Strong acid — Direction of equilibrium → 35 Weak acid

R = H or alkyl

Figure 7.17 Reaction of an alcohol with NaH.

Once an alkoxide ion has been obtained, the Williamson ether synthesis can proceed via an S_N2 mechanism by treating the alkoxide ion with an alkyl halide (Figure 7.18). The best yields are obtained if both the alkoxide ion and the electrophilic carbon atom of the alkyl halide have *fewer* alkyl groups, minimizing steric hindrance.

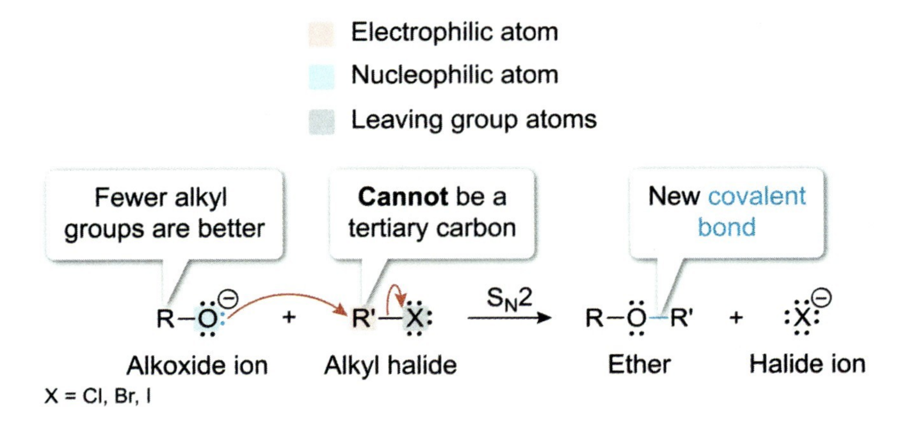

Figure 7.18 Limitations of the Williamson ether synthesis.

Many ethers can be prepared in two ways using the Williamson ether synthesis (ie, by reacting R–O⁻ with R′–X *or* by reacting R–X with R′–O⁻), but some ethers—such as aryl ethers—have constraints that limit these options. If aryl ethers are prepared by the Williamson ether synthesis, the aryl R group *must* be derived from the alkoxide nucleophile because it is *not possible* to perform a Williamson ether synthesis on an aryl halide, in which the halogen atom is attached to an sp^2 hybridized carbon atom (Figure 7.19).

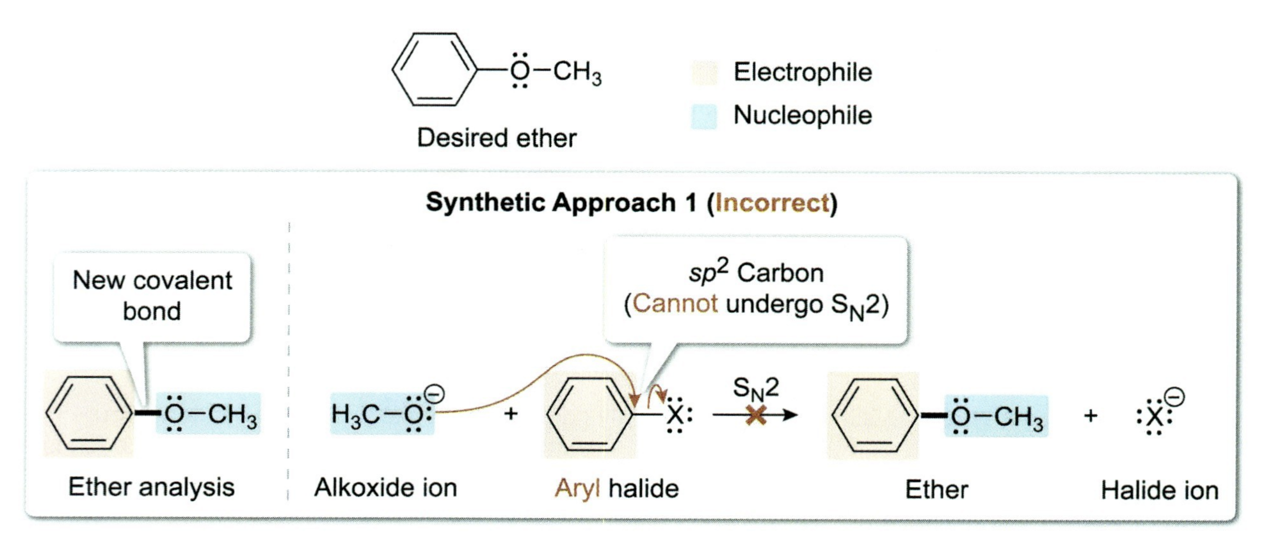

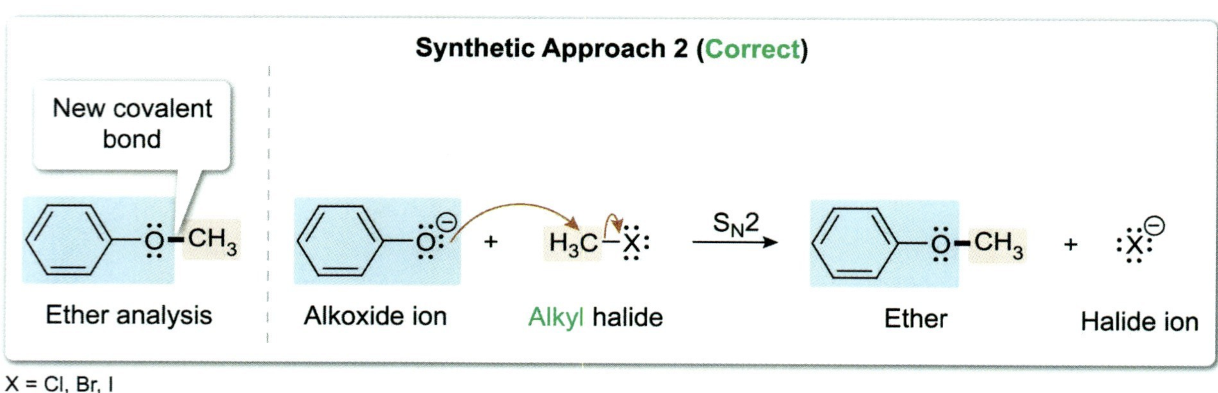

Figure 7.19 Comparison of Williamson synthetic approaches for an aryl ether.

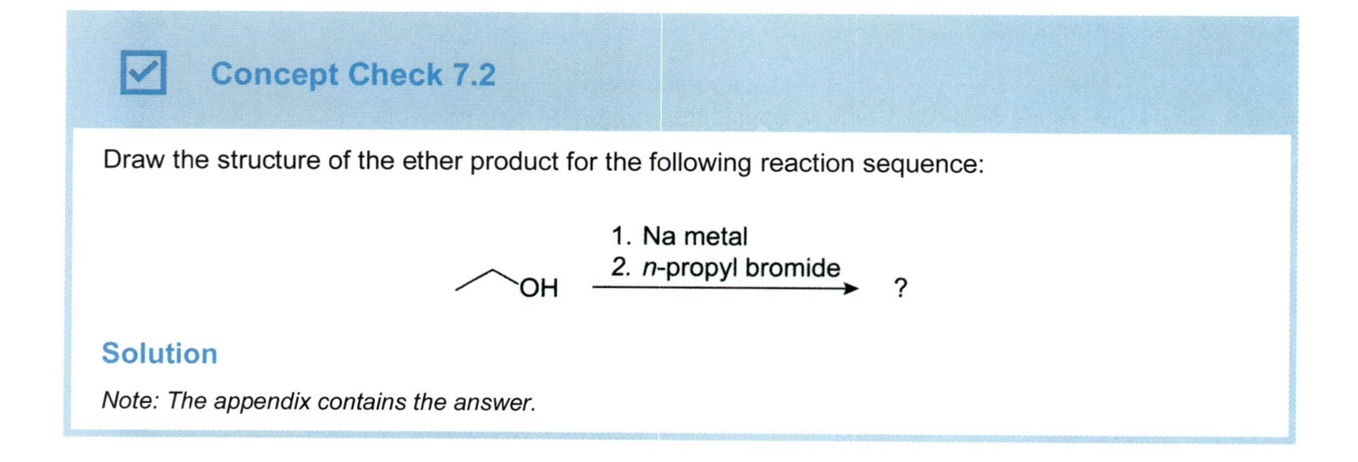

Concept Check 7.2

Draw the structure of the ether product for the following reaction sequence:

Solution

Note: The appendix contains the answer.

Lesson 7.3

Thiols, Thioethers, Thioesters, and Disulfides

Introduction

Concept 2.8.04 introduces several functional groups that contain one or more sulfur atoms (Figure 7.20). Sulfur and oxygen are both Group 6A elements, and the sulfur-containing functional groups share similar bonding patterns with their oxygen-containing counterparts:

- A thiol contains a sulfhydryl group (–SH).
- A thioether has a sulfur atom between two alkyl groups.
- A thioester has a sulfur atom between the alkyl group and a carbonyl carbon atom.
- A disulfide contains two sulfur atoms bonded to one another.

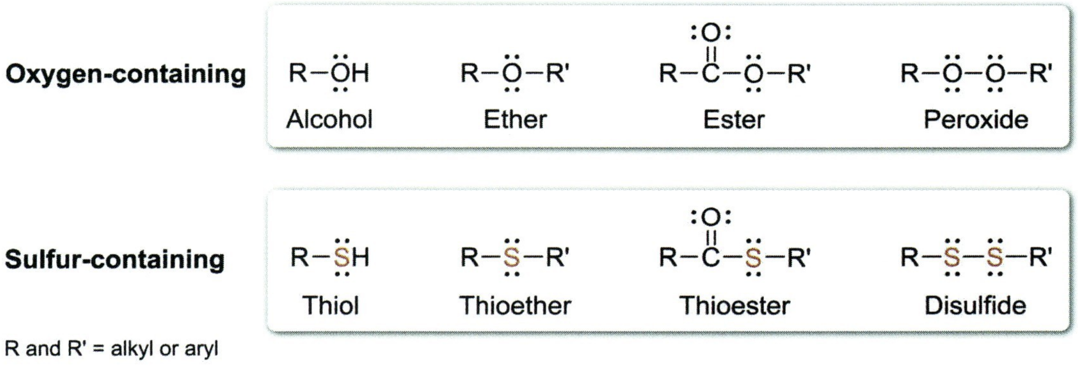

Oxygen-containing	R–OH Alcohol	R–O–R' Ether	R–C(=O)–O–R' Ester	R–O–O–R' Peroxide
Sulfur-containing	R–SH Thiol	R–S–R' Thioether	R–C(=O)–S–R' Thioester	R–S–S–R' Disulfide

R and R' = alkyl or aryl

Figure 7.20 Comparison of oxygen- and sulfur-containing functional groups.

Oxygen and sulfur atoms have an identical number of valence electrons, and *many* parallels can be drawn between the physical and chemical properties of oxygen- and sulfur-containing functional groups. This lesson focuses on concepts regarding the chemical properties and reactivity of sulfur-containing functional groups that are most relevant to the exam.

7.3.01 Physical Properties of Sulfur-Containing Functional Groups

In terms of physical properties, thiols can be considered representative of all the sulfur-containing functional groups, which have physical properties similar to their corresponding oxygen-containing functional groups.

Boiling Point

Although sulfur is more electronegative than hydrogen and a sulfur-hydrogen bond has a dipole moment oriented toward the sulfur atom, it is a *smaller* dipole moment relative to the oxygen-hydrogen bond of an alcohol (ie, thiols are *less polar* than alcohols, Figure 7.21).

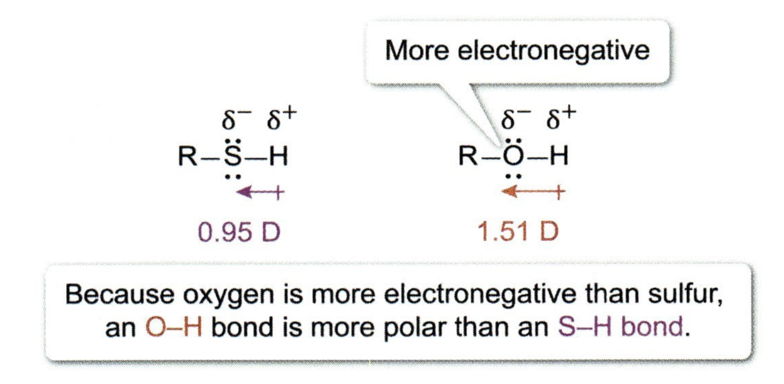

Figure 7.21 Comparison of the bond polarity of thiols and alcohols.

Consequently, thiol boiling points are *between* those of alkanes (no dipole-dipole forces) and alcohols (stronger dipole-dipole forces) of comparable molecular weight (Table 7.4).

Table 7.4 Boiling points of alkanes, thiols, and alcohols.

Molecular weight range (g/mol)	Alkane	Boiling point (°C)	Thiol	Boiling point (°C)	Alcohol	Boiling point (°C)
44–48	$CH_3CH_2CH_3$	−42	CH_3SH	6	CH_3CH_2OH	78
58–62	$CH_3(CH_2)_2CH_3$	−1	CH_3CH_2SH	35	$CH_3(CH_2)_2OH$	97
72–76	$CH_3(CH_2)_3CH_3$	36	$CH_3(CH_2)_2SH$	68	$CH_3(CH_2)_3OH$	118

Increasing boiling point — Increasing boiling point — Increasing boiling point

For a similar weight, the boiling points of thiols are greater than those of alkanes and less than those of alcohols.

Alternatively, this difference can also be viewed as a consequence of hydrogen bonding. Although the thiol sulfur-hydrogen bond is polar, sulfur is *not* capable of true hydrogen bonding. Therefore, the *decrease* in the boiling point of a thiol relative to an alcohol could be viewed as a result of the *absence* of hydrogen bonding.

The usual boiling point trends for organic molecules are also observed for thiols:

- Boiling points *increase* with molecular weight due to a *greater* number of electrons, a *larger* surface area, and a corresponding *increase* in London dispersion forces.
- Branching *decreases* the surface area, causing a *decrease* in the boiling point.

Solubility

Molecules with sulfur-containing functional groups have a solubility profile similar to those of other organic compounds in nonpolar to moderately polar organic solvents. However, thiols tend to be *less soluble* in water than their corresponding alcohols due to their reduced polarity.

Density

In general, sulfur-containing molecules remain less dense than water (1 g/mL), even though the density of a molecule with sulfur-containing functional groups tends to be *slightly greater* than the density of many other organic molecules.

Odor

Most sulfur-containing compounds tend to have a characteristic foul odor. The human sense of smell is especially sensitive to the odors of sulfur-containing compounds and can often detect very low concentrations of sulfur as an adaptation that allows humans to avoid rotting food and other harmful environmental elements.

7.3.02 Acidity and Nucleophilicity of Thiols

Among the sulfur-containing functional groups, thiols have two unique chemical properties that are relevant to the exam: acidity and nucleophilicity.

The sulfur-hydrogen bond of a thiol is less polar than the oxygen-hydrogen bond of an alcohol. Given the definition of a Brønsted-Lowry acid, it might seem reasonable to conclude that a *more polar* bond to an acidic hydrogen would be *easier* to break. However, this is not the case; a thiol is typically several orders of magnitude *more acidic* than an alcohol (Figure 7.22).

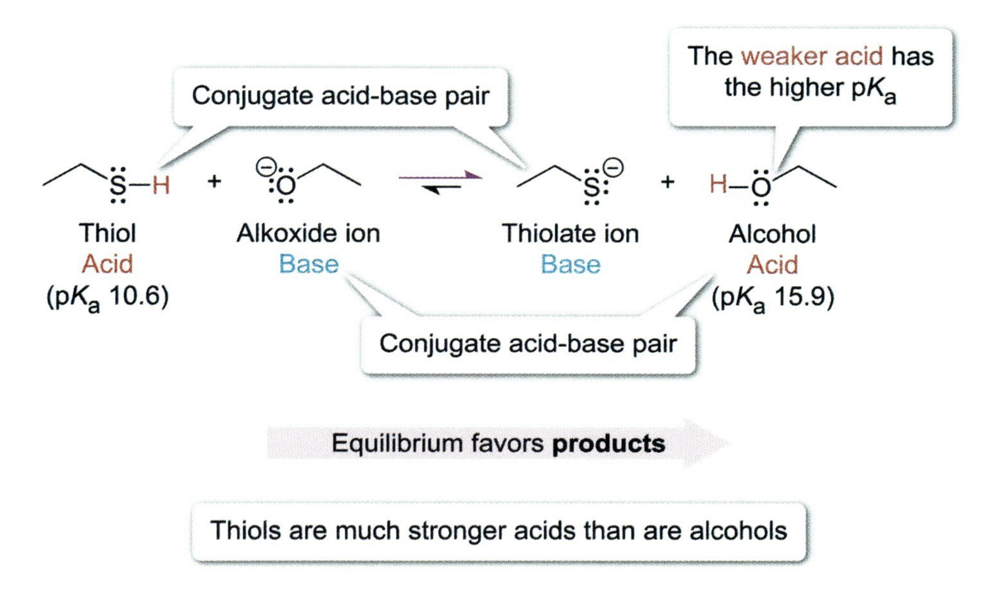

Figure 7.22 Relative acidity of a thiol and an alcohol.

A sulfur atom is *more* polarizable (Concept 5.1.05) than an oxygen atom, and therefore the conjugate base of a thiol (a **thiolate ion**, R–S$^-$) has a *greater* ability to stabilize a negative charge than an alkoxide ion. The pK_a difference between thiols and alcohols is large enough that hydroxide bases or alkoxide ions are sufficient to fully deprotonate a thiol to its thiolate ion.

The lower electronegativity and higher polarizability also makes the sulfur atom of a thiol a *much* stronger nucleophile than the oxygen atom of an alcohol (Figure 7.23). Similarly, the thiolate conjugate base is more nucleophilic than the thiol conjugate acid (Concept 5.3.01). Sulfur nucleophiles play an essential role in the action of many key enzymes and biochemical processes.

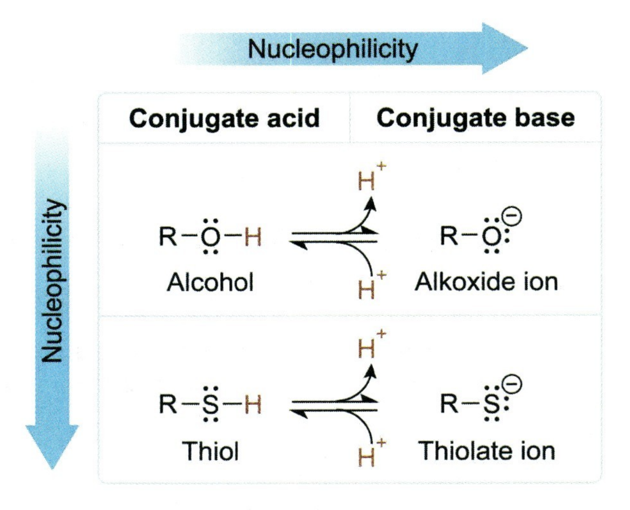

Figure 7.23 Comparative nucleophilicity of alcohols, alkoxides, thiols, and thiolates.

7.3.03 Disulfide Bonds

A disulfide bond joins two sulfur atoms through a sulfur-sulfur single bond. Each sulfur atom in a disulfide group is typically attached to a single alkyl group, and the disulfide has the general structure of R–S–S–R′ (Concept 2.8.04).

A disulfide group is formed when two thiols undergo an oxidation reaction by the removal of an equivalent of hydrogen gas (H_2) (Figure 7.24). This type of net change often takes place through the loss of both a hydride ion ($H:^-$) and a proton (H^+). However, this mechanism is beyond the scope of the exam.

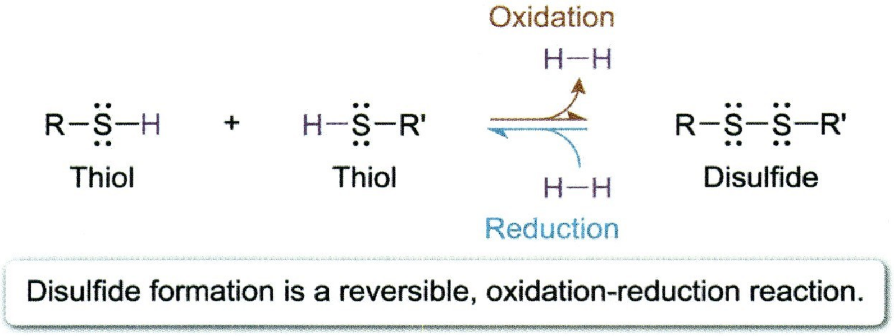

Oxidation
H–H
R–S–H + H–S–R' ⇌ R–S–S–R'
Thiol Thiol Disulfide
H–H
Reduction

Disulfide formation is a reversible, oxidation-reduction reaction.

Figure 7.24 Dynamic equilibrium between two thiols and a disulfide.

The formation of a disulfide group is a *reversible* process; a disulfide can be cleaved into two thiols through a complementary reductive process. The interconversion of two thiols and a disulfide is an example of a dynamic equilibrium.

Disulfide bonds play an essential role in the structure of proteins, in which they are formed through the oxidative coupling of thiol groups on the sidechain of the amino acid cysteine.

Lesson 8.1

Structure and Physical Properties of Alcohols

Introduction

An **alcohol** is a functional group containing an alkyl group attached to a hydroxyl group (–OH). This lesson focuses on the structural and physical characteristics of alcohols.

8.1.01 Structural Characteristics of Alcohols

The oxygen atom of an alcohol is bonded to *two* atoms (R and H) and has *two* pairs of nonbonding electrons, has four electron domains, and is sp^3 hybridized (Figure 8.1). Therefore, this oxygen atom has a tetrahedral electron geometry, a bent molecular geometry, and bond angles of approximately 109.5°.

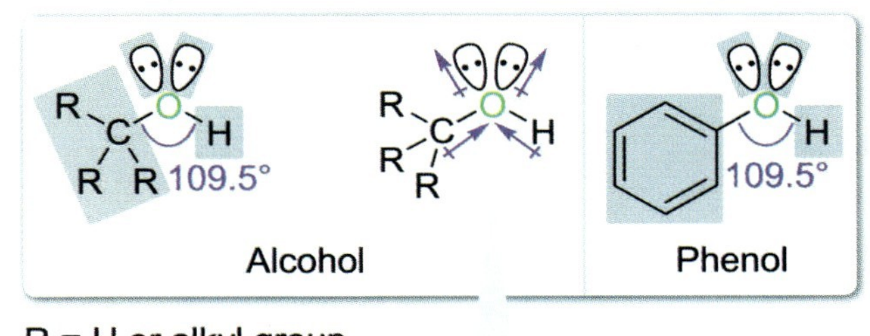

Figure 8.1 Structural characteristics of alcohols and phenols.

An alcohol contains a polar **hydroxyl group** (ie, –OH) and a nonpolar alkyl group (R). Combining the vector sum of the C–O and O–H bond dipole moments with the dipole moments from the nonbonding electrons on the oxygen creates a molecular dipole moment oriented toward the lone pairs. The polar nature of an alcohol facilitates interaction with other polar molecules and impacts its physical properties (see Concept 8.1.03).

Phenols are a subclass of alcohols that have a hydroxyl group attached to a carbon atom on a benzene ring. Although the oxygen atom of a phenol has a *similar* orbital structure as other alcohols, the *reactivity* of phenols is *different* due to the influence of the adjacent aromatic ring (see Lesson 8.3).

8.1.02 Classification of Alcohols

An alcohol can be categorized by the number of alkyl groups on the carbon bearing the hydroxyl group (Figure 8.2). The hydroxyl group (–OH) of **methyl alcohol** (CH_3OH) is attached to a carbon bonded to *zero* alkyl groups. In a **primary (1°) alcohol**, the –OH group is attached to a 1° carbon atom, whereas the –OH group on a **secondary (2°) alcohol** is attached to a 2° carbon. The –OH group on a **tertiary (3°) alcohol** is bonded to a 3° carbon.

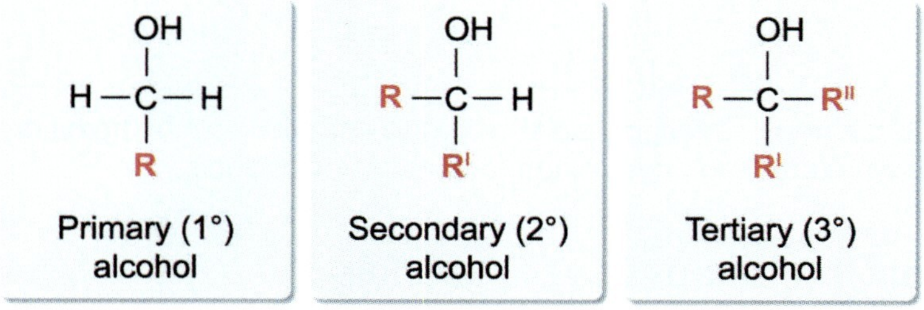

OH | H—C—H | R	OH | R—C—H | R'	OH | R—C—R'' | R'
Primary (1°) alcohol	Secondary (2°) alcohol	Tertiary (3°) alcohol

Figure 8.2 Classification of alcohols.

8.1.03 Physical Properties of Alcohols

Boiling Point

In addition to London dispersion forces and dipole-dipole interactions, an alcohol has *both* a hydrogen bond donor and acceptor, allowing for hydrogen bonding with other alcohols (Concept 2.3.03). As a result, alcohols have *higher* boiling points than either alkanes or ethers (Table 8.1).

Table 8.1 Boiling points of alkanes, ethers, and alcohols.

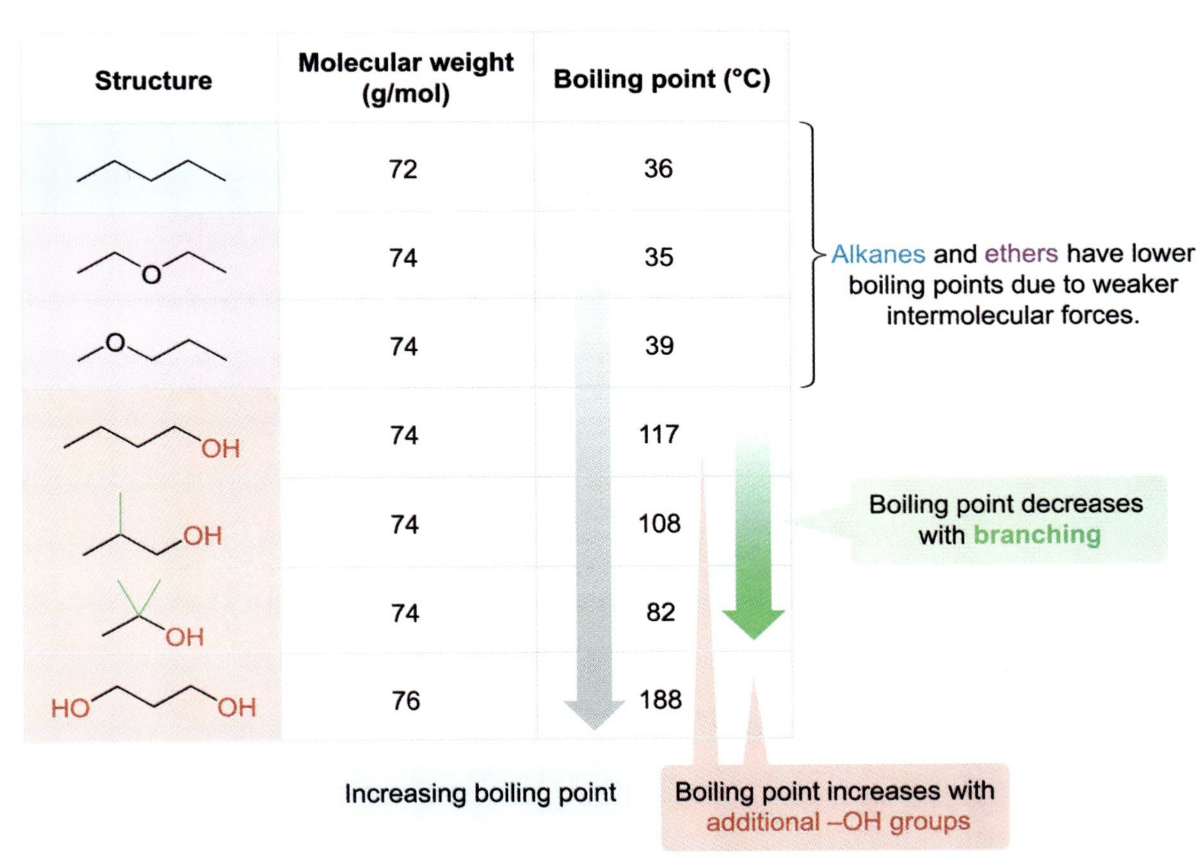

Structure	Molecular weight (g/mol)	Boiling point (°C)
	72	36
O	74	35
O	74	39
OH	74	117
OH	74	108
OH	74	82
HO OH	76	188

As observed for other functional groups, the boiling point of an alcohol *decreases* with branching. Molecules with more than one hydroxyl group (eg, diols, triols) have even higher boiling points, due to the presence of a larger number of hydrogen bond donors and acceptors.

Solubility

Based on the principle of "like dissolves like" (Concept 2.4.02), the *polar* hydroxyl group of an alcohol interacts favorably with *polar* water and is *hydrophilic*, whereas the *nonpolar* alkyl group does *not* interact favorably with water and is *hydrophobic*.

Because alcohols and water can form hydrogen bonds, alcohol molecules integrate well into the hydrogen bonding network of an aqueous solution (Figure 8.3). A single polar hydroxyl group typically allows a linear alkyl group chain with up to three carbon atoms to become fully solvated by surrounding water molecules. Alcohols with three or fewer carbon atoms are **miscible** in water and can dissolve (or mix) in *any* ratio.

Alcohols with more than one hydroxyl group are more polar and tend to have higher solubility. For example, because methanol and ethylene glycol both have a 1:1 C:O atom ratio, they have similar *ratios* of polar to nonpolar character and similar aqueous solubility.

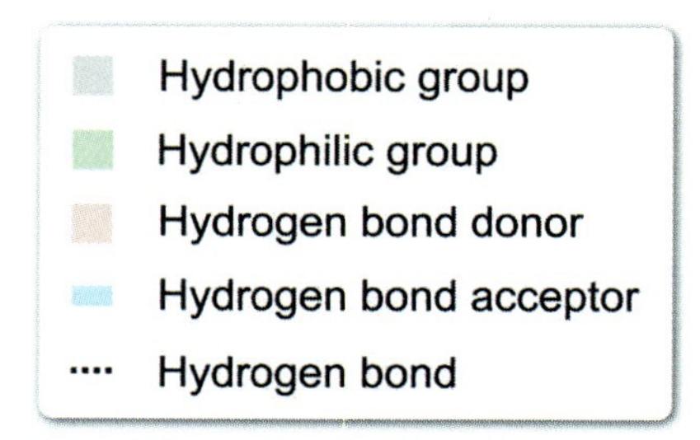

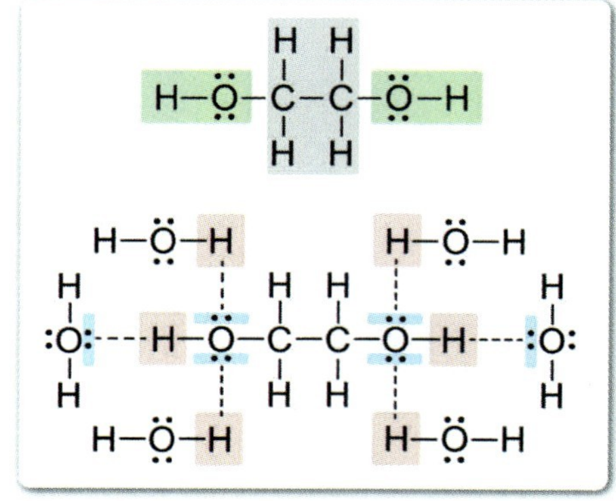

Alcohol solubility trend

$H_3C–OH$	Miscible
OH	9.1% soluble
OH	0.6% soluble

- Alcohol water solubility results from hydrogen bonding through the hydrophilic group
- Each –OH groups increases hydrogen bonding capacity

Increasing alkyl chain length *decreases* water solubility

Figure 8.3 Ethylene glycol hydrogen bonding with water and the solubility trend of alcohols.

Aqueous solubility of an alcohol *decreases* for alkyl chain lengths greater than four carbon atoms, whereas longer chains have lower solubility. Alcohols with a *branched* alkyl group have a smaller surface area and *greater* water solubility. For example, *tert*-butanol (a branched chain alcohol) is miscible in water, whereas 1-butanol (a linear chain alcohol) has 9.1% water solubility at 25 °C.

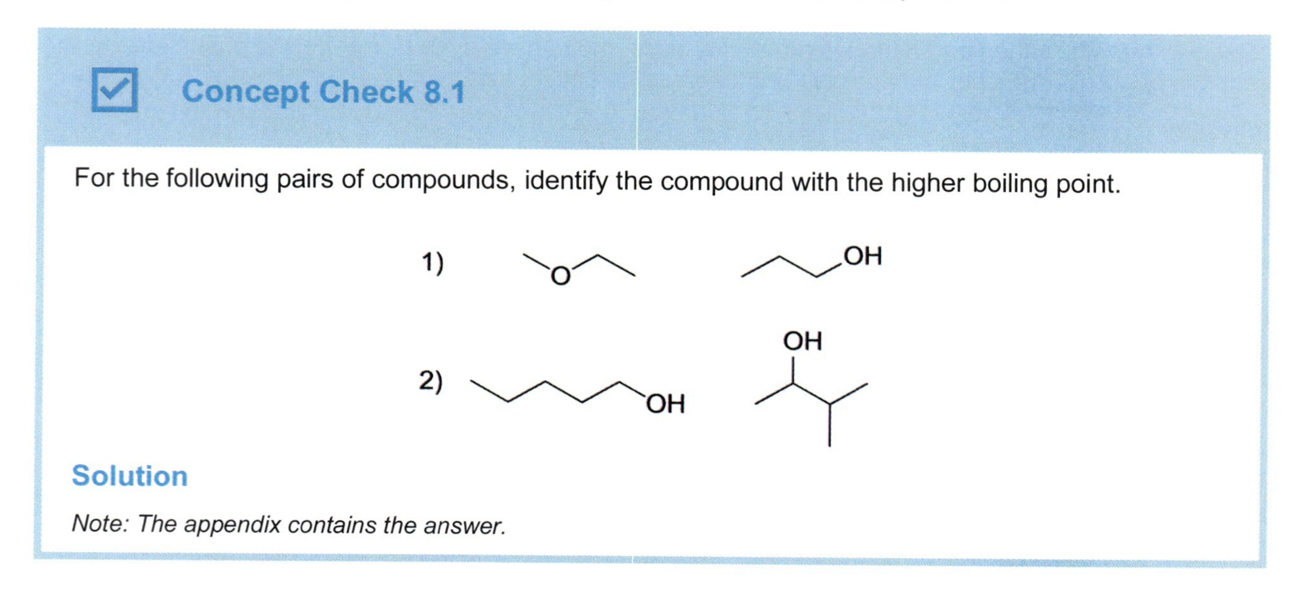

Concept Check 8.1

For the following pairs of compounds, identify the compound with the higher boiling point.

Solution

Note: The appendix contains the answer.

Lesson 8.2

Synthesis of Alcohols

Introduction

Because alcohols undergo an array of chemical reactions and are highly useful for organic synthesis (Lesson 8.3), reactions that result in the *formation* of alcohols are equally useful. This lesson covers reactions that are used to synthesize alcohols:

- Substitution reactions
- The Grignard reaction with an aldehyde or ketone
- The reduction of a carbonyl or carboxylic acid derivative

8.2.01 Alcohols through Substitution Reactions

An alcohol can be prepared by either an S_N1 reaction (Concept 5.4.02) or an S_N2 reaction (Concept 5.4.03) with an alkyl halide. The classification of the alkyl halide generally determines the classification of the resultant alcohol and the type of substitution reaction that should be used:

- Methanol or 1° alcohols are prepared by S_N2 reactions.
- 3° alcohols are prepared by S_N1 reactions.
- 2° alcohols can be prepared by *either* an S_N1 or an S_N2 reaction.

Synthesis of Methanol and 1° Alcohols by S_N2 Reactions

Because the S_N2 reaction is bimolecular, *both* the alkyl halide and the nucleophile impact the reaction rate (Figure 8.4). Sterically unhindered alkyl halides and sterically unhindered nucleophiles in a polar aprotic solvent are ideal for an S_N2 reaction (see Concept 5.4.03). To form methanol or a 1° alcohol, an S_N2 reaction requires the hydroxide ion (HO^-), a sterically unhindered strong nucleophile. The hydroxide nucleophile approaches the substrate from the opposite side of the C–X bond (ie, a backside attack), resulting in a **Walden inversion**.

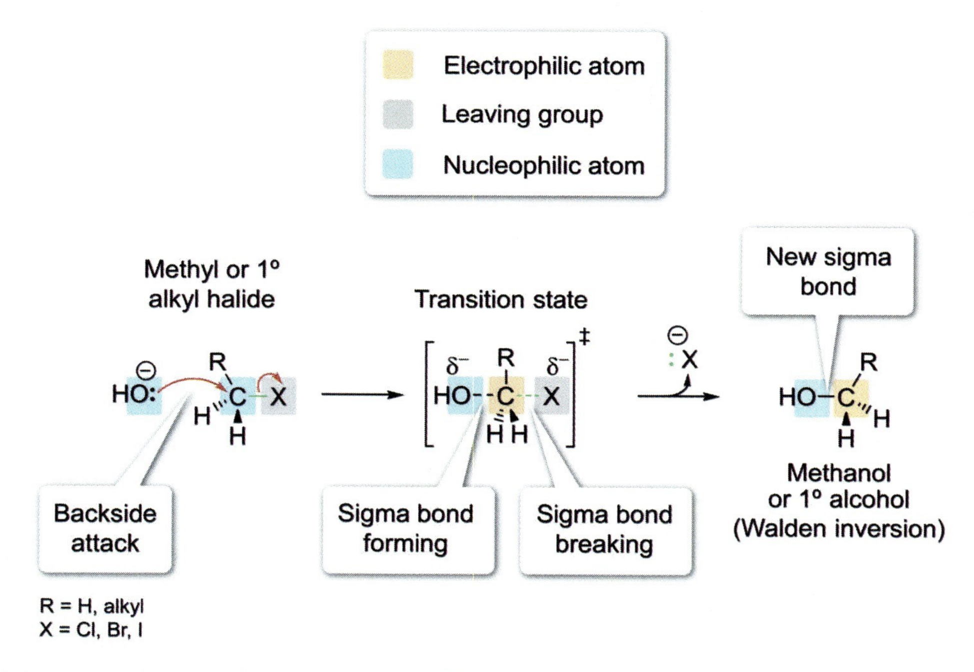

Figure 8.4 S_N2 reaction to synthesize methanol or a 1° alcohol.

Synthesis of 3° Alcohols by S_N1 Reactions

Because the S_N1 reaction is unimolecular, *only* the alkyl halide substrate impacts the reaction rate (Figure 8.5). In these reactions, water is typically both the reaction solvent and the weak nucleophile used to generate an alcohol product. Highly substituted electrophiles (eg, 3°) are the ideal substrates for an S_N1 reaction due to the inductive effect and hyperconjugation (see Concept 5.4.02).

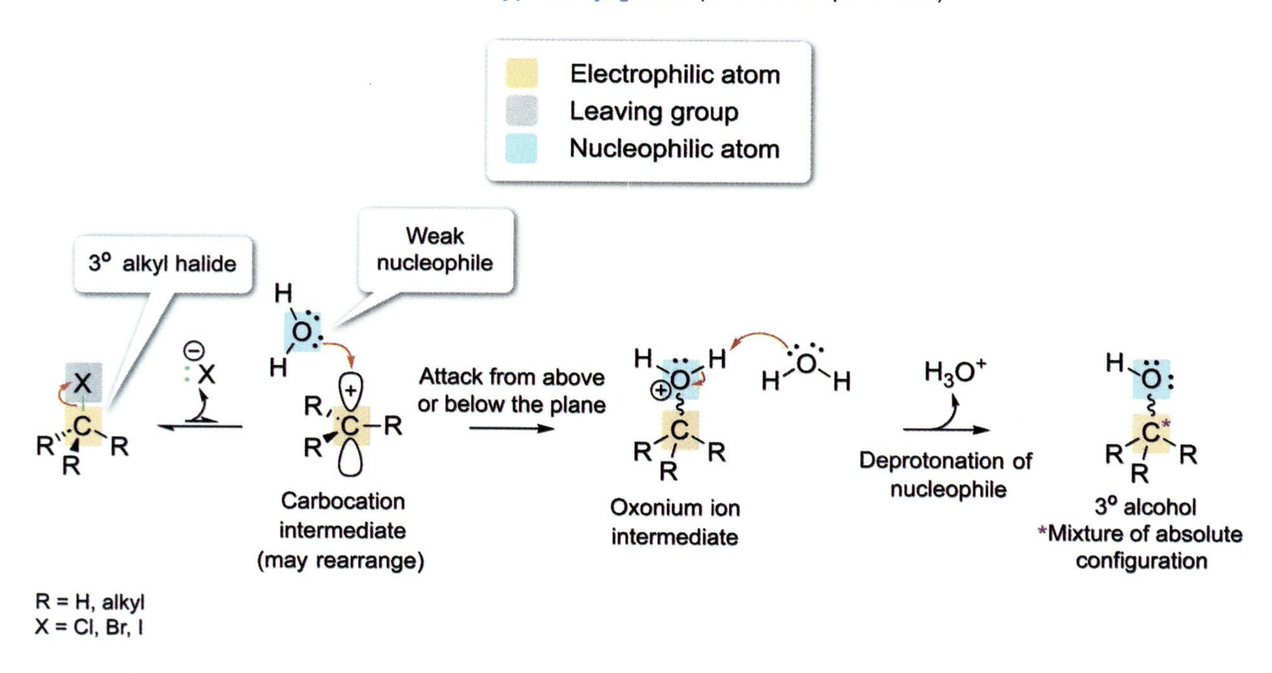

Figure 8.5 S_N1 reaction to synthesize a 3° alcohol.

The carbocation formed in an S_N1 reaction may undergo a carbocation rearrangement via a 1,2-hydride shift or a 1,2-methyl shift—processes that would *change* the position of the hydroxyl group in the product relative to the halogen in the reactant. When water attacks the carbocation to form an oxonium ion intermediate, the trigonal planar structure of the carbocation allows water to attack from above or below the plane, resulting in a *mixture* of absolute configurations.

Synthesis of 2° Alcohols by Substitution Reactions

Although 2° alcohols can be prepared using *either* an S_N2 or an S_N1 reaction, both approaches have limitations. In an S_N2 reaction, a 2° alkyl halide has greater steric hindrance, and there is usually greater competition with elimination reactions. In contrast, reaction of a 2° alkyl halide with water can be slow under S_N1 conditions and may result in a lower yield of product.

8.2.02 Grignard Reaction with an Aldehyde or Ketone

A **Grignard reagent** is an organometallic compound with the general formula R–Mg–X, in which R is an alkyl or aryl group and X is Cl, Br, or I. Concept 7.1.03 explains how a Grignard reagent is formed through the reaction of an alkyl halide and magnesium metal in an ether solvent. The R group in the Grignard reagent is strongly nucleophilic and reacts like $R:^- Mg^+X$.

The nucleophilic addition of a Grignard reagent to an aldehyde or ketone (a Type 3 electrophile) is known as a **Grignard reaction** and generates an alkoxide ion intermediate. Protonation of the alkoxide during an acidic workup results in the formation of an alcohol and an inorganic byproduct (Figure 8.6).

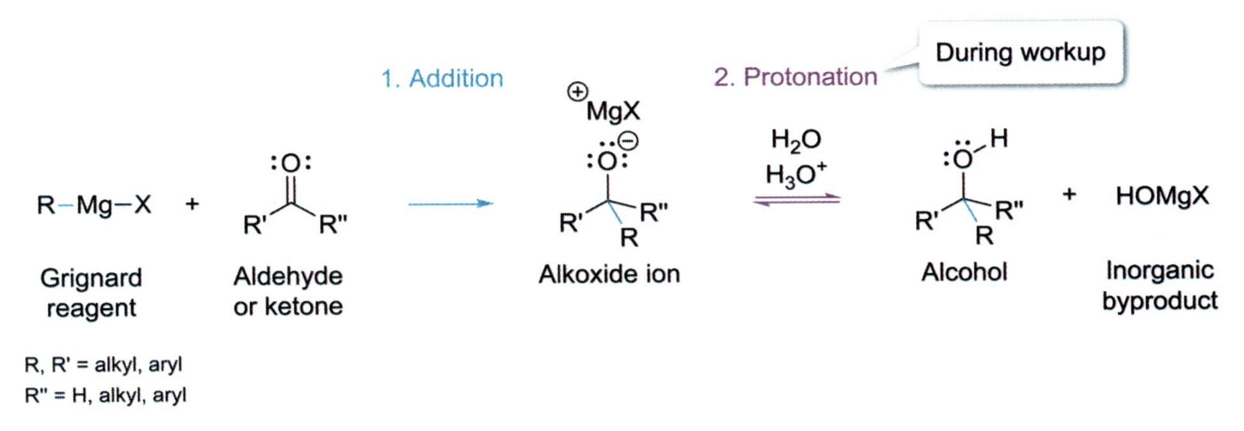

Figure 8.6 General Grignard reaction to form an alcohol.

The classification of alcohol formed in a Grignard reaction depends on the electrophile used in the reaction and is determined by the *sum* of R groups on the aldehyde or ketone and the Grignard reagent (Table 8.2):

- 1° alcohol (1 R group) = formaldehyde (0 R groups) + Grignard reagent (1 R group)
- 2° alcohol (2 R groups) = aldehyde (1 R group) + Grignard reagent (1 R group)
- 3° alcohol (3 R groups) = ketone (2 R groups) + Grignard reagent (1 R group)

Table 8.2 Formation of alcohols using a Grignard reaction.

Nucleophile	Electrophile	Addition product
	Formaldehyde 0 R groups	1° Alcohol 1 R group
R–Mg–X Grignard reagent 1 R group	Aldehyde 1 R group	2° Alcohol 2 R groups
	Ketone 2 R groups	3° Alcohol 3 R groups

Note: R, R', R" = alkyl, aryl

The Grignard reaction mechanism is a nucleophilic addition (Figure 8.7). First, the nucleophilic Grignard reagent attacks the electrophilic carbonyl carbon atom, generating an alkoxide ion. The alkoxide ion is protonated by water during workup with a dilute acid, forming the alcohol product.

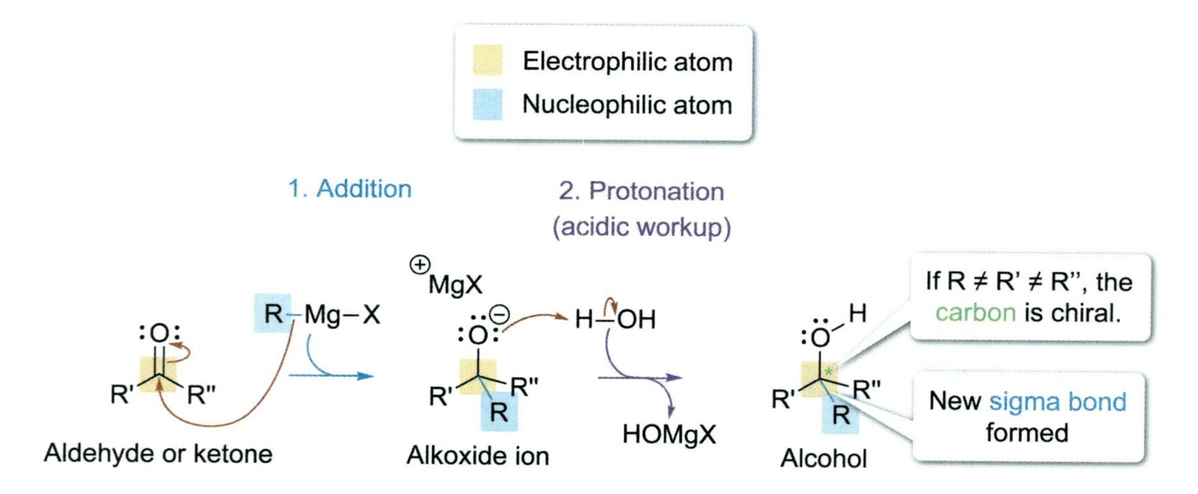

Figure 8.7 Mechanism of the Grignard reaction.

The acidic workup of a Grignard reaction *must* be performed after complete formation of the alkoxide salt. Because the Grignard reagent (a strong base) quickly reacts with water and acid, prematurely performing an acidic workup destroys the Grignard reagent and prohibits further Grignard reaction.

Because the electrophilic carbon of an aldehyde or a ketone is sp^2 hybridized and has a trigonal planar molecular geometry, the Grignard reagent can attack from *either* above or below the plane. Therefore, if the carbon bearing the hydroxyl group is chiral in the alcohol product, it will have a 1:1 mixture of absolute configurations. If there are no other chiral centers in the molecule, the alcohol product is racemic.

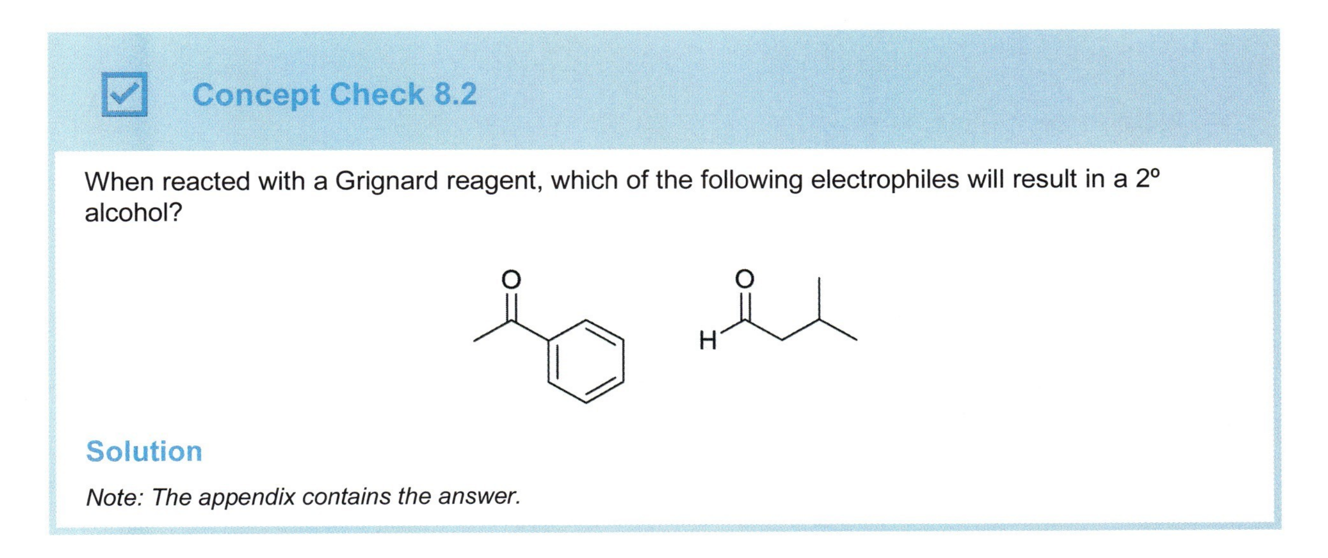

Concept Check 8.2

When reacted with a Grignard reagent, which of the following electrophiles will result in a 2° alcohol?

Solution

Note: The appendix contains the answer.

8.2.03 Reduction of Carbonyl-Containing Groups

An alcohol can be formed through reduction of a carbonyl-containing molecule. Each equivalent of the hydride ion that is used represents a two-electron reduction of the carbonyl carbon atom. The classification of alcohol formed through reduction is determined by the functional group that contains the carbonyl.

Reductions that Form 1° Alcohols

The reagents $LiAlH_4$ and $NaBH_4$ are both sources of the hydride ion and among the most common reducing agents for carbonyl-containing compounds. Reaction of an aldehyde with *either* $LiAlH_4$ or $NaBH_4$, followed by an acidic workup, forms a 1° alcohol (Figure 8.8).

O
R H
Aldehyde
R = alkyl, aryl

1. $LiAlH_4$
or $NaBH_4$
2. H_3O^+

H H
R OH
1° Alcohol

+ MH_3 + M^+ salt
M = Al or B
M = Li or Na

Figure 8.8 Hydride reduction of an aldehyde to form a 1° alcohol.

Treatment of certain carboxylic acid derivatives (eg, acid chlorides, anhydrides, esters) or a carboxylic acid with $LiAlH_4$ results in a *double* reduction (ie, two 2 e^- reductions) to form a 1° alcohol (Table 8.3). The most electrophilic carboxylic acid derivatives (ie, acid chloride, anhydride) are reduced by $NaBH_4$, whereas esters and carboxylic acids are *not* generally reduced by $NaBH_4$.

Table 8.3 Hydride reduction of an acid chloride, an anhydride, an ester, and a carboxylic acid.

Carbonyl-containing compound	Reactant	Reaction sequence	Reduced product
Acid chloride	RC(=O)Cl	1. 2 $LiAlH_4$ 2. H_3O^+ →	1° Alcohol (RCH_2OH)
		1. 2 $NaBH_4$ 2. H_3O^+ →	1° Alcohol (RCH_2OH)
Anhydride	RC(=O)OC(=O)R'	1. 2 $LiAlH_4$ 2. H_3O^+ →	1° Alcohol (RCH_2OH)
		1. 2 $NaBH_4$ 2. H_3O^+ →	1° Alcohol (RCH_2OH)
Ester	RC(=O)OR'	1. 2 $LiAlH_4$ 2. H_3O^+ →	1° Alcohol (RCH_2OH)
Carboxylic acid	RC(=O)OH	1. 2 $LiAlH_4$ 2. H_3O^+ →	1° Alcohol (RCH_2OH)

Cannot be reduced by $NaBH_4$

Note: R, R' = alkyl, aryl

A 1° alcohol can also be formed through the reduction of a carboxylic acid with borane (BH_3) (see Concept 10.3.03).

Reductions that Form 2° Alcohols

A 2° alcohol is formed through the reduction of a *ketone* with either $LiAlH_4$ or $NaBH_4$ (Figure 8.9). If the ketone is unsymmetrical (ie, it has two different alkyl groups), the hydride ion can attack the carbon atom from either side of the plane and the product will contain a 1:1 mixture of absolute configurations at this carbon atom.

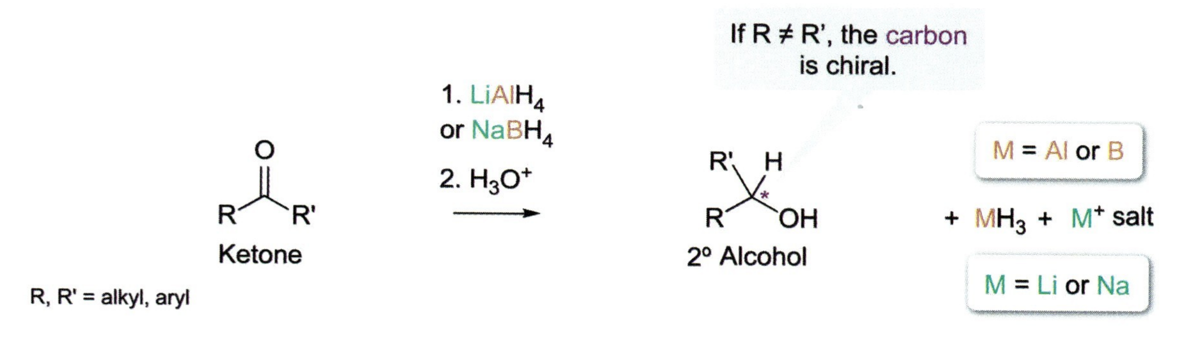

Figure 8.9 Hydride reduction of a ketone to form a 2° alcohol.

Reduction that Forms 3° Alcohols

Because a 3° alcohol has *three* alkyl groups on a carbon atom bearing a hydroxyl group, it is not possible to prepare a 3° alcohol using a hydride reduction of a carbonyl-containing compound.

Lesson 8.3

Reactions of Alcohols

Introduction

Alcohols are among the most important functional groups for chemical synthesis. This lesson describes several key reactions that can be performed with an alcohol:

- Oxidation into carbonyl-containing functional groups
- Deprotonation of the acidic hydroxyl group
- Reactions in which an alcohol is a nucleophile
- Esterification reactions
- Preparation and reactions of sulfonate esters
- Reaction with hydrohalic acids
- Reaction with phosphorus tribromide (PBr_3)
- Acidic dehydration of an alcohol
- Protection of alcohols with protecting groups

8.3.01 Oxidation of Alcohols

An alcohol may be oxidized through redox reactions. Most alcohol oxidation reactions represent a two-electron oxidation of the carbon atom bearing the hydroxyl group, in which the number of carbon-oxygen bonds increases by one. The outcomes of an alcohol oxidation *differ* depending on the alcohol classification.

Oxidation of Methanol and 1° Alcohols

Methanol (CH_3OH) and 1° alcohols (RCH_2OH) can be treated with an array of oxidizing agents to yield oxidized products (Table 8.4). Treatment with a *strong* oxidizing agent (eg, chromium-derived agents, MnO_4^-) results in a *double* ($2 \times 2\ e^- = 4\ e^-$) oxidation to a carboxylic acid; the aldehyde intermediate *cannot* be isolated using these reagents because an aldehyde tends to be *easier* to oxidize than the starting alcohol.

Table 8.4 Oxidation reactions of methanol and primary alcohols.

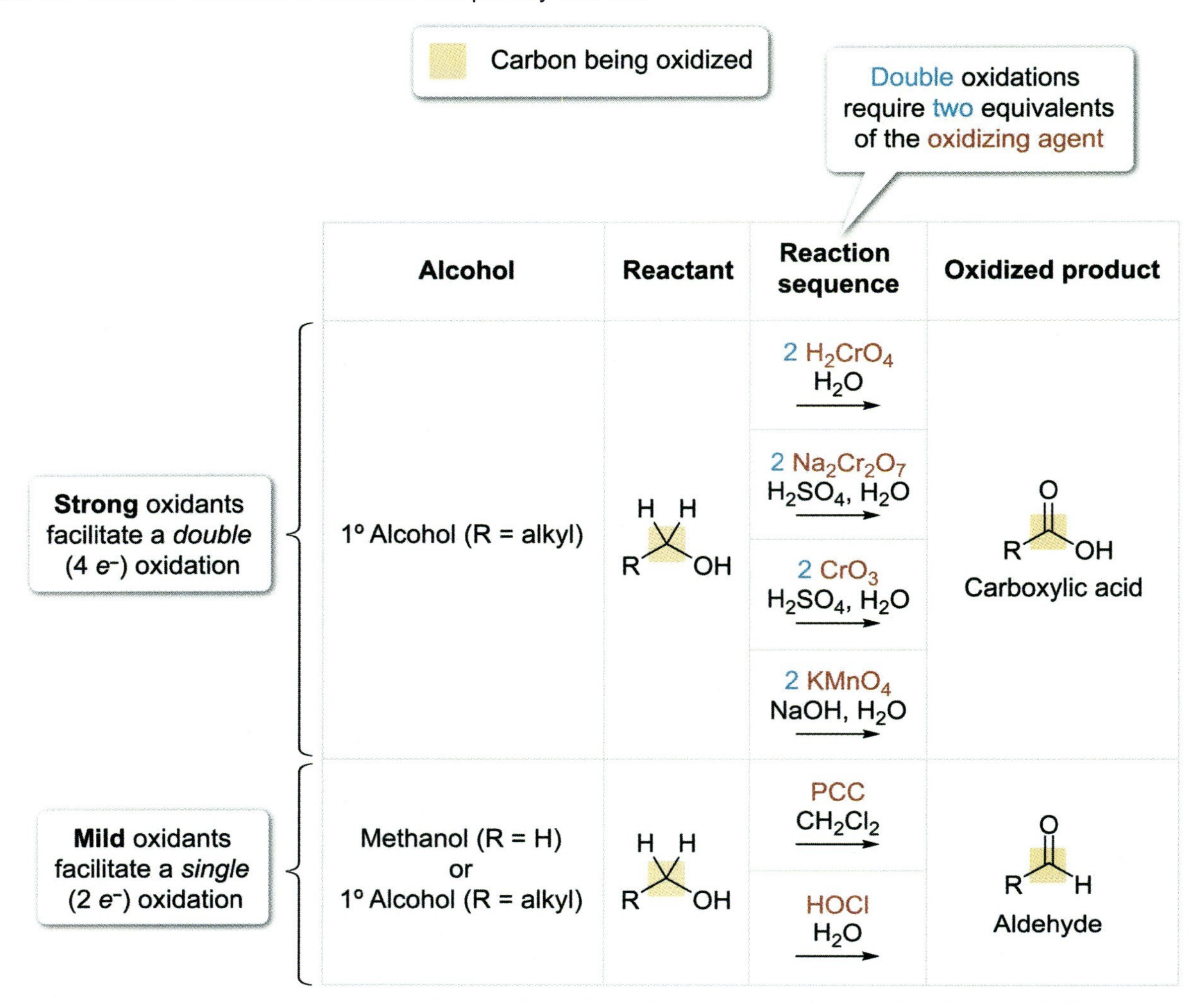

Alcohol	Reactant	Reaction sequence	Oxidized product
1° Alcohol (R = alkyl)	RCH_2OH	2 H_2CrO_4, H_2O →	Carboxylic acid ($RCOOH$)
		2 $Na_2Cr_2O_7$, H_2SO_4, H_2O →	
		2 CrO_3, H_2SO_4, H_2O →	
		2 $KMnO_4$, NaOH, H_2O →	
Methanol (R = H) or 1° Alcohol (R = alkyl)	RCH_2OH	PCC, CH_2Cl_2 →	Aldehyde ($RCHO$)
		HOCl, H_2O →	

Note: $Na_2Cr_2O_7$ is the anhydride salt of chromic acid and CrO_3 (ie, Jones reagent) is dehydrated chromic acid; both sets of reagents generate H_2CrO_4 in the presence of aqueous acid (H_2SO_4).

To recover an aldehyde product, a *milder* oxidizing agent is required. Treatment of methanol or a 1° alcohol with either pyridinium chlorochromate (PCC) or hypochlorous acid (HOCl) causes a *single* two-electron oxidation to yield an aldehyde. These agents allow for *selective* oxidation of functional groups that are prone to multiple successive oxidation reactions.

The mechanism for the oxidation of a 1° alcohol to a *carboxylic acid* by chromic acid (H_2CrO_4) is shown in Figure 8.10. First, the electron-deficient Cr (VI) atom is attacked by a lone pair from the hydroxyl group of the alcohol, and a pi bond becomes a lone pair on an oxygen atom.

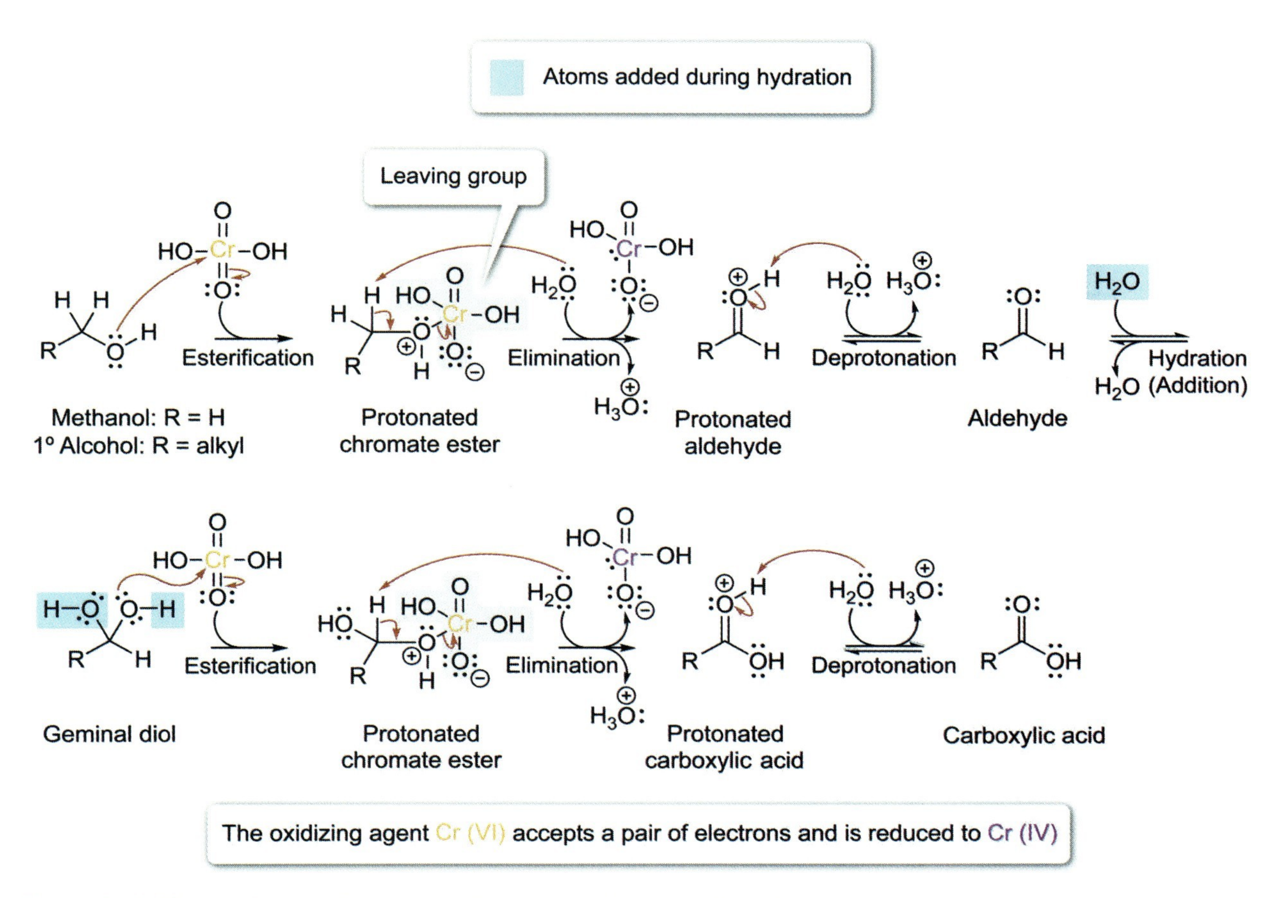

Figure 8.10 The mechanism for the chromic acid oxidation of methanol or a 1° alcohol.

Water acts as a base and removes one of the hydrogen atoms from the carbon being oxidized. The electrons from the broken carbon-hydrogen bond become a pi bond between carbon and oxygen (ie, a carbonyl group), and the bond between the alcohol-derived oxygen atom and the chromium atom breaks; in this step, the chromium accepts the lone pair and is reduced from Cr (VI) to Cr (IV). The acidic proton on the carbonyl is lost to the solvent to provide the aldehyde intermediate.

Under acidic conditions, the aldehyde intermediate undergoes a hydration reaction with a molecule of water to provide a geminal diol (ie, an aldehyde hydrate, Concept 9.3.2). The geminal diol undergoes a *similar* series of steps as the first oxidation with H_2CrO_4: one of the alcohols attacks H_2CrO_4, a base facilitates the removal of a hydrogen atom and the reduction of Cr (VI) to Cr (IV), and the protonated carboxylic acid loses an acidic hydrogen atom to the solvent to provide the carboxylic acid product.

Oxidation of 2° Alcohols

Secondary alcohols can be oxidized to ketones (Table 8.5). *All* oxidizing agents that react with methanol or a 1° alcohol (see Table 8.4) also facilitate the conversion of a 2° alcohol to a ketone. Under most conditions, it is *not possible* to oxidize a ketone to a carboxylic acid because the reaction would require a mechanism that breaks a carbon-carbon bond.

Table 8.5 Oxidation reactions of secondary alcohols.

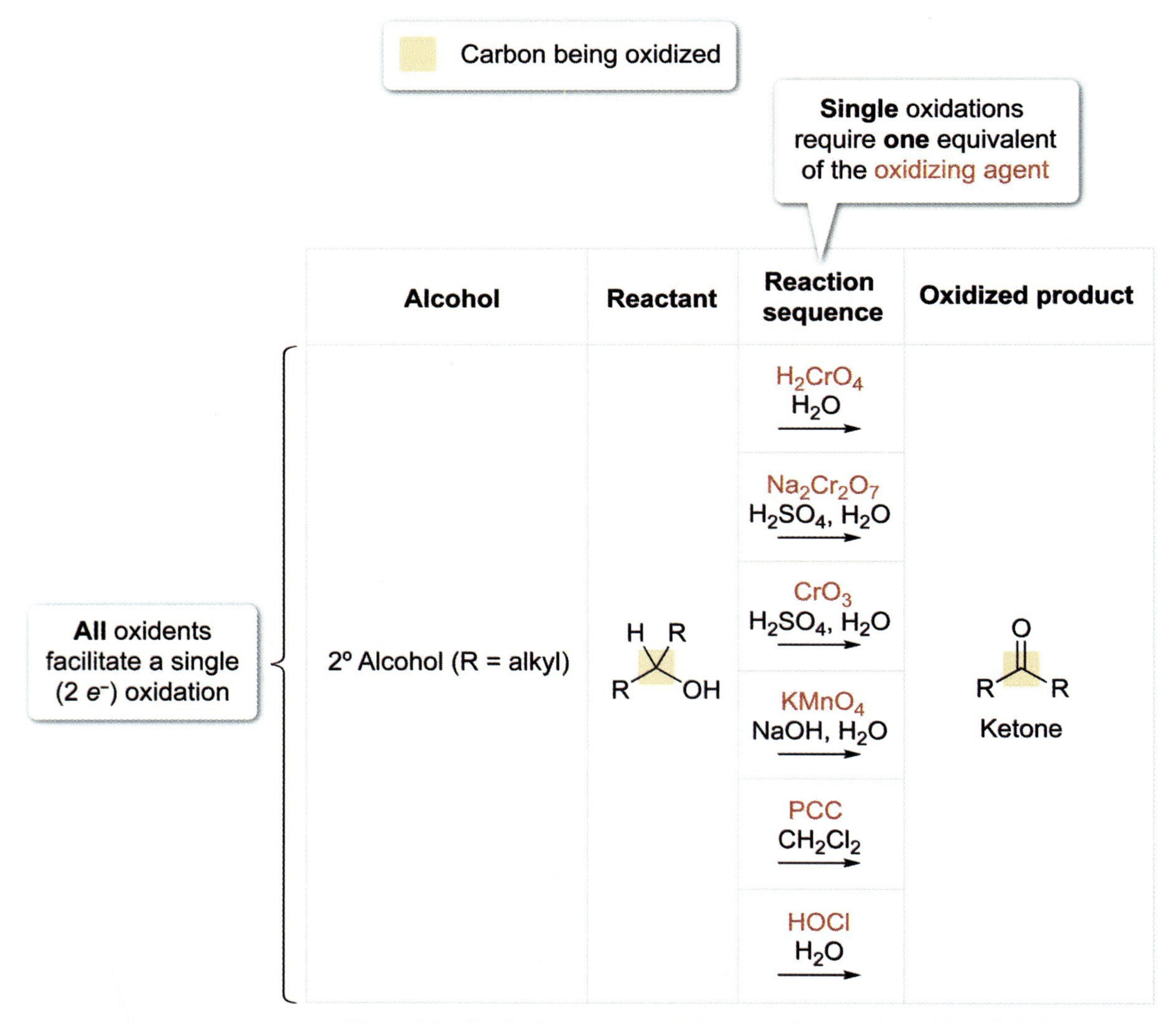

Note: $Na_2Cr_2O_7$ is the anhydride salt of chromic acid and CrO_3 (ie, Jones reagent) is dehydrated chromic acid; both sets of reagents generate H_2CrO_4 in the presence of aqueous acid (H_2SO_4).

Oxidation of 3° Alcohols

Under most conditions, it is *not possible* to oxidize a 3° alcohol. Such a reaction would require a mechanism that breaks one or more of the carbon-carbon bonds between an R group and the carbon atom bearing the hydroxyl group.

Biochemical Oxidation of Alcohols

Alcohols are frequently oxidized in biochemical systems. The oxidizing agent for biochemical oxidations is typically nicotinamide adenine dinucleotide (NAD^+). The nicotinamide ring in NAD^+ accepts a lone pair of electrons (in the form of the hydride ion at the 4-position on its nicotinamide ring, in a mechanistic fashion similar to the reduction of Cr (VI) to Cr (IV) in H_2CrO_4. The reduced form of NAD^+ is NADH, which serves as a mobile carrier of the electron pair from the accepted hydride ion (Figure 8.11).

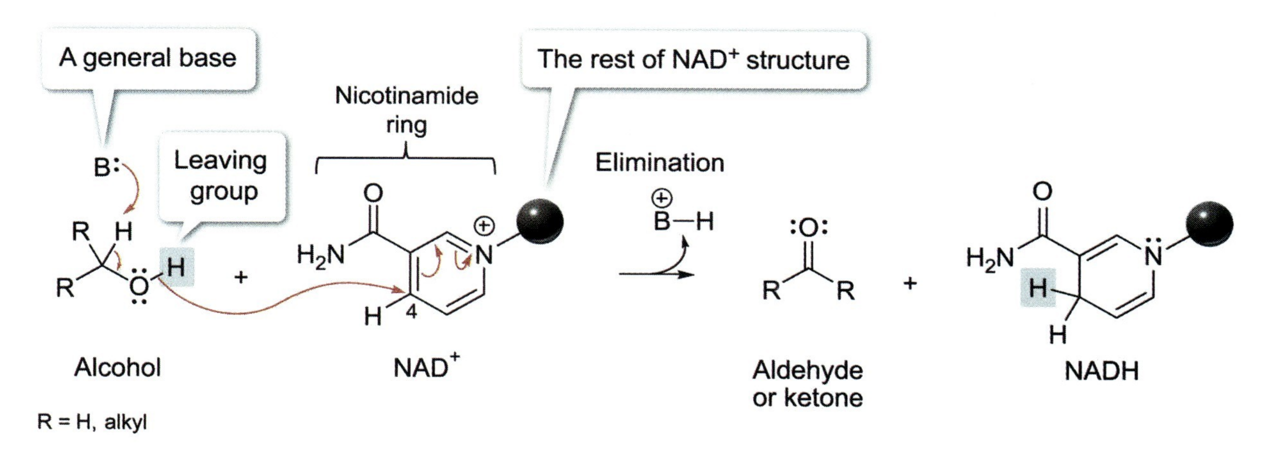

Figure 8.11 The mechanism for the biochemical oxidation of an alcohol with NAD^+.

Oxidation of Hydroquinones

Although it is usually challenging to oxidize an aromatic alcohol (eg, phenol), some aromatic *diols* (eg, *o*- and *p*-hydroquinone) undergo oxidation reactions. Figure 8.12 illustrates the oxidation reactions for the *para* system of molecules, which is generally the most frequently encountered. Unlike most of the redox chemistry discussed in this concept, *p*-hydroquinone undergoes *one*-electron oxidation reactions, up to a maximum of a *two*-electron oxidation. Each oxidation (or reduction) reaction is associated with a proton (H^+), which maintains the charge balance.

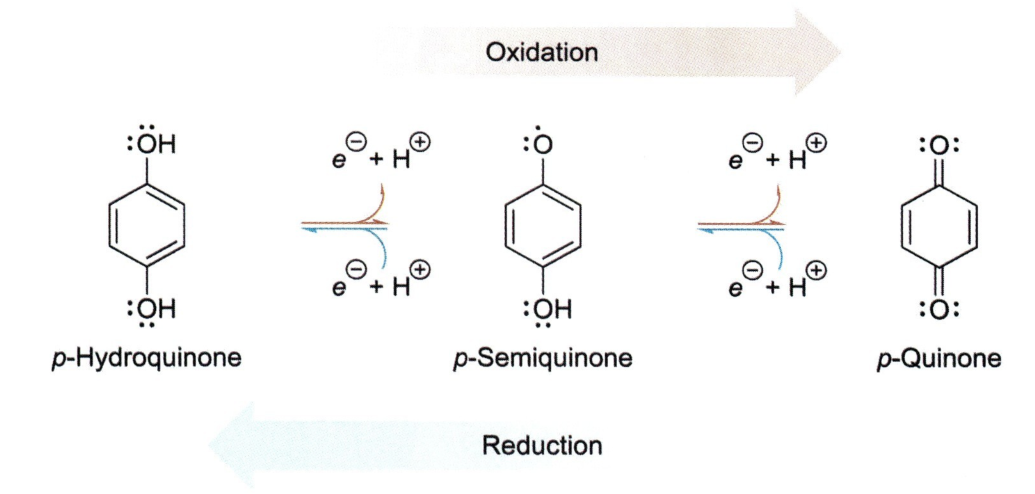

Figure 8.12 Redox reactions between a hydroquinone and a quinone.

The first one-electron oxidation of *p*-hydroquinone forms ***p*-semiquinone**, which contains an unpaired electron and is a free radical. A second one-electron oxidation of *p*-semiquinone forms *p*-quinone, whose structure contains carbonyl groups in place of the hydroxyl groups of *p*-hydroquinone. The net change of the two-electron oxidation process is a loss of 2 e^- and 2 H^+, the molecular equivalent of hydrogen gas (H_2) and one of the organic definitions of oxidation. Oxidation reactions between *p*-hydroquinone and *p*-quinone are reversible; *p*-quinone can be reduced to *p*-hydroquinone through two one-electron reduction reactions.

8.3.02 Acidity of Alcohols

Because of the polarity of the O–H bond, the hydroxyl hydrogen atom of an alkyl alcohol is a moderate Brønsted-Lowry acid with pK_a values ranging from 16 to 18 (Table 8.6). The alcohol's classification is a major determinant that correlates its structure to its pK_a. The conjugate base of an alcohol is an alkoxide ion, which bears a negative formal charge on its oxygen atom.

Table 8.6 Acidity of alcohols and basicity of alkoxide ions.

Sigma electron-donating group
Sigma electron-withdrawing group

	Conjugate acid		Conjugate base	
Alcohol name	**Structure**	**pK_a**	**Structure**	**pK_b**
1,1-Dimethylethanol (3° alcohol)	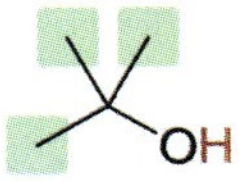 OH	18	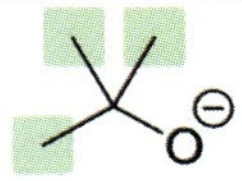 $O^{\ominus}$	–4
1-Methylethanol (2° alcohol)	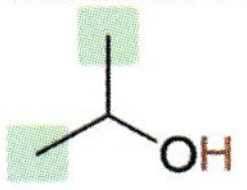 OH	16.5	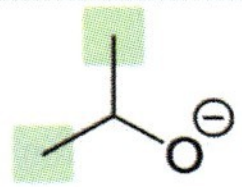 $O^{\ominus}$	–2.5
Ethanol (1° alcohol)	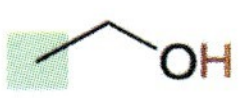 OH	15.9	 $O^{\ominus}$	–1.9
Water	H–OH	15.7	H–$O^{\ominus}$	–1.7
Methanol (methyl alcohol)	H_3C–OH	15.5	H_3C–$O^{\ominus}$	–1.5
2-Chloroethanol	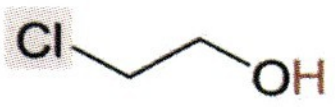 Cl, OH	14.3	 Cl, $O^{\ominus}$	–0.3
2,2-Dichloroethanol	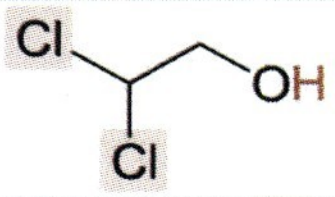 Cl, Cl, OH	13.1	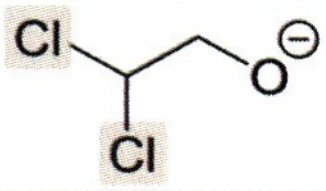 Cl, Cl, $O^{\ominus}$	0.9
2,2,2-Trichloroethanol	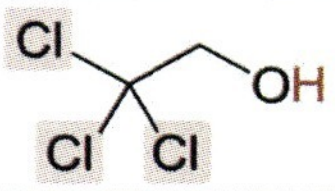 Cl, Cl, Cl, OH	12.2	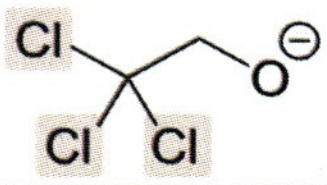 Cl, Cl, Cl, $O^{\ominus}$	1.8
Phenol	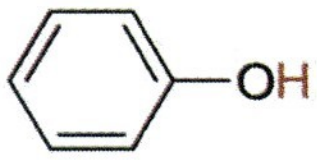 OH	10	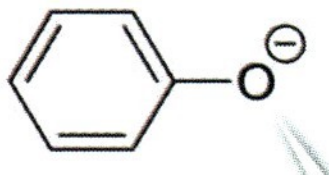 $O^{\ominus}$	4

More acidic

More basic

Delocalized by resonance

Sigma electron-donating groups **decrease** acidity
Sigma electron-withdrawing groups **increase** acidity

The stability of an alkoxide ion is influenced by the inductive effect, by which sigma electron-donating groups (eg, alkyl groups) *destabilize* the negative charge and sigma electron-withdrawing groups (eg, electronegative atoms, halogens) *stabilize* the negative charge (see Concept 6.5.03). Because an alkoxide ion with *greater* stability is more likely to be present in a conjugate acid-base equilibrium, the acidity of an alcohol *increases* with:

- *Lower* numbers of alkyl groups on the carbon bearing the hydroxyl group (ie, methyl > 1° > 2° > 3°)
- *Higher* numbers of halogen atoms in *closer* proximity to the hydroxyl group

A phenol is an alcohol that has a hydroxyl group attached to a carbon atom of an aromatic benzene ring. The conjugate base of phenol is a phenoxide ion. A phenol is *much* more acidic (pK_a 10) than an alkyl alcohol due to the ability of a phenoxide ion to delocalize its negative charge through resonance.

Generation of Alkoxide Ions and Phenoxide Ions

The pK_a range of an alkyl alcohol (pK_a 16–18) is similar to that of water (pK_a 15.7); therefore, a hydroxide ion (HO^-) is generally *not* a strong enough base to fully deprotonate an alcohol and convert it to an alkoxide ion. Concept 7.2.03 describes how treatment of an alcohol with an alkali metal (eg, sodium [Na] or potassium [K]) or sodium hydride (NaH) generates the conjugate alkoxide ion. However, the hydroxide ion *can* be used to remove the acidic proton in phenol (pK_a 10) due to the larger difference in pK_a values.

8.3.03 Alcohols as Nucleophiles

The oxygen atom in either an alcohol or an alkoxide ion has lone pairs and can react as a nucleophile in a substitution reaction (Figure 8.13). Because an alcohol is a weak nucleophile, it is more likely to undergo an S_N1 reaction. Because an alkoxide ion is a strong base, it almost always reacts in an S_N2 reaction.

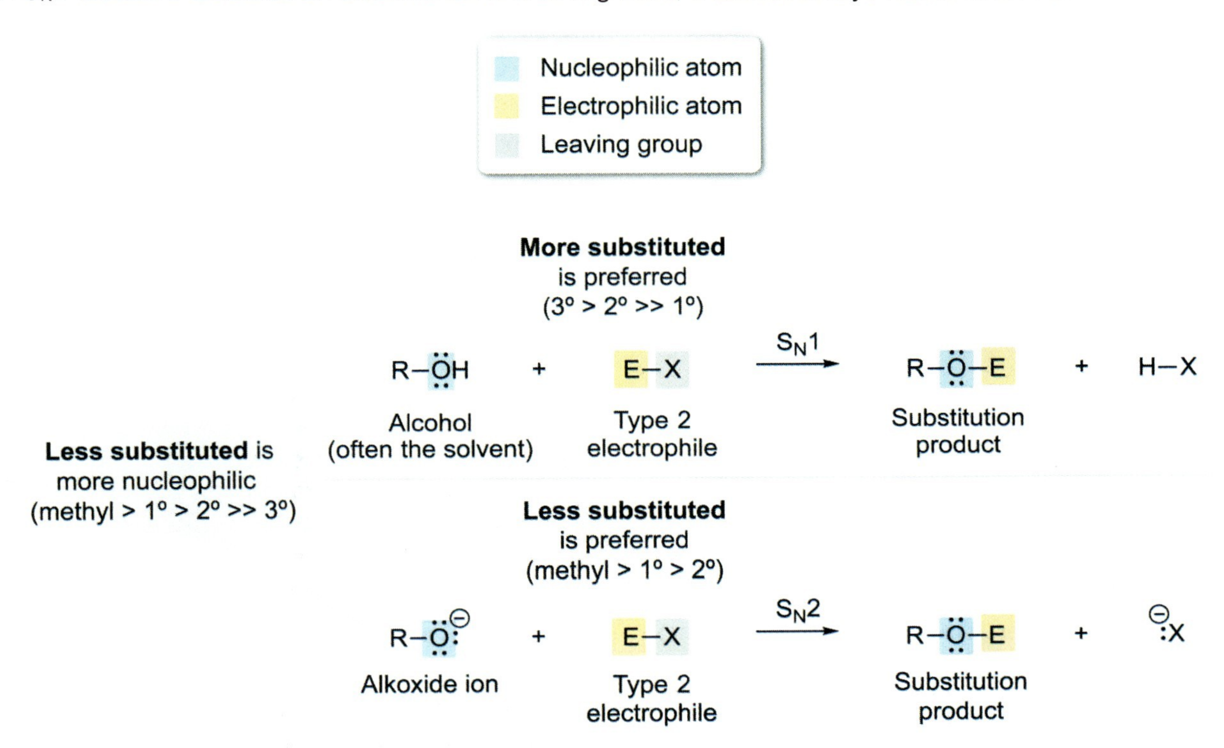

Figure 8.13 Alcohol and alkoxide ion nucleophiles in substitution reactions.

For both mechanisms, nucleophiles with low steric hindrance (ie, methyl, 1°) are ideally suited for substitution reactions. *Bulky* alcohol or alkoxide nucleophiles (ie, 2°, 3°) have a *slower* rate of reaction and, particularly for bulky alkoxide ions, a higher likelihood of reacting as a base (eg, to promote an E2 reaction) (see Concept 5.6.03).

Alcohol nucleophiles in an S_N1 reaction react with Type 2 electrophiles that are either highly substituted (ie, 3°, 2°) or stabilized by resonance (eg, allylic or benzylic). In contrast, alkoxide nucleophiles in an S_N2 reaction react fastest with electrophiles that are either less substituted (eg, methyl, 1°) or have transition states that are stabilized by conjugation (eg, allylic, benzylic).

8.3.04 Esterification

If no other context is provided, the term ester is usually assumed to be the product of a condensation reaction between an alcohol and a *carboxylic* acid (Table 8.7). Other types of esters can be formed through a reaction of an alcohol with different types of acids, including sulfonate esters, nitrate esters, and phosphate esters (eg, phosphoesters).

Table 8.7 Types of esters.

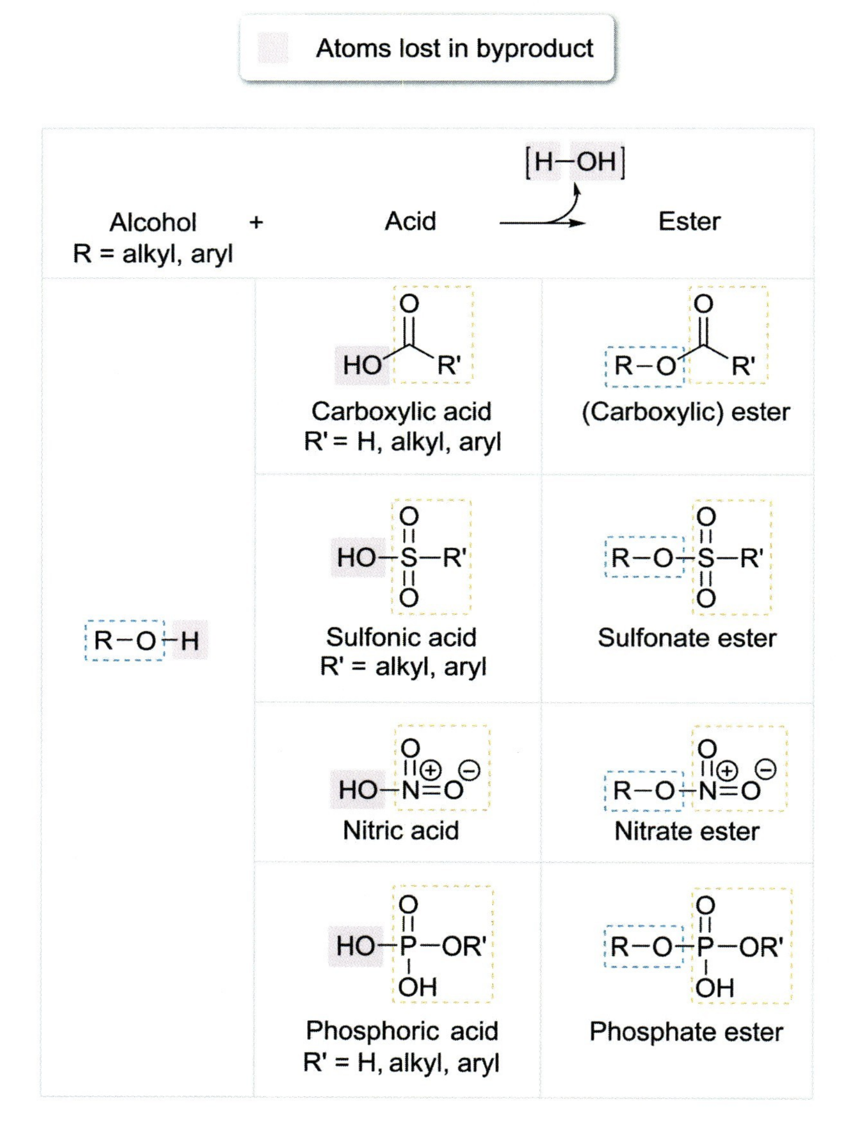

An **esterification reaction** is a chemical reaction that results in the formation of an ester and occurs between either an alcohol or an alkoxide ion and an acid or acid derivative.

8.3.05 Preparation and Reaction of Sulfonate Esters

Conversion of an Alcohol to a Sulfonate Ester

The oxygen atom of an alcohol or alkoxide ion is nucleophilic and reacts with Type 2 electrophiles through either S_N1 or S_N2 reactions (Concept 8.3.03). However, although the carbon atom bearing the alcohol or alkoxide ion functional group is electron deficient due to the electronegative oxygen atom, it is a *poor* electrophile because both the hydroxide ion and the oxide ion are *poor* leaving groups (Figure 8.14).

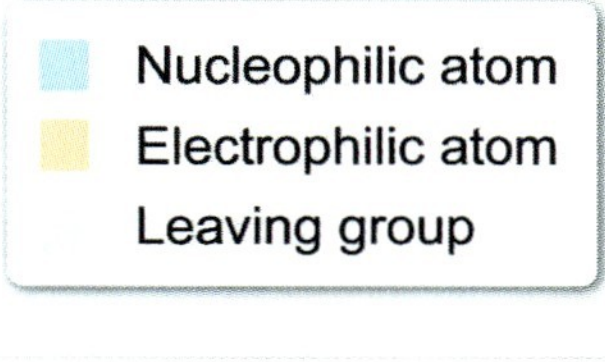

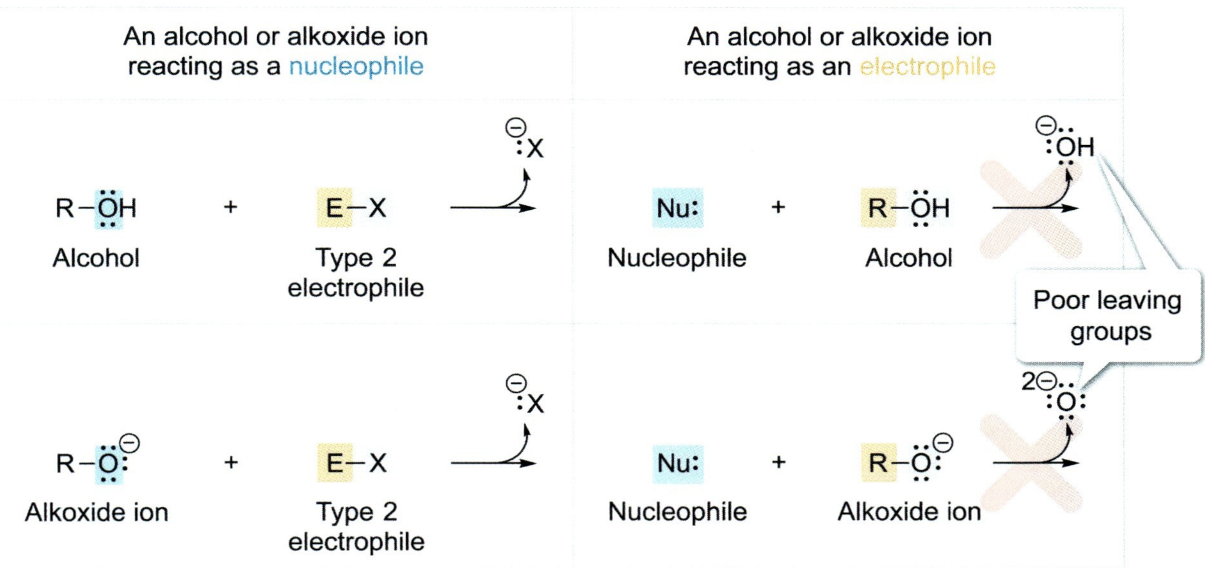

Figure 8.14 Comparison of alcohols and alkoxide ions as nucleophiles and electrophiles.

One way that chemists turn alcohols into *good* leaving groups involves the use of sulfonate esters. A **sulfonate ester** is formed through the condensation of an alcohol with a sulfonic acid (or a sulfonate acid derivative) (Figure 8.15). A sulfonate ester contains a *fantastic* leaving group, a sulfonate ion, an even *better* leaving group than a highly polarizable iodide ion (Concept 5.3.03)—after the bond to the electrophile is broken (becoming a lone pair on the sulfonate), the sulfonate ion delocalizes the negative formal charge across all three of its oxygen atoms.

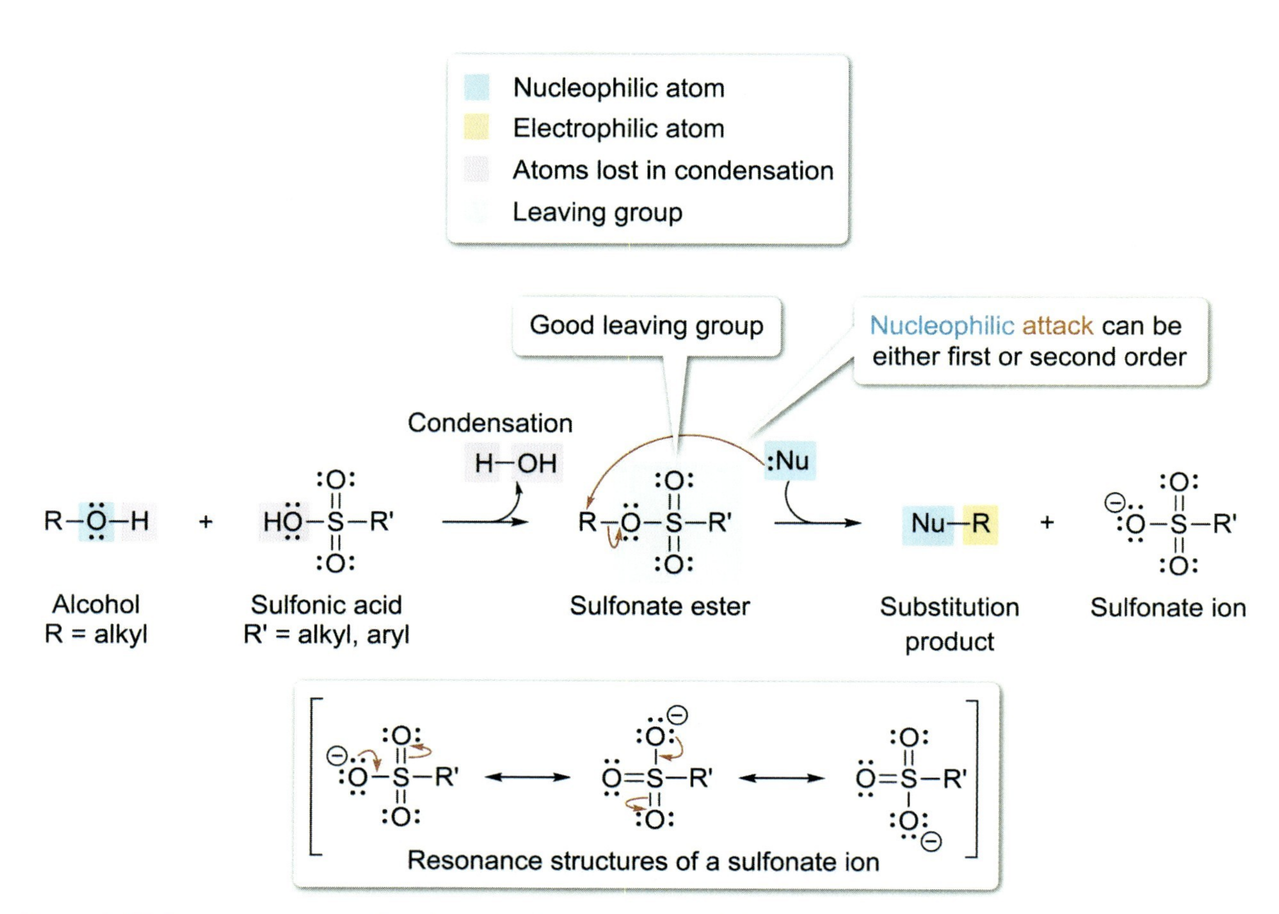

Figure 8.15 Structure and reactivity of a sulfonate ester as a good leaving group.

Figure 8.16 shows how to convert an alcohol into a sulfonate ester through a condensation reaction with a **sulfonyl chloride**, a functional group with similar reactivity as an acid chloride (Concept 11.2.03). Two common sulfonyl chlorides are ***p*-toluenesulfonyl chloride** (tosyl chloride, TsCl) and **methanesulfonyl chloride** (mesyl chloride, MsCl), which react with alcohols to form **tosylate esters** or **mesylate esters**, respectively. Because the condensation of alcohols with sulfonyl chlorides generates an equivalent of hydrochloric acid, these reactions are typically performed using a non-nucleophilic organic base (eg, pyridine) that ensures that the reaction maintains basic conditions and assists in the shuttling of acidic protons.

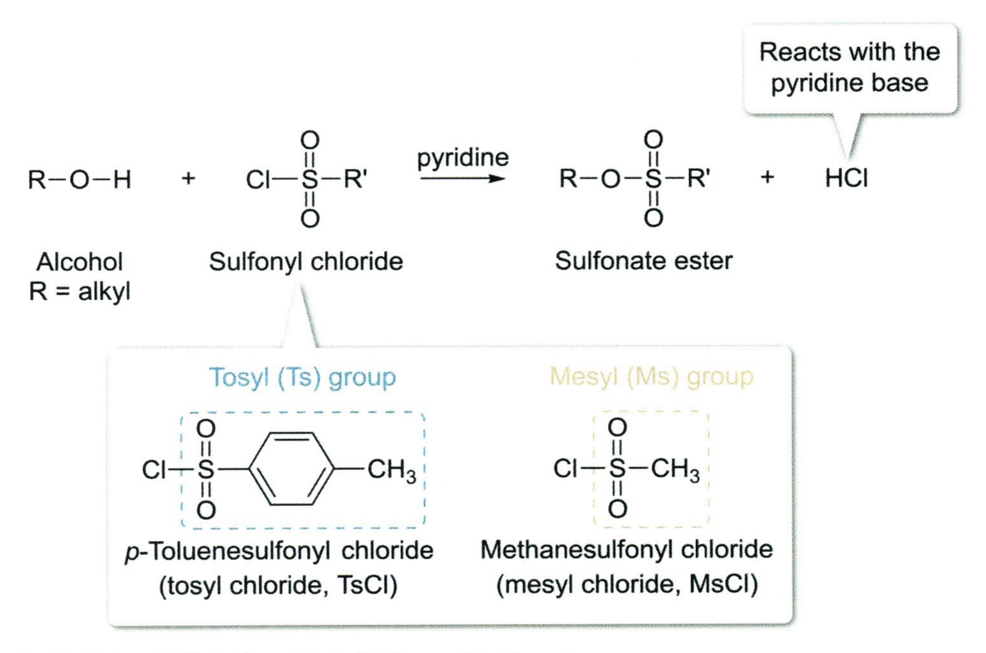

Figure 8.16 General conversion of an alcohol into a sulfonate ester.

The mechanism for the tosylation or mesylation (ie, *sulfonation*) of an alcohol is similar to a nucleophilic acyl substitution with a highly electronegative (sulfonic) acid derivative (see Concept 11.2.02) (Figure 8.17). The nucleophilic oxygen atom of the alcohol attacks the sulfur atom of the sulfonyl chloride, forming an intermediate. Then the pi bond reforms and a chloride ion is eliminated. An organic base (eg, pyridine) deprotonates the acidic proton of the oxonium ion to form the sulfonate ester.

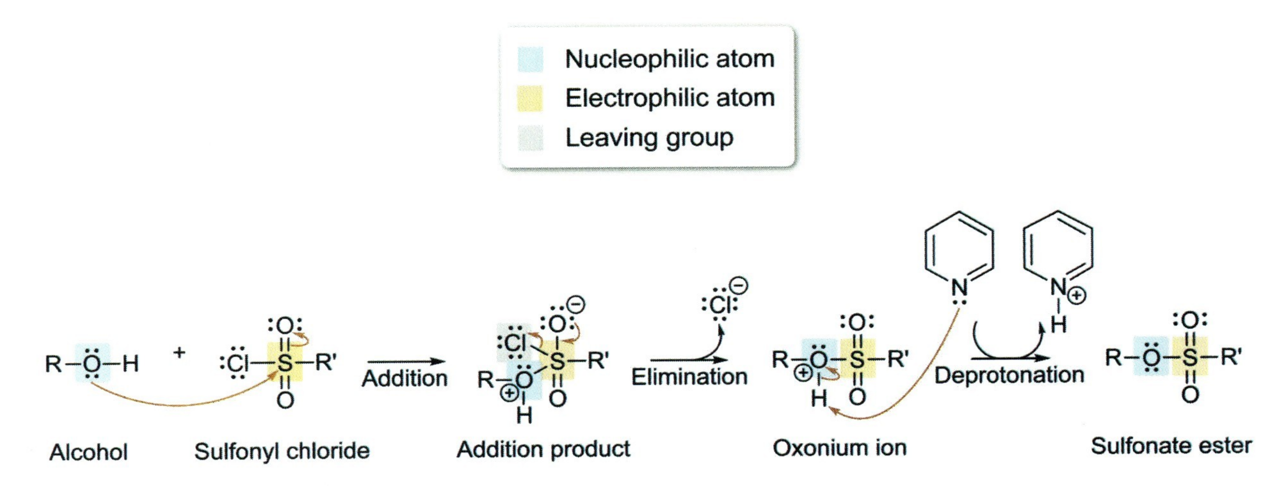

Figure 8.17 The mechanism for the conversion of an alcohol to a sulfonate ester.

Reactions of Sulfonate Esters

A sulfonate ester has a similar reactivity profile as an alkyl iodide (R–I) and undergoes many of the same chemical reactions (Figure 8.18). If the R group of a sulfonate ester is methyl, 1°, or 2°, the sulfonate ester can undergo an S_N2 reaction with a strong nucleophile (eg, HO^-, NC^-, Br^-, I^-, RO^-). By contrast, if the R group is 2° or 3°, the sulfonate ester can undergo an S_N1 reaction with a weak nucleophile (eg, H_2O, ROH).

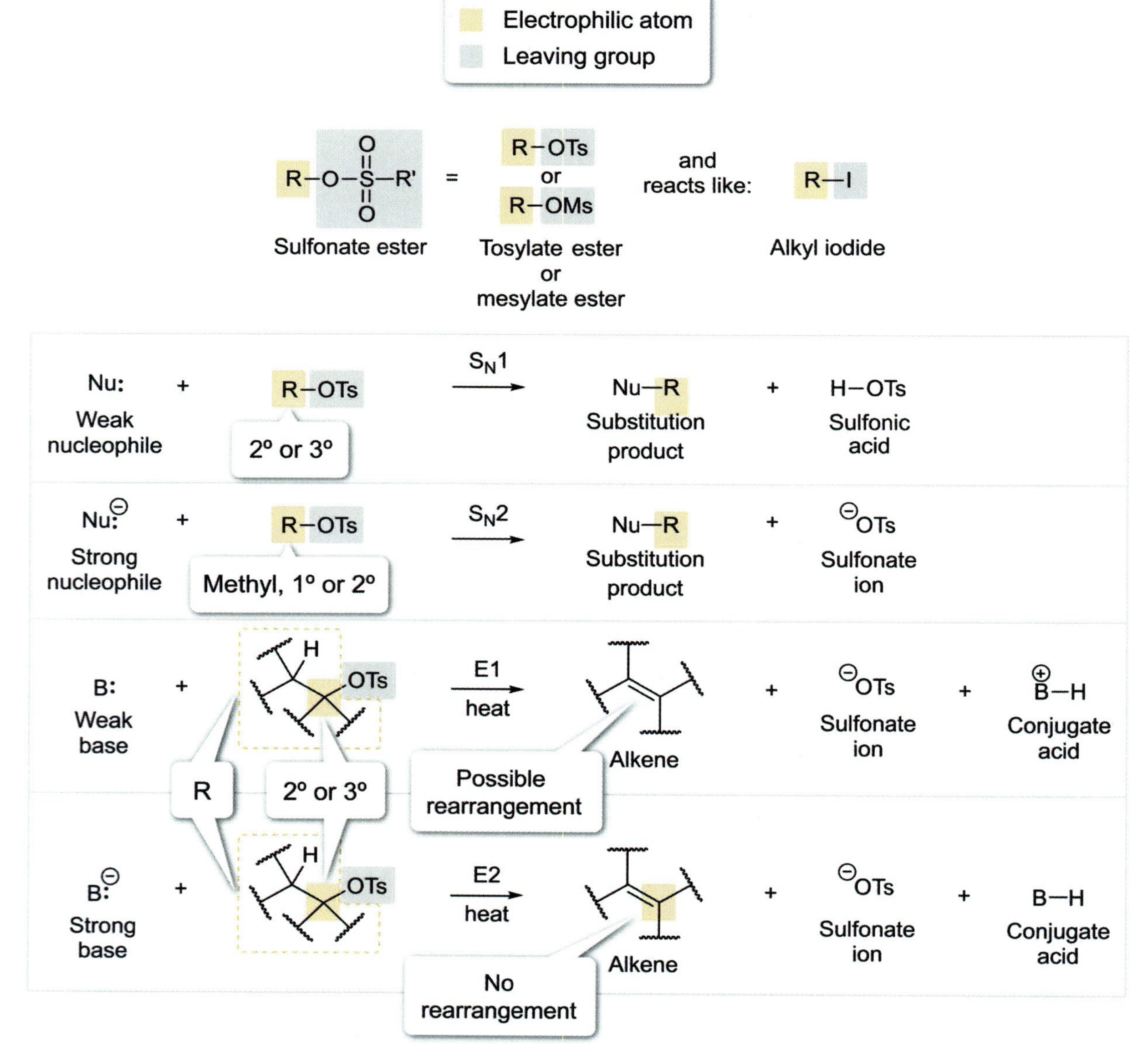

Figure 8.18 Reactions involving sulfonate esters.

A highly substituted sulfonate ester (ie, R = 2° or 3°) can also undergo an elimination reaction if treated with a base and heat. Weak bases tend to proceed through an E1 mechanism and form a carbocation intermediate, whereas strong bases proceed through an E2 mechanism and have a concerted mechanism.

Sulfonate esters react with lithium aluminum hydride ($LiAlH_4$), a powerful reducing agent, to displace the sulfonate ester with a nucleophilic hydride ion (Figure 8.19). Converting an alcohol to a sulfonate ester, followed by reaction with $LiAlH_4$, has the net effect of replacing the hydroxyl group with a hydrogen atom and is one of the few ways to reduce an alcohol to an alkane. Because the $LiAlH_4$ step proceeds through an S_N2 mechanism and *requires* a backside attack of the hydride nucleophile, it *cannot* be performed on sulfonate esters derived from 3° alcohols.

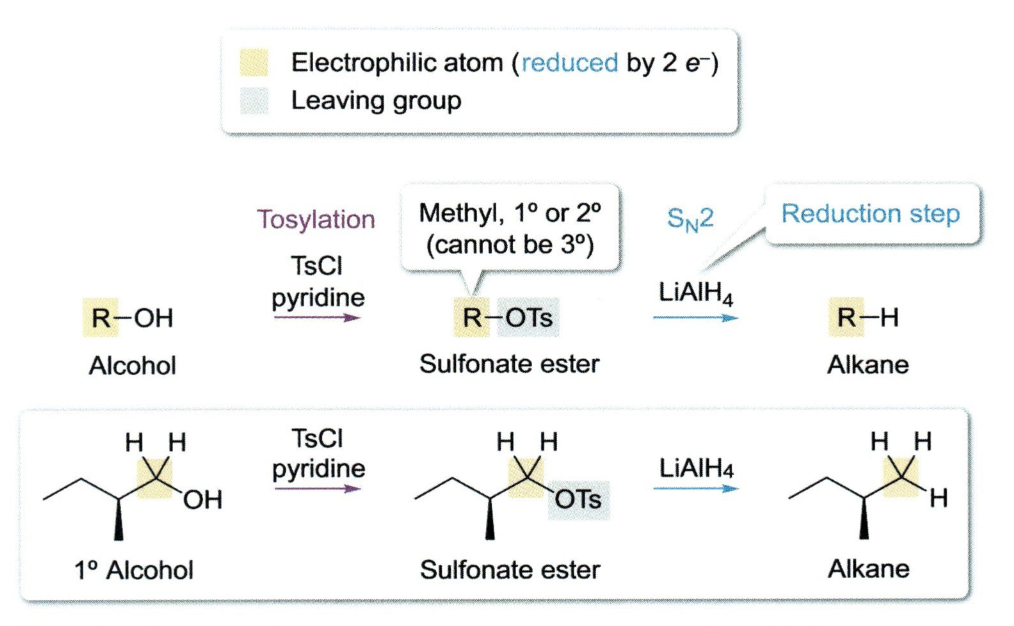

Figure 8.19 Reduction of an alcohol to an alkane through tosylation and reaction with $LiAlH_4$.

8.3.06 Reaction with Hydrohalic Acids

A second way to convert an alcohol into a good leaving group is to treat the alcohol with a strong acid. The strong acid protonates one of the lone pairs on the oxygen atom of the hydroxyl group and generates an oxonium ion (Concept 5.3.03). Then a nucleophile can displace the oxonium ion through a substitution reaction.

Treatment of an alcohol with **hydrohalic acids** (eg, HCl, HBr, HI) converts an alcohol into an alkyl halide with the halogen atom of the chosen hydrohalic acid (Figure 8.20). This reaction proceeds through two mechanistic stages:

1. Reaction of the hydroxyl group with an acid (either Brønsted-Lowry or Lewis) to generate an oxonium ion
2. Substitution of the oxonium ion with a halide ion (eg, Cl^-, Br^-, I^-)

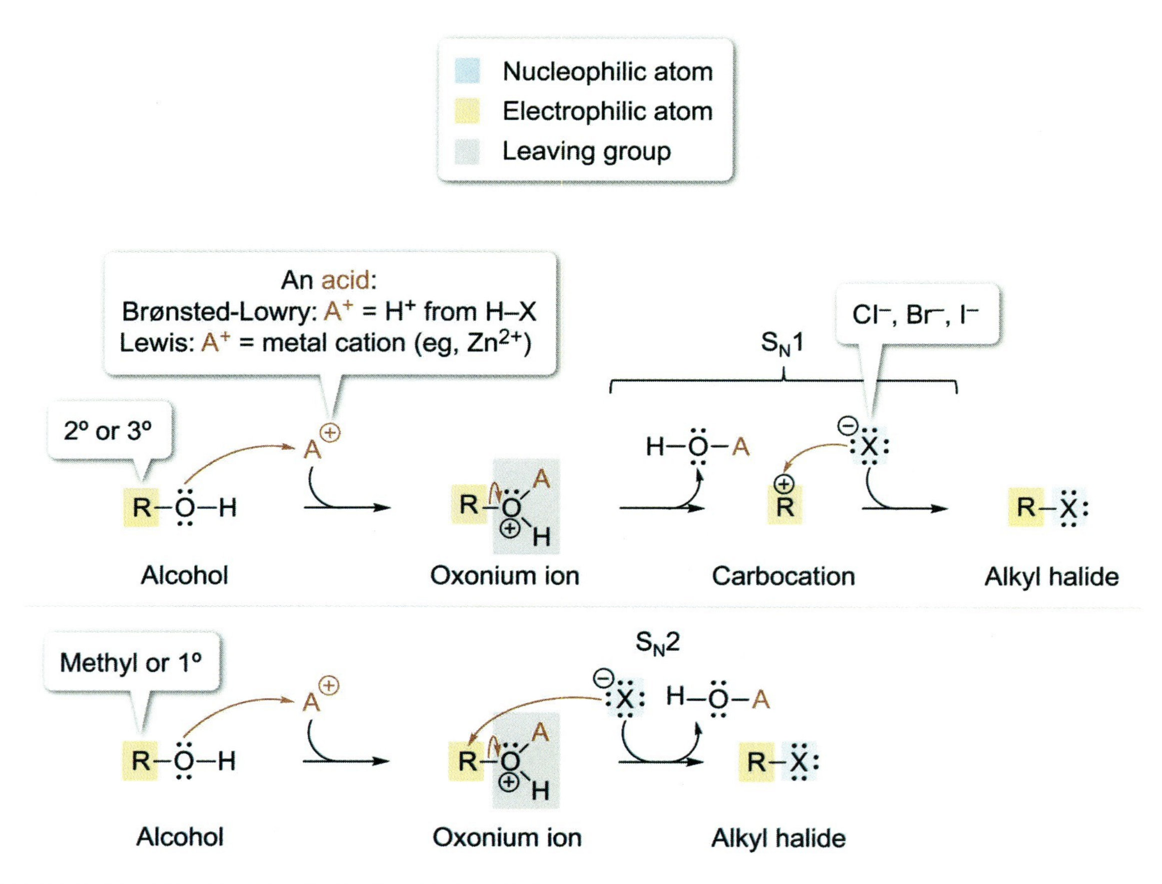

Figure 8.20 Reaction of alcohols with hydrohalic acids.

Depending on the alcohol's classification, substitution can proceed through either an S_N1 mechanism (with 3° or 2° alcohols) or an S_N2 mechanism (with methanol or 1° alcohols). A potential limitation of the S_N1 mechanism is the possibility of carbocation rearrangement (Concept 5.2.01). A second limitation is competing elimination reactions, which result in an alkene product (Concept 8.3.08).

Reaction with Hydrochloric Acid

Although the general mechanism is the same for the treatment of an alcohol with HCl, HBr, or HI, reaction with HCl has a couple of subtle nuances. Because Cl^- is a *much* weaker nucleophile than either Br^- or I^-, reaction with HCl requires a *more powerful* acid to generate a *more electrophilic* oxonium ion (ie, one that contains an *even better* leaving group).

The Lewis acid zinc (II) chloride ($ZnCl_2$) is typically used for this purpose (Figure 8.21). Treatment of an alcohol with an aqueous solution of HCl and $ZnCl_2$ allows the oxygen atom of the alcohol to attack the Lewis acidic zinc ion. The resulting alcohol-zinc complex contains an oxonium ion and is the leaving group for the substitution reaction. Displacement generates $(ZnCl_2OH)^-$, an insoluble, colorless (ie, white) solid that precipitates from the solution.

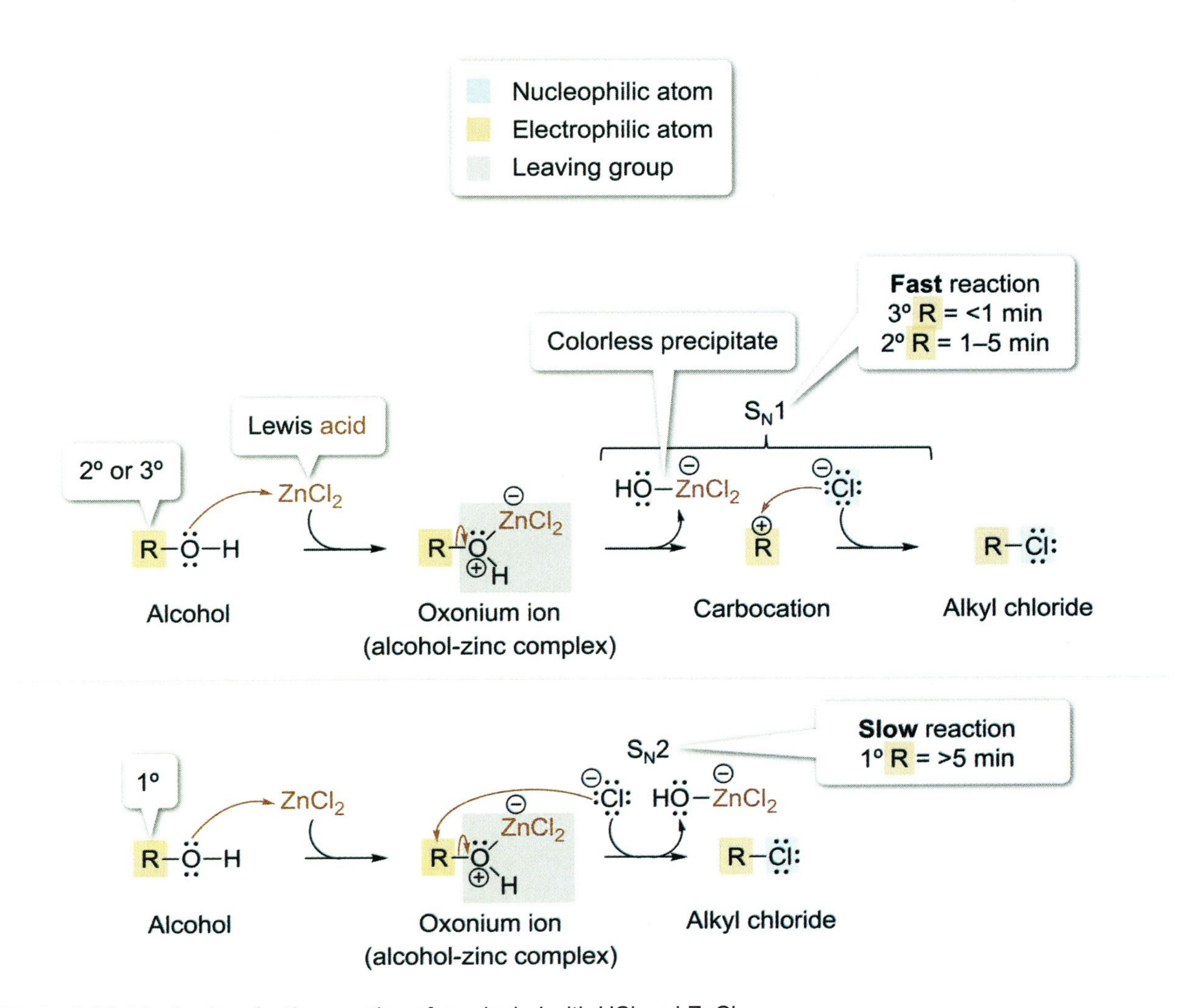

Figure 8.21 Mechanism for the reaction of an alcohol with HCl and $ZnCl_2$.

The reaction of an alcohol with HCl and $ZnCl_2$ is also known as the **Lucas test**, a means of *qualitatively* determining whether a substance contains an alcohol functional group. If the addition of an aqueous solution of HCl and $ZnCl_2$ (a clear solution) generates a cloudy solution (the $[ZnCl_2OH]^-$ precipitate) over time, an alcohol functional group is likely present.

The length of time required to form a cloudy solution can indicate the classification of the alcohol present. Tertiary alcohols proceed through a rapid S_N1 mechanism and form a precipitate in less than 1 minute. Secondary alcohols react more slowly through an S_N1 mechanism and generate a precipitate in 1–5 minutes. Primary alcohols react very slowly through an S_N2 mechanism and require *at least* 5 minutes for a precipitate to form.

8.3.07 Reaction with PBr_3

An alcohol can also be converted into an alkyl bromide in high yield through reaction with **phosphorus tribromide** (PBr_3) (Figure 8.22).

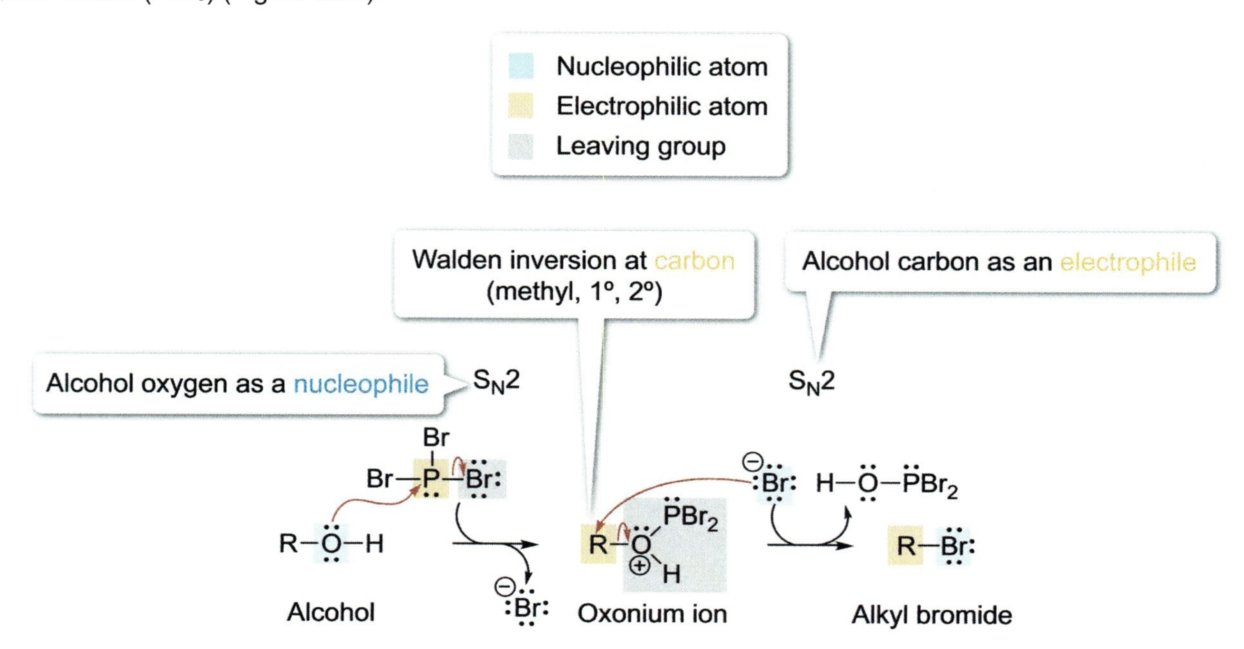

Figure 8.22 The mechanism for the reaction of an alcohol with phosphorus tribromide.

In the first step of the reaction mechanism, a lone pair on the oxygen atom of the alcohol attacks the electrophilic phosphorus atom and displaces a bromide ion through an S_N2 mechanism. The resulting intermediate is a type of oxonium ion and a good leaving group. Then the bromide ion undergoes a *second* S_N2 reaction on the carbon atom bearing the oxonium ion. For this reason, *only* methyl, 1°, and 2° alcohols react with PBr_3 to form an alkyl bromide.

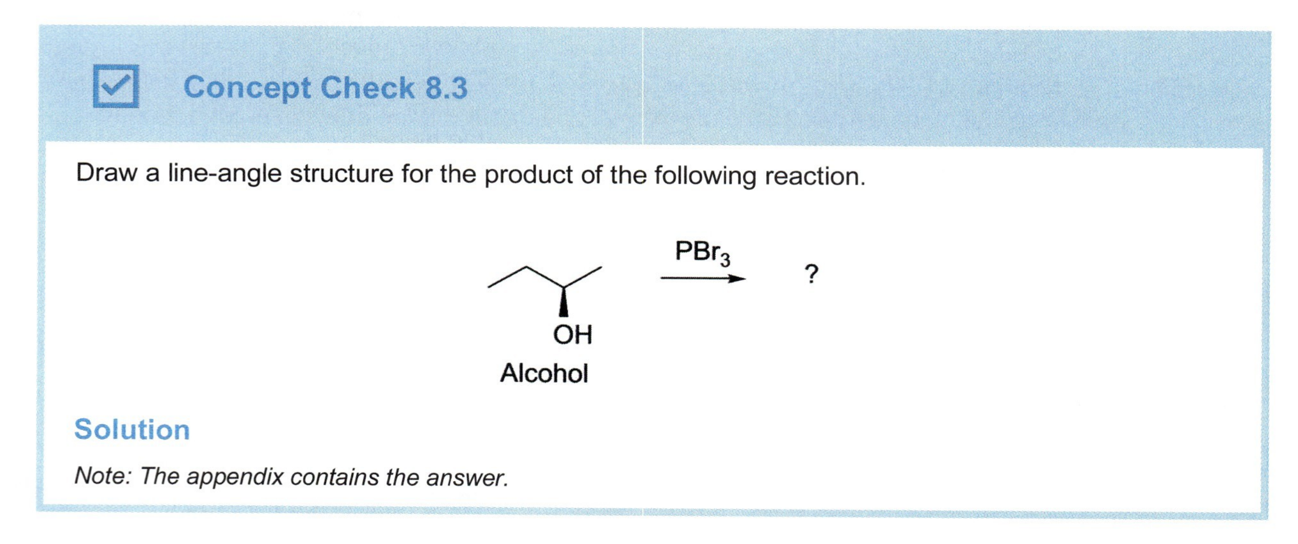

Concept Check 8.3

Draw a line-angle structure for the product of the following reaction.

Solution

Note: The appendix contains the answer.

8.3.08 Acidic Dehydration of an Alcohol

Often, substitution reactions and elimination reactions are competing processes for a given reactant and set of experimental conditions. A common way to promote an elimination reaction over a competing substitution reaction is to add heat.

The hydroxyl group of an alcohol can become protonated by a strong acid to generate an oxonium ion, a good leaving group (Concept 8.3.06). Heating an oxonium ion can promote an elimination reaction that removes a molecule of water (ie, a **dehydration reaction**) and results in the formation of an alkene (Figure 8.23). A dehydration reaction is the *reciprocal* process to a condensation reaction, which *generates* a molecule of water as the small molecule byproduct.

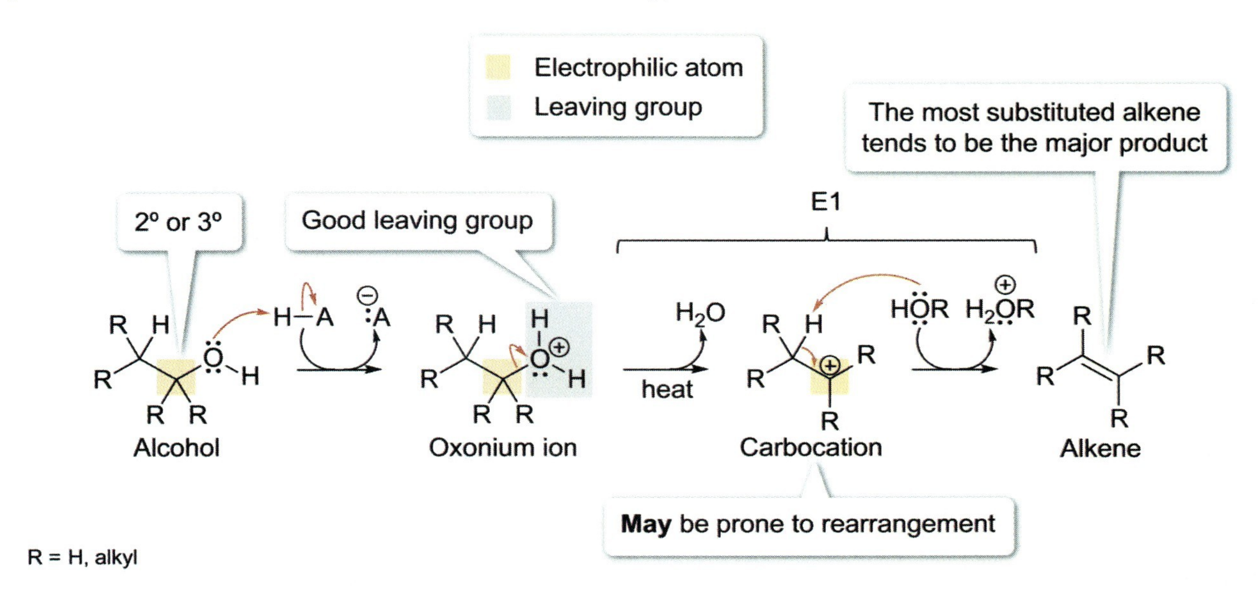

Figure 8.23 The E1 dehydration of an alcohol to an alkene.

Although the dehydration of an alcohol could occur through either an E1 or E2 mechanism, the E1 mechanism is more common because an alcohol is a *weak* base and an E2 mechanism requires a *strong* base. The E1 dehydration of an alcohol proceeds through a carbocation intermediate (ie, which may be prone to rearrangement), where the best alcohol substrates are either 2° or 3° and form the most stable initial carbocations. As a general guideline, the alkene product with the highest substitution (ie, greatest number of R groups attached to the alkene carbon atoms) tends to be formed in the highest concentration.

8.3.09 Protecting Groups for Alcohols

In organic chemistry, it is common for the same reagent or reaction conditions to react with more than one type of functional group. Because a single organic molecule may contain *several* functional groups, the broad reactivity profile of certain reagents (eg, $LiAlH_4$) can lead to difficulty ensuring that a reagent reacts *only* with a *desired* subset of functional groups for a transformation. In the absence of *selectivity*, many chemical transformations generate a large mixture of desired and undesired products. Undesired products of a chemical reaction (ie, **side products**) decrease the yield of the desired product of a reaction.

A **protecting group** is a set of atoms (usually a type of functional group) that can be *reversibly* added to a molecule to make *another* functional group unreactive to a chemical transformation (Figure 8.24). The addition of a protecting group (ie, a **protection** step) is later reversed through its removal (ie, a **deprotection** step). The reactivity profile of a protecting group should be **orthogonal** (ie, opposite or complementary) to the nature of the key reaction from which the molecule is protected (eg, oxidizing, reducing, acidic, basic).

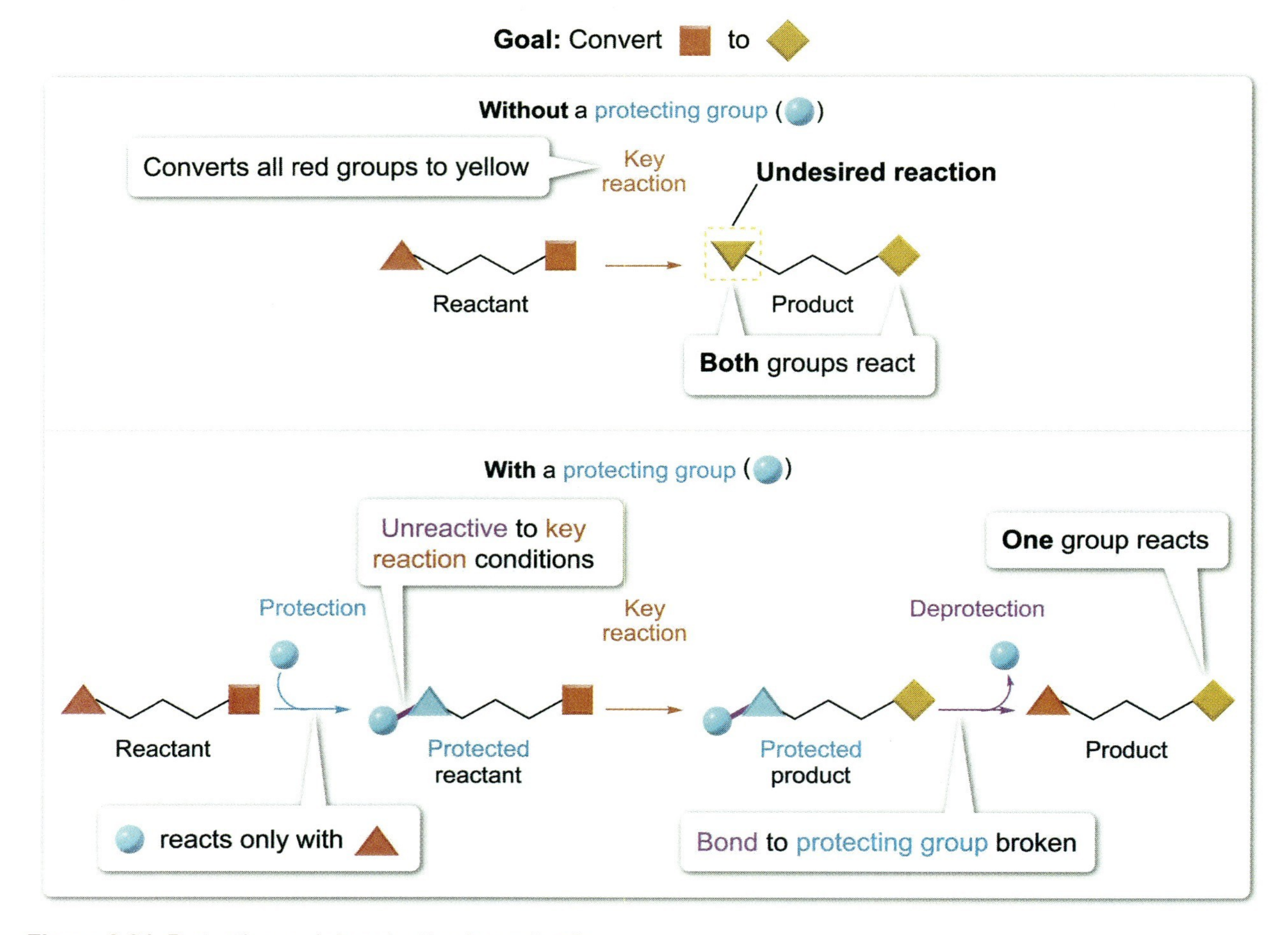

Figure 8.24 Protection and deprotection in synthesis.

Because alcohols are nucleophilic and prone to oxidation, alcohols are frequently a target for protection in chemical synthesis. Alcohols are frequently protected as a **silyl ether**, a functional group containing silicon (Si) in a Si-O bond (Figure 8.25). Silyl protecting groups typically have three alkyl or aryl groups on the silicon atom. Silyl ethers no longer have the acidic hydrogen atom of an alcohol and have increased stability to basic reaction conditions. A silyl ether is *reactive* (ie, deprotected) under acidic conditions.

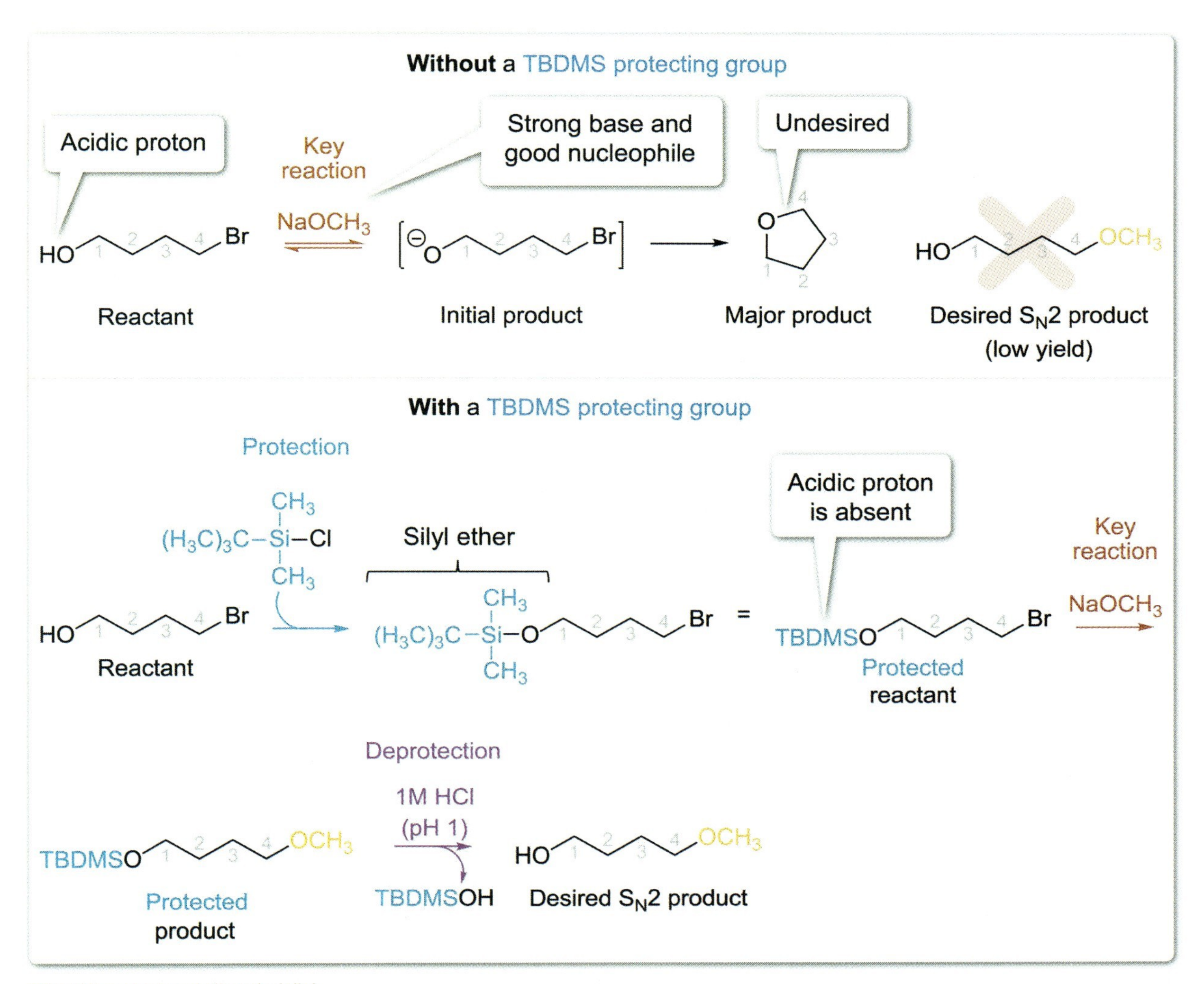

TBDMS = *tert*-butyl-dimethylsilyl

Figure 8.25 The use of a silyl ether as an alcohol protecting group.

An example of a silyl protecting group for an alcohol is a *tert*-butyldimethylsilyl (TBDMS) group. Alcohols can be converted to a TBDMS ether through reaction with TBDMS chloride, which proceeds through an S_N2 mechanism. TBDMS-protected alcohols have increased stability to basic (eg, NaH, pyridine) or nucleophilic (eg, alkoxide ion) conditions and can tolerate aqueous solutions up to approximately pH 12. However, TBDMS groups are readily removed under acidic conditions (pH <4), which restores the original alcohol.

Lesson 9.1

Structure and Physical Properties of Aldehydes and Ketones

Introduction

The carbonyl group (C=O) is central to organic chemistry and is present in several functional groups, including aldehydes, ketones, carboxylic acids, and carboxylic acid derivatives. The next few chapters in this book concern the structure, physical properties, and reactions of carbonyl-containing functional groups.

This lesson discusses the general structural and physical characteristics of aldehydes and ketones.

9.1.01 Structural Characteristics of Aldehydes and Ketones

Aldehydes and **ketones** are two functional groups that contain a **carbonyl group** (C=O) (see Concept 2.8.02). An aldehyde has the general formula R–CH(O), in which the carbonyl carbon atom is attached to *one* alkyl or aryl group (R) and one hydrogen atom. A ketone has the general formula R–C(O)–R′, in which the carbonyl carbon is attached to two R groups, which may or may not be the same.

The carbon atom of an aldehyde or ketone is attached to *three* atoms and has *zero* nonbonding electrons (Figure 9.1). Therefore, the carbonyl carbon has three electron domains, is sp^2 hybridized (Concept 1.2.03 and Concept 2.1.02), and has a trigonal planar molecular geometry with 120° bond angles.

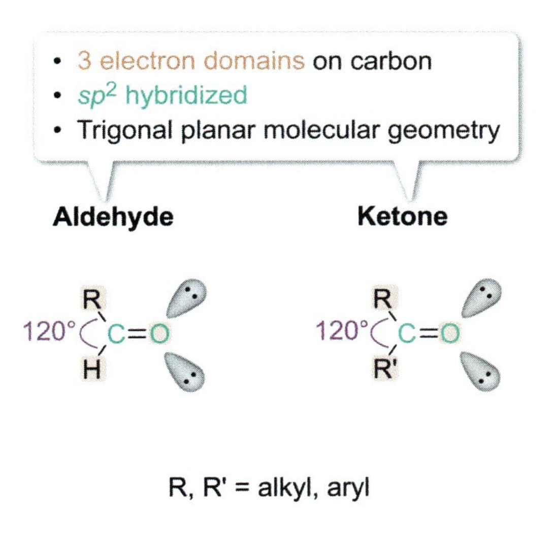

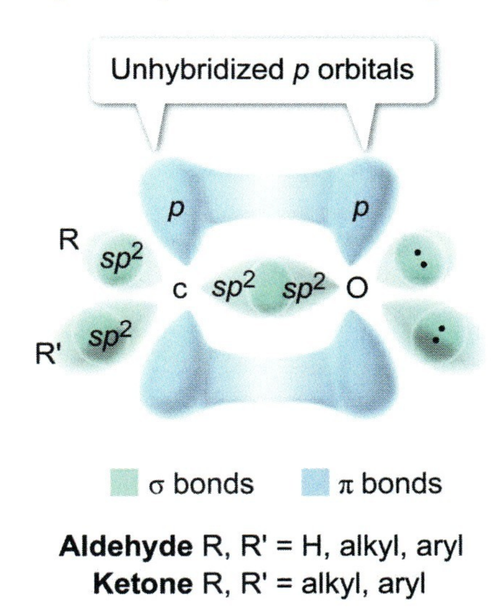

Figure 9.1 Structure of an aldehyde and a ketone.

The carbon-oxygen double bond of a carbonyl is composed of one σ bond and one π bond. The σ bond is formed by head-to-head overlap of sp^2 hybrid orbitals, and the π bond is formed by side-to-side overlap of their unhybridized *p* orbitals.

Because the carbonyl carbon atom has a trigonal planar geometry, the carbon and the three atoms attached to it are all located within the same plane (ie, are coplanar). The π bond of the carbonyl has electron density above and below this plane, giving the carbonyl two distinct faces where reactions can occur.

The carbon-oxygen double bond of a carbonyl is polar due to the difference in electronegativity between oxygen and carbon. The dipole moment is also enhanced by resonance (Figure 9.2):

- **Major resonance structure:** Each atom has a complete octet.
- **Minor resonance structure:** The electrons from the carbonyl π bond become a lone pair on the oxygen atom. This shift in electrons leads to charge separation and a positively charged carbon atom with an incomplete octet.

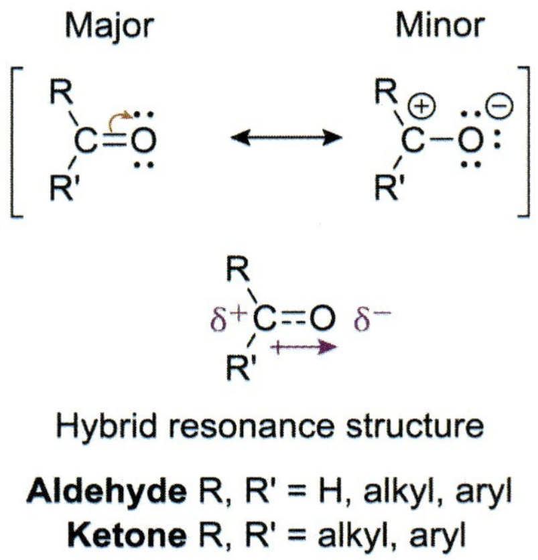

Major Minor

Hybrid resonance structure

Aldehyde R, R' = H, alkyl, aryl
Ketone R, R' = alkyl, aryl

Figure 9.2 Resonance forms of a carbonyl.

9.1.02 Physical Properties of Aldehydes and Ketones

The physical properties of aldehydes and ketones depend on the intermolecular forces present in a pure sample. The most significant intermolecular forces are dipole-dipole interactions resulting from the polar carbonyl group. Because pure aldehyde or ketone samples lack a hydrogen bond donor, a *pure* aldehyde or ketone *cannot* hydrogen bond. However, because they have hydrogen bond acceptors (the lone pairs on the oxygen atom of the carbonyl), they can hydrogen bond with *other* molecules that have a hydrogen bond donor (eg, water, alcohols).

Boiling Point

Aldehydes and ketones have *higher* boiling points than alkanes (due to their dipole-dipole interactions) and *lower* boiling points than alcohols (because pure aldehydes and ketones *cannot* hydrogen bond) (Table 9.1). Aldehydes and ketones of comparable molecular weights have similar boiling points due to similar magnitudes of dipole-dipole interactions. The boiling points of aldehydes and ketones *decrease* with branching of their alkyl chains.

Table 9.1 Boiling points of alkanes, aldehydes, ketones, and alcohols.

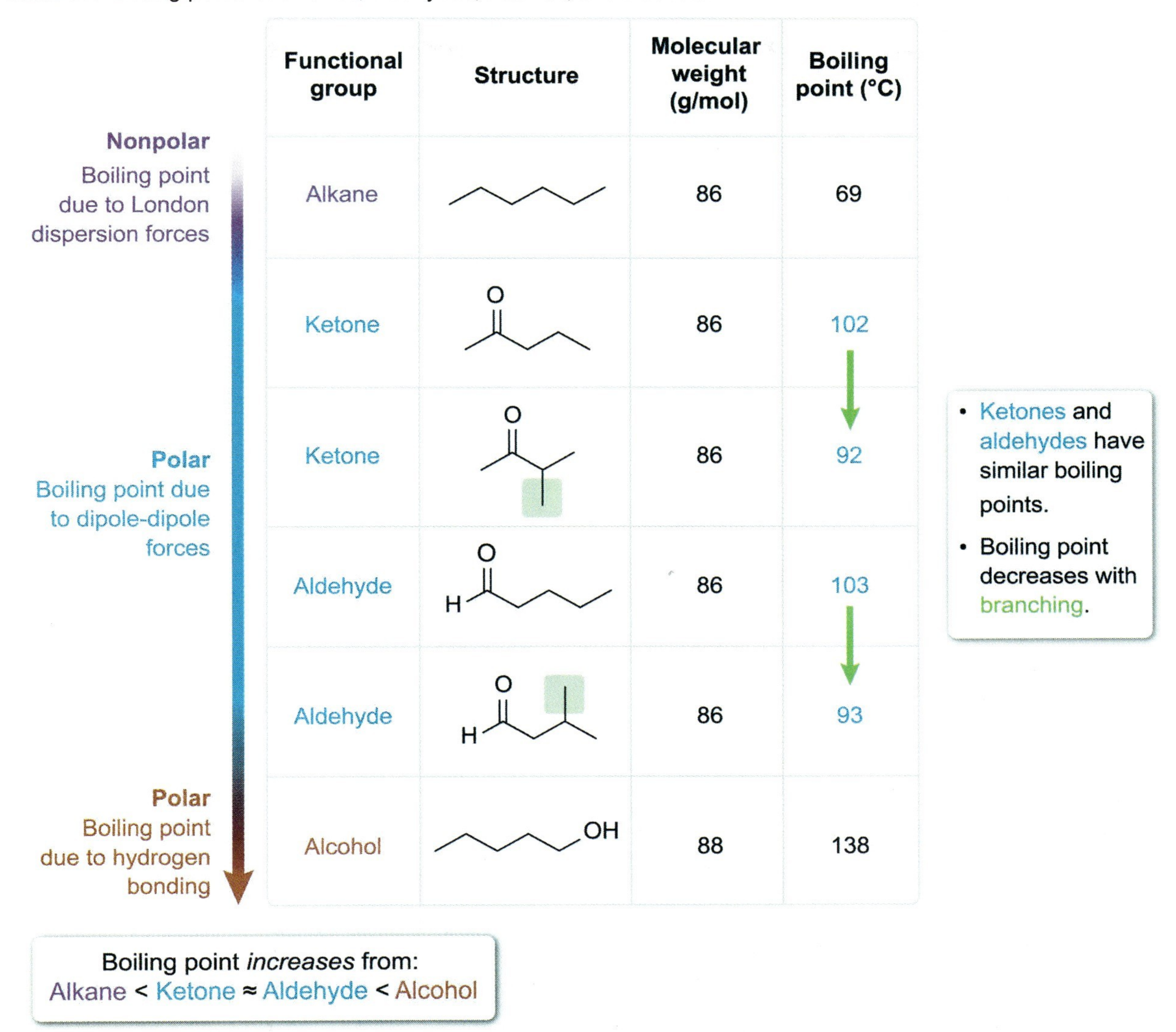

Functional group	Structure	Molecular weight (g/mol)	Boiling point (°C)
Alkane		86	69
Ketone	O	86	102
Ketone	O	86	92
Aldehyde	O, H	86	103
Aldehyde	O, H	86	93
Alcohol	OH	88	138

Solubility

Aldehydes and ketones contain both a polar carbonyl group and at least one nonpolar R group. Based on the principle of "like dissolves like," the electronegative oxygen atom of the carbonyl group is *hydrophilic* and can solvate a limited number of nonpolar R group carbon atoms. The ability of an aldehyde or ketone to accept hydrogen bonds from water *greatly* increases their water solubility. Aldehydes and ketones containing up to a total of four carbon atoms are *highly soluble* in water; acetaldehyde and acetone are *miscible* with water.

Another feature that impacts the solubility of aldehydes and ketones is their ability to undergo a hydration reaction to form a geminal diol (see Concept 9.3.02). This chemical reaction is an equilibrium process in which a portion of aldehyde and ketone molecules exposed to water are in the geminal diol form. The two hydroxyl groups of a geminal diol provide physical properties similar to those of alcohols (see Concept 8.1.03).

Lesson 9.2

Synthesis of Aldehydes and Ketones

Introduction

A traditional organic chemistry curriculum covers several ways to synthesize an aldehyde and a ketone. This lesson's discussion of aldehyde and ketone preparation is focused on methods most likely to appear on the exam:

- Oxidation of alcohols
- Reduction of carboxylic acids and carboxylic acid derivatives

9.2.01 Oxidation of Alcohols

Reaction of an alcohol with an oxidizing agent *increases* the oxidation state of the carbon bearing the hydroxyl group. The classification of the alcohol determines the functional group product.

When methanol (CH_3OH) and 1° alcohols (RCH_2OH) are treated with a *mild* oxidizing agent (eg, PCC, HOCl), they undergo a two-electron oxidation to generate an aldehyde (Figure 9.3). If an aldehyde is the desired product, it is important *not* to use a strong oxidizing agent (eg, chromium-derived agents, MnO_4^-); these agents promote a *four*-electron oxidation (*two* 2-electron oxidations) of the carbon atom to produce a *carboxylic acid*. The mechanism for the oxidation of an alcohol to an aldehyde or ketone is discussed in Concept 8.3.01.

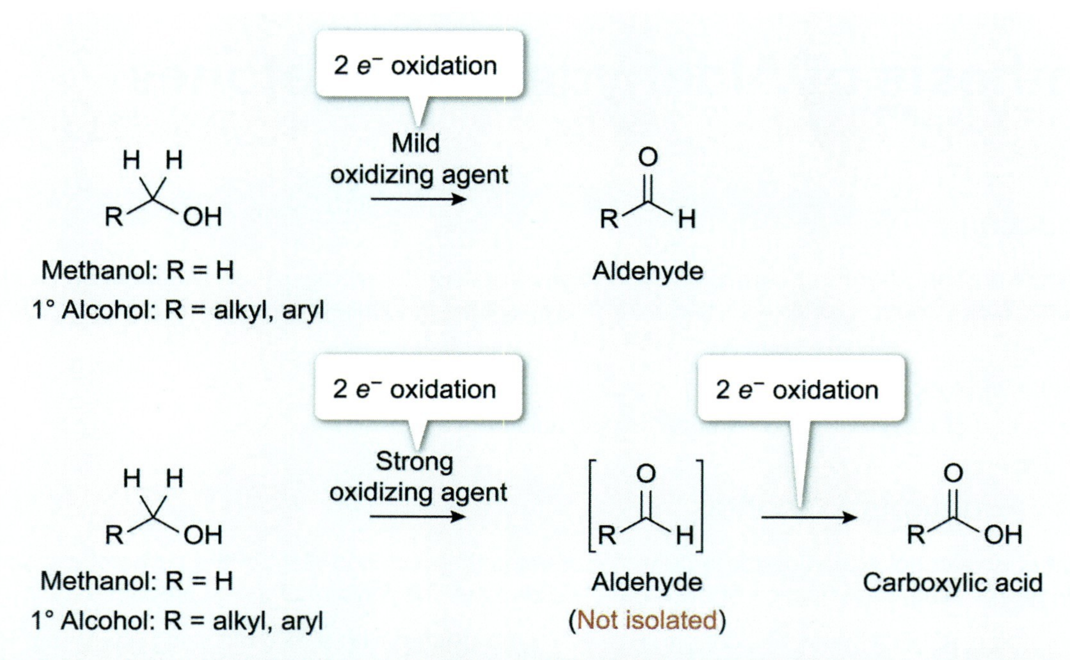

Oxidation of methanol and 1° alcohols with a **mild oxidizing agent** generates an **aldehyde**.

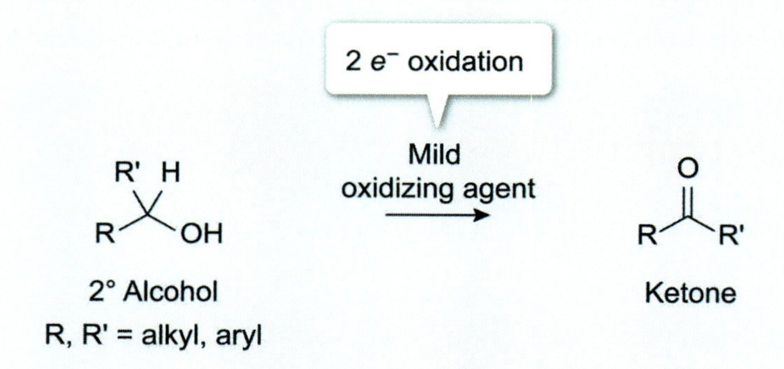

Oxidation of 2° alcohols with a **strong or mild oxidizing agent** generates a **ketone**.

Figure 9.3 Formation of an aldehyde or a ketone by the oxidation of an alcohol.

Treatment of 2° alcohols with *either* a strong *or* mild oxidizing agent facilitates a two-electron oxidation and generates a ketone. Overoxidation is not usually a concern in the oxidation of 2° alcohols. The biochemical oxidation of an alcohol with NAD^+ is discussed in Concept 8.3.01.

Concept Check 9.1

If 3 moles of Compound A are oxidized to Compound B, how many moles of PCC are needed for all of Compound A to be oxidized?

OH
HO
PCC
O O
H
Compound A
Compound B

Solution

Note: The appendix contains the answer.

9.2.02 Reduction of Carboxylic Acids and Carboxylic Acid Derivatives

An aldehyde can be synthesized from a carboxylic acid or carboxylic acid derivative via a two-electron reduction. Because an aldehyde is *more* reactive than the carboxylic acid or acid derivative, overreduction *often* occurs and the aldehyde intermediate is further reduced to a 1° alcohol (Concept 10.3.03 and Concept 11.3.03).

However, an acid chloride can be *selectively* reduced to an aldehyde using the sterically hindered hydride reducing agent, $LiAlH(OtBu)_3$ (see Concept 11.3.03).

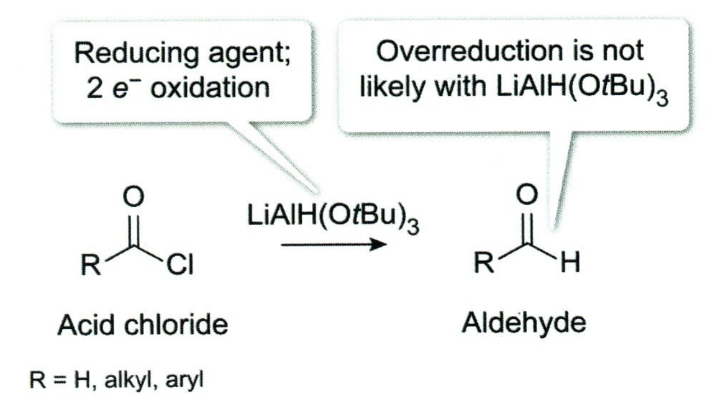

Figure 9.4 Synthesis of an aldehyde by selective reduction of an acid chloride.

Ketones *cannot* be prepared using a hydride reduction because the structure of a ketone inherently *lacks* a hydrogen atom on its carbonyl group (ie, the hydrogen atom added through a hydride ion reduction).

Lesson 9.3

Reactions of Aldehydes and Ketones

Introduction

This lesson focuses on the reactivity of aldehydes and ketones and the way these reactions are used to generate other functional groups. The first six concepts describe the reactivity of aldehydes and ketones as Type 3 electrophiles that undergo nucleophilic addition reactions:

- Acidic and basic mechanisms of a nucleophilic addition
- Hydration of aldehydes and ketones
- Formation of cyanohydrins
- Formation of imines and imine variants
- Formation of hemiacetals and acetals
- The use of acetals as a protecting group for aldehydes and ketones

The final two concepts describe the redox reactivity of aldehydes and ketones:

- Oxidation of aldehydes
- Reduction of aldehydes and ketones

9.3.01 Nucleophilic Addition

Concept 5.5.02 introduces the nucleophilic addition reaction, in which a nucleophile and a hydrogen atom add across the pi bond of a Type 3 electrophile (Figure 9.5). All nucleophilic addition reactions in this book proceed through *at least* two mechanistic steps in which the nucleophile and the hydrogen atom are added to the Type 3 electrophile in *separate* mechanistic steps. The typical regiochemical outcome has the nucleophile attached to the electrophilic carbon atom and the hydrogen atom attached to the electronegative heteroatom (X). For the reactions of aldehydes and ketones, the electronegative heteroatom is oxygen.

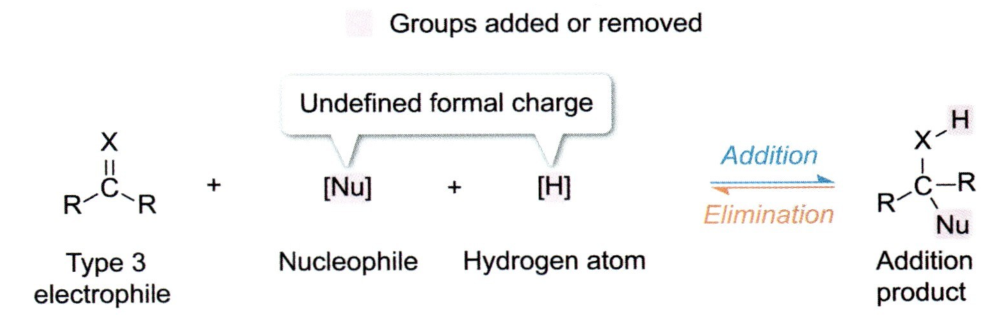

Figure 9.5 Reversible nature of a nucleophilic addition reaction.

The nucleophilic addition mechanism is the *primary* reaction mechanism for aldehydes and ketones and can occur under acidic or basic conditions. Unlike some other reactions, the individual steps of nucleophilic addition are often in dynamic equilibrium, meaning both the forward (addition) and reverse (elimination) processes are possible.

The Acidic Nucleophilic Addition Mechanism

The **acidic nucleophilic addition mechanism** (Figure 9.6) has the following *three* mechanistic steps:

1. Protonation
2. Addition of the nucleophile
3. Deprotonation

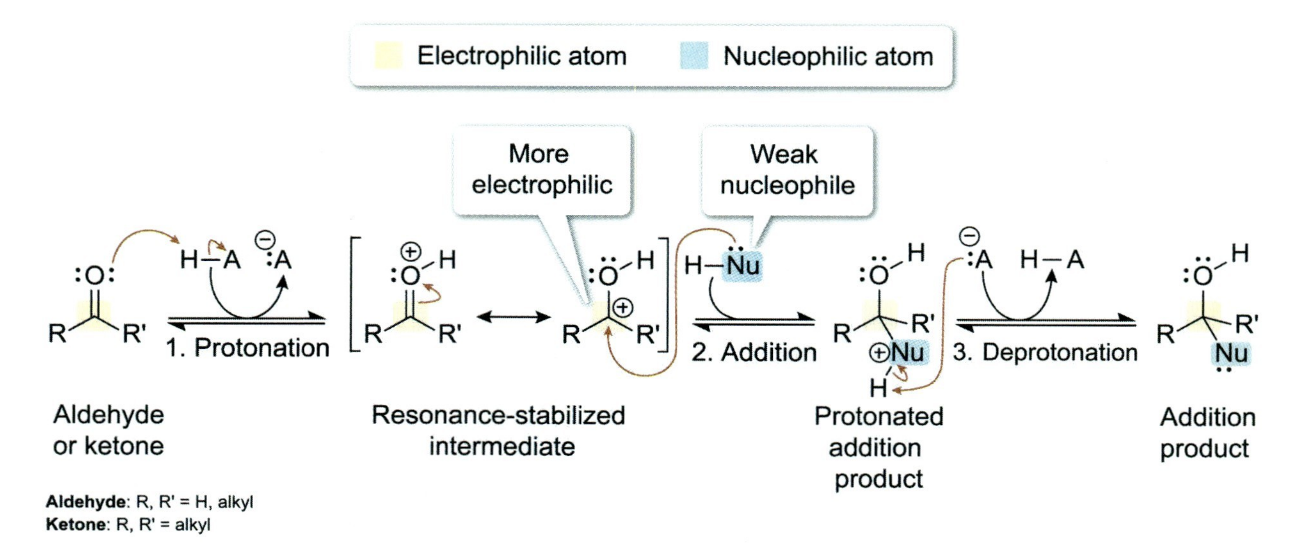

Figure 9.6 A general acidic nucleophilic addition mechanism.

Because the nucleophile in an acidic mechanism tends to be weak (eg, water, alcohols), the carbonyl oxygen atom must first be protonated by a strong acid; protonation makes the carbonyl *carbon* atom a stronger electrophile. The nucleophile then undergoes an *addition* with the carbonyl carbon atom, and the pi bond becomes a lone pair on the carbonyl oxygen atom. Because the nucleophile in an acidic mechanism is typically neutral, the final step is deprotonation of the acidic proton that was generated following nucleophilic attack.

The Basic Nucleophilic Addition Mechanism

In contrast, a **basic nucleophilic addition mechanism** (Figure 9.7) has *two* mechanistic steps:

1. Addition of the nucleophile
2. Protonation

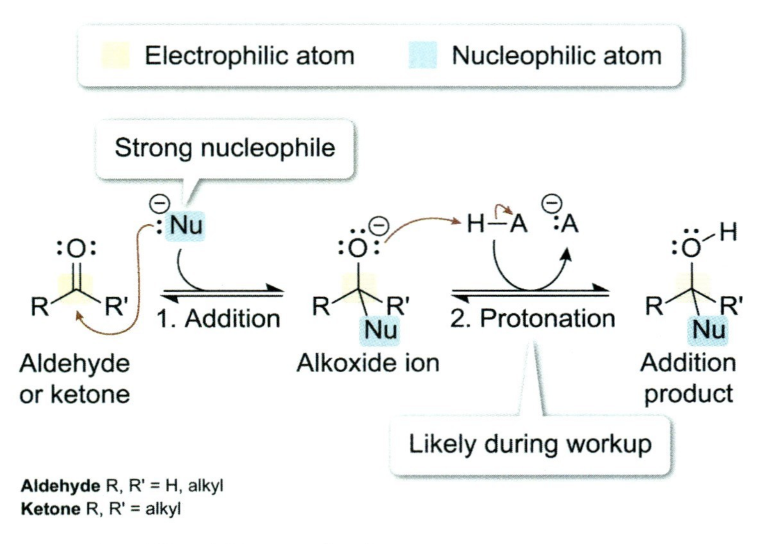

Figure 9.7 A general basic nucleophilic addition mechanism.

A basic nucleophilic addition mechanism uses a strong nucleophile that directly attacks the electrophilic carbon atom. During the *addition*, the π bond of the carbonyl becomes a lone pair on the oxygen atom and forms an alkoxide ion. Then the alkoxide ion is protonated by a Brønsted-Lowry acid.

9.3.02 Hydration of Aldehydes and Ketones

The carbonyl group of an aldehyde or ketone can undergo a nucleophilic addition with a molecule of water (ie, a **hydration reaction**) to form an **aldehyde hydrate** or a **ketone hydrate**, respectively (Figure 9.8). The hydration product is a **geminal diol**, a molecule with two hydroxyl groups attached to the *same* carbon atom.

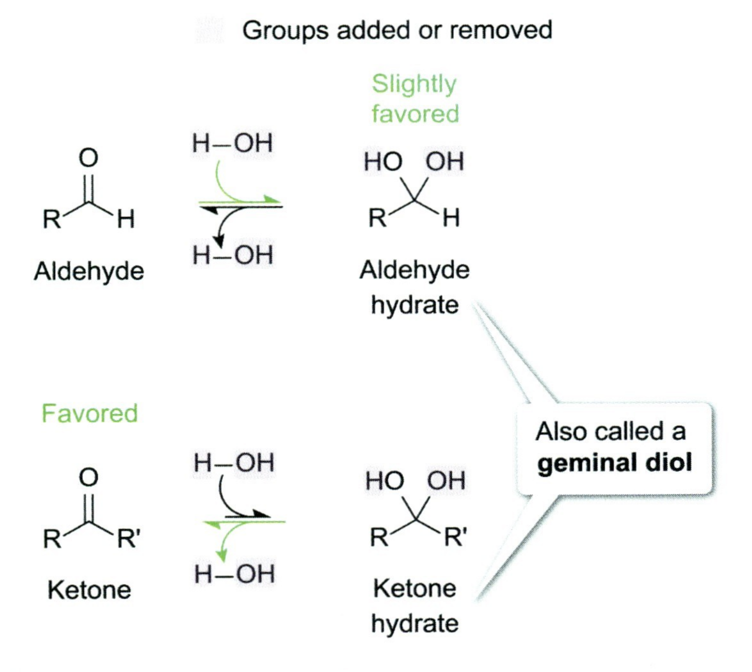

Figure 9.8 Formation of a carbonyl hydrate (geminal diol).

The forward (hydration) and reverse (dehydration) reactions form a dynamic equilibrium. Aldehydes are *much* more likely than ketones to form a stable hydrate; the hydration equilibrium constant (K_{eq}) for an aldehyde is generally about 1, with formaldehyde having an atypically high K_{eq} of 40. These values (ie, $K_{eq} \geq 1$) indicate that the hydrate is slightly favored. In contrast, ketones give a K_{eq} value of 10^{-4} to 10^{-2}, and the hydration equilibrium favors the carbonyl form.

Carbonyl hydration reactions require water and *either* an acid or a base catalyst. Consequently, the mechanism for hydration of an aldehyde or a ketone proceeds through *either* an acidic nucleophilic addition *or* a basic nucleophilic addition (see Concept 9.3.01).

The Acidic Mechanism of Hydration

In the acidic mechanism, the oxygen atom of the carbonyl is first protonated by the acid catalyst (H_3O^+) (Figure 9.9). Then the oxygen atom of water attacks the electrophilic carbon atom, generating an oxonium ion intermediate. Finally, a molecule of water deprotonates the oxonium ion to regenerate the acid catalyst and provide the hydrate product.

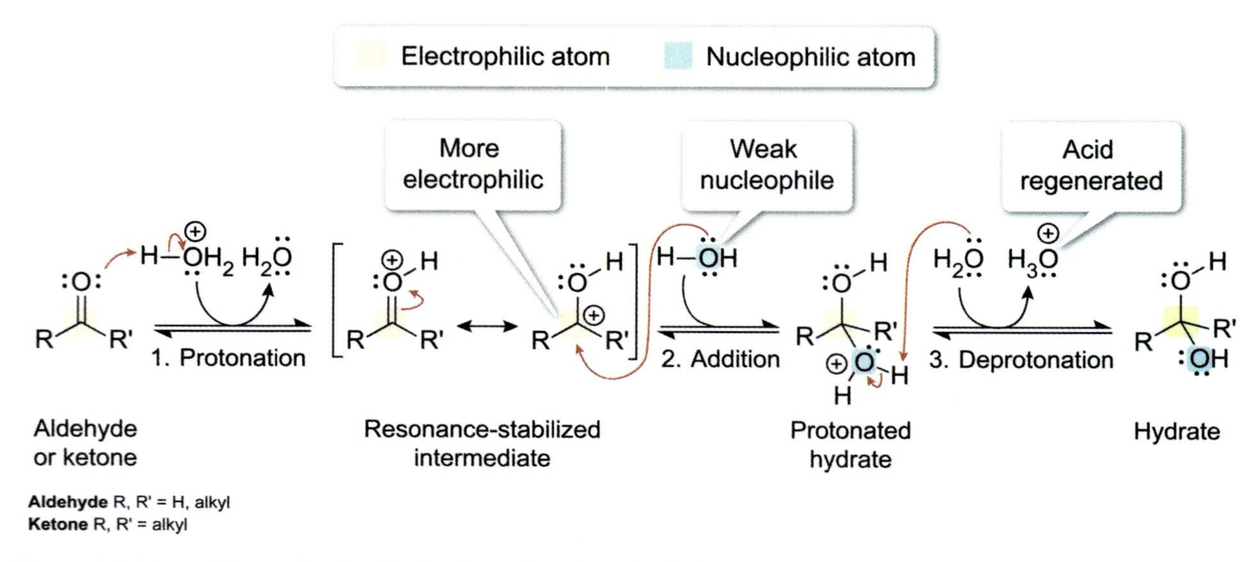

Figure 9.9 The acidic mechanism for the hydration of an aldehyde or a ketone.

The Basic Mechanism of Hydration

In the basic mechanism, the carbonyl carbon atom is directly attacked by a nucleophilic hydroxide ion (Figure 9.10). Then the resulting alkoxide ion is protonated by a molecule of water to regenerate the base catalyst and form the hydrate product.

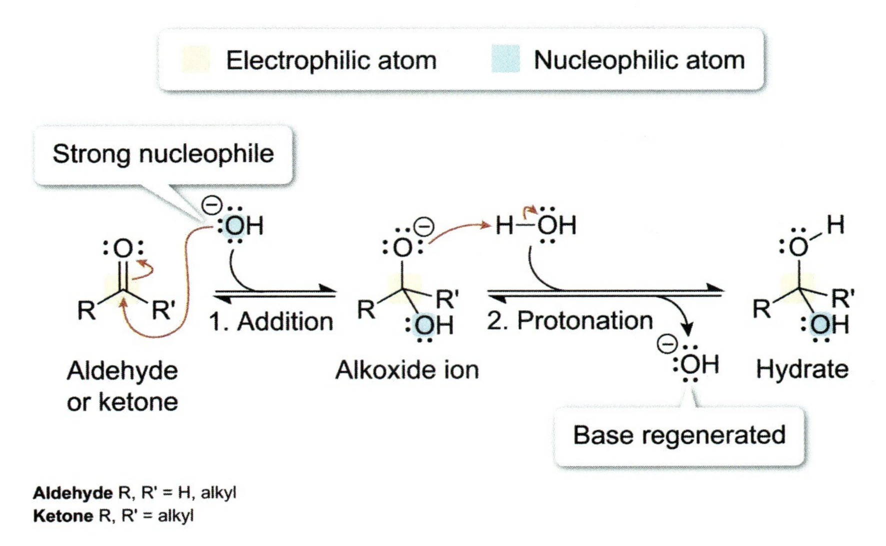

Figure 9.10 The basic mechanism for the hydration of an aldehyde or a ketone.

9.3.03 Formation of Cyanohydrins

An aldehyde or a ketone reacts with a cyanide ion (NC^-) through a nucleophilic addition reaction to form a **cyanohydrin**, a functional group that contains a nitrile and a hydroxyl group attached to the same carbon atom (Figure 9.11).

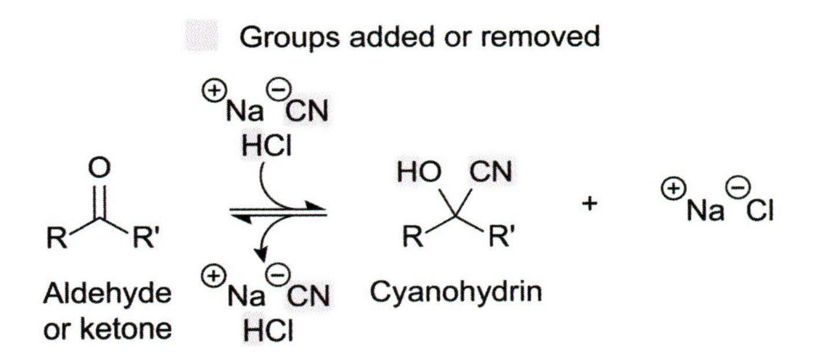

Figure 9.11 General reaction to form a cyanohydrin.

Because the cyanide ion is a strong nucleophile, the formation of a cyanohydrin follows the basic nucleophilic addition mechanism (Figure 9.12). First, the cyanide ion directly attacks the electrophilic carbon atom of the carbonyl to form an alkoxide ion. Then the alkoxide ion is protonated by the strong acid to form the cyanohydrin.

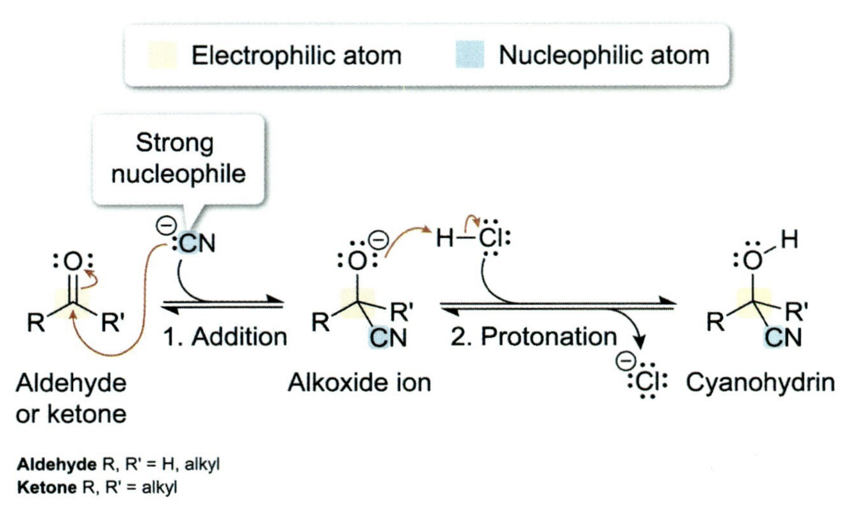

Figure 9.12 The mechanism to form a cyanohydrin.

9.3.04 Formation of Imines and Related Compounds

An electrophilic aldehyde or ketone reacts with the nucleophilic nitrogen atom of either ammonia (NH_3) or an amine in the presence of a weak acid to form stable products. Although these reactions follow the same general mechanism, the names and structures of the products *differ* based on the starting nucleophile (Figure 9.13). Tertiary (3°) amines are *unable* to form a stable product with an aldehyde or ketone.

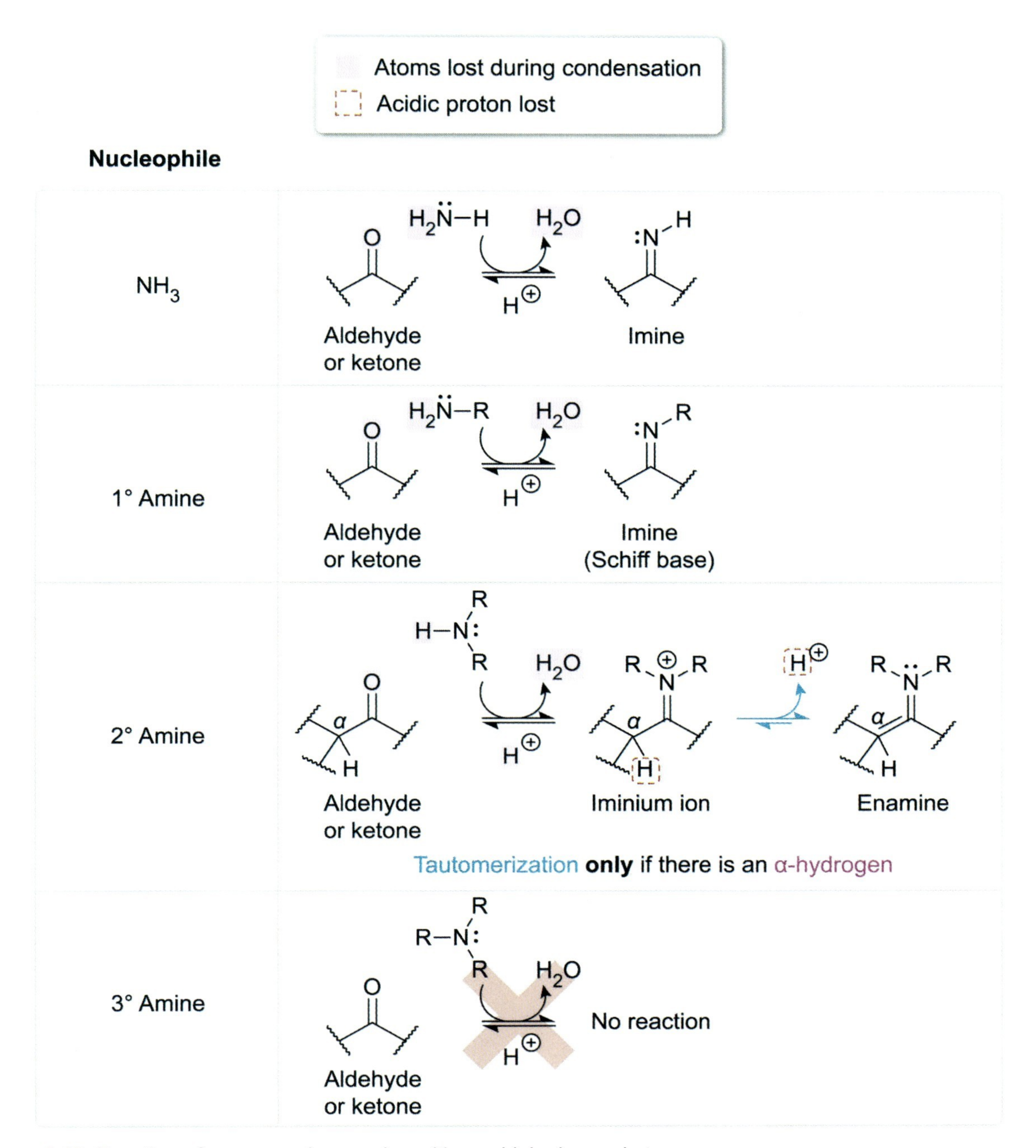

Figure 9.13 Reaction of an ammonia or amine with an aldehyde or a ketone.

In these reactions, the nitrogen-containing molecule undergoes a condensation reaction with the aldehyde or ketone carbonyl to form a product containing a carbon-nitrogen double bond (ie, C=N) in place of the carbon-oxygen double bond. The carbonyl oxygen and two hydrogen atoms are lost as a molecule of water. Condensation with either ammonia or a 1° amine forms a functional group called an **imine**; an imine derived from a 1° amine is sometimes called a **Schiff base**.

Condensation with a 2° amine initially forms a functional group called an **iminium ion**, which contains a C=N bond and has marginal stability. If an iminium ion *lacks* hydrogen atoms at either α position (ie, the carbons *adjacent* to the C=N), the iminium ion tends to equilibrate back to the stable aldehyde or ketone. However, if the iminium ion *does* have one or more α-hydrogen atoms, it will undergo tautomerization to form a stable product called an **enamine** (see Concept 9.4.06).

The conversion of an aldehyde or ketone to an imine or iminium ion has two stages with three steps each (a total of six steps, each in dynamic equilibrium) (Figure 9.14). The first stage is acidic nucleophilic addition between the nitrogen-containing nucleophile and the carbonyl group, and the second stage is acidic dehydration of the alcohol (see Concept 8.3.08). The six mechanistic steps for this reaction are:

1. Protonation
2. Addition of the nucleophile
3. Deprotonation
4. Protonation
5. Elimination of the leaving group
6. Deprotonation

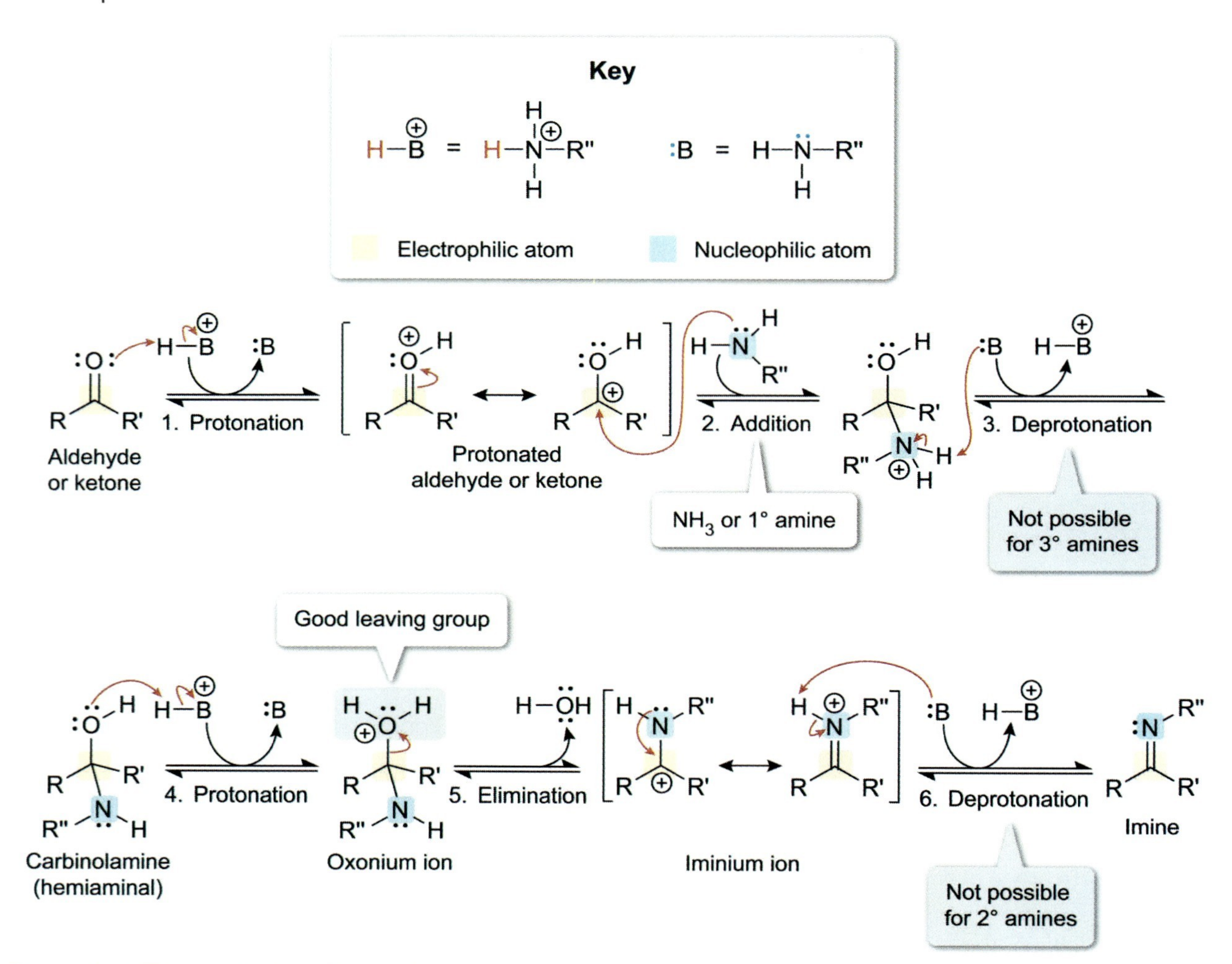

Figure 9.14 The mechanism of imine formation.

First, the carbonyl oxygen atom is protonated by weak acid—strong acids *cannot* be used because they would *also* protonate the nucleophilic nitrogen atom, *consuming* its lone pair. After protonation, the nitrogen nucleophile undergoes *addition* with the carbonyl carbon atom. Next, the acidic proton on the nitrogen atom is deprotonated to form a **carbinolamine** (or **hemiaminal**) intermediate, a functional group with a hydroxyl group attached to the same carbon as a nitrogen atom.

In the second stage of the mechanism, the hydroxyl oxygen is protonated by acid to form an oxonium ion, a good leaving group. Formation of an oxonium ion is the *primary* reason why an acid catalyst is needed for the overall reaction. Then a molecule of water is eliminated to form an iminium intermediate, which is stabilized through resonance with the nitrogen atom lone pair. Finally, the iminium intermediate is deprotonated to form the imine product. The final deprotonation step is *not possible* when a 2° amine nucleophile is used—there would be *no* hydrogen atoms on the nitrogen to remove—leading to a final product of either an iminium ion or an enamine (see Concept 9.4.06).

The conversion of an aldehyde or a ketone into an imine has several related variants that differ in the groups attached to the nucleophilic nitrogen atom (Table 9.2).

Table 9.2 Functional group variants for the condensation of a nitrogen nucleophile with an aldehyde or a ketone.

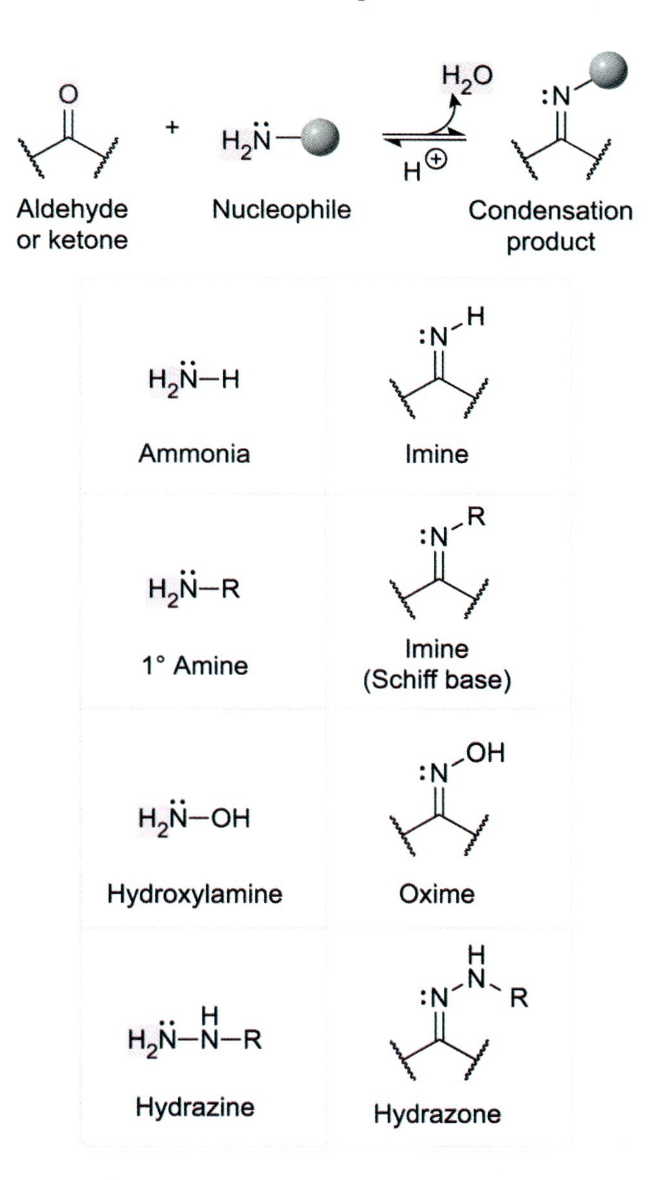

9.3.05 Formation of Hemiacetals and Acetals

In the presence of an acid catalyst, electrophilic aldehydes or ketones can react with the nucleophilic oxygen atom of alcohols to form stable products. The first stable intermediate formed is a **hemiacetal**, which has a hydroxyl group (–OH) and an alkoxy group (–OR) attached to the same carbon atom.

Continued reaction with a *second* equivalent of the alcohol eliminates a molecule of water to form an **acetal**, a functional group with *two* alkoxy groups attached to the same carbon atom. The products of a reaction with a *ketone* may also be described using the terms **hemiketal** or **ketal** (Figure 9.15).

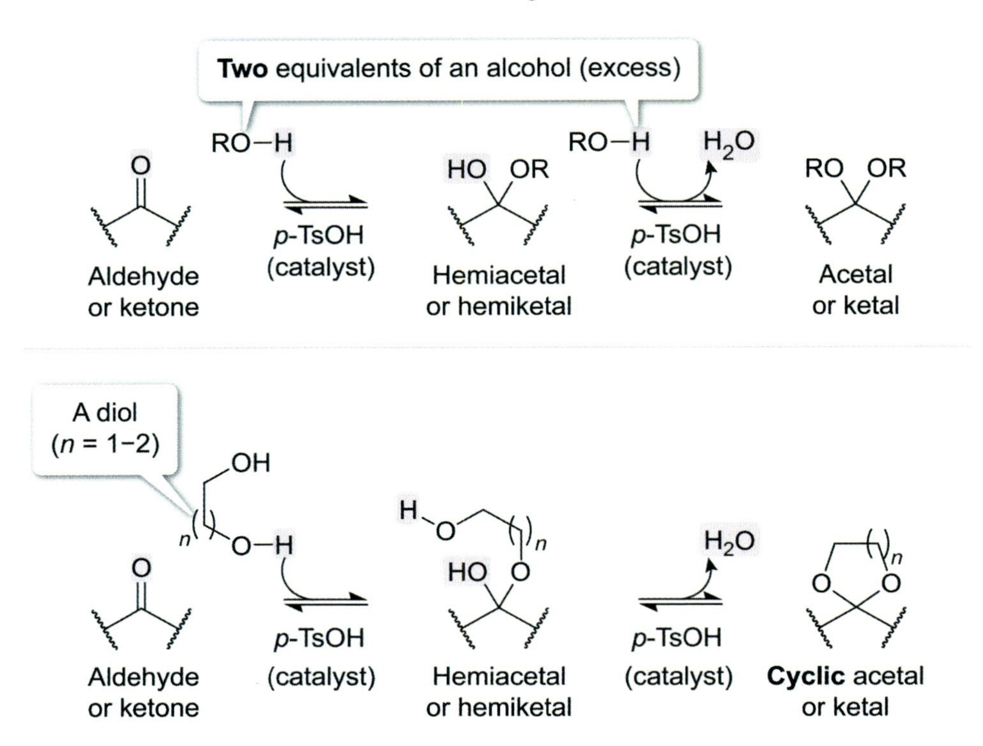

Figure 9.15 General reaction of an alcohol with an aldehyde or ketone.

Alternately, it is possible to form an acetal from the reaction of an aldehyde or ketone with a diol, which has two hydroxyl groups in the *same* molecule. In these situations, the product is a *cyclic* acetal, in which the alkyl R groups of the acetal connect through the *same* alkyl chain.

The formation of a hemiacetal or acetal is a *reversible* process that requires a catalytic amount of a Brønsted-Lowry acid—commonly, ***p*-toluene sulfonic acid** (**tosic acid**, *p*-TsOH). Typically, an *excess* of alcohol is used to shift the equilibrium toward increased formation of the acetal.

The conversion of an aldehyde or ketone to an acetal has two stages and a total of seven mechanistic steps, *all* of which are dynamic equilibria (Figure 9.16). The first stage is an acidic nucleophilic addition between the alcohol and the carbonyl group; the second stage is an acid-catalyzed substitution of an alcohol (see Concept 8.3.06). The seven mechanistic steps for this reaction are:

1. Protonation
2. Addition of the nucleophile
3. Deprotonation
4. Protonation of the leaving group
5. Elimination of the leaving group
6. Nucleophilic attack
7. Deprotonation

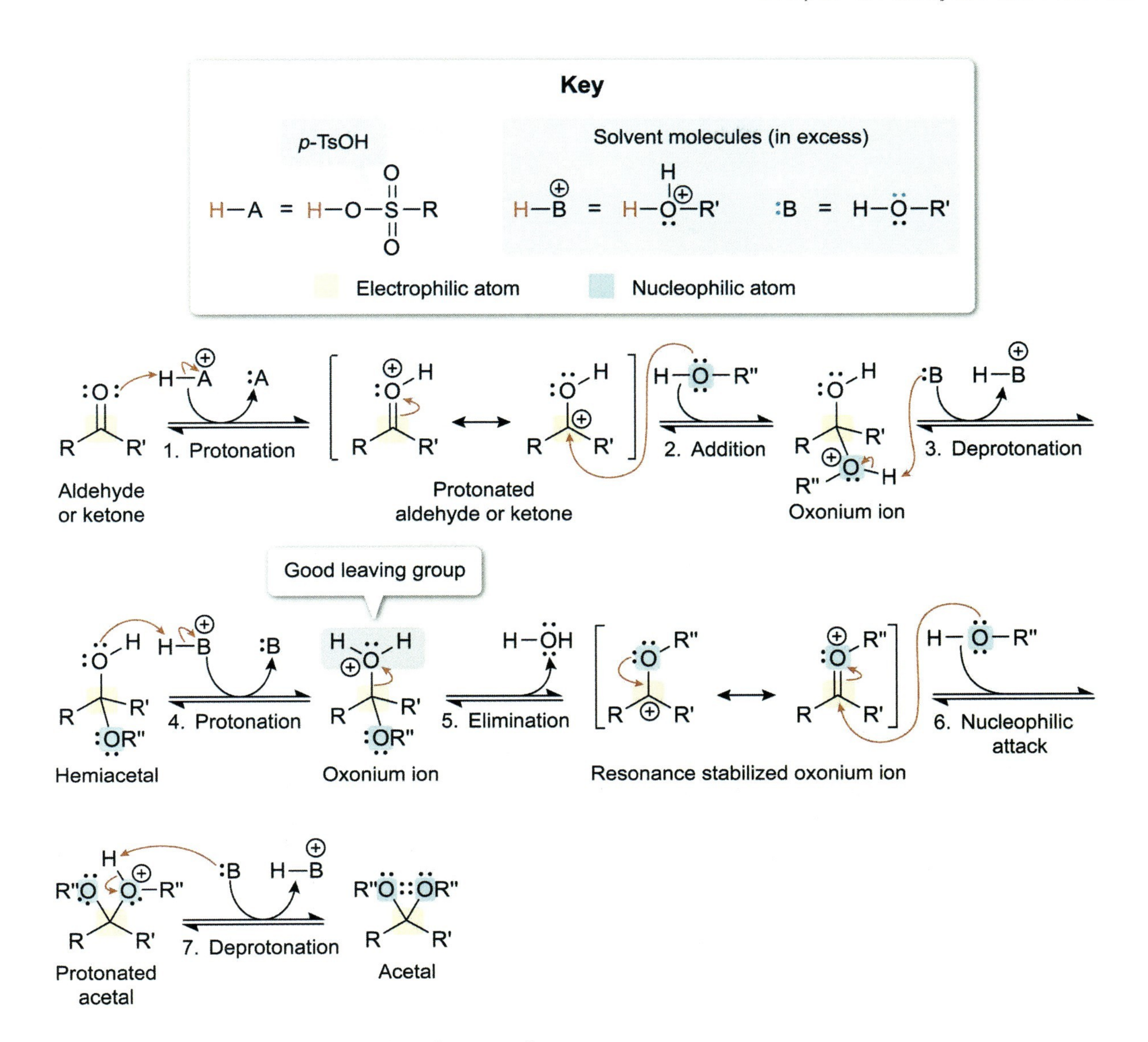

Figure 9.16 Mechanism for the formation of an acetal.

First, the carbonyl oxygen atom is protonated by the acid catalyst. Then an alcohol undergoes an *addition* with the carbonyl carbon atom. Next, an acidic proton on the oxonium ion intermediate is deprotonated to form a hemiacetal intermediate.

To begin the second stage of the mechanism, the oxygen atom of the hydroxyl group is protonated by an acid, forming a *different* oxonium ion to serve as a good leaving group. The leaving group leaves (as a molecule of water) forming a resonance-delocalized intermediate. Next, the *second* alcohol molecule attacks the electrophilic carbon atom. Finally, an acidic proton is removed from the oxonium ion intermediate to form the acetal product.

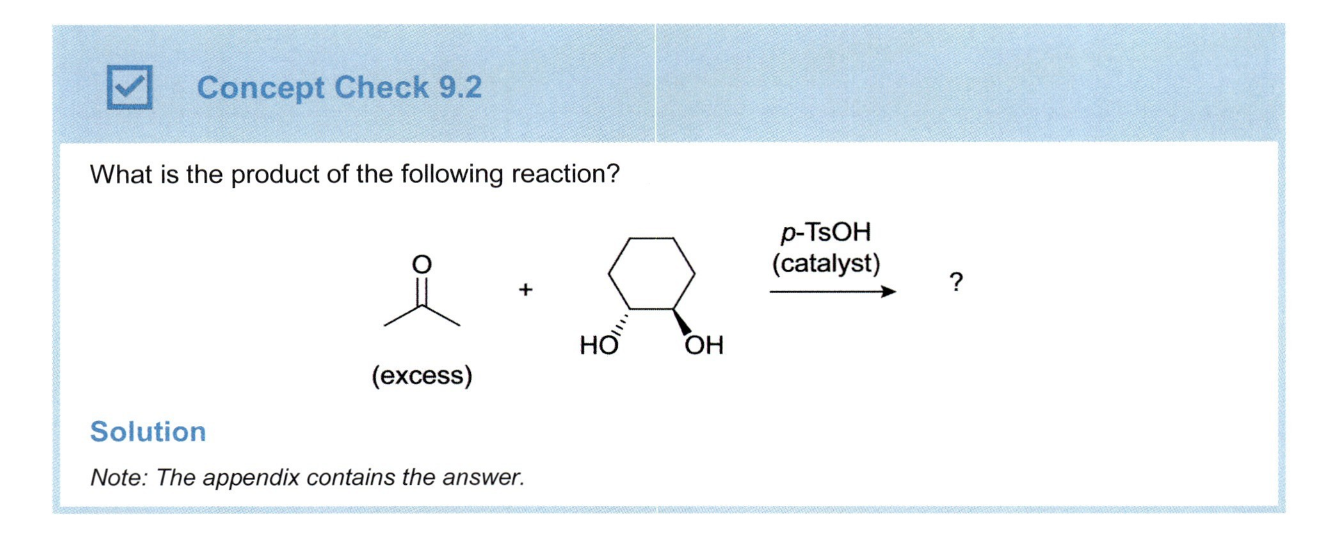

9.3.06 Protecting Groups for Aldehydes and Ketones

One of the ways that an aldehyde or ketone can be reversibly protected is as an acetal (Figure 9.17). The acetal **protecting group** can be formed either through reaction with two equivalents of alcohol or, more commonly, through reaction with a 1,2- or 1,3-diol (eg, ethane-1,2-diol, propane-1,3-diol) to form a *cyclic* acetal. The protection step is typically accomplished by dissolving the aldehyde or ketone in the desired alcohol (ensuring that the alcohol is in excess) and heating the solution with a catalytic amount of strong acid (ie, *p*-TsOH).

Unreactive to nucleophilic conditions

p-TsOH; HO–(CH₂)ₙ–OH → Protection (– H_2O) → Key reaction (S_N2) with :Nu⁻ (– Br⁻) → Deprotection (1M HCl, THF) → HO–(CH₂)ₙ–OH

Aldehyde or ketone (reactant) | Acetal-protected reactant | Acetal-protected S_N2 product | S_N2 Product

Figure 9.17 The use of an acetal as a protecting group for a ketone or aldehyde.

In general, an acetal protecting group is *unreactive* under strongly basic or nucleophilic conditions, and *reactive* (ie, reversibly deprotected) under acidic conditions. An acetal protecting group can typically be removed by reacting the protected compound with an aqueous solution of strong acid (eg, 1M HCl) mixed with a water-miscible solvent (eg, tetrahydrofuran [THF]).

Aldehydes are generally more electrophilic than ketones due to reduced steric hindrance; therefore, it is sometimes possible to *selectively* protect an aldehyde as an acetal in the presence of a ketone. With the aldehyde group protected, a compound that reacts with carbonyl groups will then *specifically* target the ketone on the structure. Figure 9.18 shows a reaction scheme resulting in the *selective* reduction of a ketone to a 2° alcohol using $LiAlH_4$, a nucleophilic reducing agent (see Concept 9.3.08). Removal of the acetal protecting group then yields a product containing both an aldehyde and a 2° alcohol.

Without an acetal protecting group

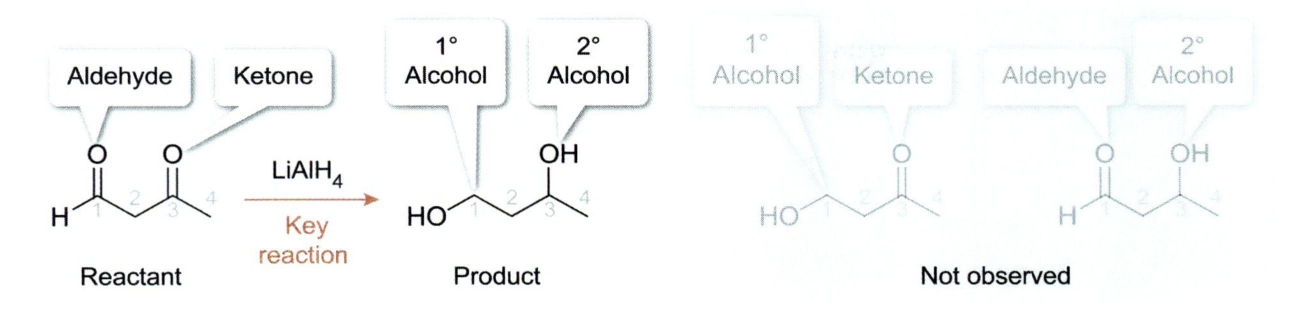

With an acetal protecting group

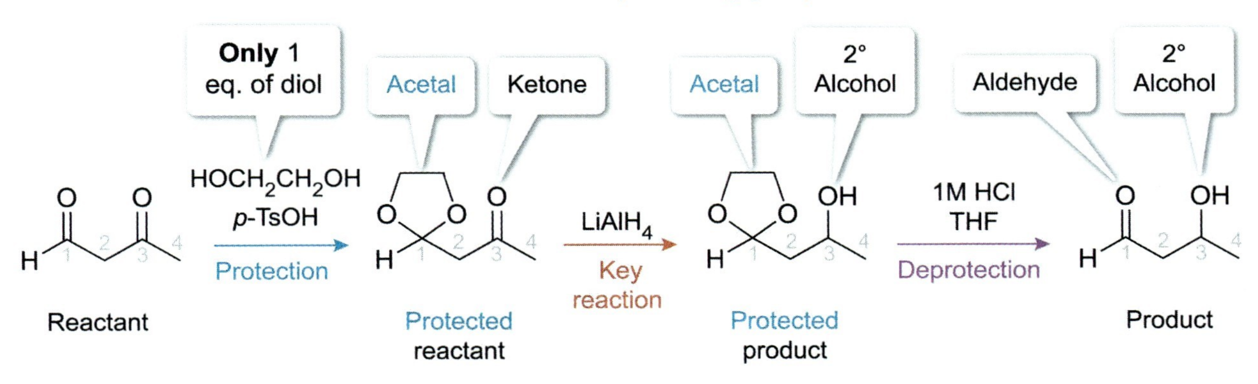

Figure 9.18 An acetal can selectively protect an aldehyde.

Alternately, an acetal can be used as a protecting group for a 1,2- or 1,3-diol (Figure 9.19). To achieve this, a low molecular weight ketone (eg, acetone [$CH_3C(O)CH_3$]) is used in excess with a catalytic amount of *p*-TsOH to convert the 1,2- or 1,3-diol into an acetal. This type of chemistry is frequently used for carbohydrates, which contain many alcohol functional groups. On deprotection, the ketone is easily removed with acid to provide the deprotected diol as the isolated product.

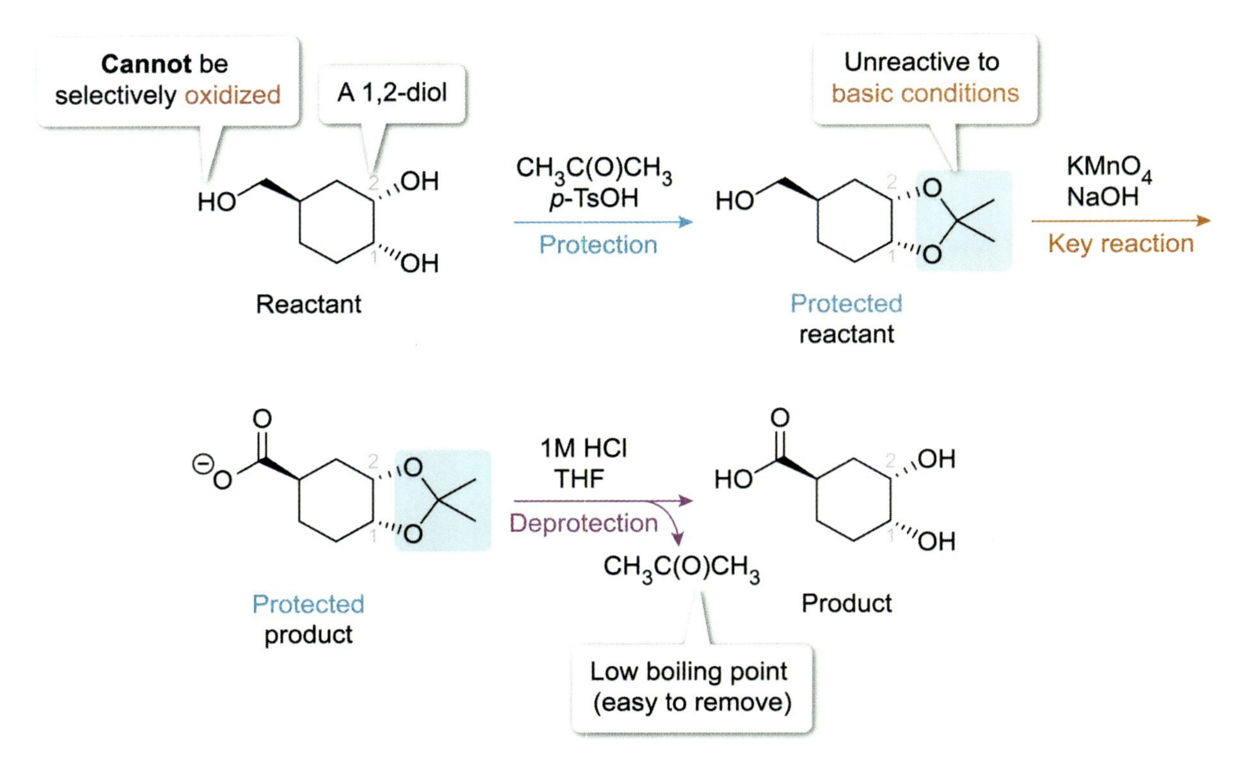

Figure 9.19 The use of an acetal as a protecting group for a 1,2-diol.

9.3.07 Oxidation of Aldehydes

Treatment of an aldehyde with an oxidizing agent promotes a two-electron oxidation to form a carboxylic acid (Table 9.3). In general, an aldehyde is among the functional groups *most* prone and *easiest* to oxidize. Various oxidizing agents (eg, H_2CrO_4, $Na_2Cr_2O_7$, CrO_3, $KMnO_4$, Ag^+) are sufficient to produce a carboxylic acid (see Concept 8.3.01).

Table 9.3 Oxidation reactions of aldehydes.

Carbon being oxidized

Single oxidations require only **one** equivalent of a 2 e^- oxidizing agent

	Functional group	Reactant	Reactant sequence	Oxidized product
All oxidants facilitate a single (2 e^-) oxidation	Aldehyde (R = alkyl)	RCHO	H_2CrO_4, H_2O →	RCOOH Carboxylic acid
			$Na_2Cr_2O_7$, H_2SO_4, H_2O →	
			CrO_3, H_2SO_4, H_2O →	
			$KMnO_4$, NaOH, H_2O →	
			2 Ag^+ (Tollens's test) →	

As a **1** e^- oxidizing agent, **two** equivalents are required.

Note: $Na_2Cr_2O_7$ is the anhydride salt of chromic acid and CrO_3 (ie, Jones reagent) is dehydrated chromic acid; both sets of reagents generate H_2CrO_4 in the presence of aqueous acid (H_2SO_4).

The mechanism for the oxidation of an aldehyde to a carboxylic acid using chromic acid (H_2CrO_4) is shown in Figure 9.20. Under acidic conditions, an aldehyde undergoes a hydration reaction with a molecule of water to produce a geminal diol (ie, an aldehyde hydrate) (see Concept 9.3.02). One of the hydroxyl groups of the geminal diol attacks H_2CrO_4 to form a protonated chromate ester intermediate. Then a base facilitates removal of a hydrogen atom and the reduction of Cr (VI) to Cr (IV). Finally, the protonated carboxylic acid loses an acidic hydrogen atom to the solvent to yield the carboxylic acid product.

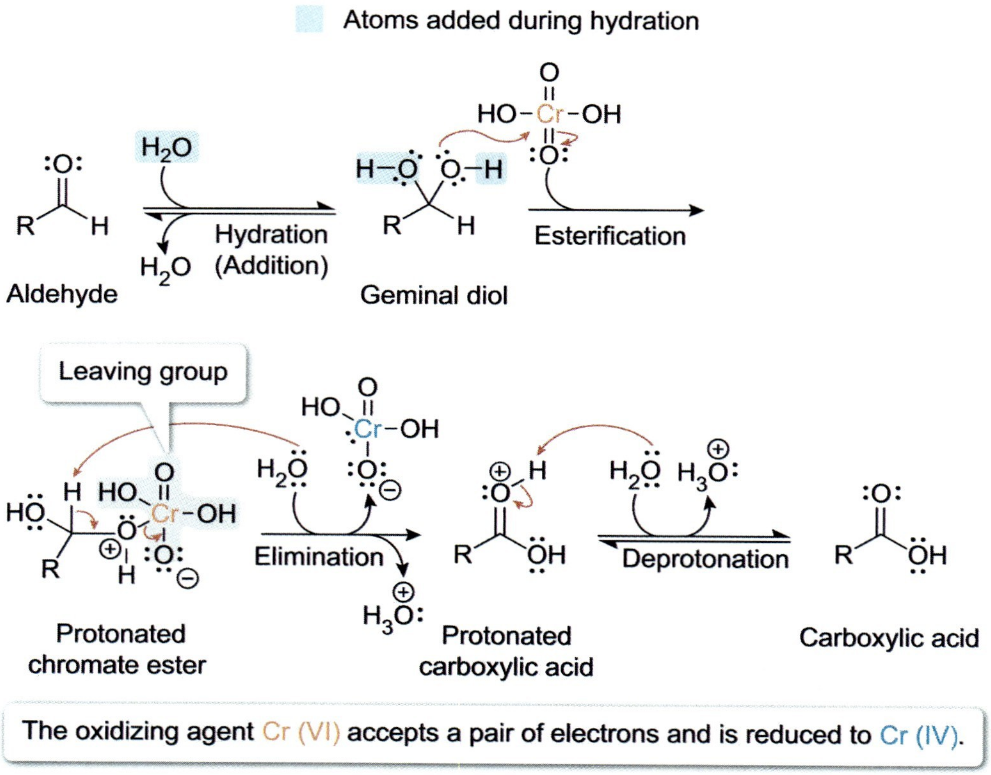

Figure 9.20 The mechanism for the oxidation of an aldehyde by chromic acid.

Tollens's Test

The silver (I) ion (Ag^+) is a mild oxidizing agent used in **Tollens's test**, which tests for the presence of an aldehyde functional group. To perform Tollens's test, a solution of the Ag^+-containing oxidizing agent ($[Ag(NH_3)_2]^+$, **Tollens's reagent**) is prepared in a basic solution of $AgNO_3$ with NaOH and NH_4OH (Figure 9.21). Then the sample to be tested is added to Tollens's reagent and the mixture heated in a test tube or similar vessel.

A *positive* Tollens's test results in the formation of a *silver mirror*, formed by the precipitation of elemental silver (Ag^0) on the surface of the test tube, indicating that an aldehyde functional group was present in the sample. In some specific cases, a *ketone* functional group can *also* give a positive Tollens's test if the ketone can undergo successive base-catalyzed keto-enol tautomerization reactions into an aldehyde (eg, carbohydrates).

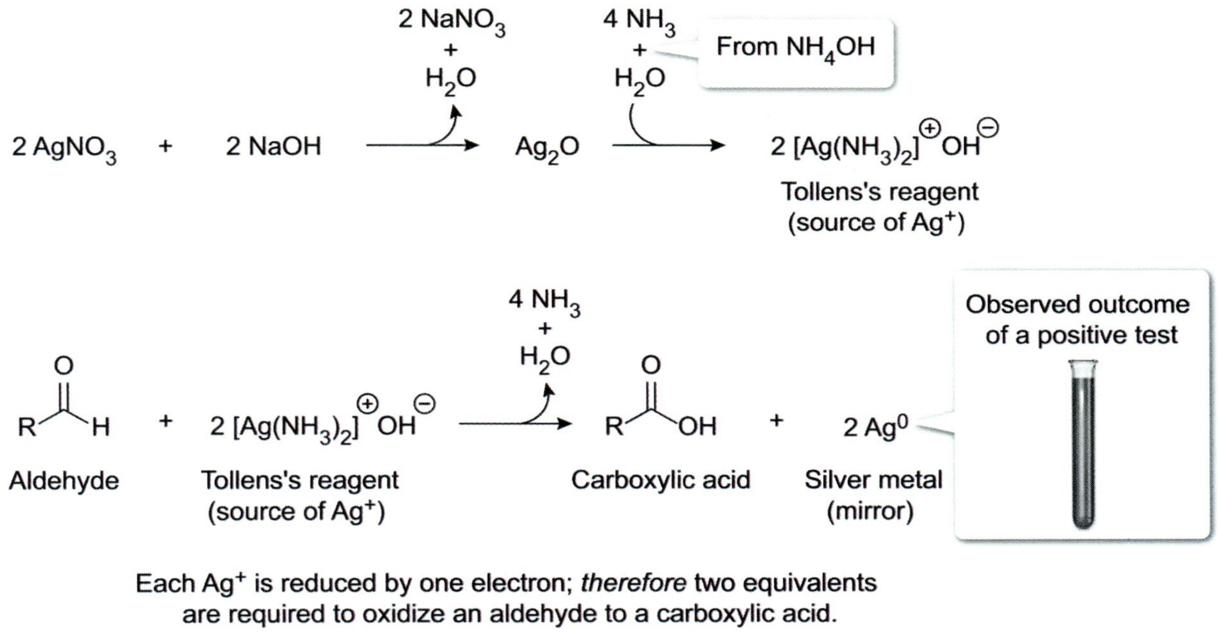

$$2\,AgNO_3 + 2\,NaOH \xrightarrow{} Ag_2O \;(+\; 2\,NaNO_3 + H_2O) \xrightarrow{4\,NH_3 + H_2O} 2\,[Ag(NH_3)_2]^{\oplus}OH^{\ominus}$$

From NH_4OH

Tollens's reagent (source of Ag^+)

$$RCHO + 2\,[Ag(NH_3)_2]^{\oplus}OH^{\ominus} \xrightarrow{} RCOOH + 2\,Ag^0 \;(+\; 4\,NH_3 + H_2O)$$

Aldehyde — Tollens's reagent (source of Ag^+) — Carboxylic acid — Silver metal (mirror)

Observed outcome of a positive test

Each Ag^+ is reduced by one electron; *therefore* two equivalents are required to oxidize an aldehyde to a carboxylic acid.

Figure 9.21 Oxidation of an aldehyde by Tollens's reagent.

Tollens's test is a redox reaction. Ag^+ (in the $[Ag(NH_3)_2]^+$ complex) is the oxidizing agent and is *reduced* to Ag^0. Because the reduction of Ag^+ to Ag^0 is a *one*-electron process, *two* equivalents of $[Ag(NH_3)_2]^+$ are required to oxidize an aldehyde to a carboxylic acid.

Biological Oxidation of Aldehydes

Oxidation reactions of aldehydes are common in biochemical settings (Figure 9.22). First, the aldehyde undergoes an addition reaction with a nucleophilic heteroatom (from an enzyme or another substrate). Then the tetrahedral intermediate collapses, eliminating a hydride ion. The oxidizing agent NAD^+ accepts the hydride ion at the four-position on its nicotinamide ring and forms its reduced derivative, NADH.

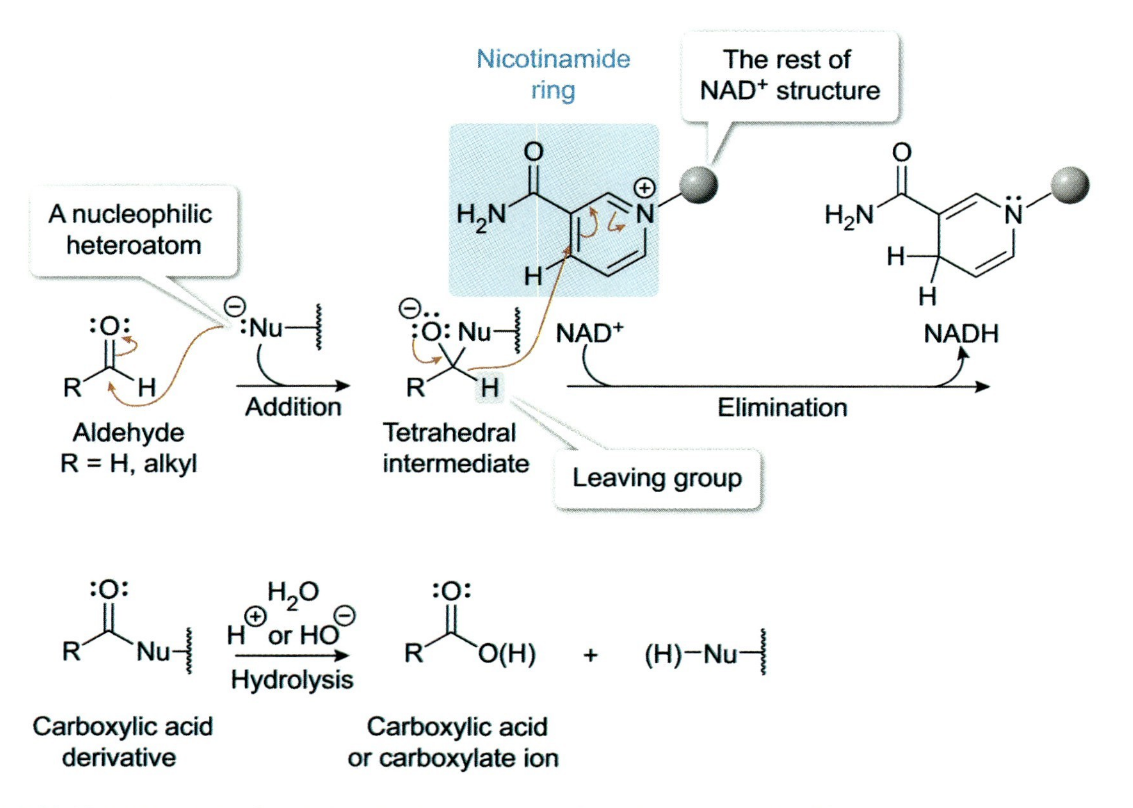

Figure 9.22 The mechanism for the biochemical oxidation of an aldehyde with NAD^+.

The product of this reaction is a carboxylic acid derivative (eg, ester, amide), which is hydrolyzed (see Concept 11.3.01) to produce a carboxylic acid or carboxylate ion.

Oxidation of Ketones

Unlike an aldehyde, a ketone *cannot* be oxidized. Such a reaction would require a mechanism that breaks one or more of the carbon-carbon bonds between an R group and the carbon atom of the ketone carbonyl.

9.3.08 Reduction of Aldehydes and Ketones

An aldehyde or a ketone can be reduced in several ways. Most reduction reactions are two-electron reductions that reduce an aldehyde to a 1° alcohol and a ketone to a 2° alcohol (Figure 9.23).

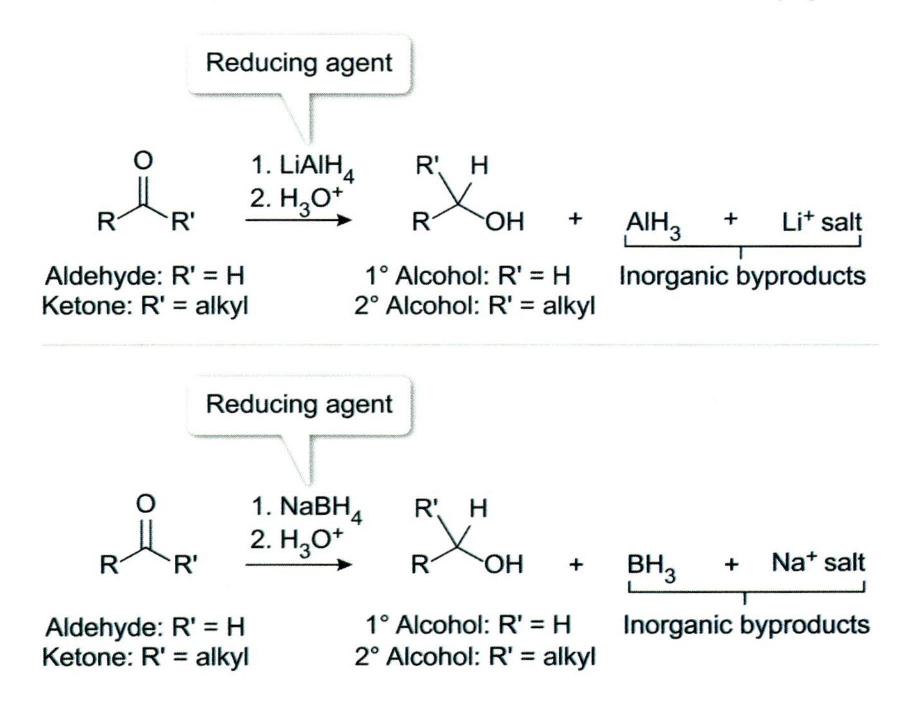

Figure 9.23 The reduction of an aldehyde or ketone to an alcohol using $LiAlH_4$ or $NaBH_4$.

The reducing agents lithium aluminum hydride ($LiAlH_4$) and sodium borohydride ($NaBH_4$) can generally be used interchangeably to facilitate this conversion because both are sources of the nucleophilic hydride ion. If other functional groups are present (ie, carboxylic acids, carboxylic acid derivatives), it may be preferable to use $NaBH_4$, which is a *milder* reducing agent and does not generally reduce functional groups aside from aldehydes and ketones.

Because both $LiAlH_4$ and $NaBH_4$ generate a *basic* reaction environment, a second step involving an *acidic* aqueous workup is required. For a $LiAlH_4$ reaction, the inorganic byproducts of the reduction are AlH_3 and lithium salts, whereas the inorganic byproducts for $NaBH_4$ are BH_3 and sodium salts.

The mechanism for the $LiAlH_4$ reduction of an aldehyde or ketone is a basic nucleophilic addition, followed by an acidic workup (Figure 9.24). The MH_4^- complex ion of each reducing agent (eg, AlH_4^-, BH_4^-) reacts like $H:^-$ + MH_3; one of the M–H bonds (acting like $H:^-$) donates its electron density to the electrophilic carbon atom of the aldehyde or ketone. The resulting alkoxide ion becomes protonated during the acidic workup to generate the alcohol product.

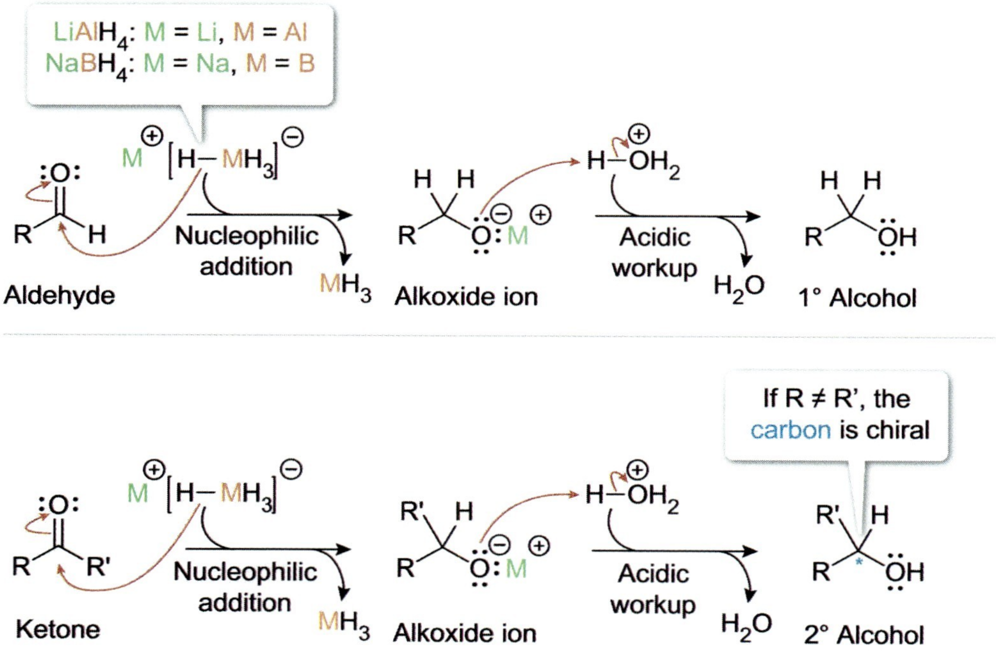

Figure 9.24 The mechanism for the reduction of an aldehyde or ketone by $LiAlH_4$ or $NaBH_4$.

If the ketone is unsymmetrical and has two *different* alkyl groups, the 2° alcohol product will contain a new chiral carbon atom. Because the electrophilic carbon atom of a ketone is planar (see Concept 9.1.01), the hydride ion can attack the electrophilic carbon atom from *either* side of the plane. Consequently, the 2° alcohol product contains a 1:1 mixture of absolute configurations at its new chiral carbon atom. If this chiral carbon atom is the *only* stereocenter in the 2° alcohol, the product is racemic.

Lesson 9.4

Alpha Reactions of Aldehydes and Ketones

Introduction

Lesson 9.3 discusses the reactivity of aldehydes and ketones as Type 3 *electrophiles* (Figure 9.25). Nucleophilic attack at the carbonyl carbon shifts the electrons in the pi bond to a lone pair on the oxygen atom.

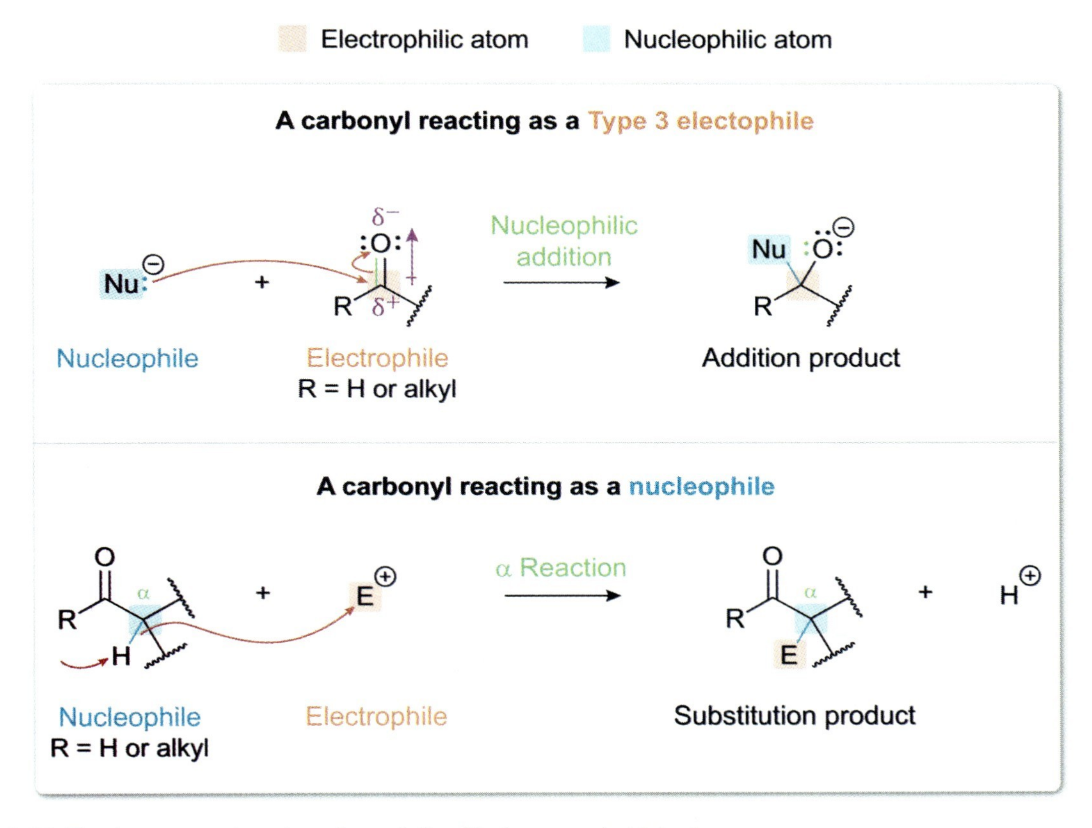

Figure 9.25 The two general modes of reactivity of ketones and aldehydes.

Carbonyl compounds (specifically the α-carbons of a carbonyl) can also react as *nucleophiles* in a class of reactions broadly described as **alpha (α) reactions**. This lesson covers concepts related to α reactions of aldehydes and ketones.

9.4.01 Enol and Enolate Functional Groups

Enols and enolates are two functional groups that are combinations of other functional groups. The name "**enol**" is derived from the terms for an alkene (*en*) and an alcohol (*ol*) and describes a molecule in which a hydroxyl group is *directly* attached to one of the carbon atoms of a double bond (Figure 9.26). An **enolate** is an enol that has lost the acidic proton of its hydroxyl group (ie, an enolate is the anionic conjugate base of an enol).

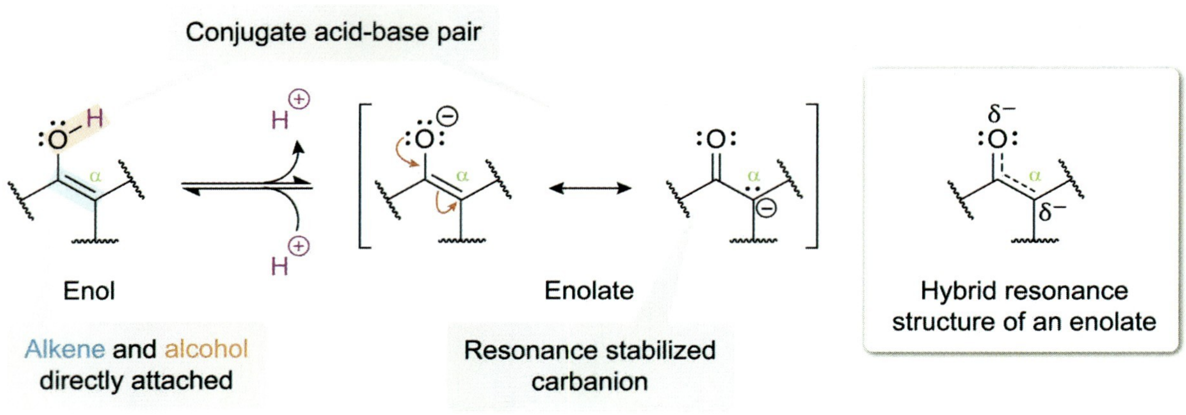

Conjugate acid-base pair

Enol

Alkene and alcohol directly attached

Enolate

Resonance stabilized carbanion

Hybrid resonance structure of an enolate

Figure 9.26 Relationship between an enol and an enolate.

Unlike enols, some enolate electrons are inherently delocalized through resonance. Delocalization leads to the other resonance form of an enolate, which contains a lone pair on the carbon atom adjacent to a carbonyl group (the **α-carbon**). The hybrid resonance structure of an enolate depicts a delocalized π system with partial negative charges on the oxygen atom and the α-carbon. However, the resonance form with the negative charge on the carbon atom best illustrates the reactivity profile of the α-carbon of an enolate as a nucleophile (see Concept 9.4.03).

9.4.02 Tautomerization

Tautomers are constitutional isomers (see Lesson 3.1) that differ in two ways (Figure 9.27):

- One hydrogen differs in position by two atoms (ie, it moves to an atom in a 1,3-relationship with the original position).
- The relative position of a π bond differs by one atom.

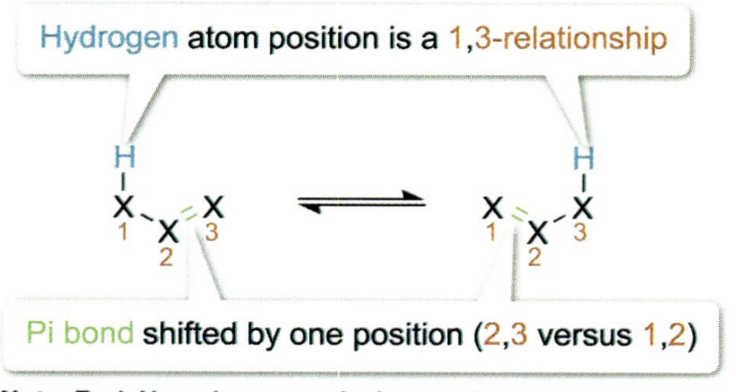

Figure 9.27 General definition of tautomers.

Tautomers differ from resonance structures in several important ways. Tautomers are *different* molecules, often with *different* functional groups, because they have *different* sigma bond connectivity of atoms. The interconversion between tautomers (ie, **tautomerization**) is a chemical reaction in dynamic equilibrium. In contrast, resonance structures are different electronic representations of the *same molecule* (the hybrid resonance structure). The sigma bond connectivity of atoms *cannot* change between resonance structures.

The first tautomeric relationship that students encounter in organic chemistry is **keto-enol tautomerization** (Figure 9.28). The keto (carbonyl) tautomer contains a carbon-oxygen double bond and a hydrogen at its α position, whereas the enol tautomer contains a carbon-carbon double bond and a hydrogen attached to the oxygen atom (as a hydroxyl group). For most keto-enol tautomers, the position of the equilibrium (ie, K_{eq}) tends to favor the keto tautomer.

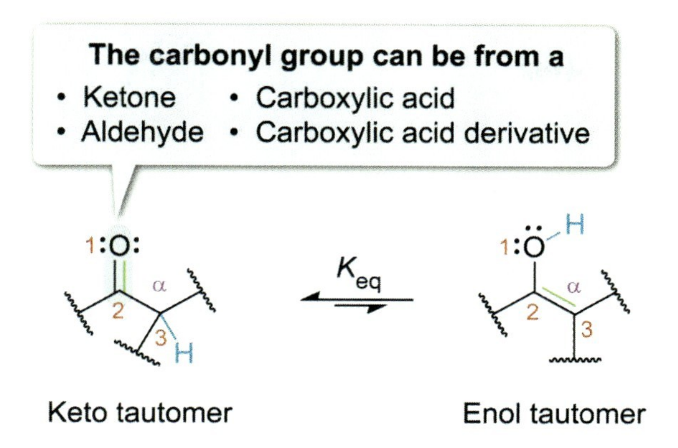

Figure 9.28 Keto-enol tautomerization.

Like many of the reactions discussed in Lesson 9.3, keto-enol tautomerization can occur with either an acid catalyst or a base catalyst.

In the acid-catalyzed mechanism, the acid (HA) protonates the carbonyl oxygen atom to create a resonance-stabilized oxonium intermediate (Figure 9.29). Then, like an E1 reaction (see Concept 5.6.02), the conjugate base (A^-) removes a hydrogen atom from the α-carbon (regenerating the acid catalyst) and the electrons from its sigma bond shift to create the new pi bond of the enol tautomer.

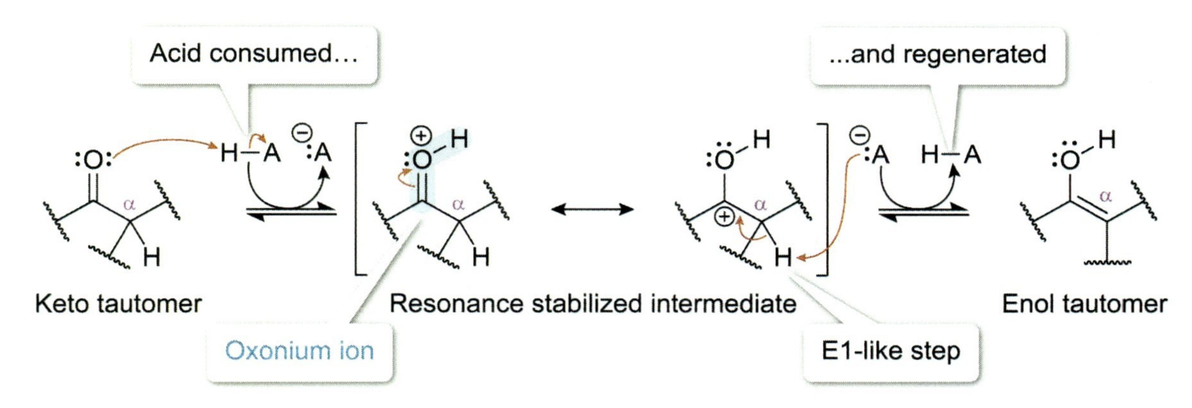

Figure 9.29 Acid-catalyzed keto-enol tautomerization.

In the base-catalyzed mechanism, the base catalyst (A^-) removes a proton from the α-carbon to create a resonance-stabilized enolate ion intermediate (Figure 9.30). After resonance-delocalization, protonation of the enolate by the conjugate acid (HA) regenerates the base catalyst and yields the enol tautomer.

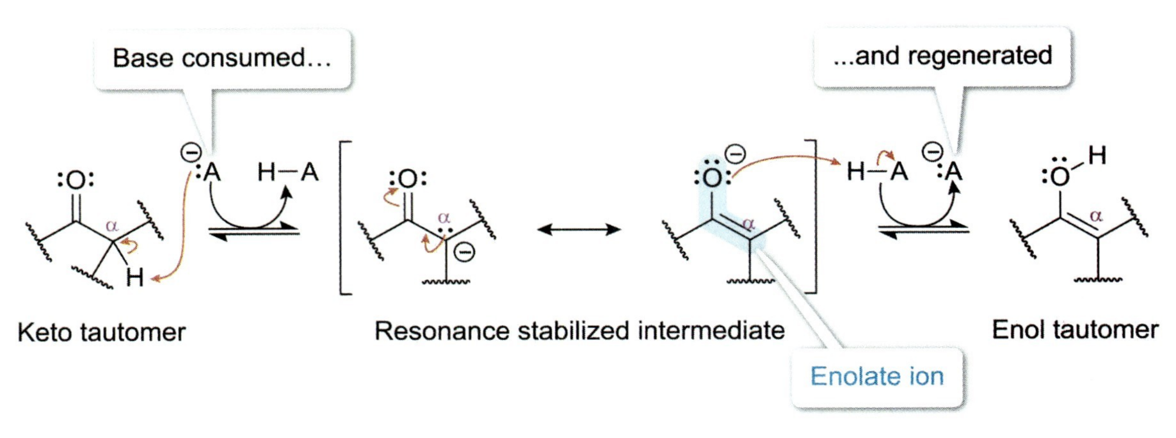

Figure 9.30 Base-catalyzed keto-enol tautomerization.

Keto-enol tautomerization is an example of a dynamic equilibrium; therefore, individual reactions between the keto tautomer and the enol tautomer continue to occur, even after a reaction reaches a state of equilibrium (ie, no net change in concentrations). A potential implication of the dynamic tautomerization process is that if any molecule with a chiral center α to a carbonyl group is exposed to *either* catalytic acid *or* base, racemization of the absolute configuration at the α-carbons (Figure 9.31) can occur.

This occurs because tautomerization causes the hybridization of the α-carbon to change from *sp*^{3} (and potentially chiral) to *sp*^{2}. The *sp*^{2} α-carbon is now *achiral*, meaning that its stereochemical information is lost. During tautomerization *back* to the keto tautomer, the proton can be added to *either* the top face *or* the bottom face of the planar enol, leading to racemization.

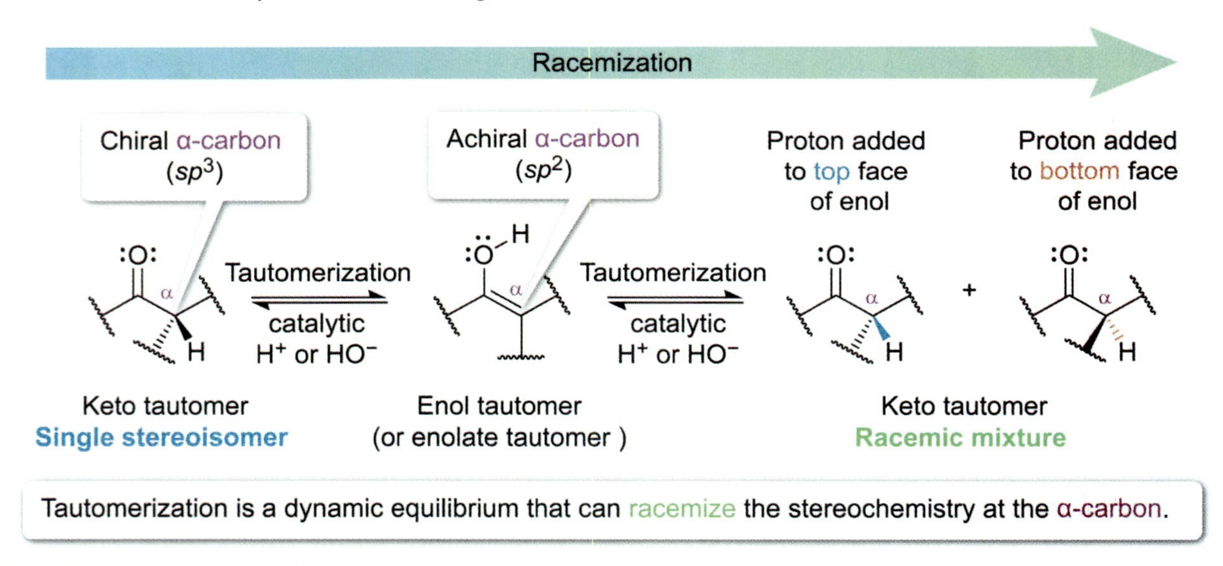

Figure 9.31 Racemization of an α-carbon through dynamic tautomerization.

Concept Check 9.3

For each of the following pairs of compounds, determine if they meet the criteria of tautomers.

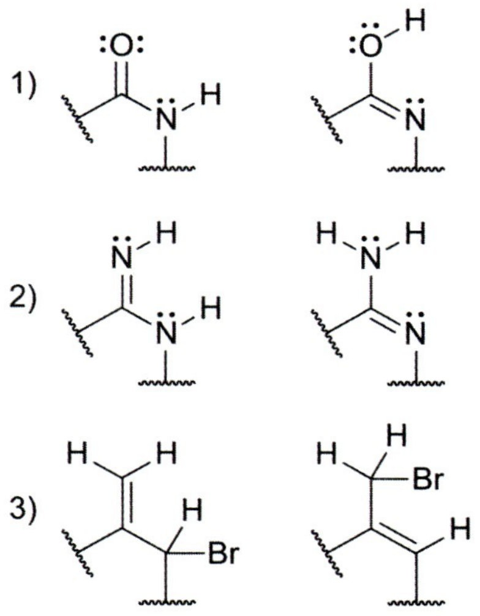

1) :O: ... N–H (amide) / :O–H ... =N (imidic acid)

2) N–H ... N–H (amidine) / H–N–H ... =N

3) H, H ... H, Br / H, H, Br ... H

Solution

Note: The appendix contains the answer.

9.4.03 General Features of Alpha Reactions

Alpha reactions involve a nucleophilic α-carbon and therefore *require* either an enol or enolate ion intermediate to generate them (Figure 9.32).

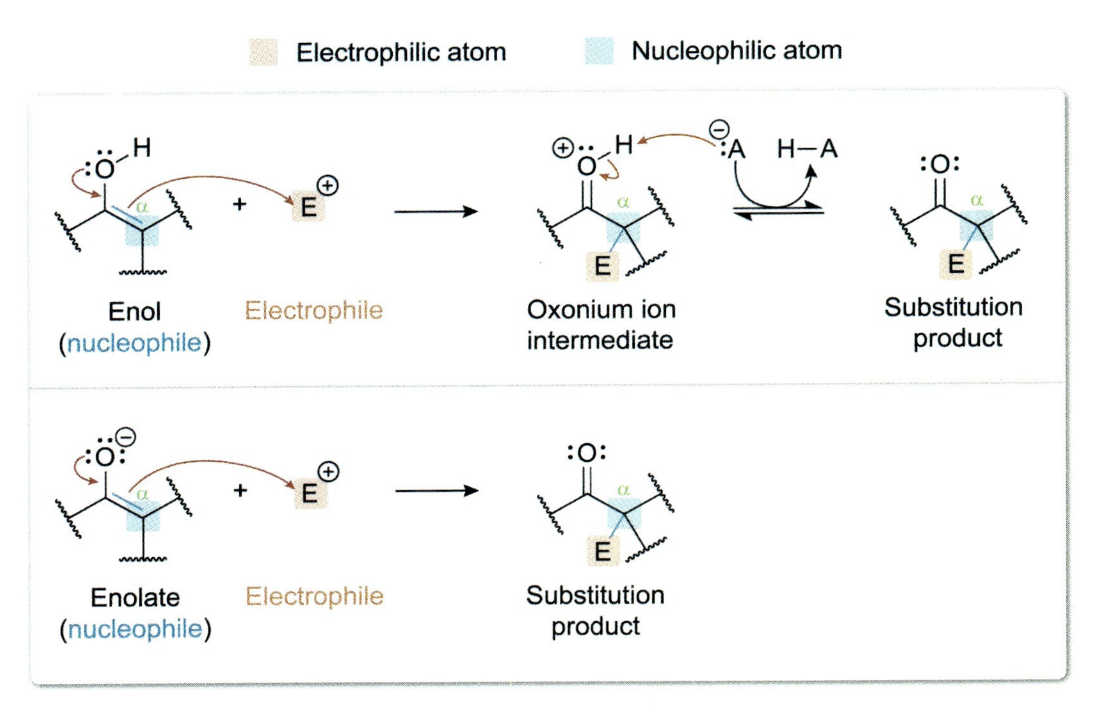

Figure 9.32 General mechanism of an enol and an enolate ion as nucleophiles.

In these general mechanisms, a lone pair on the enol or enolate oxygen atom becomes a pi bond, whereas the carbon-carbon pi bond electrons act as a *nucleophile* to attack an electrophile. The result is a carbonyl group with a new sigma bond between the α-carbon and the electrophilic atom.

To tautomerize into an enol or enolate ion, a carbonyl-containing functional group must have *at least* one α-hydrogen atom (Concept 9.4.02). This hydrogen becomes substituted with an incoming electrophile (Figure 9.33); for this reason, alpha reactions are often called **α-substitution reactions**.

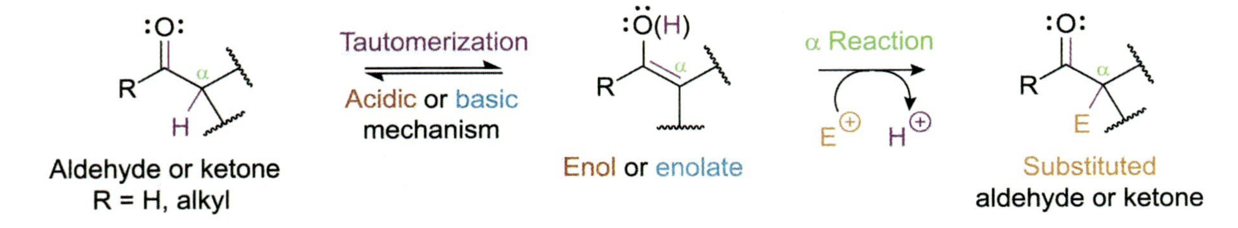

Figure 9.33 Alpha substitution through sequential tautomerization and alpha reaction.

9.4.04 The Acidity of Alpha Protons

An α-proton to a carbonyl group is a Brønsted-Lowry acid (Figure 9.34). Although protons on most sp^3 carbon atoms are typically *extremely* weak acids (pK_a ~50), protons attached to the α-carbon are *significantly* more acidic (pK_a 20). The massive (10^{30}-fold) increase in acidity occurs because the conjugate base of an acidic α-proton (the enolate ion) is stabilized by resonance. Because of *increased* delocalization of their conjugate base enolate ions, the alpha protons of a **1,3-dicarbonyl** group are even *more acidic* than the alpha protons of an aldehyde or ketone.

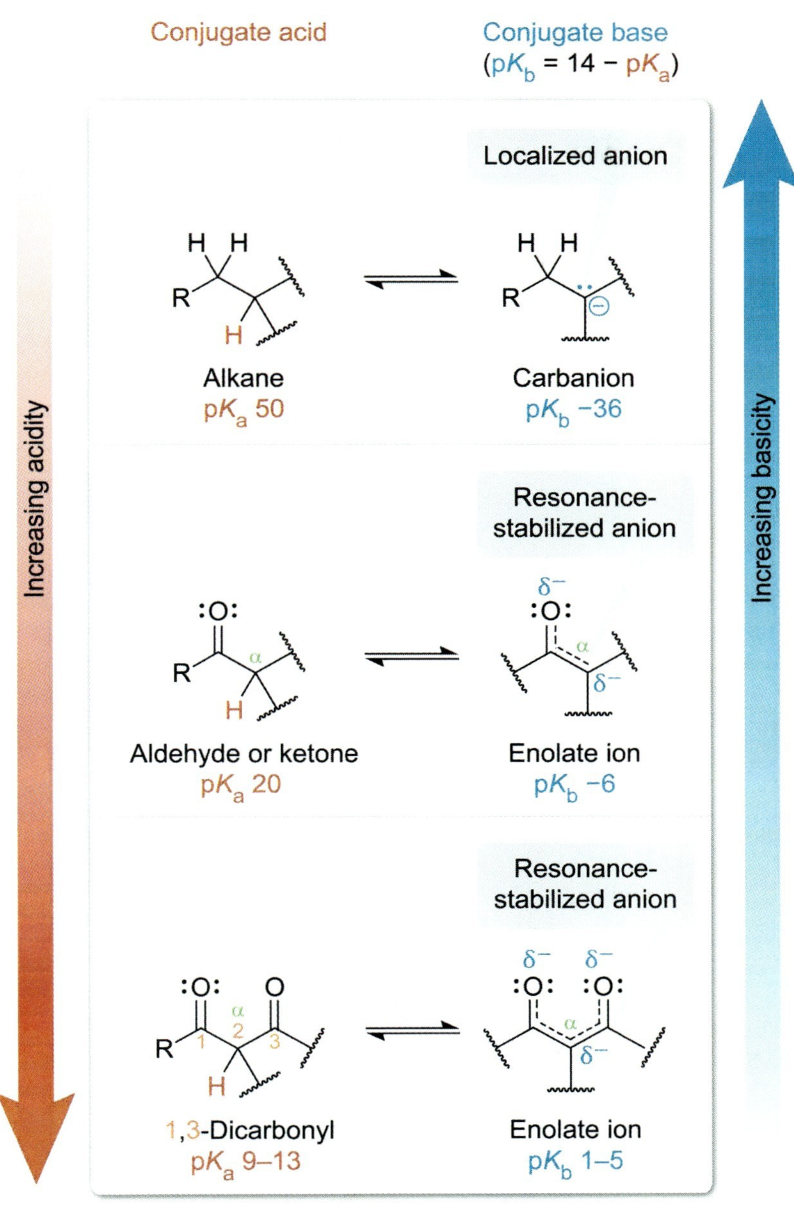

Figure 9.34 The acidity of alpha protons.

Although some alpha reactions take place under acidic conditions, under which the effective nucleophile of an aldehyde or ketone reactant is an enol, it is more common for an alpha reaction to take advantage of α-proton acidity and occur under basic conditions. In these cases, one of several bases can be used to generate the necessary enolate ion nucleophile.

The Equilibrium Enolate Approach

Treatment of an aldehyde or ketone with either a hydroxide ion (HO^-) or alkoxide ion (RO^-) establishes an *equilibrium* with its conjugate enolate ion (Figure 9.35). Because of the reciprocal relationship between the strength of conjugate acids and bases, hydroxide and alkoxide ions (pK_b −2 to −4) are slightly *weaker* bases than an enolate ion (pK_b −6). Consequently, the equilibrium of this reaction favors the *reactants* and only a *small* concentration of enolate ion is present at equilibrium.

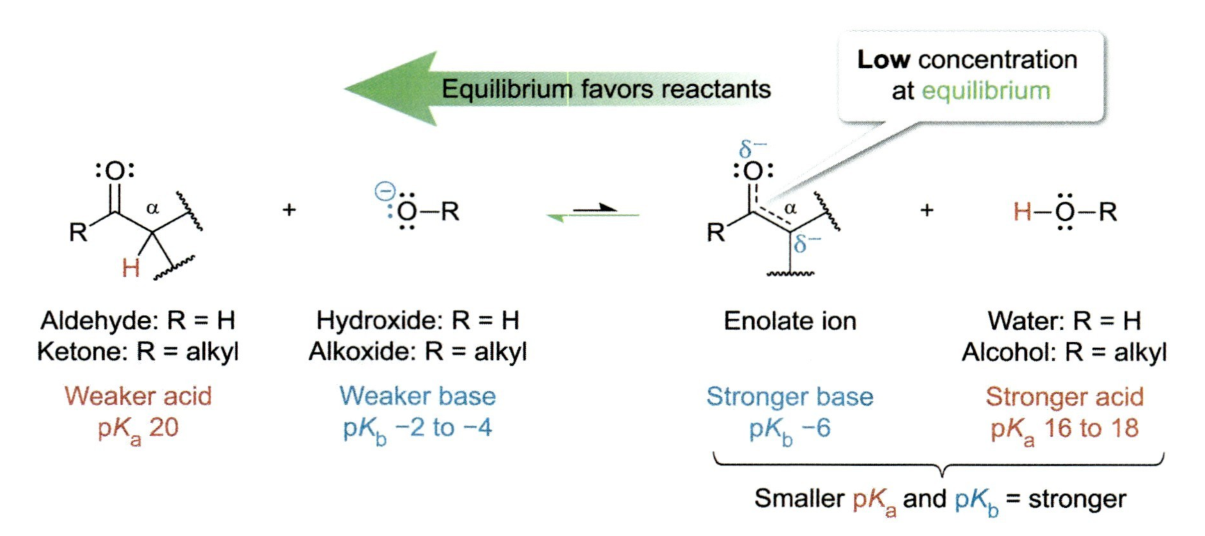

Figure 9.35 Generation of an enolate ion through an equilibrium.

Despite the low equilibrium enolate concentration, alpha reactions can result in a high (although not 100%) yield of substitution product (Figure 9.36). According to Le Châtelier's principle, after disturbance of the tautomerization equilibrium—such as the *removal* of enolate through its *irreversible* reaction with an electrophile—the system acts to *restore* equilibrium by *producing more enolate*. The additional enolate ions can then react with the electrophile, producing more substitution product.

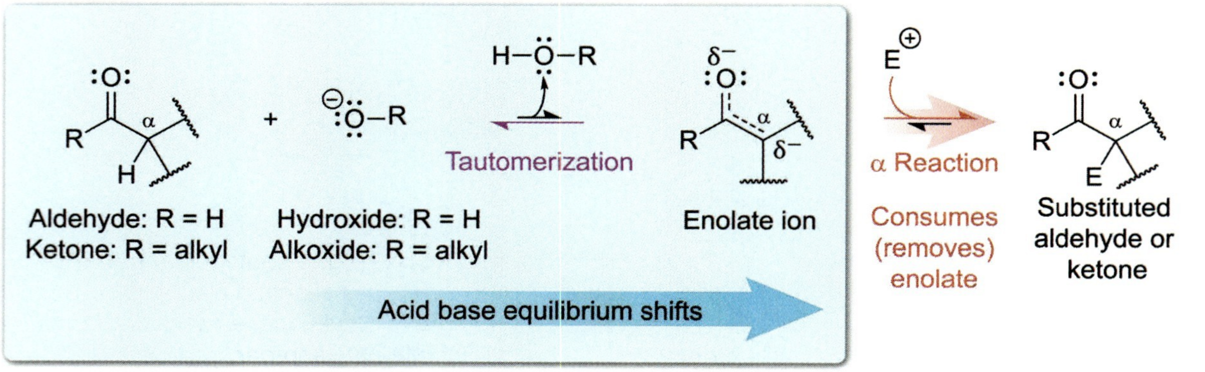

Aldehyde: R = H
Ketone: R = alkyl
Hydroxide: R = H
Alkoxide: R = alkyl
Tautomerization
Enolate ion
α Reaction
Consumes (removes) enolate
Substituted aldehyde or ketone
Acid base equilibrium shifts

Although tautomerization using $^{-}$OR generates low [enolate], α reaction consumes [enolate] and the acid-base equilibrium shifts to increase [enolate].

Figure 9.36 Impact of an alpha reaction on the tautomerization equilibrium.

Through this iterative process, most of the aldehyde or ketone reactant can be converted to alpha reaction product. However, the equilibrium enolate approach is viable only if *both steps* (ie, acid-base reaction, alpha reaction) take place in the *same* reaction vessel. Consequently, the equilibrium enolate approach is limited to situations in which the alpha reaction electrophile is unreactive with a hydroxide ion or alkoxide ion, bases that can *also* act as strong nucleophiles.

The Quantitative Enolate Approach

For situations in which *complete* conversion of a ketone or aldehyde to an enolate ion is required, a *much* stronger base is necessary (Figure 9.37). Sodium hydride (NaH, pK_b −21) and lithium diisopropylamide (LDA, pK_b −26) are stronger bases than enolate ions (pK_b −6), so the acid-base equilibrium greatly favors formation of the enolate ion product.

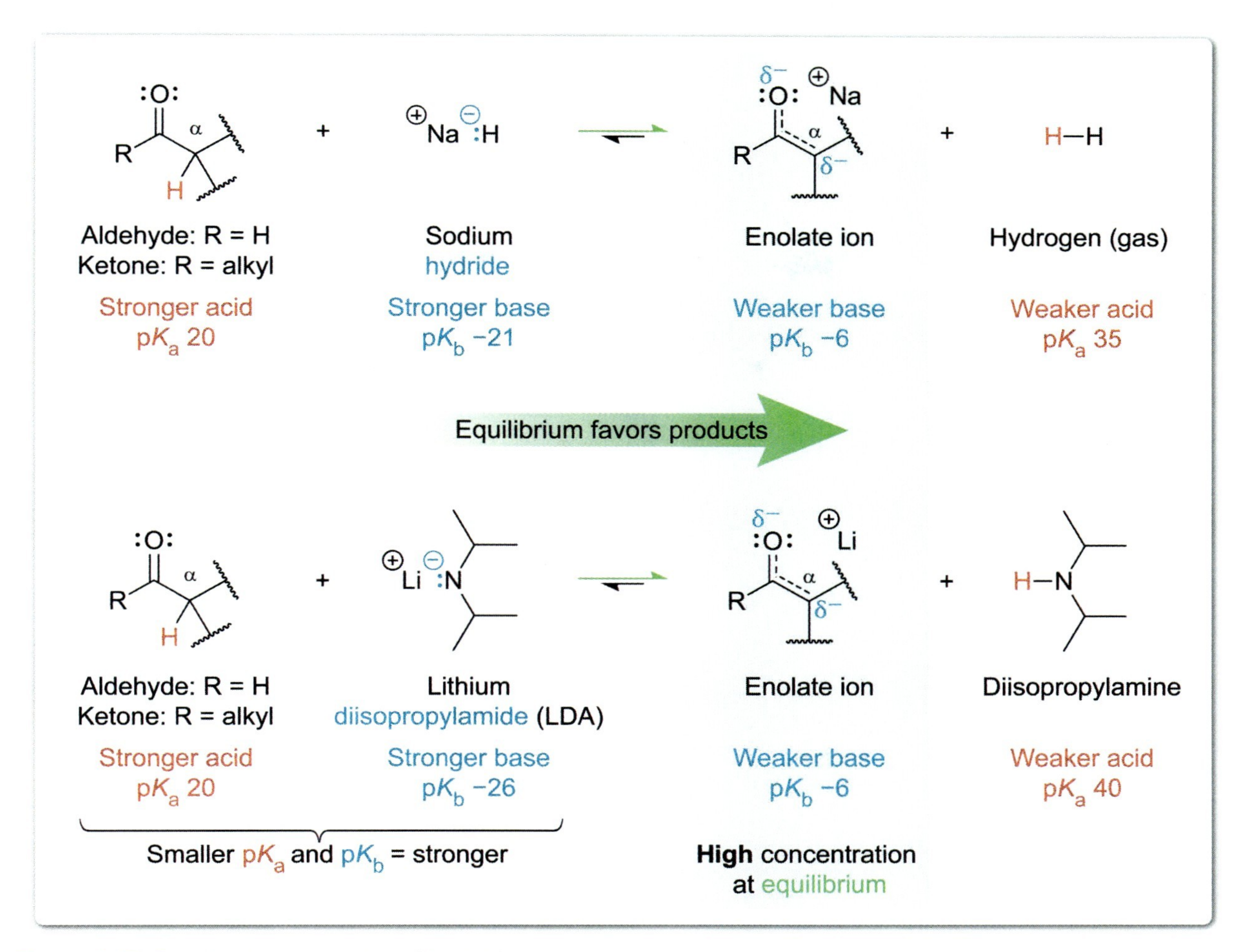

Figure 9.37 Quantitative generation of an enolate ion.

An advantage of the quantitative enolate approach is that the formation of the enolate and the alpha reaction can take place as *separate*, sequential steps. Because the alpha reaction electrophile is not in direct reaction contact with the base used for tautomerization (ie, NaH, LDA), these reagents do not need to be compatible with one another. Consequently, the quantitative enolate approach can accommodate a wider variety of electrophiles for alpha reactions.

9.4.05 Kinetic and Thermodynamic Enolates

The base used to generate an enolate ion for an alpha reaction has another important impact on the reaction outcome. As introduced in Concept 5.6.02, regioisomers are positional isomer products, a mixture of which may be produced from a chemical reaction. If a ketone contains *two different* α-carbons (each with at least one α-hydrogen), *two different* enolate ions (ie, two regioisomers) are possible (Figure 9.38).

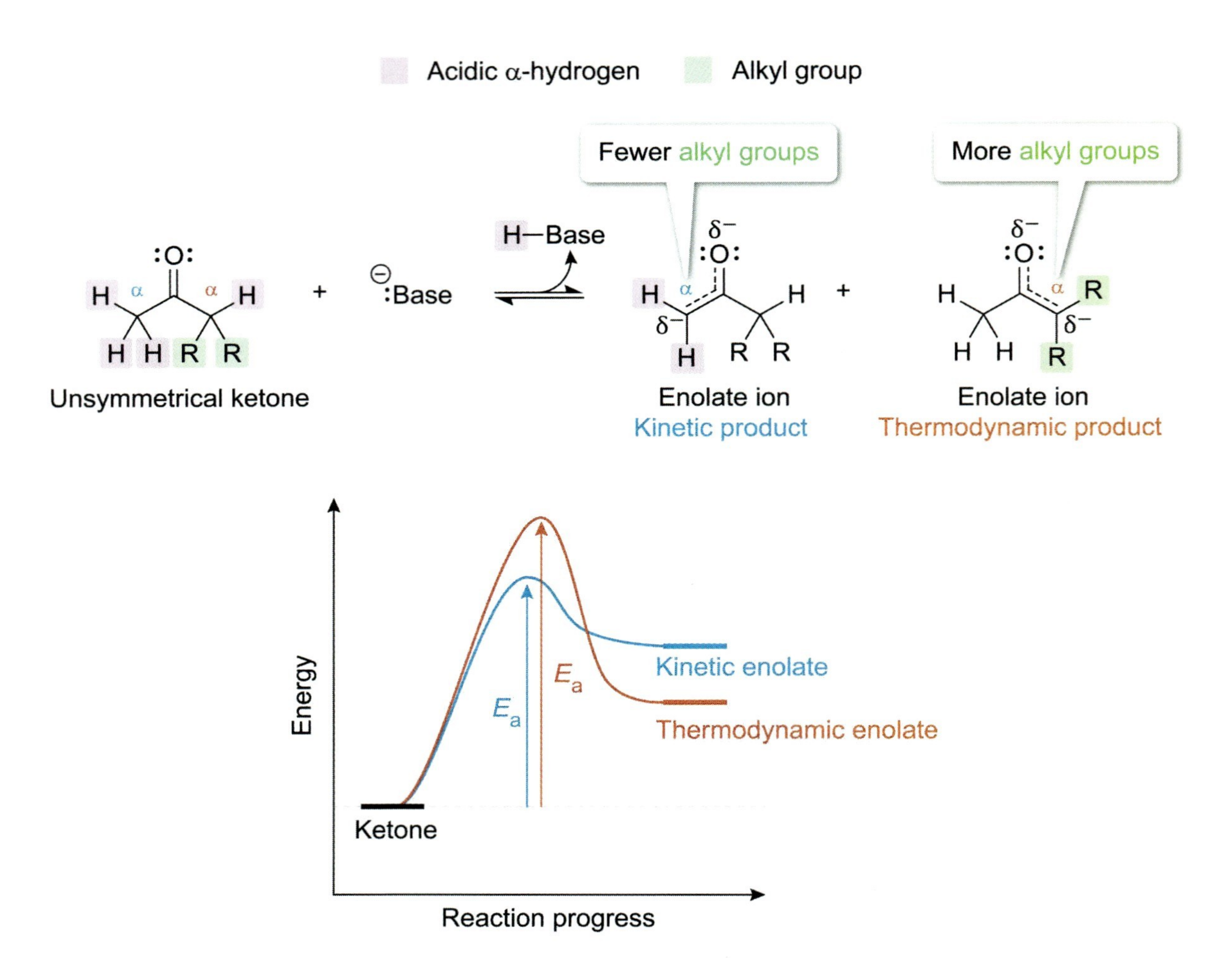

Figure 9.38 Kinetic and thermodynamic enolate ions.

The two regioisomers are the kinetic product and the thermodynamic product. The **kinetic product** is the product that forms *faster*. Typically, the kinetic enolate results from deprotonation of the α-carbon bonded to *fewer* alkyl groups. With fewer alkyl groups, bases can approach more easily due to *less* steric hindrance. Less steric hindrance also means a lower-energy transition state, a lower activation energy (E_a), and—given that reaction rate is inversely related to E_a—a *faster* rate of formation.

In contrast, the **thermodynamic product** is the *product* that has the *lower energy* (ie, greater stability). Given that an enolate ion contains a carbon-carbon pi bond (like an alkene), enolate ion stability increases with an *increasing* number of α-carbon R groups (Zaitsev's rule, see Concept 5.6.02). Therefore, the thermodynamic enolate product is the product involving the α-carbon with *more* alkyl groups.

The hydride ion from NaH is small and is minimally affected by steric hindrance (Figure 9.39). Therefore, the hydride ion *easily* approaches an acidic α-hydrogen surrounded by a high number of alkyl groups. Consequently, treatment of a ketone with NaH generates the *more stable* (ie, thermodynamic) enolate product.

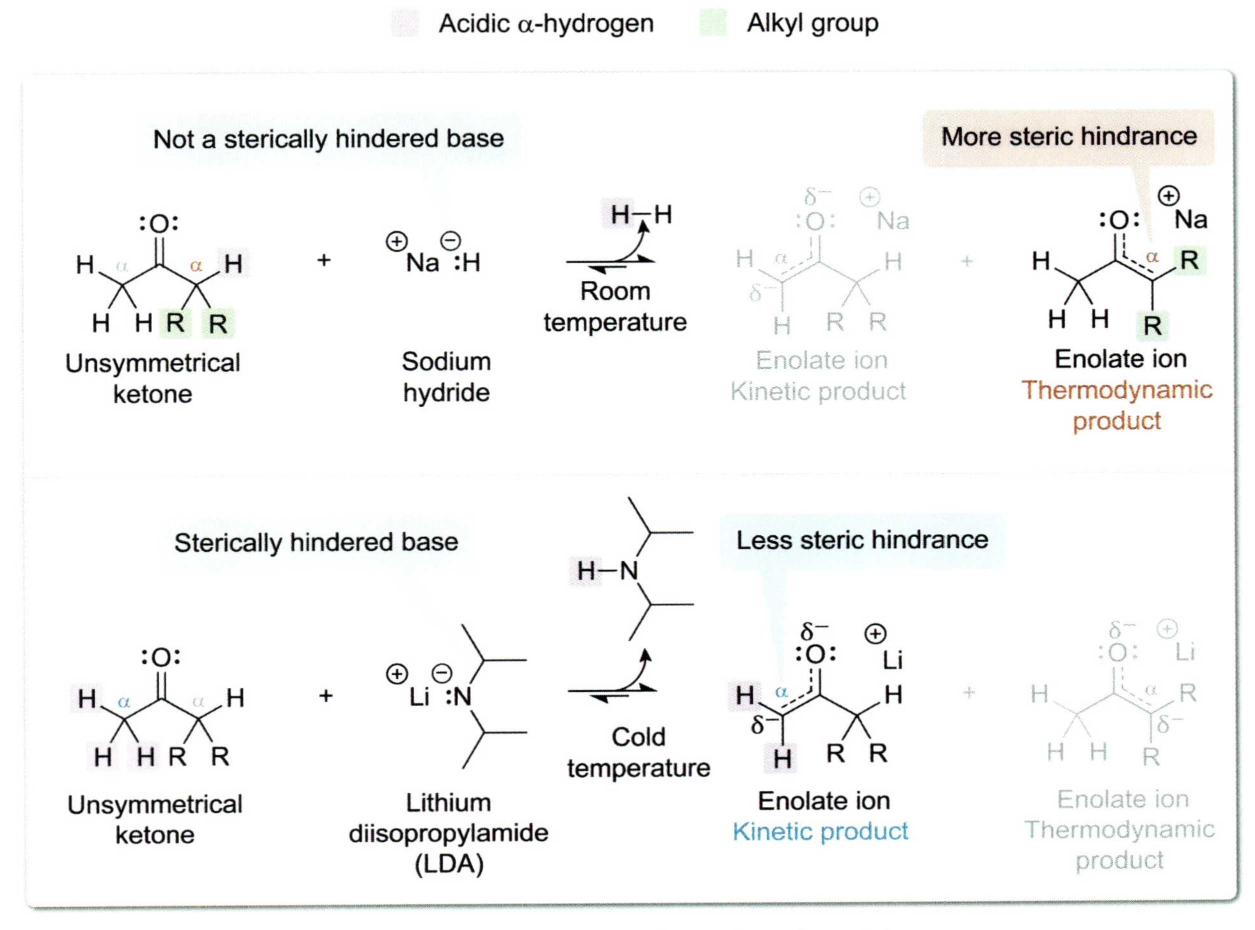

Figure 9.39 Preferential generation of a kinetic enolate or a thermodynamic enolate.

In contrast, the basic nitrogen atom in LDA is attached to two bulky isopropyl groups, *increasing* steric hindrance when interacting with reactants and *limiting* its ability to approach acidic protons. Therefore, it is easier (ie, faster) for LDA to remove an acidic α-hydrogen surrounded by *fewer* alkyl groups. Consequently, treatment of a ketone with LDA will generate the *less stable* (ie, kinetic) enolate product.

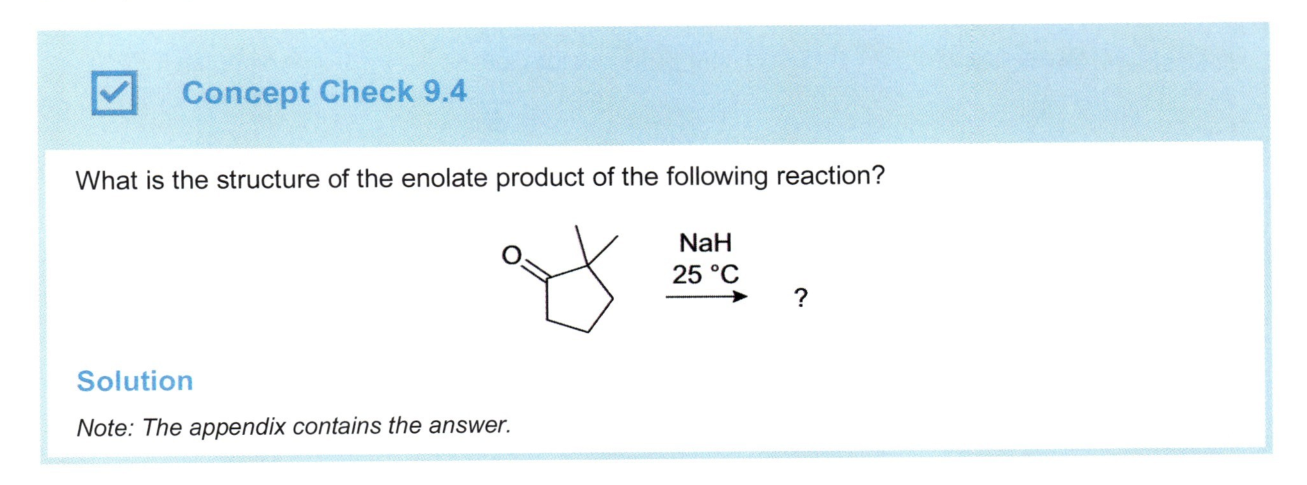

Concept Check 9.4

What is the structure of the enolate product of the following reaction?

Solution

Note: The appendix contains the answer.

9.4.06 Alpha Alkylation Reactions

Alkylation reactions are reactions that effectively *add* an alkyl group to an existing molecule. In an α-alkylation reaction, a nucleophilic enolate ion reacts with an alkyl halide, adding an alkyl group (R) to the α-carbon (Figure 9.40).

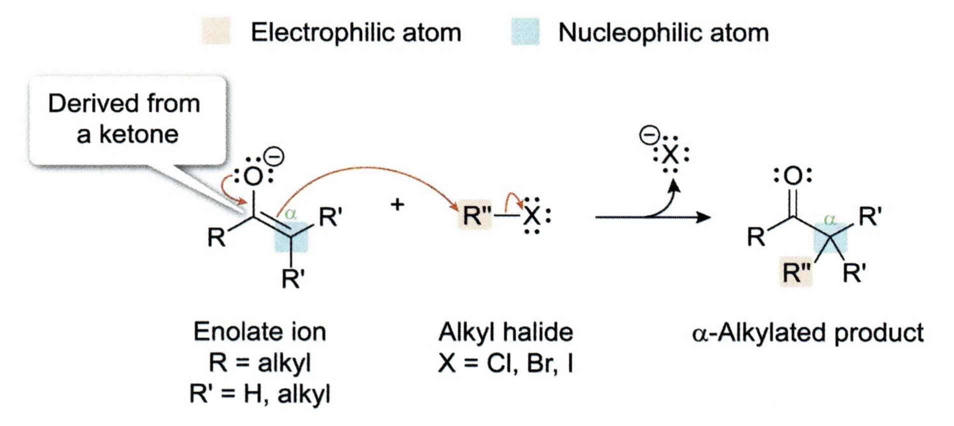

Figure 9.40 α-Alkylation of an enolate ion.

Because alkylation occurs through a concerted S_N2 mechanism, alkyl halides with minimal steric hindrance (eg, methyl, 1°) provide the best results (Concept 5.3.02). To reduce the risk of over-alkylation (if there are *multiple* acidic α-hydrogen atoms), a limiting amount of base that facilitates the quantitative formation of enolate (Concept 9.4.04) is used.

As a sterically unhindered strong base, NaH promotes the formation of the more-stable (ie, more-substituted) thermodynamic enolate ion, whereas the sterically hindered strong base LDA promotes the formation of the less-stable (ie, less-substituted) kinetic enolate ion (Figure 9.41).

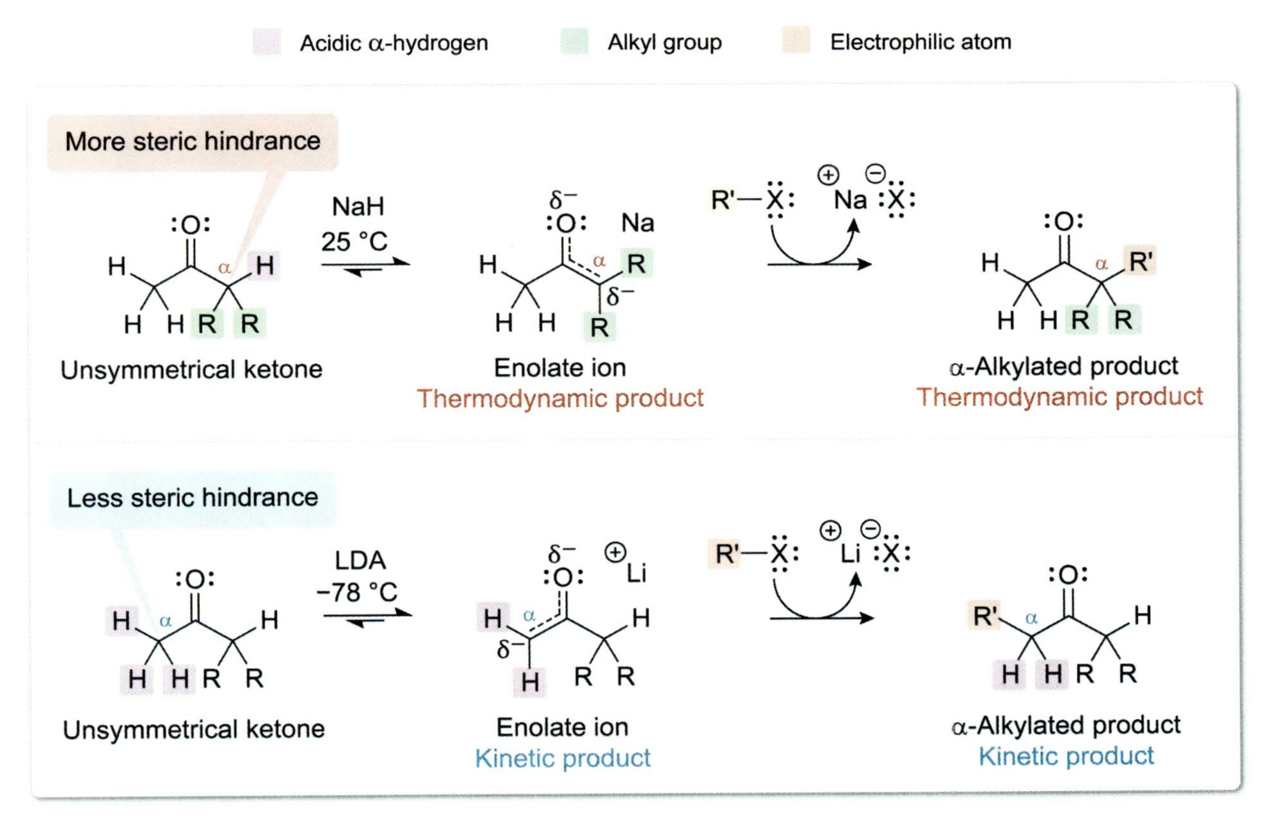

Figure 9.41 Regioselective production of α-alkylation products.

Alternately, α-alkylation can involve an enamine intermediate. An **enamine** is a functional group in which the nitrogen atom of an amine is covalently attached to a carbon atom of an alkene (Figure 9.42). After condensation between a ketone and a secondary amine (eg, pyrrolidine) to form an iminium ion (see Concept 9.3.04), the iminium undergoes tautomerization and deprotonation to form an enamine. If the starting ketone is unsymmetrical, the thermodynamic (ie, more-substituted) enamine is preferred.

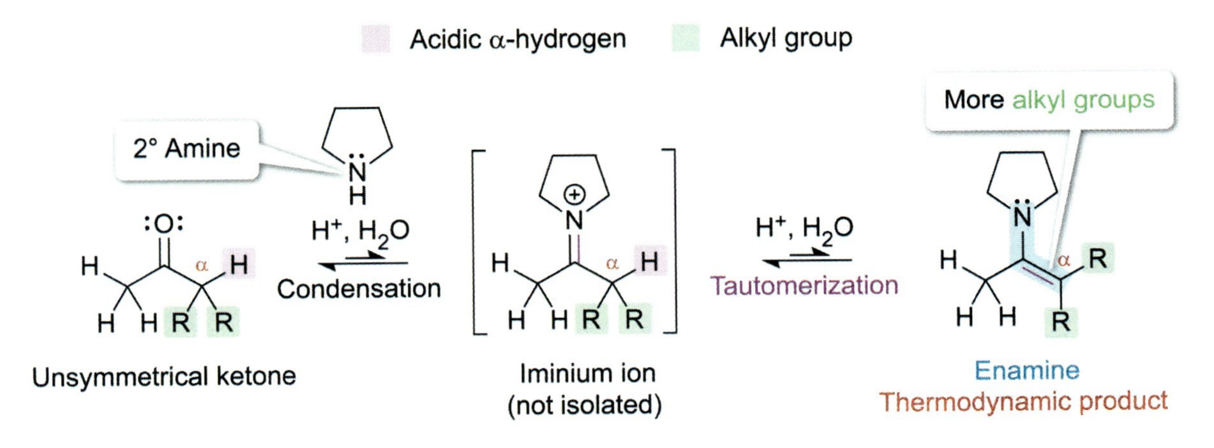

Figure 9.42 Synthetic preparation of an enamine.

Enamines react like enols and are nucleophilic at their α-carbon atom. The α-alkylation reaction of an enamine with an alkyl halide is called the **Stork reaction** and proceeds through an S_N2 mechanism (Figure 9.43). Acidic hydrolysis of the iminium ion restores the original ketone functional group (see Concept 9.3.04), such that the three-step process results in an α-alkylated ketone product.

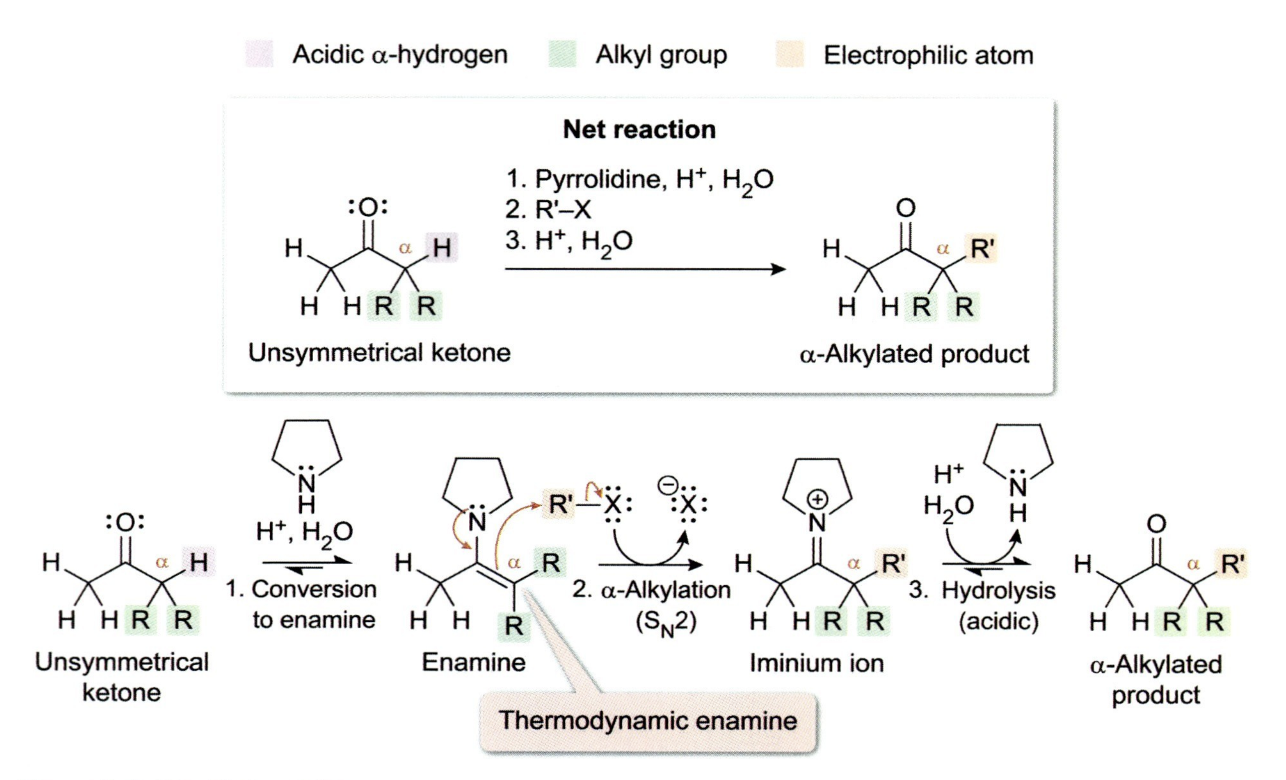

Figure 9.43 The Stork reaction.

9.4.07 Alpha Halogenation Reactions

Halogenation reactions replace a hydrogen atom in a molecule with a halogen atom (Figure 9.44).

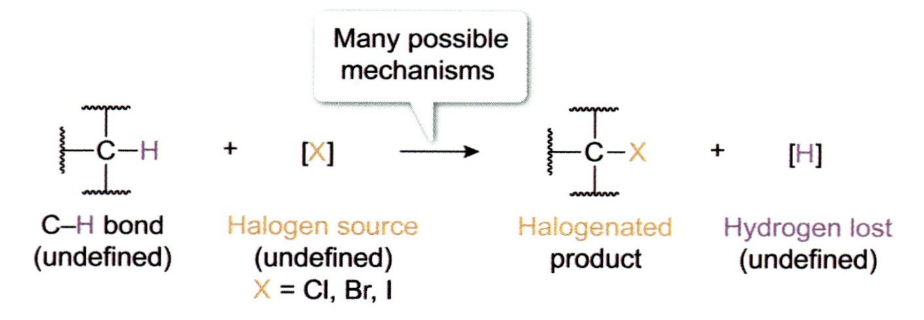

Figure 9.44 A general halogenation reaction.

The source of the halogen atom (X) can vary. Often, the halogen atom comes from a molecule of diatomic halogen (X_2, Figure 9.45). Cl_2, Br_2, and I_2 can be used to add Cl, Br, or I to the substrate, respectively; fluorination reactions are generally too reactive for most laboratory settings.

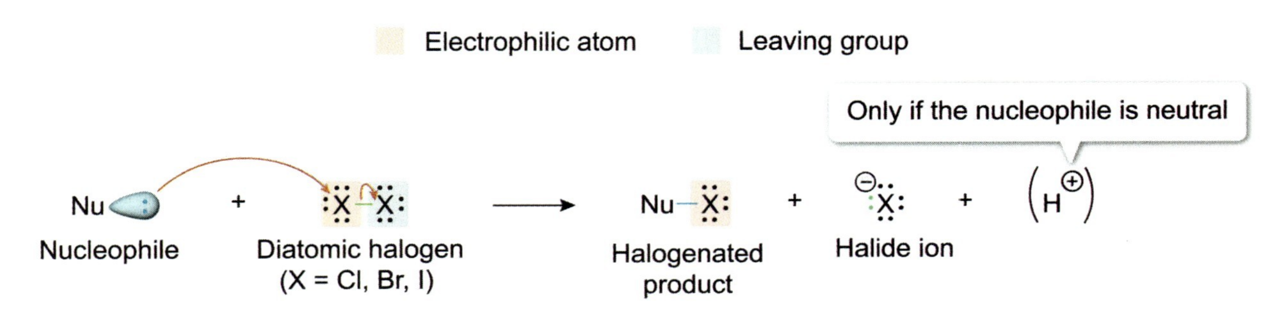

Figure 9.45 The general reactivity of a diatomic halogen (X_2) as a Type 2 electrophile.

In an **α-halogenation reaction**, a nucleophilic enol or enolate ion reacts with a molecule of X_2 to add a halogen atom (X) to the α-carbon (Figure 9.46). The halogenation step occurs through a concerted S_N2 mechanism through either basic (enolate intermediate) or acidic (enol intermediate) approaches.

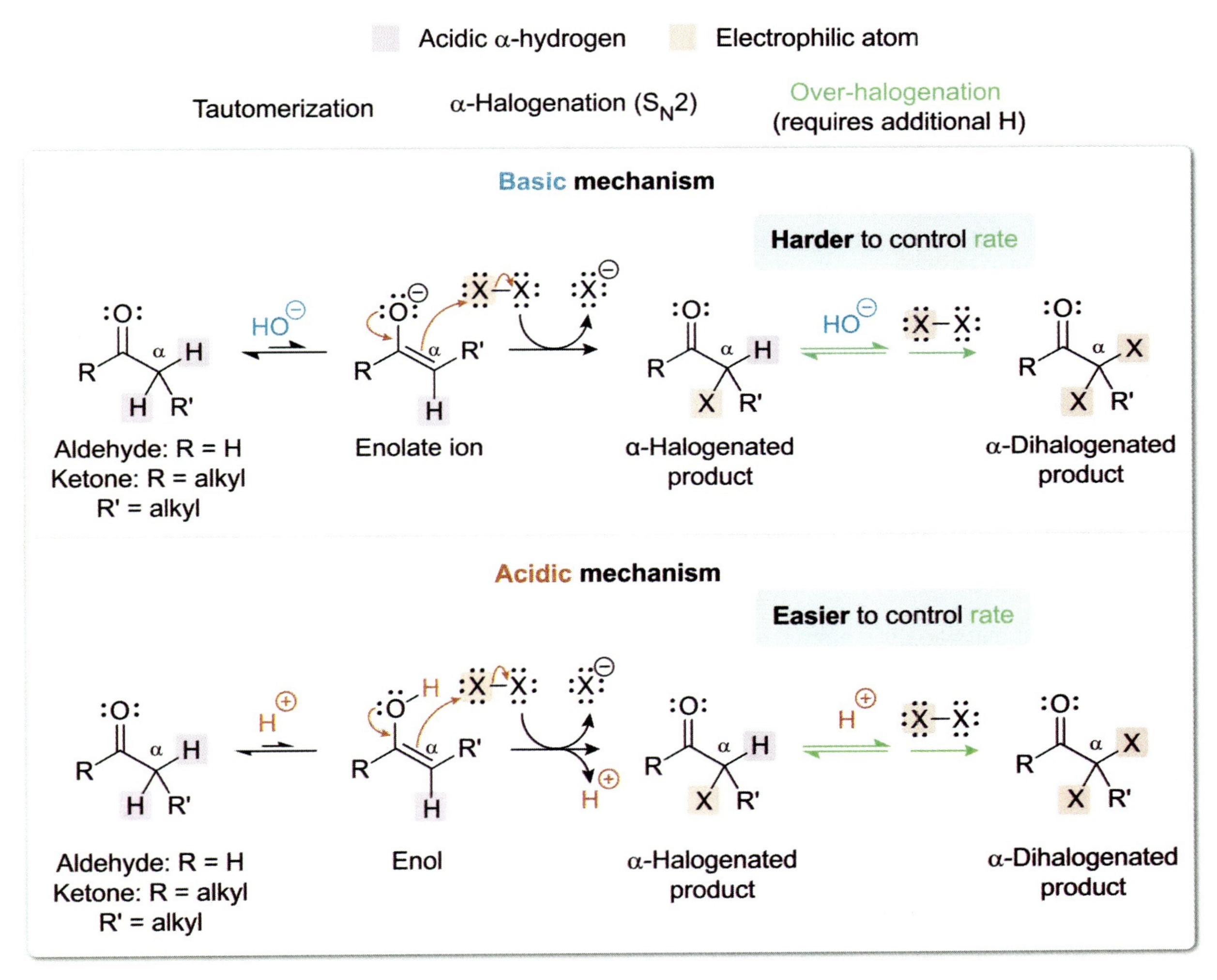

Figure 9.46 Basic and acidic α-halogenation reactions.

Because an enolate ion is more nucleophilic than an enol (see Concept 5.3.01), basic α-halogenation tends to occur faster than the acidic α-halogenation. Furthermore, the inductive effect causes halogens to stabilize the enolate anion, meaning basic α-halogenation generates products that are *more reactive* than the reactant. Consequently, the risk of *poly*halogenation is *very high* under basic conditions.

By contrast, acidic α-halogenation is easier to control; acidic reaction mechanisms involve an *enol* intermediate, which reacts slower than an enolate ion, resulting in a lower risk of unintended over-halogenation.

9.4.08 Aldol Condensations

Condensation reactions occur when two smaller molecules are combined (condensed) into a larger molecule (Figure 9.47). Condensation reactions also produce a byproduct consisting of atoms lost during the bond-forming process (eg, water, HBr).

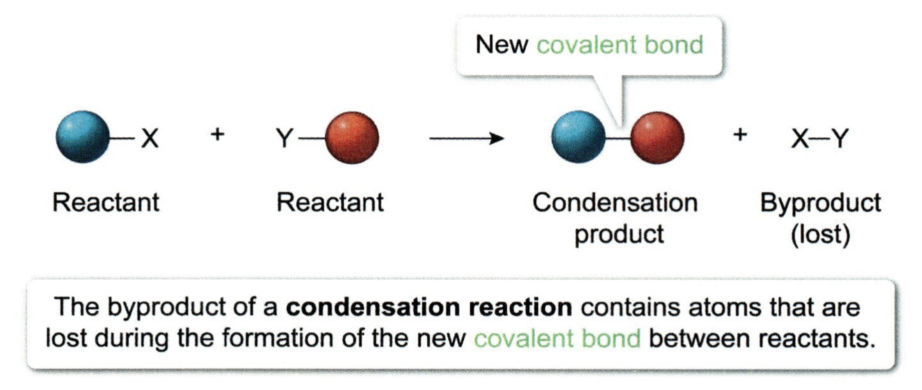

Figure 9.47 A general condensation reaction.

The Aldol Addition, the Retro-Aldol Reaction, and the Aldol Condensation

One of the most biochemically relevant condensation reactions is the aldol condensation (Figure 9.48). The aldol condensation consists of two major stages: aldol *addition*, then dehydration. An **aldol addition** is an α reaction involving two molecules (often of the same species) undergoing a nucleophilic addition (Concept 5.5.02). In other words, the α-carbon of an enol or enolate ion performs a nucleophilic attack on the electrophilic carbonyl of the second molecule. The result of aldol addition is a new carbon-carbon bond in an **aldol** product (either a β-hydroxy ketone or a β-hydroxy aldehyde).

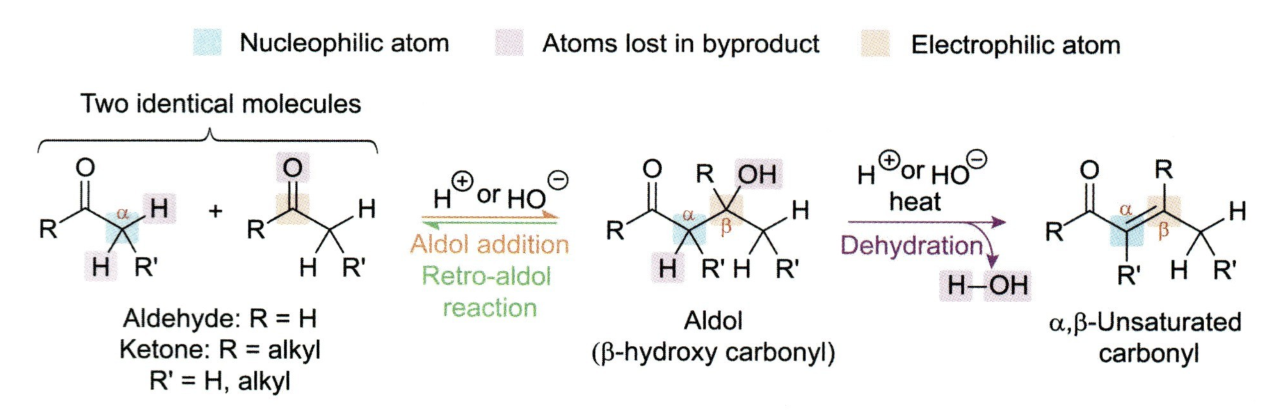

Figure 9.48 A general aldol condensation.

In most cases, the aldol addition is one direction of a reversible, dynamic equilibrium. The reverse process is called the **retro-aldol reaction**.

Prolonged experimental exposure or heating of an aldol reaction can promote **dehydration** of the aldol (ie, elimination of water) to give a conjugated product: the **α,β-unsaturated carbonyl**. Dehydration *requires* a second, acidic α-hydrogen atom, which is eliminated alongside the β-hydroxy group to form the pi bond. The combined sequence of an aldol addition followed by dehydration is known as the **aldol condensation**.

The complete aldol condensation is a sequence of three reaction stages that can be performed under *either* acidic or basic conditions:

1. Tautomerization
2. Nucleophilic addition
3. Elimination

The acidic aldol mechanism begins with acidic tautomerization of the ketone or aldehyde to an enol intermediate (Figure 9.49). Then acidic protonation of the carbonyl oxygen atom of the electrophilic reactant creates an activated oxonium intermediate, which is attacked by the nucleophilic enol. Finally, the β-hydroxyl group of the aldol intermediate is protonated by acid, and a molecule of water is eliminated through an E1 mechanism.

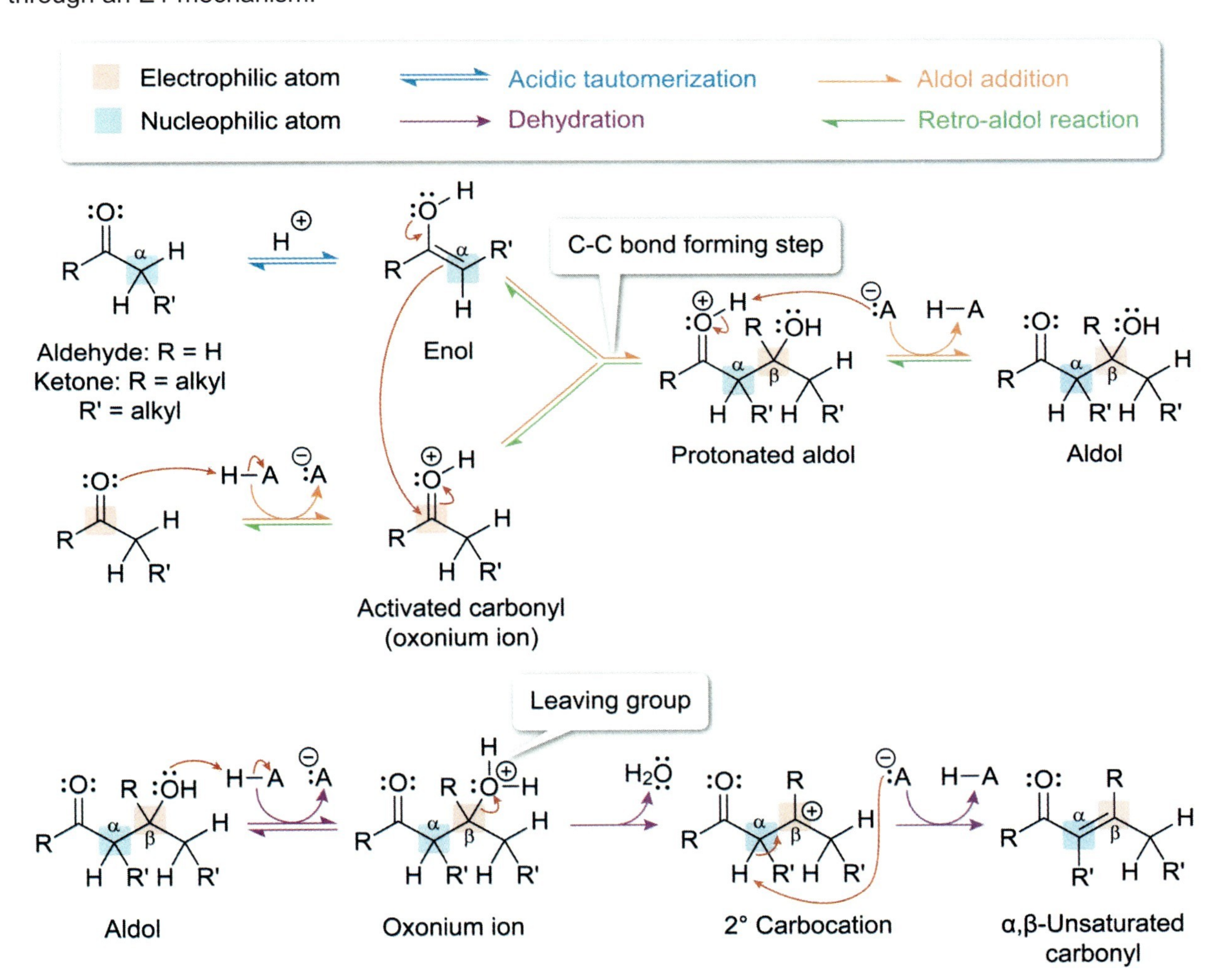

Figure 9.49 The aldol condensation mechanism under acidic conditions.

Under basic conditions, the ketone or aldehyde undergoes basic tautomerization to an enolate ion intermediate (Figure 9.50). Because an enolate ion is more nucleophilic than an enol (see Concept 5.3.01), its α-carbon directly attacks the electrophilic carbonyl carbon to generate an alkoxide ion. Protonation of the alkoxide regenerates a molecule of base and the aldol intermediate.

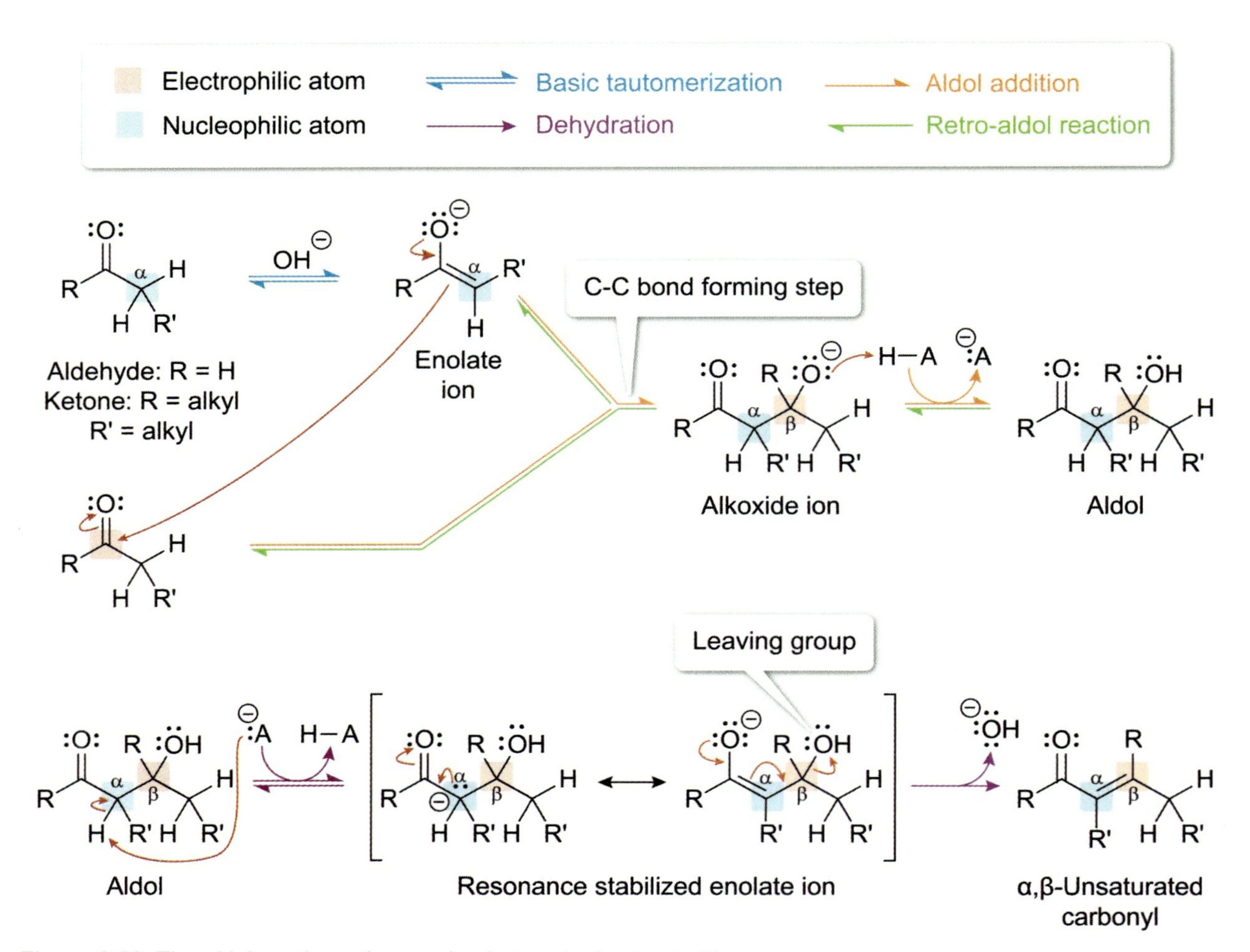

Figure 9.50 The aldol condensation mechanism under basic conditions.

Although the hydroxide ion is normally a poor leaving group for a *concerted* reaction (eg, S_N2, E2), it is sometimes possible to eliminate poor leaving groups if the elimination mechanism takes place over *multiple* steps. A molecule of base removes an acidic α-hydrogen atom, and the sigma bond electrons become delocalized into the adjacent carbonyl. Then the resonance-stabilized intermediate forms a new pi bond between the α-carbon and the β-carbon, eliminating a hydroxide ion in the process.

The dehydration reaction (an elimination) becomes *irreversible* when the reaction is *heated*. Heat increases the evaporation of water produced by the dehydration stage, effectively *removing* water from the reaction and shifting the dehydration equilibrium toward increased formation of the α,β-unsaturated carbonyl product. To isolate the aldol *addition* product (ie, the aldol itself) the reaction should be performed at *cold* temperatures (Figure 9.51).

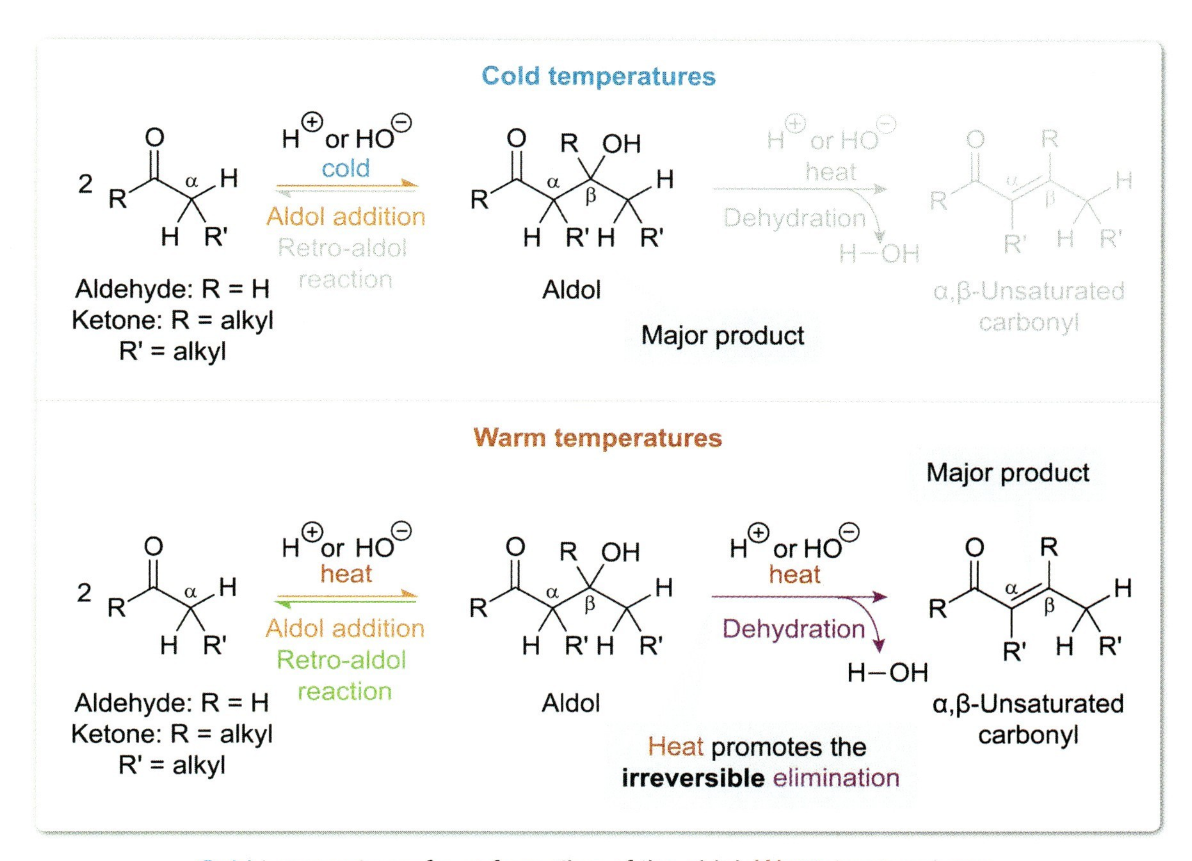

Cold temperatures favor formation of the aldol. Warm temperatures favor dehydration and formation of the α,β-unsaturated carbonyl.

Figure 9.51 The role of temperature in the formation of an aldol or an α,β-unsaturated carbonyl.

The Crossed Aldol Condensation

The **crossed aldol condensation** is an aldol condensation between two molecules of *different* substances. A crossed aldol condensation requires greater planning than reacting an equimolar mixture of two different aldehyde or ketone reactants—without such experimental considerations, a crossed aldol condensation between two different aldehydes can yield up to four different α,β-unsaturated aldehyde products (Figure 9.52).

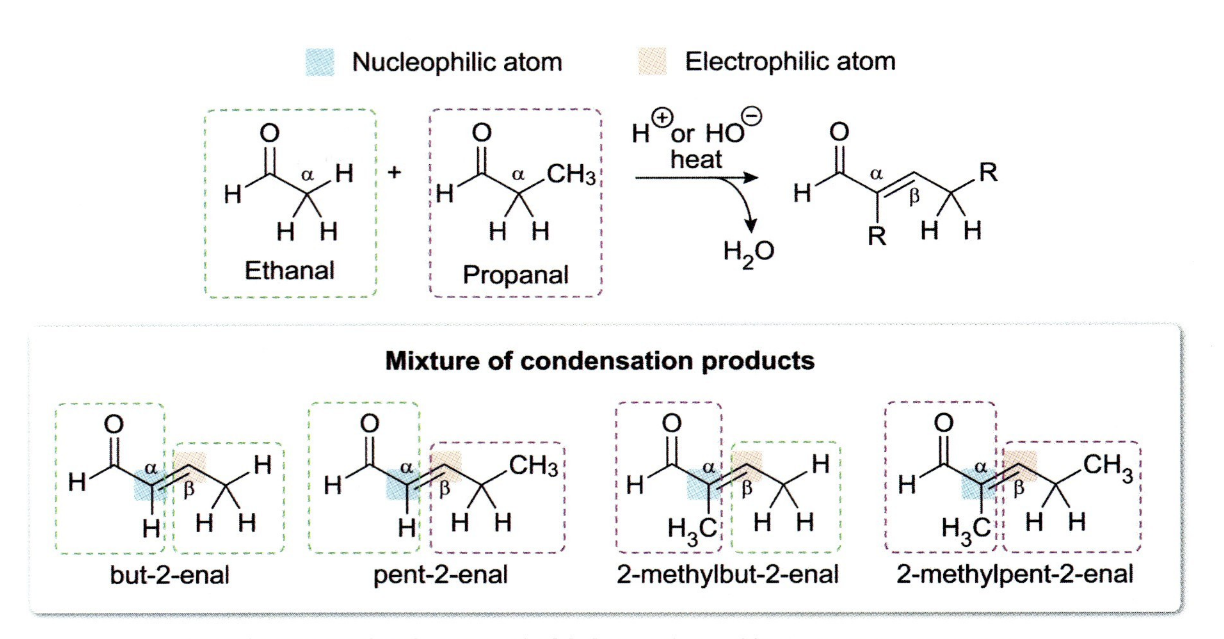

In an unconstrained **crossed aldol reaction**, either reactant can act as the nucleophile or the electrophile to yield a mixture of products.

Figure 9.52 Condensation products from an unconstrained crossed aldol reaction.

A crossed aldol reaction achieves the highest yield of a *single* α,β-unsaturated carbonyl product when (Figure 9.53):

1. The *nucleophilic* reactant is the only one with acidic α-hydrogen atoms.
2. The *electrophilic* reactant is present in excess and is *more* electrophilic than the nucleophilic reactant.
3. Reactants are introduced in a specific **order of addition**, such that the nucleophile is added slowly to a basic solution of the electrophile.

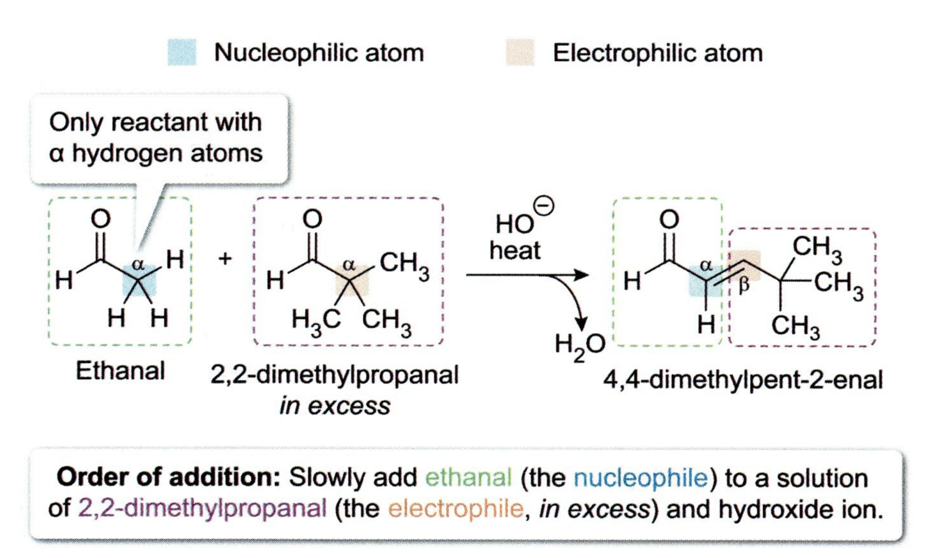

Order of addition: Slowly add ethanal (the nucleophile) to a solution of 2,2-dimethylpropanal (the electrophile, *in excess*) and hydroxide ion.

Figure 9.53 A selective crossed aldol condensation.

The Intramolecular Aldol Condensation

An aldol condensation can also occur between two ketone or aldehyde functional groups *within the same molecule* (ie, an **intramolecular aldol condensation**). An intramolecular aldol condensation can often be classified as a type of crossed aldol condensation, in which the nucleophile and electrophile happen to be within the same molecule. The product of an intramolecular aldol condensation is a *cyclic* α,β-unsaturated carbonyl (Figure 9.54).

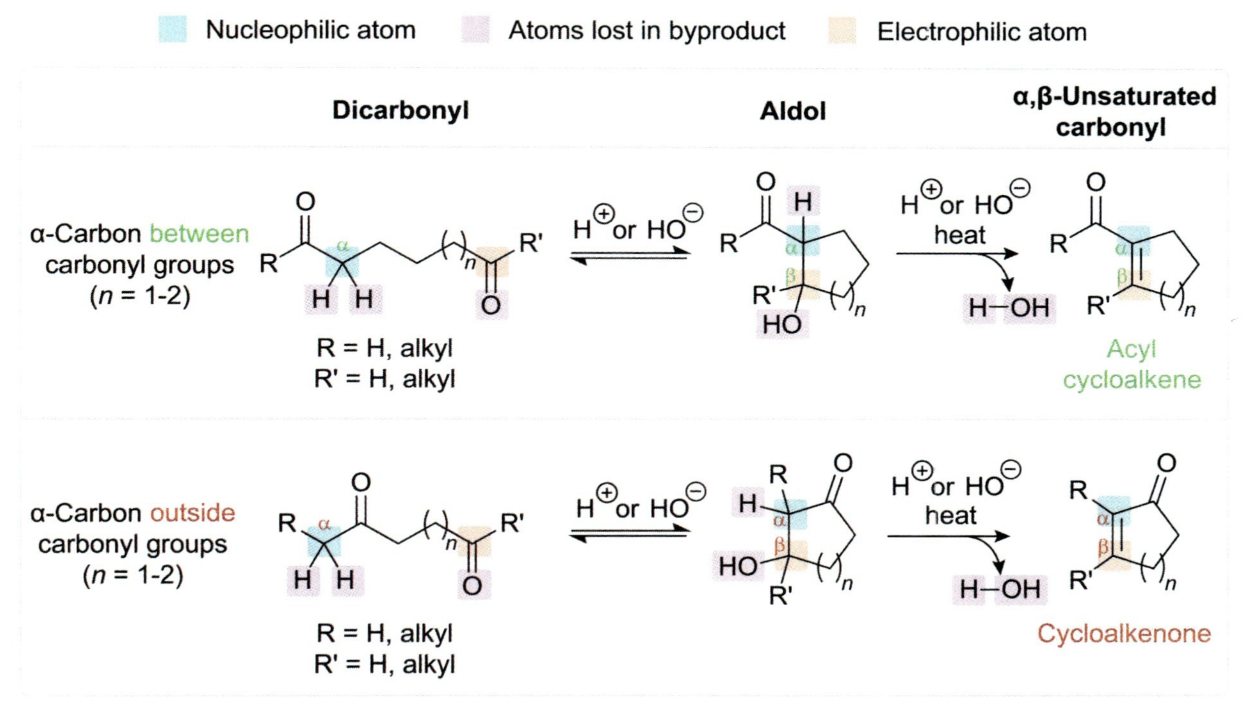

In an **intramolecular aldol condensation,** a carbonyl nucleophile reacts with a carbonyl electrophile within the same molecule to give five- and six-carbon cyclic products.

Figure 9.54 A general intramolecular aldol condensation.

Because an aldol addition is an equilibrium reaction, the most stable (thermodynamic) products tend to be formed in higher concentrations. Because an intramolecular aldol reaction forms a *new* ring, products with the most stable ring sizes (ie, smallest ring strain) are preferred. Consequently, the major products of most intramolecular aldol condensations contain five- and six-membered rings.

To identify the major products of an intramolecular aldol condensation, it is often necessary to first identify each of the *possible* product permutations (Figure 9.55). For the critical carbon-carbon bond forming step of the aldol condensation mechanism:

- Either of the carbonyl carbon atoms could act as the *electrophilic* atom.
- Any α-carbon atoms containing *two* acidic α-hydrogen atoms could act as the *nucleophilic* atom.

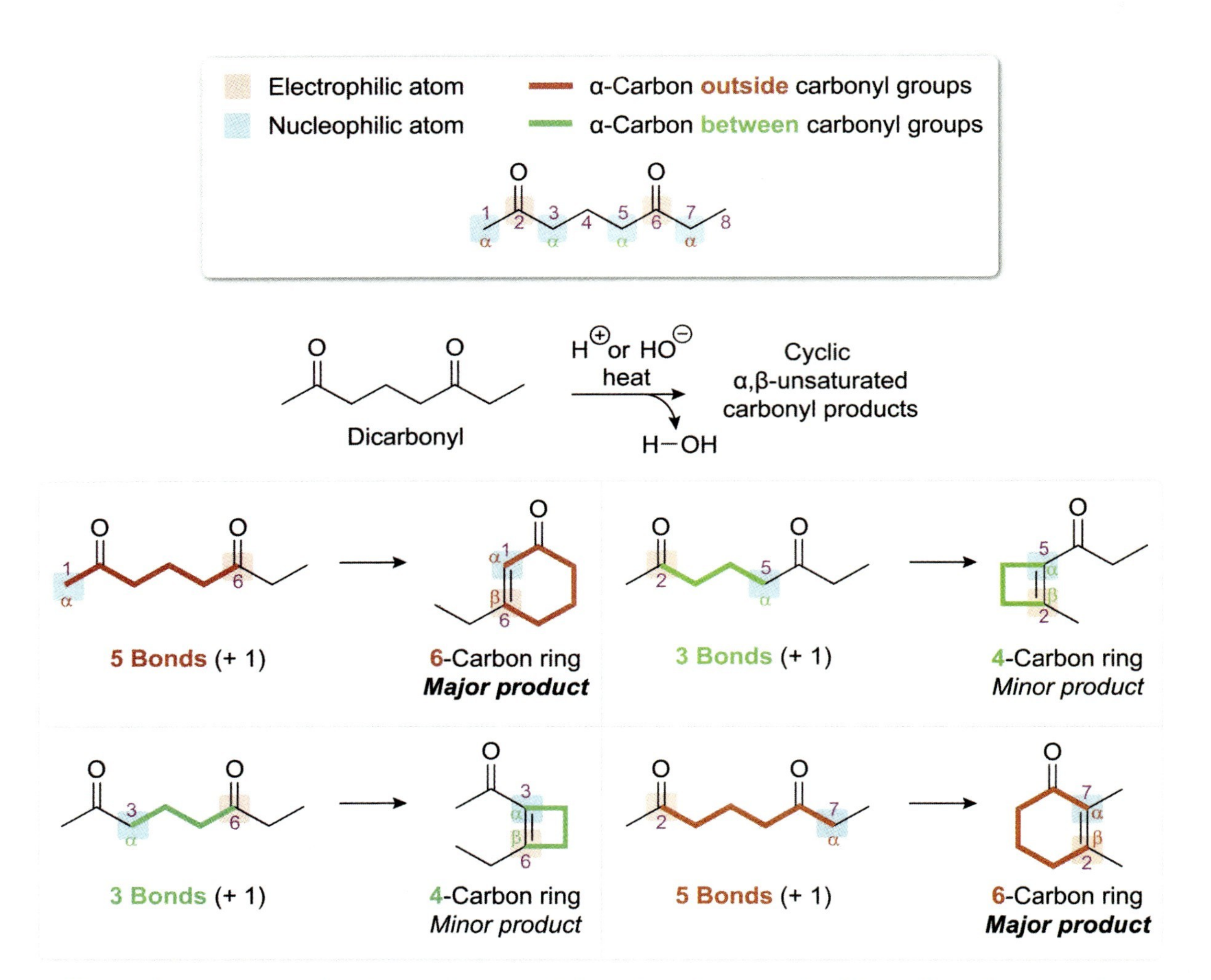

The ***major products*** of an intramolecular aldol condensation contain either a 5- or 6-carbon ring.

Figure 9.55 Determining the major products of an intramolecular aldol condensation.

To form a five- or six-membered ring, the electrophilic atom and the nucleophilic atom must be separated by either four or five sigma bonds (one bond in the final cyclic product is formed through the addition reaction). A single dicarbonyl substrate (with a maximum of *four* α-carbons) can produce up to *four* cyclic α,β-unsaturated carbonyl products.

Concept Check 9.5

Propose a viable synthetic approach that would allow the following molecule to be prepared through an aldol condensation:

O

CH_3

Solution

Note: The appendix contains the answer.

9.4.09 Michael Addition

1,2-Addition and 1,4-Addition to an α,β-Unsaturated Carbonyl

An α,β-unsaturated carbonyl is a type of conjugated molecule (see Concept 6.5.01). Because π electron density is distributed *throughout* the conjugated system, the polar nature of the carbon-oxygen double bond leads to *uneven* electron distribution of the *carbon-carbon* double bond (Figure 9.56). The hybrid resonance structure of an α,β-unsaturated carbonyl contains *two* electron-deficient (δ^+) carbon atoms: the carbonyl carbon atom and the β-carbon atom.

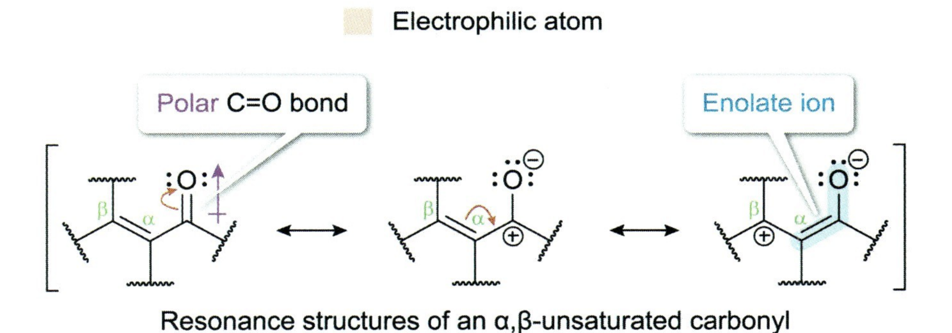

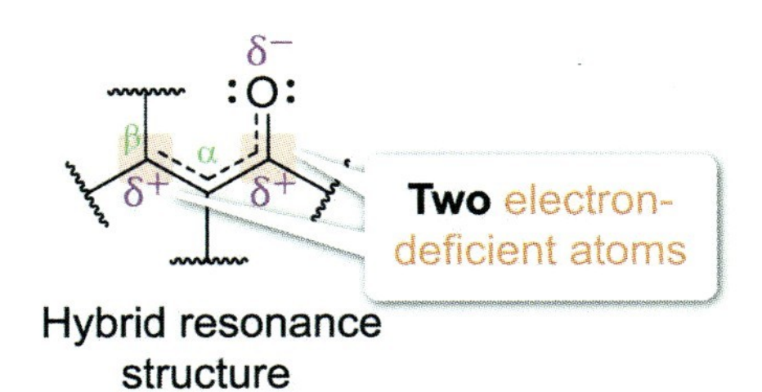

Figure 9.56 Conjugated nature of an α,β-unsaturated carbonyl.

Consequently, *either* atom can act as an electrophilic atom in a reaction with a nucleophile (Figure 9.57). Attack at the carbonyl carbon atom is a nucleophilic addition that results in an alkoxide ion product, which can later be protonated. Because of their relative positions, this type of addition is called a **1,2-addition**.

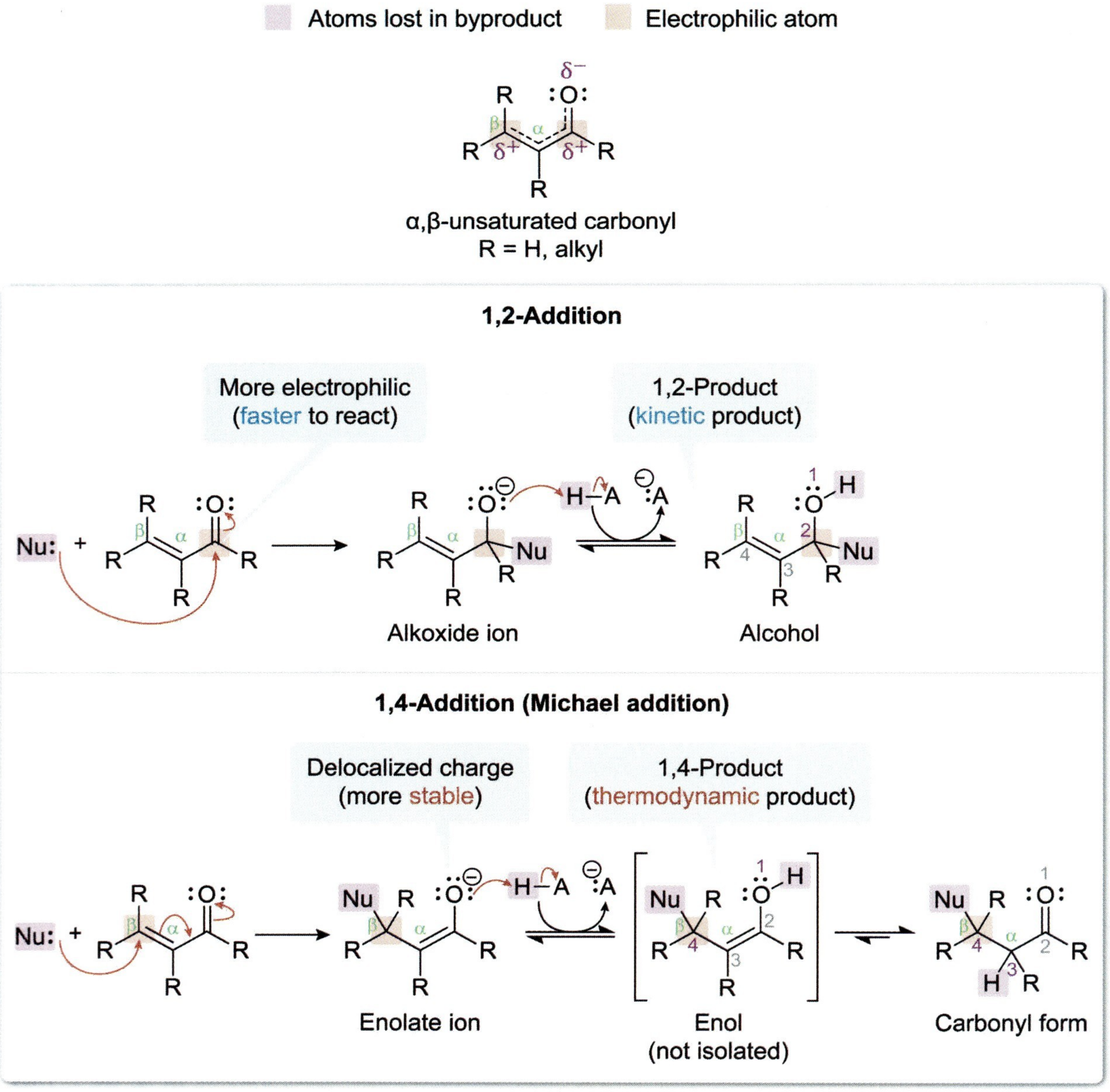

Atoms lost in byproduct   Electrophilic atom

α,β-unsaturated carbonyl
R = H, alkyl

1,2-Addition

More electrophilic (faster to react)

1,2-Product (kinetic product)

Alkoxide ion

Alcohol

1,4-Addition (Michael addition)

Delocalized charge (more stable)

1,4-Product (thermodynamic product)

Enolate ion

Enol (not isolated)

Carbonyl form

Figure 9.57 The mechanisms for a 1,2-addition and 1,4-addition to an α,β-unsaturated carbonyl.

Nucleophilic attack at the β-carbon atom shifts the carbon-carbon π bond one position and the carbon-oxygen π electrons become a lone pair. Because the oxygen anion is an enolate ion, protonation generates an enol. This type of addition, which adds new groups at positions 1 and 4 of the conjugated system, is called a **1,4-addition**, a **conjugate addition,** or a **Michael addition**. The enol tautomerizes to its carbonyl form, which is typically the isolated final product following a 1,4-addition.

Michael Addition

In the Michael addition, a nucleophile is added to the electrophilic β-carbon of an α,β-unsaturated carbonyl (Figure 9.58). Following tautomerization, the net result is the addition of a nucleophile and a hydrogen atom across the α,β-carbon-carbon double bond, producing a β-substituted carbonyl derivative (a **Michael adduct**).

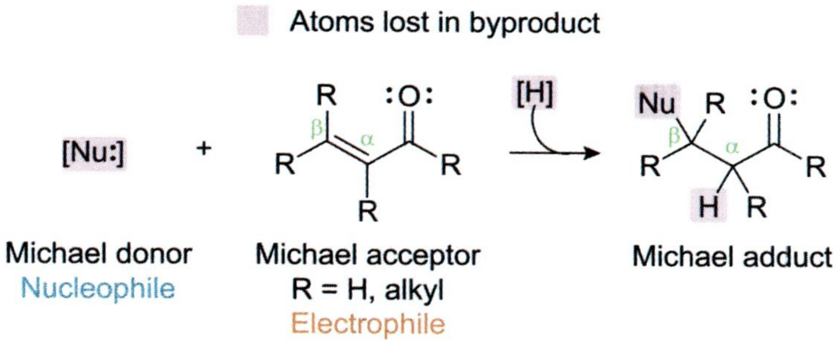

Atoms lost in byproduct

[Nu:] + (R)(R)C$_\beta$=C$_\alpha$(R)–C(=O:)–R —[H]→ Nu–C$_\beta$(R)(R)–C$_\alpha$(H)(R)–C(=O:)–R

Michael donor
Nucleophile

Michael acceptor
R = H, alkyl
Electrophile

Michael adduct

Figure 9.58 A general Michael addition.

Only a subset of all nucleophiles and electrophiles participate in the Michael addition (Table 9.4). A **Michael donor** is a *nucleophile* that can undergo the Michael addition, whereas a **Michael acceptor** contains an α,β-unsaturation attached to a pi electron-withdrawing group.

Table 9.4 Michael donors and Michael acceptors.

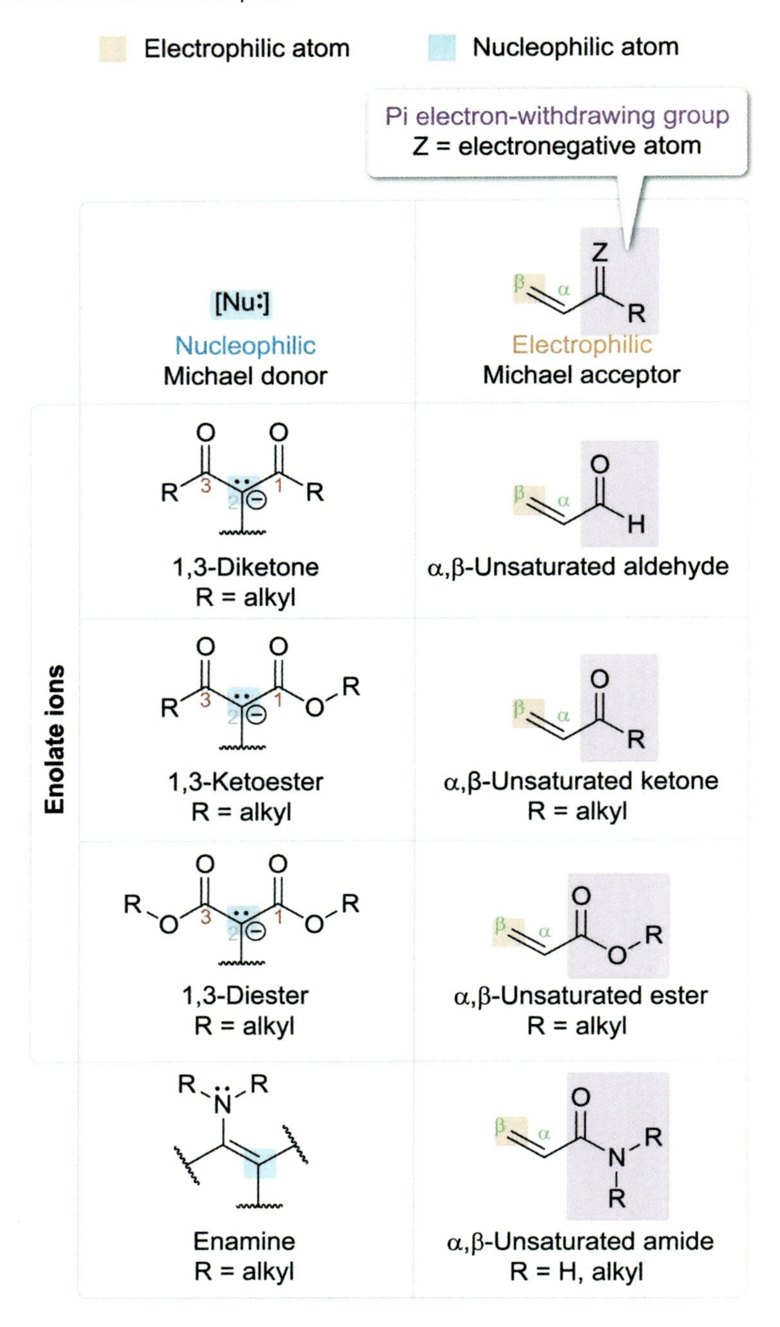

A regular enol or enolate ion (ie, one that is derived from only a single carbonyl group) is *not* typically a good Michael donor. However, enolate ions generated from a 1,3-dicarbonyl group (see Concept 9.4.04) or enamines (see Concept 9.4.06) are both excellent Michael donors.

9.4.10 Robinson Annulation

The Robinson annulation is an alpha reaction with a longer, more-complex reaction mechanism composed of a sequence of foundational mechanisms. An **annulation reaction** is a reaction that forms a *new* ring of atoms. The **Robinson annulation** involves the basic reaction of an enolate ion of a 1,3-dicarbonyl (ie, a Michael donor) with an α,β-unsaturated ketone (ie, a Michael acceptor), resulting in the formation of a new cyclohexenone ring (Figure 9.59).

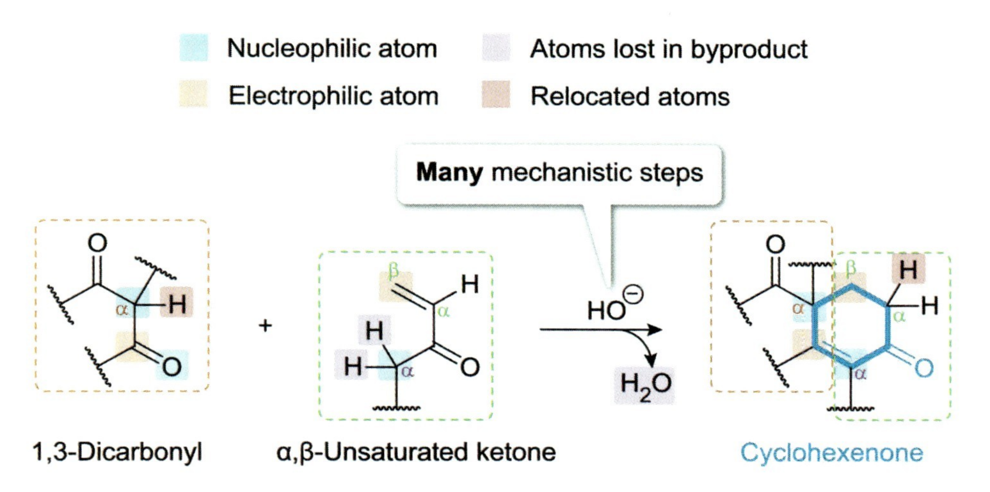

Figure 9.59 A general Robinson annulation.

The mechanism of the Robinson annulation is a sequence of the following foundational mechanisms (stages):

1. Formation of an enolate ion (see Concept 9.4.04)
2. A Michael addition (see Concept 9.4.09)
3. *Two* basic tautomerization reactions (see Concept 9.4.02)
4. An intramolecular aldol condensation (see Concept 9.4.08)

The Robinson annulation mechanism is depicted in Figure 9.60. In the first stage, a base removes an acidic α-hydrogen atom from the α-carbon atom between the two carbonyl groups to form an enolate that is reactive as a Michael donor. In the second stage, the nucleophilic enolate reacts with the electrophilic β-carbon of an α,β-unsaturated ketone (a Michael addition) to produce the initial enolate adduct.

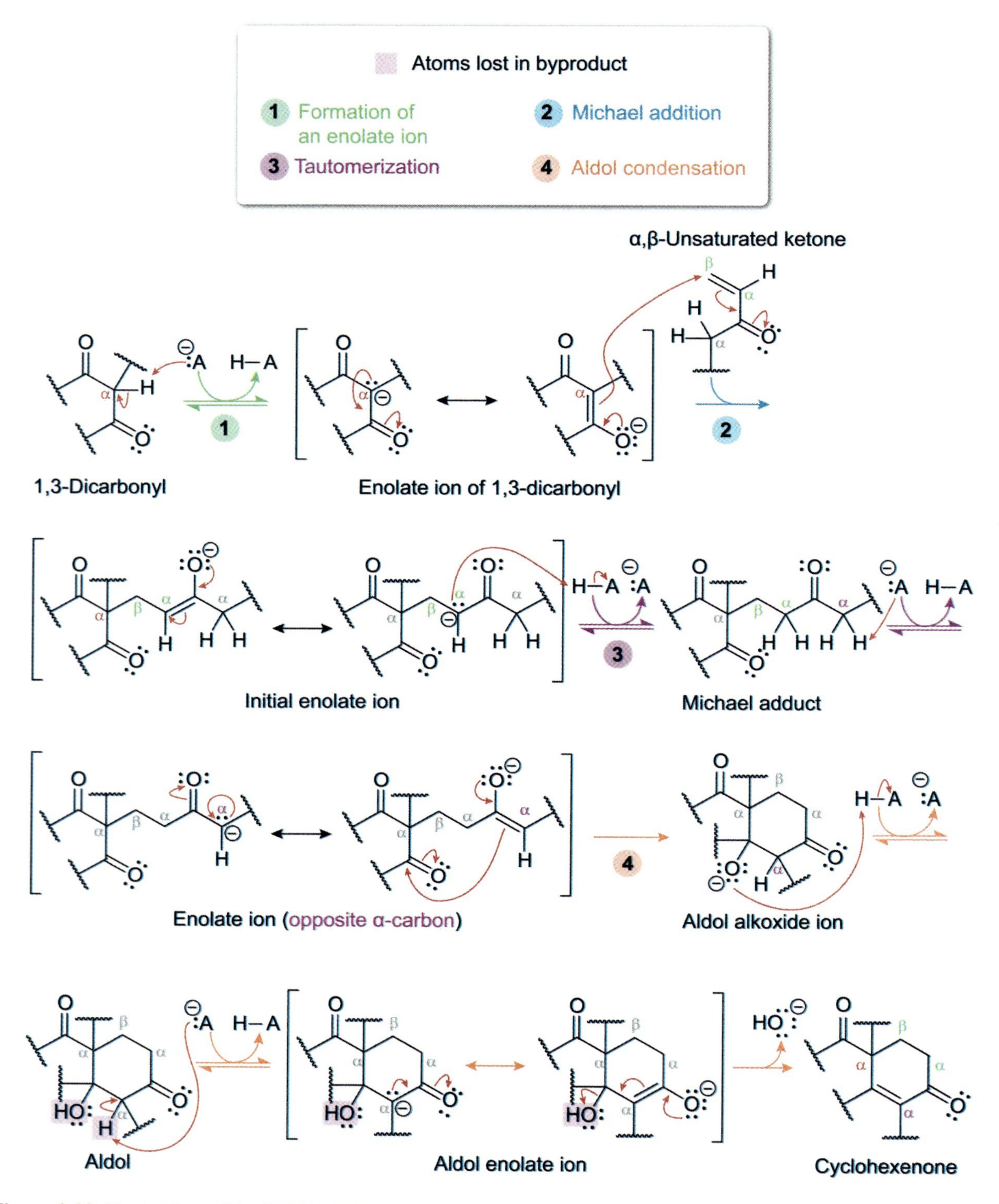

Figure 9.60 Mechanism of the Robinson annulation.

The initial enolate ion after the Michael addition tautomerizes to its more stable ketone form (the *first* of two successive tautomerizations within Stage 3). For a Robinson annulation, the ketone product must *also* contain two acidic α-hydrogen atoms on the *other* α-carbon (ie, the α-carbon that was not originally part of the α,β-unsaturation). The ketone intermediate then undergoes a *second* tautomerization to generate an enolate ion involving *that* α-carbon, ending Stage 3.

Stage 4 is an intramolecular aldol condensation (Concept 9.4.08) between the Stage 3 enolate and a carbonyl from the original 1,3-dicarbonyl reactant. After the final dehydration step, the net result of the Robinson annulation is a substituted cyclohexenone.

Lesson 10.1

Structure and Physical Properties of Carboxylic Acids

Introduction

A carboxylic acid has the general formula R–C(O)OH, in which R is an alkyl group or an aryl group. Although a carboxylic acid contains a carbonyl group, the properties and reactivity of a carboxylic acid are different than those of an aldehyde or a ketone. This lesson discusses the structure and physical properties of carboxylic acids.

10.1.01 Structural Characteristics of Carboxylic Acids

A **carboxylic acid** is defined by the presence of a **carboxyl group** [–C(O)OH], a functional group in which a hydroxyl group is attached to a carbonyl carbon. The carbon atom of a carboxyl group is attached to *three* atoms and has *zero* nonbonding electrons (Figure 10.1); therefore, this carbon atom is sp^2 hybridized (Concept 1.2.03 and Concept 2.1.02), has a trigonal planar molecular geometry, and has bond angles of 120°. The oxygen atom in the carbonyl of a carboxylic acid is *also* sp^2 hybridized and has a trigonal planar electron domain geometry.

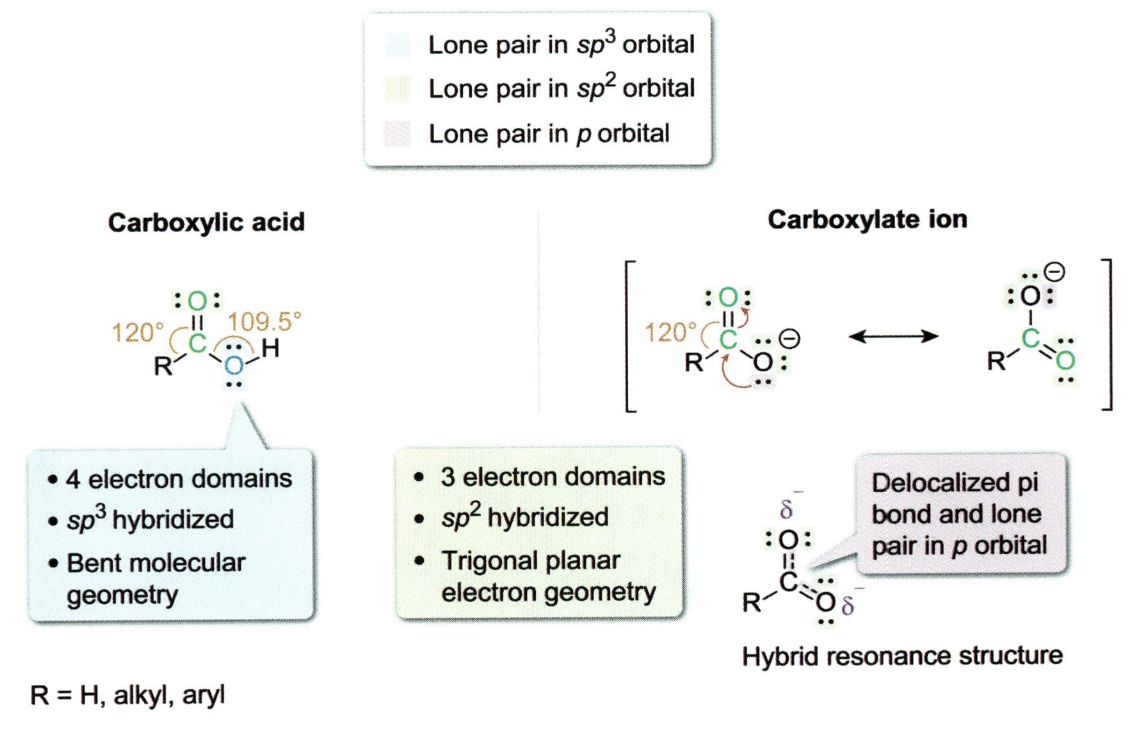

Figure 10.1 Structural characteristics of a carboxylic acid and a carboxylate ion.

The oxygen atom in the hydroxyl group is attached to *two* atoms and has *two* pairs of nonbonding electrons. Therefore, the oxygen atom of the hydroxyl group has four electron domains and is sp^3 hybridized, with a tetrahedral electron group geometry and a bent molecular geometry.

When a carboxylic acid acts as a Brønsted-Lowry acid and donates its acidic hydrogen atom to a base (Concept 10.1.03), the oxygen atom in the hydroxyl group gains a negative formal charge, forming a carboxylate ion. The negative charge becomes delocalized through resonance with the adjacent carbonyl group; therefore, the oxygen atom of the former hydroxyl group *must* be sp^2 hybridized (Concept 2.7.04).

10.1.02 Physical Properties of Carboxylic Acids

The most significant intermolecular force that impacts a carboxylic acid's physical properties is hydrogen bonding. The lone pairs of both oxygen atoms in the carboxyl group are hydrogen bond acceptors, and the hydrogen atom of the hydroxyl group is a hydrogen bond donor.

Hydrogen bonding between two carboxylic acid molecules forms a **dimer** instead of an extended hydrogen bond network (Concept 8.1.03). The prevalence of a carboxylic acid dimer effectively *doubles* the molecular weight of a carboxylic acid because individual molecules are very likely to be present in a *noncovalent* dimer pair (Figure 10.2).

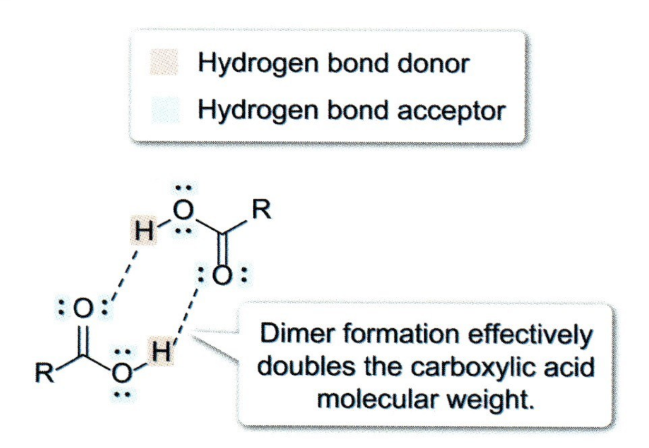

Figure 10.2 Structure of a carboxylic acid dimer.

Boiling Point

Carboxylic acids have much higher boiling points than alkanes or alcohols due to more extensive intermolecular forces (Table 10.1). Because the molecular weight of the carboxylic acid effectively *doubles* when it becomes a dimer, the energy required to disrupt the intermolecular forces between the molecules in the dimer is *larger*, resulting in *much* higher boiling points.

Table 10.1 Boiling points of alkanes, alcohols, and carboxylic acids.

Hydroxyl group | Carboxyl group | Branching

Functional group	Structure	Molecular weight (g/mol)	Boiling point (°C)
Alkane		114	125
Alcohol	OH	116	175
Carboxylic acid	O, OH	116	205
Carboxylic acid	O, OH	116	197
Carboxylic acid	O, OH	116	191
Dicarboxylic acid	HO, O, OH, O	118	235

Nonpolar Boiling point due to London dispersion forces (Alkane)

Polar Boiling point due to hydrogen bonding (Alcohol)

Polar Boiling point due to dimer formation (Carboxylic acids)

Boiling point *decreases* with branching.

A second carboxyl group further increases the boiling point

A dicarboxylic acid has a *second* carboxyl group, which *further increases* its capacity for hydrogen bonding. Consequently, dicarboxylic acids have even higher boiling points than carboxylic acids. Like other functional groups, the boiling point of a carboxylic acid is *lower* if its alkyl chain is branched (Concept 2.4.01).

Solubility

Carboxylic acids are **amphiphilic** molecules with *both* a polar hydrophilic carboxyl group and a nonpolar hydrophobic alkyl chain. The properties of an amphiphilic molecule are influenced by *both* sets of molecular properties. The nonpolar character of a carboxylic acid *increases* for *longer* carbon chains because the hydrophobic portion of the molecule has a greater relative effect on the molecular properties.

Because a carboxylic acid has both hydrogen bond acceptors and donors, a carboxylic acid can form hydrogen bonds with water, greatly increasing the water solubility of a carboxylic acid. Based on the principle of "like dissolves like," the polar carboxyl group interacts favorably with polar water. Carboxylic acids containing up to four carbon atoms are miscible with water (ie, they fully dissolve in *any* ratio). The water solubility of a carboxylic acid *decreases* with carbon chain length because its hydrophobic region *increases* and no longer can be fully solvated. Carboxylic acids with 12 or more carbon atoms have limited water solubility.

Carboxylic acids are also typically soluble in alcohols because of their ability to form hydrogen bonds. Longer-chain carboxylic acids are more soluble in alcohols than in water because alcohols are less polar than water and can better solvate the nonpolar alkyl chain.

10.1.03 Acidity of Carboxylic Acids

A **carboxylic acid** is an example of a **Brønsted-Lowry acid**. In comparison to strong mineral acids (eg, HCl, HBr), a carboxylic acid is considered a weak acid. However, among organic functional groups, a carboxylic acid is one of the stronger acidic groups.

The increased acidity of carboxylic acids is illustrated through a comparison with alcohols (Figure 10.3). Although both functional groups have an acidic O–H, the *stabilities* of the resulting conjugate bases are *different*. The conjugate base of an alcohol (ie, an alkoxide ion) has its negative formal charge *localized* to its oxygen atom (see Concept 8.3.02), whereas the conjugate base of a carboxylic acid (ie, a carboxylate ion) has its negative formal charge *delocalized* by resonance between *two* oxygen atoms.

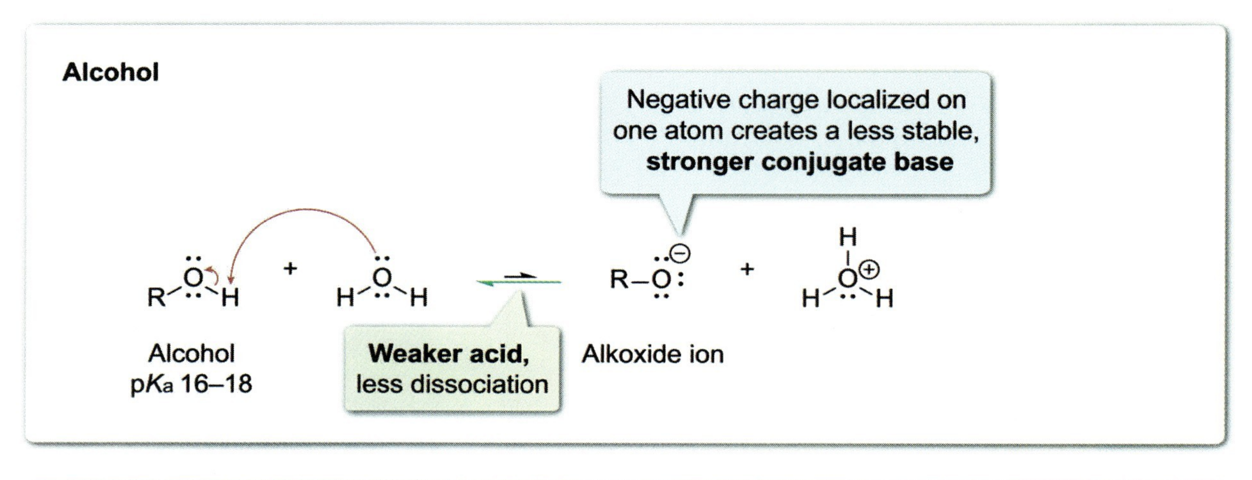

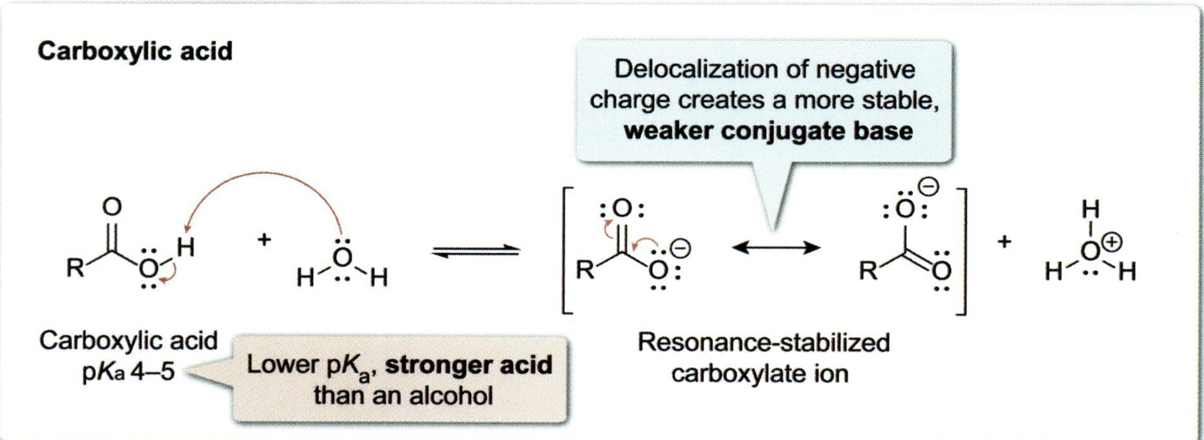

Figure 10.3 The acidities of alcohols and carboxylic acids.

A product stabilized by resonance is *more* likely to predominate in an equilibrium (Concept 2.7.04). Because the carboxylate ion is more stable than an alkoxide ion, it is more likely to be formed through an acid-base equilibrium. Consequently, a carboxylic acid (pK_a 4–5) is a *stronger* acid than an alcohol (pK_a 16–18), and a carboxylate ion is a *weaker* base than an alkoxide ion.

The stability of the carboxylate ion is also impacted by nearby groups through the inductive effect. Sigma electron–donating groups (eg, alkyl groups) add electron density, *destabilize* the negative charge of the carboxylate ion, and *decrease* the acidity of the corresponding carboxylic acid (ie, pK_a *increases*) (Table

10.2). The impact of the inductive effect is cumulative; a *greater number* of sigma electron–donating groups leads to a further decrease in acidity.

Table 10.2 The effect of sigma electron–withdrawing groups and sigma electron–donating groups on the acidity of carboxylic acids.

Sigma electron–donating groups | Sigma electron–withdrawing groups

	Conjugate acid		Conjugate base	
Carboxylic acid name	**Structure**	**pK_a**	**Structure**	**pK_b**
2,2-Dimethylpropanoic acid	O, OH	5.0	O, O⊖	9.0
Propanoic acid	O, OH	4.9	O, O⊖	9.1
Acetic acid	O, OH	4.7	O, O⊖	9.3
Chloroacetic acid	Cl, O, OH	2.9	Cl, O, O⊖	11.1
Trichloroacetic acid	Cl, Cl, Cl, O, OH	0.6	Cl, Cl, Cl, O, O⊖	13.4

Sigma electron–donating groups **decrease** acidity

Sigma electron–withdrawing groups **increase** acidity

More acidic

More Basic

Sigma electron–withdrawing groups (eg, halogens, electronegative atoms) help *stabilize* the negative charge of a carboxylate ion, *increasing* the acidity of the corresponding carboxylic acid (ie, pK_a *decreases*). The impacts of the inductive effect are cumulative; a *greater number* of sigma electron–withdrawing groups lead to further increases in acidity. In addition, the *closer* proximity of sigma electron–withdrawing groups to the carboxyl group further *increases* acidity (Table 10.3).

Table 10.3 The effect of proximity of a sigma electron–withdrawing group on the acidity of a carboxylic acid.

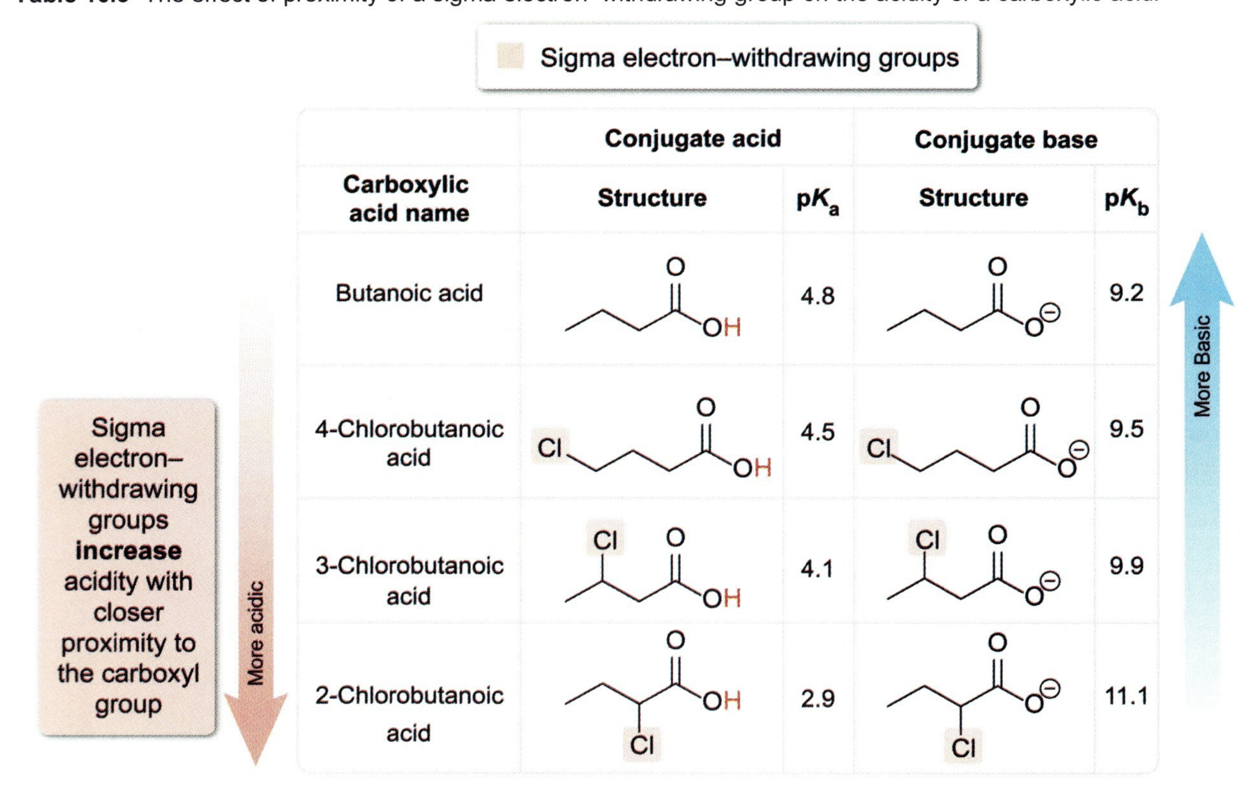

Sigma electron–withdrawing groups

Carboxylic acid name	Conjugate acid: Structure	pK_a	Conjugate base: Structure	pK_b
Butanoic acid		4.8		9.2
4-Chlorobutanoic acid		4.5		9.5
3-Chlorobutanoic acid		4.1		9.9
2-Chlorobutanoic acid		2.9		11.1

The acidity of substituted benzoic acids is affected by the presence of pi electron–withdrawing groups or pi electron–donating groups on the aromatic ring (Table 10.4). An *ortho* or *para* pi electron–withdrawing group (eg, $-NO_2$) *removes* electron density from the benzoate ion, *stabilizing* it and *increasing* its acidity. Conversely, a pi electron–donating group (eg, $-OCH_3$) *adds* electron density to the benzoate ion, *destabilizing* it and *decreasing* the acidity of the benzoic acid.

Table 10.4 The relative acidity of substituted benzoic acids and basicity of benzoate ions.

Pi electron–donating group | Pi electron–withdrawing group

Carboxylic acid name	Conjugate acid: Structure	pK_a	Conjugate base: Structure	pK_b
p-Methoxybenzoic acid		4.5		9.5
Benzoic acid		4.2		9.8
p-Nitrobenzoic acid		3.4		10.6

Pi electron–donating groups **decrease** acidity

Pi electron–withdrawing groups **increase** acidity

More acidic

More basic

Dicarboxylic Acids

Although simple dicarboxylic acids are symmetrical, the acidities of the two carboxylic acid groups are *different* (ie, these molecules have *two* pK_a values, one for each acidic proton). This effect is caused by electrostatic repulsion, in which similarly charged ions repel each other. Because a carboxylate ion already has a negative formal charge, it is *more difficult* to remove the second acidic proton from a dicarboxylic acid because the resulting dicarboxylate ion would have *two* negatively charged groups that repel each other.

Electrostatic repulsion depends on the *distance* between the two groups with similar charges; therefore, the relative acidity of the first acidic proton (pK_{a1}) and the second acidic proton (pK_{a2}) varies with the number of carbon atoms separating the carboxylic acid groups (Table 10.5). The first acidic proton (pK_{a1}) typically falls within the range for a carboxylic acid (pK_a 4–5). The second acidic proton (pK_{a2}) is approximately 10 times *less* acidic than pK_{a1} (ie, a difference of about 1 pK_a unit). The magnitude of this difference (pK_{a2} − pK_{a1}) slightly *decreases* with each additional carbon atom separating the two groups.

Table 10.5 The relative acidity of dicarboxylic acids.

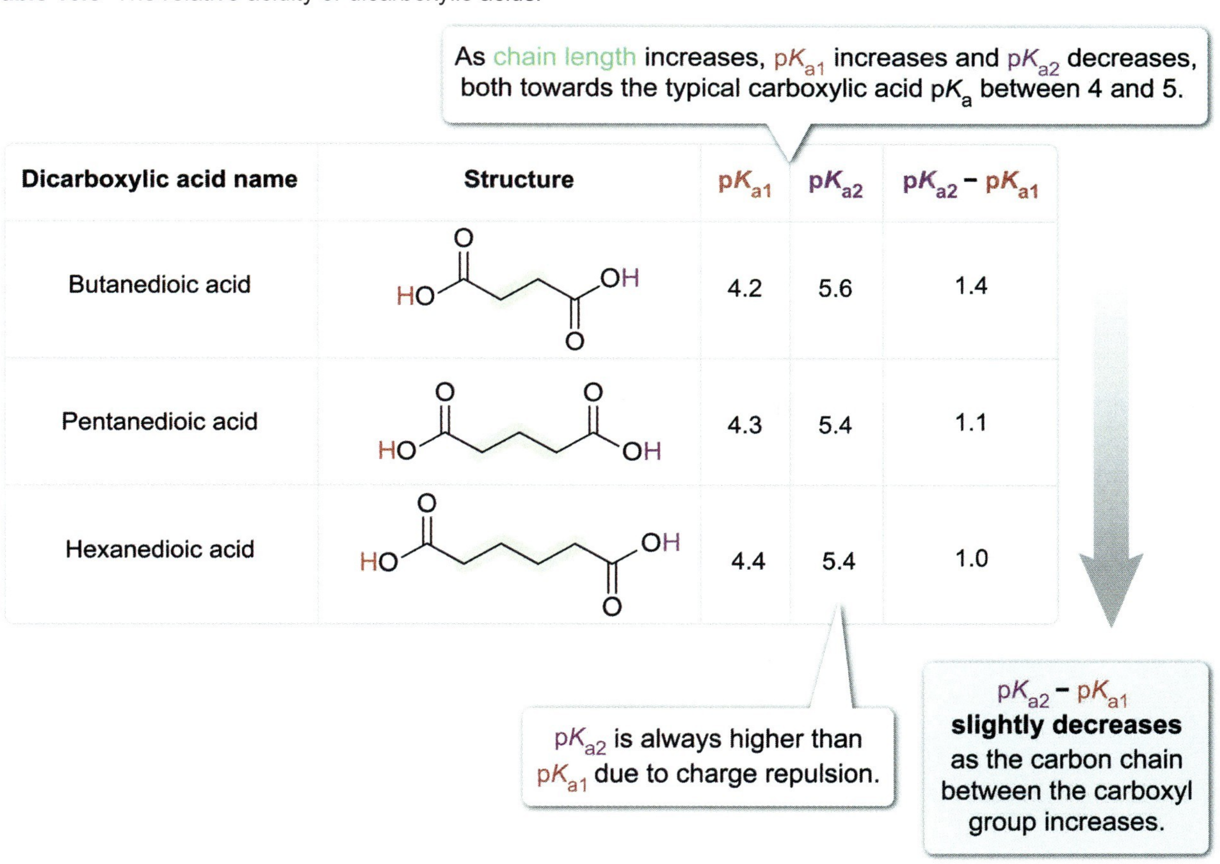

Dicarboxylic acid name	Structure	pK_{a1}	pK_{a2}	pK_{a2} - pK_{a1}
Butanedioic acid		4.2	5.6	1.4
Pentanedioic acid		4.3	5.4	1.1
Hexanedioic acid		4.4	5.4	1.0

As chain length increases, pKa1 increases and pKa2 decreases, both towards the typical carboxylic acid pKa between 4 and 5.
HO
OH
O
pKa2 is always higher than pKa1 due to charge repulsion.
pKa2 - pKa1 slightly decreases as the carbon chain between the carboxyl group increases.

Concept Check 10.1

In each of the following pairs of carboxylic acids, which molecule is the stronger acid and why?

1) FCH_2COOH CF_3COOH

2) $HCOOH$ CH_3CH_2COOH

Solution

Note: The appendix contains the answer.

Lesson 10.2

Synthesis of Carboxylic Acids

Introduction

This lesson focuses on a selection of reactions used to synthetically prepare a carboxylic acid, including the oxidation of alcohols and aldehydes, the hydrolysis of carboxylic acid derivatives, and the malonic ester synthesis.

10.2.01 Oxidation of Alcohols and Aldehydes

A common way to prepare a carboxylic acid is through oxidation of *either* a 1° alcohol or an aldehyde (Table 10.6). Although the *same* oxidizing agents are used for both types of reactions, the number of *equivalents* of oxidizing agent are different. Oxidation of a 1° alcohol to a carboxylic acid is a four-electron process and requires *two* equivalents of a two-electron oxidizing agent; oxidization of an *aldehyde* requires only *one* equivalent.

Table 10.6 Oxidation reactions that form a carboxylic acid.

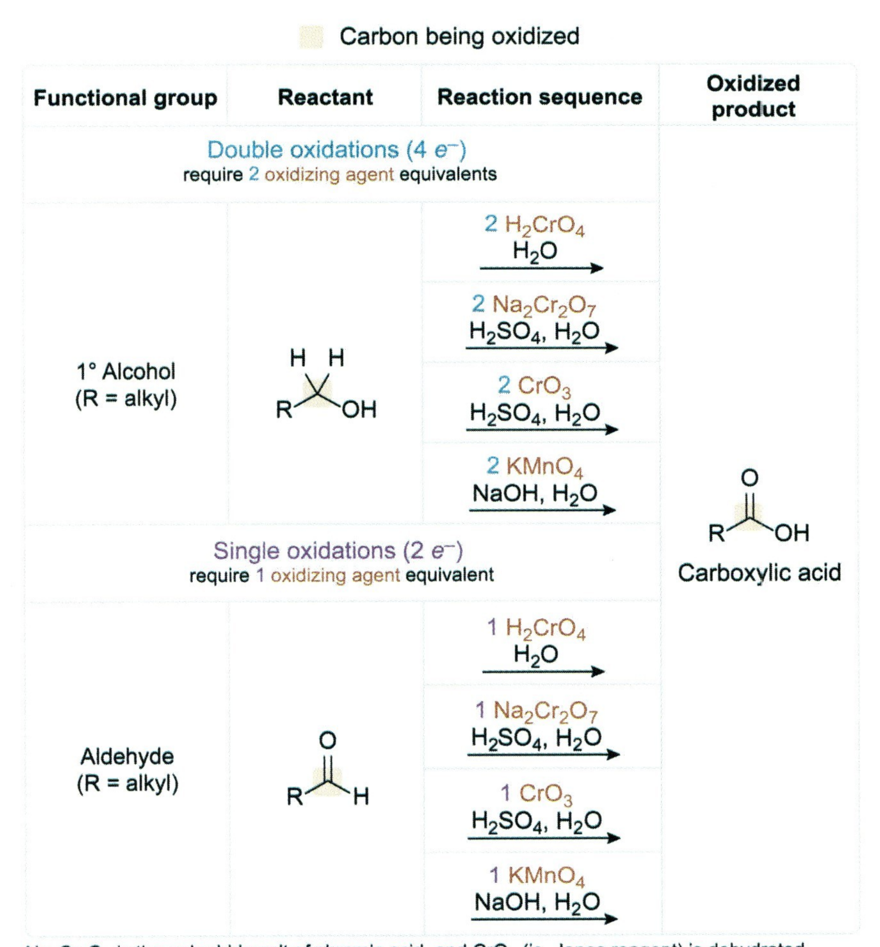

Carbon being oxidized

Functional group	Reactant	Reaction sequence	Oxidized product
Double oxidations (4 e^-) require 2 oxidizing agent equivalents			
1° Alcohol (R = alkyl)	RCH_2OH	2 H_2CrO_4, H_2O →	Carboxylic acid ($RCOOH$)
		2 $Na_2Cr_2O_7$, H_2SO_4, H_2O →	
		2 CrO_3, H_2SO_4, H_2O →	
		2 $KMnO_4$, NaOH, H_2O →	
Single oxidations (2 e^-) require 1 oxidizing agent equivalent			
Aldehyde (R = alkyl)	$RCHO$	1 H_2CrO_4, H_2O →	
		1 $Na_2Cr_2O_7$, H_2SO_4, H_2O →	
		1 CrO_3, H_2SO_4, H_2O →	
		1 $KMnO_4$, NaOH, H_2O →	

$Na_2Cr_2O_7$ is the anhydride salt of chromic acid, and CrO_3 (ie, Jones reagent) is dehydrated chromic acid; both sets of reagents generate H_2CrO_4 in the presence of aqueous acid (H_2SO_4).

10.2.02 Hydrolysis of Carboxylic Acid Derivatives

Carboxylic acid derivatives (ie, acid halides, anhydrides, esters, amides) contain an acyl group attached to a heteroatom (eg, oxygen, nitrogen, halogens). Carboxylic acids can be produced through the aqueous hydrolysis of carboxylic acid derivatives (Table 10.7). In these reactions, the R group of the carboxylic acid derivative is retained, and the variable Z group is replaced with a hydroxyl group.

Table 10.7 Preparation of a carboxylic acid through the hydrolysis of carboxylic acid derivatives.

Acyl group | Z group (attached through a heteroatom)

Reactivity of the acyl group (increasing from amide to acid halide)

Carboxylic acid derivative	Reactant	Reaction sequence	Hydrolysis product
Acid halide (R = alkyl; X = Cl, Br, I)	RC(=O)X	H_2O	RC(=O)OH Carboxylic acid
Anhydride (R, R' = alkyl)	RC(=O)OC(=O)R'	H_2O	
Ester (R, R' = alkyl)	RC(=O)OR'	H_2O, $H^{\oplus}$ or $HO^{\ominus}$	
Amide (R, R', R" = H, alkyl)	RC(=O)NR'R"	H_2O, $H^{\oplus}$ or $HO^{\ominus}$, heat	

Reactions performed under basic conditions (eg, HO^-) initially generate a carboxylate ion and require an acidic workup to form a carboxylic acid.

The most reactive carboxylic acid derivatives (ie, acid halides, anhydrides) (see Concept 11.2.01), react directly with water to form a carboxylic acid. The hydrolysis of less-reactive carboxylic acid derivatives (ie, esters, amides) generally requires heat and either an acid or a base catalyst (see Concept 11.3.01). Hydrolysis takes place through a nucleophilic acyl substitution mechanism, with nuances related to the type of catalyst used and the relative reactivity of the carboxylic acid derivative (see Concept 11.2.01).

10.2.03 Malonic Ester Synthesis

Functional groups that contain a carbonyl (C=O) can often undergo α-alkylation reactions (Concept 9.4.06). The **malonic ester synthesis** provides the means to prepare α-substituted (eg, α-alkylated) derivatives of acetic acid from the reactant diethyl malonate, a diester of three-carbon dicarboxylic acid malonic acid (Figure 10.4).

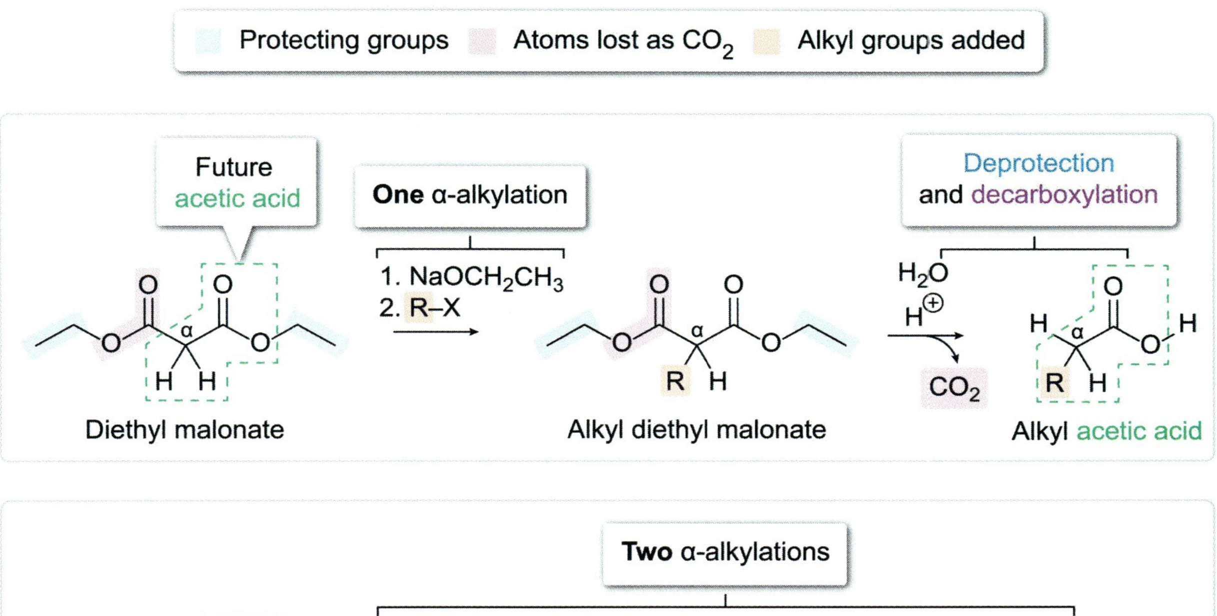

Figure 10.4 The general malonic ester synthesis of an alkyl acetic acid or a dialkyl acetic acid.

Treatment of diethyl malonate with a base, followed by an S_N2-amenable alkyl halide (eg, methyl, 1°, or 2° R–X) provides an alkylated diethyl malonate intermediate. This process can be *repeated*, potentially with a *different* alkyl halide (R′–X), to provide a dialkylated diethyl malonate intermediate. Finally, treatment with aqueous acid produces a substituted acidic acid product.

Although it is usually more challenging to perform an α-alkylation reaction *directly* on a carboxylic acid reactant, the structure of diethyl malonate is more amenable for several reasons. First, the ethyl groups in diethyl malonate serve as protecting groups for the two carboxylic acid groups in malonic acid.

Second, because diethyl malonate contains a 1,3-dicarbonyl group, its α-hydrogen atoms are *more* acidic than a standard α-hydrogen atom and are *quantitatively* removed using an alkoxide ion base. The malonic ester synthesis *always* uses the ethoxide ion (eg, sodium ethoxide) as the base because its alkyl group is identical to that of diethyl malonate, so any basic transesterification reactions that occur will not impact the structure of the diester (see Concept 11.3.02).

The 1,3-relationship of the ester groups in diethyl malonate is also important for promoting a necessary decarboxylation reaction in the final step of the malonic ester synthesis. A β-carboxy carbonyl is unstable and spontaneously loses a molecule of carbon dioxide (CO_2) (see Concept 10.3.07). In this way, the 1,3-relationship in diethyl malonate is first used to facilitate *easier* α-alkylation steps, then later used to promote the removal of one of the two carboxylic acid groups, resulting in a product containing a *single* carboxylic acid functional group.

The mechanism of the malonic ester synthesis has the following general steps (Figure 10.5):

1. Deprotonation
2. Alkylation (S_N2)
3. Acidic hydrolysis and decarboxylation (simultaneously)

Note: Steps 1 and 2 can be repeated to yield a *di*alkylated acetic acid product.

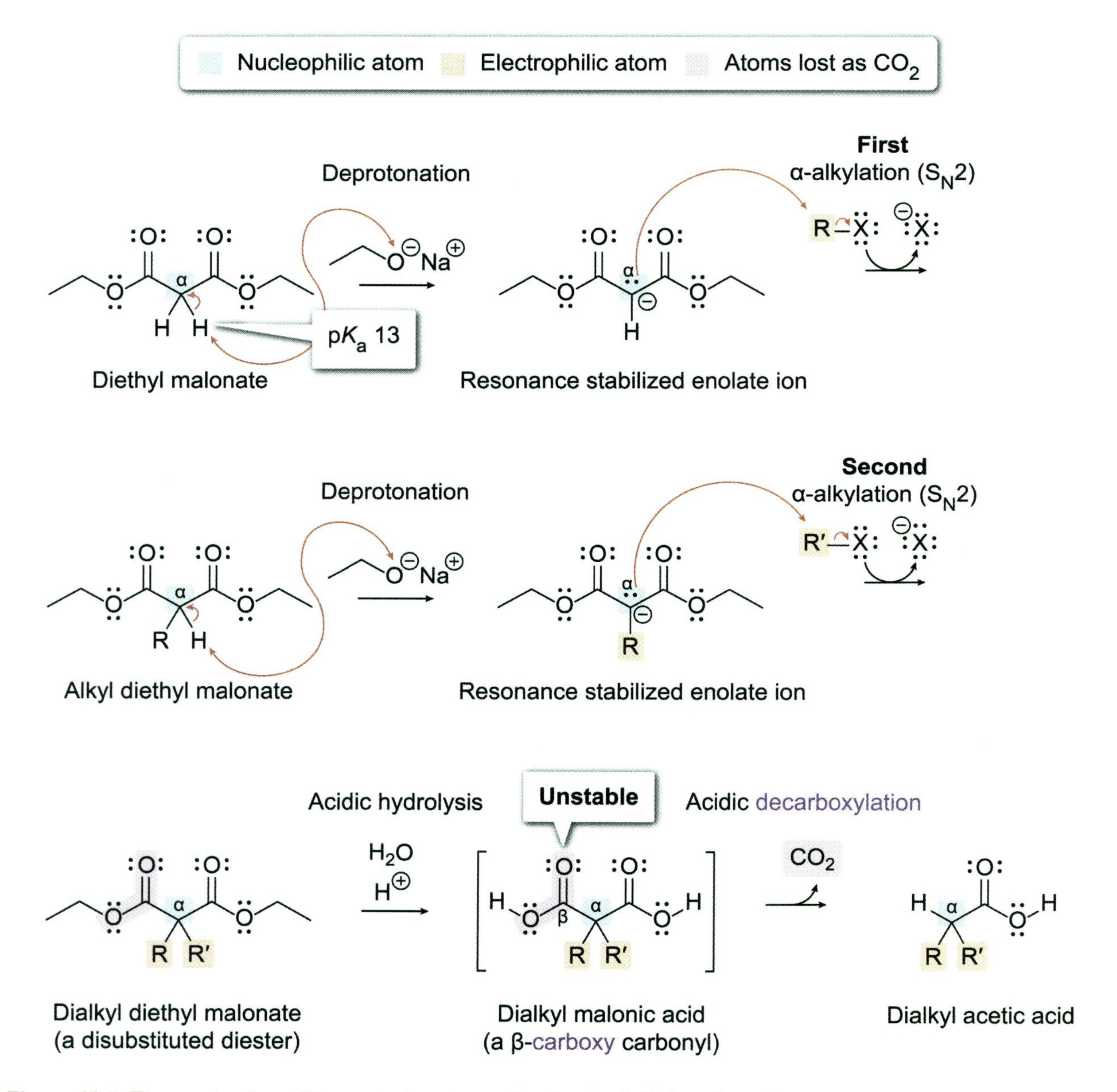

Figure 10.5 The mechanism of the malonic ester synthesis of a dialkyl acetic acid.

Diethyl malonate is initially treated with one equivalent of sodium ethoxide. Because the pK_a of the α-hydrogen atoms (pK_a 13) is several orders of magnitude *lower* (ie, *more* acidic) than the conjugate acid of the ethoxide ion (ethanol, pK_a 15.9), the resonance-stabilized enolate ion is generated quantitatively (see Concept 9.4.04). The addition of an alkyl halide (R–X) promotes an α-alkylation through an S_N2

mechanism to generate an alkyl diethyl malonate intermediate. The only restriction to the alkyl halide is that it must be amenable to an S_N2 reaction (ie, it must be methyl, 1°, or 2°). If a *dialkylated* product is desired (as shown in Figure 10.5), these two steps can be repeated a second time with (potentially) a *different* alkyl halide (R′–X).

Finally, the two ester groups in the alkylated diethyl malonate intermediate are hydrolyzed with aqueous acid. The resulting alkylated malonic acid intermediate is a β-carboxy carboxylic acid (a type of β-carboxy carbonyl) that spontaneously undergoes an acidic decarboxylation to evolve a molecule of carbon dioxide and provide the alkylated acetic acid product.

Concept Check 10.2

Can the following carboxylic acid be prepared using only a malonic ester synthesis? Why or why not?

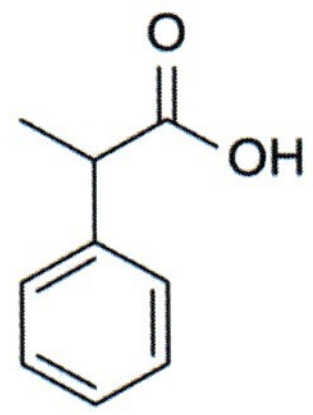

O
OH

Solution

Note: The appendix contains the answer.

Lesson 10.3

Reactions of Carboxylic Acids

Introduction

This lesson discusses the ways that carboxylic acids can be converted to other substances, which is related to the general interconversion of carboxylic acid derivatives (see Lesson 11.2). The following reactions are discussed:

- Acidic and basic mechanisms of a nucleophilic acyl substitution
- Fischer esterification
- Reduction of a carboxylic acid
- Conversion of a carboxylic acid to an acid chloride
- Cyclization of a carboxylic acid to a lactone
- The Hell-Volhard-Zelinsky reaction
- Decarboxylation of β-carboxy carbonyls

10.3.01 Nucleophilic Acyl Substitution

Nucleophilic acyl substitution is the *primary* reaction mechanism for carboxylic acid reactants and requires *at least* two steps. First, a nucleophile undergoes an *addition* to the π bond of the carbonyl group, forming a tetrahedral intermediate. Then the leaving group is *eliminated* as the π bond of the carbonyl is reformed. Consequently, a nucleophilic acyl substitution is sometimes called an **addition-elimination reaction**. The nucleophilic acyl substitution mechanism has acidic and basic variants.

The Acidic Nucleophilic Acyl Substitution Mechanism

The **acidic nucleophilic acyl substitution mechanism** (Figure 10.6) has *six* mechanistic steps; four involve proton transfers (ie, protonation, deprotonation):

1. Protonation
2. Addition of the nucleophile
3. Deprotonation
4. Protonation
5. Elimination of the leaving group
6. Deprotonation

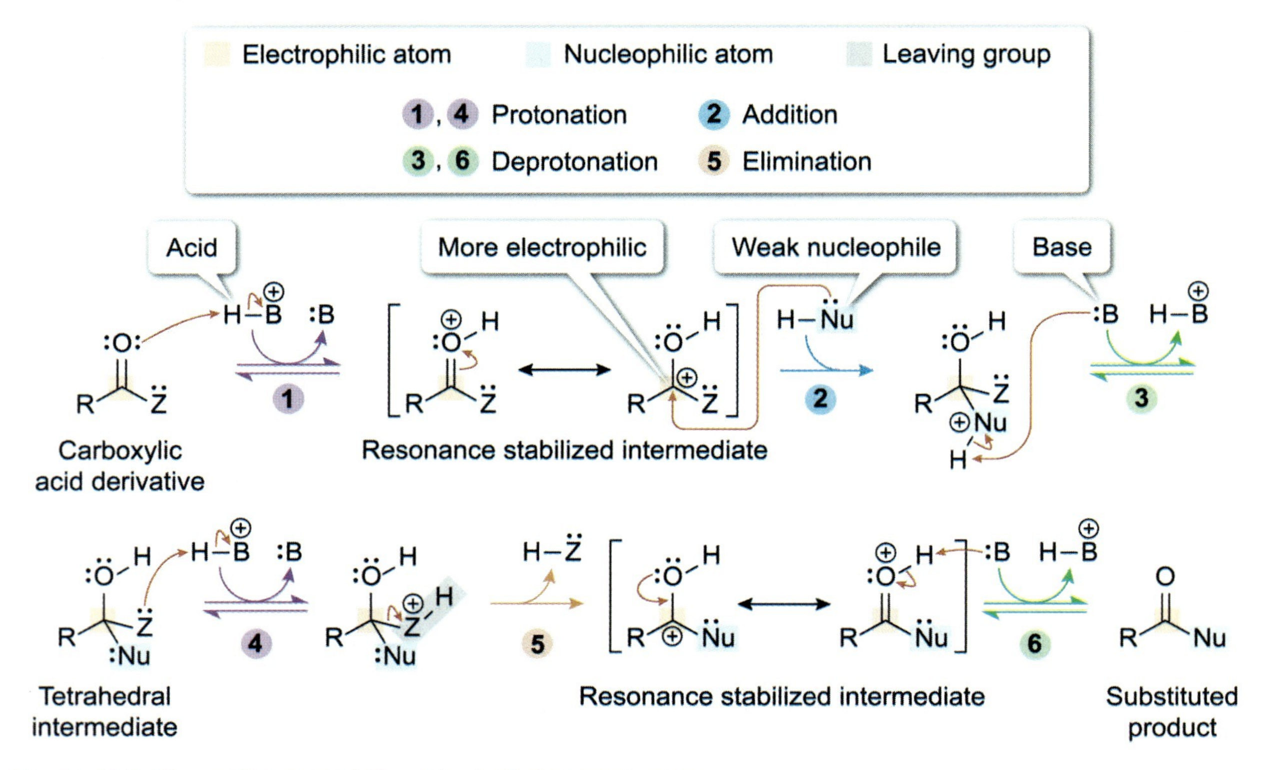

Figure 10.6 The acidic nucleophilic acyl substitution mechanism.

Because the nucleophile in an acidic mechanism tends to be weak (eg, water, an alcohol), the first step is the protonation of the carbonyl oxygen by a strong acid, increasing the electrophilicity of the carbonyl carbon atom. Then the nucleophile undergoes an *addition* with the carbonyl carbon atom. Because the nucleophile in an acidic mechanism is typically neutral, the next step is deprotonation of an acidic proton.

The second half of the acidic nucleophilic acyl substitution mechanism also requires an acidic catalyst. The Z group is protonated by a strong acid to turn it into a better leaving group. Then the leaving group is eliminated to generate a resonance stabilized intermediate. Finally, the acidic proton on the carbonyl oxygen atom is removed, yielding the substituted product.

The Basic Nucleophilic Acyl Substitution Mechanism

A **basic nucleophilic acyl substitution mechanism** (Figure 10.7) has only *two* mechanistic steps:

1. Addition of the nucleophile
2. Elimination of the leaving group

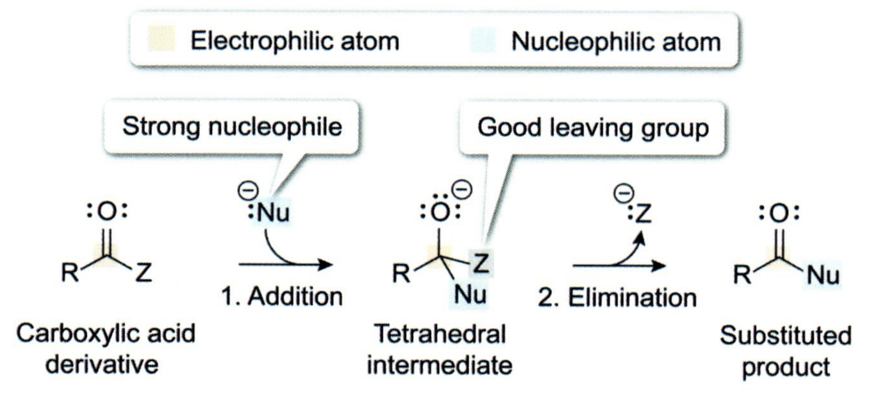

Figure 10.7 A general basic nucleophilic acyl substitution mechanism.

In a basic nucleophilic acyl substitution, a strong nucleophile directly attacks the electrophilic carbon atom of the carboxylic acid derivative. During the *addition*, the pi bond of the carbonyl becomes a lone pair on the oxygen atom. Because the anionic tetrahedral intermediate contains a good leaving group, the pi bond reforms and the leaving group is *eliminated*.

10.3.02 The Fischer Esterification

The **Fischer esterification** is a reaction between a carboxylic acid and an alcohol in the presence of a strong acid catalyst to form an ester and a molecule of water (Figure 10.8). In the *absence* of an acid catalyst, the Fischer esterification does *not* proceed at a reasonable rate. Because the alcohol functions as the nucleophile, the reaction is *most* effective with alcohols *lacking* steric hindrance (eg, methanol, 1° alcohols).

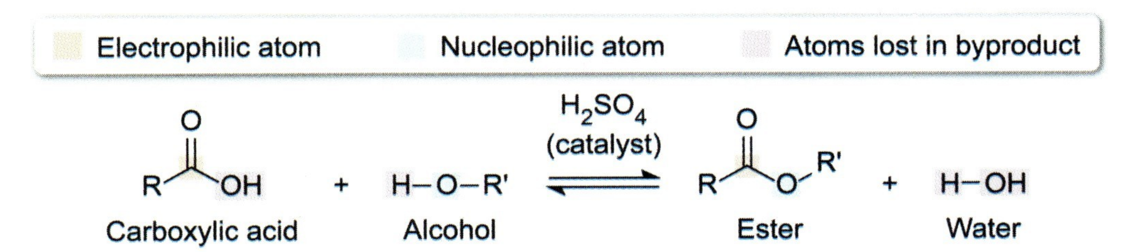

Figure 10.8 A general Fischer esterification reaction.

The Fischer esterification is an equilibrium process with a K_{eq} that typically favors the reactants (a carboxylic acid and an alcohol). At equilibrium, only a small amount of ester is normally present. To improve the yield of ester, one or more of the following approaches using Le Châtelier's principle are used to drive the equilibrium toward increased ester production:

- Use of an *excess* of alcohol reactant.
- *Removal* of the water product.

The Fischer esterification mechanism is an acidic nucleophilic acyl substitution and follows the same sequence of six mechanistic steps (see Concept 10.3.01) (Figure 10.9). The initial source of acidic hydrogen atoms is the strong acid (eg, H_2SO_4), which protonates one of the lone pairs on the carbonyl oxygen atom. Resonance delocalization of the pi bond generates an electrophilic carbon atom, which is attacked by the alcohol oxygen atom. The resulting oxonium ion intermediate contains an acidic hydrogen atom that is removed by a molecule of the alcohol solvent (HOR′).

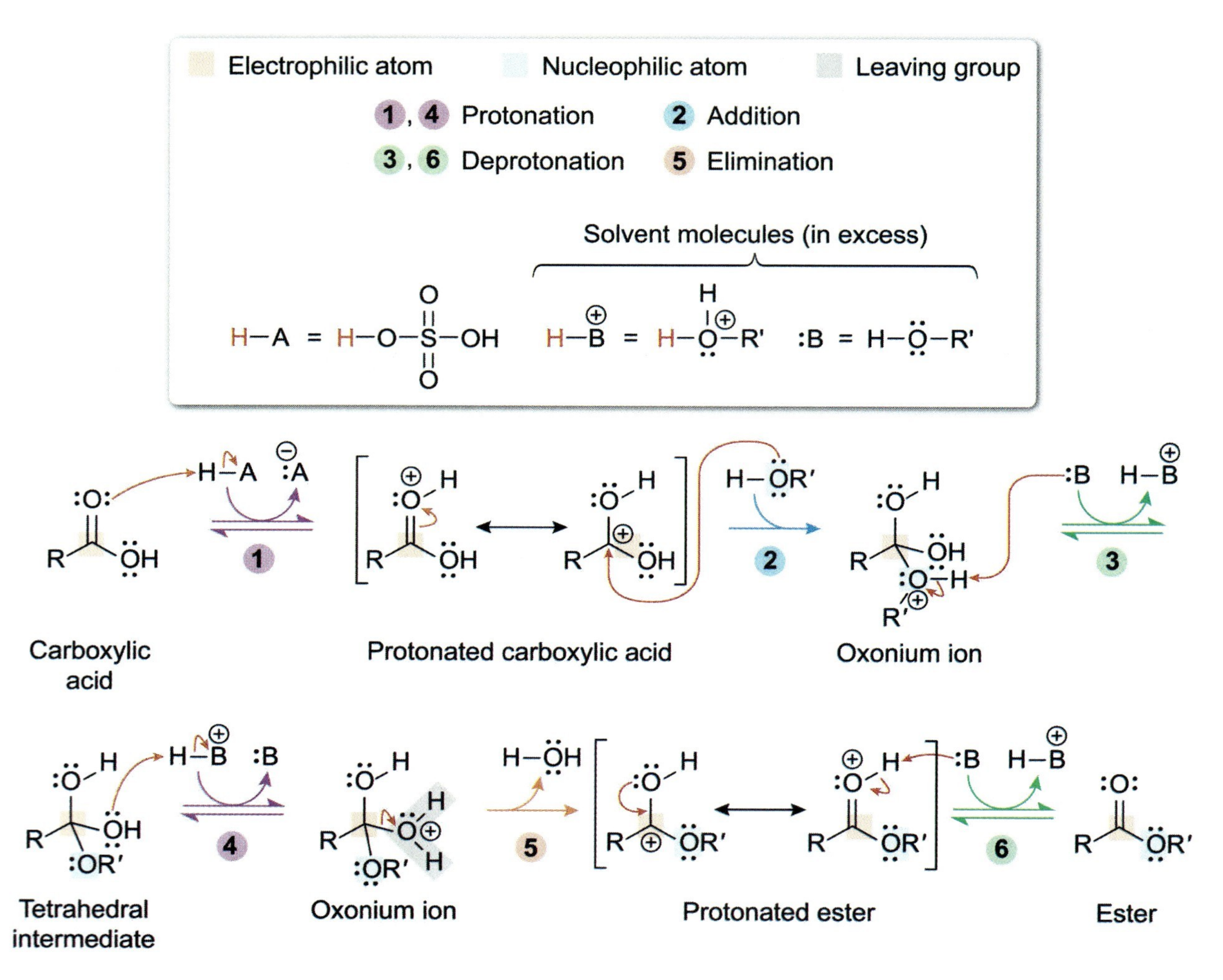

Figure 10.9 The Fischer esterification mechanism.

The tetrahedral intermediate in the Fischer esterification contains two hydroxyl groups attached to the original electrophilic carbon atom. *Either* hydroxyl group is then protonated by an acid to generate an oxonium ion, which is a good leaving group. Following the elimination of a water molecule, the resonance-stabilized protonated ester is deprotonated by the solvent to yield the ester product.

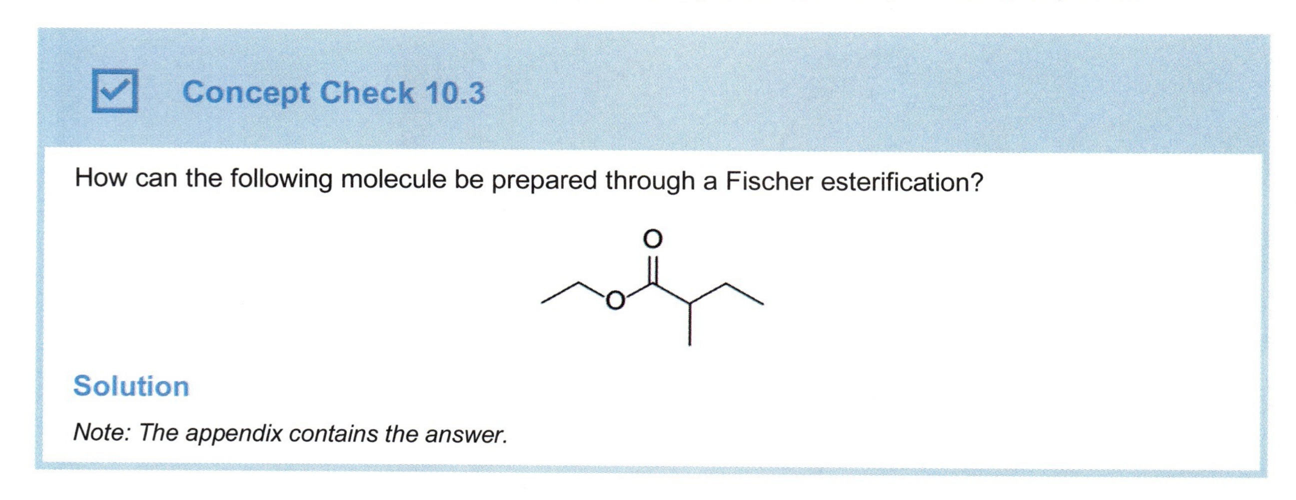

Concept Check 10.3

How can the following molecule be prepared through a Fischer esterification?

Solution

Note: The appendix contains the answer.

10.3.03 Reduction of a Carboxylic Acid

A carboxylic acid can be reduced by a reducing agent via several methods. Most successful reduction reactions are sequential two-electron reductions that proceed through an aldehyde intermediate en route to a 1° alcohol. This concept discusses two complementary methods that reduce a carboxylic acid to a 1° alcohol:

- Reduction by lithium aluminum hydride (LAH, $LiAlH_4$)
- Reduction by borane (BH_3)

Lithium Aluminum Hydride Reduction

Treating a carboxylic acid with $LiAlH_4$ results in a *double* ($2 \times 2e^- = 4e^-$) reduction to a 1° alcohol (Figure 10.10). A carboxylic acid contains an acidic hydrogen atom that requires an additional equivalent of a hydride ion to act as a base, meaning that *three* equivalents of $LiAlH_4$ are needed in total. Because of the basic environment, a second step involving acidic aqueous workup is also required. The inorganic byproducts of the reduction are hydrogen gas, aluminum complexes, and lithium salts.

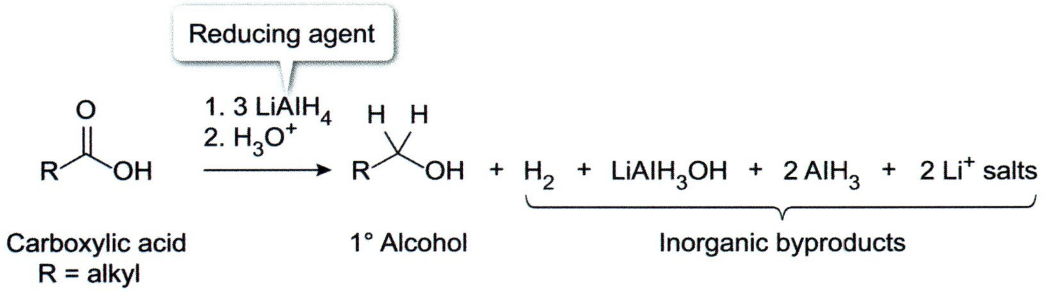

Reducing agent

$$RCOOH \xrightarrow[2.\ H_3O^+]{1.\ 3\ LiAlH_4} RCH_2OH + H_2 + LiAlH_3OH + 2\ AlH_3 + 2\ Li^+ \text{ salts}$$

Carboxylic acid
R = alkyl

1° Alcohol

Inorganic byproducts

Figure 10.10 The reduction of a carboxylic acid to a 1° alcohol using lithium aluminum hydride ($LiAlH_4$).

The mechanism for the $LiAlH_4$ reduction of a carboxylic acid has three stages, each requiring one equivalent of $LiAlH_4$ (Figure 10.11):

1. Deprotonation
2. Nucleophilic acyl substitution (basic mechanism)
3. Nucleophilic addition (basic mechanism)
   Note: Acidic workup neutralizes the anionic salts and excess $LiAlH_4$.

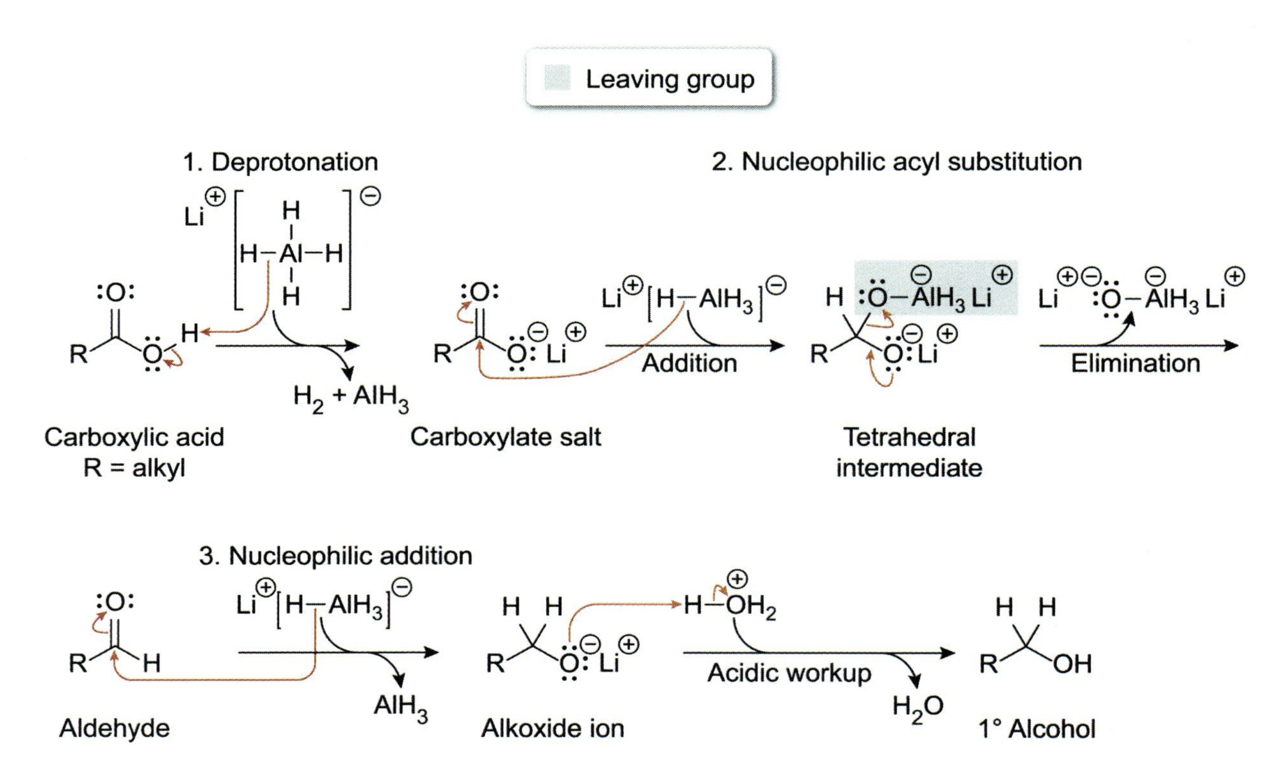

Figure 10.11 The mechanism for the reduction of a carboxylic acid using $LiAlH_4$.

The AlH_4^- complex ion reacts like $H:^-$ + AlH_3; one of the Al–H bonds donates its electron density to *either* an acidic hydrogen atom or an electrophilic carbon atom. In the deprotonation stage, the first equivalent of $LiAlH_4$ removes the acidic hydrogen atom on the carboxylic acid, generating a carboxylate ion salt. Then the second equivalent of $LiAlH_4$ undergoes a nucleophilic acyl substitution with the carboxylate ion salt. One of the Al–H bonds attacks the electrophilic carbon atom of the carboxylic acid. The tetrahedral intermediate collapses to reform the carbonyl pi bond, eliminating the good leaving group $LiAlH_3OH$.

In the final stage, a third molecule of $LiAlH_4$ undergoes a nucleophilic addition with the carbonyl of the aldehyde intermediate. The resulting alkoxide ion becomes protonated during the acidic workup to generate the 1° alcohol product.

The $LiAlH_4$ (present in excess) will *also* reduce other carbonyl-containing functional groups (eg, esters, amides, ketones, aldehydes) in the reactant (Figure 10.12).

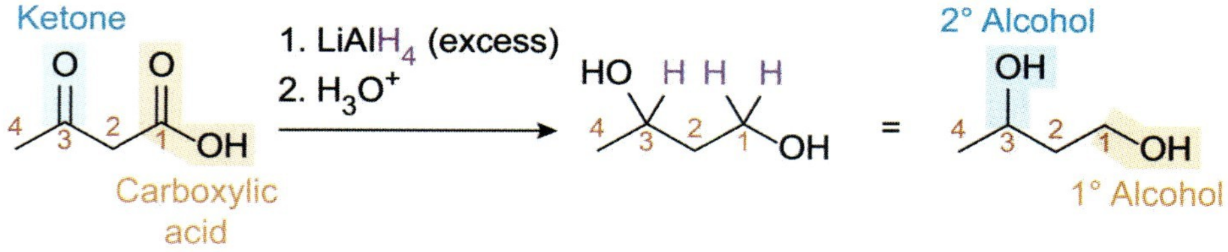

Figure 10.12 $LiAlH_4$ promotes the reduction of all carbonyl-containing functional groups.

Borane Reduction

It is possible to *selectively* reduce a carboxylic acid to a 1° alcohol through treatment with borane (BH_3) (Figure 10.13). Although this process is also a *double* (ie, $4e^-$) reduction, it can be performed in the presence of other carbonyl-containing functional groups (eg, esters, amides, ketones, aldehydes), which do *not* react with BH_3. However, BH_3 *does* react with the pi bonds in alkenes and alkynes through an electrophilic addition, which can become problematic if the carboxylic acid reactant also contains either of these functional groups.

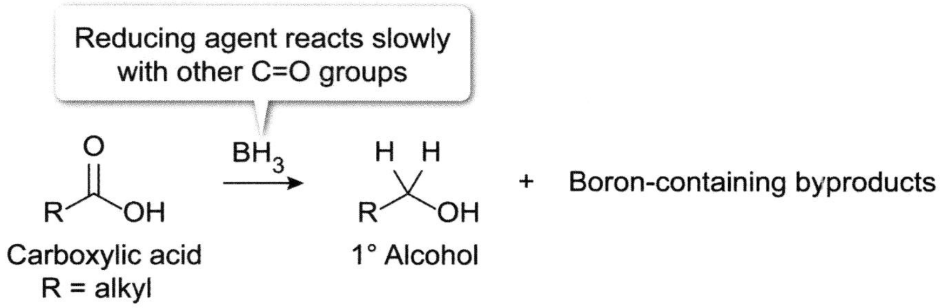

Reducing agent reacts slowly with other C=O groups

BH_3

Carboxylic acid
R = alkyl

1° Alcohol

+ Boron-containing byproducts

Figure 10.13 The reduction of a carboxylic acid to a 1° alcohol using borane (BH_3).

The byproducts of this reduction are several boron-containing salts and complexes, which are removed on workup. Although it is not possible to directly perform a *single* 2 e^- reduction of a carboxylic acid and isolate the aldehyde product in high yield, synthetic methods have been developed to selectively reduce a *carboxylic acid derivative* to an aldehyde (see Concept 11.3.03).

10.3.04 Conversion to an Acid Chloride

Carboxylic acid derivatives have different heteroatom groups (Z) in place of the hydroxyl group of the original carboxylic acid (Figure 10.14). Acid halides play an important role in the interconversion between carboxylic acids and their derivatives.

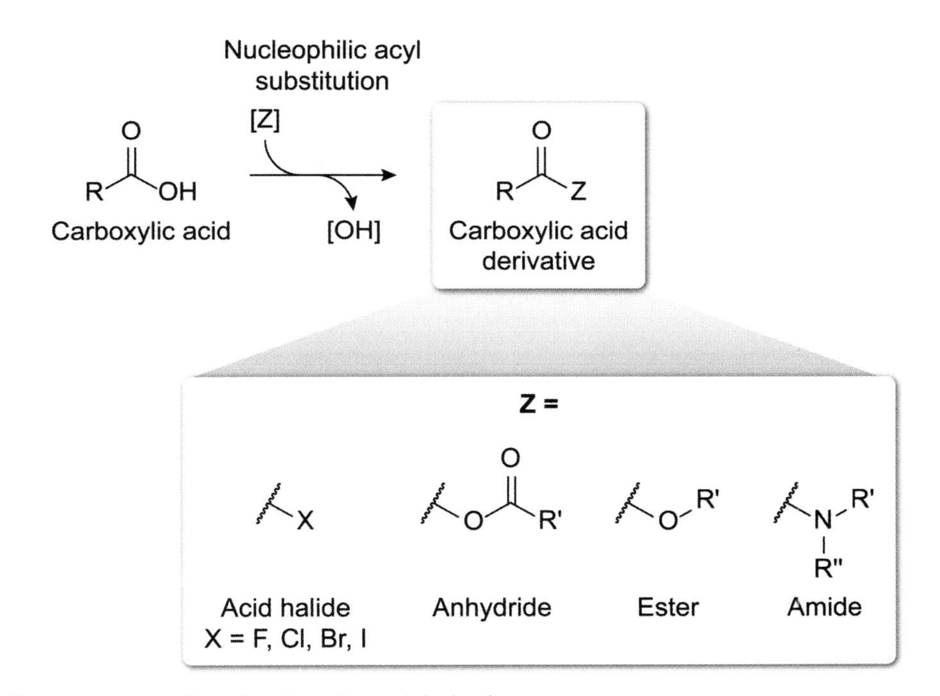

Figure 10.14 General preparation of carboxylic acid derivatives.

Acid halides, as the *most reactive* carboxylic acid derivatives (see Concept 11.2.01), are used to directly prepare *all* other carboxylic acid derivatives. The most commonly used acid halide is an **acid chloride**, which is prepared through the reaction of a carboxylic acid with a chlorinating agent such as thionyl chloride ($SOCl_2$) (Figure 10.15).

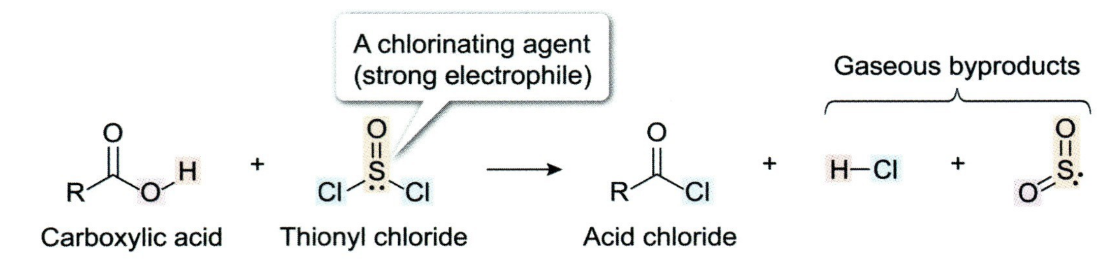

Figure 10.15 Conversion of a carboxylic acid to an acid chloride.

Thionyl chloride contains a sulfur-oxygen double bond (ie, a **thionyl group**) and two chlorine atoms attached to the sulfur. The sulfur atom in thionyl chloride is *strongly* electrophilic and reacts with most nucleophiles, including water. Treatment of a carboxylic acid with thionyl chloride produces acid chloride in high yield. The only byproducts of this reaction are gases that are easily removed.

The mechanism of the reaction between a carboxylic acid and thionyl chloride proceeds through a *modified* nucleophilic acyl substitution reaction. Although the reaction does not involve the use of a Brønsted-Lowry acid, thionyl chloride is a strong electrophile (a Lewis *acid*), and the reaction is most similar to the *acidic* nucleophilic acyl substitution mechanism (Concept 10.3.01). The mechanism proceeds through two successive stages:

1. Nucleophilic acyl (thionyl) substitution at the *sulfur* atom of $SOCl_2$
2. Nucleophilic acyl substitution at the *carbon* atom of the original carboxylic acid

In the first stage, the *carbonyl* oxygen atom of the carboxylic acid acts as a nucleophile and attacks the electrophilic sulfur atom of thionyl chloride (Figure 10.16)—the *addition* step of the first nucleophilic acyl (thionyl) substitution reaction. The tetrahedral intermediate collapses, reforming the S=O pi bond and *eliminating* a chloride ion. Then a lone pair on the hydroxyl group of the original carboxylic acid is delocalized through resonance into the adjacent C=O pi bond, changing which oxygen atom bears the positive formal charge. The result of this delocalization is a protonated (ie, activated) mixed anhydride.

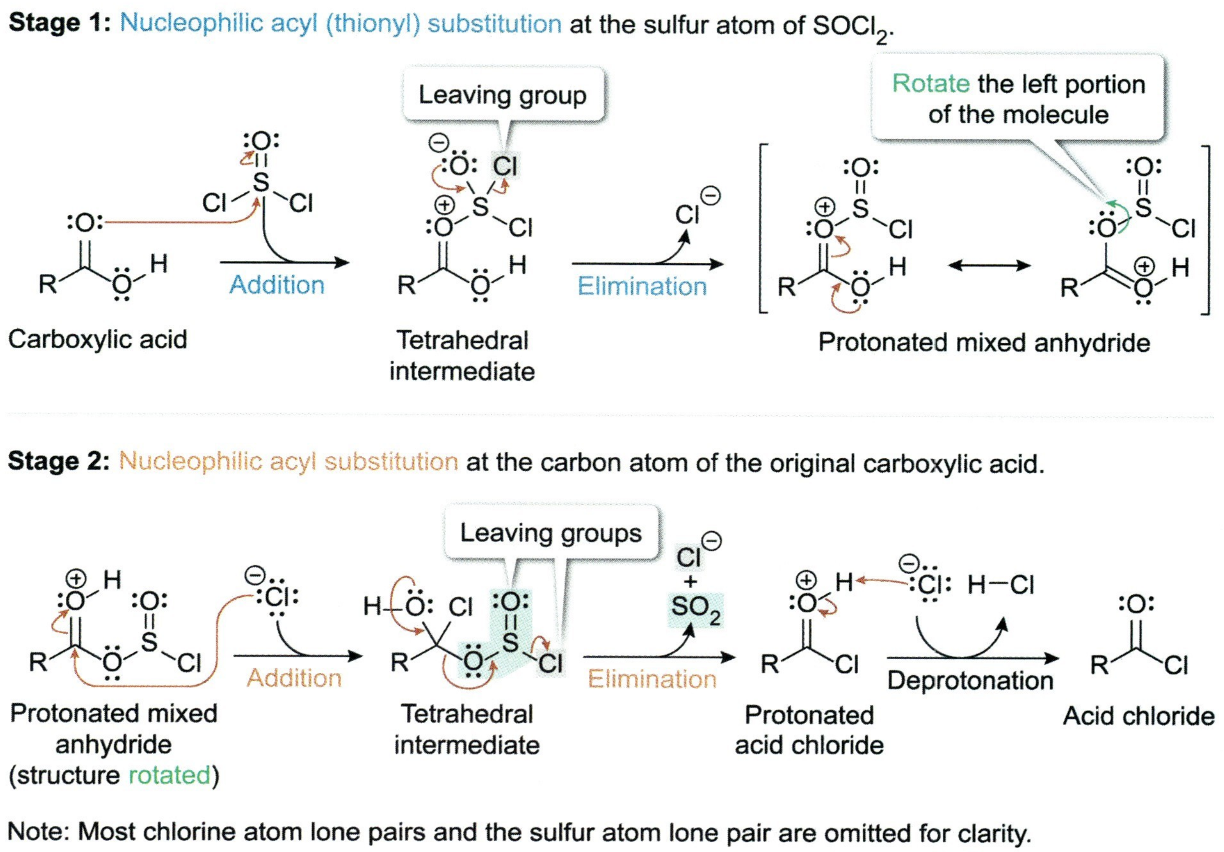

Stage 1: Nucleophilic acyl (thionyl) substitution at the sulfur atom of $SOCl_2$.

Leaving group

Rotate the left portion of the molecule

Addition

Elimination

Carboxylic acid

Tetrahedral intermediate

Protonated mixed anhydride

Stage 2: Nucleophilic acyl substitution at the carbon atom of the original carboxylic acid.

Leaving groups

Addition

Elimination

Deprotonation

Protonated mixed anhydride (structure rotated)

Tetrahedral intermediate

Protonated acid chloride

Acid chloride

Note: Most chlorine atom lone pairs and the sulfur atom lone pair are omitted for clarity.

Figure 10.16 The mechanism of the reaction between a carboxylic acid and thionyl chloride ($SOCl_2$).

In the second stage, a chloride ion acts as a nucleophile and attacks the *carbon* atom of the protonated carbonyl group, forming a tetrahedral intermediate. The tetrahedral intermediate then collapses to reform the carbon-oxygen pi bond and eliminate the rest of the mixed anhydride as *two* leaving groups:

- The electrons in the broken carbon-oxygen bond become a pi bond between the oxygen and sulfur atoms; this portion becomes a molecule of gaseous SO_2 (a good leaving group).
- The chlorine atom attached to the sulfur atom is lost as a chloride ion (a good leaving group).

In the final step of the reaction mechanism, a chloride ion removes the acidic proton on the carbonyl group, forming the acid chloride product.

10.3.05 Cyclization to a Lactone or Lactam

A lactone is typically prepared from a single molecule with *both* an alcohol functional group *and* a carboxylic acid functional group, separated by at least two carbon atoms (Figure 10.17). Similarly, a lactam is generally prepared from a single molecule containing *both* an amine and a carboxylic acid, separated by at least two carbon atoms.

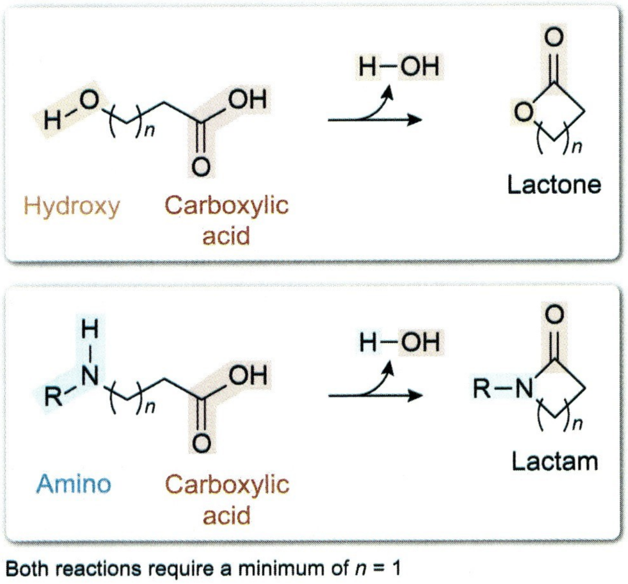

H−OH
Lactone
Hydroxy Carboxylic acid
H−OH
R−N
Lactam
Amino Carboxylic acid

Both reactions require a minimum of $n = 1$

Figure 10.17 Preparing a lactone or a lactam.

Creation of a lactone or lactam requires a specialized reagent such as **dicyclohexylcarbodiimide (DCC)** to facilitate the *selective* and *mild* intramolecular (ie, within the same molecule) condensation between either an alcohol or an amine and a carboxylic acid (Figure 10.18).

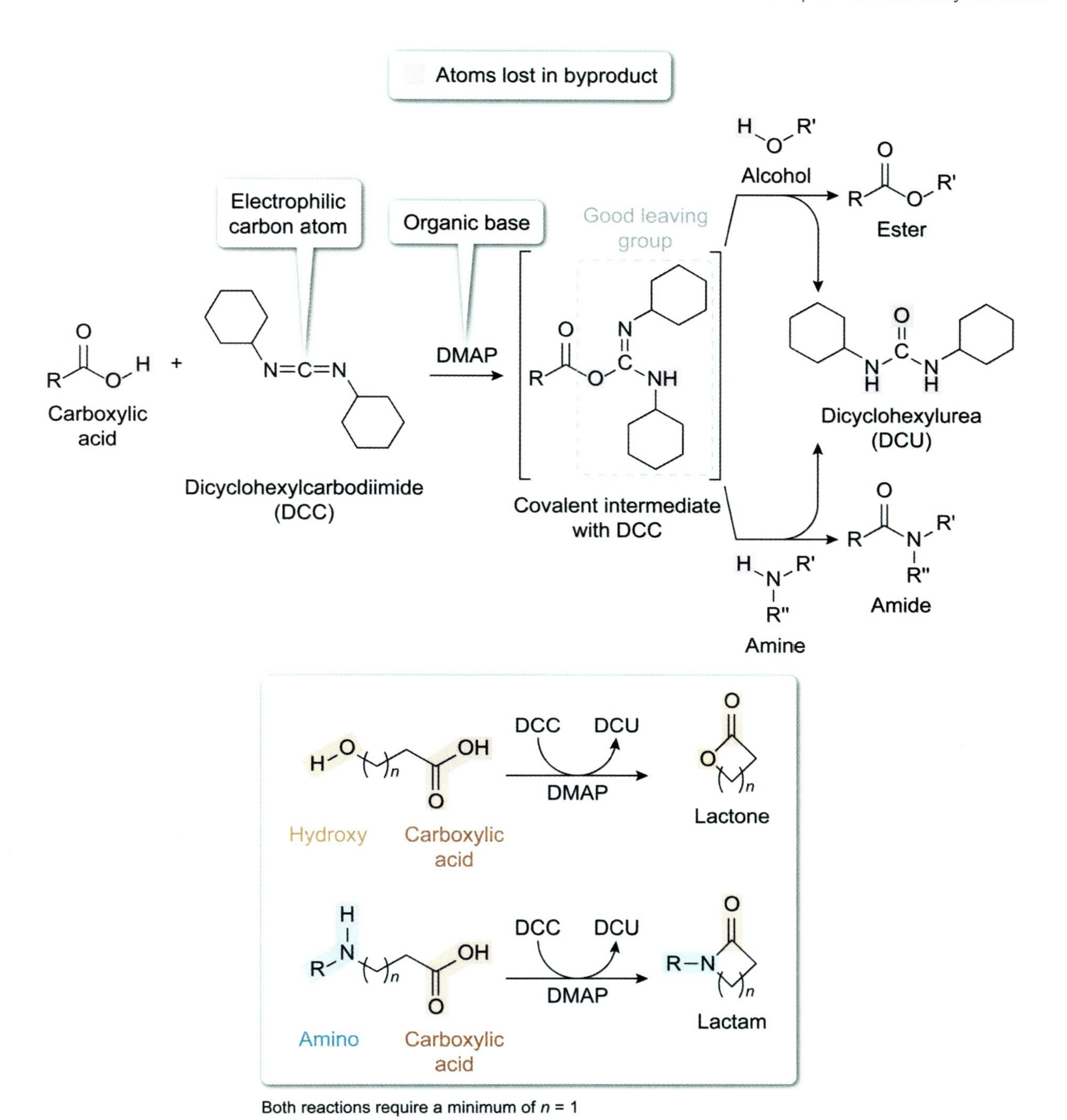

Figure 10.18 The use of DCC to promote formation of a lactone and a lactam.

DCC contains a strongly electrophilic carbon atom between two electronegative nitrogen atoms and serves to *activate* the carboxylic acid for nucleophilic attack by the alcohol or amine. DCC reacts with the carboxylic acid to create a covalent intermediate, in which the DCC-derived portion is a good leaving group. Then an alcohol or an amine (within the same molecule) can undergo nucleophilic acyl substitution with the activated carboxylic acid to form the lactone or lactam, respectively.

The reaction with DCC also requires a sterically hindered (ie, bulky) organic base (eg, 4-(*N*,*N*-dimethylamino)pyridine [DMAP]) to promote the proton-transfer steps required for the coupling mechanism. The byproduct of a DCC-promoted lactonization reaction is dicyclohexylurea (DCU), which contains the atoms of DCC *and* the two hydrogen atoms and one oxygen lost during the condensation.

Concept Check 10.4

Draw the structure of the lactone product that is formed by the following reaction.

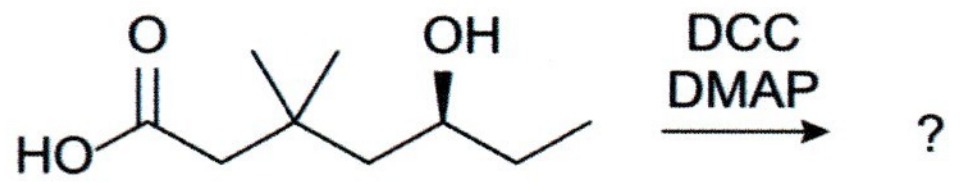

Solution

Note: The appendix contains the answer.

10.3.06 Alpha Bromination (Hell-Volhard-Zelinsky Reaction)

Although ketones and aldehydes are typically the functional groups most prone to α reactions, other carbonyl-containing functional groups, such as carboxylic acids and some carboxylic acid derivates (see Concept 10.2.03), can also undergo α reactions through a similar mechanistic process.

In the Hell-Volhard-Zelinsky (HVZ) reaction, a carboxylic acid is treated with a mixture of phosphorus tribromide (PBr_3) and bromine (Br_2) to provide an initial α-halogenated product, an α-bromo acid bromide (Figure 10.19). Treatment of the α-bromo acid bromide with water hydrolyzes the acid bromide and provides the final product of an HVZ reaction, an α-bromo carboxylic acid.

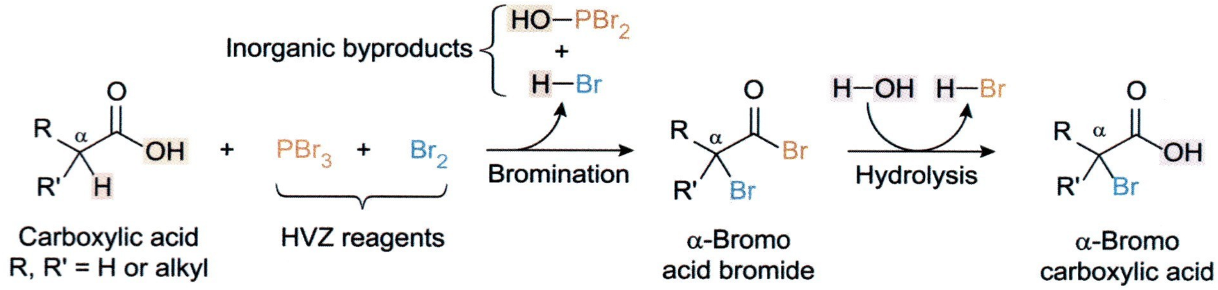

Figure 10.19 A general Hell-Volhard-Zelinsky (HVZ) reaction.

The mechanism of an HVZ reaction (and subsequent hydrolysis) takes place over four stages (Figure 10.20):

1. Conversion of the carboxylic acid to an acid bromide with PBr_3 (an S_N2 reaction followed by a nucleophilic acyl substitution reaction)
2. Acidic tautomerization to an enol of an acid bromide
3. α-Halogenation
4. Hydrolysis (a nucleophilic acyl substitution reaction)

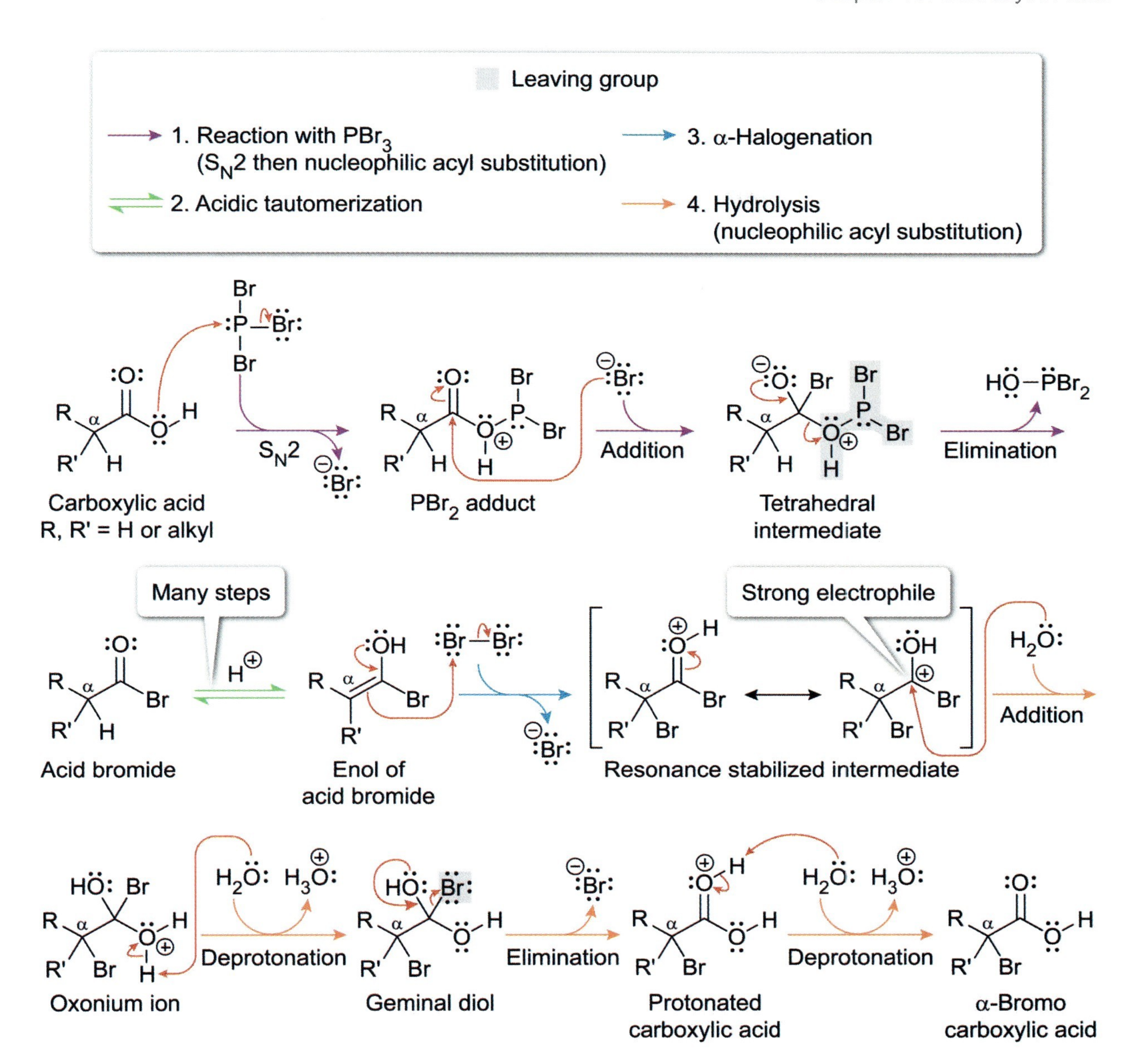

Figure 10.20 The mechanism of an HVZ reaction.

In the first stage, the hydroxyl group of the carboxylic acid attacks the electrophilic PBr_3 through an S_N2 reaction. The displaced bromide ion attacks the electrophilic carbon atom of the PBr_2 adduct to form a tetrahedral intermediate during the addition step of a nucleophilic acyl substitution. Because the $-O(H)PBr_2$ group is a good leaving group, it is eliminated to form the acid bromide.

The second stage of the HVZ reaction requires the acid bromide to have at least one α-hydrogen atom to undergo acidic tautomerization to its enol form (see Concept 9.4.02). In the third stage, the nucleophilic enol is α-brominated by a molecule of Br_2 to provide a resonance-stabilized intermediate, the protonated (ie, activated) form of an α-bromo acid bromide (see Concept 9.4.07).

In the final stage, water acts as a weak nucleophile and is added to the carbon atom of the resonance-stabilized intermediate. The resulting oxonium ion is deprotonated to form a geminal diol, which then eliminates a bromide ion to form a protonated carboxylic acid intermediate. Deprotonation yields the final α-bromo carboxylic acid product.

10.3.07 Decarboxylation

In organic chemistry, specific combinations and spatial relationships between functional groups can lead to molecules with unique chemical properties. One example of this effect is observed in molecules containing a carboxylic acid or carboxylate ion at the β position of *another* carbonyl group. The carbonyl group can be a part of many types of functional groups (eg, ketone, carboxylic acid, ester, amide). Some molecule classes that fulfill this functional group relationship include β-ketoacids and β-diacids. Broadly, molecules that belong to one of these classes are described as **β-carboxy(late) carbonyls**.

A β-carboxy(late) carbonyl is an *unstable* molecule that spontaneously undergoes a decarboxylation reaction (Figure 10.21). In a **decarboxylation reaction**, a reactant loses a carboxylic acid or carboxylate ion functional group, usually as carbon dioxide (CO_2). Because the evolved CO_2 is typically in the gas phase, this transformation is *irreversible*.

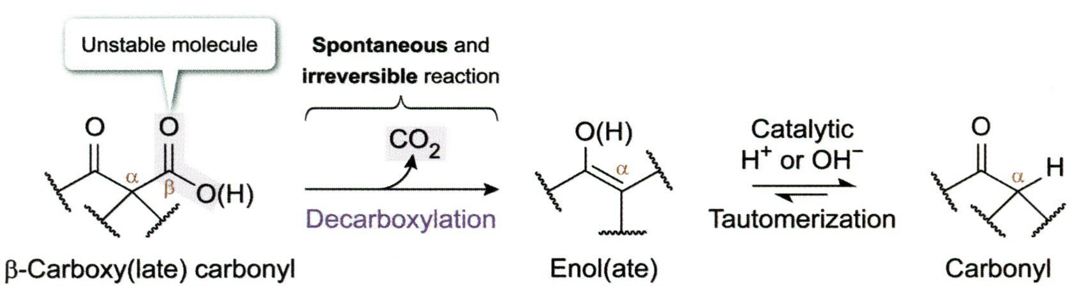

Figure 10.21 The general decarboxylation of a β-carboxy(late) carbonyl.

When a β-carboxy(late) carbonyl undergoes a decarboxylation reaction, the initial product is an enol or enolate ion. Because most reactions contain a catalytic concentration of either an acid or a base, the enol or enolate ion undergoes keto-enol tautomerization to reform the initial carbonyl. In this way, the net reaction of decarboxylation is the replacement of a β-carboxylic acid or β-carboxylate ion with a hydrogen atom. The mechanism of decarboxylation has different steps depending on whether the reaction takes place under acidic (ie, carboxylic acid) or basic (ie, carboxylate ion) conditions.

Acidic decarboxylation of a β-carboxy carbonyl is shown in Figure 10.22. Although the first mechanistic step is *concerted* (ie, all electron flow takes place simultaneously), it is often easier to conceptualize the mechanism in the following order:

1. The carbonyl oxygen atom attacks the acidic hydrogen atom, breaking the O–H bond of the carboxylic acid.
2. The electrons from the broken O–H bond become a pi bond between oxygen and the β-carbon, which in turn breaks the bond between the α-carbon and the β-carbon.
3. The electrons from the broken α-β bond become a pi bond between the α-carbon and the carbonyl carbon atom, and electrons from the carbonyl pi bond shift to a lone pair on the oxygen atom.

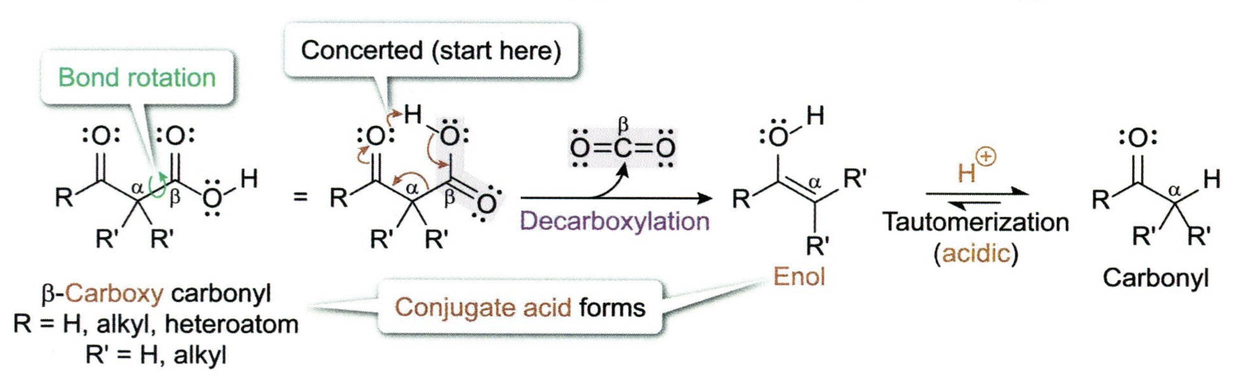

Figure 10.22 The acidic decarboxylation mechanism of a β-carboxy carbonyl.

Basic decarboxylation of a β-carboxylate carbonyl is shown in Figure 10.23. A lone pair on the anionic oxygen atom becomes a pi bond with the β-carbon, eliminating a resonance-stabilized enolate (a type of good leaving group). The enolate ion intermediate then undergoes tautomerization to the carbonyl product.

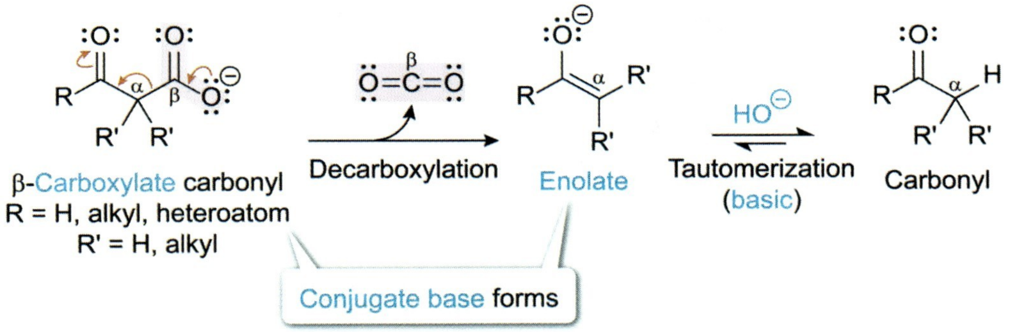

Figure 10.23 The basic decarboxylation mechanism of a β-carboxylate carbonyl.

Lesson 11.1

Structure and Physical Properties of Carboxylic Acid Derivatives

Introduction

This chapter focuses on carboxylic acid derivatives, most of which contain a carbonyl bonded to an alkyl group and either oxygen, nitrogen, or a halogen—the only exception is nitriles, which do not contain a carbonyl. Carboxylic acid derivatives are useful molecules in organic synthesis; not only can they be prepared from one another (ie, interconversion) (Lesson 11.2), but they can also be converted into other functional groups (Lesson 11.3).

This lesson discusses the structural features and physical properties of carboxylic acid derivatives.

11.1.01 Structural Characteristics of Carboxylic Acid Derivatives

The carbonyl-containing carboxylic acid derivatives have the general formula R–C(O)–Z, in which the Z group contains a heteroatom (ie, halogen, oxygen, nitrogen) directly attached to the carbonyl.

Acid Halides, Anhydrides, and Esters

Acid halides have the general formula R–C(O)–X, in which X is a halogen and R represents an alkyl group, an aryl group, or a hydrogen atom (Figure 11.1). **Anhydrides** are represented by the general formula R–C(O)–O–C(O)–R′, in which R and R′ may or may not be the same. **Esters** have the general formula R–C(O)–OR″, where R″ *cannot* be a hydrogen atom (otherwise, the functional group would be a carboxylic acid). A cyclic variant of an ester is a lactone in which the R group and the R″ group are connected (Concept 4.3.06).

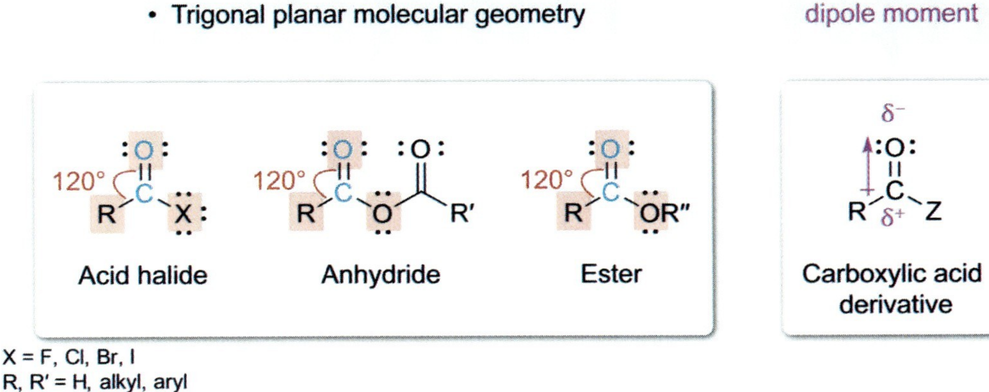

Figure 11.1 Structural characteristics of acid halides, anhydrides, and esters.

In each functional group, the carbonyl carbon and oxygen atoms each have three electron domains, are sp^2 hybridized (Concept 1.2.03 and Concept 2.1.02), and have a trigonal planar electron geometry with 120° bond angles.

Like aldehydes, ketones (Concept 9.1.01), and carboxylic acids (Concept 10.1.01), the carbon-oxygen double bond of a carboxylic acid derivative carbonyl is polar due to the difference in electronegativity between oxygen and carbon, resulting in a molecular dipole. The polar nature of the carbonyl influences the physical properties of acid halides, anhydrides, and esters.

Amide

Amides are carboxylic acid derivatives with the general formula R–C(O)–N(R′)$_2$, in which the R′ groups on the N atom may or may not be the same (Figure 11.2). The carbonyl carbon and oxygen atoms of an amide have the same structural characteristics as the previously discussed carbonyl-containing compounds: three electron domains, *sp*^{2} hybridization, trigonal planar electron geometry, and 120° bond angles.

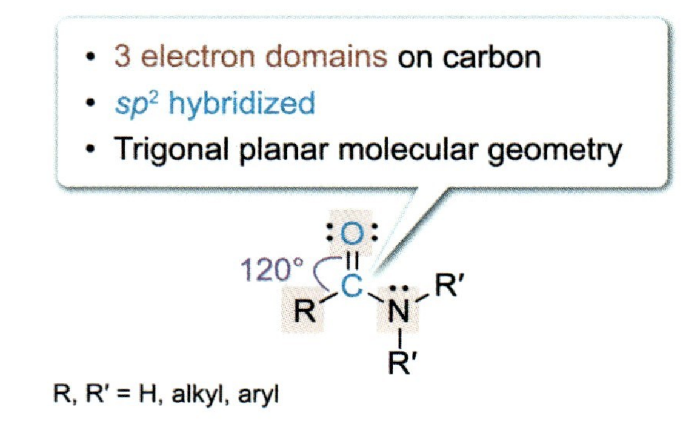

Figure 11.2 Structural characteristics of amides.

Hybridization of the amide nitrogen atom is significantly impacted by resonance. In Figure 11.3, all the atoms of Resonance structure 1 have a complete octet, and the nitrogen atom of the amide *appears* to have four electron domains and therefore may *appear* to be *sp*^{3} hybridized.

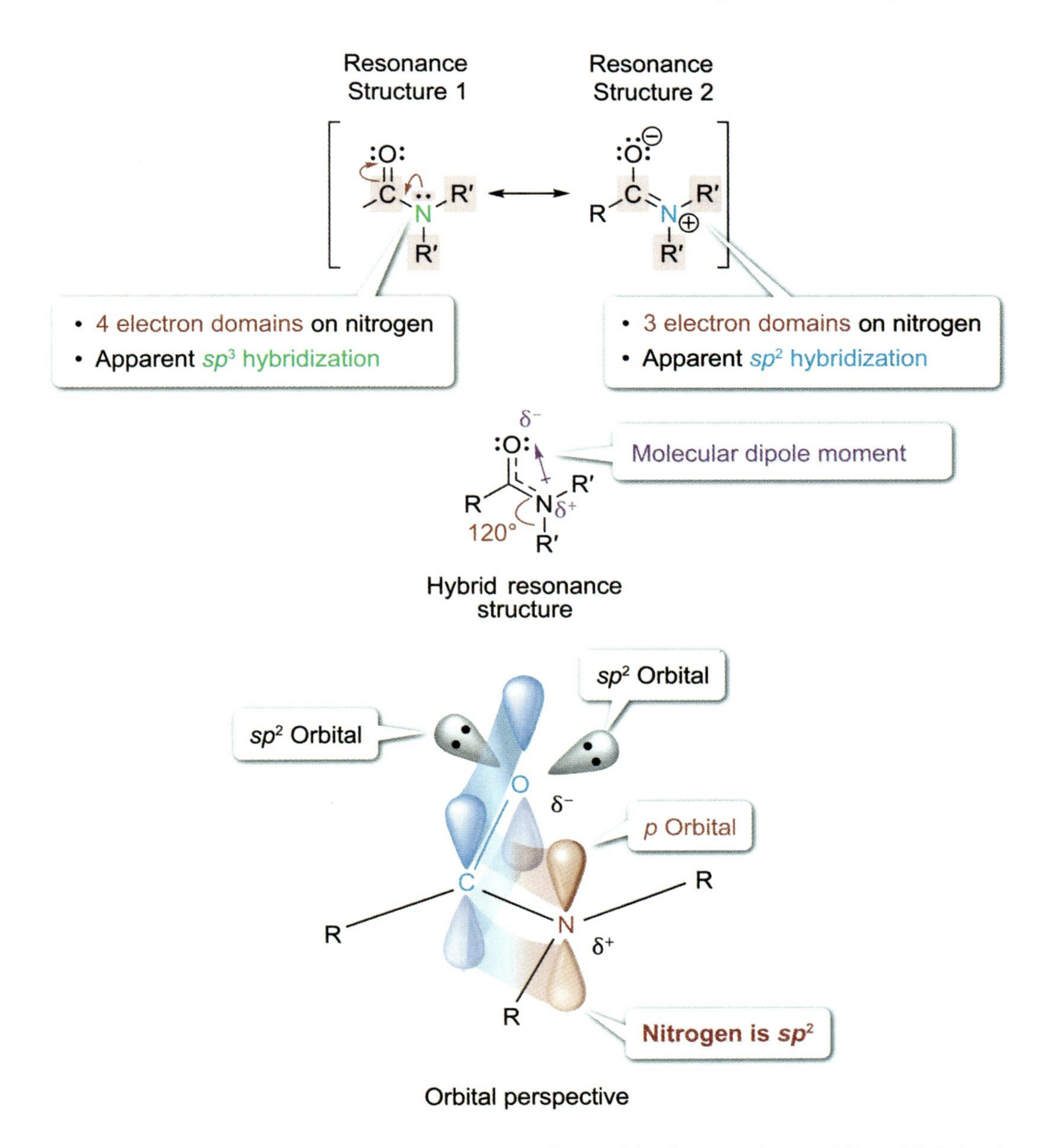

Figure 11.3 The impact of resonance on the hybridization of an amide nitrogen atom and the orbital structure of an amide.

However, the *p* orbital of the carbon has side-to-side overlap with the orbital containing the lone pair electrons of the nitrogen atom. Therefore, the nitrogen atom lone pair is conjugated with the π system of the adjacent carbonyl group (Concept 6.5.01) by resonance delocalization. This is shown in Resonance structure 2, in which the lone pair on the nitrogen atom becomes a pi bond and the electrons from the carbonyl pi bond become a lone pair on the oxygen atom.

This shift in electrons leads to a charge separation. Because the nitrogen atom of an amide is *less* electronegative than the heteroatoms of other carboxylic acid derivatives, an amide can more easily delocalize its lone pair and is more stable bearing a positive formal charge.

Because an amide is resonance stabilized, the C–N bond has partial double bond character, and **the amide nitrogen is *sp*² hybridized**. The amide nitrogen atom has three electron domains and trigonal planar geometry with approximately 120° bond angles. The resonance stabilization of an amide and the molecular dipole moment resulting from the *partial* separation of charges (ie, δ^+, δ^-) also impact its physical properties and reactivity (Concept 11.2.01).

An amide can be classified by the number of nonhydrogen groups attached to the N atom (Concept 12.1.02). A cyclic variant of an amide is a lactam, in which the R group is directly attached to the nitrogen atom (Concept 4.4.02).

Nitrile

Nitriles are unlike other carboxylic acid derivatives because they contain a carbon-nitrogen triple bond (known as a **cyano group**) and no carbonyl group. Figure 11.4 shows that a nitrile has the general formula R″–C≡N. The carbon and nitrogen atoms each have *two* electron domains, are *sp* hybridized, and have linear molecular geometry with a 180° bond angle.

- 2 electron domains on carbon
- *sp* hybridized
- Linear molecular geometry

Figure 11.4 Structural characteristics of nitriles.

In nitriles, the carbon-nitrogen triple bond is polar and the electrons between the carbon atom and nitrogen atom are not shared equally due to the electronegativity difference between carbon and nitrogen. The resulting partial charges and molecular dipole moment impact the physical properties of nitriles.

11.1.02 Physical Properties of Carboxylic Acid Derivatives

The most significant intermolecular force affecting the physical properties of the carboxylic acid derivatives is dipole-dipole interactions resulting from their polar molecular character (Figure 11.5). An amide has the strongest dipole-dipole forces due to the partial charge separation (ie, δ^+, δ^-) resulting from its hybrid resonance structure.

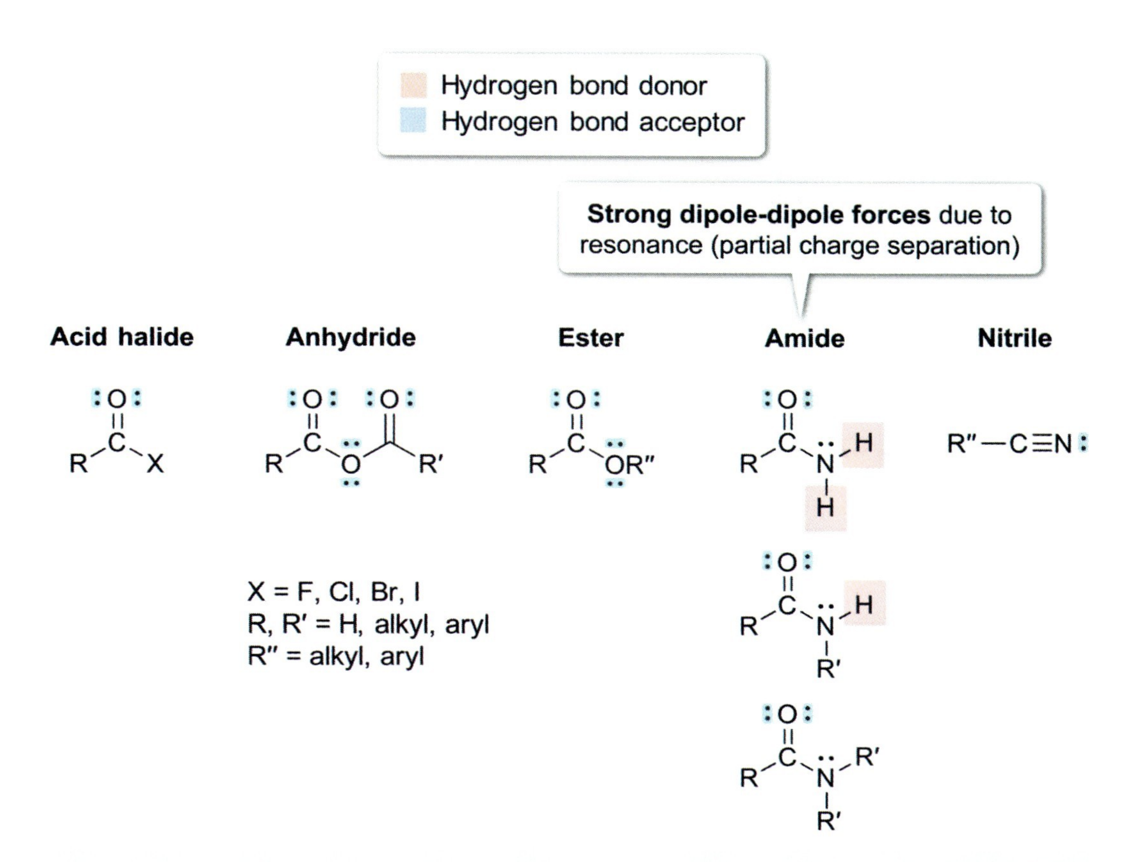

Figure 11.5 The intermolecular forces of carboxylic acid derivatives that most significantly impact their physical properties.

The lone pairs on the oxygen atoms of acid halides, anhydrides, esters, and amides and the lone pair on the nitrogen atom of nitriles are hydrogen bond acceptors. Therefore, each carboxylic acid derivative can form hydrogen bonds. Because acid halides, anhydrides, esters, and nitriles lack a hydrogen bond donor, pure samples of molecules that contain only these functional groups *cannot* hydrogen bond. These groups can only participate in hydrogen bonds if *another* functional group is present to contribute a donor.

The lone pair of an amide nitrogen atom is *not* a hydrogen bond acceptor due to the resonance delocalization of the lone pair. If an amide nitrogen atom is attached to *at least* one hydrogen atom, a pure sample of the amide *can* participate in hydrogen bonds.

Boiling Point

Molecules with more extensive intermolecular forces have higher boiling points than molecules with weaker intermolecular forces. The data in Table 11.1 show the boiling point trend for the carboxylic acid derivatives and how their boiling points compare to other functional groups. Molecules with similar molecular weights are compared to eliminate differences due to London dispersion forces.

Table 11.1 Boiling points of alkanes, alcohols, carboxylic acids, and carboxylic acid derivatives.

Hydrogen bond donor

Nonpolar: Boiling point due solely to London dispersion forces

Acid chloride and ester boiling points are closer to alkanes despite being moderately polar molecules.

Nitriles and alcohols have similar boiling points.

Boiling point *increases*: alkane < acid chloride ≈ ester < nitrile ≈ alcohol < carboxylic acid < amide

Amide boiling points *increase* as the number of hydrogen bond donors increases.

Functional group	Structure	Boiling point (C°)
Alkane		36
Acid chloride	O, C, Cl	52
Ester	O, C, OCH_3	57
Nitrile	C≡N	117
Alcohol	OH	118
Carboxylic acid	O, OH	141
3° Amide	O, C, H, N	153
2° Amide	O, C, N, H	202
1° Amide	O, C, N, H, H	213

* The molecular weight range is from 69-79 g/mol.

Although the carboxylic acid derivatives (except for nitriles) contain a polar carbonyl group, the boiling point of carboxylic acid derivatives may not be significantly impacted. For example, acid halides and esters, which are two of the more reactive carboxylic acid derivatives (Concept 11.2.01), have boiling points closer to that of an alkane than other carboxylic acid derivatives despite being moderately polar molecules. This is an example of how physical properties (eg, boiling point) and chemical properties (eg, reactivity) are independent of one another.

Nitriles and alcohols have similar boiling points, and both are higher than those of esters and acid halides. A nitrile experiences strong dipole-dipole interactions, whereas an alcohol participates in hydrogen bonding (Concept 8.1.03).

Carboxylic acids form stable *dimers* through hydrogen bonding (Concept 10.1.02), resulting in boiling points higher than those of alcohols or nitriles.

Amides are the *highest*-boiling carboxylic acid derivative. Their high boiling point is due to *both* hydrogen bonding and the strong dipole-dipole interactions resulting from charge separation due to resonance. The extent of hydrogen bonding is *greatest* for an amide with two N–H bonds, resulting in a boiling point *higher* than that of an amide with fewer N–H bonds.

Solubility

Low-molecular-weight esters and amides are water soluble, but solubility decreases with increasing carbon chain length. Similarly, low-molecular-weight nitriles are water soluble. Acetonitrile (CH_3CN) is miscible in water, whereas nitriles containing three or four carbon atoms are water soluble.

Lesson 11.2

Interconversion of Carboxylic Acid Derivatives

Introduction

Many of the chemical reactions of carboxylic acid derivatives involve preparing them from carboxylic acids or from one another (ie, interconversion). This lesson discusses the following aspects of the interconversion of carboxylic acid derivatives:

- The relative reactivity of carboxylic acid derivatives
- Mechanisms of interconversion reactions
- Interconversion reactions between carboxylic acid derivatives

11.2.01 Relative Reactivity of Carboxylic Acid Derivatives

Carboxylic acid derivatives generally react as Type 3 electrophiles (Concept 5.3.02) as the carbonyl carbon atom is electrophilic and capable of reaction with nucleophiles. Consequently, the electrophilicity of the carbonyl carbon atom is the central factor that determines the relative reactivity of the carboxylic acid derivatives.

Because the *difference* between the carbonyl-containing carboxylic acid derivatives lies in the identity of the group attached to the carbonyl carbon atom (ie, the Z group) (Figure 11.6), this group is primarily responsible for the *difference* in reactivity among the carboxylic acid derivatives. The Z group influences the electrophilicity of the carbonyl carbon atom in three ways:

1. The inductive effect
2. Resonance stabilization
3. Basicity of the leaving group

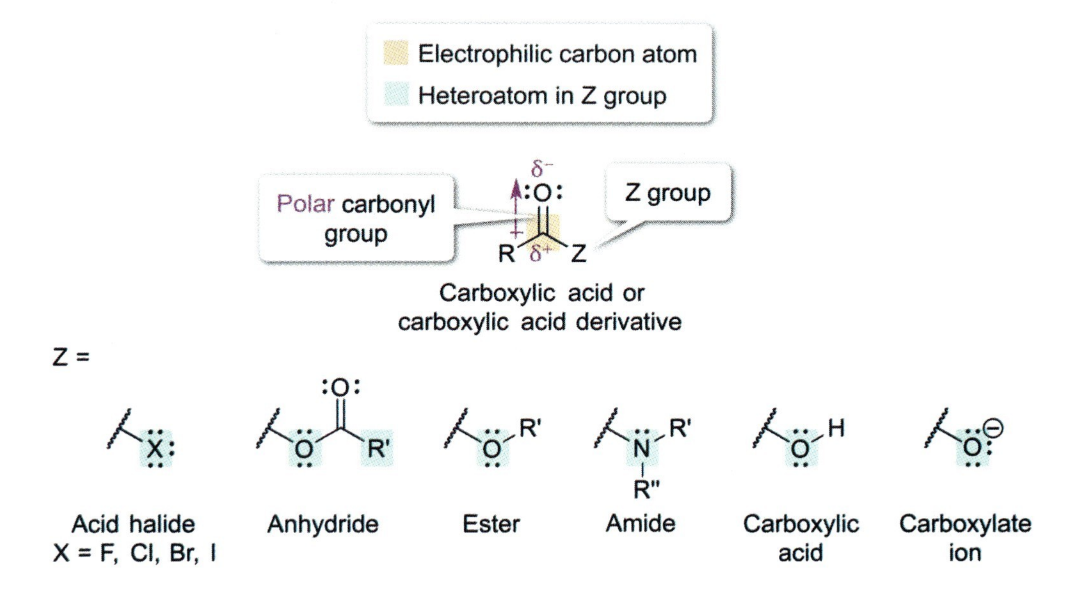

Figure 11.6 Structure of a carboxylic acid and the carbonyl-containing carboxylic acid derivatives.

The Inductive Effect

A carboxylic acid derivative has a Z group containing a heteroatom directly attached to the carbonyl carbon atom. *All* the heteroatoms within a Z group are *more* electronegative than carbon and participate in a polar carbon-heteroatom bond (Figure 11.7). Due to the inductive effect, the polar carbon-heteroatom bond *increases* the electrophilicity of the carbonyl carbon atom.

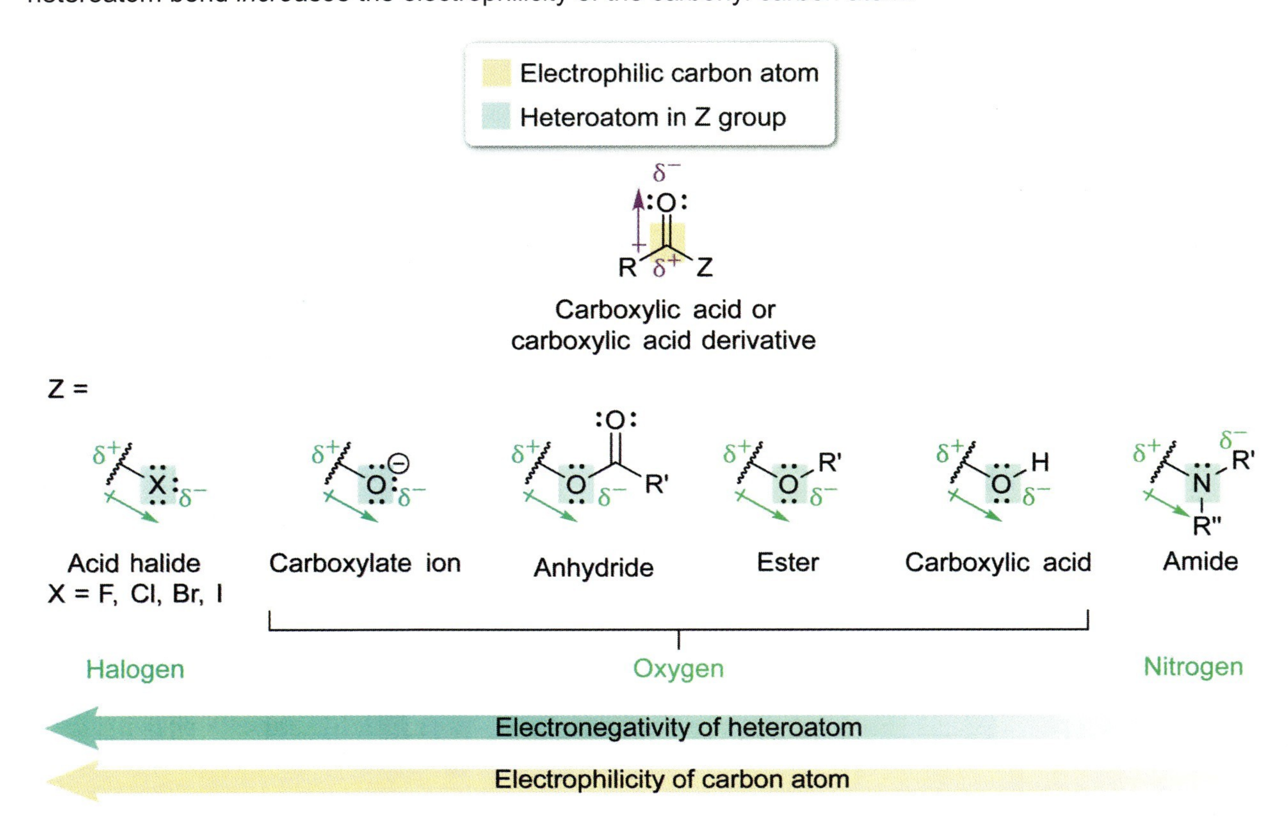

Figure 11.7 Impact of the inductive effect on the electrophilicity of carboxylic acid derivatives.

Based *solely* on the inductive effect, a Z group containing a halogen should be the most reactive, followed by Z groups containing an oxygen, and then Z groups containing a nitrogen. The additional lone pair on the oxygen atom in a carboxylate ion further increases the magnitude of its inductive effect.

Resonance Stabilization from Lone Pairs

The heteroatoms within the Z groups of carboxylic acid derivatives each contain at least one lone pair that is delocalized into the pi bond of the carbonyl through resonance (Figure 11.8). The *greater* the resonance stabilization of a carboxylic acid derivative, the *less* likely it will be to react as a Type 3 electrophile. *More* electronegative heteroatoms are *more* likely to hold onto their lone pairs, have *less* resonance stabilization, and are therefore *more* reactive. Based on this relationship, acid halides should have the highest reactivity, followed by anhydrides, esters, carboxylic acids, carboxylate ions, and then amides.

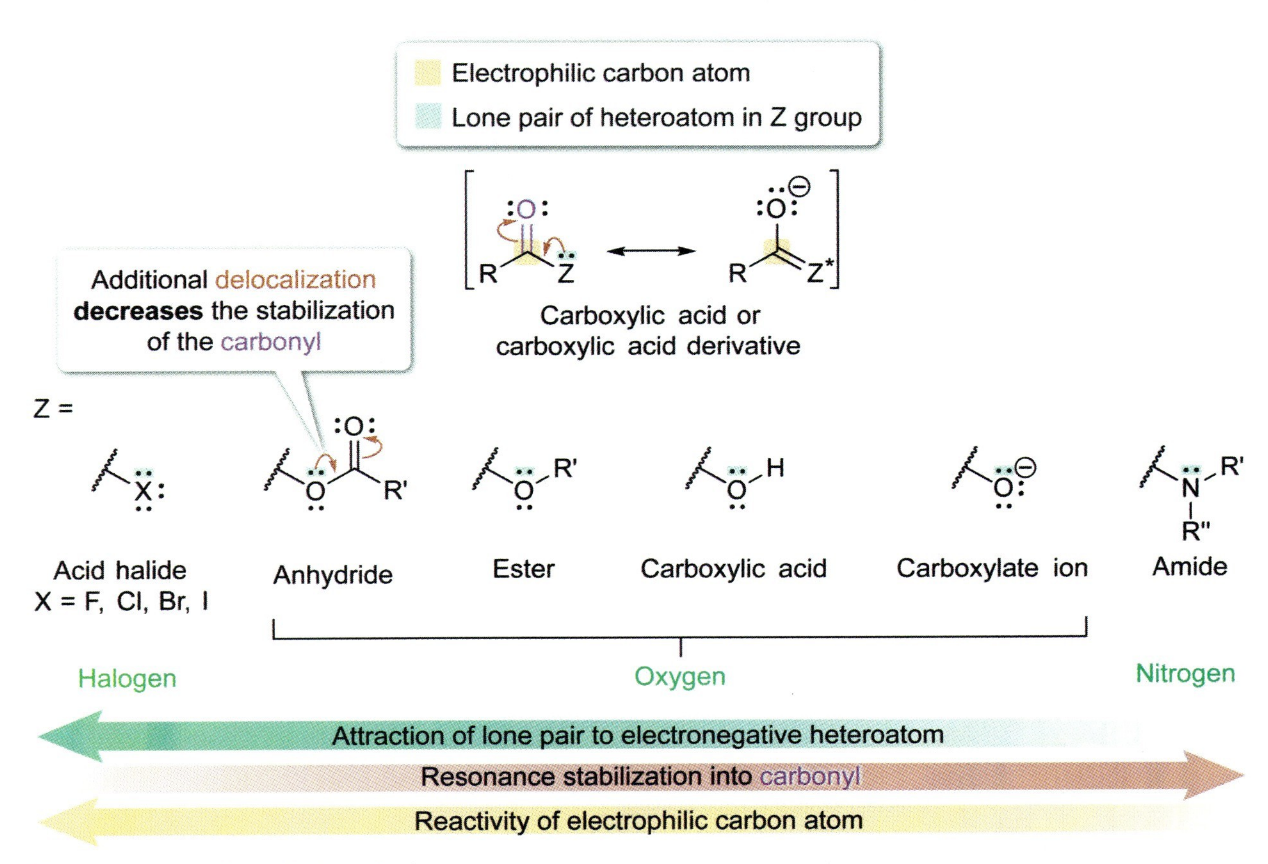

Figure 11.8 Impact of resonance stabilization on the reactivity of carboxylic acid derivatives.

The Basicity of the Leaving Group

One of the most important factors that influence the relative reactivity of the carboxylic acid derivatives is the ability of the Z group to act as a good leaving group in a nucleophilic acyl substitution reaction. As introduced in Concept 5.3.03, the best leaving groups are *weak* conjugate bases of strong acids. The strength of a base is quantified by its pK_b. Consequently, Z groups with a *higher* pK_b are the *weakest* bases, the *best* leaving groups, and have *higher* relative reactivity as carboxylic acid derivatives (Figure 11.9).

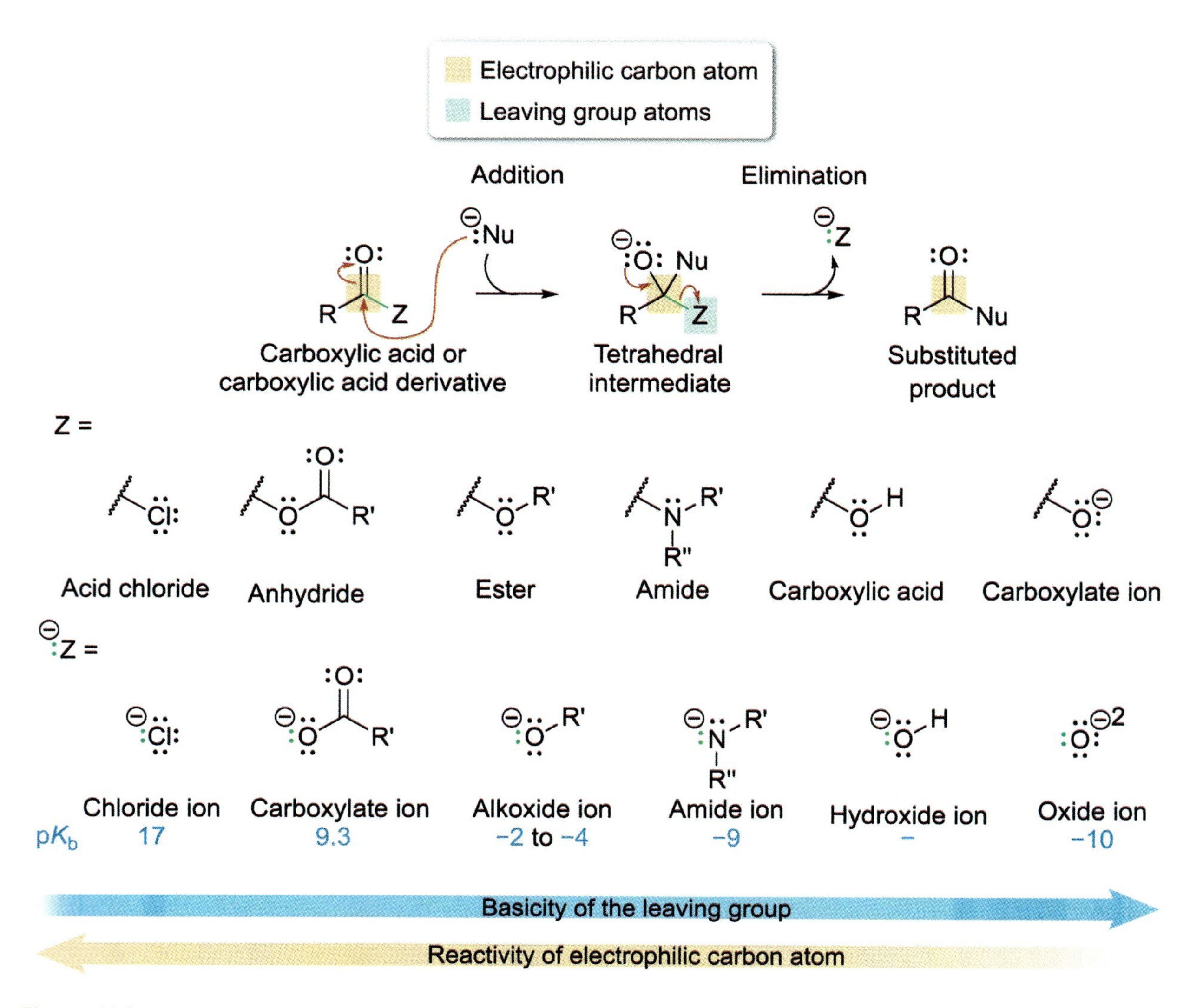

Figure 11.9 Impact of leaving group basicity on the reactivity of a carboxylic acid derivative.

Summary of the Relative Reactivity of Carboxylic Acid Derivatives

After factoring in the impact of the inductive effect, resonance stabilization, and the basicity of the leaving group, the reactivity trend of carboxylic acid derivatives is revealed (Figure 11.10). The most reactive carboxylic acid derivatives are acid halides. Although anhydrides have nearly the same reactivity profile as acid halides, anhydrides are much less frequently encountered in chemical synthesis due to the added steps required for their preparation (Concept 11.2.03). The next most reactive carboxylic acid derivative are the esters, which are significantly more reactive than amides due to all three factors.

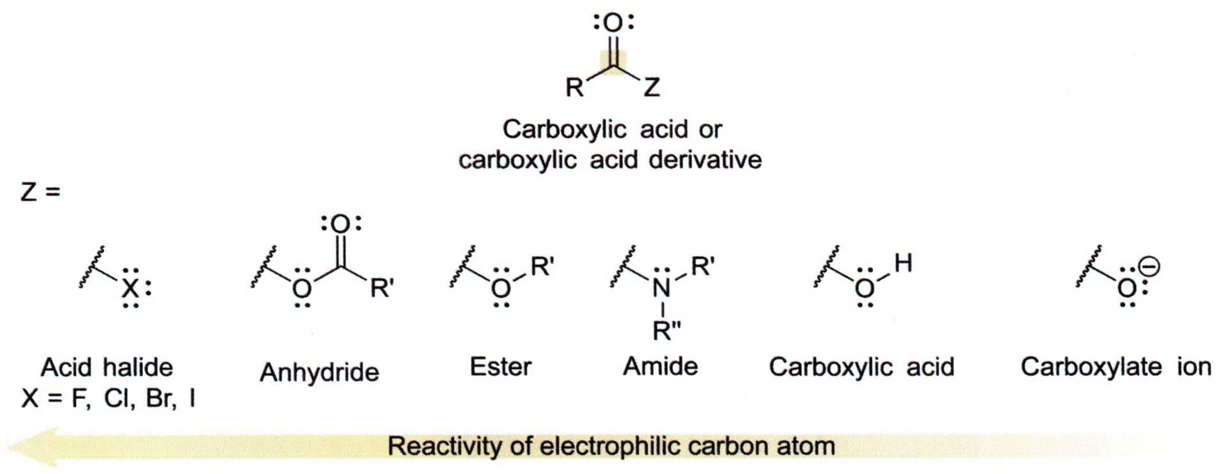

Carboxylic acid or carboxylic acid derivative

Z =

Acid halide X = F, Cl, Br, I | Anhydride | Ester | Amide | Carboxylic acid | Carboxylate ion

Reactivity of electrophilic carbon atom

Figure 11.10 Relative reactivity (electrophilicity) of carboxylic acid derivatives.

The most reactive carboxylic acid derivative that remains stable to the aqueous environment of a biological system is the ester (or its sulfur derivative, a thioester). Acid halides and anhydrides are typically too electrophilic and rapidly react with water.

11.2.02 Nucleophilic Acyl Substitution Revisited

As introduced in Concept 10.3.01, the nucleophilic acyl substitution mechanism is the *primary* reaction mechanism of carboxylic acids and carboxylic acid derivatives and involves different steps under acidic conditions and basic conditions. Although these general mechanisms are applicable to the reactions of carboxylic acid derivatives, additional nuances must be considered related to the relative electrophilicity of the carboxylic acid derivatives.

Highly electrophilic carboxylic acid derivatives generally react through a *basic* mechanism (Table 11.2). These derivatives do *not* require an activation event (eg, addition of a strong acid) before nucleophilic attack. If a neutral nucleophile (eg, water, an alcohol) reacts with one of these derivatives, the basic mechanism typically requires a final deprotonation step immediately after the elimination step. Furthermore, reactions with highly electrophilic derivatives do not usually require heat to speed up the rate of reaction.

Table 11.2 Nucleophilic acyl substitution reaction parameters for carboxylic acid derivatives.

Electrophilicity	Carboxylic acid derivative	Acidic or basic mechanism	Requires heat
High	Acid halide Anhydride	Basic*	No
Intermediate	Ester	Acidic	Yes
		Basic	No
Low	Amide	Acidic or basic	Yes

* If reacted with a neutral nucleophile, an additional deprotonation step may be required following the elimination step.

Esters (intermediate electrophilicity) and amides (low electrophilicity) undergo nucleophilic acyl substitution reactions through either an acidic or basic mechanism. Weak nucleophiles typically follow acidic mechanisms, whereas stronger nucleophiles typically follow basic mechanisms. For esters, acidic conditions often require the addition of heat, and most reactions involving amides require heat and react slowly.

Although *many* specific nucleophilic acyl substitution reactions of carboxylic acid derivatives are discussed in Concept 11.2.03 and Lesson 11.3, few specific mechanisms are depicted. Instead, *patterns* of the described reactions related to both the *general* mechanisms (acidic and basic) and the parameters described in this concept are discussed.

11.2.03 Interconversion Reactions Between Carboxylic Acid Derivatives

Figure 11.11 provides a summary of the reactions that interconvert the carboxylic acid derivatives and carboxylic acids. Carboxylic acids and acid chlorides play *central* roles in the interconversion process, as the reaction of a carboxylic acid with thionyl chloride to produce an acid chloride is the *only* reaction that converts a functional group with *lower* reactivity to a functional group with *higher* reactivity. All other interconversion reactions are *only* favorable if a higher reactivity functional group is converted to a lower reactivity functional group.

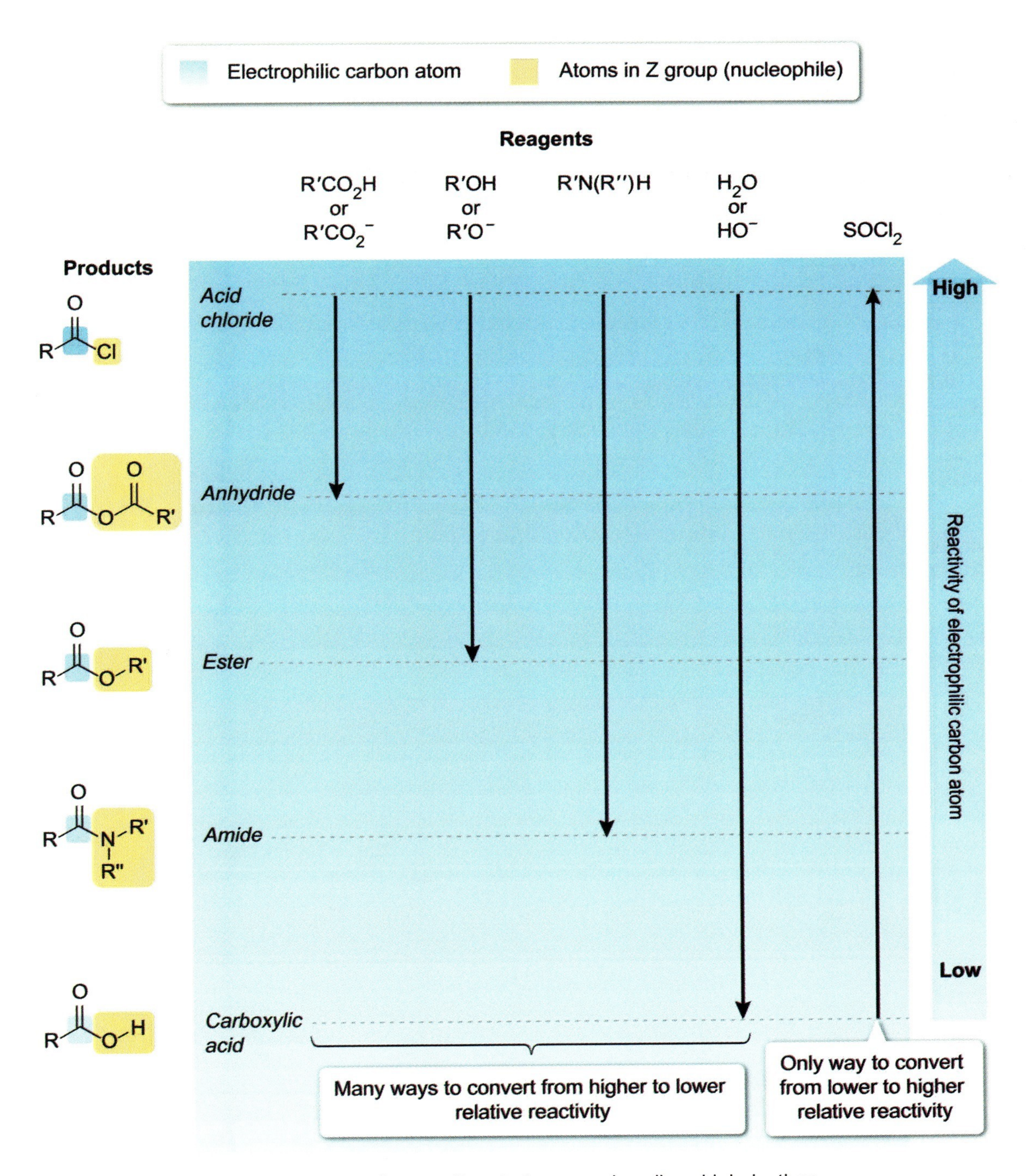

Figure 11.11 Summary of the interconversion reactions between carboxylic acid derivatives.

Although the steps of the nucleophilic acyl substitution reaction mechanism may differ, many of the interconversion reactions can be performed with either the conjugate acid or conjugate base form of the nucleophile. The only exceptions are conversions to an amide, which are often performed using an amine (the conjugate acid), rather than an amide ion (the conjugate base).

Synthesis of Anhydrides

Anhydrides are commonly formed by first converting a carboxylic acid to an acid chloride using thionyl chloride, and then reacting the acid chloride with either a carboxylic acid or a carboxylate ion (Figure 11.12). If a carboxylic acid is used for the second step, a molecule of HCl, a strong acid, is released and a sterically hindered, non-nucleophilic base (eg, a tertiary amine like triethylamine, $(CH_3CH_2)_2NCH_2CH_3$) is typically added to react with the evolved acidic proton.

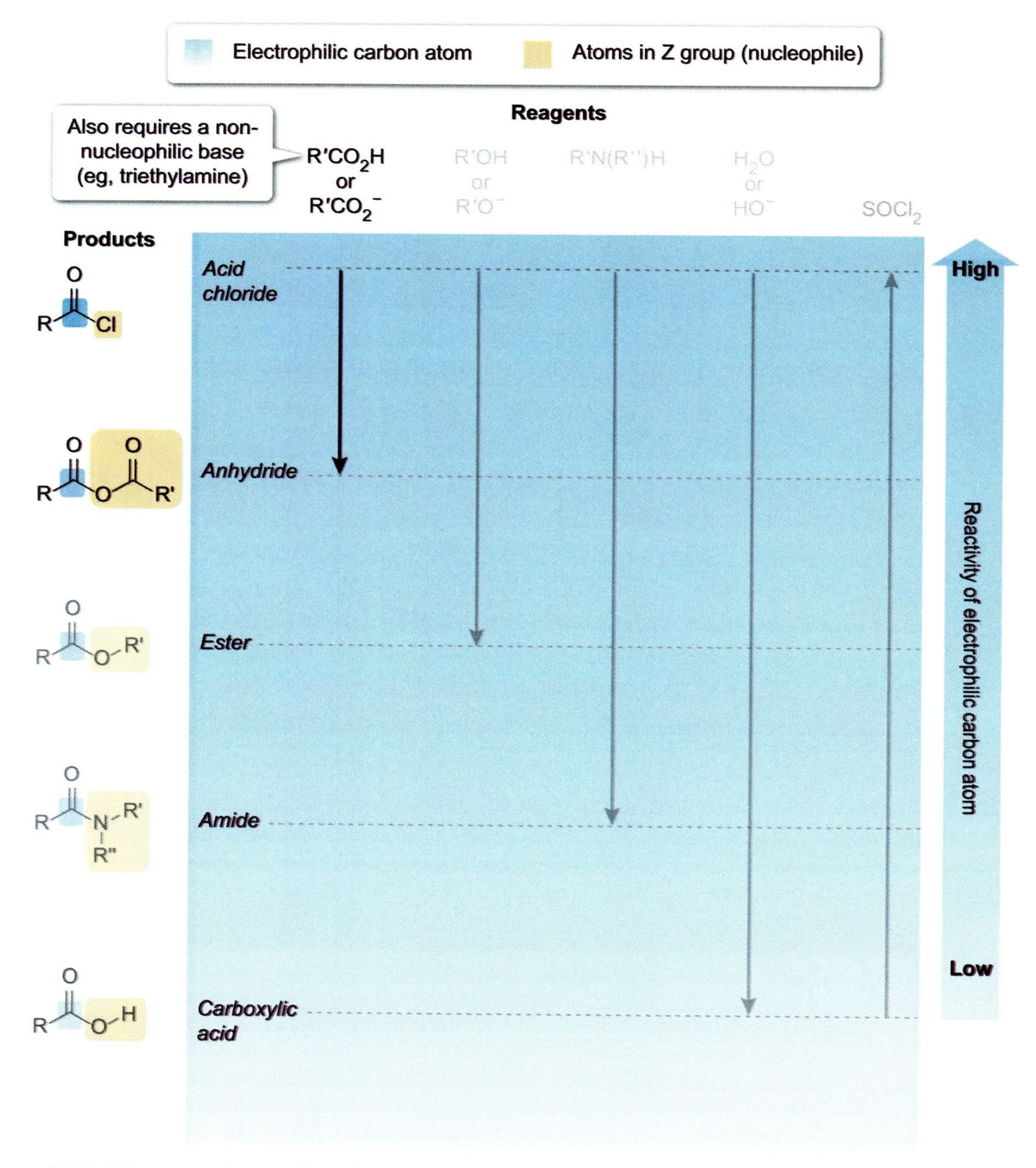

Figure 11.12 Interconversion reactions that form an anhydride.

Synthesis of Esters

An ester can be prepared through the reaction of either an acid chloride or an anhydride with a nucleophilic alcohol or alkoxide ion (Figure 11.13). The nucleophilic acyl substitution mechanism starts with nucleophilic attack of the electrophilic carbon atom (ie, the basic mechanism). If an alcohol is the nucleophile, a non-nucleophilic base (eg, triethylamine) is often added to react with the acidic proton from the oxonium ion intermediate.

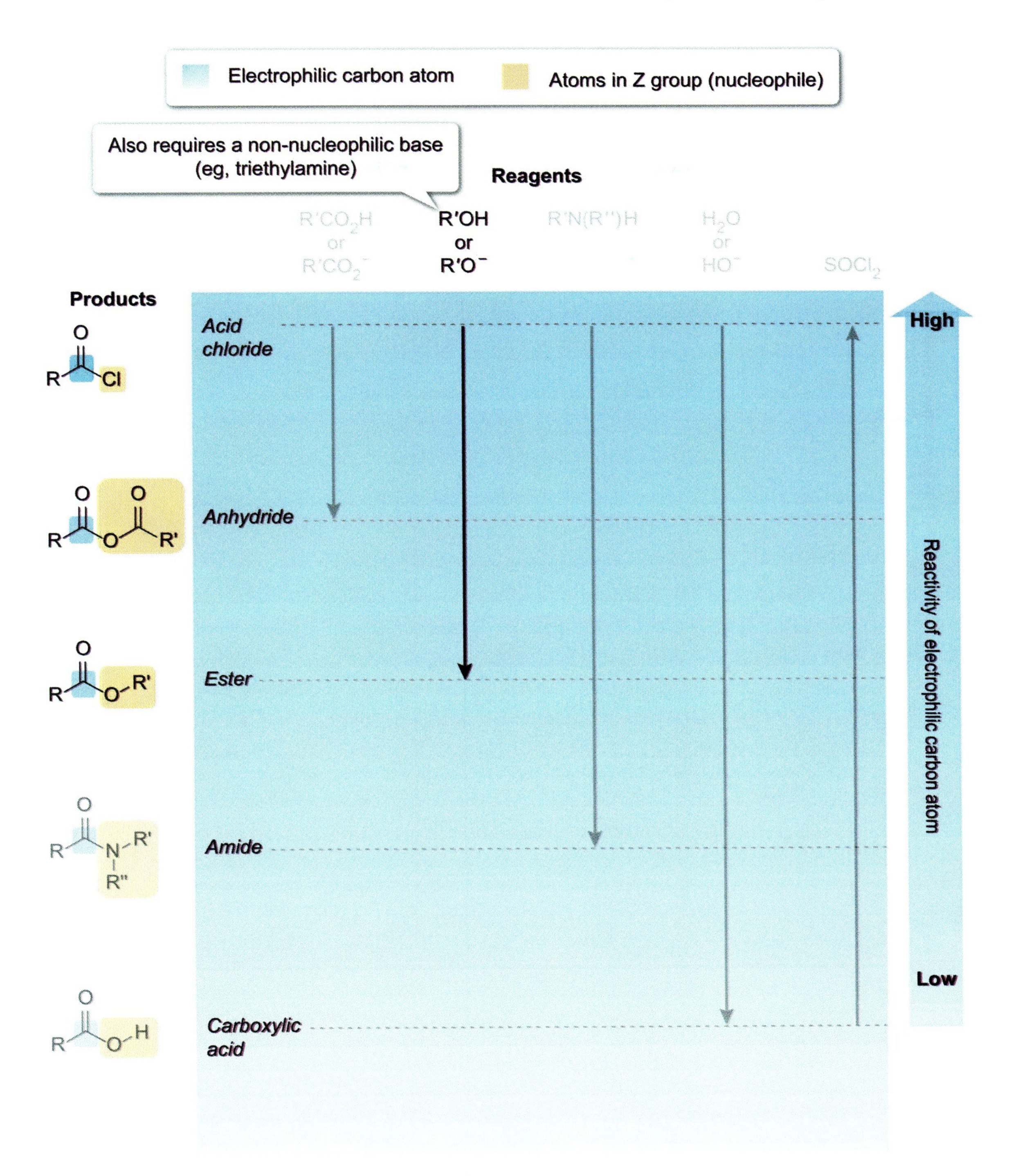

Figure 11.13 Interconversion reactions that form an ester.

The synthesis of an ester is sometimes called an **acyl transfer reaction**, particularly if the alcohol or alkoxide ion constitutes the greater portion of the final ester structure relative to the carbonyl-derived portion (eg, ethanoyl [acetyl], $CH_3C(O)–$). Acyl transfer reactions are frequently encountered within biochemistry.

Synthesis of Amides

As the *least* electrophilic carboxylic acid derivative, an amide can be prepared from any other carboxylic acid derivative (Figure 11.14). The nucleophile in each of these interconversions is an amine. If an acid chloride or anhydride is the electrophile, the reaction proceeds through a basic mechanism with a final deprotonation step and does not generally require heat. Conversely, the reaction of an amine with an ester (an **ammonolysis**) is a *slower* process that is often aided with heat.

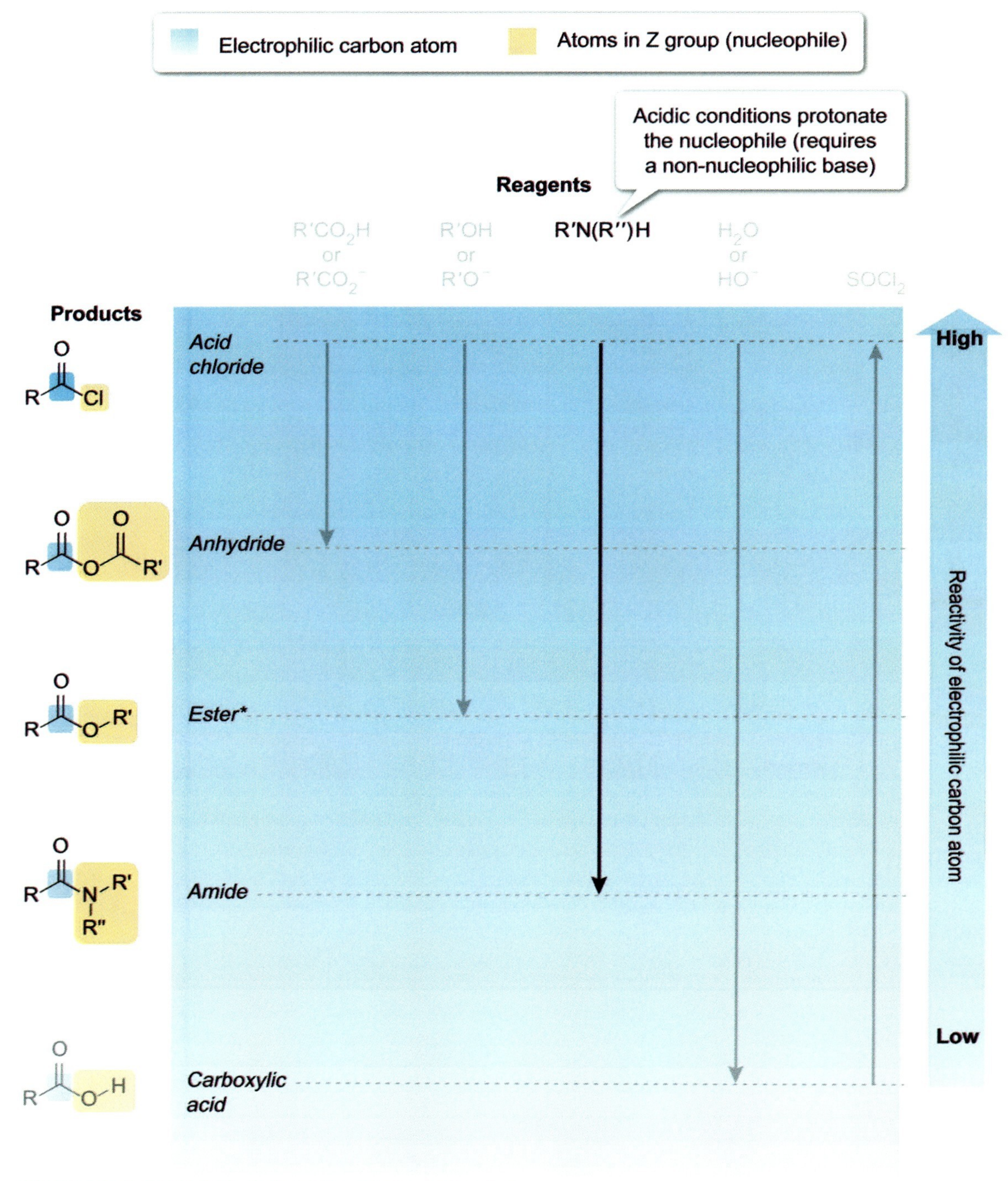

Figure 11.14 Interconversion reactions that form an amide.

Reactions to prepare amides generally *cannot* be conducted in the presence of acid, as the acid is likely to protonate the amine to form a non-nucleophilic ammonium ion. Rather, most reactions to prepare amides require either a sacrificial, non-nucleophilic base (eg, triethylamine) or an excess of amine reactant to promote proton transfers that keep the reaction at a basic pH.

Concept Check 11.1

Is the following reaction likely to produce the indicated product in high yield? Why or why not?

$$\text{N-ethylbenzamide} \xrightarrow[\text{25 °C}]{H_3C{-}O^{\ominus}\ Na^{\oplus},\ H_3C{-}OH} H_3C{-}O{-}C(=O){-}C_6H_5$$

Solution

Note: The appendix contains the answer.

Lesson 11.3

Other Reactions of Carboxylic Acid Derivatives

Introduction

In addition to interconversion reactions, carboxylic acid derivatives can also undergo a variety of other chemical transformations, including hydrolysis, transesterification, and reduction reactions.

11.3.01 Hydrolysis to a Carboxylic Acid

All carboxylic acid derivatives can be hydrolyzed back to the parent carboxylic acid. A **hydrolysis reaction** is a reaction that consumes a molecule of water to *break* (ie, lyse) a sigma bond (Figure 11.15). Typically, the atoms in water are added to each side of the broken sigma bond. A hydrolysis reaction is the reverse process of a condensation reaction, which *evolves* a molecule of water as a byproduct.

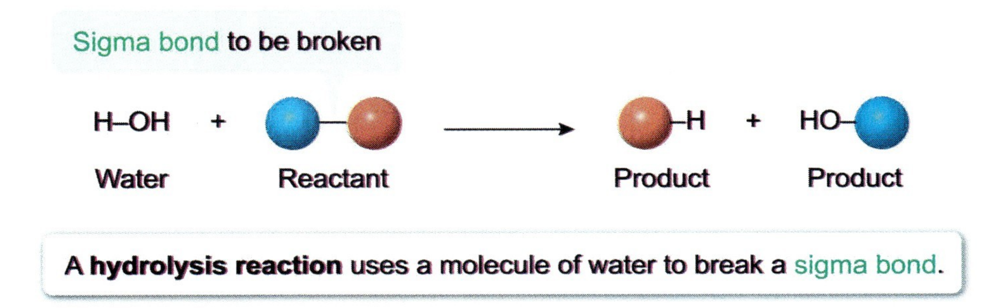

Figure 11.15 A general hydrolysis reaction.

Hydrolysis of a carboxylic acid derivative breaks the sigma bond between the carbonyl carbon atom and the Z group through a nucleophilic acyl substitution mechanism (either acidic or basic variation) (Figure 11.16). Acidic hydrolysis involves water and either a strong Brønsted-Lowry acid (eg, HCl, H_2SO_4) or a strong Lewis acid (eg, BF_3) catalyst, whereas basic hydrolysis reaction involves water and a source of the hydroxide ion. When the hydroxide ion is used, the initial hydrolysis product is a carboxylate ion, as the excess base deprotonates the carboxylic acid. Aqueous workup using acid (eg, 1M HCl) produces the final carboxylic acid, which is typically the isolated product.

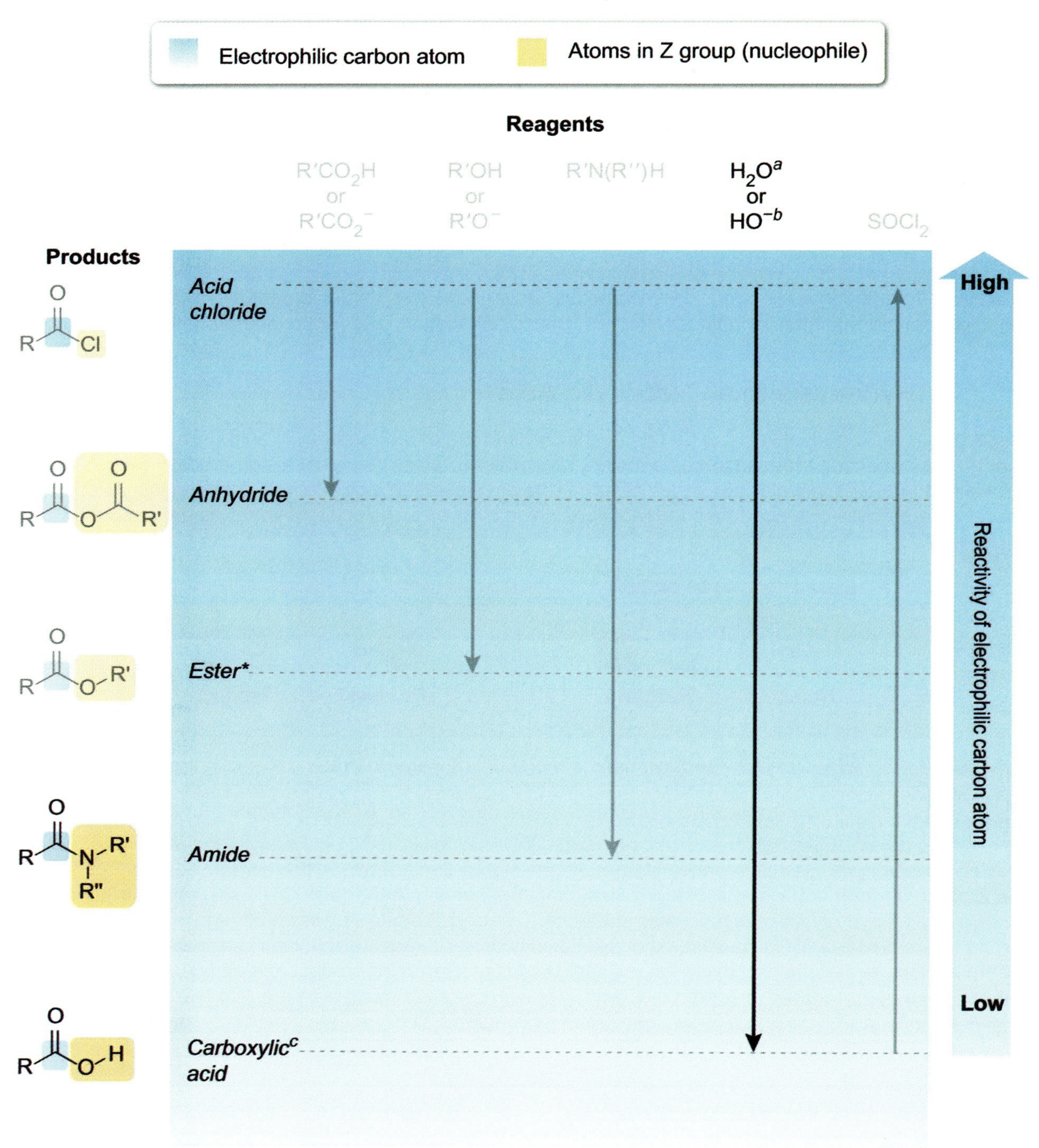

[a] Esters and amides require both heat and an acid catalyst (H^+).
[b] Amides require heat.
[c] If the hydroxide ion is used, the reaction requires an acidic workup

Figure 11.16 Hydrolysis of carboxylic acid derivatives to a carboxylic acid.

Because acid chlorides and anhydrides are the most electrophilic carboxylic acid derivatives, their hydrolysis reactions proceed through a basic mechanism in which a molecule of water or hydroxide ion directly attacks the electrophilic carbon atom; a strong acid is not required for additional activation of the carbonyl group toward attack.

Hydrolysis of an ester or amide can proceed under acidic or basic conditions. Acidic hydrolysis of an ester is the reverse mechanism of the Fischer esterification and requires both a strong acid catalyst and heat. Basic hydrolysis of an ester is typically faster and may not require heating. An amide is less electrophilic than an ester; therefore, both the acidic hydrolysis and basic hydrolysis of an amide proceed at much slower rates and generally require prolonged heating.

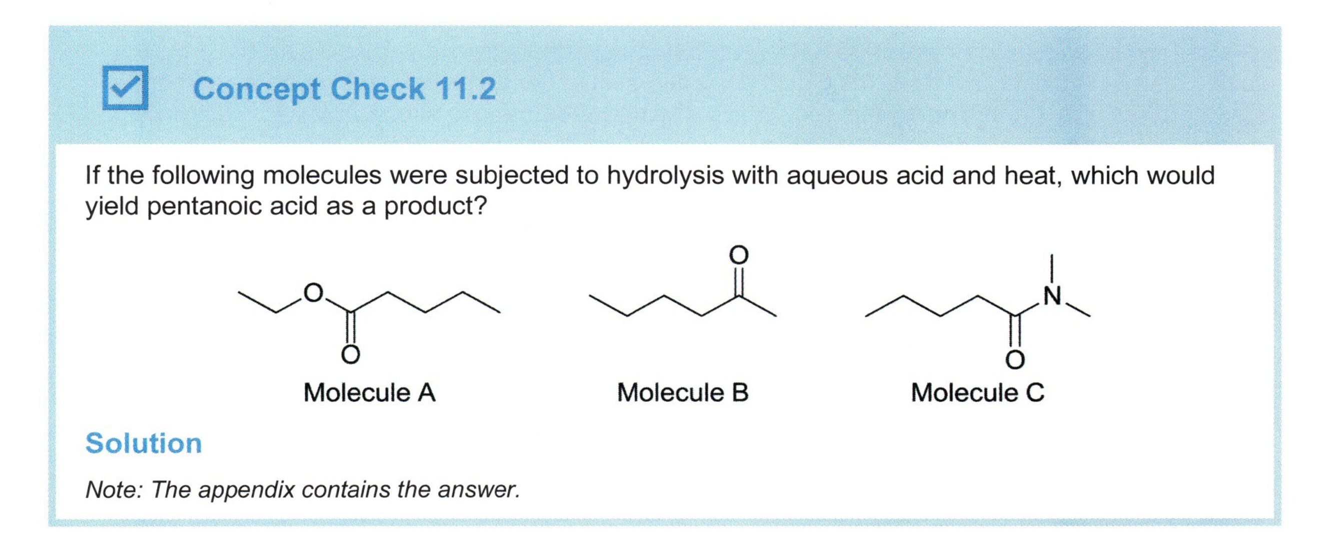

Concept Check 11.2

If the following molecules were subjected to hydrolysis with aqueous acid and heat, which would yield pentanoic acid as a product?

Solution

Note: The appendix contains the answer.

11.3.02 Transesterification

Transesterification is a chemical process in which the alkoxy group (–OR') of an ester is substituted for a *different* alkoxy group (ie, an ester reactant becomes a *different* ester product). A transesterification reaction is a type of acyl transfer reaction (see Concept 11.2.03).

A generalized transesterification reaction occurs when an ester reacts with either a nucleophilic alcohol (the conjugate acid form) or alkoxide ion (the conjugate base form) (Figure 11.17) to form a *different* ester and a *different* alcohol or alkoxide ion. In both cases, the transesterification reaction is a dynamic equilibrium; an *excess* of alcohol or alkoxide ion is required to shift the equilibrium toward increased formation of product.

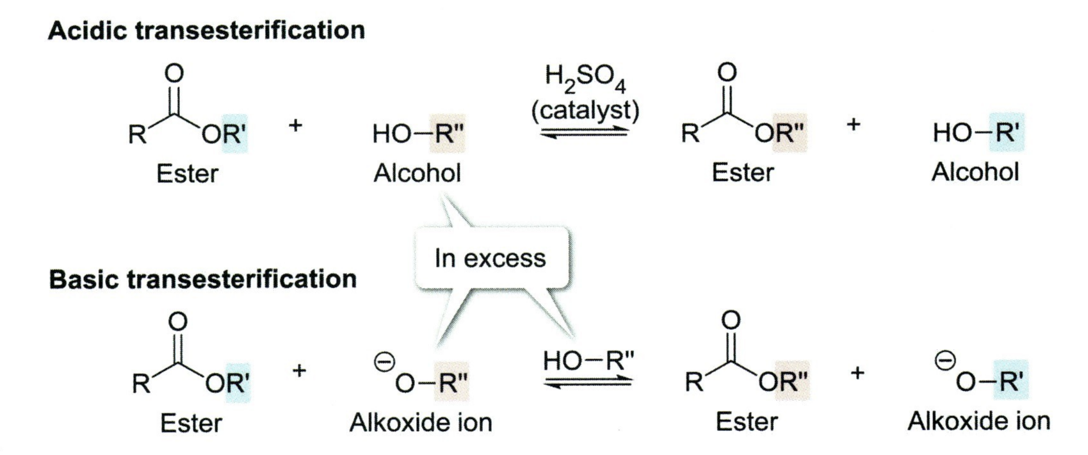

Figure 11.17 General transesterification reactions under acidic conditions and basic conditions.

The mechanism of transesterification occurs through a nucleophilic acyl substitution reaction, which can proceed under *either* acidic or basic conditions. The acidic transesterification mechanism is *nearly identical* to the mechanism for the Fischer esterification and requires a strong acid catalyst (see Concept

10.3.02). If an alkoxide ion is used as the nucleophile to promote a transesterification reaction, the transesterification mechanism proceeds through a basic nucleophilic acyl substitution mechanism.

11.3.03 Reduction of Carboxylic Acid Derivatives

Lithium Aluminum Hydride and Sodium Borohydride Reductions

Like a carboxylic acid (see Concept 10.3.03), carboxylic acid derivatives can also be reduced by a reducing agent (Table 11.3). Carboxylic acid derivatives have the same oxidation state as carboxylic acids; therefore, most reducing agents (eg, $LiAlH_4$, $NaBH_4$) promote *two* successive two-electron reductions that effectively add *two* equivalents of hydride ion to the reactant.

Table 11.3 The reduction of carboxylic acid derivatives by $LiAlH_4$ or $NaBH_4$.

Atoms lost during the reduction

Only the **most electrophilic** carboxylic acid derivatives are reduced by $NaBH_4$ (acid chloride, anhydride)

Results from a **difference** in the mechanism of reduction (amide)

Carboxylic acid derivative	Reactant	Reaction sequence	Reduced product
Acid chloride	R–C(=O)–Cl	1. 2 $NaBH_4$ 2. H_3O^+	1° Alcohol (R–CH₂–OH)
		1. 2 $LiAlH_4$ 2. H_3O^+	1° Alcohol
Anhydride	R–C(=O)–O–C(=O)–R'	1. 2 $NaBH_4$ 2. H_3O^+	1° Alcohol
		1. 2 $LiAlH_4$ 2. H_3O^+	1° Alcohol
Ester	R–C(=O)–O–R'	1. 2 $LiAlH_4$ 2. H_3O^+	1° Alcohol
Amide	R–C(=O)–N(R')–R'	1. 2 $LiAlH_4$	Amine (R–CH₂–N(R')–R')

All carboxylic acid derivatives can be reduced by the powerful reducing agent, lithium aluminum hydride ($LiAlH_4$). However, only the most electrophilic carboxylic acid derivatives (ie, acid chlorides, anhydrides) can be reduced by the milder reducing agent sodium borohydride ($NaBH_4$); esters and amides *cannot* easily be reduced by $NaBH_4$. Reduction of an acid chloride, anhydride, or ester provides a primary alcohol, whereas reduction of an amide provides an amine.

Although acid chlorides, anhydrides, and esters all lose the atoms that constitute their Z group when reduced, reduction of an amide results in the loss of only the oxygen atom of the carbonyl. The difference in the atoms lost during reduction is due to a difference in the mechanisms of reduction. An acidic workup is not typically required for the reduction of an amide by $LiAlH_4$ because the nitrogen in the amine product is basic and would become protonated under acidic conditions.

The mechanisms of reduction by $LiAlH_4$ and $NaBH_4$ are *identical* (aside from the source of the hydride ion); consequently, the mechanisms shown in this concept will use $LiAlH_4$ as a representative example. The mechanism for the $LiAlH_4$ reduction of acid chlorides, anhydrides, or esters has two stages (each requires one equivalent of $LiAlH_4$) (Figure 10.18):

1. Nucleophilic acyl substitution (basic mechanism)
2. Nucleophilic addition (basic mechanism)
   Note: Acidic workup neutralizes the anionic salts and excess $LiAlH_4$.

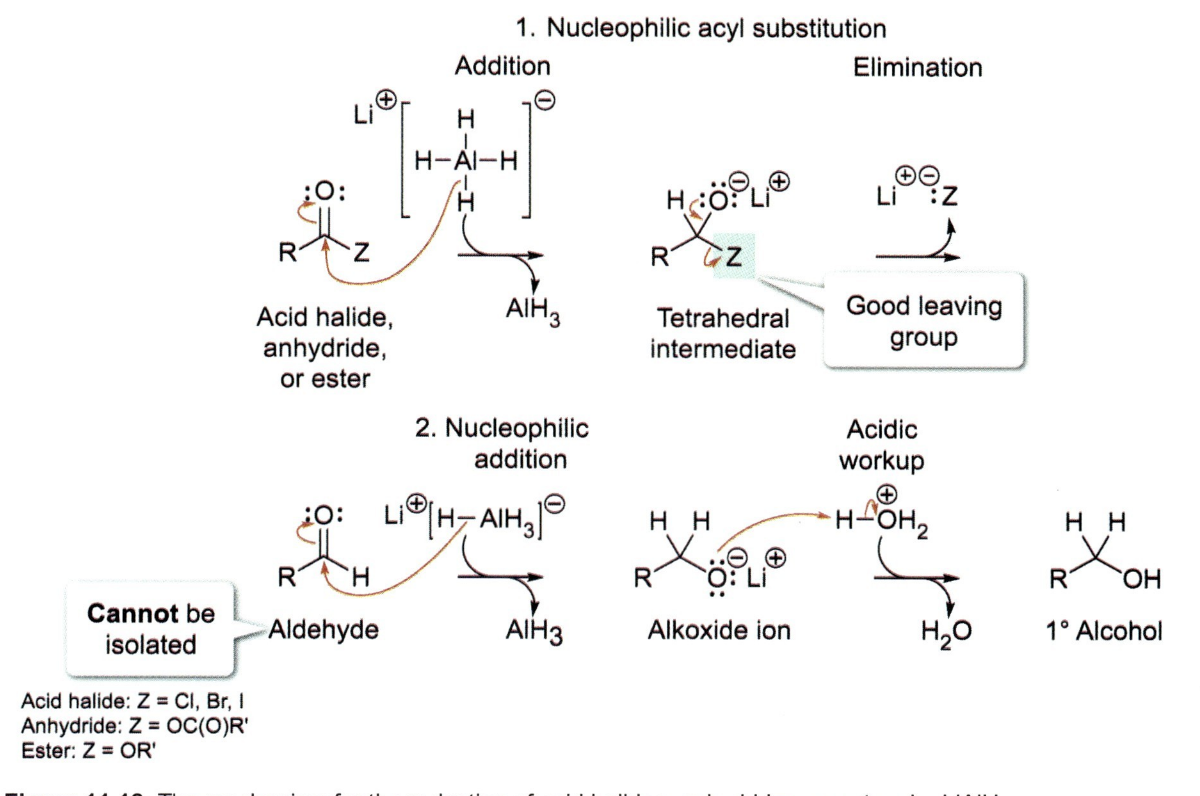

Figure 11.18 The mechanism for the reduction of acid halides, anhydrides, or esters by $LiAlH_4$.

One of the Al–H bonds donates its electrons to the electrophilic carbon atom in an addition step. Then, the tetrahedral intermediate reforms the pi bond and eliminates the Z group, a good leaving group under basic reaction conditions. A second molecule of $LiAlH_4$ undergoes a nucleophilic addition with the carbonyl of the aldehyde intermediate. The resulting alkoxide ion becomes protonated during the acidic workup to generate the 1° alcohol product.

Although the reduction of an amide with $LiAlH_4$ also proceeds through successive nucleophilic acyl substitution and nucleophilic addition reactions, the mechanism for an amide reduction features several key differences (Figure 11.19). Following the addition of hydride, the AlH_3 associates with the *oxygen* atom of the carbonyl, which turns the complex into a good leaving group. The lone pair on the nitrogen atom forms a pi bond and eliminates $LiOAlH_3^-$ to form an iminium ion intermediate.

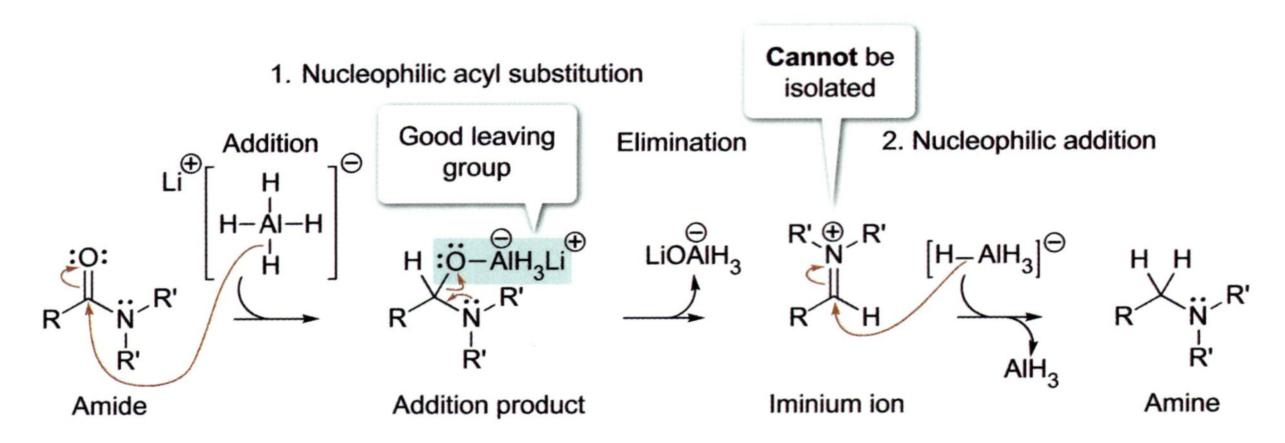

Figure 11.19 The mechanism for the reduction of an amide by $LiAlH_4$.

The iminium ion undergoes a nucleophilic addition with a second equivalent of $LiAlH_4$ to form the amine product. Through this reaction, 1°, 2°, and 3° amides are reduced to 1°, 2°, and 3° amines, respectively.

Selective Reduction with Lithium Tri(*tert*-butoxy)aluminum Hydride

The reducing agent lithium tri(*tert*-butoxy)aluminum hydride ($LiAlH(OtBu)_3$) allows for the *selective* reduction of certain carboxylic acid derivatives (eg, acid chloride) to aldehydes (Figure 11.20). A molecule of $LiAlH(OtBu)_3$ contains just *one* Al–H bond that can participate in the nucleophilic acyl substitution stage of the reduction mechanism (see Figure 11.19).

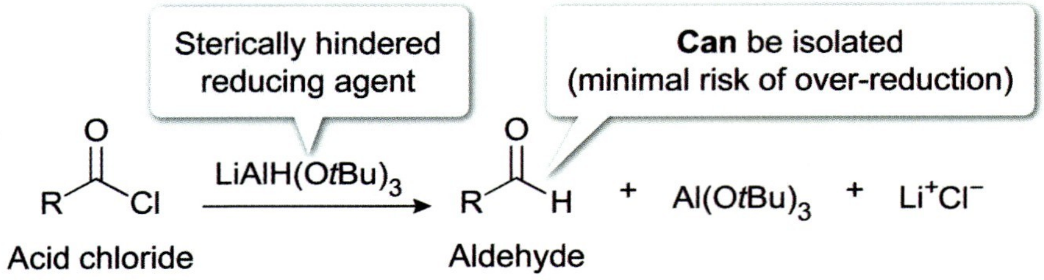

Sterically hindered reducing agent

Can be isolated (minimal risk of over-reduction)

$LiAlH(OtBu)_3$

+ $Al(OtBu)_3$ + Li^+Cl^-

Acid chloride

Aldehyde

Figure 11.20 The selective reduction of an acid chloride by $LiAlH(OtBu)_3$.

The three *tert*-butoxy groups are sterically hindered and *decrease* the rate of reaction with the strongly electrophilic acid chloride. Through careful control of both the reaction temperature and number of equivalents of $LiAlH(OtBu)_3$ added to the reaction, overreduction of the aldehyde intermediate to the 1° alkoxide ion can be minimized.

Lesson 12.1

Structure and Physical Properties of Amines and Amides

Introduction

As introduced in Concept 2.8.03, an **amine** is an organic derivative of ammonia (NH_3) in which one or more of the hydrogen atoms is replaced by an alkyl or aryl group. In contrast, an **amide** is an NH_3 derivative in which *one* of the hydrogen atoms is replaced by an acyl group.

This lesson discusses the general characteristics of amines and amides.

12.1.01 Structural Characteristics of Amines and Amides

An amine consists of a nitrogen atom attached to *three* other groups through σ bonds, with *at least one* carbon-nitrogen bond (see Concept 2.8.03). Because the nitrogen atom of an amine also has one pair of nonbonding electrons, the nitrogen atom has a total of four electron domains and is typically sp^3 hybridized (Figure 12.1).

However, when the nitrogen atom lone pair is conjugated with a π system (see Concept 6.5.01), the amine nitrogen atom will typically be sp^2 hybridized because the lone pair is delocalized within the π system through resonance (see Concepts 2.7.04 and 6.5.03). For example, the nitrogen atom of aniline has *mostly* sp^2 character.

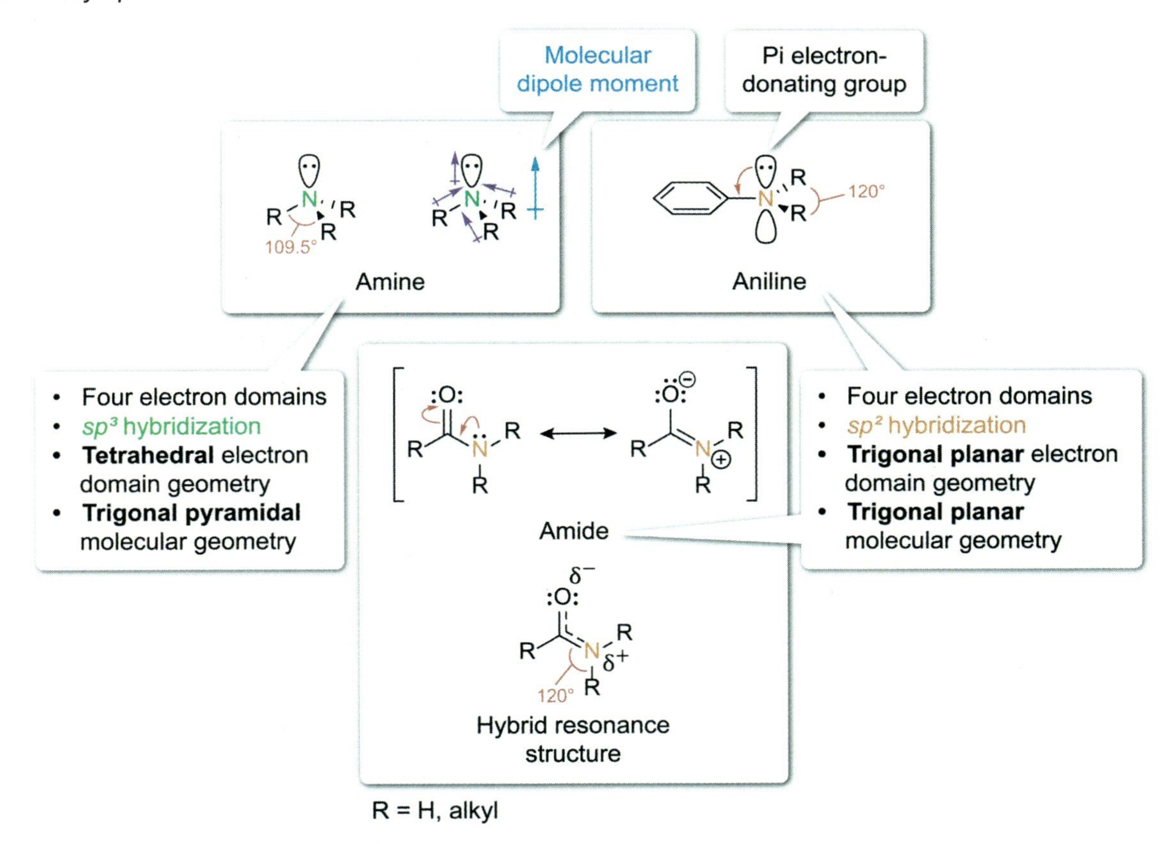

Figure 12.1 Structural characteristics of amines and amides.

Although the nitrogen atom of an amide's major resonance contributor may *appear* to have four electron domains, its lone pair is *inherently* conjugated with the adjacent carbonyl group (C=O). For this reason, the *delocalized* lone pair does *not* contribute an electron domain, resulting in the nitrogen atom of an amide being sp^2 hybridized and trigonal planar (see Concept 2.7.04).

Because nitrogen is less electronegative than oxygen and more electronegative than carbon or hydrogen, an amine is a *moderately* polar functional group with a molecular dipole moment oriented toward the lone pair of the amine. In contrast, the resonance forms of an amide create a hybrid resonance structure that contains a *partial* separation of charges (eg, δ^+, δ^-). Consequently, an amide is a *significantly* more polar functional group.

12.1.02 Classification of Amines and Amides

Amines can be further classified by the number of alkyl groups attached to the nitrogen atom, as shown in Figure 12.2. An ammonium ion is the only amine variant in which the nitrogen atom does not have a lone pair, as those electrons are required to form the fourth σ bond. Note that **ammonia (NH_3)** is an *inorganic* compound that contains *zero* alkyl groups and is *not* an amine.

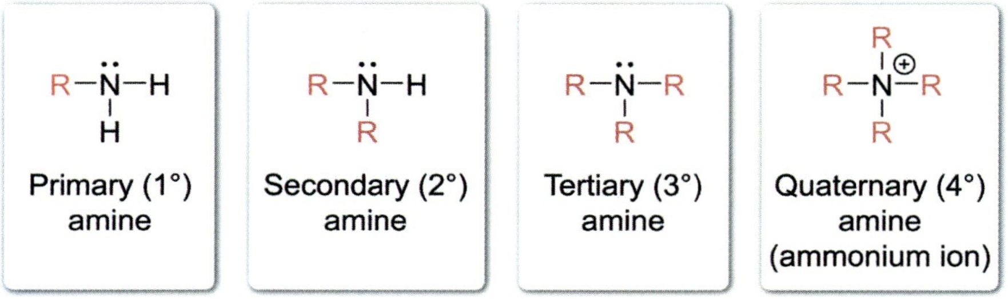

Primary (1°) amine | Secondary (2°) amine | Tertiary (3°) amine | Quaternary (4°) amine (ammonium ion)

Figure 12.2 Classification of amines.

The classification of amides follows a similar pattern as amines with the carbon atom of the *required* acyl group counting as one of the nonhydrogen groups attached to the nitrogen atom (Figure 12.3). The lone pair on the nitrogen atom is delocalized through resonance into the adjacent carbonyl group, making it challenging to use the nitrogen lone pair in an amide to form a fourth σ bond. For this reason, a quaternary (4°) amide is not possible.

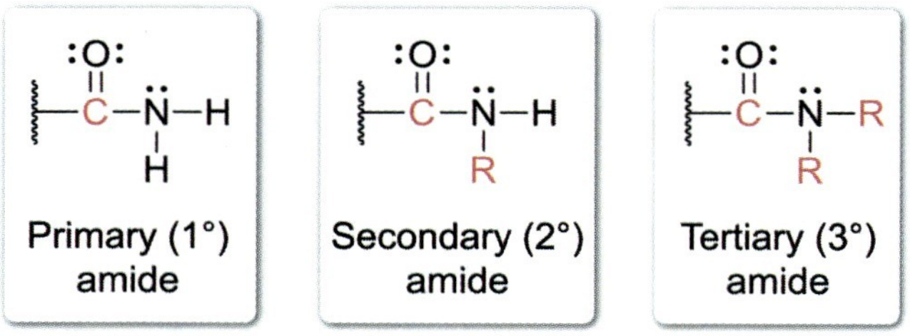

Primary (1°) amide | Secondary (2°) amide | Tertiary (3°) amide

Figure 12.3 Classification of amides.

12.1.03 Physical Properties of Amines and Amides

The physical properties of amines and amides are generally a function of the intermolecular forces present in a pure sample (see Lesson 2.3). The most influential intermolecular forces for these functional groups are dipole-dipole interactions resulting from their polar molecular character (Figure 12.4).

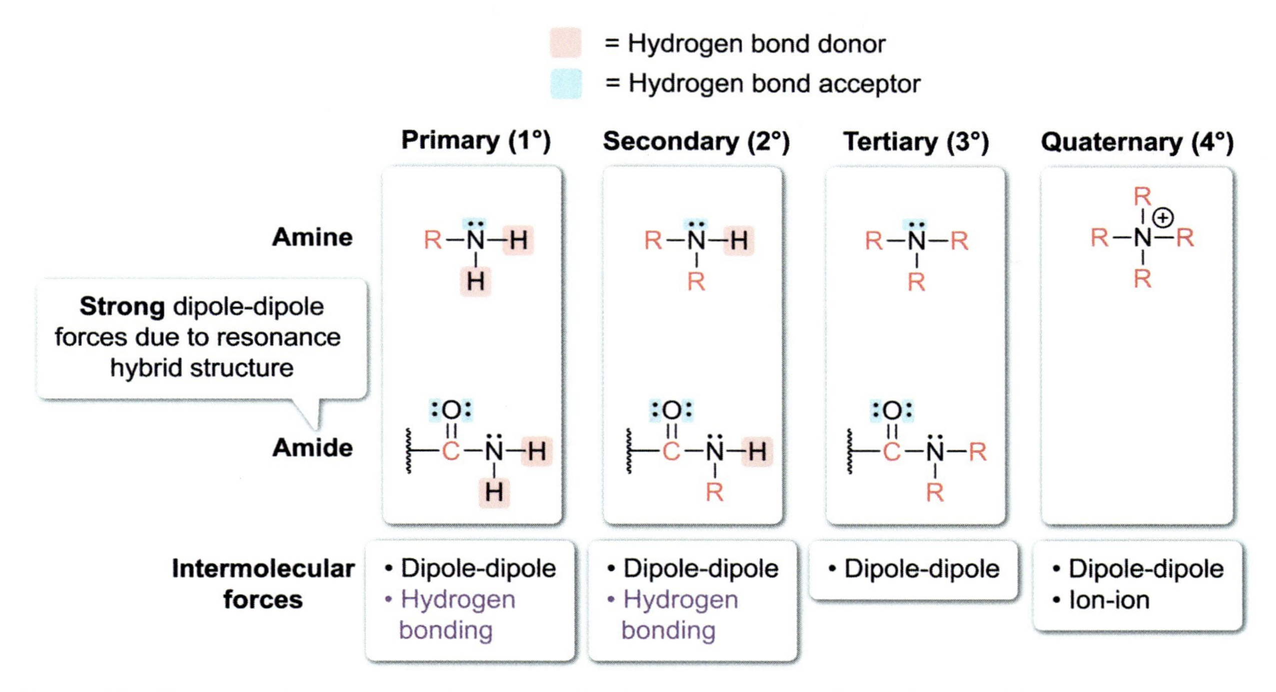

Figure 12.4 The intermolecular forces of amines and amides that most significantly impact their physical properties.

Certain classes of amines and amides can participate in a particularly strong type of dipole-dipole interaction called hydrogen bonding (see Concept 2.3.03). Primary and secondary amines and amides contain *both* hydrogen bond donors (ie, N–H bonds) and hydrogen bond acceptors (ie, the N of an amine or the O of an amide). The amide nitrogen atom *does not act* as a hydrogen bond acceptor due to the resonance delocalization of its *one* lone pair.

In contrast, 3° amines and amides have *zero* hydrogen bond donors and are *unable* to hydrogen bond. Likewise, a quaternary (4°) amine (eg, NR_4^+) also lacks the ability to hydrogen bond (it contains *zero* hydrogen bond donors *or* acceptors), but the nitrogen atom bears a *permanent* positive formal charge. For this reason, 4° amines demonstrate repulsive ionic interactions with other molecules of 4° amine and attractive interactions with oppositely charged counterions.

Although an amine is moderately polar, amides demonstrate much stronger dipole-dipole interactions due to resonance-induced separation of charge. Consequently, an amide is among the *most polar* organic functional groups, and the intermolecular forces in a sample of an amide tend to be *much* stronger than the intermolecular forces in a sample of an amine.

Boiling Point

The boiling point of a substance is proportional to the energy required to disrupt its liquid-phase intermolecular forces. Consequently, molecules with stronger intermolecular forces have higher boiling points.

In general, an amine is *less* polar than an alcohol due to the *lower* electronegativity of nitrogen relative to oxygen. Likewise, an amide is *more* polar than an alcohol due to the partial charges of its hybrid resonance structure. The data in Table 12.1 demonstrate that the boiling point *increases* from amine to alcohol to amide within the same classification level (eg, 1°, 2°, 3°).

Table 12.1 Boiling points of alkanes, amines, alcohols, and amides of comparable molecule weight.

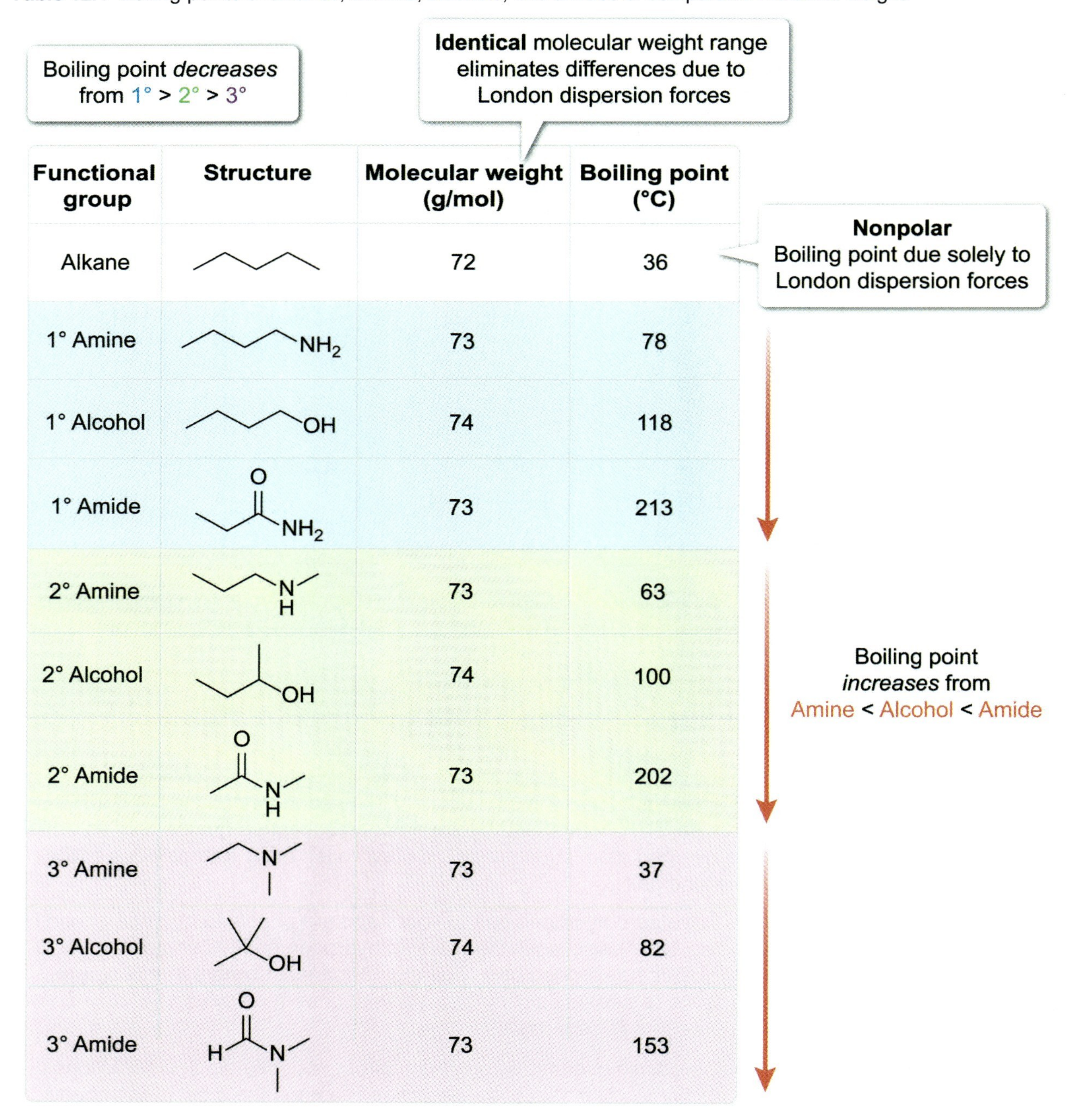

Functional group	Structure	Molecular weight (g/mol)	Boiling point (°C)
Alkane		72	36
1° Amine		73	78
1° Alcohol		74	118
1° Amide		73	213
2° Amine		73	63
2° Alcohol		74	100
2° Amide		73	202
3° Amine		73	37
3° Alcohol		74	82
3° Amide		73	153

Because 1° amines and amides have the *greatest* number of hydrogen bond donors (ie, N–H bonds), the extent of hydrogen bonding is *greater* and these functional groups have the *highest* relative boiling points. Secondary amines and amides have *fewer* hydrogen bond donors and therefore lower boiling points, whereas 3° amines and amides *are unable* to hydrogen bond and have the lowest boiling points. For the alcohols, the extent of hydrogen bonding is *unchanged,* and the decrease in boiling point from 1° to 2° to 3° is the result of decreased surface area (ie, more branching) resulting in weaker London dispersion forces (see Concept 8.1.03).

Solubility

As discussed in Concept 12.1.01, the N or O heteroatoms of an amine or amide are polar and the alkyl groups are nonpolar. Following the principle of "like dissolves like" (see Concept 2.4.02), *polar* bonds interact favorably with *polar* water and are *hydrophilic*. In contrast, *nonpolar* alkyl groups do not interact favorably with *polar* water and are *hydrophobic*.

Except for 4° amines, which are cationic and interact with water through an ion-dipole interaction, *all* classes of amines and amides have at least one hydrogen bond acceptor (ie, the nitrogen lone pair for amines and the oxygen lone pairs for amides) and have the capacity to hydrogen bond *with* water. Consequently, small amines and amides (ie, those containing six or fewer total carbon atoms) are soluble in aqueous solution but are *not* fully miscible in water. Aqueous solubility *decreases* with *increasing* carbon chain length, as the nonpolar character of larger alkyl groups increasingly overcomes the polar amine or amide functional group.

Odor

Amines tend to have a characteristic unpleasant odor, which can be described as the smell of rotting fish. In contrast, most amides and 4° amines *lack* the characteristic odor, in part due to their decreased volatility and higher boiling points.

Lesson 12.2

Synthesis of Amines and Amides

Introduction

The previous lesson discusses amines and amides in the context of structural features (Concept 12.1.01), classification terminology (Concept 12.1.02), and physical properties (Concept 12.1.03).

The focus of this lesson is a selection of reactions that are used to synthetically prepare amines and amides, with an emphasis on those most likely to be represented on the exam.

12.2.01 Amine Synthesis through Direct Substitution

An amine can be prepared through a bimolecular nucleophilic substitution reaction (ie, an S_N2 reaction) between a nitrogen atom nucleophile and an alkyl halide electrophile—a type of **alkylation reaction** (Figure 12.5). As with any S_N2 reaction, the electrophile *cannot* be a 3° alkyl halide.

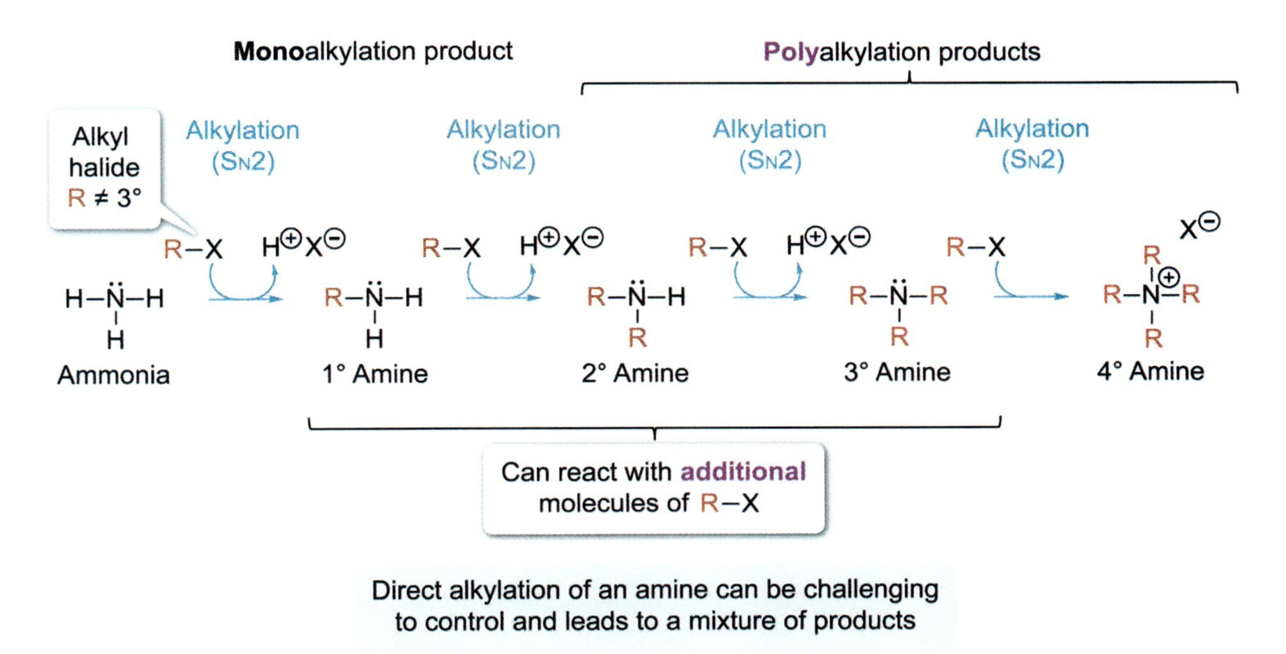

Figure 12.5 Direct alkylation of a nitrogen nucleophile with an alkyl halide.

Even if the number of reactant equivalents is carefully controlled, the direct alkylation of a nitrogen nucleophile tends to generate a *mixture* of products due to **polyalkylation** (ie, more than one alkylation reaction). For example, the S_N2 reaction of ammonia (NH_3) with an alkyl halide initially generates a 1° amine, which can itself act as a nucleophile for a *second* alkylation reaction to generate a 2° amine, and so on.

For these reasons, there are two high-yield ways to prepare an amine by *direct* S_N2 reaction (Figure 12.6). Both methods involve using an excess of *one* reactant. The first option is to use a large excess of the electrophilic alkyl halide to promote *polyalkylation* and the exclusive formation of the most highly substituted 4° amine.

The second option is to use a large excess of NH_3 to promote *monoalkylation* and the formation of a 1° amine. Excess NH_3 can be removed much more easily than can excess 1° or 2° amine reactant following the reaction. For this reason, this methodology is not applicable for the production of 2° or 3° amines from 1° or 2° amines, respectively.

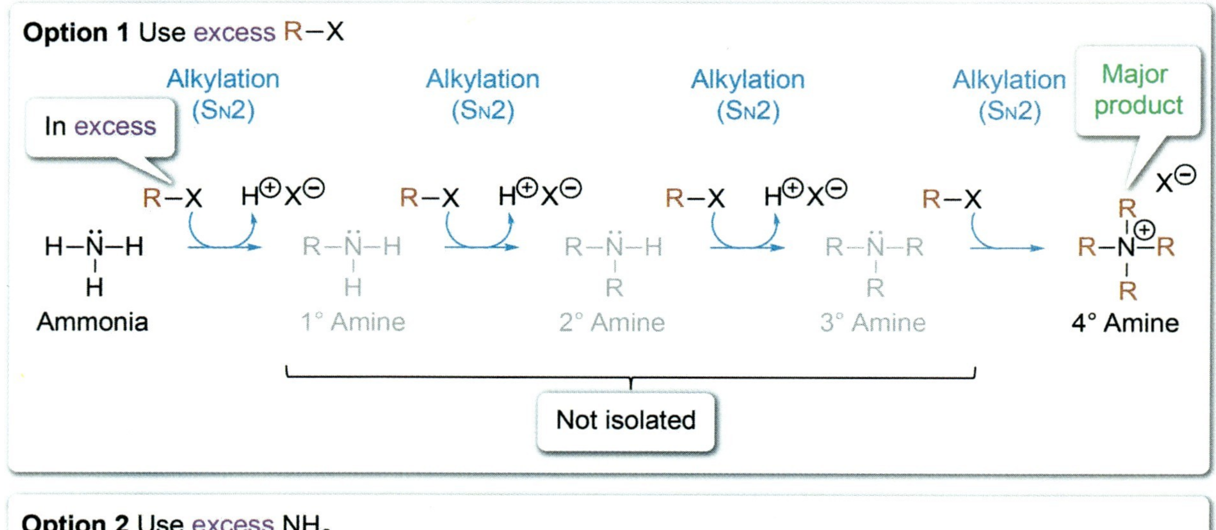

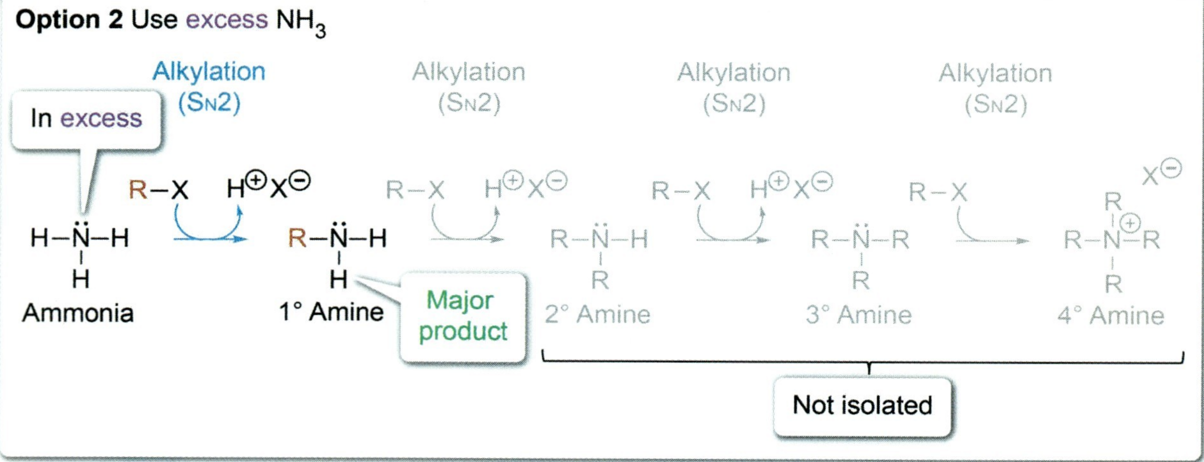

Figure 12.6 Useful direct alkylation reactions to prepare amines.

The Gabriel amine synthesis is an alternative option to avoid using excess reactant when preparing a 1° amine using an S_N2 reaction. This alternative is discussed in Concept 12.2.02.

12.2.02 The Gabriel Amine Synthesis

The Gabriel amine synthesis is one approach that circumvents both polyalkylation and the need to use *excess* nucleophile in the production of a 1° amine (as discussed in Concept 12.2.01). The nitrogen atom nucleophile for the Gabriel amine synthesis is the **phthalimide ion**, which is usually administered as a potassium ion salt (ie, potassium phthalimide) (Figure 12.7). The nitrogen atom in phthalimide is *much* more acidic than a normal amide N–H group because the negative charge of its conjugate base (ie, phthalimide anion) is delocalized through resonance with *both* carbonyl groups.

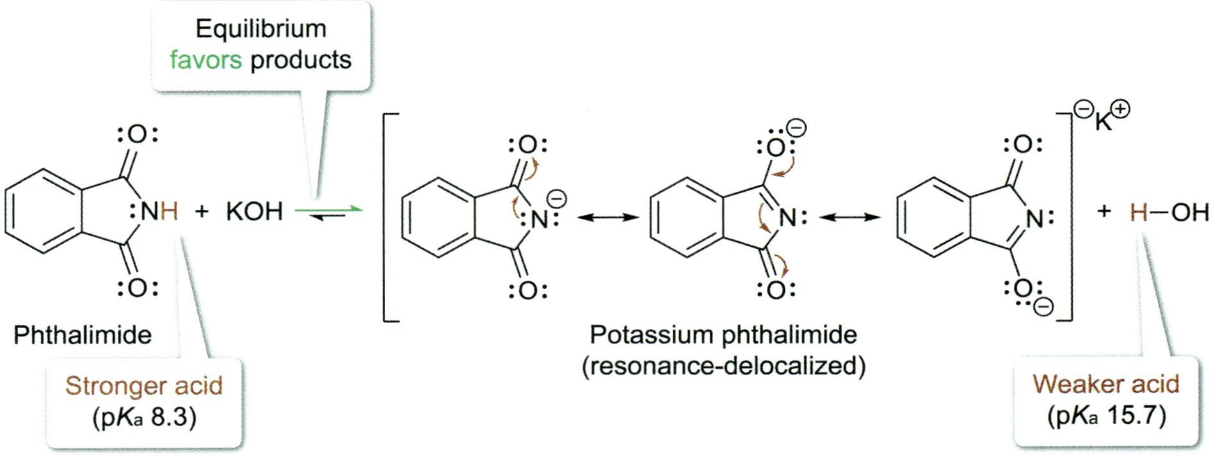

Equilibrium favors products

Phthalimide + KOH ⇌ Potassium phthalimide (resonance-delocalized) + H–OH

Stronger acid (pK_a 8.3)

Weaker acid (pK_a 15.7)

Figure 12.7 Generation of a phthalimide ion.

In the **Gabriel amine synthesis**, an alkyl halide is treated with potassium phthalimide, which is a strong nucleophile at its nitrogen atom (Figure 12.8). Because the phthalimide ion undergoes an S_N2 reaction with the alkyl halide, the structure of the alkyl halide cannot be 3°. The phthalimide ion serves as a protected derivative of a nitrogen atom nucleophile that is *not* susceptible to polyalkylation reactions.

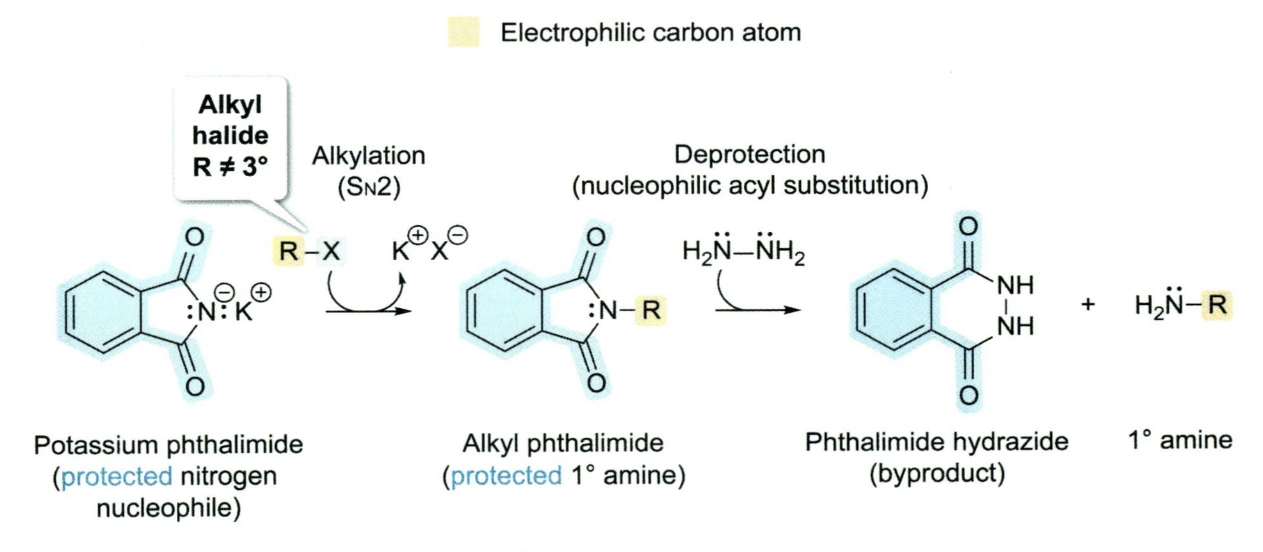

Figure 12.8 The Gabriel amine synthesis of a 1° amine.

In the second step of the Gabriel amine synthesis, the alkyl phthalimide is heated with **hydrazine** (H_2N-NH_2). The *two* nucleophilic amino groups of hydrazine react with the electrophilic carbonyl groups of the phthalimide through *two* successive nucleophilic acyl substitution reactions. These reactions deprotect the nitrogen atom and displace the 1° amine product, leaving phthalimide hydrazide. The phthalimide hydrazide byproduct typically has low solubility in the reaction solvent and precipitates from the reaction as a colorless solid.

12.2.03 Amine Synthesis through Reductive Amination

One of the most versatile ways to prepare an amine is through a process called **reductive amination**, which *combines* the condensation reaction of a nitrogen nucleophile with an aldehyde or ketone (see Concept 9.3.04) and the reduction of a polar pi bond by a hydride ion reducing agent (see Concept

9.3.08). Although the condensation reaction step is *reversible*, the subsequent reduction reaction is *irreversible*, helping drive formation of the amine product (Figure 12.9).

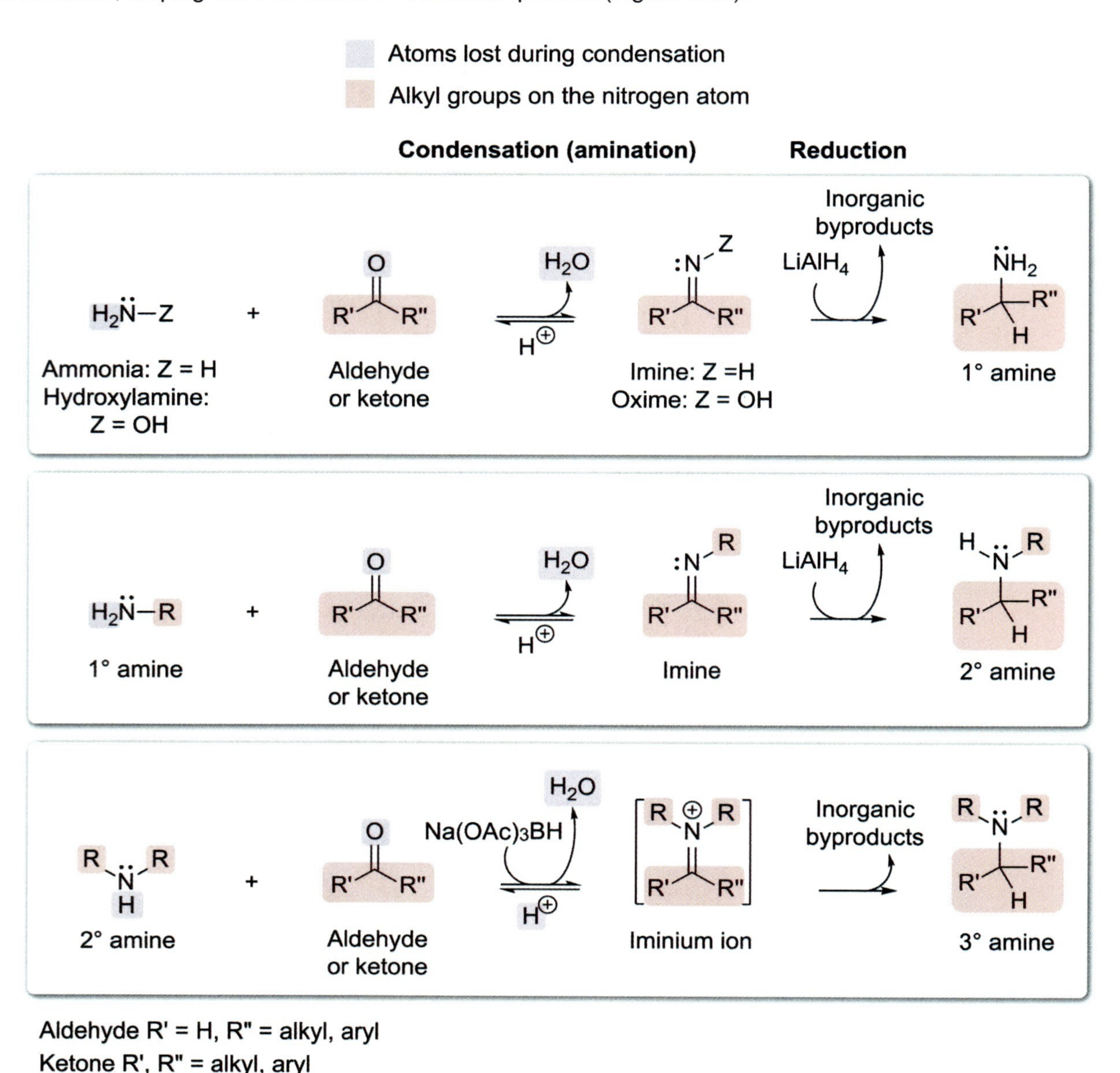

Figure 12.9 The general use of reductive amination to prepare a 1°, 2°, or 3° amine product.

The net outcome of a reductive amination is to add exactly *one* additional alkyl group to a nucleophilic nitrogen atom. Consequently, an NH_3, 1° amine, or 2° amine reactant can be converted to a 1° amine, 2° amine, or 3° amine product, respectively. The only limitation of reductive amination is that the added alkyl group *must* have at least one hydrogen atom at the carbon atom directly attached to the nitrogen.

Because reductive amination is a sequential process, each stage has its own reaction mechanism. The initial amination stage occurs through an acidic nucleophilic addition mechanism (see Concept 9.3.04). The reduction stage occurs through a basic nucleophilic addition mechanism (see Concept 9.3.08).

Although the term reductive amination implies the same general steps and features, there may be minor modifications for practical reasons depending on the class of amine being formed (Table 12.2). If the desired product is a 1° amine, it is much more common to use hydroxylamine (ie, H_2NOH) as the nitrogen nucleophile rather than ammonia (NH_3), which forms imine intermediates that are more unstable and difficult to isolate in high yield. Then, during the reduction step with $LiAlH_4$, *both* the carbon-nitrogen double bond *and* the N–O bond are reduced to give a 1° amine product.

Table 12.2 Summary of the differences in the reductive amination processes.

Nitrogen atom nucleophile	Intermediate	Reducing agent	Concurrent amination and reduction	Product
Hydroxylamine	Oxime	$LiAlH_4$	No	1° Amine
1° Amine	Imine	$LiAlH_4$	No	2° Amine
2° Amine	Iminium ion	$Na(OAc)_3BH$	Yes	3° Amine

The preparation of 2° amines does not feature any modifications to the reaction process, and both steps are performed as originally described. After the condensation of a 1° amine with an aldehyde or ketone, treatment with $LiAlH_4$ reduces the carbon-nitrogen double bond of the imine intermediate to a single bond.

The preparation of 3° amine product through reductive amination requires the reaction of a 2° amine with an aldehyde or ketone. The condensation product is an iminium ion, which is generally challenging to isolate. For this reason, the highest reductive amination yields are achieved when the condensation *and* reduction steps take place *concurrently* under the same conditions within the same reaction vessel. However, the reducing agent $LiAlH_4$ *cannot* be used to reduce an iminium ion as it *also* reacts with the acid catalyst *and* the aldehyde or ketone reactant (prior to condensation with the 2° amine).

Therefore, an *alternative* hydride ion reducing agent, sodium triacetoxyborohydride ($Na(OAc)_3BH$), is used to prepare a 3° amine through reductive amination. $Na(OAc)_3BH$ *selectively* reduces an iminium ion to a 3° amine in the presence of *both* the acid catalyst and an aldehyde or ketone. As such, all reactants *and* $Na(OAc)_3BH$ are all added to the reaction vessel together to generate the 3° amine in one synthetic step, *without* the need to isolate the iminium ion intermediate.

Concept Check 12.1

What is the structure of the aldehyde or ketone that would be used to complete the following synthesis involving reductive amination?

NH_2 + ? 1. $H_3O^{\oplus}$ 2. $LiAlH_4$ → H N

Aldehyde or ketone

Solution

Note: The appendix contains the answer.

12.2.04 Amide Synthesis through Nucleophilic Acyl Substitution

An amide, one of the least reactive carboxylic acid derivatives, can be prepared from the reaction of a nitrogen nucleophile (eg, NH_3, 1° amine, 2° amine) with a *more* reactive carboxylic acid derivative (eg, acid chloride, anhydride, ester) (see Concept 11.2.03). Because the rate of interconversion *decreases* for less reactive (ie, less electrophilic) carboxylic acid derivatives (see Concept 11.2.01), the most common way to prepare an amide is through the reaction of a nitrogen nucleophile with a highly reactive acid chloride in a non-nucleophilic and basic organic solvent, such as pyridine (Figure 12.10).

Because these conversions effectively transfer the acyl group of the carboxylic acid derivative to the nucleophilic nitrogen atom, they can be described as **acylation reactions**.

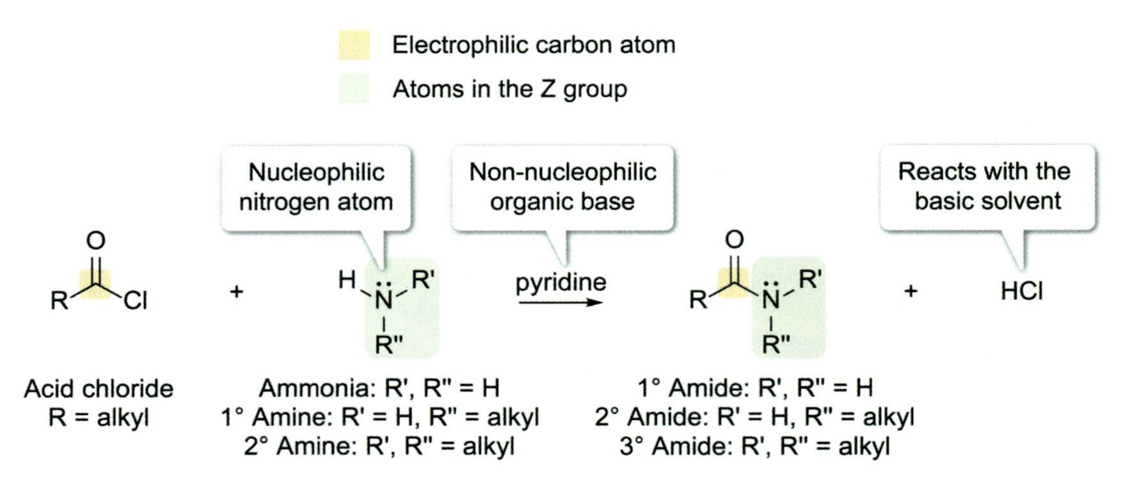

Figure 12.10 The general synthesis of an amide through acylation of a nitrogen nucleophile.

In the reaction mechanism, the nucleophilic nitrogen atom attacks the electrophilic carbon atom of the acid chloride and forms a tetrahedral intermediate. The intermediate then reforms the π bond of the carbonyl group, eliminating a chloride ion as the leaving group. The basic solvent (eg, pyridine) removes one of the acidic protons from the nitrogen atom to form the amide product.

Like reductive amination, the acylation of a nitrogen nucleophile *increases* the number of carbon-containing groups by exactly one:

- NH_3 nucleophiles (0 alkyl groups) yield a 1° amide (1 acyl group).
- 1° amine nucleophiles (1 alkyl group) yield a 2° amide (1 acyl group + 1 alkyl group).
- 2° amine nucleophiles (2 alkyl groups) yield a 3° amide (1 acyl group + 2 alkyl groups).

12.2.05 Amine Synthesis through Acylation-Reduction

Amines can alternatively be prepared through a complementary, flexible process called **acylation-reduction**, which refers to a two-stage series of reactions (Figure 12.11) in which:

1. A nitrogen nucleophile is acylated by a highly electrophilic carboxylic acid derivative (eg, acid chloride) to form an amide intermediate (see Concept 12.2.04).
2. The resulting amide is treated with $LiAlH_4$ to effectively reduce the amide carbonyl (C=O) to a methylene (CH_2) group (see Concept 11.3.03).

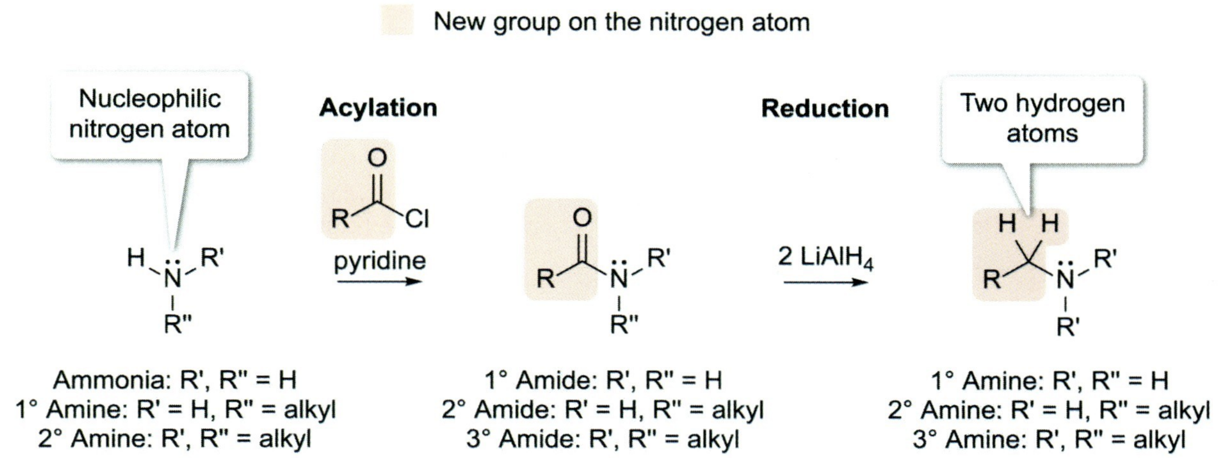

Figure 12.11 The general use of acylation-reduction to prepare a 1°, 2°, or 3° amine product.

The process of acylation-reduction is similar to reductive amination (see Concept 12.2.03) in several ways. Both processes add one *new* alkyl group to a nucleophilic nitrogen atom of ammonia or 1° and 2° amines. Therefore, 1°, 2°, or 3° amines can be *iteratively* prepared through acylation-reduction, each acylation-reduction cycle adding one alkyl group. Like reductive amination has a structural requirement that the newly added alkyl group *must* contain at least *one* hydrogen atom on the carbon attached to the nitrogen (from the *single* reduction of the double bond), acylation-reduction *requires* the same carbon atom to have *two* hydrogen atoms (due to the *double* reduction of an amide by $LiAlH_4$).

Because the lone pairs on the nitrogen atom of an amide are delocalized through resonance into the adjacent carbonyl group, the nitrogen atom of an amide is a poor nucleophile. Consequently, much like reductive amination, there is a very low risk of polyacylation (ie, more than one successive acylation reaction) during the acylation step of acylation-reduction.

Lesson 12.3

12.3 Reactions of Amines and Amides

Introduction

Although this chapter has presented amines and amides together to draw parallels or identify differences between the two nitrogen-containing functional groups, the chemical reactivities of amines and amides are *very* different. Consequently, the chemical reactivity of amides is discussed in the context of the other carboxylic acid derivatives (see Lesson 11.3).

This lesson focuses on the chemical reactivity of amines as bases (Concept 12.3.01) and as nucleophiles (Concept 12.3.02).

12.3.01 Basicity of Amines

The lone pair on the nitrogen atom of an *alkyl* amine is generally a moderate Brønsted-Lowry base. Because an alkyl group is a sigma electron–donating group that increases the electron density on the nitrogen atom, it would be expected that each additional alkyl group would *increase* the basicity of an amine. However, each additional nonpolar alkyl group also *decreases* the aqueous solvation of its conjugate ammonium ion, a feature that would be expected to *decrease* the basicity of the amine. Because these two effects somewhat cancel each other out, alkyl amines have a relatively *narrow* pK_b range between 2.5 and 3.5, somewhat independent of the amine classification (Table 12.3).

Table 12.3 The relative basicity of alkyl amines and acidity of alkyl ammonium ions.

	Conjugate base		Conjugate acid	
Name of base	**Structure**	**pK_b**	**Structure**	**pK_a**
Ammonia	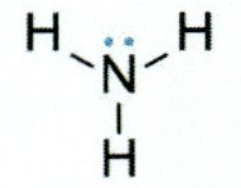 H–N̈(–H)–H	4.7	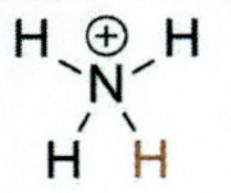 H–N⊕(H)(H)–H	9.3
Ethylamine (1° amine)	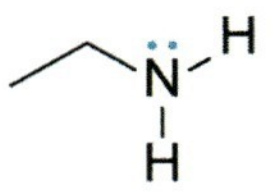 Et–N̈(H)–H	3.4	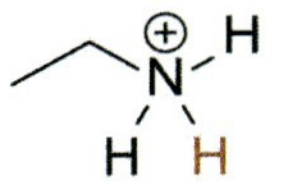 Et–N⊕(H)(H)–H	10.6
Triethylamine (3° amine)	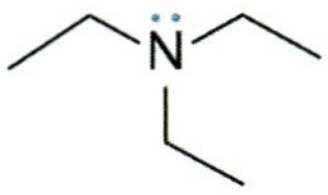 Et₃N:	3.2	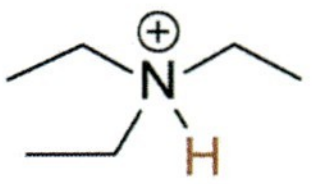 Et₃N⊕–H	10.8
Diethylamine (2° amine)	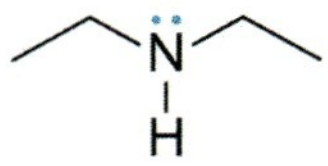 Et–N̈(–H)–Et	3.0	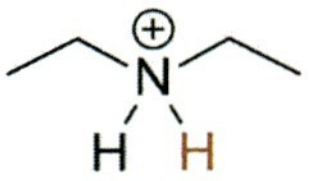 Et–N⊕(H)(H)–Et	11.0
Piperidine	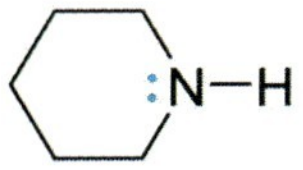 :N–H	2.9	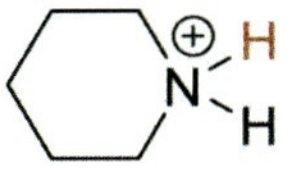 N⊕(H)–H	11.1
Pyrrolidine	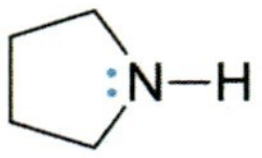 :N–H	2.7	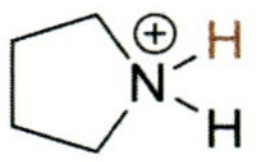 N⊕(H)–H	11.3
Hydroxide ion	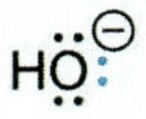 HO^{-}	-1.7	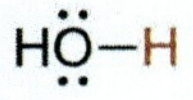 HO–H	15.7

Similar basic strength

More basic

More acidic

In contrast, the lone pair on the nitrogen atom of an *aromatic* amine has more variable basicity (Table 12.4). Because the lone pair in aniline is conjugated with the delocalized π system of its adjacent phenyl ring, it is *less* available to abstract an acidic proton and is therefore a *weaker* base (pK_b 9.4) than an alkyl amine (pK_b 2.5–3.5).

The presence of *ortho* or *para* pi electron–donating groups (eg, $-OCH_3$) *increases* the electron density of the nitrogen atom and *increases* its basicity (pK_b 8.7), whereas pi electron-withdrawing groups (eg, $-NO_2$) have the opposite effect and *decrease* its basicity (pK_b 13). These effects are not generally observed for *meta*-substituted isomers.

Table 12.4 The relative basicity of aromatic amines and acidity of aromatic ammonium ions.

Pi electron-donating group | Pi electron-withdrawing group

Name of base	Conjugate base Structure	pK_b	Conjugate acid Structure	pK_a
Pyrrole	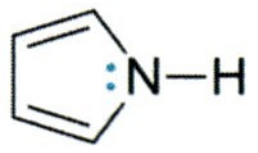 :N–H (pyrrole)	14	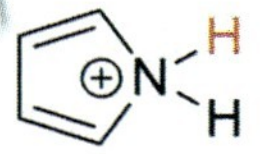 ⊕NH₂ (protonated pyrrole; no longer aromatic)	0
p-Nitroaniline	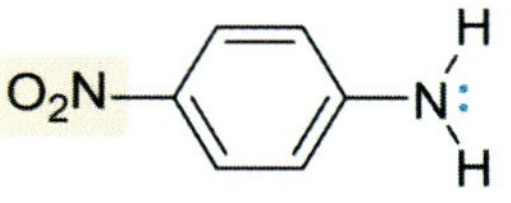 O_2N–C₆H₄–NH_2	13	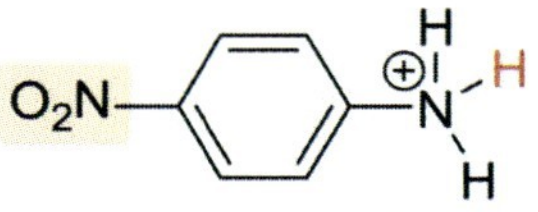 O_2N–C₆H₄–NH_3^+	1.0
Aniline	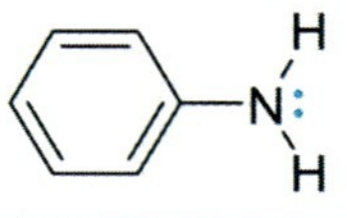 C₆H₅–NH_2	9.4	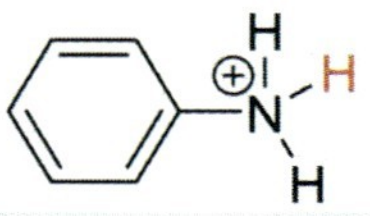 C₆H₅–NH_3^+	4.6
Pyridine	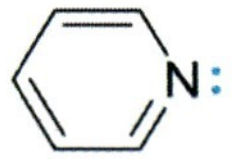 pyridine N:	8.8	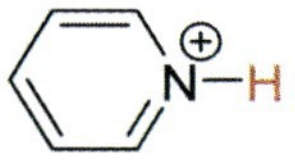 pyridinium N⊕–H	5.2
p-Methoxyaniline	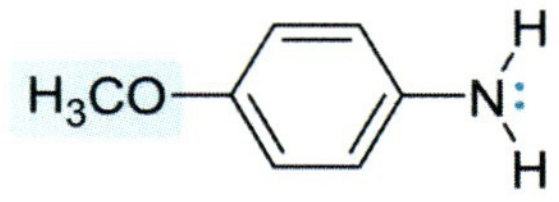 H_3CO–C₆H₄–NH_2	8.7	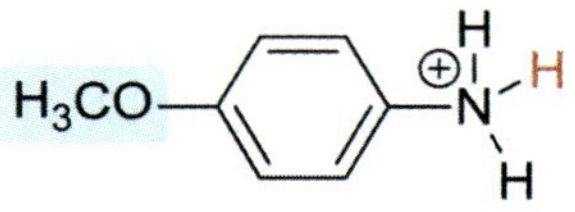 H_3CO–C₆H₄–NH_3^+	5.3
Ammonia	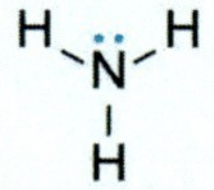 NH_3	4.7	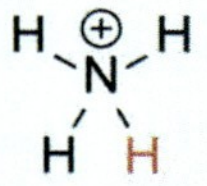 NH_4^+	9.3

More basic ↓ | More acidic ↑

The basicity of nitrogen atoms within heterocyclic aromatic rings (eg, pyrrole, pyridine) depends on the orbital structure of the nitrogen atom (see Concept 6.5.03). The nitrogen lone pair in pyridine is located within an sp^2 orbital, which is *perpendicular* to the *p* orbitals of the aromatic ring and *unable* to participate in conjugation with the aromatic ring. Therefore, it is available to act as a base *without* disrupting the aromaticity of the ring.

In contrast, the nitrogen lone pair in pyrrole is in an unhybridized *p* orbital that *contributes* to ring aromaticity. Because protonation of pyrrole creates a conjugate acid, thereby *breaking* the stability afforded by aromaticity, this process is *not* as favorable, making pyrrole an exceptionally *weak* nitrogen base (pK_b 14).

Reactions in Which Amines Act as Bases

A traditional sophomore organic chemistry curriculum contains *numerous* examples of reactions in which amines participate as bases. However, within the exam-relevant body of organic transformations described within this book, an amine reacts as a base in the following reactions:

- The use of LDA to generate a quantitative enolate ion (Concept 9.4.04) or kinetic enolate ion (Concept 9.4.05)
- The use of a non-nucleophilic organic base during nucleophilic acyl substitution reactions of an acid chloride or anhydride (see Concepts 11.2.03, 12.2.04, and 12.2.05)

Generation of Ammonium Ions

Due to the position of the acid-base equilibrium (see Concept 5.1.03), the formation of an ammonium ion is favored when an amine reacts with an acid that is *stronger* (ie, lower pK_a) than the resultant ammonium ion. Consequently, alkyl ammonium ions (pK_a 10.5–11.5) are favored to form in reactions with an acid of pK_a < 10.5, and aromatic or resonance-delocalized ammonium ions (eg, anilinium ion, pyridinium ion) are favored to form in reactions with an acid of pK_a < 4.

12.3.02 Amines as Nucleophiles

The nitrogen atom of an amine is often considered a strong nucleophile. Some exceptions to this generalization include nitrogen atoms with:

- Four bonds (ie, quaternary amines, ammonium ions), which *lack* the required lone pair for reactivity as nucleophiles
- Resonance-delocalized lone pairs in a π system (eg, anilines, enamines)
- Lone pairs that are *required* for the aromaticity of a ring (eg, pyrrole)
- Steric hindrance due to bulky R groups (eg, *tert*-butyl, LDA)

Previous lessons have introduced chemical reactions that involve an amine acting as a nucleophile. Each of these classes of reactions are summarized in this concept.

Amines in Nucleophilic Substitution Reactions

Although it is unlikely for an amine nucleophile to participate in a unimolecular (S_N1) substitution reaction mechanism (see Concepts 5.4.02 and 12.2.01), amines are good nucleophiles and readily undergo alkylation reactions with unhindered electrophiles through an S_N2 mechanism (see Concept 5.4.03), as shown in Figure 12.12. The net result of an alkylation reaction adds *one* new alkyl (R) group to the amine nitrogen atom in place of *either* a hydrogen atom or its lone pair.

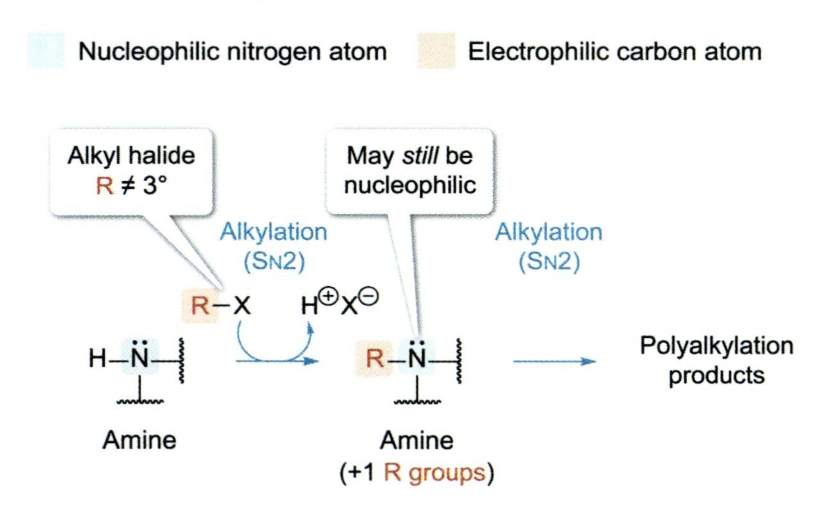

Figure 12.12 Alkylation of an amine nucleophile.

Because the product of the alkylation of an amine is *also* an amine and is often still nucleophilic, the risk of polyalkylation can be a significant limitation to this type of transformation. However, this can be mediated either by adapting the experimental stoichiometry of the reactants (Concept 12.2.01) or by using a protected nitrogen nucleophile (Concept 12.2.02).

Amines in Nucleophilic Addition Reactions

Nucleophilic 1° or 2° amines can also undergo a reversible condensation reaction with the carbonyl group of aldehydes or ketones in the presence of a weak acid (Concept 9.3.04) as shown in Figure 12.13.

These reactions form products containing a carbon-nitrogen double bond and eliminate a molecule of water.

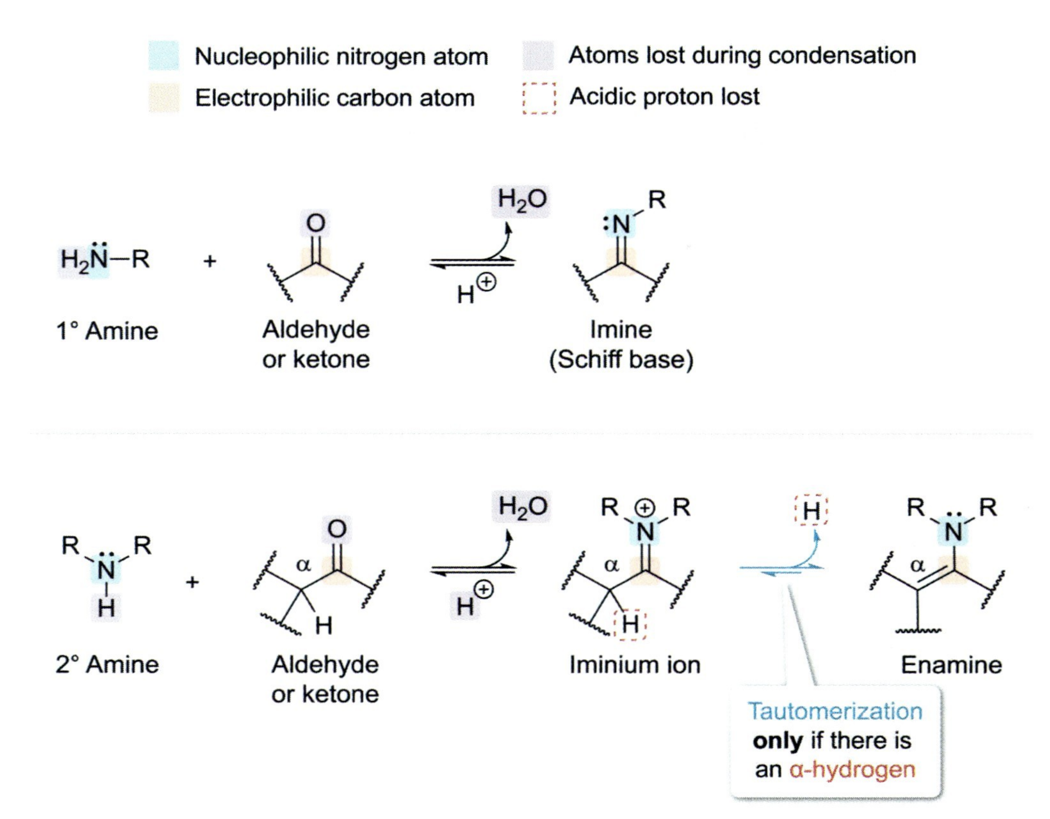

Figure 12.13 Condensation of an amine nucleophile with the carbonyl group of an aldehyde or ketone.

The mechanism for these condensation reactions is an acidic nucleophilic addition followed by an acid-catalyzed dehydration of an alcohol to form either an **imine** (from a 1° amine nucleophile) or **iminium ion** (from a 2° amine nucleophile) condensation product. Although an iminium ion is generally unstable, iminium ions containing at least one acidic α-hydrogen atom may tautomerize to form a stable **enamine** product (see Concept 9.4.06). Together, imines and imine derivatives can be employed in the controlled synthesis of substituted amines through reductive amination (see Concept 12.2.03).

Amines in Nucleophilic Acyl Substitution Reactions

A nucleophilic amine can also react with a carboxylic acid derivative to form an amide product (Figure 12.14). Acid halides, anhydrides, and esters are all more electrophilic than the resultant amide, such that their reaction with an amine is favorable. Each electrophilic carboxylic acid derivative contains a leaving group (the Z group), which is displaced by the amine through a nucleophilic acyl substitution mechanism.

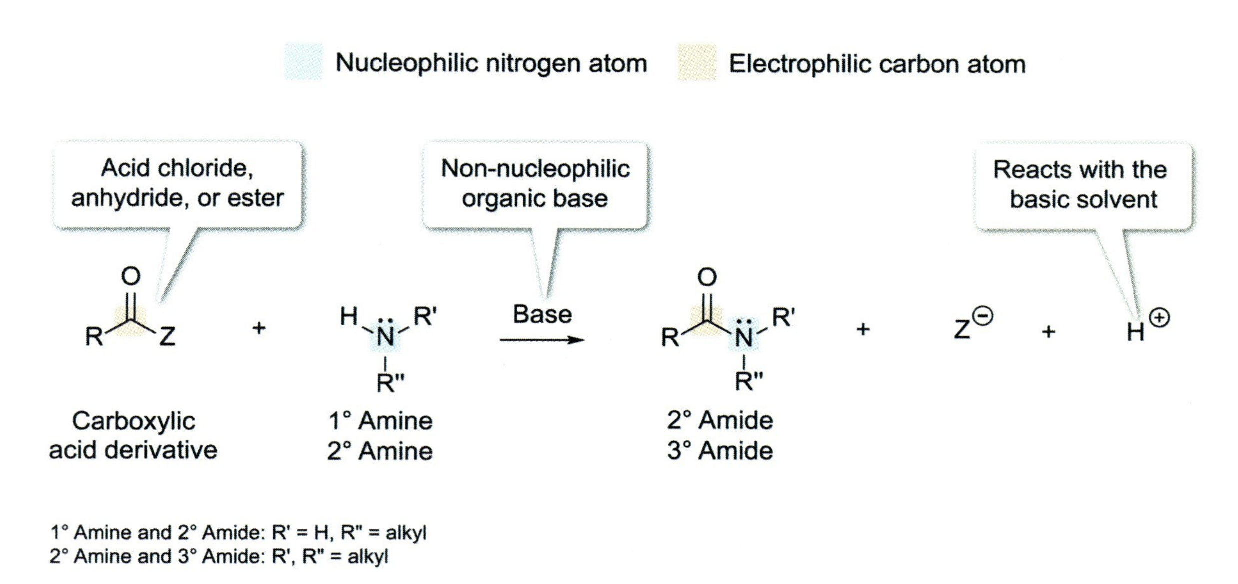

1° Amine and 2° Amide: R' = H, R" = alkyl
2° Amine and 3° Amide: R', R" = alkyl

Figure 12.14 Acylation of an amine nucleophile.

Alternatively, this transformation may be described as an acylation of an amine (see Concept 12.2.04). Like an alkylation, the acylation process effectively adds one additional carbon-containing group to the nucleophilic nitrogen atom of the amine. Unlike alkylation, the amide product is far *less nucleophilic* than the amine reactant so there is *minimal* risk of polyacylation.

Amine reactions with carboxylic acid derivatives typically proceed by a modified basic nucleophilic acyl substitution mechanism and do not require the addition of heat except with less reactive derivatives such as esters (see Concept 11.2.03).

END-OF-UNIT MCAT PRACTICE

Congratulations on completing **Unit 2: Functional Groups and Their Reactions**.

Now you are ready to dive into MCAT-level practice tests. At UWorld, we believe students will be fully prepared to ace the MCAT when they practice with high-quality questions in a realistic testing environment.

The UWorld Qbank will test you on questions that are fully representative of the AAMC MCAT syllabus. In addition, our MCAT-like questions are accompanied by in-depth explanations with exceptional visual aids that will help you better retain difficult MCAT concepts.

TO START YOUR MCAT PRACTICE, PROCEED AS FOLLOWS:

1) Sign up to purchase the UWorld MCAT Qbank
   IMPORTANT: You already have access if you purchased a bundled subscription.
2) Log in to your UWorld MCAT account
3) Access the MCAT Qbank section
4) Select this unit in the Qbank
5) Create a custom practice test

Unit 3 Separation Techniques, Spectroscopy, and Analytical Methods

Chapter 13 Separation and Purification Methods

13.1 Laboratory Techniques Utilizing Solubility

13.1.01 General Principles of Solubility
13.1.02 Solid-Liquid Separations
13.1.03 Liquid-Liquid Separations

13.2 Distillation

13.2.01 General Principles of Distillation
13.2.02 Simple Distillation
13.2.03 Vacuum Distillation
13.2.04 Fractional Distillation

13.3 Chromatography

13.3.01 General Principles of Chromatography
13.3.02 Gas Chromatography
13.3.03 Introduction to Liquid-Solid Chromatography
13.3.04 Paper Chromatography
13.3.05 Thin-Layer Chromatography (TLC)
13.3.06 Column Chromatography
13.3.07 High-Performance Liquid Chromatography (HPLC)

Chapter 14 Spectroscopy and Analysis

14.1 Laboratory Analysis

14.1.01 Methods to Analyze pH

14.2 Polarimetry

14.2.01 Plane-Polarized Light
14.2.02 Relationship between Polarimetry and Chirality
14.2.03 Observed Rotation and Specific Rotation

14.3 The Electromagnetic Spectrum

14.3.01 Introduction to the Electromagnetic Spectrum
14.3.02 Spectroscopy and the Electromagnetic Spectrum

14.4 Mass Spectrometry

14.4.01 Introduction to Mass Spectrometry
14.4.02 Molecular Ionization
14.4.03 The Mass Spectrometer
14.4.04 The Mass Spectrum
14.4.05 Isotope Signatures in Mass Spectrometry
14.4.06 Fragmentation of Radical Cations

14.5 UV-Vis Spectroscopy

14.5.01 Electronic Transitions
14.5.02 The Beer-Lambert Law
14.5.03 The UV-Vis Spectrophotometer
14.5.04 The UV Spectrum
14.5.05 The Visible Spectrum

14.6 Infrared Spectroscopy

14.7 ^{1}H NMR Spectroscopy

Lesson 13.1

Laboratory Techniques Utilizing Solubility

Introduction

The previous two units in this book cover the foundational principles of organic chemistry (Unit 1) and the chemical reactions of the functional groups (Unit 2). In general, these topics encompass the traditional *lecture* component of the sophomore organic chemistry sequence, in the context of concepts most relevant to the exam.

The final unit of this book focuses on the broad themes of separation techniques and spectroscopy, which are typically emphasized in the *laboratory* component of sophomore organic chemistry. Chapter 13 discusses three methodological approaches that chemists employ to separate the components in a mixture: solubility-based methods (Lesson 13.1), distillation (Lesson 13.2), and chromatography (Lesson 13.3).

This lesson introduces how solubility (ie, the amount of a substance that can dissolve in another substance) can be used to separate materials in a mixture. Concept 13.1.01 introduces solubility-based methodologies, with a focus on the relationship between compound structure, intermolecular forces, molecular polarity, and solubility patterns. The remaining concepts each focus on one type of separation methodology.

13.1.01 General Principles of Solubility

In Concept 2.4.02, the **solubility** of a compound is defined as the maximum mass of a substance (the **solute**) that can dissolve in a quantity of a liquid (the **solvent**). Often, solubility is reported in units of grams of solute per 100 mL solvent; larger values indicate a more soluble compound.

In general, the solubility of organic compounds is described by the phrase "like dissolves like"; that is, compounds with similar polar or nonpolar character are soluble in one another. At a molecular level, the solubility of a substance is impacted by a combination of molecular polarity and other intermolecular forces in which it can participate. Compounds achieve the greatest solubility when *both* their polarity and intermolecular forces are similar to those of the solvent.

Although chemists regularly use the term **insoluble** to describe substances with *poor* solubility in a solvent, this term can be misleading. *All* substances are at least *somewhat* soluble in a solvent, even if the solubility is very, very low. For this reason, insoluble substances are never *fully* insoluble. In contrast, some substances demonstrate infinite *solubility* (ie, no upper solubility limit) in a solvent; these solute-solvent pairs are described as being **miscible**.

In organic chemistry, the solubility of an organic substance needs to be considered with regard to *both* water and organic solvents (Table 13.1). As a polar solvent, water tends to have favorable interactions with polar solutes (eg, molecules with ionic- or dipole-associated intermolecular forces). Solutes with a greater number of hydrogen bond donors and acceptors tend to have greater water solubility.

Table 13.1 Differences in the characteristics of aqueous and organic solvents.

Feature	Aqueous solvent (water)	Organic solvent
Molecular polarity	Polar	Nonpolar to Polar
Major intermolecular forces for solvation	Ion-associated and dipole-associated	Induced dipole–associated and dipole-associated
Prevalence of hydrogen bond donors and acceptors	More prevalent	Less prevalent

In contrast, the solubility characteristics of organic solvents are more variable. Some organic solvents have predominantly nonpolar character, favorable interactions with nonpolar solutes, and *fewer* hydrogen bond donors and acceptors. Other organic solvents are much more polar and are amenable to interactions with polar solutes.

The polarity range of organic solvents is closely associated with the concepts of miscibility and **immiscibility** (ie, the *inability* to fully dissolve two substances together) (Table 13.2). Organic solvents can be broadly divided into two categories: those that are immiscible (ie, insoluble) with water and those that are miscible with water. Organic solvent polarity is described as a gradient rather than a binary (ie, have or have not). Organic solvents with greater polarity are more likely to be miscible with water, a highly polar liquid.

Table 13.2 Polarity and miscibility of common organic solvents.

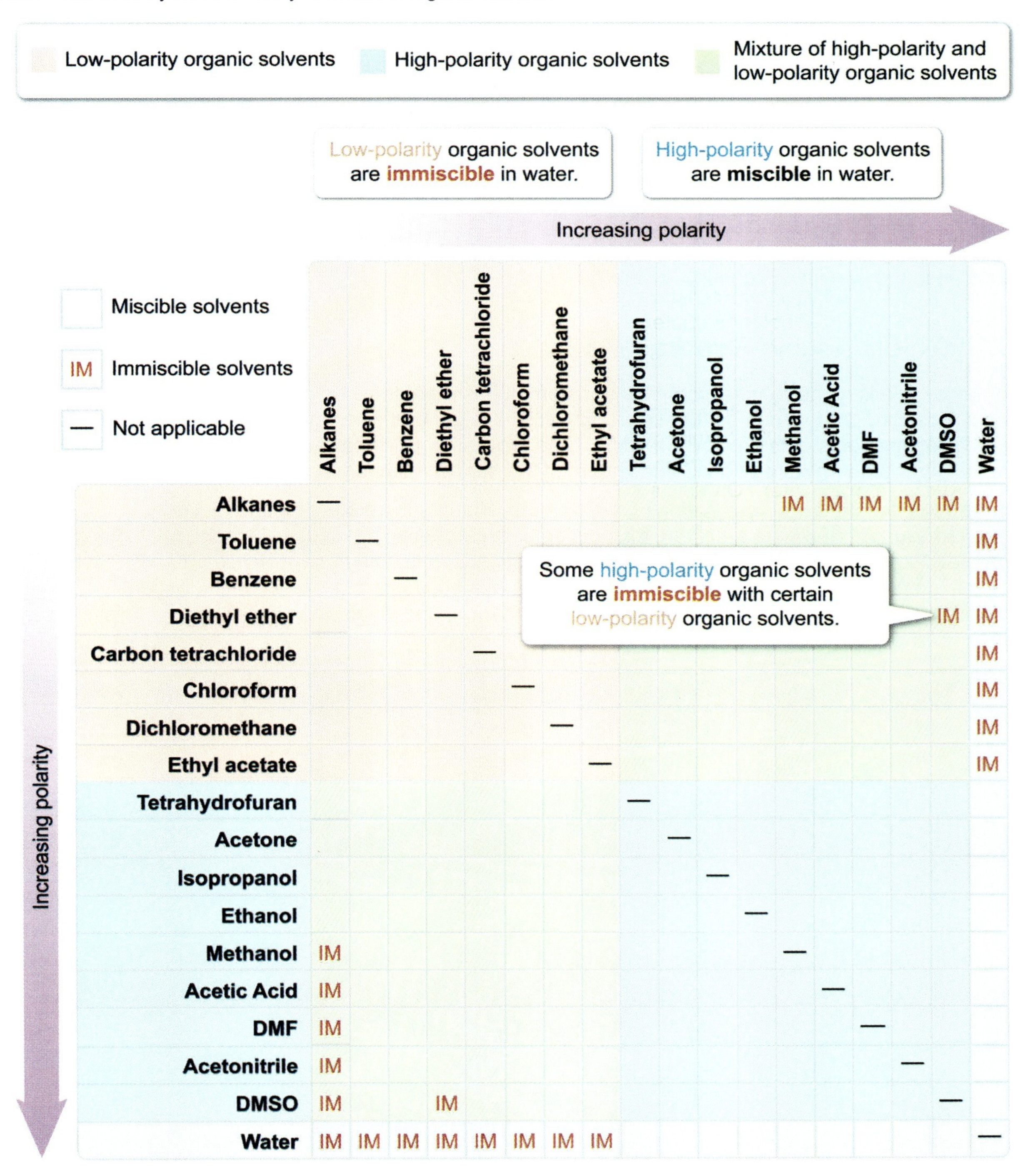

	Alkanes	Toluene	Benzene	Diethyl ether	Carbon tetrachloride	Chloroform	Dichloromethane	Ethyl acetate	Tetrahydrofuran	Acetone	Isopropanol	Ethanol	Methanol	Acetic Acid	DMF	Acetonitrile	DMSO	Water
Alkanes	—												IM	IM	IM	IM	IM	IM
Toluene		—																IM
Benzene			—															IM
Diethyl ether				—													IM	IM
Carbon tetrachloride					—													IM
Chloroform						—												IM
Dichloromethane							—											IM
Ethyl acetate								—										IM
Tetrahydrofuran									—									
Acetone										—								
Isopropanol											—							
Ethanol												—						
Methanol	IM												—					
Acetic Acid	IM													—				
DMF	IM														—			
Acetonitrile	IM															—		
DMSO	IM			IM													—	
Water	IM	IM	IM	IM	IM	IM	IM	IM										—

Organic solvents that are immiscible in water are generally considered to be either nonpolar or only moderately polar (ie, **low-polarity organic solvents**). Low-polarity organic solvents include the hydrocarbons (eg, hexane, pentane), aromatic solvents (eg, benzene, toluene), ethers, halogenated hydrocarbons (eg, CCl_4, $CHCl_3$, CH_2Cl_2), and esters (eg, ethyl acetate). In contrast, organic solvents that are miscible in water are typically **high-polarity organic solvents**. High-polarity organic solvents include the cyclic ether tetrahydrofuran (THF), acetic acid, short-chain alcohols (eg, isopropanol, ethanol, methanol), and the polar aprotic solvents (see Concept 5.3.01).

Most organic solvents are miscible with one another. The few exceptions to this statement are observed when organic solvents from the *extreme* ends of the polarity spectrum are mixed. For example, a *highly* nonpolar organic solvent (eg, alkanes, linear ethers) may be immiscible with some of the *highest* polarity organic solvents (eg, acetic acid, methanol, DMF, DMSO, acetonitrile). It is unlikely that these exceptions would be directly tested on the exam, as mixtures of two or more low-polarity organic solvents *or* two or more high-polarity organic solvents are almost always miscible.

Solubility is a physical property that is temperature dependent. Although there are exceptions, the quantitative solubility of a solid solute typically *increases* with the solvent temperature, up to the boiling point of the solvent.

These topics concerning solubility are most relevant to chromatography (Lesson 13.3) and the methods described in the remaining concepts in this lesson.

13.1.02 Solid-Liquid Separations

In organic chemistry, performing chemical transformations represents only half of the synthesis process. Even if a desired product can be formed in high yield, it is generally necessary to isolate (separate) it from the other components of the reaction mixture. The process of **purification** is the removal of undesirable components from a mixture and the enrichment of the desired component.

A **mixture** is defined as a combination of two or more unique substances (Figure 13.1). A **homogeneous mixture** has a *uniform* distribution of components and is usually more challenging to separate. Examples of homogeneous mixtures include solutions and miscible liquids (see Concept 13.1.01). In contrast, a **heterogeneous mixture** has a *nonuniform* distribution of components. Examples of heterogeneous mixtures include suspensions, emulsions, and immiscible liquids.

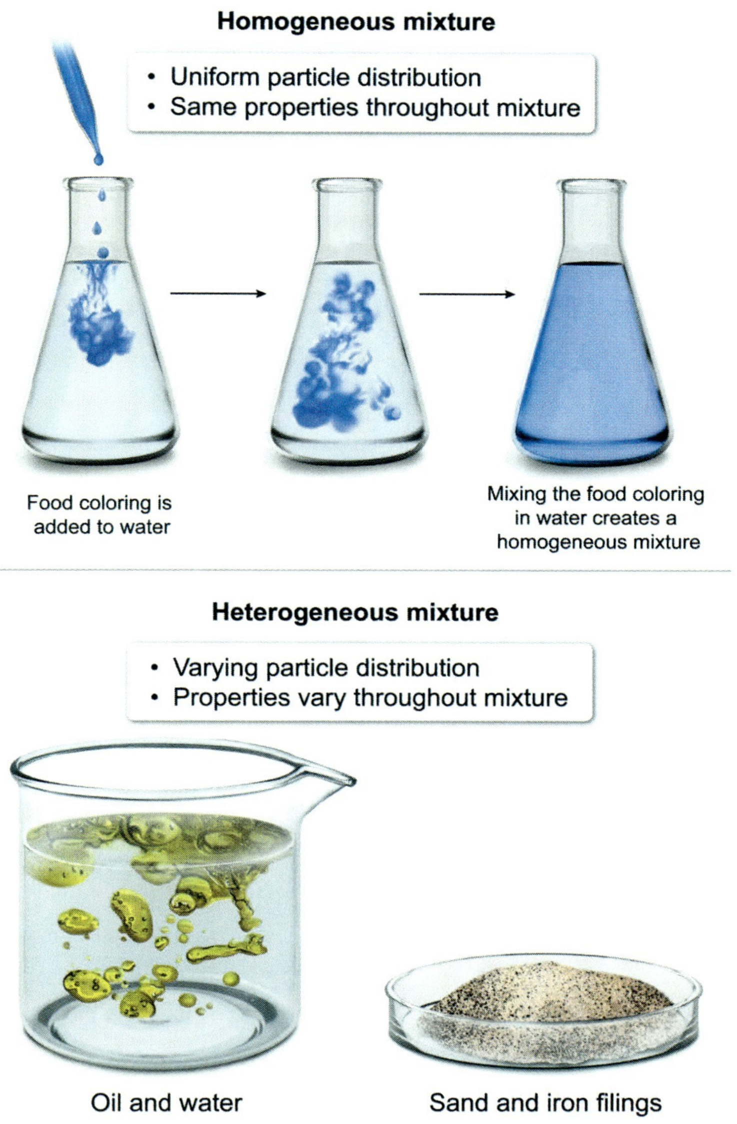

Homogeneous mixture

- Uniform particle distribution
- Same properties throughout mixture

Food coloring is added to water

Mixing the food coloring in water creates a homogeneous mixture

Heterogeneous mixture

- Varying particle distribution
- Properties vary throughout mixture

Oil and water

Sand and iron filings

Figure 13.1 Homogeneous and heterogeneous mixtures.

One of the most common general approaches to separation and purification involves the creation of a heterogeneous mixture containing an insoluble solid suspended in a liquid. The following laboratory techniques all result in the formation of a **solid-liquid mixture** for the purposes of separation and purification:

- Precipitation
- Trituration
- Recrystallization

Precipitation

A physical or chemical process that transforms a homogeneous mixture (ie, a solution) into a heterogeneous mixture of a solid and a liquid is called a **precipitation**. In general chemistry, precipitation reactions are introduced in the context of ionic double-replacement chemical reactions (see General Chemistry Concept 2.4.01). In organic chemistry, precipitations can occur through *either* a chemical *or* a physical process.

Physical precipitation exploits changes in solubility due to physical changes (eg, changes in temperature). For example, consider a solution containing a saturating or near-saturating amount of solute. Decreasing the temperature of the solution can *decrease* the solubility of the solute to below the solute's concentration (ie, the solute concentration exceeds the solubility limit, and the solution is **supersaturated**). At this point, the excess solute typically precipitates from the solution as a solid (Figure 13.2).

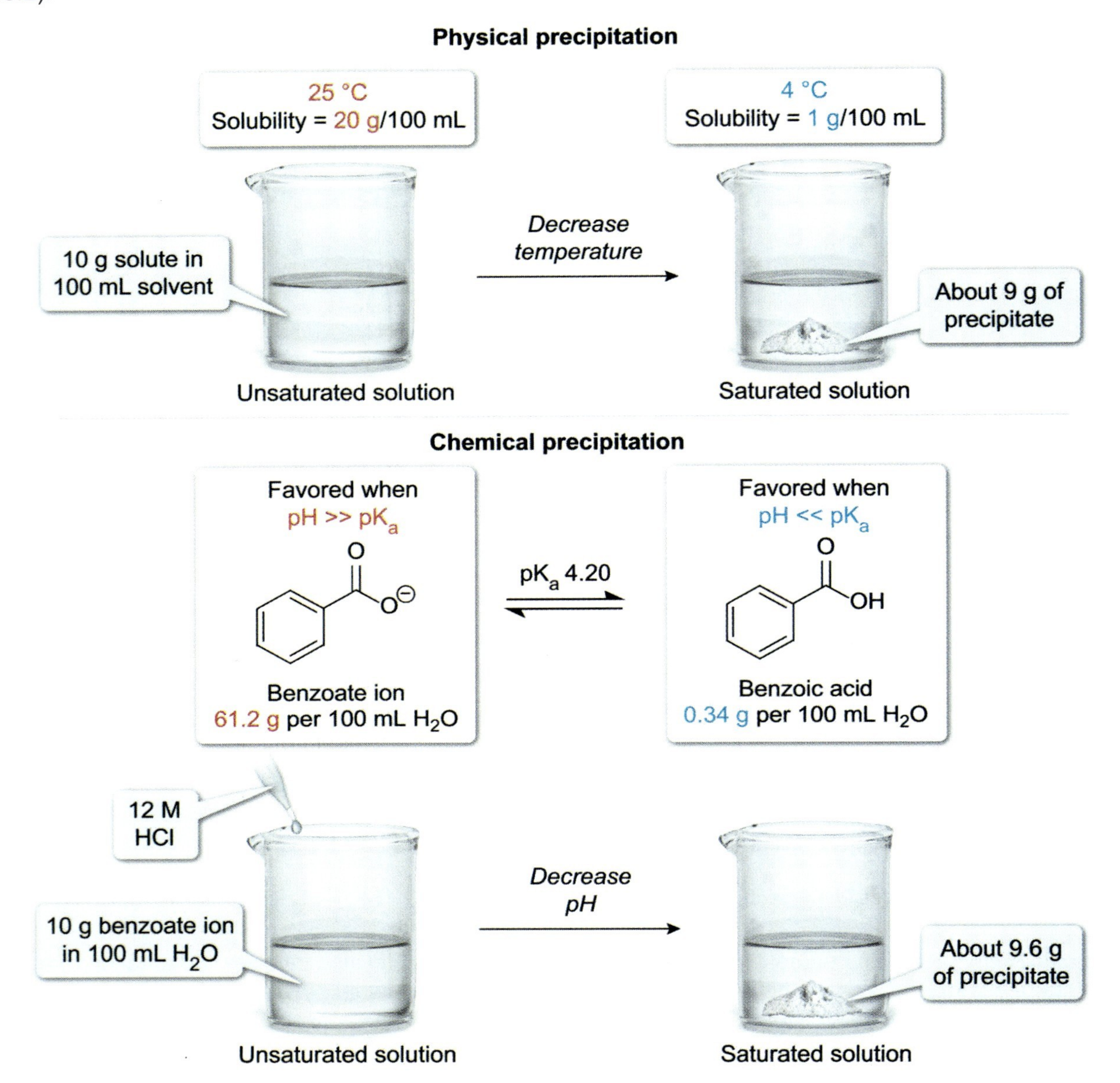

Figure 13.2 Physical and chemical precipitation.

Chemical precipitation exploits a solubility change that occurs upon a chemical reaction. Acid-base reactions are a common (and reversible) way to chemically modify the structure (and therefore solubility) of a solute. As introduced in Concept 13.1.01, ionized organic compounds typically have higher solubility in water than their neutral counterparts. As an example, the benzoate ion contains a negatively charged carboxylate group and is *more* soluble in water (61.2 g/100 mL) than its conjugate acid, benzoic acid (0.34 g/100 mL).

If an unsaturated aqueous solution of benzoate ion is treated with a strong acid to lower the pH below benzoic acid's pK_a (pK_a 4.20), the benzoate ion is quantitatively converted to benzoic acid. Because benzoic acid has a much lower aqueous solubility than the benzoate ion, any benzoic acid that exceeds the solubility limit precipitates from the solution as a solid.

Trituration

In **trituration**, a solvent is added in which the desired substance is *insoluble* but impurities are soluble. In other words, the solvent dissolves the impurities but leaves the desired compound as a solid (Figure 13.3). The solid can then be isolated from the impure solution through filtration.

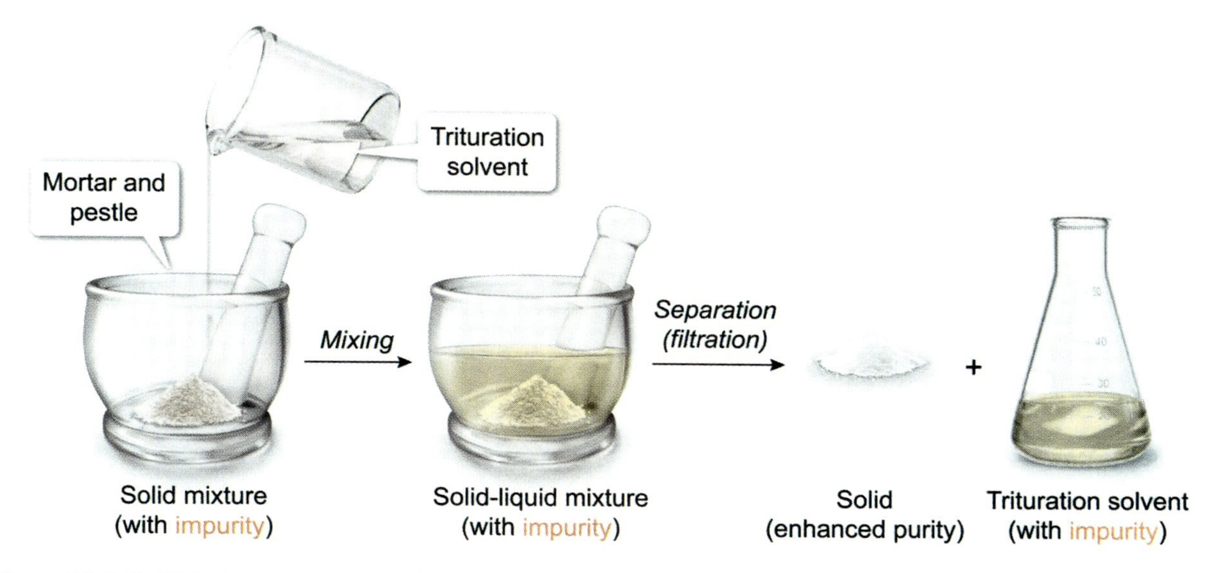

Figure 13.3 Purification through trituration.

Recrystallization

Recrystallization is a laboratory technique that utilizes solubility differences under different temperatures, solvent ratios, or other conditions to *enhance* the purity of a solid. In the most common method of recrystallization, an *impure* solid is dissolved in a *minimum* amount of a boiling solvent to create an initial *saturated* solution (Figure 13.4). Then, the saturated solution is *slowly* cooled to *decrease* the solubility of the desired solid and thereby induce precipitation.

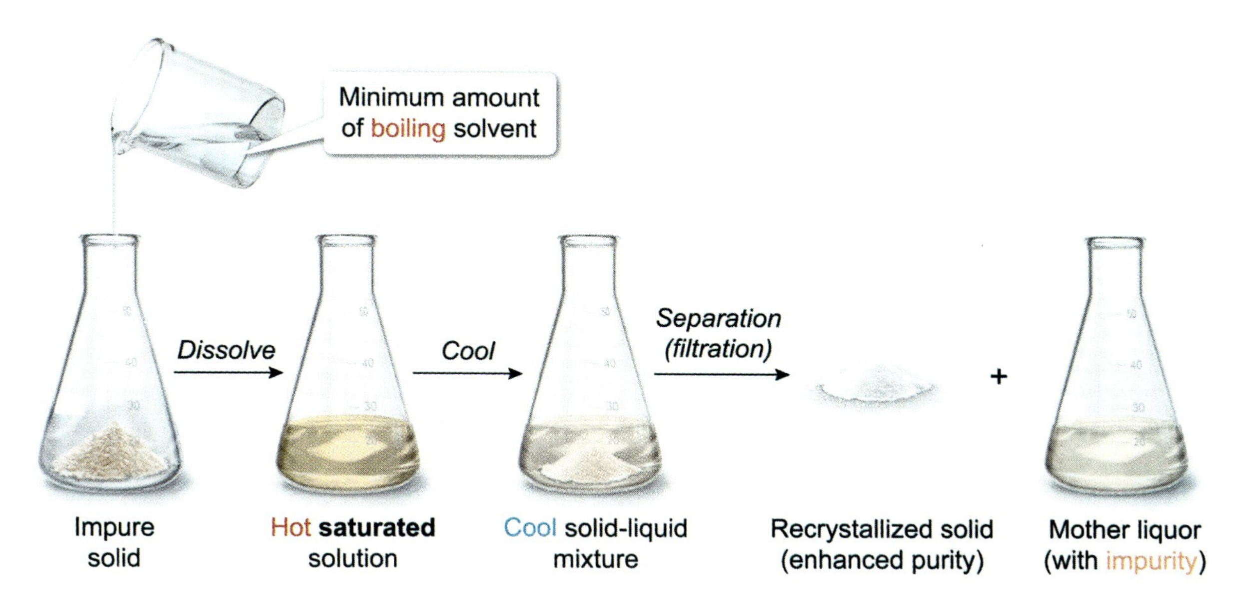

Figure 13.4 Recrystallization of a solid.

In recrystallization, the impurities are intended to enter the *opposite* phase of the desired material and *either*:

- Remain *undissolved* in the *hot* solution, such that they can be removed by filtration
- Remain *dissolved* in the cooled recrystallization solvent

Because a portion of the desired solute remains dissolved even at low temperatures, it is *not possible* to recover all of the desired solid. Sometimes, it may be necessary to recover the lost solute by concentrating (ie, evaporating) the separated recrystallization solvent, now called the **mother liquor**.

An ideal recrystallization solvent typically has a large solubility difference for the desired solute at different temperatures (ie, very soluble at high temperatures and minimally soluble at low temperatures). Common solvents for recrystallization include benzene, dichloromethane, ethyl acetate, alcohols (eg, methanol, ethanol), and water.

In some cases, a mixture of solvents is used for the recrystallization procedure (Figure 13.5). In a **mixed solvent recrystallization**:

1. The impure solid is first dissolved in a *minimum* amount of a boiling solvent in which it has *high* solubility.
2. Then, a second *low* solubility solvent is added to the hot solution until the mixture reaches the **cloud point** and begins to form a precipitate.

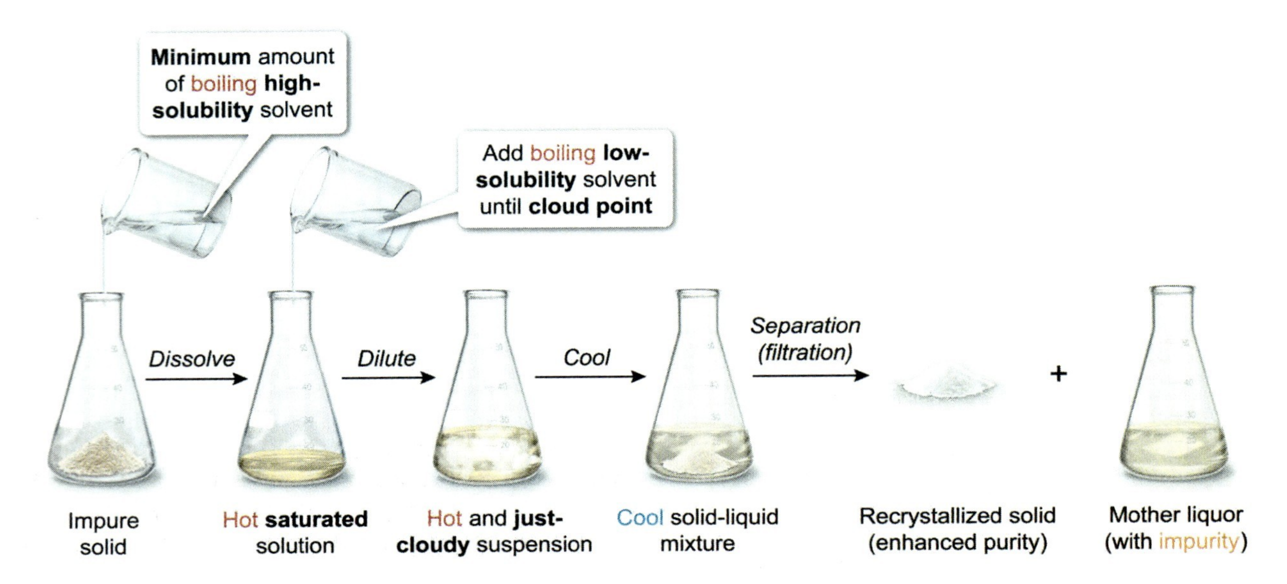

Figure 13.5 Mixed solvent recrystallization of a solid.

A significant advantage of mixed solvent recrystallization is that the solubility profile of the solvent mixture can be specifically tailored to a particular compound. Normally, the two (or more) solvents in a mixed solvent recrystallization are miscible with one another, where the *more polar* solvent is the *higher* solubility (ie, first) solvent, and the *less polar* solvent is the *lower* solubility (ie, second) solvent. Some common mixed solvent recrystallization solvent pairs include ethyl acetate (high solubility) and hexane (low solubility) for nonpolar organic molecules, or methanol (high solubility) and dichloromethane (low solubility) for polar organic molecules.

Filtration

Once a mixture of a solid and a liquid has been generated, they need to be physically separated from one another. The most common way to separate a solid and a liquid is through **filtration**. In a filtration, the solid (ie, the **precipitate**) is retained on the filter, whereas the liquid (ie, the **filtrate**) passes through it.

The filter needs to be unreactive with both the precipitate and the filtrate, porous (ie, having small holes to allow passage of the filtrate), and easily cleaned or dried.

The simplest form of filtration is **gravity filtration** (Figure 13.6), which uses a piece of thick filter paper (eg, a coffee filter) that is often **fluted** (ie, accordion-folded) to increase the surface area for passage of the filtrate. The fluted filter paper is placed in a funnel and the solid-liquid mixture is poured into it. The liquid filtrate (containing the solvent and any dissolved substances) passes through the fluted filter paper, whereas the solid precipitate does not.

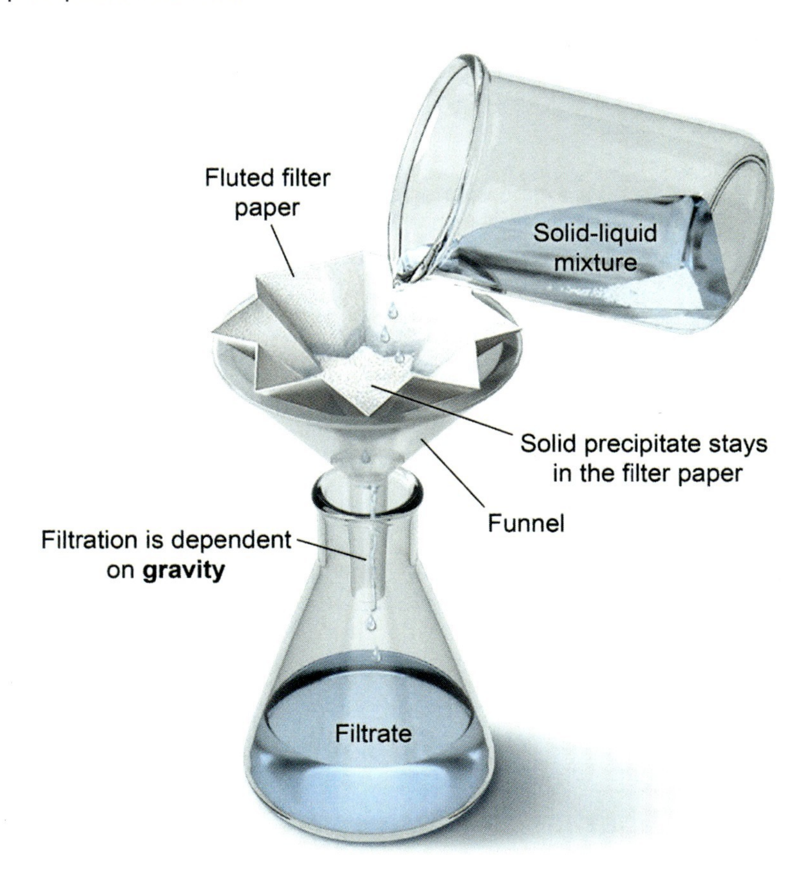

Figure 13.6 Gravity filtration.

Vacuum filtration, another form of filtration, uses a vacuum source to draw the filtrate through the filter (Figure 13.7). A specialized funnel called a **Büchner funnel** is typically used for vacuum filtration. The filter for vacuum filtration is usually either a circular piece of filter paper or a porous glass structure called a **frit**.

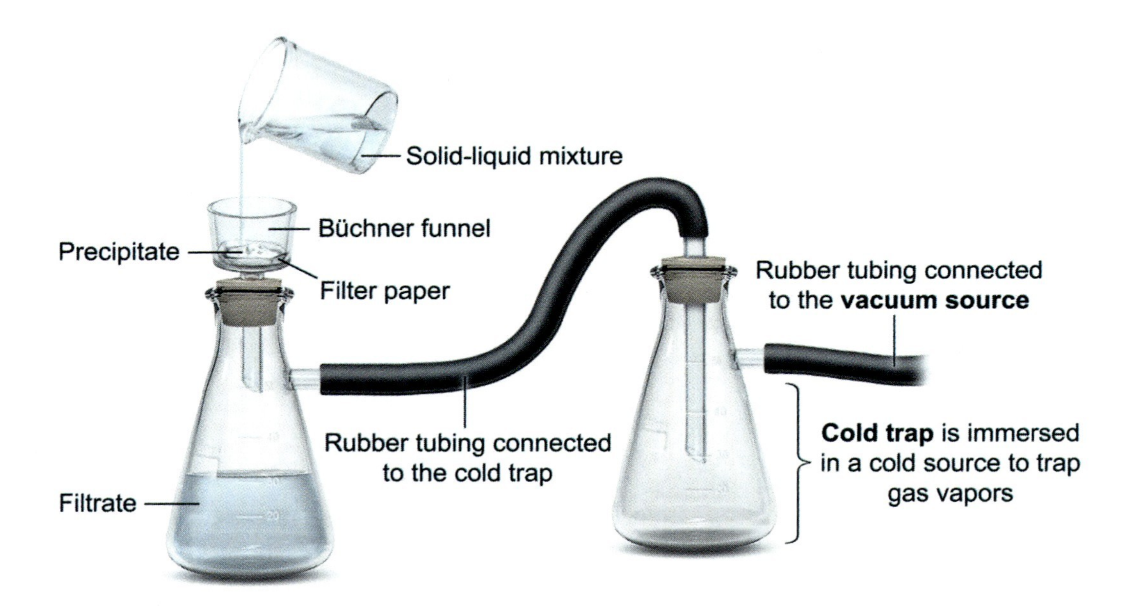

Figure 13.7 Vacuum filtration.

In gravity filtration or vacuum filtration, the desired substances can be present in the precipitate, the filtrate, or both. If the precipitate contains the desired substance, a small amount of *cold* filtrate solvent can be passed over the precipitate to **wash** the solid and remove any residual impurities. Finally, the solid can be **dried** of residual solvent either with mild heat, vacuum, or the passage of air.

If the filtrate contains the desired substance, the solvent can be evaporated to recover the dissolved solute. The recovered solute can then be dried using one of the methods described previously.

13.1.03 Liquid-Liquid Separations

Liquid-liquid separation is a frequently employed method for separating the components of a mixture. In **liquid-liquid separation**, a mixture of solutes is separated based on differences in solute solubility between two *immiscible* solvents. Typically, one solvent is aqueous and the other is organic. The **immiscible solvents** are mixed in the same container to create a two-phase (ie, liquid-liquid) system, in which the less-dense liquid is the top layer.

Most solute molecules typically have a significantly *higher* solubility in one of the two immiscible solvents. The relative solubility of a substance (A) in water compared to its solubility in an organic solvent is quantified by the **partition coefficient (*P*)**, which is a type of dynamic equilibrium constant (*K*) (Figure 13.8):

$$P = \frac{\text{Concentration of A in organic solvent}}{\text{Concentration of A in water}} = \frac{[\text{A}]_{\text{organic}}}{[\text{A}]_{\text{water}}}$$

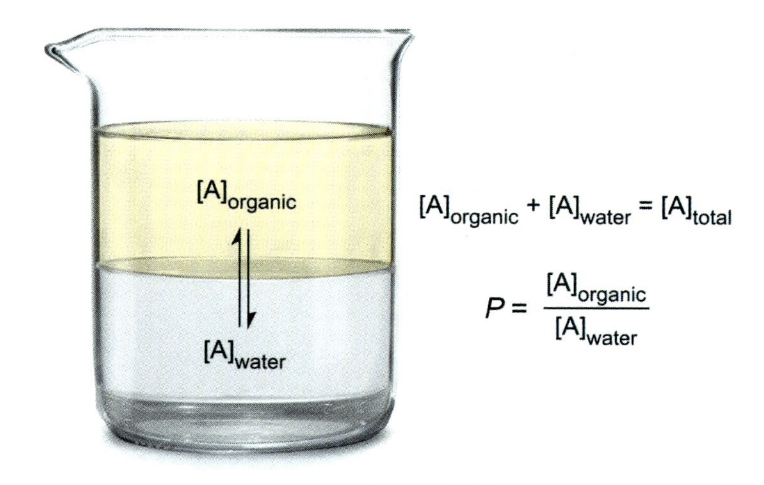

The **partition coefficient (*P*)** describes the solubility equilibrium of a solute [A] between an equivalent volume of an organic phase and an aqueous phase.

Figure 13.8 The partition coefficient.

Compounds with a *large* partition coefficient are more soluble in the organic solvent, are typically nonpolar and hydrophobic, and tend *not* to be ionized. In contrast, compounds with a *small* partition coefficient are more soluble in *water*, are typically polar and hydrophilic, and may be ionized.

During liquid-liquid separations, an **extraction** occurs when a *desired* substance is removed from its original solution by dissolving (ie, partitioning) into the newly added liquid layer. In contrast, a **wash** occurs when an *undesired* substance (eg, byproduct, impurity) is dissolved into the newly added liquid layer (Table 13.3).

Table 13.3 Comparison of liquid-liquid separation approaches.

Liquid-liquid separation	Original solution	Added solvent layer	Location of desired substance after mixing
Organic extraction	Aqueous	Organic	Organic
Organic wash	Aqueous	Organic	Aqueous
Aqueous extraction	Organic	Aqueous	Aqueous
Aqueous wash	Organic	Aqueous	Organic

In an **extraction**, the desired substance is *transferred* to the added solvent layer, while in a **wash**, the desired substance *remains* in the original solvent layer.

Performing a Liquid-Liquid Separation

Liquid-liquid separations are typically performed in a **separatory funnel** (Figure 13.9).

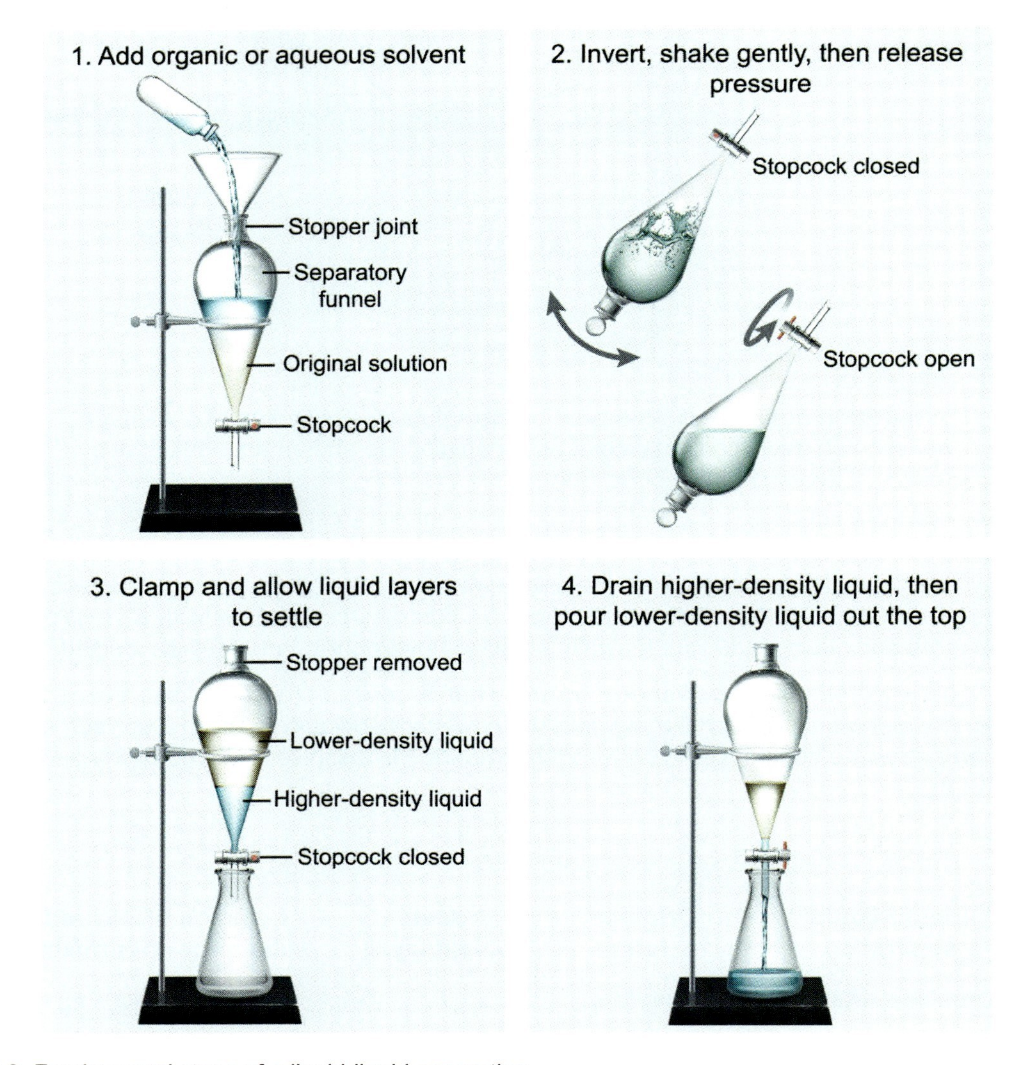

Figure 13.9 Fundamental steps of a liquid-liquid separation.

The solute and the two immiscible liquids are poured into the separatory funnel, and the mixture is thoroughly mixed to allow the solute to partition into its preferred solvent. After mixing, the liquids then separate into different layers. The top layer contains the less-dense liquid, and the bottom layer contains the denser liquid.

Depending on the identities of the solvents used, either layer may be the aqueous layer or the organic layer. Most organic solvents are less dense than water and are therefore the *top* layer in the separatory funnel; however, two notable exceptions are the halogenated solvents dichloromethane (CH_2Cl_2) and chloroform ($CHCl_3$), which are denser than water, making them the *bottom* layer in the separatory funnel (see Concept 7.1.01).

Table 13.4 lists the general steps of a typical liquid-liquid separation.

Table 13.4 General steps in the workup of an organic reaction.

Step	Technique	Description
1	Organic extraction	The nonpolar substances in the crude reaction mixture are **extracted** into an organic solvent. Repeated 3 times.
2	Aqueous wash	The combined organic extracts (ie, organic layer) are **washed** with water to remove residual polar components.
3	Brine wash	The organic layer is **washed** with saturated aqueous NaCl (eg, brine) to remove most water molecules from the organic layer.
4	Drying, filtration	The organic layer is **dried** over an anhydrous salt (eg, Na_2SO_4 or $MgSO_4$) to remove remaining water molecules, then **filtered**.
5	Concentration	The organic solvent is **evaporated** to generate the organic solutes from the reaction.

pH-Dependent Liquid-Liquid Separations

Molecules with **ionizable** functional groups (eg, carboxylic acids, phenols, amines) may have different aqueous solubilities depending on the pH (see Concept 13.1.01). Although the *neutral* forms of these functional groups have a higher solubility in an organic solvent, their *ionized* forms generally have a higher solubility in an aqueous solvent.

A sequence of liquid-liquid separations can be utilized to separate an ionizable organic molecule from a mixture of other organic substances. First, the organic mixture is dissolved in a low-polarity organic solvent that is immiscible with water (eg, CH_2Cl_2, ethyl acetate). Then, this organic solution is extracted by an aqueous solution with *either* an acidic or basic pH (Figure 13.10).

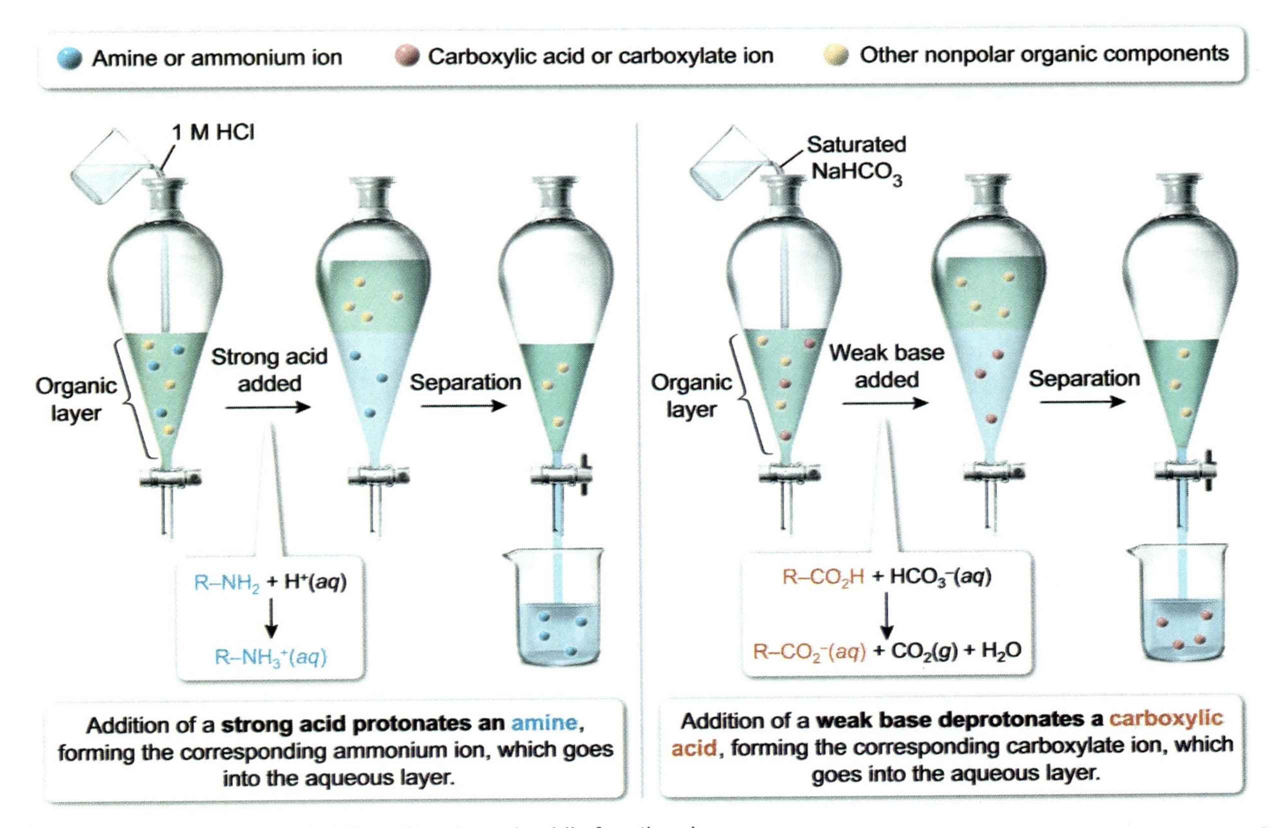

Figure 13.10 Aqueous extraction of basic and acidic functional groups.

Aqueous extraction of an *acidic* functional group (eg, carboxylic acid) requires an aqueous solution with a *basic* pH (eg, saturated $NaHCO_3$, 1M NaOH), whereas aqueous extraction of a *basic* functional group (eg, amine) requires an aqueous solution with an *acidic* pH (eg, 1M HCl). In each case, the addition of the pH-adjusted aqueous solution results in the ionization of the functional groups and the movement of the molecules from the organic phase into the newly added aqueous phase.

After extraction, the pH of the collected aqueous extract can be adjusted to the *opposite* pH extreme, converting the charged compound back to its uncharged form. This allows the combined aqueous solution to be **back-extracted** with an organic solvent to transfer the *neutralized* organic solute into a fresh organic layer, without impurities (Figure 13.11). Finally, the combined organic extracts can be concentrated through evaporation to isolate the ionizable organic molecule.

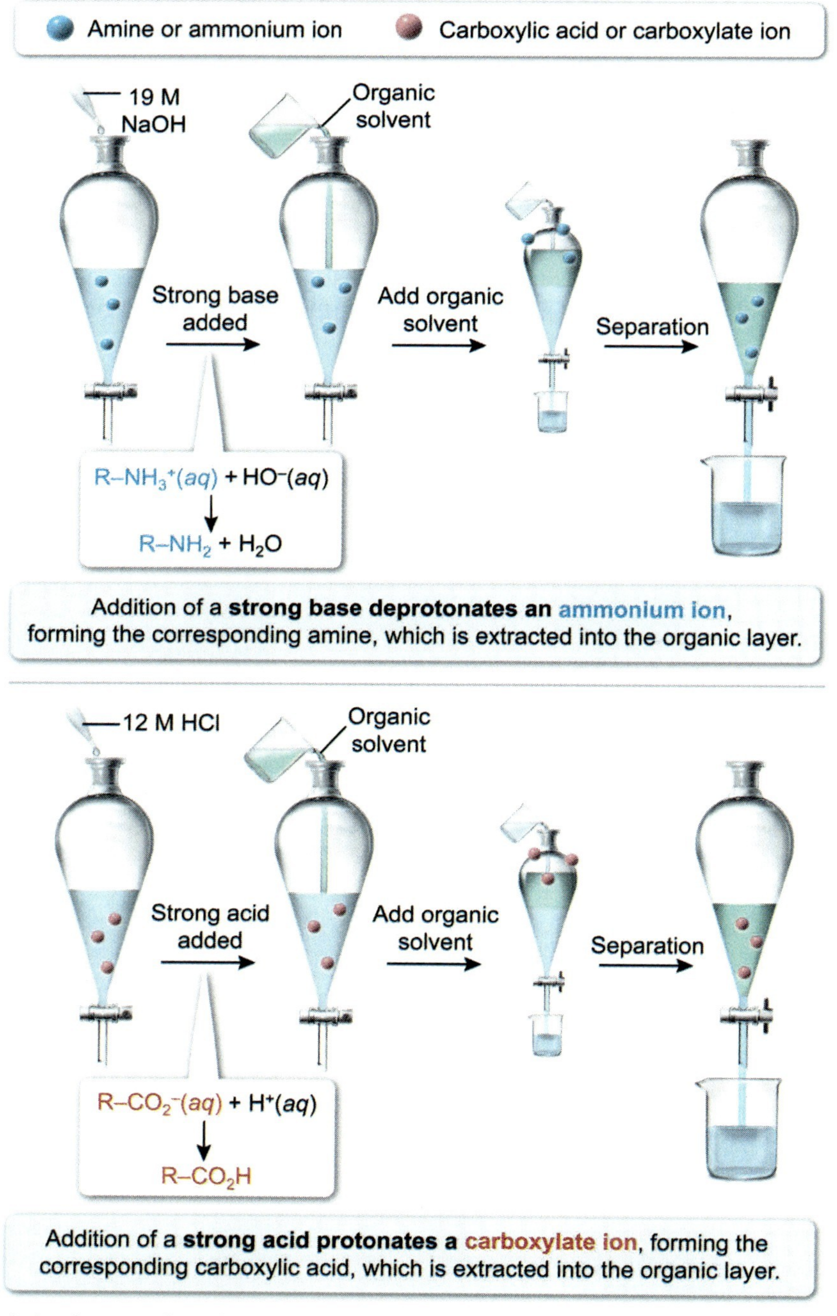

Figure 13.11 Organic back-extraction of basic and acidic functional groups.

One potential limitation of acid-base extractions is that other similarly ionizable substances may *also* be extracted. For example, it would be very challenging to selectively isolate one specific carboxylic acid from a mixture of several carboxylic acid molecules.

Concept Check 13.1

How could liquid-liquid extraction be employed to separate a mixture of the following two compounds?

OH
H_3C

O
OH
H_3C

Solution

Note: The appendix contains the answer.

Lesson 13.2

Distillation

Introduction

Lesson 13.1 describes laboratory methods used to separate or purify mixtures of substances based on differences in solubility. Heterogeneous solid-liquid mixtures can be separated using precipitation followed by filtration (see Concept 13.1.02), whereas heterogeneous liquid-liquid mixtures (involving immiscible solvents) can be separated based on the partitioning of a solute between one of the two liquid phases (see Concept 13.1.03).

This lesson describes how a homogeneous mixture of *miscible* liquids can be separated through a process called **distillation**.

13.2.01 General Principles of Distillation

Concept 2.4.01 defines the boiling point (bp) of a substance as the temperature at which either

- Its vapor pressure is equivalent to the ambient pressure of the system
- The substance undergoes a rapid phase transition from liquid to gas

For a *pure* liquid A in a closed container, the gaseous vapor above the liquid has a composition *identical* to that of the liquid (ie, both are pure, 100% substance A). However, for a *mixture* of liquids, the vapor has a composition *different* from that of the liquid mixture.

This is because a given molar amount of the lower-bp substance (ie, the substance that has a *higher* vapor pressure) in a liquid mixture *contributes more vapor molecules* than the same amount of the higher-bp substance.

Consequently, if a mixture of liquids is heated, the vapors collected, and the collected vapors condensed *back to a liquid* (ie, one **vaporization-condensation cycle**), the composition of the condensed liquid mixture is *enriched in the lower-bp liquid* and deficient in the higher-bp liquid (Figure 13.12). In contrast, the solution in the original container, having *lost* more lower-bp molecules, is now enriched in the *higher-bp* substance. The *larger* the boiling point difference between the components of the liquid mixture, the *greater* the enrichment (ie, separation).

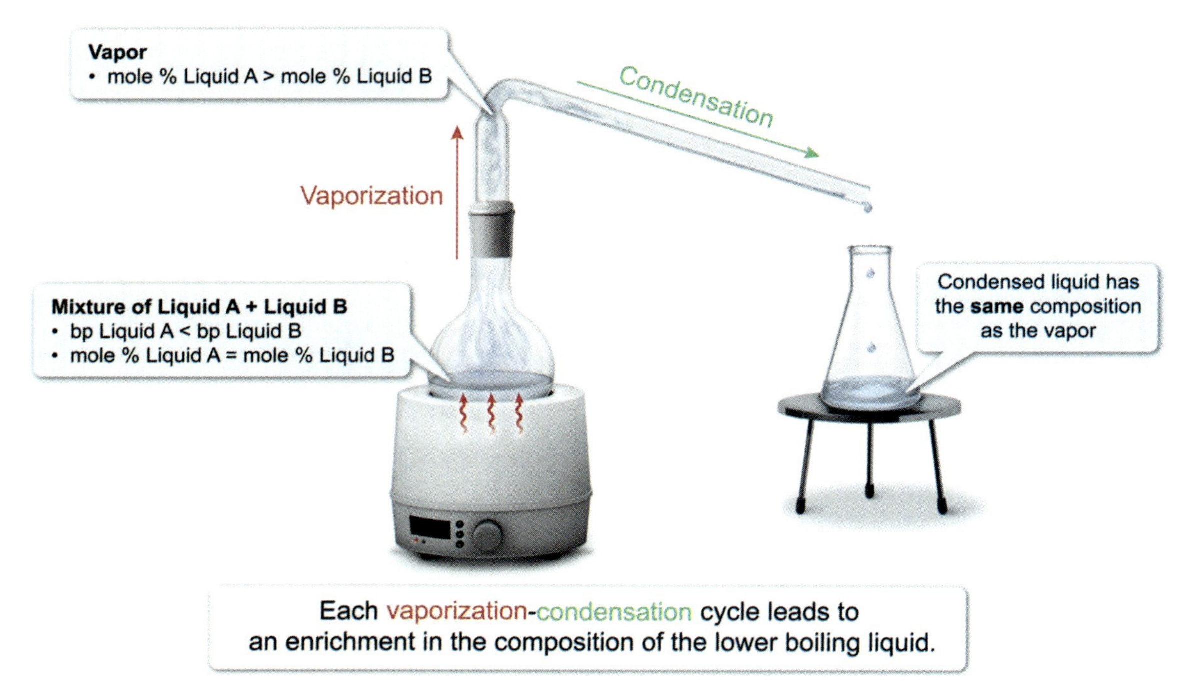

Figure 13.12 The separation of a mixture of volatile liquids through a single vaporization-condensation cycle.

Distillation is a laboratory procedure in which the components in a mixture are separated based on a difference in boiling points using vaporization-condensation cycles. At least one of the components must be volatile. The initial liquid mixture is called the **distilland**, and the resultant liquid after distillation is called the **distillate** (Figure 13.13).

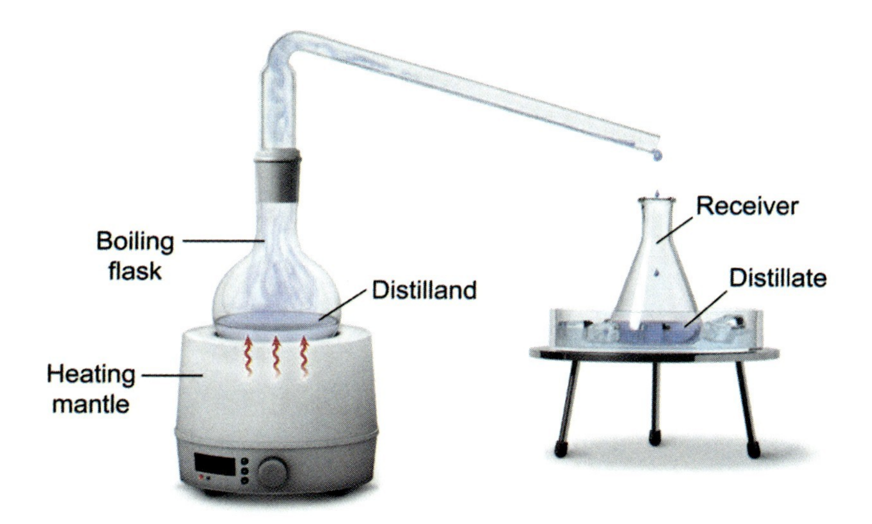

Figure 13.13 Required components for a distillation.

To ensure a safe and steady-paced distillation, either boiling chips or a stir bar is placed inside the boiling flask with the distilland (Figure 13.14). **Boiling chips** are made of a porous material and provide surfaces with nucleation sites where small bubbles can form, while the action of a **magnetic stir bar** can be controlled by a magnetic stirrer, usually located below the heating mantle. These items both disrupt the surface tension of the distilland and allow it to boil evenly, reducing the risk of **superheating** (ie, heating a liquid *above* its boiling point) and **bumping** (ie, vigorous formation of bubbles from a superheated liquid).

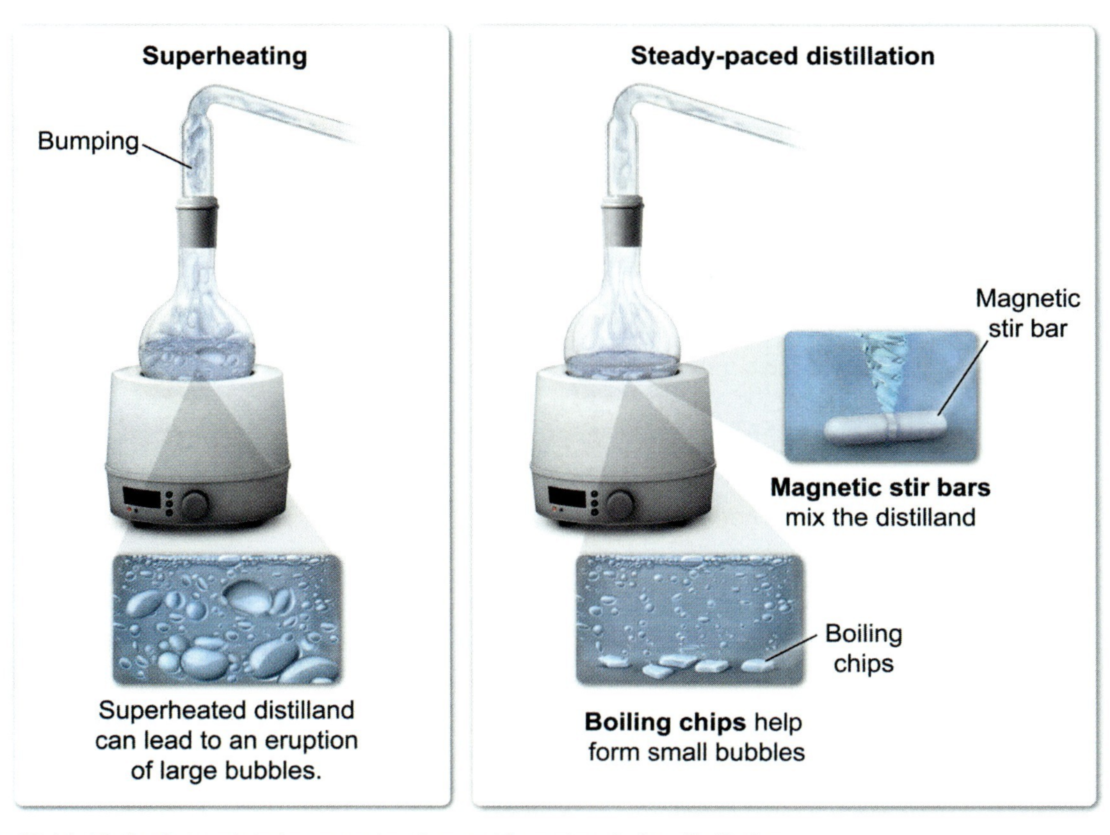

Figure 13.14 Methods to minimize superheating and bumping during distillation.

To monitor a laboratory distillation, a **thermometer** is connected to the top of the **distillation adapter** and lowered until its bulb is *just below* the shortest path vapors could take between the boiling flask and the receiver. Proper thermometer bulb placement ensures that the temperature of vapors is accurately measured during distillation (Figure 13.15).

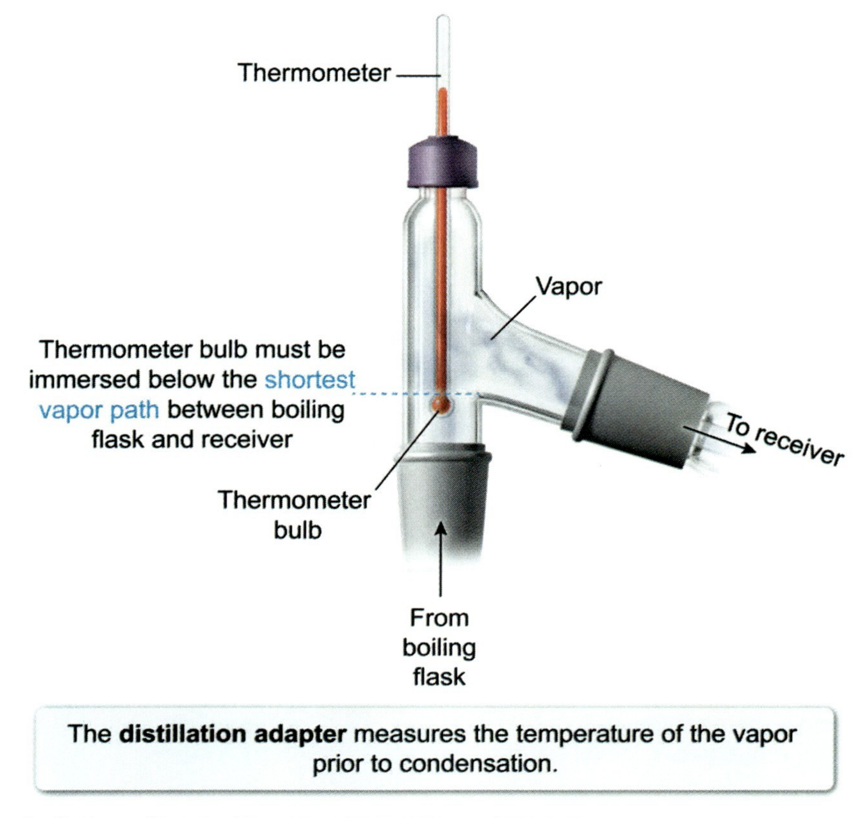

Figure 13.15 The distillation adapter with proper thermometer placement.

Many types of distillation use a condenser to increase the efficiency of vapor condensation (Figure 13.16). A **condenser** is a jacketed glass tube (ie, a tube within a tube) that allows a cooling agent to surround the vapor path. The cool inner surface facilitates the condensation of vapors to form the distillate. Typically, the cooling agent (often water) is directed into the *lower* condenser port and exits the *higher* condenser port. This way, the flow of water through the condenser runs against gravity and ensures the greatest cooling efficiency by maximizing the fill of the jacket and by operating in a countercurrent fashion to the condensed vapor flow.

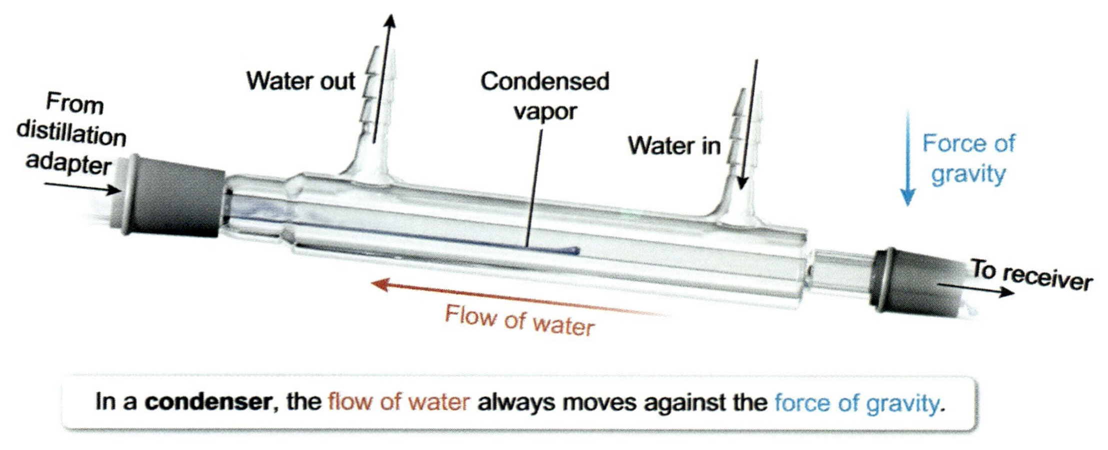

Figure 13.16 Use of a condenser in distillation.

Toward the end of the distillation apparatus, the **receiver adapter** may be connected to either an inert gas line (eg, N_2 if the components are sensitive to oxygen) or a vacuum source (for vacuum distillation

[Concept 13.2.03]). If neither of these items are required, the receiver adapter port is either unused (ie, open to the air) or connected to a **drying tube**, which prevents contamination by ambient water vapor (Figure 13.17).

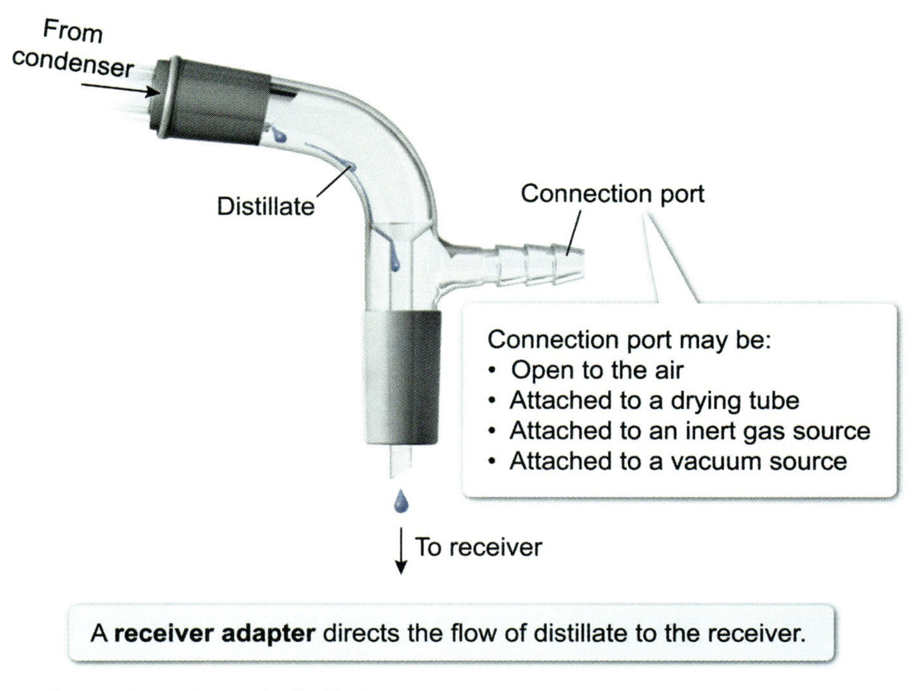

Figure 13.17 Use of a receiver adapter in distillation.

Each of the remaining concepts in this lesson focus on one variation of distillation, with emphasis on the applications of each method.

13.2.02 Simple Distillation

The simplest form of distillation is appropriately called a simple distillation. A **simple distillation** has exactly *one* vaporization-condensation cycle, in which vaporization occurs *only* in the boiling flask and condensation takes place in the condenser (Figure 13.18).

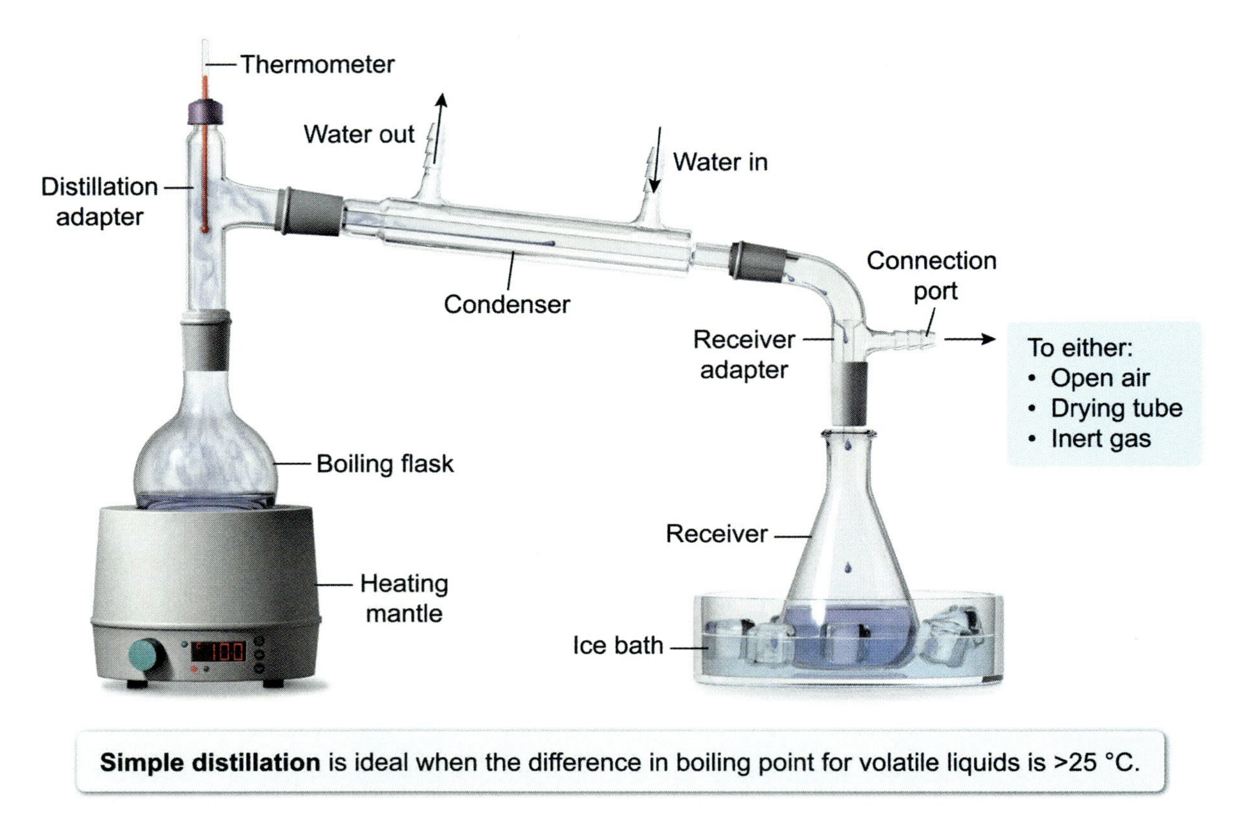

Figure 13.18 Simple distillation.

Simple distillation is used to separate substances with **large boiling point differences**. Ideally, it is used to purify a volatile liquid with *small amounts* of liquid contaminants that have boiling points *at least* 25 °C above that of the volatile liquid component. Contaminants that have boiling points *less* than 25 °C above that of the major liquid component may not be fully removed by simple distillation. In addition, for the reasons outlined in Concept 13.2.03, simple distillation is best performed for liquids with moderate to low boiling points (ie, boiling points *lower* than 150 °C), which includes many types of organic compounds.

13.2.03 Vacuum Distillation

As discussed in Concept 2.4.01 and Concept 13.2.01, the boiling point of a substance is defined as the temperature at which its vapor pressure is equivalent to the ambient pressure of the system. For a system that is open to the surrounding air, a substance boils when its vapor pressure equals *atmospheric pressure*—approximately 1 atm or 760 torr.

Consequently, a distillation conducted at *reduced* pressure will have boiling points at a *lower* temperature. In **vacuum distillation**, the *interior* of the distillation apparatus is placed under reduced pressure, *decreasing* the boiling point of its contents (Figure 13.19). A vacuum distillation is ideal when the atmospheric boiling point of the distilland is **greater than 150 °C**. The vacuum *lowers* the observed boiling point, which decreases the time required to perform the distillation and reduces the risk of heat-promoted chemical reactions.

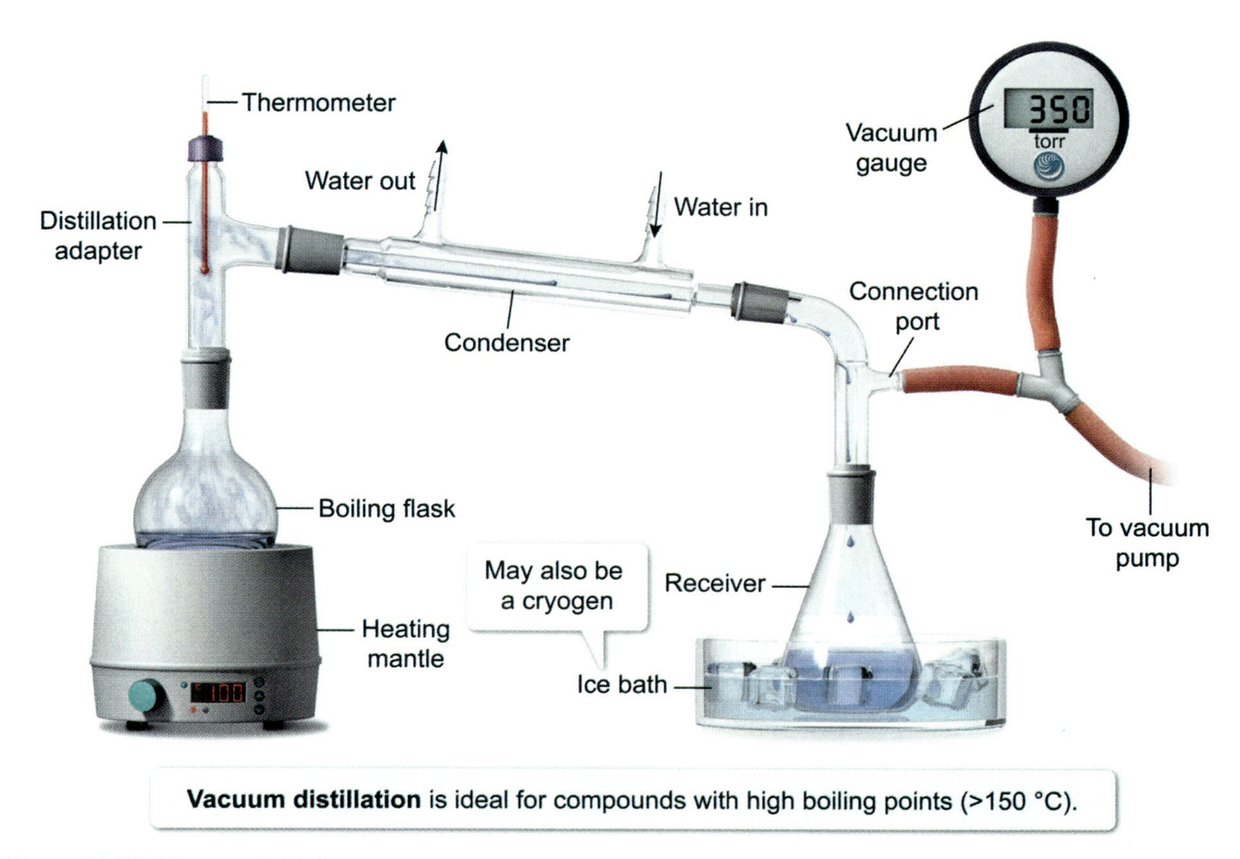

Figure 13.19 Vacuum distillation.

Note that vacuum distillations typically use a magnetic stir bar instead of boiling chips to prevent bumping. Under reduced pressure, the air pockets inside boiling chips are removed, which decreases their ability to prevent superheating.

13.2.04 Fractional Distillation

Concept 13.2.02 explains that simple distillation, which involves only a single vaporization-condensation cycle, is most effective when boiling points differ by *more* than 25 °C.

However, the amount of separation provided by a vaporization-condensation cycle is *cumulative*, and *multiple* vaporization-condensation cycles in sequence result in *greater* separation and enrichment of the lower-boiling component within the distillate. Figure 13.20 depicts one way in which multiple vaporization-condensation cycles could be accomplished. A *sequence* of simple distillations, in which the distillate from the previous simple distillation becomes the distilland for the subsequent simple distillation, increasingly enriches the lower-boiling liquid component.

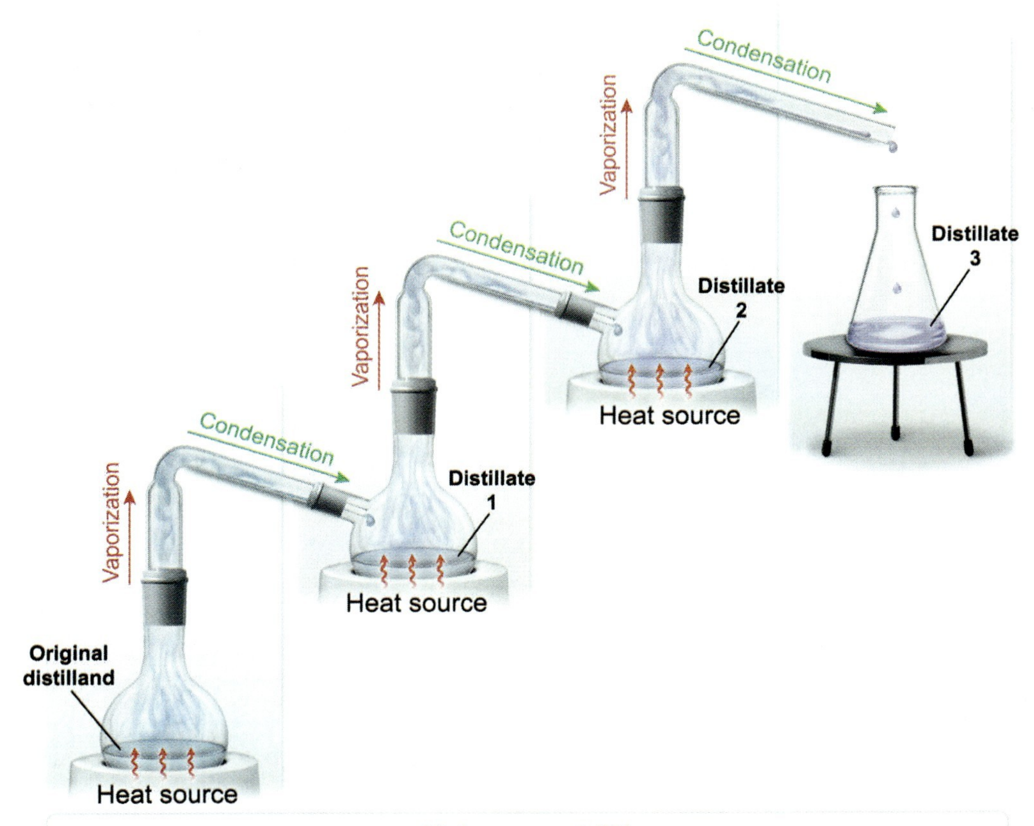

	Mole percent (%)			
Component	**Original distilland**	**Distillate 1**	**Distillate 2**	**Distillate 3**
Liquid A	50	81	95	99
Liquid B	50	19	5	1

Note: The bp of Liquid A < the bp of Liquid B.

Each *successive* vaporization-condensation cycle leads to an enrichment in the composition of the lower boiling liquid.

Figure 13.20 The effect of sequential simple distillations on distillate composition.

Alternatively, a distilling column may be used to cause the same effect (Figure 13.21). A **distilling column** usually contains a packing material (eg, beads, glass chips) or an irregular interior surface that provides *additional* surface area for distilled vapors to condense. Vapors condense within the distilling column and may then be revaporized to *further enrich* the composition of the low-boiling liquid component. Ultimately, the vapor leaving a distilling column has undergone **more than one** vaporization-condensation cycle and is *more* enriched in the low-boiling component than the vapor from a simple distillation.

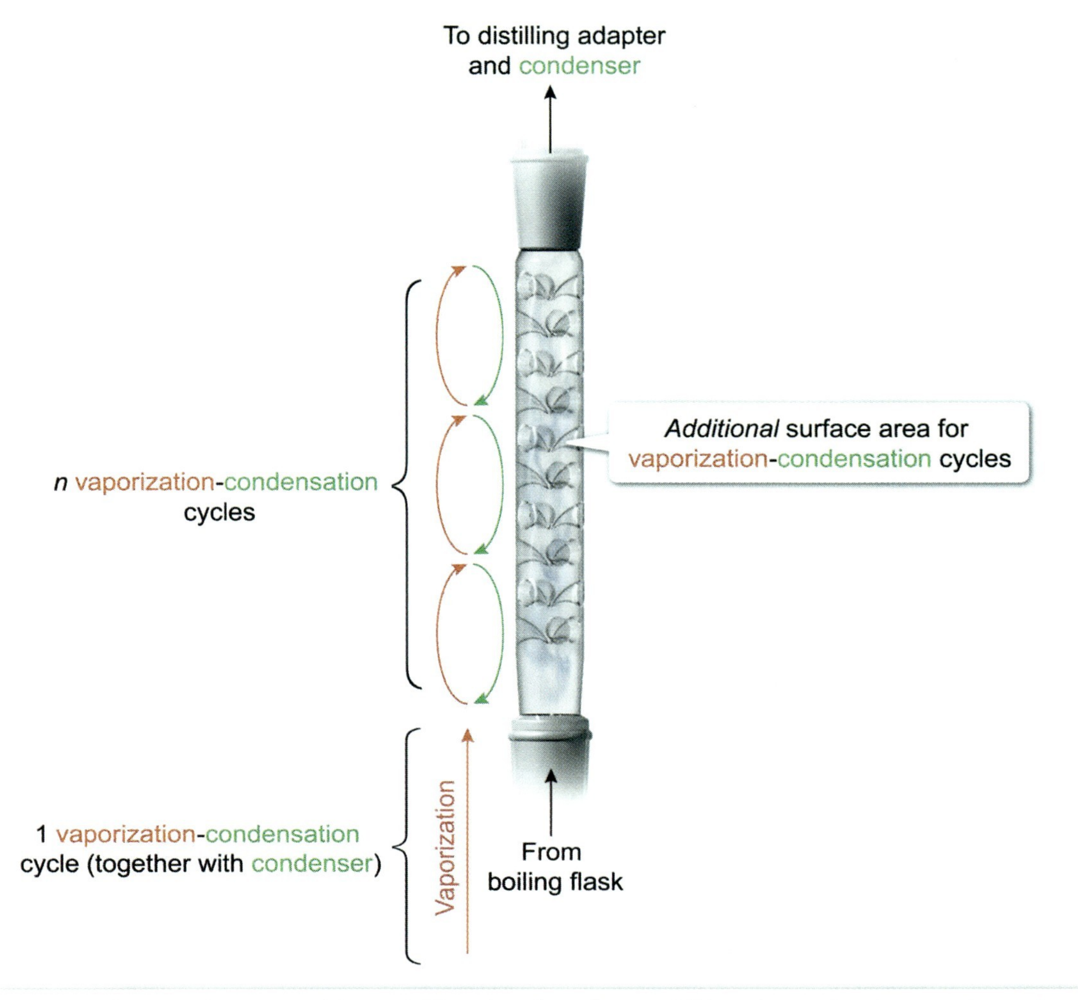

Figure 13.21 Function of a distilling column.

The efficiency of a distilling column is proportional to the average number of *additional* vaporization-condensation cycles it provides to a distillation. Distillations that utilize a distilling column are called **fractional distillations** (Figure 13.22). Because of their capacity for additional enrichment, fractional distillations are typically used to separate mixtures of volatile liquids that have a *difference* in boiling points of **less than 25 °C**.

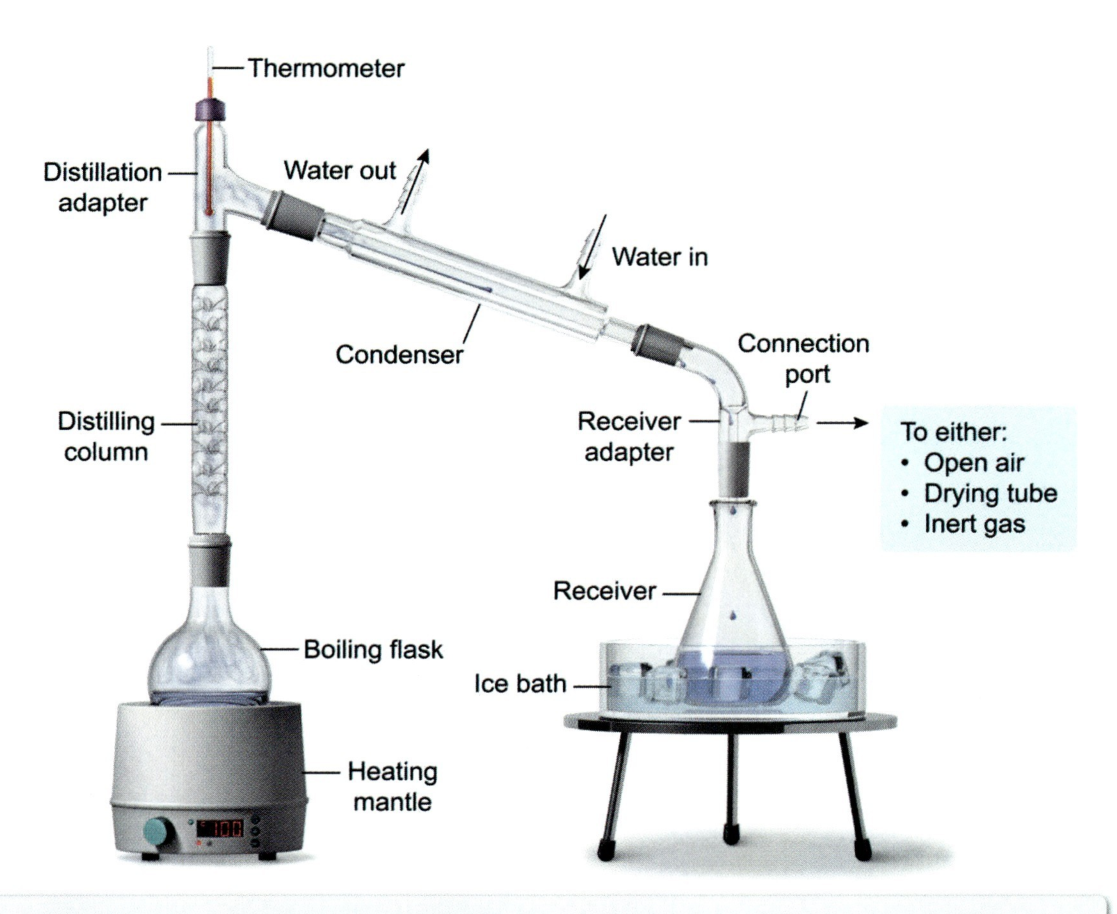

Figure 13.22 Fractional distillation.

Concept Check 13.2

What distillation method would be most appropriate to separate a 1:1 molar mixture of formamide (bp = 210 °C) and acetanilide (bp = 304 °C)?

Solution

Note: The appendix contains the answer.

Lesson 13.3

Chromatography

Introduction

This lesson describes how a mixture of substances can be separated through a process called **chromatography**. Concept 13.3.01 introduces the topic of chromatography, the experimental data produced through chromatography, and general considerations for performing effective chromatographic separations. The remaining concepts discuss specific types of chromatography.

13.3.01 General Principles of Chromatography

Chromatography is a laboratory method that separates molecules based on the *differential* interactions between the components of a mixture with a two-phase system consisting of a *mobile* (ie, moving) phase and a *stationary* (ie, not moving) phase (Figure 13.23). As the **mobile phase** flows with respect to the **stationary phase**, substances that *preferentially* interact with the mobile phase become spatially separated from substances that *preferentially* interact with the stationary phase.

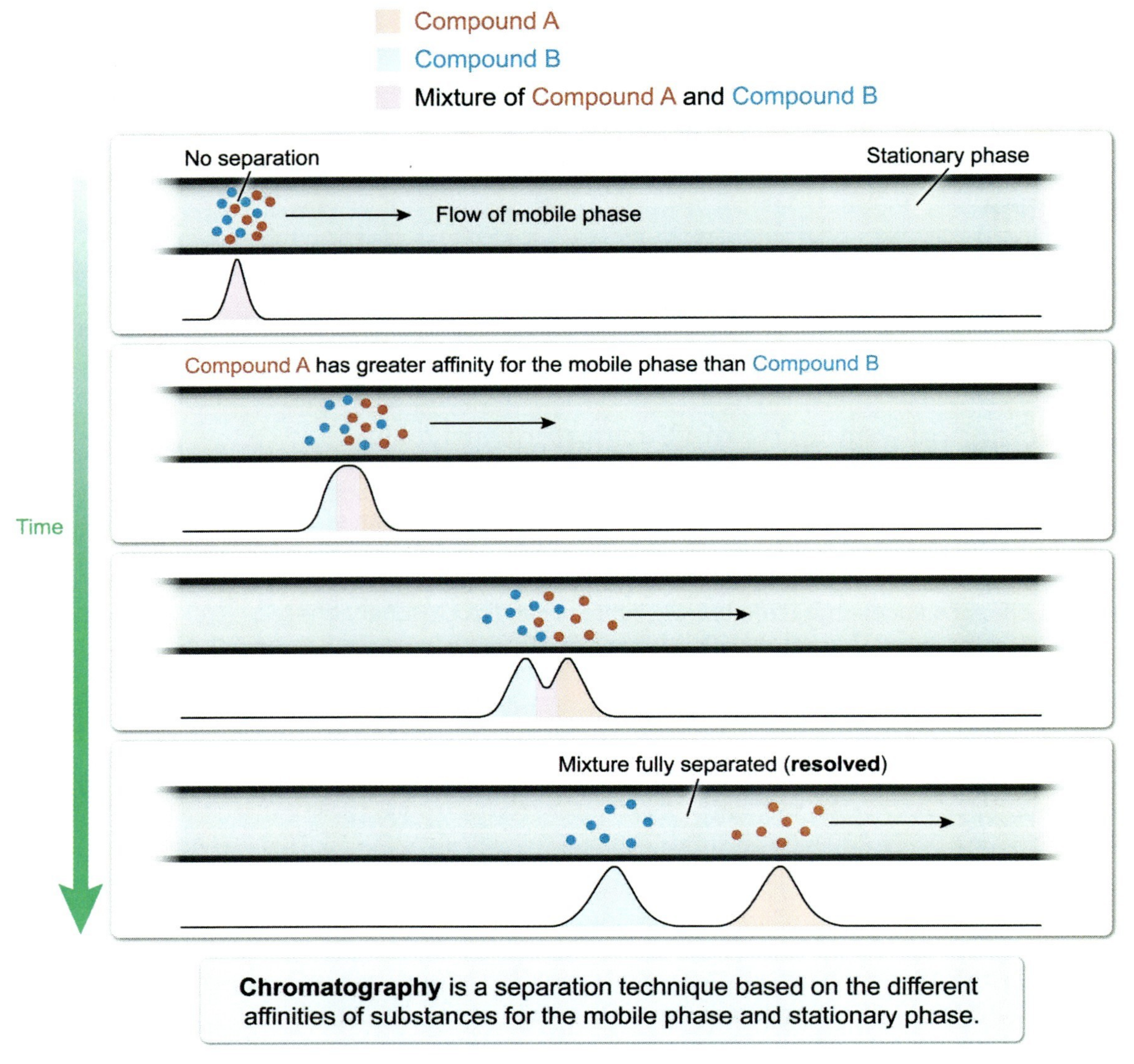

Figure 13.23 A general chromatography-based separation.

Given that most substances demonstrate *some* extent of interaction with *both* the mobile and stationary phases, chromatographic separation is based on the *relative difference* in interaction strength. When there are *large* differences in preference (eg, component A preferentially interacts with the mobile phase while component B preferentially interacts with the stationary phase) separation tends to occur *faster*. In contrast, when there are *small* differences in preference (eg, components A and B have similar preferences for the mobile or the stationary phase), separation tends to occur *more slowly*.

Chromatographic **resolution** (ie, the separation of the components of a mixture) is also dependent on the length of the stationary phase. *Longer* stationary phases provide *greater* resolution, and *shorter* stationary phases have *reduced* resolution. This is because longer stationary phases allow more *time* for separations to occur.

Consequently, the data produced through chromatography (called a **chromatogram**) display the measured signal intensity (*y*-axis) as a function of time (*x*-axis) (Figure 13.24). Different chromatography methods measure different types of signals over time.

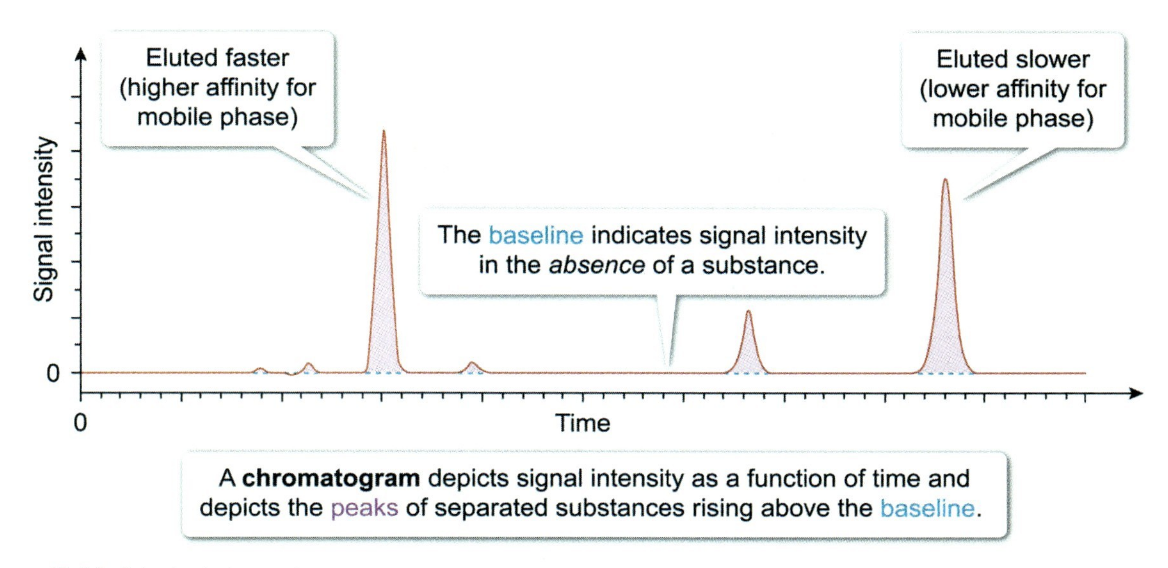

Figure 13.24 A typical chromatogram.

Within a chromatogram, the **baseline** represents the *absence* of detected sample and typically displays a *linear* relationship over time. In most chromatograms, the baseline is flat and has a slope of zero. However, it is possible for some chromatograms to exhibit a sloped baseline, such as in methods that involve a mobile phase gradient (Concept 13.3.07). In either scenario, regions of the chromatogram where the signal *increases* above the baseline are called **peaks** and represent the presence of one or more substances from the original mixture.

A chromatogram peak is generally characterized by its retention time, dimensions, and shape. The **retention time** is the point on the *x*-axis where the signal has the greatest intensity, which is usually at or near the midpoint of the peak (Figure 13.25). Retention times have units that match the time units of the *x*-axis (eg, seconds, minutes, hours).

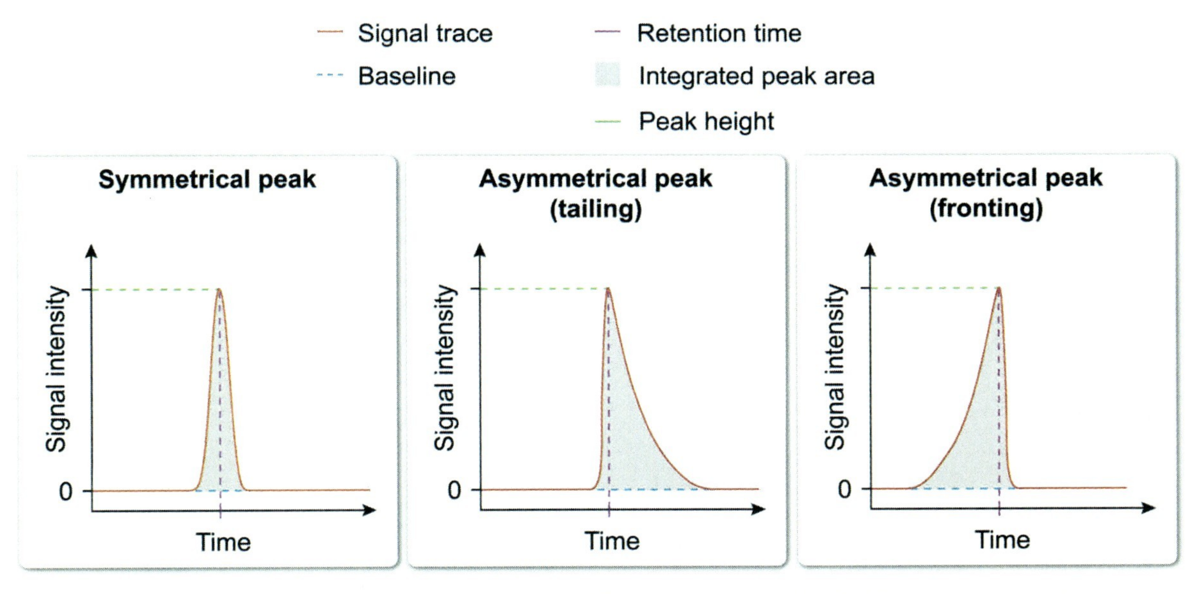

Figure 13.25 Examples of various chromatogram peaks. The retention time is identified as the timepoint corresponding to the highest intensity.

Since signal detection occurs after separation, a *lower* retention time indicates a substance that had a *higher affinity for the mobile phase* and therefore moved away from the stationary phase (ie, **eluted**) *faster*. Likewise, a *higher* retention time indicates a substance that had a *higher affinity for the stationary phase* and was *slower* to elute. Consequently, retention times are *directly* related to the amount of time a substance is retained by the stationary phase, and they are *inversely* related to the preferential interaction with the mobile phase.

The dimensions of a peak (eg, height, width, area) relate to the **amount of substance** detected. The **peak area** is the area between the signal trace (ie, data) and the baseline and is typically calculated by an instrument through integration. Peaks with *larger* areas generally indicate a *larger* amount of substance detected. Sometimes, **peak height** (ie, the distance from the baseline to the highest intensity point) is used instead of peak area if integration is not possible.

In the context of chromatogram peaks, the components in a mixture are fully separated (ie, resolved) when peaks have a region of baseline between them. A *larger* region of baseline separation generally indicates *greater* resolution of substances and a more effective separation.

13.3.02 Gas Chromatography

Gas chromatography (GC) is a chromatography method in which the mobile phase is a stream of an inert *gas* such as helium (He), argon (Ar), or nitrogen (N_2). The stationary phase is a narrow-diameter tube that is *either* packed with small particles of a solid (as in gas-solid chromatography) or lined with a liquid polymer (as in **gas-liquid chromatography**). In either case, the stationary phase is typically called a gas chromatography column.

A typical gas chromatography experiment has the following essential procedural steps (Figure 13.26):

1. The sample to be separated is **injected** into the column inlet (ie, entry point), often as a solution in a solvent with a low boiling point (eg, methanol, ethanol, hexane).
2. The sample is **heated** to vaporize its components.
3. The carrier gas (ie, the mobile phase) carries the sample components through the column (containing the stationary phase) and the components are **separated**. Usually, the column is located within a heated chamber (ie, oven) to maintain an elevated temperature throughout the separation.
4. A **detector** measures the materials that elute from the column over time, generating a chromatogram.

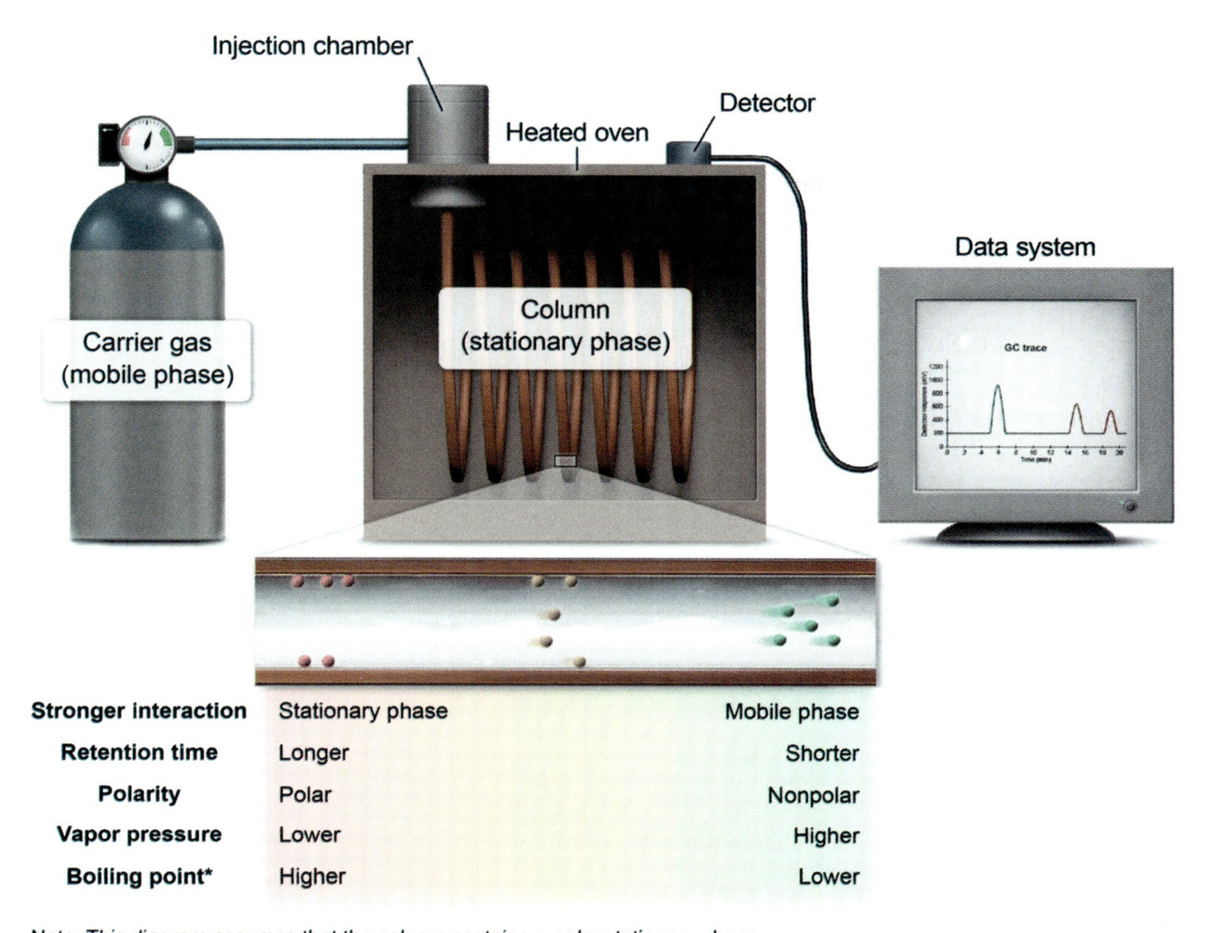

Note: This diagram assumes that the column contains a *polar* stationary phase
*While boiling point and vapor pressure are *often* synonymous, vapor pressure is generally the better metric.

In **gas chromatography**, passage of a carrier gas through a column separates the components of a mixture based on a *combination* of polarity and vapor pressure.

Figure 13.26 The essential components of gas chromatography.

Samples that are best suited for gas chromatography should be heat stable and should *not* readily decompose. Since GC requires substances to be vaporized for separation, an organic compound must have appreciable vapor pressure between 150 to 300 °C.

While a GC experiment can utilize many different types of detectors, one of the most frequently utilized detectors is a mass spectrometer (see Lesson 14.4). In a **gas chromatography–mass spectrometry (GC-MS)** experiment, the GC column *separates* the components of the mixture, and each separated component is *analyzed* by mass spectrometry to construct its mass spectrum. In this way, GC-MS generates a mass spectrum for *each* retention time within the gas chromatogram, providing a rich source of information about the components in the original mixture.

Order of Elution

The order of elution of substances from a GC column is dependent on two factors (Table 13.5):

- The relative vapor pressures (and therefore relative boiling points) of the substances being separated
- The polarity of the material within the GC column

Table 13.5 Influence of variables on the order of elution in gas chromatography.

Variable	Order of elution (Increasing retention time)
Vapor pressure	Nonpolar → Polar
Polar stationary phase (GC column)	Nonpolar → Polar
Nonpolar stationary phase (GC column)	Polar → Nonpolar

Less common (Nonpolar stationary phase)

Note: Assumes substances have a similar molecular weight.

While the vapor pressure and stationary phase *both* influence the retention time of substances in gas chromatography, in most situations, **nonpolar substances have a lower retention time than polar substances.**

Because the mobile phase in GC is a gas, molecules that more easily transition to the gas phase tend to elute more quickly. This corresponds to *volatile* molecules, which have *higher vapor pressures* and **lower boiling points**. In contrast, less volatile molecules can temporarily deposit onto the surface of a solid stationary phase or dissolve into a liquid stationary phase, slowing their travel. Therefore, less volatile molecules (ie, lower vapor pressures, higher boiling points) elute *more slowly*. In this way GC can be thought of as an *analytical* complement (see Concept 13.3.03) to the *preparative* method of distillation (Lesson 13.2).

This general relationship between elution time and boiling point is often accurate and is likely sufficient to answer most exam questions. However, the complete picture of GC order of elution, like other forms of chromatography, also involves specific molecule–stationary phase interactions and therefore is more complex.

For example, a GC column containing a *polar* stationary phase tends to interact favorably with *polar* substances and may retain them longer than nonpolar components. In some cases, low boiling point molecules that are small and highly polar have strong enough interactions that they elute *after* larger nonpolar molecules with *high* boiling points.

In nonpolar GC columns, interactions between nonpolar molecules tends to be weak, and separation is based *mostly* on vapor pressure and boiling point. However, when comparing the retention times of two molecules with *similar boiling points*, polar molecules tend to elute *before* nonpolar molecules. This result may be explained through the principle of "like dissolves like": the nonpolar molecules dissolve and are therefore retained longer, whereas polar molecules are not.

Unless otherwise stated, however, it is reasonable to correlate GC elution times with a molecule's vapor pressure and boiling point.

13.3.03 Introduction to Liquid-Solid Chromatography

In **liquid-solid chromatography** (often called simply *liquid chromatography*), a *liquid* mobile phase moves through a stationary phase composed of small *solid* particles. Throughout a liquid chromatography experiment, solutes dissolved in the mobile phase separate based on the solutes' relative affinity for the mobile versus the stationary phase.

Normal-Phase and Reversed-Phase Approaches

All liquid-solid chromatography methods are described as either a normal-phase or a reversed-phase separation (Figure 13.27). In **normal-phase (NP)** liquid-solid chromatography, the liquid mobile phase is *nonpolar* (eg, hexanes) and the solid stationary phase is *polar* (eg, silica gel). Conversely, in **reversed-phase (RP)** liquid-solid chromatography, the liquid mobile phase is *polar* and the solid stationary phase is *nonpolar*.

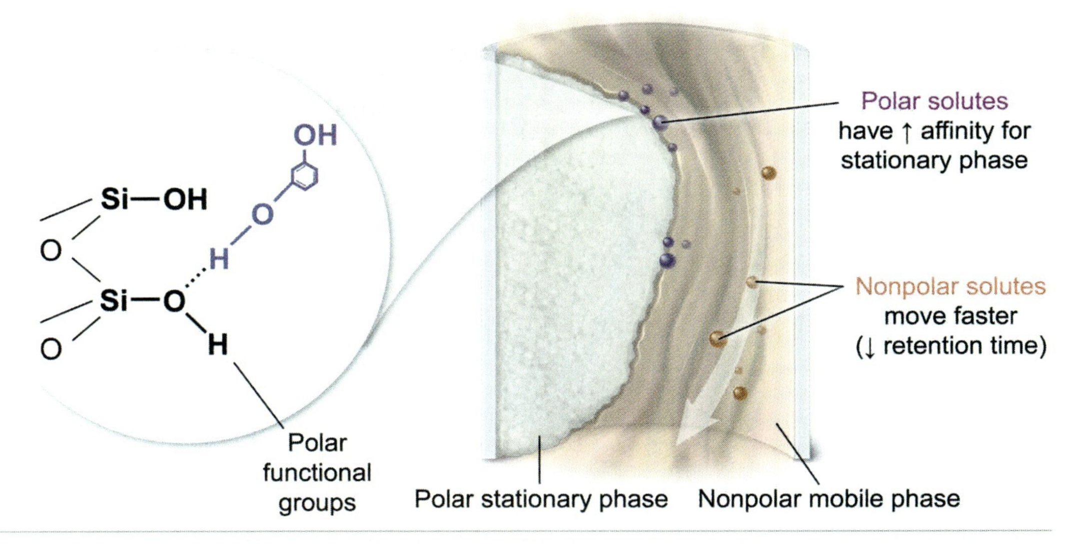

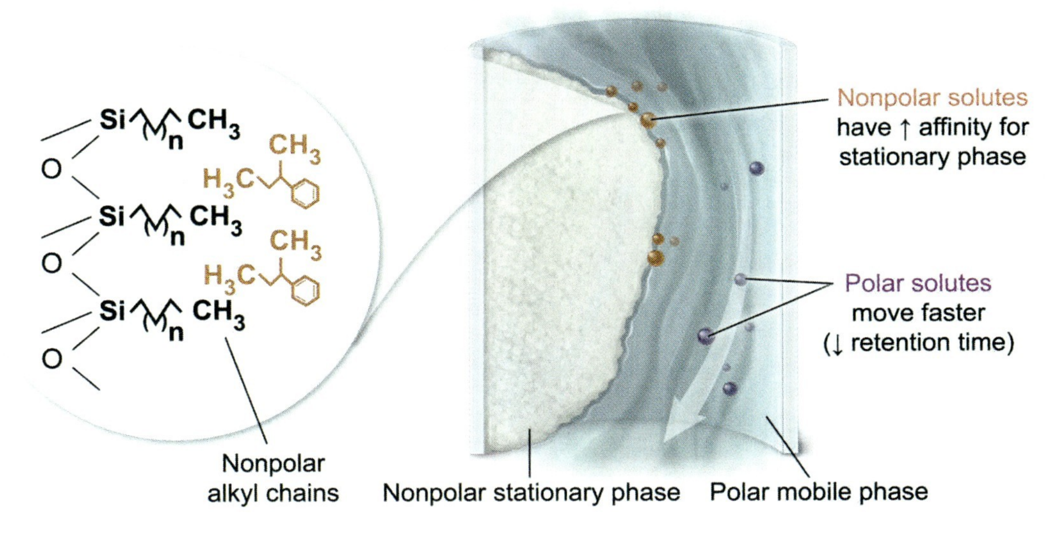

Figure 13.27 Normal- and reversed-phase liquid-solid chromatography.

The relationship between solute molecules, the mobile phase, and the stationary phase is derived from the principle of "like dissolves like" (see Concept 2.4.02 and Concept 13.1.01) and the intermolecular forces. In normal-phase chromatography, *nonpolar* solute molecules have a greater affinity for the nonpolar *mobile phase* and elute from the stationary phase *faster* (ie, they have a shorter retention time). In contrast, *polar* solute molecules have a greater affinity for the *polar stationary phase* and are eluted *slower* (ie, they have a longer retention time).

In reversed-phase chromatography, *polar* solute molecules have a greater affinity for the polar mobile phase and are eluted from the nonpolar stationary phase faster, while nonpolar solute molecules have a greater affinity for the nonpolar stationary phase and are eluted slower.

Due to the complementary nature of normal-phase and reversed-phase chromatography, these techniques are used to separate *different* mixtures of compounds. Normal-phase chromatography, which uses a nonpolar mobile phase, is typically used to isolate desired nonpolar to moderately polar organic substances from unwanted polar byproducts or contaminants. Likewise, reversed-phase chromatography, which uses a polar mobile phase, is typically used to isolate desired polar organic substances from unwanted nonpolar byproducts or contaminants.

Analytical and Preparative Chromatography

Liquid-solid chromatography methods can be used either to *analyze* the components of a mixture, to *isolate* larger quantities of one or more pure substances, or *both*. **Analytical liquid-solid chromatography** provides data (eg, a chromatogram) about the relative composition of the mixture (Table 13.6). Analytical chromatography generally requires only a small amount (1 ng to 1 mg) of the mixture. Consequently, the material used for analytical chromatography is often *not recovered* and is discarded with the experiment waste.

Table 13.6 Comparison of analytical and preparative chromatography.

Chromatography type	Quantity of material separated	Material recovered?	Experimental outcome
Analytical	Small	No	Data about mixture composition
Preparative	Small to large	Yes	At least one pure substance

In comparison, **preparative liquid-solid chromatography** *recovers* one or more components of the mixture, which can later be used for other laboratory purposes (eg, the next step in a synthesis). While preparative liquid-solid chromatography can accomplish the separation of *either* small or large (1 kg or more) quantities of the mixture, the specific approaches tend to be quite different depending on the scale. Effective preparative liquid-solid chromatography methods minimize the loss of material during separation.

Overview of Liquid-Solid Chromatography Methods

The final four concepts in this lesson each describe a commonly encountered liquid-solid chromatography method. Table 13.7 provides a summary of the following liquid-solid chromatography methods:

Table 13.7 Comparison of liquid-solid chromatography methods.

Chromatography method	Analytical or preparative?	Mobile and stationary phases	Mixtures most effectively separated
Paper chromatography	Analytical	Normal	Highly polar solutes
Thin-layer chromatography (TLC)	Analytical or preparative	Normal	Nonpolar solutes
		Reversed	Polar solutes
Column chromatography	Preparative	Normal	Nonpolar solutes
		Reversed	Polar solutes
High-performance liquid chromatography (HPLC)	Analytical or preparative	Normal	Nonpolar solutes
		Reversed	Polar solutes

13.3.04 Paper Chromatography

Paper chromatography, which uses paper as the stationary phase, employs highly polar organic solvents (eg, alcohols) or an alcohol and water mixture as the mobile phase. The paper stationary phase is composed of molecules of cellulose, a *polar* carbohydrate polymer.

While alcohols (eg, methanol, ethanol, isopropanol) are normally considered to be highly polar organic solvents (see Concept 13.1.01), alcohols are *less* polar than the cellulose in the paper stationary phase. Consequently, paper chromatography is a *normal-phase* separation method. Paper chromatography is typically used to analyze a mixture of *polar* organic substances (eg, carbohydrates, amino acids, organic dyes); it is generally *not* used as a means of preparative chromatography.

To perform paper chromatography, a small amount of sample is placed (or *spotted*) on one end of a piece of chromatography paper (Figure 13.28). The chromatography paper is placed in a closed chamber containing the mobile phase, such that the end of the paper is immersed in the mobile phase, but *not* the spot containing the sample. Capillary action draws the mobile phase upward *through* the paper and facilitates separation. The chromatography paper is removed from the chamber *before* the line of the mobile phase (ie, **solvent front**) reaches the end of the paper.

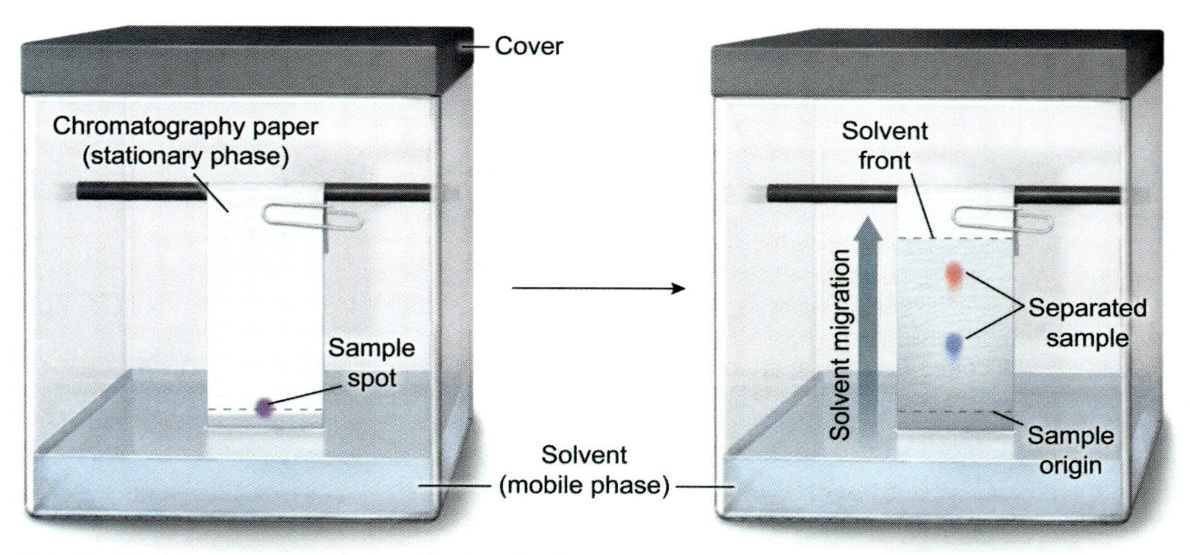

Figure 13.28 Paper chromatography.

If the separated substances have a visible color, paper chromatography data can be analyzed directly. If the separated substances do *not* have a visible color, a chemical stain is typically used to identify the location of substances. A **chemical stain** is a reagent or solution that contains chemicals that react with one or more functional groups, producing a colored product that facilitates the visual detection of otherwise colorless substances.

The Retention Factor

Since the chromatogram in paper chromatography is the paper with the separated components, the data are *static*. Consequently, the concept of retention time (see Concept 13.3.01) is modified to retention *factor*. **Retention factor (R_f)** is the ratio of the distance a *spot* traveled from the origin relative to the distance the *solvent* traveled (ie, from the origin to the solvent front).

$$R_\text{f} = \frac{\text{Distance a spot traveled from origin}}{\text{Distance the solvent traveled from origin}}$$

R_f is always between 0 and 1 (Figure 13.29). Substances with $R_f = 0$ interact very strongly with the *stationary* phase and *do not move* at all with the passage of the mobile phase. Substances with $R_f = 1$ interact very strongly with the *mobile* phase and are *not impeded* by interaction with the stationary phase.

Most substances fall between these extremes and indicate *partial* interaction with both phases during the separation.

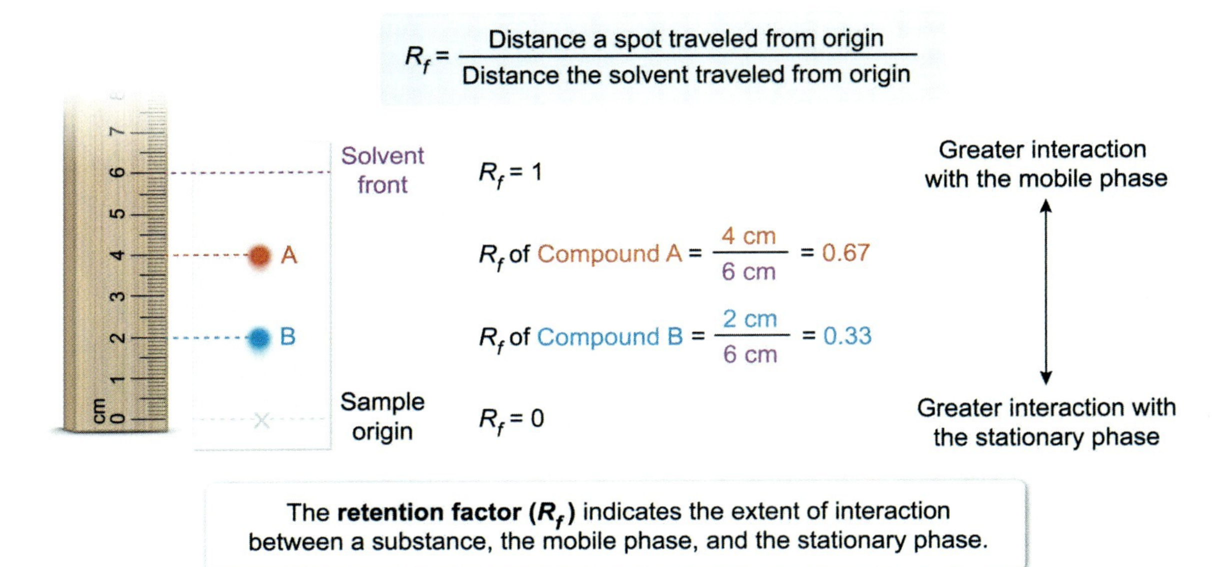

Figure 13.29 The retention factor.

The magnitude of a compound's R_f is an indication of its relative polarity. In paper chromatography, a larger R_f indicates a relatively *nonpolar* compound that had greater interaction with the (relatively) nonpolar *mobile phase* (ie, the alcohol or alcohol/water mixture). In contrast, a smaller R_f indicates a *polar* compound that did *not* travel far because of greater interaction with the highly polar *stationary phase* (ie, the paper).

Note that large R_f values correspond to *fast* travel rates. Although a paper chromatography experiment stops *before* the solvent reaches the end of the stationary phase, *if* the substances were allowed to elute, high-R_f substances would elute *first* and have a *short* elution time. Therefore, the value obtained for the retention *factor* (R_f) is *inversely correlated* to the retention *time* (R_t) value that would be obtained for a similar experiment.

The concept of R_f is also described in relation to thin-layer chromatography in Concept 13.3.05.

13.3.05 Thin-Layer Chromatography (TLC)

In a modern chemistry laboratory, paper chromatography has largely been replaced by more efficient methods, such as thin-layer chromatography. **Thin-layer chromatography (TLC)** is a liquid-solid chromatography method that employs a narrow layer of stationary phase that has been deposited onto an inert surface (eg, glass, plastic, aluminum) called a **TLC plate** (Figure 13.30).

A distinct advantage of TLC is the ability to use a wide variety of solid materials for the thin coating of stationary phase. Common stationary phases for TLC include polar materials such as silica gel and alumina (for normal-phase chromatography) and nonpolar materials such as silica gel modified with long-chain (eg, C_8, C_{18}) alkyl groups (for reversed-phase chromatography).

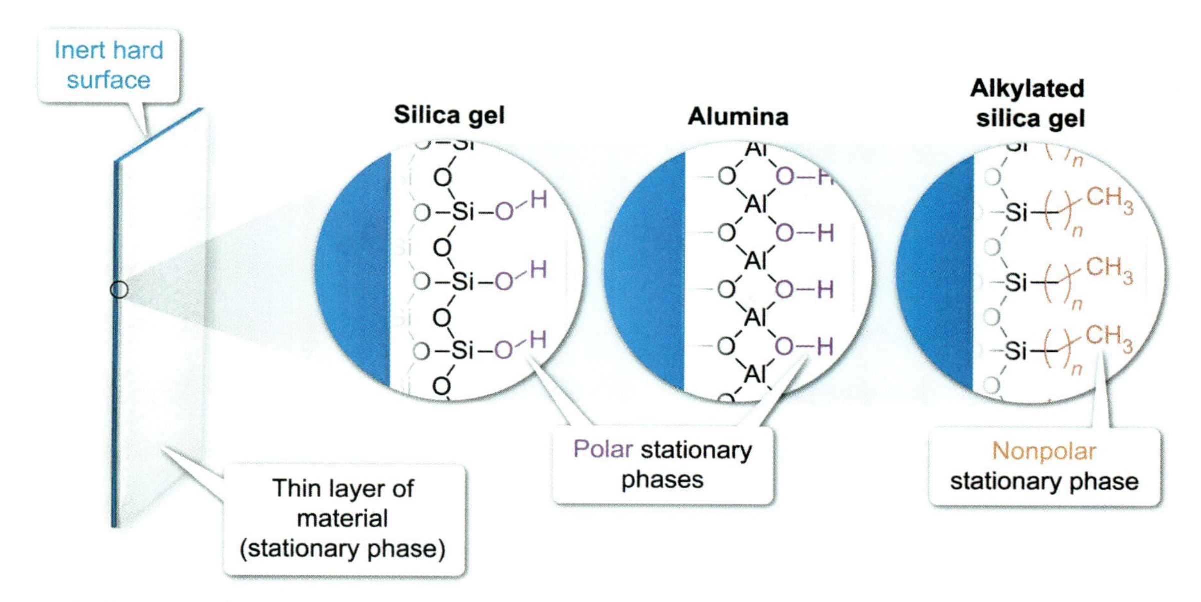

Figure 13.30 A typical TLC plate schematic.

The *thickness* of the stationary phase on a TLC plate determines whether a TLC experiment is used for analytical or preparative purposes. TLC plates with *thin* stationary phase coatings are generally used to *analyze* the components of a mixture or the eluent fractions from processes such as column chromatography (see Concept 13.3.06). TLC plates with *thick* stationary phase coatings can accommodate the separation of a *larger* amount of sample and may be used in **preparative TLC (PTLC)**, in which the purified components are extracted from the separated TLC plate.

To perform TLC, a small amount of sample is spotted on one end of a TLC plate (Figure 13.31). The TLC plate is placed in a closed chamber containing the mobile phase, such that the end of the TLC plate, but *not* the spot containing the sample, is immersed in the mobile phase. Capillary action draws the mobile phase *through* the stationary phase of the TLC plate, facilitating separation. It is important that the chamber remain sealed until the end of the experiment to minimize evaporation of the solvent as it travels. The developed (ie, separated) TLC plate is removed from the chamber *before* the line of the mobile phase (ie, **solvent front**) reaches the end of the TLC plate.

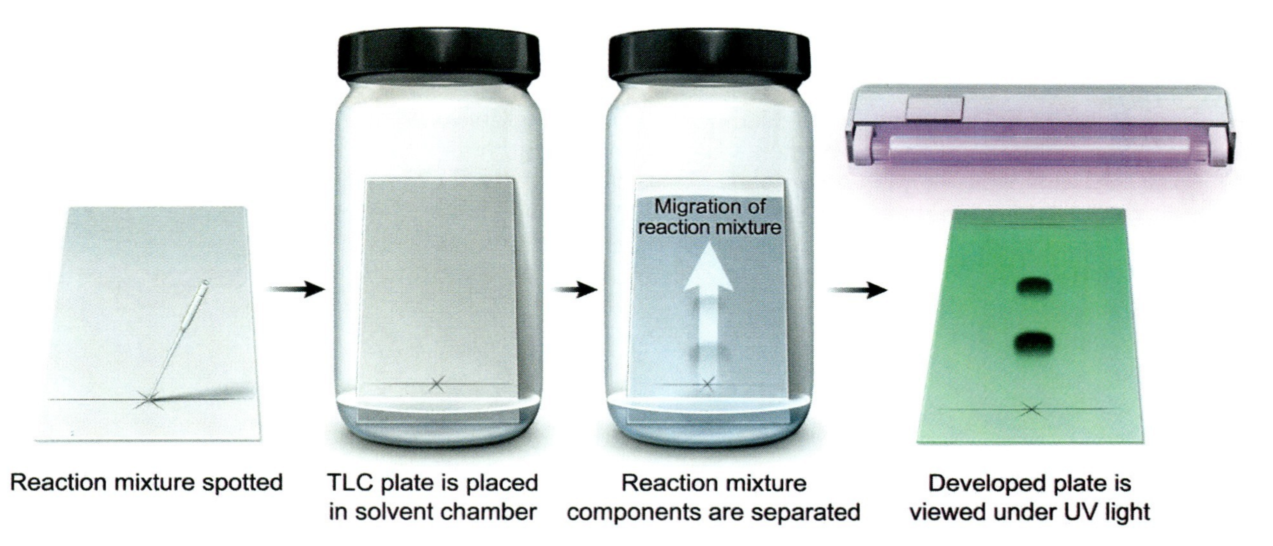

Figure 13.31 Thin-layer chromatography.

In TLC, it is *less common* for the separated substances to have a visible color and a **visualization method** is generally required to observe the spots on a TLC plate (Figure 13.32). Many TLC plates contain a fluorescent indicator embedded in the stationary phase to help identify the locations of separated spots. The separated substance in a spot *blocks* the fluorescence process, such that spots appear dark on a fluorescent green background.

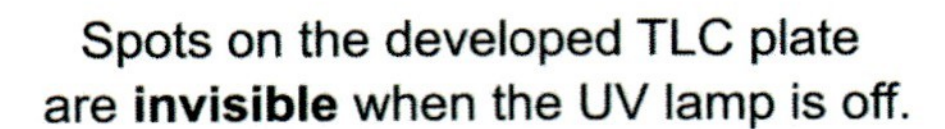

Spots on the developed TLC plate are **invisible** when the UV lamp is off.

Separated compounds (spots) **block the interaction** of UV light with the fluorescent indicator and **appear dark**.

Figure 13.32 Visualization of a developed TLC plate using UV light.

UV-active spots can be circled using a pencil when under the UV lamp to allow the location of the spots to be known in the *absence* of UV light. An advantage of using UV light to visualize a TLC plate is that the UV light is generally *nondestructive* to most chemicals. Alternatively, a **chemical stain** may also be used to visualize the spots on a TLC plate (see Concept 13.3.04). However, since a chemical stain promotes a chemical reaction with functional groups in the separated substances on the TLC plate, treatment with a chemical stain is a *destructive* process that *irreversibly changes* the chemical composition of the substances.

Once visualized, the retention factor (R_f) of the spots can be calculated (see Concept 13.3.04). The R_f value of a given spot may be used for any of the following purposes:

- To provide information about the relative polarity of the substance
- To compare to the R_f values of *known* pure substances, providing information about the possible *chemical identity* of the substance
- To aid in determining the optimum solvent ratio of the mobile phase for a desired separation (explained later in this concept)

Mobile Phase Mixtures in Thin-Layer Chromatography

Like recrystallization (Concept 13.1.02), liquid-solid chromatography can use a *mixture* of solvents to fine-tune mobile phase properties. A common solvent mixture in normal-phase chromatography is the *nonpolar* solvent hexane with the *moderately polar* solvent ethyl acetate. This mixture can be used at different ratios, depending on the separation needed. Changing the nonpolar to polar solvent ratio adjusts the overall polarity of the mobile phase and impacts the separation of substances (Figure 13.33).

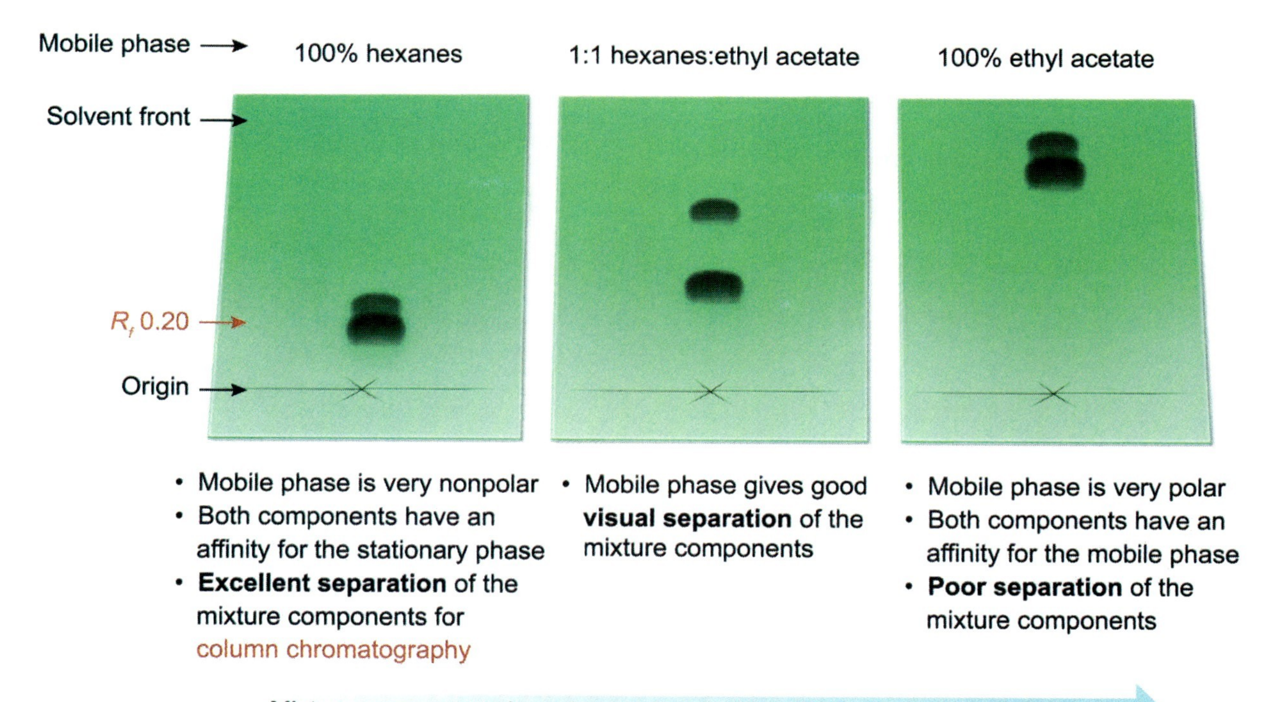

Figure 13.33 Testing different mobile phase mixtures to optimize a chromatographic separation.

Solvent mixtures which have a higher a proportion of the *nonpolar* component lead to a *lower* mobility of spots (ie, *lower* R_f values), while solvent mixtures which have a higher proportion of the *polar* component lead to a *greater* mobility of spots (ie, *higher* R_f values).

Analytical TLC experiments have the greatest resolution (ie, effective *visual* separation of components) when the developed spots are distributed across the *widest range* of the physical TLC plate. Whereas, a TLC experiment intended to *simulate* a separation using column chromatography (Concept 13.3.06) has the greatest *preparative* resolution when the substance to be isolated has an R_f of approximately 0.20.

13.3.06 Column Chromatography

The most common chromatography method for preparatively *isolating* pure components in a mixture is column chromatography. **Column chromatography** is a liquid-solid chromatography method in which the stationary phase is held in a glass tube called a **chromatography column** (Figure 13.34).

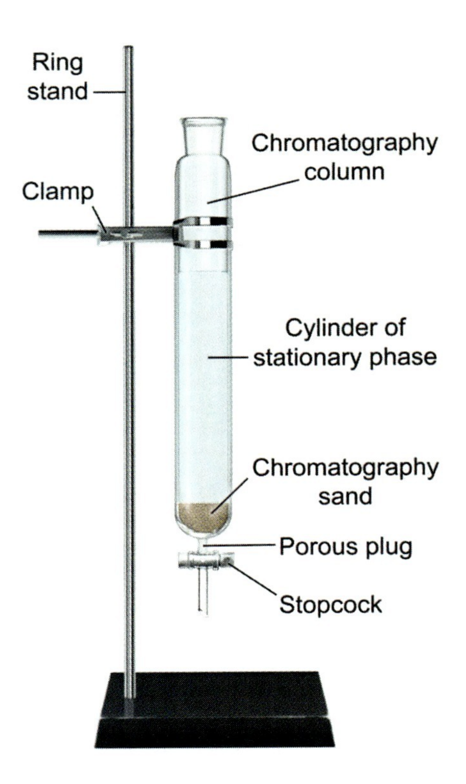

Figure 13.34 A typical chromatography column setup.

Column chromatography shares many similarities with TLC (see Concept 13.3.05). Like TLC, column chromatography can use the same variety of solid materials for the stationary phase, and the polar or nonpolar characteristics of the stationary phase determine whether a separation is normal phase (ie, a polar stationary phase and a nonpolar mobile phase) or reversed phase (ie, a nonpolar stationary phase and a polar mobile phase) (see Concept 13.3.03).

TLC is frequently employed as a model to predict the separation of compounds through column chromatography (Figure 13.35). While the flow of mobile phase in a developing TLC plate is typically *upward* (ie, bottom to top, against gravity), the flow of mobile phase in column chromatography is typically *downward* (ie, top to bottom, with gravity). Therefore, rotating a developed TLC plate 180° aligns the mobile phase flow of the TLC plate with the column and indicates the order of elution of substances from the column. Substances with a larger R_f on the TLC plate migrate down the column *faster* (ie, *small retention time*), while substances with a *smaller* R_f on the TLC plate migrate down the column *slower* (ie, *large retention time*).

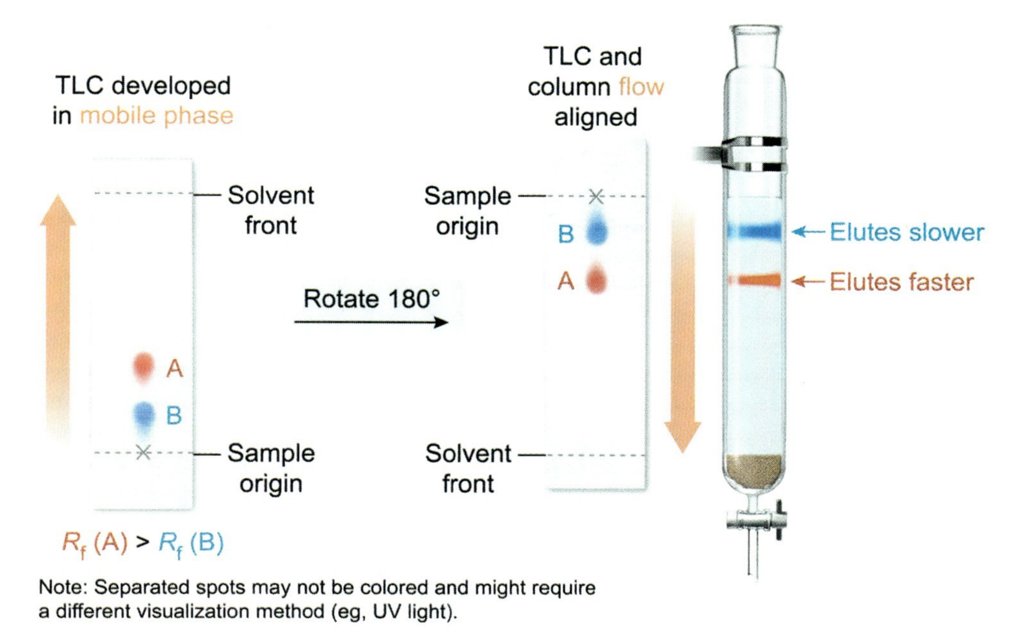

Figure 13.35 Spatial relationship between TLC and column chromatography separations.

To perform column chromatography, the sample is placed at the top of the cylinder of stationary phase within the chromatography column (Figure 13.36). Then, mobile phase is added to the top of the stationary phase. While it is possible to perform column chromatography using only gravity to draw the mobile phase through the stationary phase, the column can also be lightly *pressurized* (eg, 1 to 5 atmospheres) to *push* the mobile phase through the stationary phase. Methods that use light amounts of additional force *beyond* gravity to speed up column chromatography are sometimes called **flash chromatography** methods.

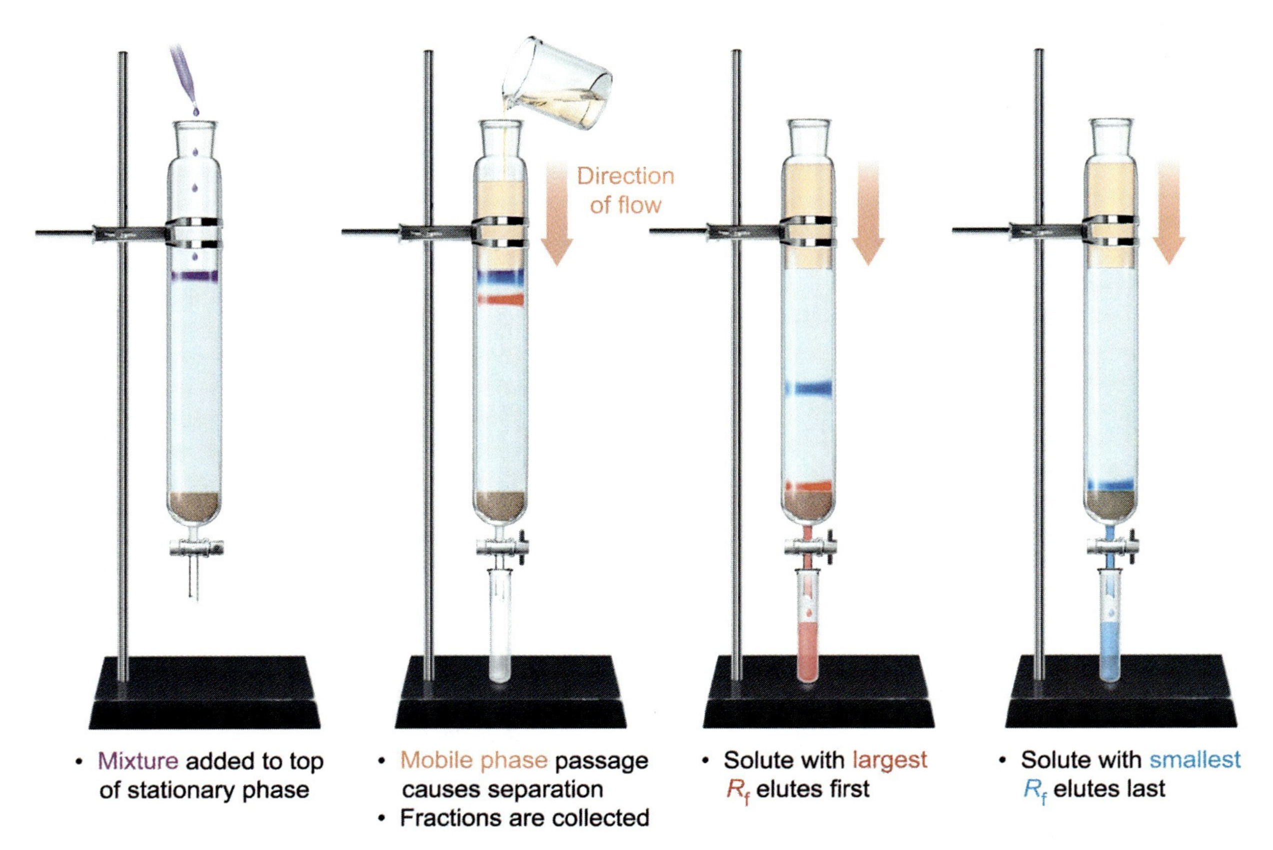

Figure 13.36 Performing a separation using column chromatography.

The flowing mobile phase that *leaves* the chromatography column through the stopcock is collected in small portions called **fractions**. Often, each column chromatography fraction is collected in a test tube.

In most cases, it is not possible to visually detect the presence of solutes in the mobile phase fractions. Instead, the mixture composition of fractions is usually analyzed by TLC to determine which fractions contain the desired substances in a purified form. The TLC R_f values of the original mixture components are compared to the observed R_f values from the fractions, with the goal of identifying the fractions containing each pure substance.

Variables in Column Chromatography

Column chromatography has experimental variables that can be manipulated to meet the needs of a given separation. For column chromatography, the two main variables are the quantity and dimensions of the stationary phase and the solvent composition of the mobile phase.

The *cross-sectional area* and *length* of the cylinder of stationary phase within a chromatography column determines the amount of material (ie, grams) that can be effectively separated (Table 13.8). Generally, the *larger* the cross-sectional area (or the *diameter* of the column), the *greater* the amount of sample that can be effectively resolved. Similarly, a *longer* cylinder of stationary phase provides a *greater* distance for the separation of components, which typically increases the *resolution* of a column chromatography separation.

Table 13.8 Impacts of variables in column chromatography.

Component	Variable	Impacts
Stationary phase	Cross-sectional area	Determines the amount of material that can be effectively separated
	Length	Influences the resolution of substances
Mobile phase	Nonpolar to polar solvent ratio	Influences the R_f values of separated substances, which can impact resolution

Concept 13.3.05 introduces how *mixtures* of solvents can be used in a chromatographic separation. In normal-phase chromatography to isolate polar compounds, a mobile phase with higher *nonpolar* character (ie, a *greater* percentage of a *nonpolar* solvent such as hexane) leads to *less* mobility of solutes, a *smaller* R_f, and *longer* elution and retention times (R_t). By comparison, a mobile phase with higher *polar* character (ie, a *greater* percentage of a more *polar* solvent such as ethyl acetate) leads to *more* mobility, *larger* R_f values, and *shorter* elution and retention times.

13.3.07 High-Performance Liquid Chromatography (HPLC)

High-performance liquid chromatography (HPLC), another variety of liquid-solid chromatography, is a separation method ideal for *small* sample sizes. Like TLC and column chromatography, HPLC is a liquid-solid chromatography method that separates mixtures based on polarity and can be performed either normal phase or reversed phase (see Concept 13.3.03).

Normal-phase HPLC uses a polar stationary phase (eg, silica gel) and a mixture of relatively nonpolar organic solvents as the mobile phase. The silica gel stationary phase contains silicon atoms bonded to hydroxyl groups, which are *polar*. Reversed-phase HPLC uses a nonpolar stationary phase (eg, silica linked to nonpolar C_{18} alkyl groups) and a highly polar mobile phase made up of a mixture of water and a miscible polar organic solvent. The most common stationary and mobile phases used in HPLC and their relative polarity are listed in Table 13.9.

Table 13.9 Commonly used stationary and mobile phases in HPLC.

	Stationary phase	Mobile phase
Normal phase	**Polar** Silica gel, alumina	**Nonpolar** Hexanes, ethyl acetate, acetone, methanol, ethanol
Reversed phase	**Nonpolar** C_{18}, C_8	**Polar** Water, acetonitrile, methanol

As discussed in Concept 13.3.03, the relationship between molecules, the stationary phase, and the mobile phase is based on the principle of "like dissolves like" and intermolecular forces. In *normal-phase HPLC*, more-polar molecules have a greater affinity for the polar stationary phase than do less-polar molecules, which have a greater affinity for the nonpolar mobile phase. As a result, the less-polar molecules elute *faster* and have a *shorter* retention time than more-polar molecules (Figure 13.37).

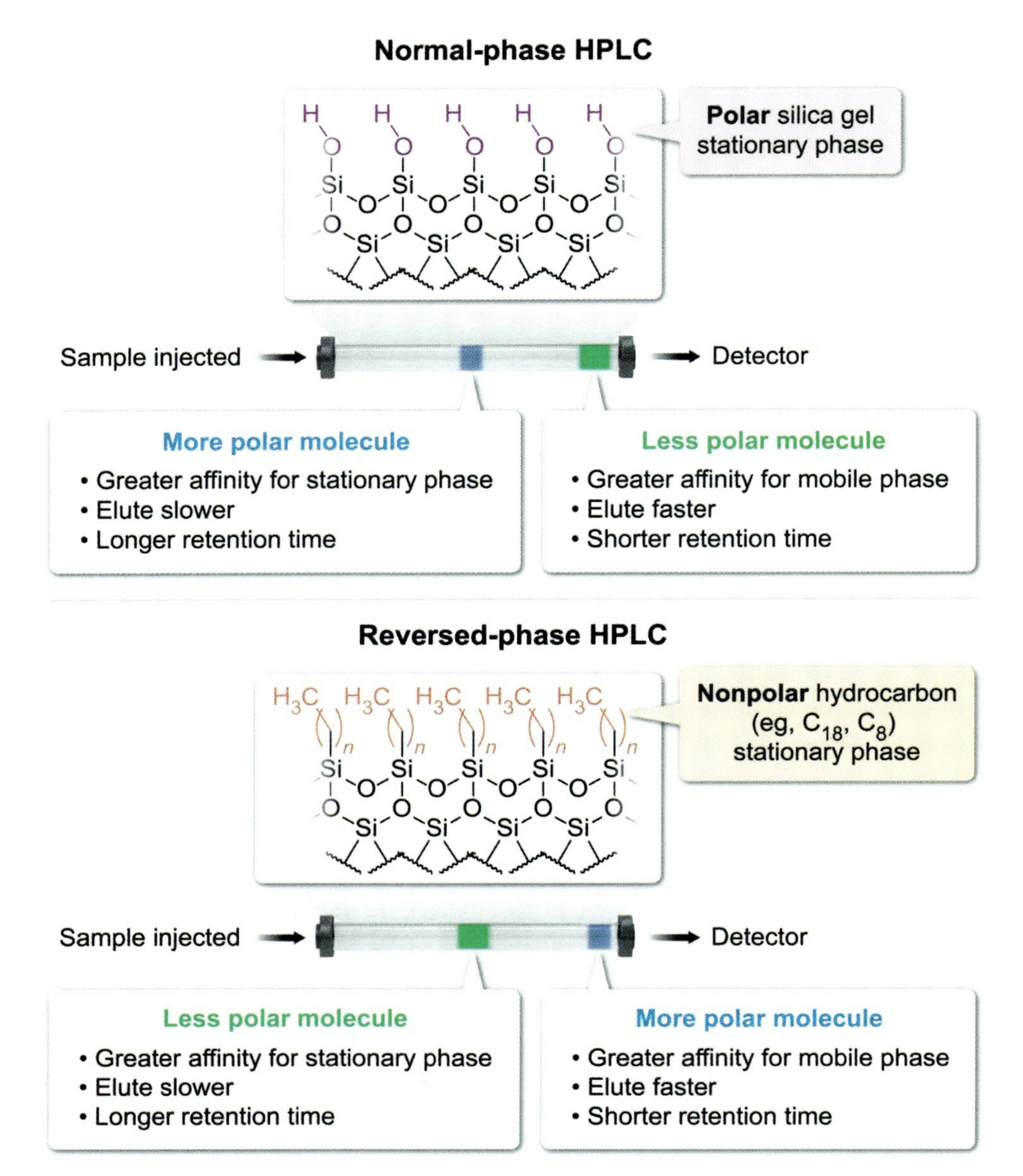

Figure 13.37 Order of elution for normal- and reversed-phase HPLC.

In *reversed-phase HPLC*, less-polar molecules have a greater affinity for the nonpolar stationary phase than do more-polar molecules, which have a greater affinity for the polar mobile phase. Therefore, the more-polar molecules elute *faster* and have a *shorter* retention time than less-polar molecules.

A typical HPLC experiment (Figure 13.38) has the following steps:

1. The solvent (mobile phase) is pumped through the system at a high pressure.
2. The sample to be separated is dissolved in a solvent and then injected into the flowing mobile phase.
3. The sample, carried by the mobile phase, travels through the column (stationary phase), where the mixture components are separated.
4. A detector monitors the solution (ie, mobile phase and any separated sample components) as they exit the column, generating a chromatogram.
5. After passing through the detector, the solution is collected (either as waste or a fraction to be saved).

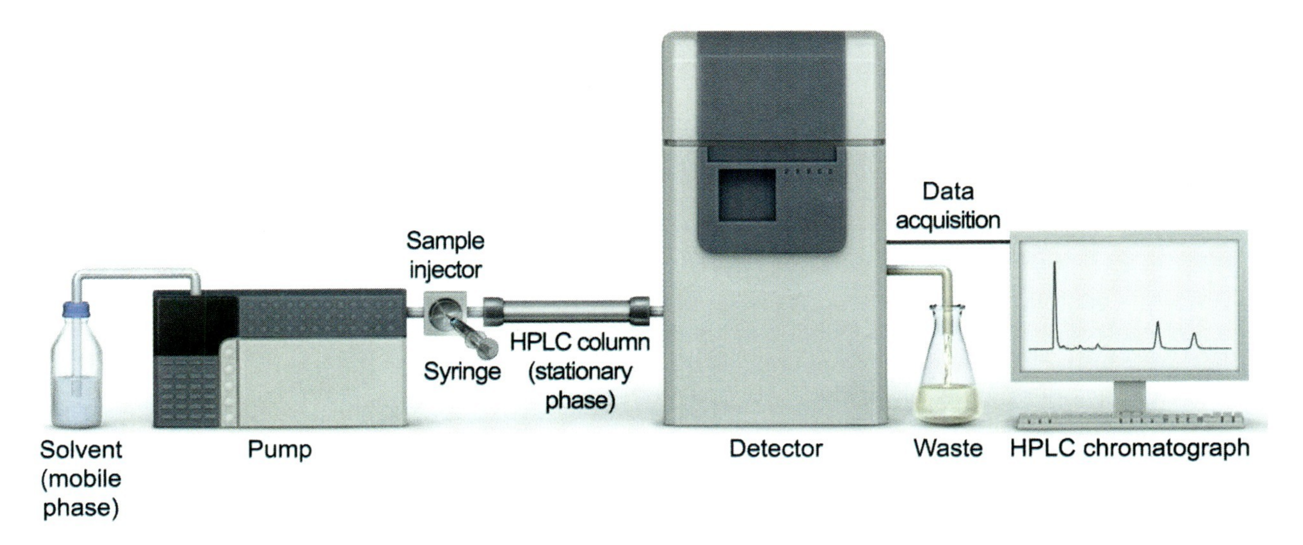

Figure 13.38 The essential components of high-performance liquid chromatography.

The most common HPLC detector is a UV detector (see Lesson 14.5). UV detectors are useful for compounds that can absorb ultraviolet light, such as conjugated or aromatic molecules. Depending on the detector, absorption at one or more wavelengths can be monitored. When a molecule is detected (ie, it absorbs UV light), a signal appears on the chromatogram as a peak. The height of the peak indicates the absorption intensity (*y*-axis) and relates both to the molecule's absorptivity and to its abundance. The retention time is indicated by the peak's position on the *x*-axis.

Variables in HPLC

HPLC has experimental variables that can be changed to meet the needs of a given separation. Aside from the choice of normal phase versus reversed phase, the three main variables for HPLC are the mobile phase flow rate, solvent composition of the mobile phase, and the stationary phase column size (Table 13.10).

Table 13.10 Impacts of variables in HPLC.

Component	Variable	Impacts
Mobile phase	Flow rate	Influences retention times and peak resolution
	Solvent ratio	Influences separation of mixture components (peak resolution)
Stationary phase	Column size	Determines the type of HPLC separation: analytical or preparative

The rate at which the solvent (ie, mobile phase) travels through the system can be adjusted based on experimental needs. A *faster flow rate* causes the mixture components to elute faster (ie, have shorter retention times) but yields poor resolution of the peaks.

As with other chromatographic methods, the ratio of the solvents in the mobile phase mixture can also be changed to improve separation. With HPLC, the composition of the mobile phase either can remain consistent throughout the separation, or the solvent ratio *can be changed* as a **gradient** *during* the separation.

HPLC has two applications—to *analyze* the components of a mixture (**analytical**) or to *isolate* larger quantities of pure substances (**preparative**) (see Concept 13.3.03). As with column chromatography (Concept 13.3.06), HPLC columns with a larger cross-sectional area (or a larger diameter) can handle more sample and therefore can be used in preparative HPLC with small scale (eg, mg scale) separation. Smaller HPLC columns are used for analytical separations, which can be performed much faster due to the smaller volumes involved.

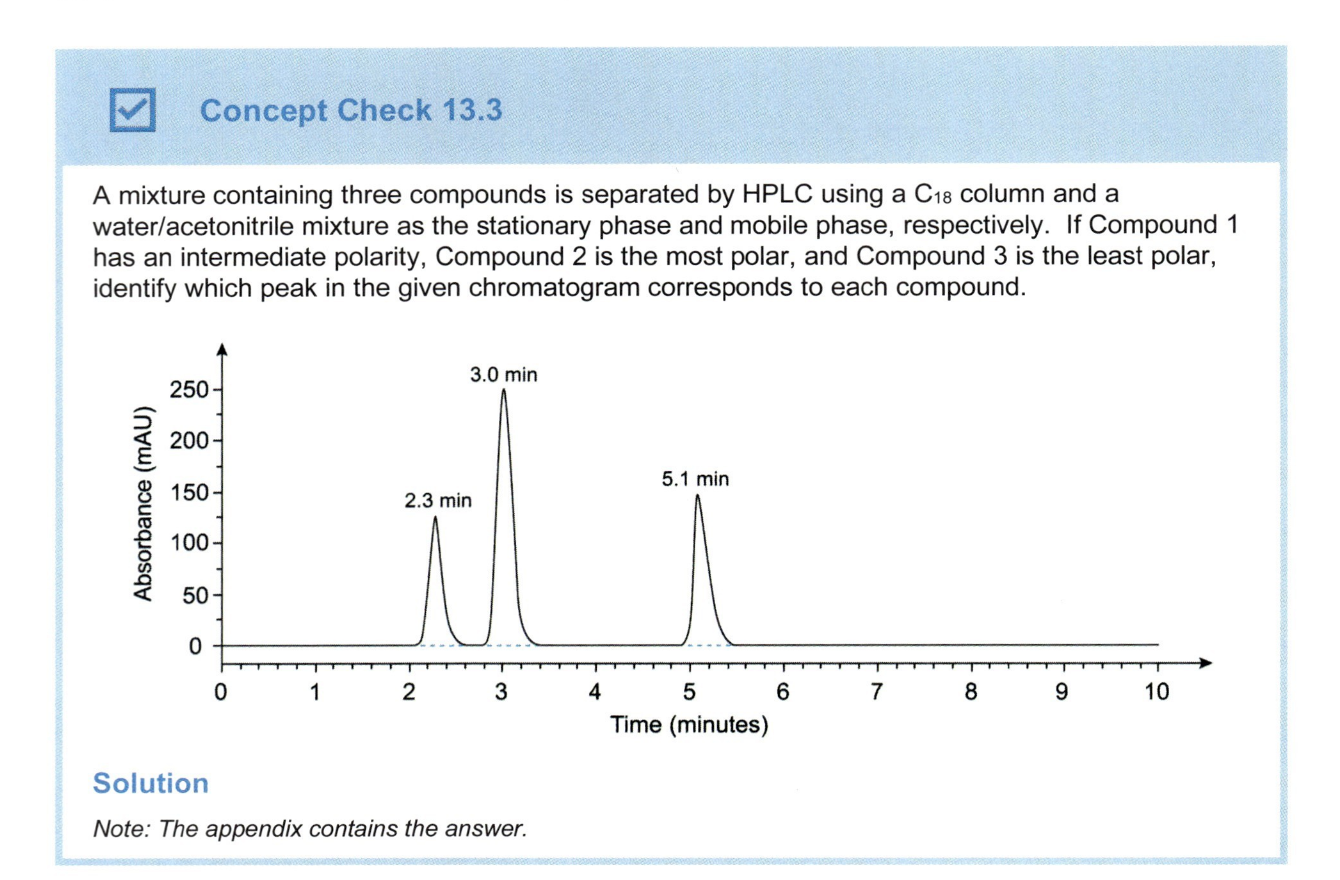

Concept Check 13.3

A mixture containing three compounds is separated by HPLC using a C_{18} column and a water/acetonitrile mixture as the stationary phase and mobile phase, respectively. If Compound 1 has an intermediate polarity, Compound 2 is the most polar, and Compound 3 is the least polar, identify which peak in the given chromatogram corresponds to each compound.

Solution

Note: The appendix contains the answer.

Lesson 14.1

Laboratory Analysis

Introduction

This chapter focuses on laboratory methods used to *analyze* chemical samples. Unlike the techniques described in Chapter 13, *none* of the approaches discussed in this chapter can enhance the purity of a substance. Rather, these techniques provide information about the physical properties, chemical properties, or structure of a substance, all of which aid in the determination or validation of a chemical's identity.

This lesson focuses on general laboratory methods used to determine the acidic or basic properties of a sample.

14.1.01 Methods to Analyze pH

As discussed in Concept 5.1.03, the pH scale provides the primary means to quantify the acidic or basic character of a solution. The pH of a solution can be determined by the equation:

$$pH = -\log[H^+]$$

where $pH < 7$ indicates an acidic solution, $pH = 7$ indicates a neutral solution, and $pH > 7$ indicates a basic solution.

Several approaches can be used to analyze the acidity or basicity of an *aqueous* solution. Although acids and bases can certainly exist in an *organic* solution (eg, carboxylic acid, amine), the analytical methods discussed in this lesson *require* the presence of water and do not typically provide valid data when used on a solely organic solution.

Litmus Paper

A common way to determine if an aqueous solution is acidic, basic, or neutral involves the use of litmus paper. **Litmus paper** is paper containing one of two dyes that change color in response to changes in the aqueous pH of a solution (Figure 14.1). **Blue litmus paper** contains a *blue* dye that changes to *red* in the presence of an *acid*, and **red litmus paper** contains a *red* dye that changes to *blue* in the presence of a *base*. In a neutral aqueous solution, the colors of *both* blue litmus paper and red litmus paper remain *unchanged*.

Figure 14.1 Litmus paper provides a qualitative measurement of the pH of an aqueous solution.

Adding a small piece of blue or red litmus paper to an aqueous solution is one of the fastest ways to *qualitatively* determine if a solution is acidic, basic, or neutral. However, due to its binary nature (ie, blue to red or red to blue), litmus paper is *unable* to *quantitatively* measure the *magnitude* of the acidity or basicity of a solution.

Red litmus paper indicates only that the solution has a pH value *below* approximately 4.5, and blue litmus paper indicates only that the solution has a pH value *above* approximately 8.3. pH values in the range of 4.5–8.3 (ie, almost a 10,000-fold range in [H^+]) do not typically result in an observable change in litmus paper color, so litmus paper cannot be used to measure solutions in this pH range with great accuracy.

Wide-Range pH Paper

Although litmus paper provides a quick method to broadly determine if an aqueous solution is significantly acidic or basic, more precise pH values are sometimes necessary. **Wide-range pH paper** provides one such semiquantitative method, in which paper embedded with several pH-sensitive organic dyes is used to differentiate pH values between 0 and 14 in approximately 1–2 pH unit intervals (Figure 14.2). pH paper variations containing several dyes with narrower pH ranges (eg, pH 0.0–1.5, pH 6.0–8.0) also exist and can be useful if more precise pH measurements are necessary.

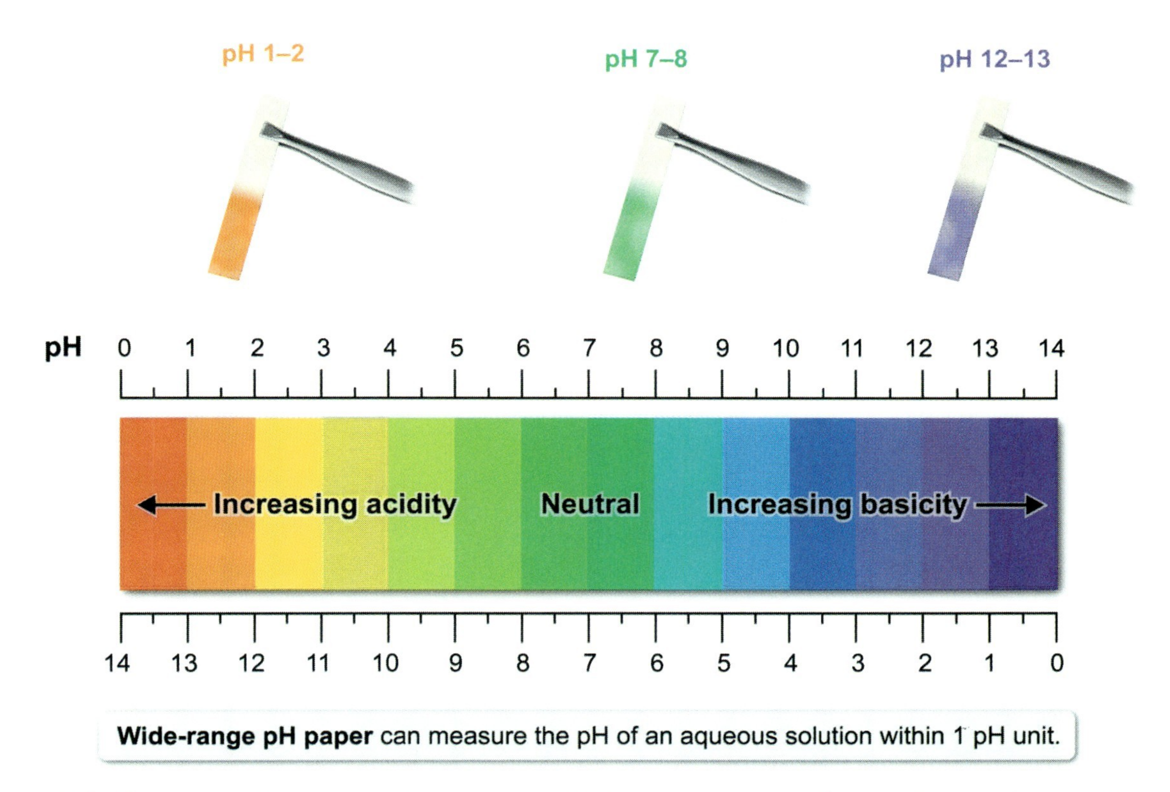

Figure 14.2 Wide-range pH paper provides a semiquantitative measurement of the pH of an aqueous solution.

pH Meters

The pH of an aqueous solution can be *quantitatively* measured using a pH meter. A **pH meter** is an electronic device connected to a pH probe that precisely measures the pH of an aqueous solution, usually to two decimal places (Figure 14.3). The **pH probe** measures the chemical potential of a solution in relation to the chemical potential of **pH standards** (reference solutions of a known pH), and the pH meter converts the voltage data into a pH measurement.

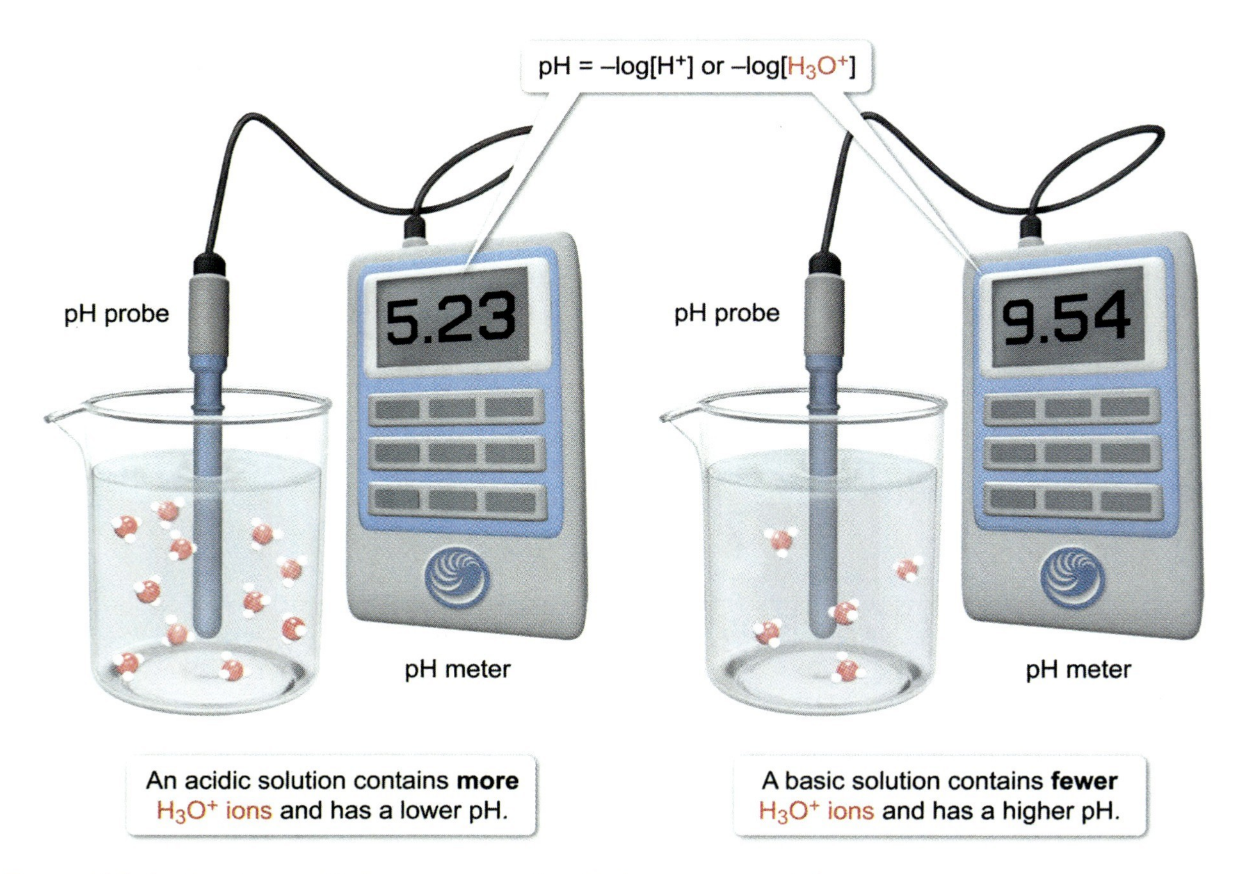

Figure 14.3 A pH meter quantitatively measures the pH of an aqueous solution.

Lesson 14.2

Polarimetry

Introduction

Enantiomers, introduced in Concept 3.3.04, have identical physical properties except for the sign of their optical activity (ie, the direction they rotate plane-polarized light). This lesson focuses on the following aspects of **polarimetry**, a method used to analyze the optical activity of a chiral molecule:

- Plane-polarized light
- The relationship between polarimetry and chirality
- Observed rotation and specific rotation

14.2.01 Plane-Polarized Light

Most light sources (eg, a light bulb, the sun) emit *unpolarized* light made of waves oscillating in all directions (ie, a mixture of light waves with varying polarization); such light has no *net* direction of oscillation. When unpolarized light passes through a polarizing filter, only light waves *aligned with* the polarizing filter can pass through (Figure 14.4). The filtered light waves are axially aligned in one orientation and are known as **plane-polarized light** (linearly polarized light).

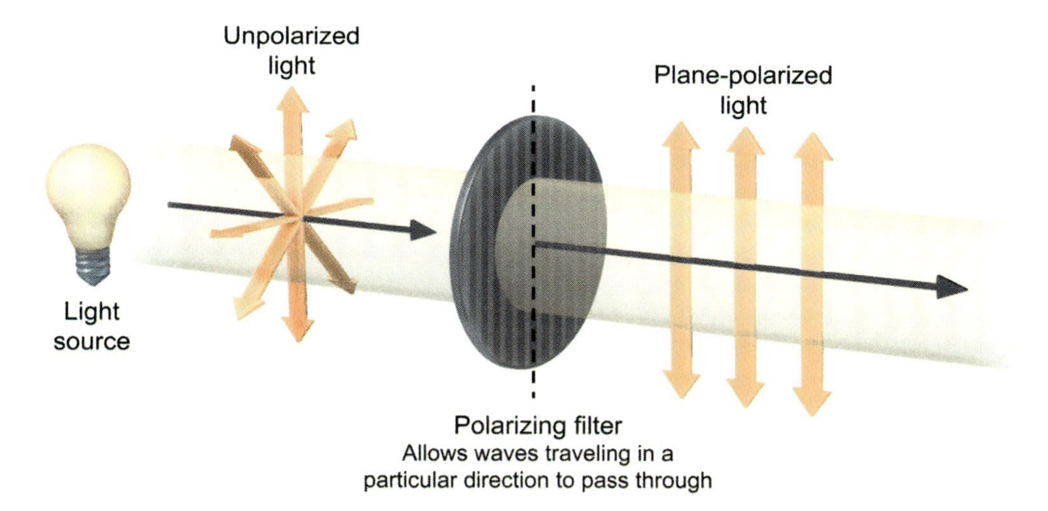

Figure 14.4 Light polarization.

When light passes through more than one polarizing filter, the orientation of the second polarizing filter (relative to the first filter) determines whether the polarized light can pass through. If the two polarizing filters are *parallel* (ie, have the same orientation), the plane-polarized light fully passes through (Figure 14.5). Conversely, if the second polarizing filter is *perpendicular* to the first polarizing filter (ie, the slits on the second filter are rotated 90° relative to the first filter), the plane-polarized light is fully blocked and does *not* pass through.

At other angles, the amount of light that passes through the second filter is *diminished*, with the amount of light passing through corresponding to the vector component in line with the second polarizer's axis.

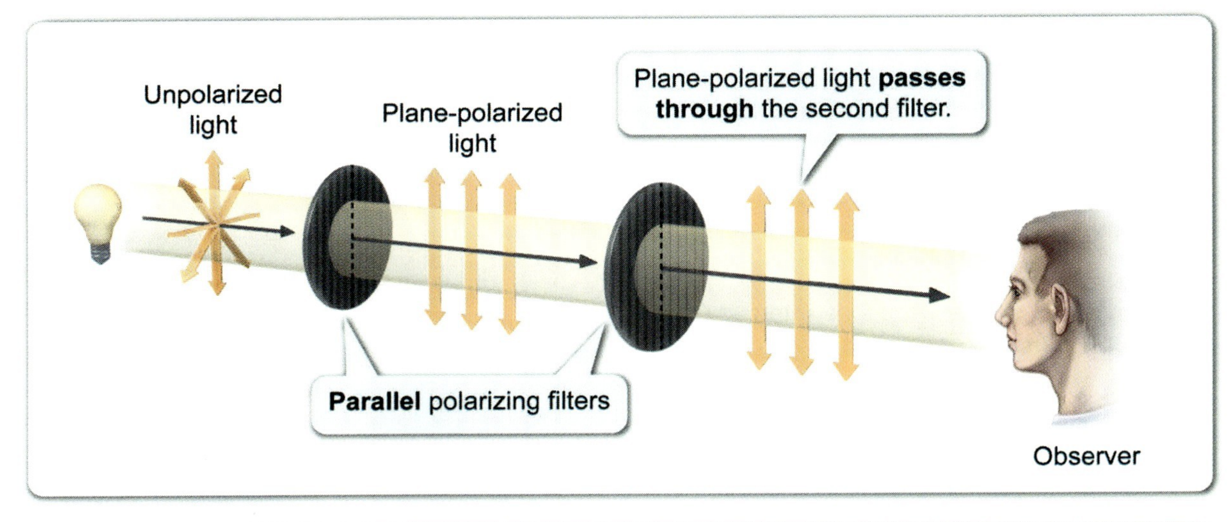

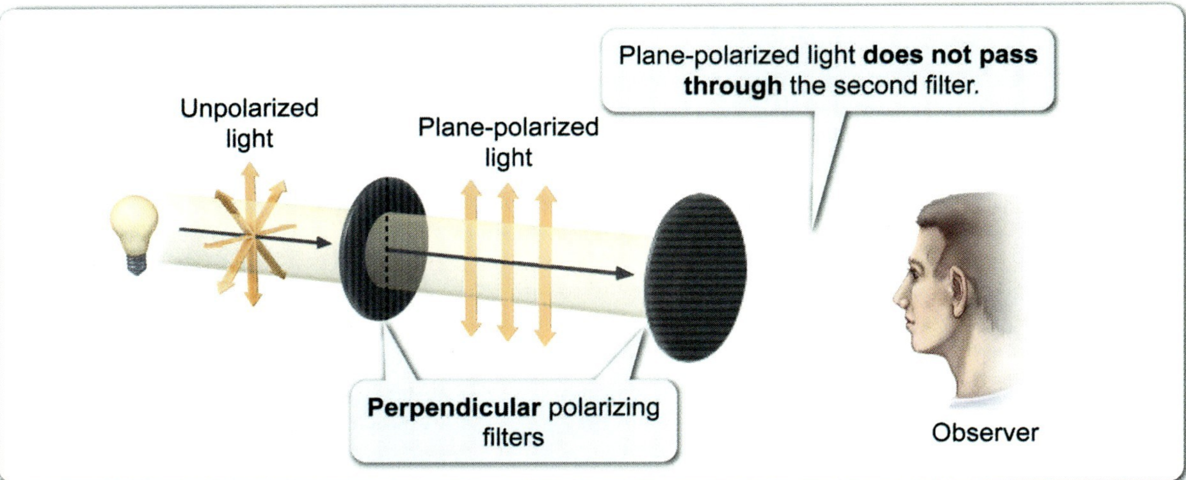

Figure 14.5 Plane-polarized light passing through two polarizing filters.

14.2.02 Relationship between Polarimetry and Chirality

A unique feature of chiral molecules is that they can rotate plane-polarized light. Therefore, chiral molecules are called **optically active**. Optical activity is a property of the chiral molecule as a whole and is independent of the number of chiral centers in the molecule (ie, a molecule with more chiral centers does *not* necessarily rotate light more than a molecule with fewer centers).

Polarimetry is an *analysis* method that measures the optical activity of a chiral molecule. The instrument used to measure the rotation of plane-polarized light by an optically active sample is called a **polarimeter**. A general diagram of the components of a polarimeter is shown in Figure 14.6.

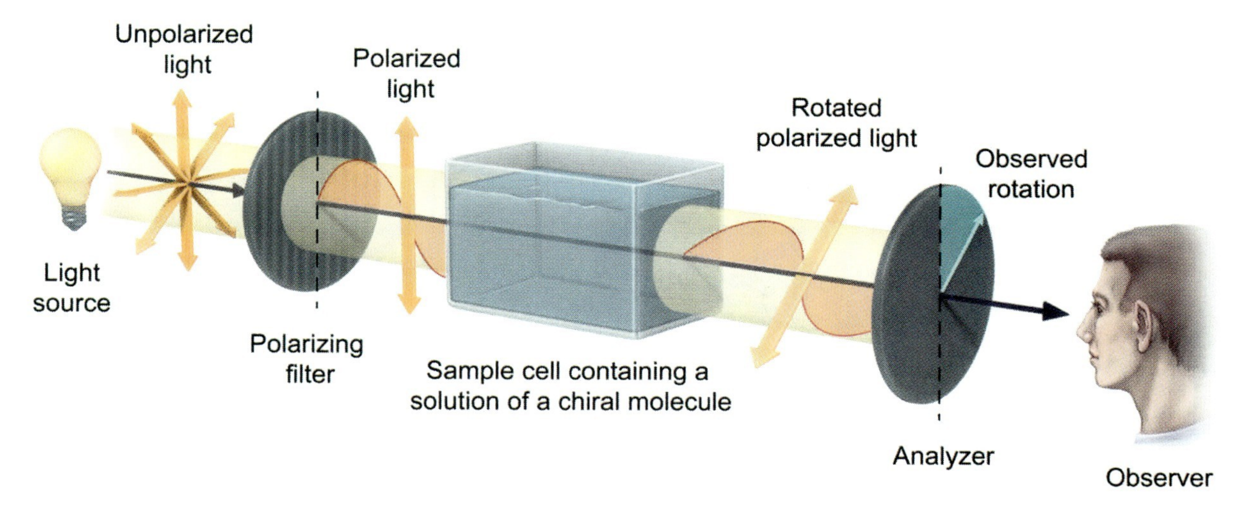

Figure 14.6 The components of a polarimeter.

The polarimeter light source is usually a sodium lamp, which emits light at 589 nm (known as the D line of sodium). The unpolarized light goes through a polarizing filter, generating plane-polarized light. As the plane-polarized light travels through the sample cell, chiral molecules *rotate* the plane-polarized light. The *concentration* of the sample solution and the path *length* of the sample cell are important because both parameters affect the extent to which the plane-polarized light is rotated (see Concept 14.2.03).

After passing through the sample, the plane-polarized light goes through an *analyzing filter*, which is a second polarizing filter used to determine how much the plane-polarized light was rotated by the chiral molecule. If the analyzer starts parallel to the first polarizer, a diminished amount of light reaches the observer because the light and the analyzer are not aligned. The observer can then rotate the analyzing filter until a *maximum* amount of light is observed.

More commonly, though, the analyzer starts *perpendicular* to the first polarizer, which should block all light from reaching the observer if not for the presence of optically active molecules. The observer can then rotate the analyzing filter until a *minimum* amount of light is observed. In either case, the measured angle difference between the chosen endpoint and the starting reference is known as the **observed rotation** (see Concept 14.2.03).

A chiral molecule that rotates plane-polarized light clockwise (ie, the top of the plane rotates to the right) is **dextrorotatory** (*d*), and the direction of rotation is denoted by a plus sign (+). A chiral molecule that rotates plane-polarized light counterclockwise (ie, to the left) is **levorotatory** (*l*), and the direction of rotation is denoted by a minus sign (−). The direction a chiral molecule rotates plane-polarized light can be designated at the beginning of the compound's IUPAC name as (+) or (−). The (+) or (−) *does not have any correlation* to the *R*/*S* absolute configuration of the chiral center(s) in a molecule (see Concept 3.3.03) or to most molecules' D/L configuration (see Concept 3.3.06).

As introduced in Concept 3.3.04, enantiomers have *identical* chemical and physical properties, *except* for the direction they rotate plane-polarized light. Enantiomers rotate plane-polarized light the *same magnitude* but in *opposite directions* (Figure 14.7). The sample solution concentration and sample cell path length must be the same to compare the rotations.

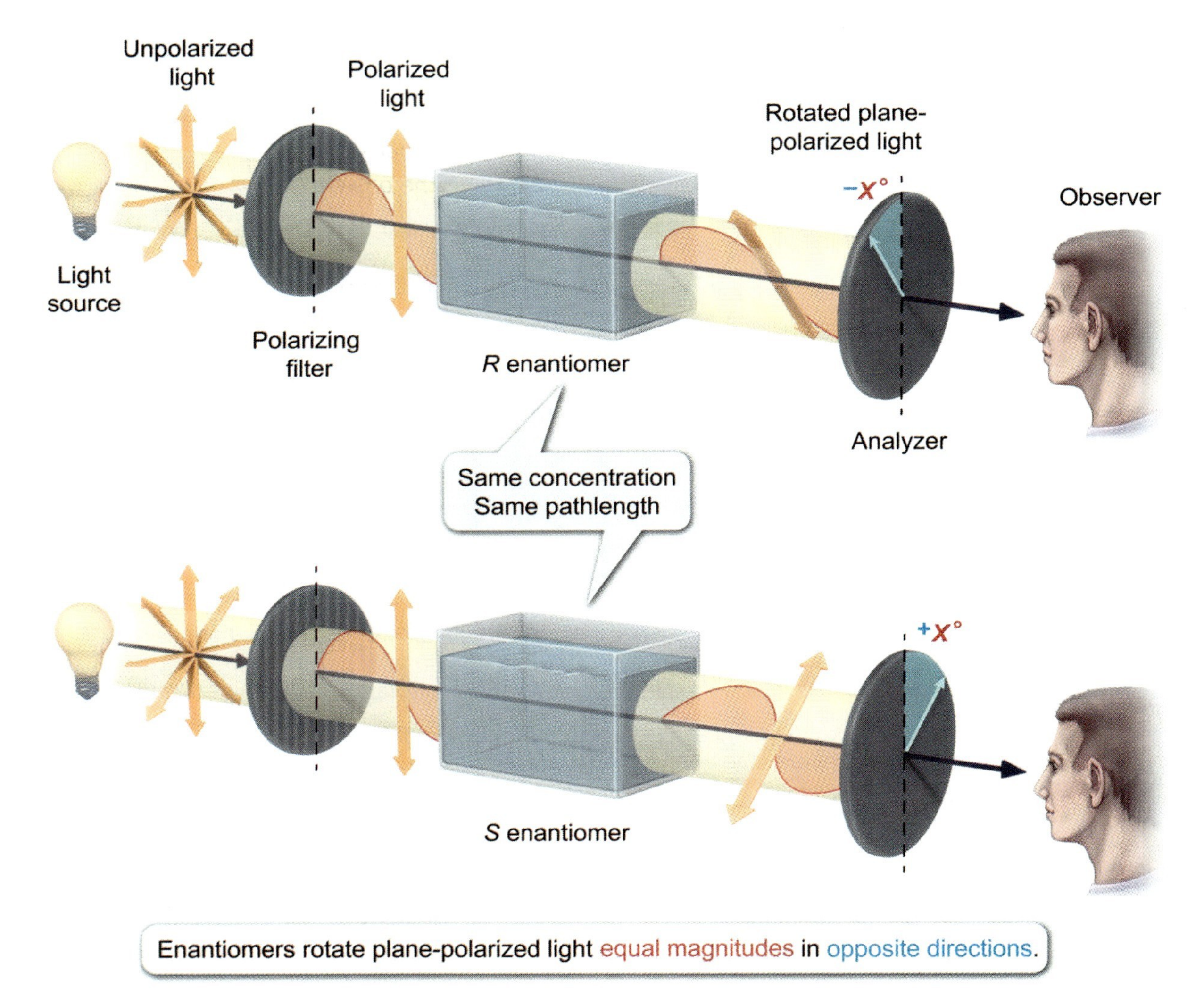

Figure 14.7 Enantiomers and optical activity.

When measuring *mixtures* of chiral compounds, the observed rotation is the weighted sum of the observed rotations for each *pure* chiral component. For example, if a pure sample of compound A has an observed rotation of −12° and a pure sample of compound B has an observed rotation of +7°, then an equimolar (ie, 1:1 molar ratio) mixture of compounds A and B (given the same [A], [B], and pathlength as the pure samples) has an observed rotation of −5°.

Similarly, if a solution is a racemic mixture (ie, contains a 50:50 mixture of enantiomers, see Concept 3.3.05), the (+) enantiomer rotates the plane-polarized light a certain number of degrees clockwise and the (−) enantiomer rotates the plane-polarized light the same number of degrees counterclockwise. The two rotations cancel each other out and a rotation of *zero* is observed. Therefore, a racemic mixture is **optically inactive**.

Because diastereomers are chiral molecules, they do rotate plane-polarized light, but there is no correlation in the magnitude or direction of the rotation of plane-polarized light for a pair of diastereomers. Although meso compounds contain chiral centers, meso compounds are *achiral* and therefore do *not* rotate plane-polarized light.

14.2.03 Observed Rotation and Specific Rotation

As introduced in Concept 14.2.02, **observed rotation**, α_{obs}, is the angle a sample of chiral molecules rotates plane-polarized light. Observed rotation is obtained by the observer from the analyzer and is dependent on the concentration (c) of the sample solution and the path length (ℓ) of the sample cell. As such, a more concentrated sample or a longer path length *increases* the observed rotation.

Because optical activity is a characteristic property of a compound (similar to boiling point, melting point), it is useful to compare the α_{obs} of different samples or compounds. However, the dependence of α_{obs} on c and ℓ requires that such comparisons be made using measurements taken under the same conditions. By convention, a set of normalized conditions (c = 1 g/mL and ℓ = 1 dm) is used to define the **specific rotation** [α], which is the observed rotation under these normalized conditions.

Sometimes a concentration of 1 g/mL is not possible (eg, small sample size, an instrument that uses a sample cell with a longer/shorter path length than 1 dm). For a sample with an unknown specific rotation, accounting for non-normalized conditions can be accomplished by using the equation in Figure 14.8 to calculate [α].

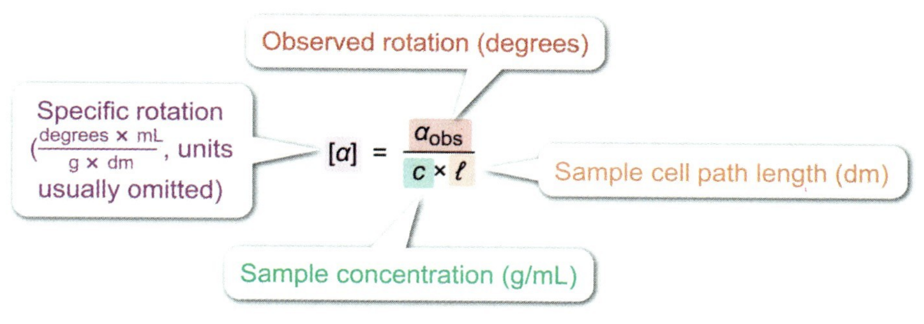

Figure 14.8 Specific rotation equation.

A specific rotation has units of $\frac{\text{degrees} \times \text{mL}}{\text{g} \times \text{dm}}$; however, the value is often reported *without* units due to the understood usage of normalized conditions. Rotations are usually measured at 20 °C using the sodium D line (589 nm). When these specifications are used, the specific rotation can be explicitly notated as $[\alpha]_D^{20}$. If other temperatures or wavelengths are used, the temperature and wavelength are typically reported with the specific rotation.

The specific rotation equation can be rearranged to solve for the expected observed rotation:

$$\alpha_{\text{obs}} = [\alpha]c\ell$$

This form of the equation is similar to the expression of the Beer-Lambert law (see Concept 14.5.02) and shows how polarimetry is related to other laboratory techniques that involve the interaction of molecules with light.

Concept Check 14.1

Sample 1 is a 0.250 g/mL solution of Compound X. Sample 1 is placed in a 2.00 dm sample cell and has an observed rotation of 20.0° counterclockwise. Sample 2 is a 0.100 g/mL solution of the same Compound X. If Sample 2 is placed in a 2.00 dm sample cell in a polarimeter, what observed rotation is expected?

Solution

Note: The appendix contains the answer.

Lesson 14.3

The Electromagnetic Spectrum

Introduction

Previous lessons in this chapter discuss laboratory methods that yield information about physical and chemical properties of substances. The remaining lessons in this chapter largely focus on **spectroscopy** (ie, analytical methods that use electromagnetic radiation to analyze the properties of a substance). This lesson introduces the general properties of electromagnetic radiation (Concept 14.3.01) and the relationship between electromagnetic radiation and spectroscopy (Concept 14.3.02). Subsequent lessons focus on specific types of spectroscopy, the data they generate, and the interpretation of spectroscopic data.

14.3.01 Introduction to the Electromagnetic Spectrum

Electromagnetic radiation is a form of energy that includes visible **light**. According to the **wave-particle duality**, light has *both* wave-like and particle-like properties. From the wave-like perspective, light is composed of transverse waves (Figure 14.9) that travel at a constant velocity of approximately 3×10^8 m/s in a vacuum (the **speed of light**, c). The distance between two equivalent points on a wave (eg, peak to peak) is called the **wavelength** (λ).

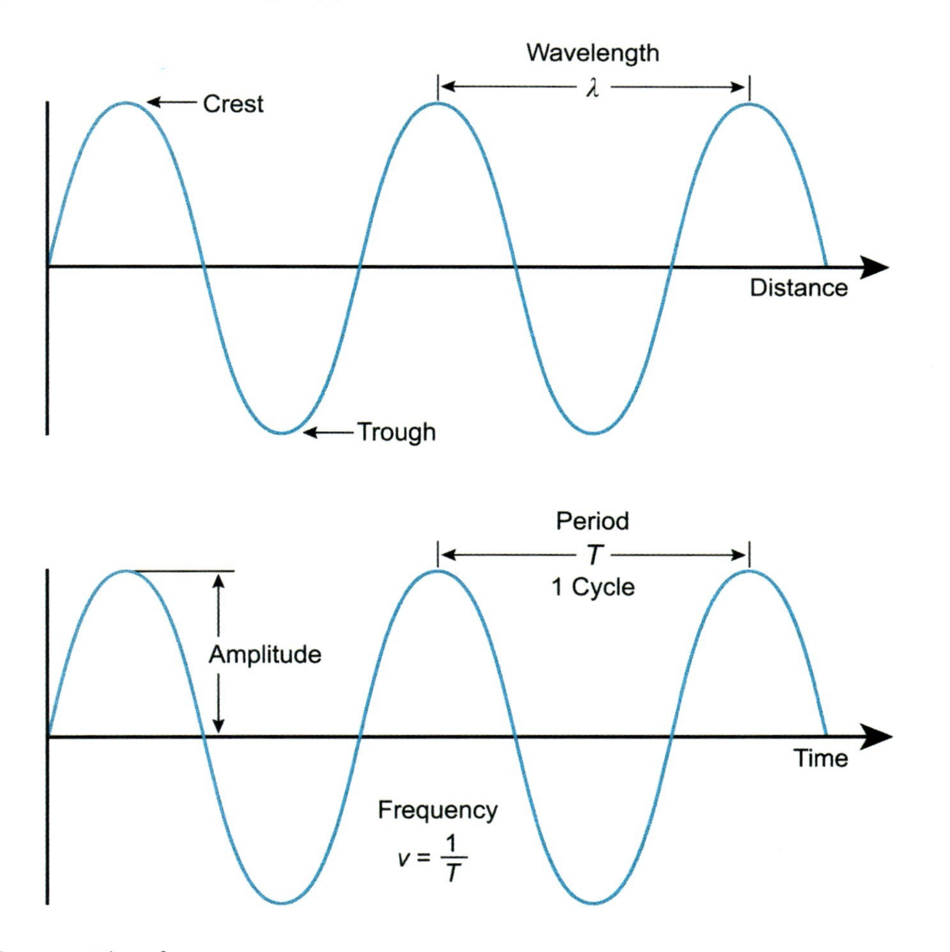

Figure 14.9 The properties of a wave.

The **frequency** (v) of a wave is defined as the number of wave cycles that pass by a point per unit of time and is typically expressed in units of hertz (Hz, cycles per second). Because the speed of light in a vacuum is *constant*, the wavelength and frequency of light are inversely proportional to one another, as described by the equation below.

$$c = \lambda v$$

From the particle-like perspective, light is composed of packets of energy called **photons**. The energy of a photon is *proportional* to its frequency and *inversely proportional* to its wavelength. For example, photons with a *higher* frequency (ie, *shorter* wavelength) have a *higher* energy. The relationship between the energy of a photon and its frequency is described by **Planck's equation**, where **Planck's constant** (h) is defined as 6.6×10^{-34} J·s:

$$E = hv = \frac{hc}{\lambda}$$

The **electromagnetic spectrum** represents the full range of wavelengths, frequencies, and energies of electromagnetic radiation. It is often convenient to describe the spectrum in terms of *energy* with named regions characterized as a subset of similar energies. Within the band of **visible light** (ie, light that can be detected by the human eye), red light has the *longest* wavelength and *lowest* energy, and purple light has the *shortest* wavelength and *highest* energy. Additional subsets (eg, radio, infrared, ultraviolet) are shown in Figure 14.10.

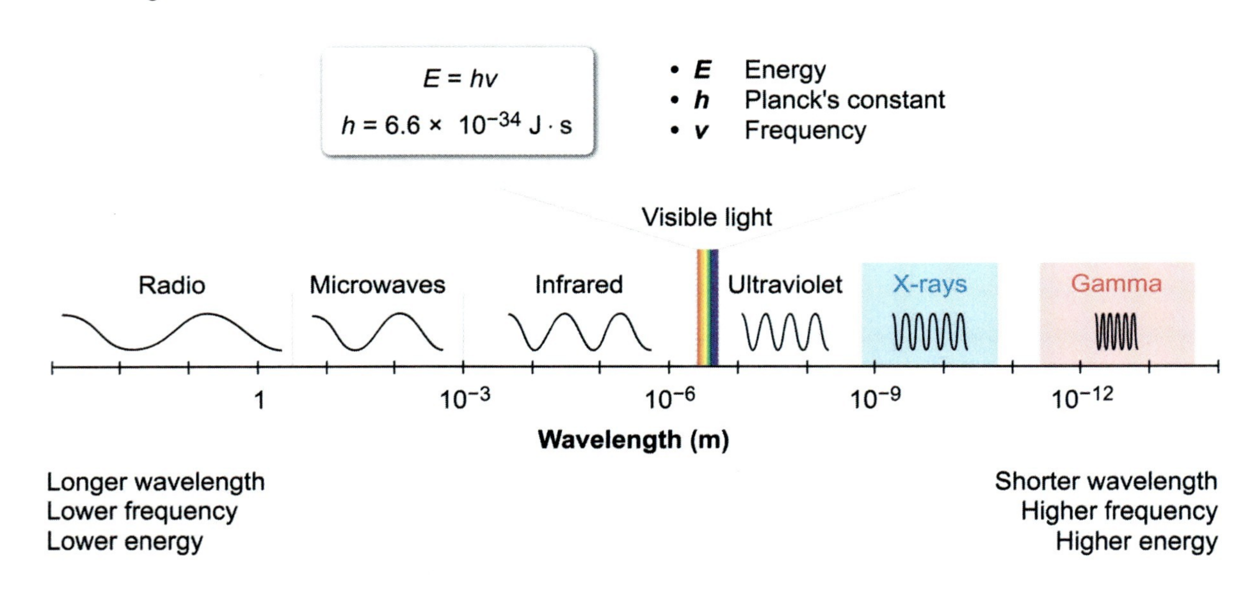

Figure 14.10 The electromagnetic spectrum.

Concept Check 14.2

A photon of light produced by a green laser pointer has a wavelength of 532 nm. What is the energy in kJ for a photon of light produced by this laser pointer?

Solution

Note: The appendix contains the answer.

14.3.02 Spectroscopy and the Electromagnetic Spectrum

The interaction of a photon with a molecule *may* lead to one of several molecular changes such as:

- Changes in molecular motion (eg, vibration, rotation)
- Electron transitions between orbitals with *specific* energy levels

In most cases, the energy of the photon must be a *specific* (ie, quantized) value that matches the *exact* energy necessary for the molecular change (Figure 14.11). If a photon has the *exact* energy needed for a molecular change, the photon is **absorbed** by the molecule and the change occurs. In general, *higher* energy photons promote *more significant* molecular changes.

A few types of molecular changes require the energy of the photon to *exceed* a **threshold** value. For example, if the energy of a photon *exceeds* the ionization energy of a substance, then the photon will cause the substance to lose an electron (ie, be **ionized**). Excess energy *beyond* the required threshold is transferred to the products, for example as kinetic energy of an ejected electron (see the **photoelectric effect**, Physics Concept 4.3.03).

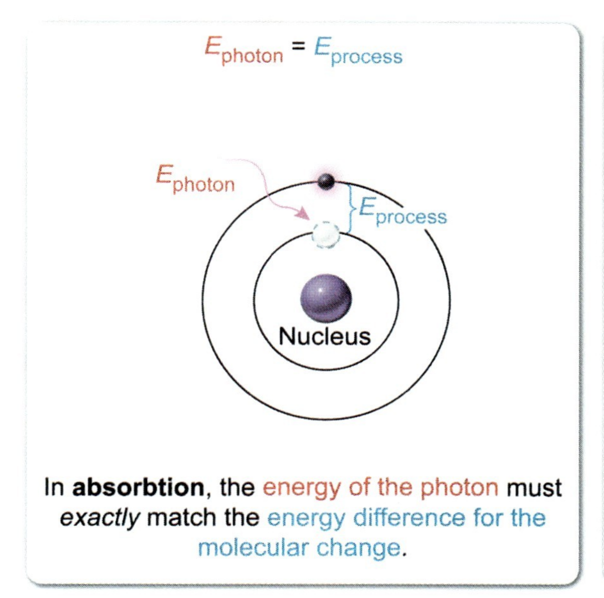

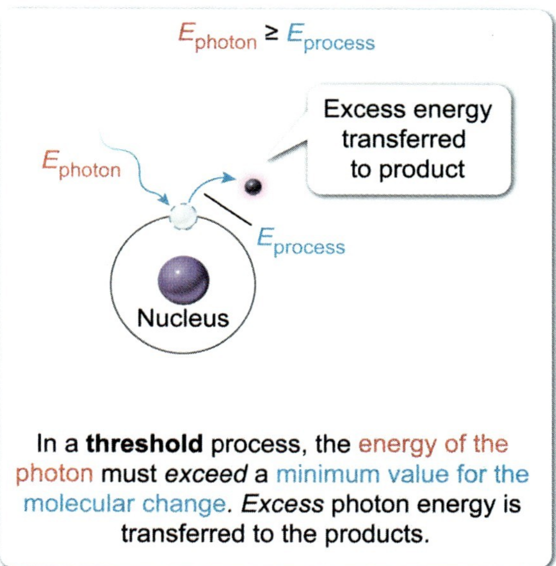

Figure 14.11 Comparison of absorption and threshold molecular interactions with a photon.

In chemistry, **spectroscopy** (ie, the study of interactions between electromagnetic radiation and matter) is used to analyze chemical substances. **Absorption spectroscopy** focuses on interactions in which electromagnetic radiation is absorbed by matter. A **spectrum** (plural: **spectra**) is a graph of the *data* collected from spectroscopic methods.

The Relationship between the Electromagnetic Spectrum and Spectroscopic Methods

Different regions of the electromagnetic spectrum correspond with specific energy ranges needed to promote *particular* types of molecular changes (Figure 14.12), which are assessed through spectroscopy.

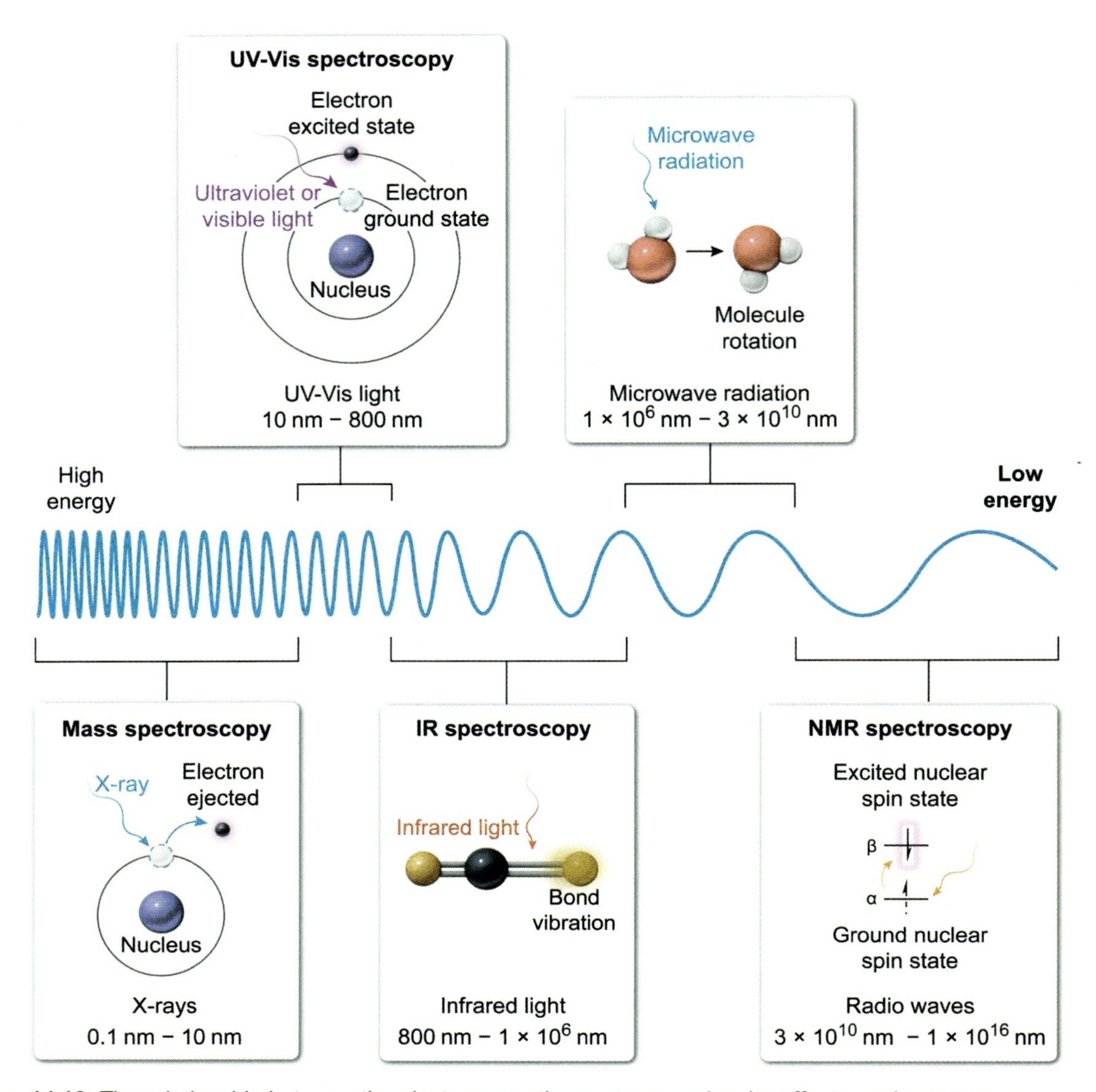

Figure 14.12 The relationship between the electromagnetic spectrum, molecular effects, and spectroscopy.

The remaining lessons in Chapter 14 discuss specific types of interactions with the electromagnetic spectrum and the associated types of spectroscopic/spectrometric experiments:

- X-rays (and other types of high-energy radiation) have sufficient energy to *ionize* a substance by ejecting an electron from an atom, which is a necessary step in **mass spectrometry** (see Lesson 14.4).
- Ultraviolet-visible (UV-Vis) light can cause the *transition* of an electron from its original orbital (ground state) to a vacant, higher-energy orbital (excited state); this transition is measured by **UV-Vis spectroscopy** (see Lesson 14.5).
- Infrared (IR) light can promote *molecular vibrations* (ie, the relative motions of atoms in a molecule), which are measured by **IR spectroscopy** (see Lesson 14.6).
- Microwave radiation promotes the *rotation* of whole molecules and can be used to heat samples or speed reactions by increasing rotational kinetic energy; the exam does not include any spectroscopic methods using microwave radiation.
- Radio waves can cause transitions in the *nuclear spin* of atoms; these transitions are measured by nuclear magnetic resonance (NMR) spectroscopy (see Lesson 14.7).

Lesson 14.5

Mass Spectrometry

Introduction

This lesson focuses on **mass spectrometry** as an experimental method and provides an overview of the ionization of analyzed samples (Concept 14.4.02) and the general function of a mass spectrometer instrument (Concept 14.4.03). A discussion of the features of a mass spectrum and the interpretation of data produced by a mass spectrometer is also given along with consideration of selected isotope signatures (Concept 14.4.05) and molecular fragmentation patterns (Concept 14.4.06).

14.4.01 Introduction to Mass Spectrometry

Mass spectrometry is an instrumental methodology that can determine the composition of a substance by analyzing the *population* of ions produced following the nonspecific (ie, somewhat random) ionization of a sample. Although nonspecific ionization (Concept 14.4.02) is a *destructive* technique that results in the sample being irreversibly destroyed, mass spectrometry is *very* sensitive and can be completed using less than a nanogram of sample.

The data produced by mass spectrometry are reported on a **mass spectrum**, which depicts the abundance of ions formed, sorted by their mass-to-charge ratio (Concept 14.4.04). A mass spectrum can often provide several key pieces of information about the molecular structure of a substance, summarized in Table 14.1.

Table 14.1 Molecular information gained through mass spectrometry.

Information gained	Description
Molecular weight	The molecular weight (or molar mass) of the intact substance is obtained from the *molecular ion.*
Presence or absence of elements	Certain elements create unique *mass patterns* (eg, N), or *isotope signatures* (eg, C, Br, Cl, S, I) in the mass spectrum.
Molecular structure	Analysis of ion fragments can help determine the *connectivity* of atoms in the substance.

14.4.02 Molecular Ionization

Broadly, **ionization** is a process that results in the formation of one or more ions from an uncharged reactant. One of the most common ways to ionize a compound for mass spectrometry is **electron ionization (EI)**, which uses a high-energy beam of free electrons (e^-) to strip away a bound electron from a molecule (Figure 14.13). For this process, the sample is first converted to the *gas phase* within a vacuum before being bombarded by the electron beam. EI *nonspecifically* removes one or more electrons from a sigma bond, pi bond, or lone electron pair of a sample molecule.

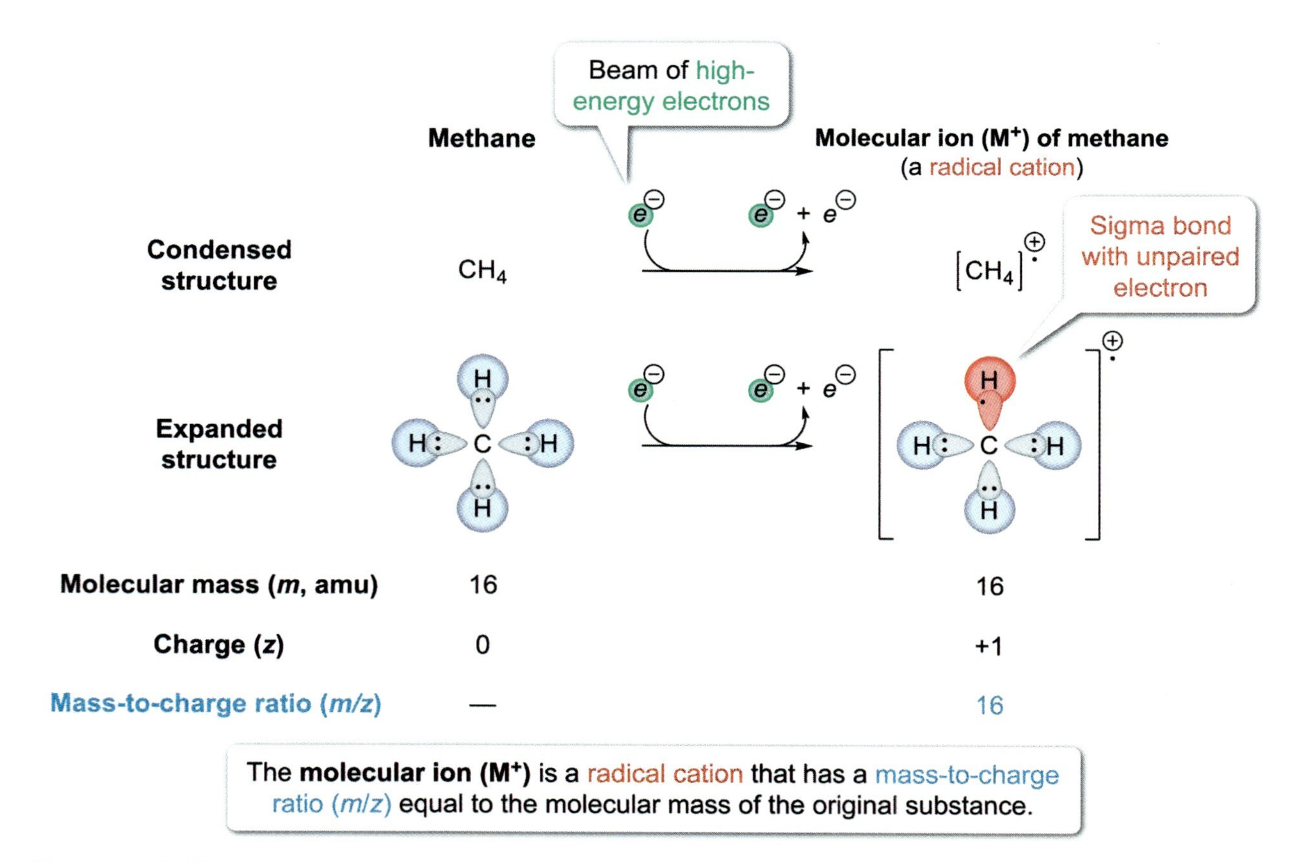

Figure 14.13 Electron ionization and the formation of a radical cation.

Because the *removed* electron has a negative charge that is *lost* by the ionized molecule, the product of EI is a reactive intermediate called a **radical cation**. Given that the mass of the lost electron is *negligible* compared to the mass of nuclear components, the mass (m) of a radical cation is *practically identical* to the mass of the original sample molecule.

If EI removes exactly *one* electron from an original sample molecule, the resultant radical cation has a charge (z) of 1+. As such, the **mass-to-charge (m/z) ratio** of a radical cation that has lost one electron has the *same value* as the molecular weight or atomic mass of the original sample molecule. This specific type of radical cation is called a **molecular ion (M^+)** because it has both the mass and the structure of the original sample molecule.

Because the molecular ion for a sample contains *two* reactive features (ie, it is *both* a radical *and* a cation), it is often unstable. As a result, a portion of *individual* molecules within the *population* of ions may spontaneously break apart into smaller fragments. The smaller, less massive fragments that still bear a charge have a *smaller* m/z than the M^+, which usually has the *highest* m/z of all ions produced through EI. Fragmentation is discussed further in Concept 14.4.06.

14.4.03 The Mass Spectrometer

The instrument used to perform mass spectrometry is a **mass spectrometer**. Although various types of mass spectrometers exist, all mass spectrometers have a:

- Sample source
- Method to *ionize* the sample
- Method to *separate* ions of different masses
- Detector to *quantify* ions

Figure 14.14 shows a general diagram of one type of mass spectrometer. The sample source enters the ionization chamber and is ionized by a high-energy electron beam. The resulting radical cations are accelerated using negatively charged plates and pass into a magnetic field that runs *perpendicular* to the path of the ions.

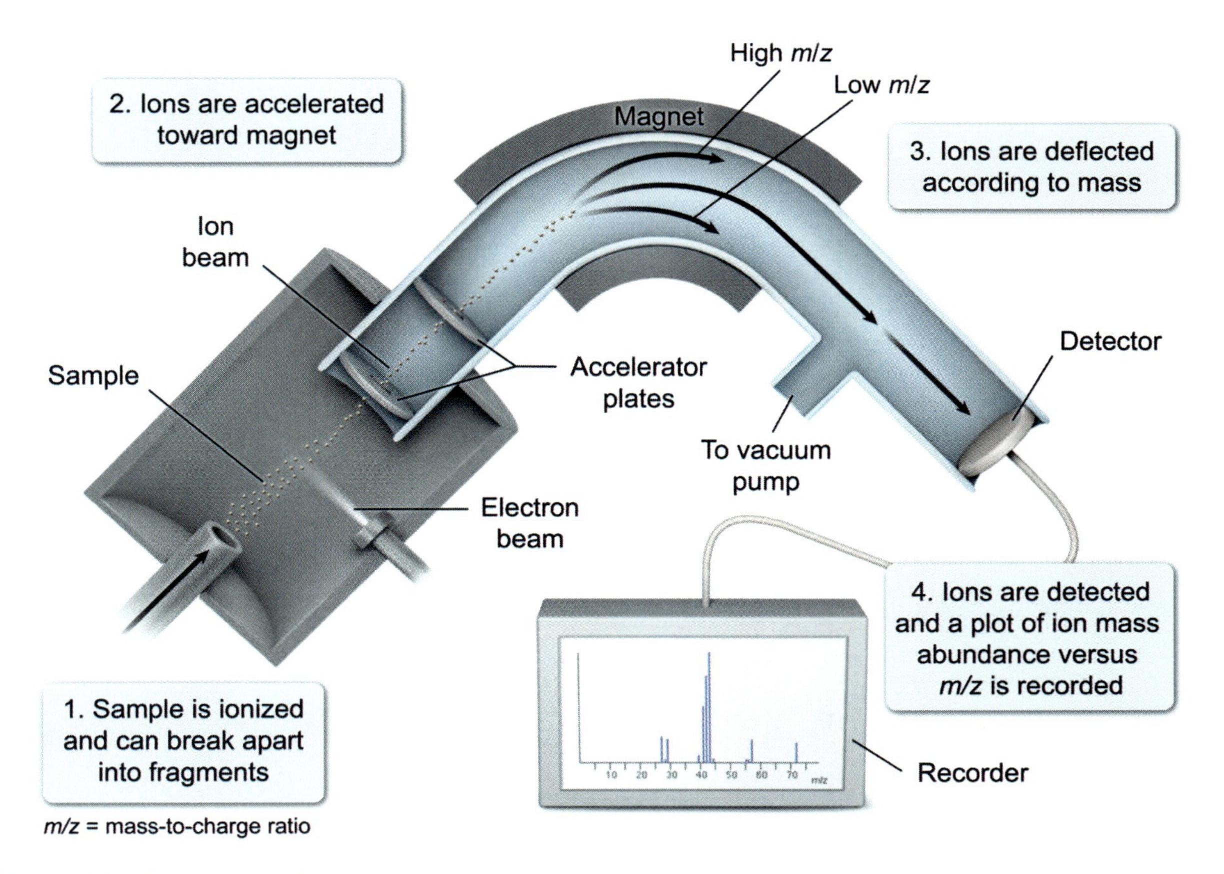

Figure 14.14 Components of a mass spectrometer.

The magnetic field results in a force that *curves* the path of the ions. In a constant magnetic field, the *magnitude* of the curvature depends on an ion's *mass* (ie, smaller ions are curved to a *greater* extent) and its *charge* (ie, larger charges are curved to a *greater* extent). In mass spectrometry, ion path curvature is usually considered in terms of the mass-to-charge ratio (*m*/*z*) (Concept 14.4.02).

Only ions with a specific *m*/*z* (depending on the extent of curvature) pass through and reach the **detector**, which counts the number of ion impacts. By varying the strength of the magnetic field, the *range* of ion *m*/*z* values quantified by the detector changes. The data from the detector are recorded in a **mass spectrum** (Concept 14.4.04), which is a graph of abundance versus *m*/*z*.

The Use of a Mass Spectrometer as a Detector in Chromatography

A mass spectrometer can be a standalone instrument that analyzes discrete (ie, single) sample sources. However, mass spectrometers may also be coupled with *other* instrumental methods that provide chromatographic separation of the components in a mixture, such as gas chromatography or high-performance liquid chromatography (Concept 13.3). In this type of coupled instrument (eg, GC-MS, HPLC-MS), the mass spectrometer analyzes the separated components of the mixture as they elute from the separation column and produces a mass spectrum for *each* peak in the chromatogram of a mixture.

14.4.04 The Mass Spectrum

A **mass spectrum** (Figure 14.15) displays the data of a mass spectrometry experiment with the abundance of ions detected (percent abundance, *y*-axis) plotted as a function of their mass-to-charge ratio (*m*/*z*, *x*-axis). Because an ionized sample contains a *population* of individual molecules, a mass spectrum displays the *population* of ions with each *m*/*z* value.

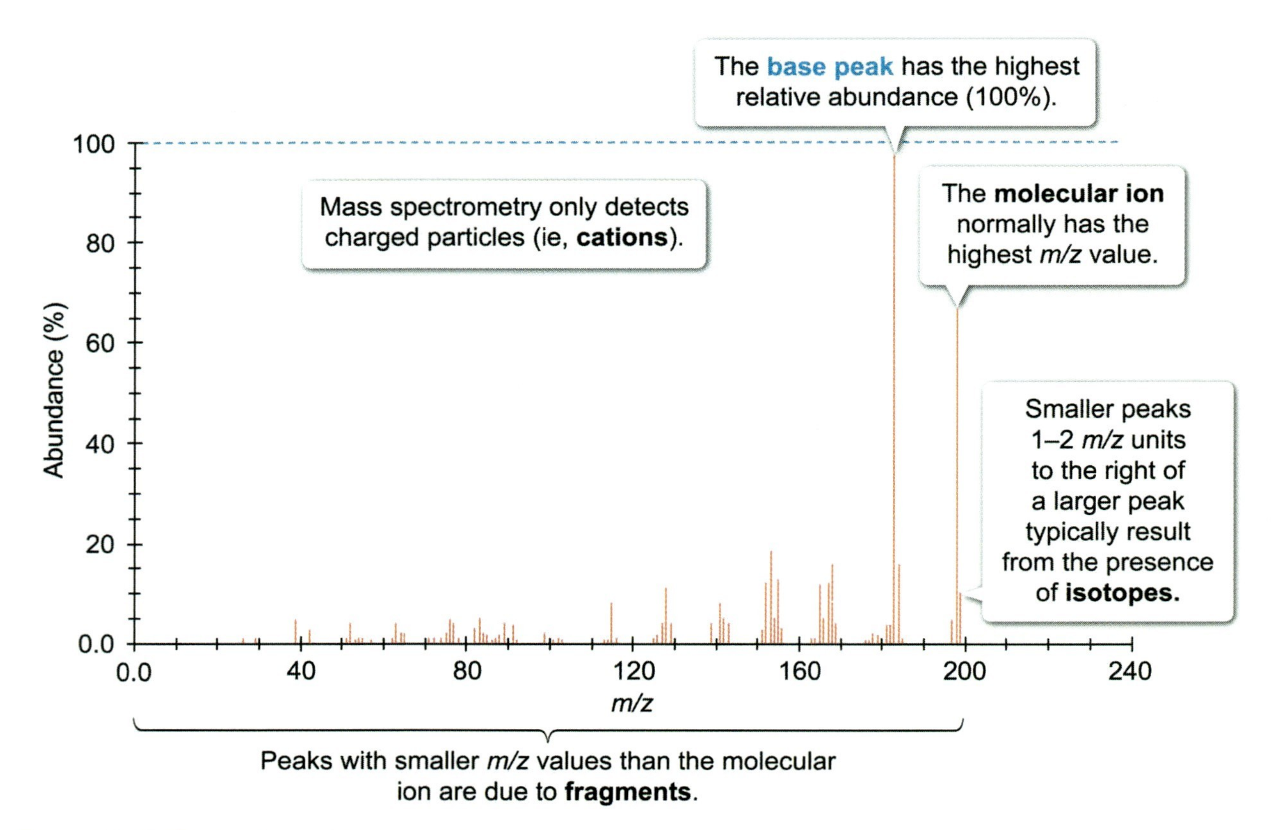

Figure 14.15 The anatomy of a mass spectrum.

The *y*-axis of a mass spectrum is displayed on a 0%–100% scale. The *m*/*z* value with the *highest* abundance (ie, the tallest peak) is labeled as having 100% abundance by convention and is called the **base peak**; all other peaks are assigned abundances relative to this peak. The base peak typically corresponds to the cation with the greatest stability and/or relative rate of formation and can provide structural clues about the identity of the original molecule.

Normally, the significant peak with the *highest m*/*z* ratio corresponds to the **molecular ion**, which has the *highest* mass (ie, the mass of the original molecule) and the *lowest* positive charge (ie, 1+). Identifying the molecular ion is a critical step toward determining the molecular mass of the original sample molecule.

The molecular ion can also break apart into a radical and a cation to produce *smaller* molecular fragments that appear at *lower m*/*z* values than the molecular ion. Fragmentation is discussed in more detail in Concept 14.4.06.

Shorter peaks are usually seen 1–2 *m*/*z* beside most of the taller, high abundance peaks, typically due to the presence of multiple naturally occurring **isotopes** for the elements within the structure of the detected cation. The isotope patterns of several elements are discussed in Concept 14.4.05.

14.4.05 Isotope Signatures in Mass Spectrometry

The elemental identity of an atom is determined by its **atomic number** (ie, the number of *protons* within the nucleus). Atoms with the same number of protons but a *different* number of neutrons are **isotopes** of the same element. Different isotopes are distinguished from each other by their **mass number** (ie, the *sum* of protons and neutrons), placed as a superscript to the left side of their atomic symbol (eg, ^{12}C, ^{13}C, and ^{14}C). Elemental masses listed on the periodic table are a weighted average of the naturally occurring isotopes for an element.

Mass spectrometry detects *individual* ions within an ionized sample population (Concept 14.4.03). Because isotopes have different masses, individual ions containing *different* isotopes are detected at *different m/z* values. The relative heights of the *m/z* peaks in a mass spectrum reflect the natural abundance of the associated isotopes.

Isotope-associated *m/z* peaks are often referenced using **M+X notation**:

- The peak for the isotope with the *lowest* atomic mass is designated as the **M+ peak**.
- Peaks for isotopes of greater mass are labeled as M+X, where X is the number of *additional* neutrons (Figure 14.16).

M+ notation should not be confused with the symbol for a molecular ion (M^+), which has a *superscript* plus symbol (Concept 14.4.02).

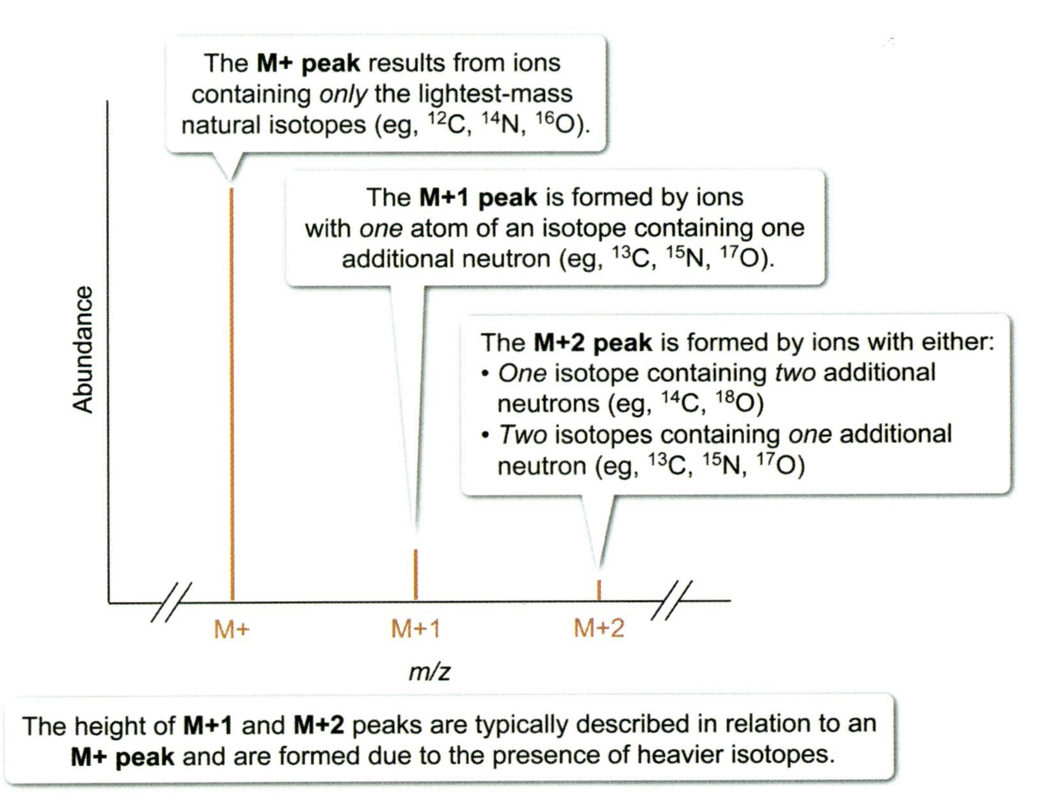

Figure 14.16 M+X notation in mass spectrometry.

In relation to an M+ peak, the **M+1 peak** is *one m/z* unit higher because the detected ion contains *one* atom of an isotope with *one* additional neutron. Similarly, the **M+2 peak** is *two m/z* units higher because either:

- *One* atom of an isotope contains *two* additional neutrons (eg, *one* ^{14}C instead of *one* ^{12}C), or
- *Two* atoms of an isotope contains *one* additional neutron (eg, *two* ^{13}C instead of *two* ^{12}C)

An organic compound being analyzed is likely to contain *several* different elements (eg, C, H, O, N), *each* with its own natural ratios of isotopes. *Any* ions that contain an isotope with additional neutrons can contribute to the M+1 or M+2 peaks. Consequently, the percent abundance of M+1 and M+2 peaks in a mass spectrum are potentially influenced by isotope ratios for *all* the elements in an ion.

Several of the common elements found in organic molecules have easily identifiable **isotope signatures** that appear in mass spectra and indicate the presence of these elements in a sample. Table 14.2 provides a summary of the isotope signatures of some selected elements; the remainder of this concept gives examples of these signatures in mass spectra.

Table 14.2 Isotope signatures of selected common elements observed in mass spectrometry.

Element	Major stable isotopes	Isotopic abundance (%)	Signature in mass spectrometry
Carbon	^{12}C	98.9	In relation to an M+ peak, the M+1 peak has a relative abundance of 1.1% per carbon atom.
	^{13}C	1.1	
Nitrogen	^{14}N	99.6	Ions with an odd number of nitrogen atoms have a molecular ion with an odd *m*/*z* value.
	^{15}N	0.4	
Bromine	^{79}Br	50.7	Bromine-containing ions have an M+ to M+2 ratio of approximately 1:1.
	^{81}Br	49.3	
Chlorine	^{35}Cl	75.8	Chlorine-containing ions have an M+ to M+2 ratio of approximately 3:1.
	^{37}Cl	24.2	
Sulfur	^{32}S	95.0	Sulfur-containing ions have an atypically large M+2 peak, approximately 4% of the M+ peak per sulfur atom.
	^{33}S	0.8	
	^{34}S	4.2	
Iodine	^{127}I	100.0	Fragmentation produces an iodine cation at *m*/*z* 127 and a mass spectrum with large gaps in detected *m*/*z* values.

Carbon Isotope Signatures

Carbon has two stable isotopes relevant to mass spectrometry: ^{12}C (98.9% abundant) and ^{13}C (1.1% abundant). Because there is a 1.1% chance that *any* carbon atom in a sample is ^{13}C, the abundance of the M+1 peak is 1.1% of the abundance of the M+ peak *per carbon atom* in the ion's structure, as seen in Figure 14.17.

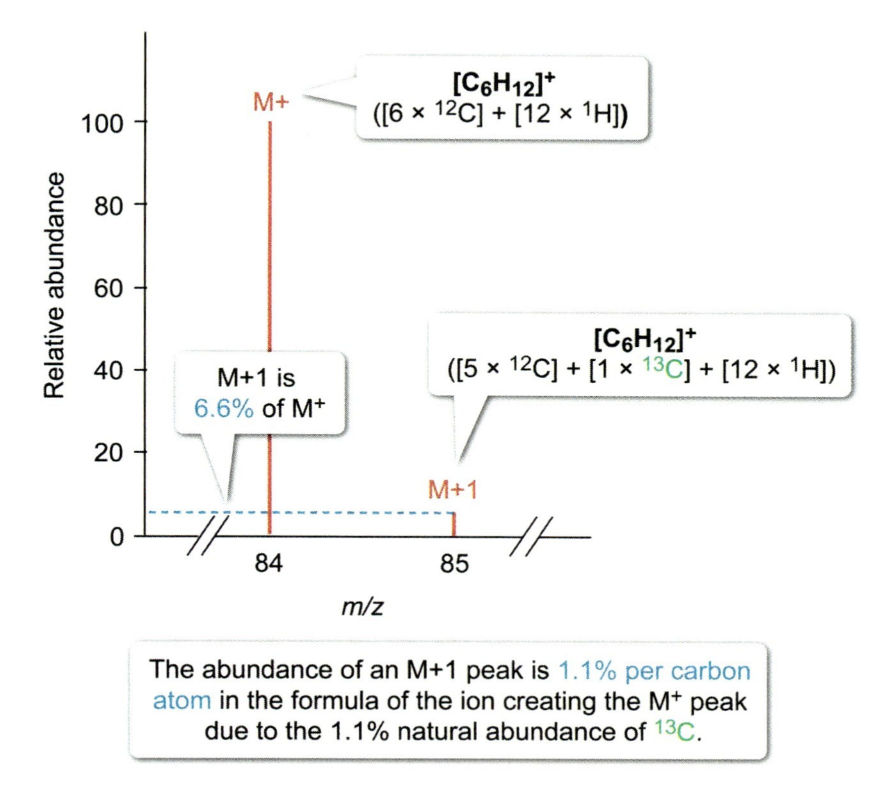

Figure 14.17 Using the M+1 peak to estimate the number of carbon atoms in an ion.

Analysis of the M+1 to M+ peak ratio is a simple way to estimate the number of carbon atoms present in the molecular ion.

$$\frac{\text{Relative abundance of (M + 1)}}{\text{Natural abundance of }^{13}\text{C}} = \frac{6.6\%}{1.1\%} = 6 \text{ Carbon atoms in structure}$$

Nitrogen Isotope Signatures

The three most common nonhydrogen elements in organic compounds are carbon, oxygen, and nitrogen (Figure 14.18). A divalent oxygen atom and a tetravalent carbon atom *both* have atom groupings that result in an M^+ peak with an *even* mass. Given that the major stable isotope of nitrogen is ^{14}N, a trivalent nitrogen atom is *unusual* in that it leads to an atom grouping with an *odd* mass ($^{14}N + {}^{1}H = 15$ amu).

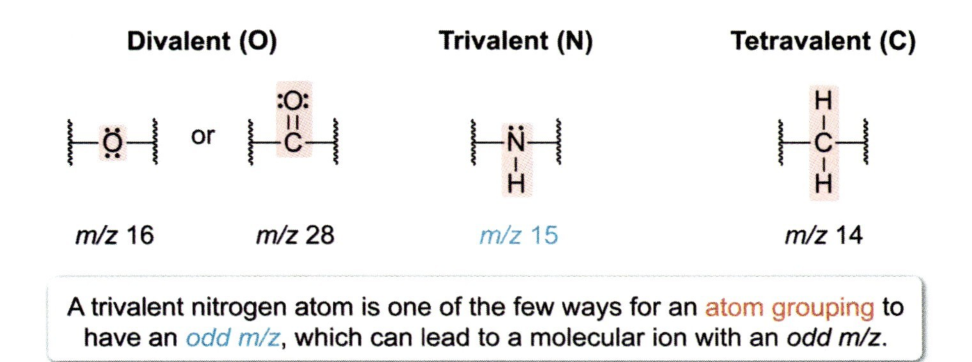

Figure 14.18 Atom groupings that generate even and odd *m/z* values in mass spectrometry.

Consequently, *most* organic compounds comprised of only carbon, hydrogen, and oxygen have a *molecular ion* with an *even m/z* value and *ion fragments* (Concept 14.4.06) with an *odd m/z* value. However, an organic molecule that also has an *odd* number of nitrogen atoms has a *molecular ion* with an *odd m/z* value and nitrogen-containing *ion fragments* with an *even m/z* value.

Bromine Isotope Signatures

Elemental bromine has two stable isotopes relevant to mass spectrometry: ^{79}Br (50.7% abundant) and ^{81}Br (49.3% abundant). Because the natural abundances of the bromine isotopes are *nearly equal*, bromine-containing ions demonstrate nearly equal M+ and M+2 peaks separated by 2 *m/z* units with a height ratio of *nearly 1:1*, as illustrated in Figure 14.19.

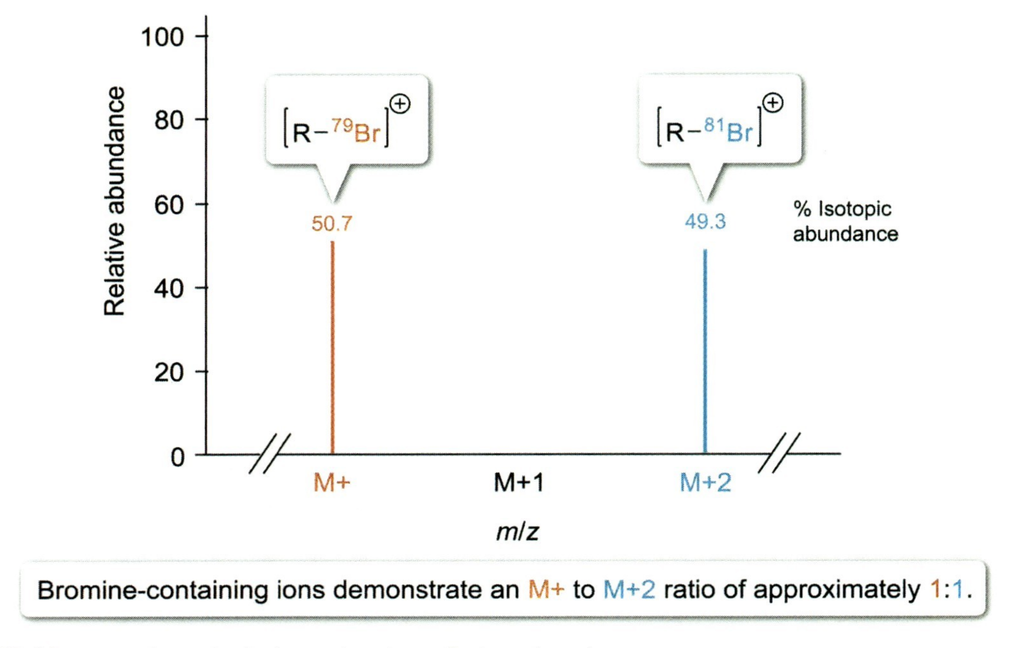

Figure 14.19 Mass spectrometry isotope signature of a bromine atom.

Chlorine Isotope Signatures

Elemental chlorine has two stable isotopes relevant to mass spectrometry: ^{35}Cl (75.8% abundant) and ^{37}Cl (24.2% abundant). Because the natural abundance of chlorine isotopes is approximately 3:1, chlorine-containing ions produce an M+ to M+2 peak ratio of *nearly 3:1* with the peaks separated by 2 *m/z* units, as shown in Figure 14.20.

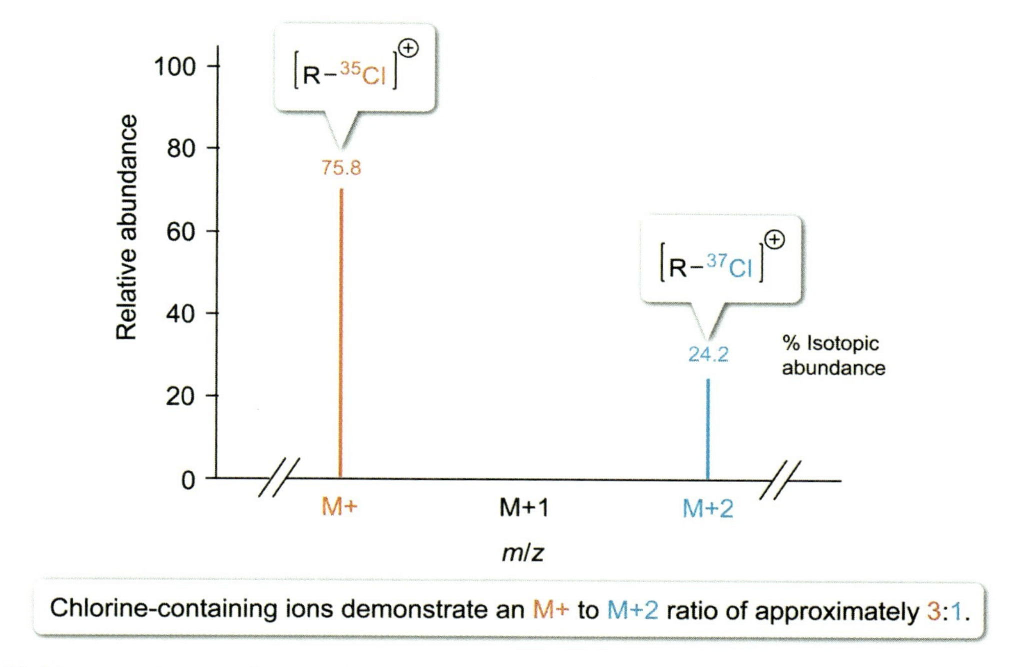

Figure 14.20 Mass spectrometry isotope signature of a chlorine atom.

Sulfur Isotope Signatures

Elemental sulfur has three stable isotopes relevant to mass spectrometry: ^{32}S (95.0% abundant), ^{33}S (0.8% abundant), and ^{34}S (4.2% abundant). Because the ^{34}S isotope has a much *larger* percent abundance than most other elements on earth (aside from bromine and chlorine), an atypically *large* M+2 peak is produced (ie, increases by approximately 4.2% per sulfur atom relative to the associated M+), as seen in Figure 14.21.

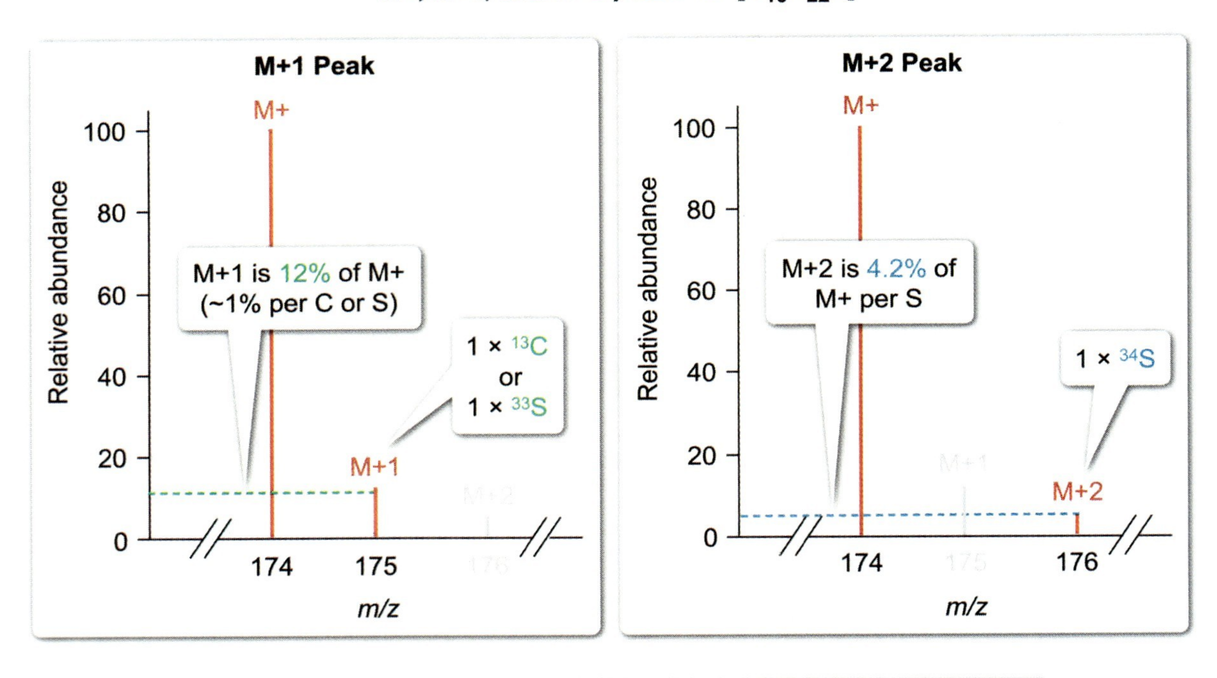

The ^{33}S isotope (0.8%) contributes to M+1 similar to ^{13}C (1.1%), and the ^{34}S isotope (4.2%) leads to a much larger M+2 than normal.

Figure 14.21 Mass spectrometry isotope signature of a sulfur atom.

Iodine Isotope Signatures

Although elemental iodine has only one stable isotope (^{127}I), there are two clues that readily indicate the presence of iodine in a sample (Figure 14.22). First, an iodine atom is a great leaving group and readily forms a stable iodine cation (I^+) with a unique *m*/*z* of 127. Second, the loss of a 127 amu group from the M^+ leads to a *large* gap in detectable ions in the mass spectrum.

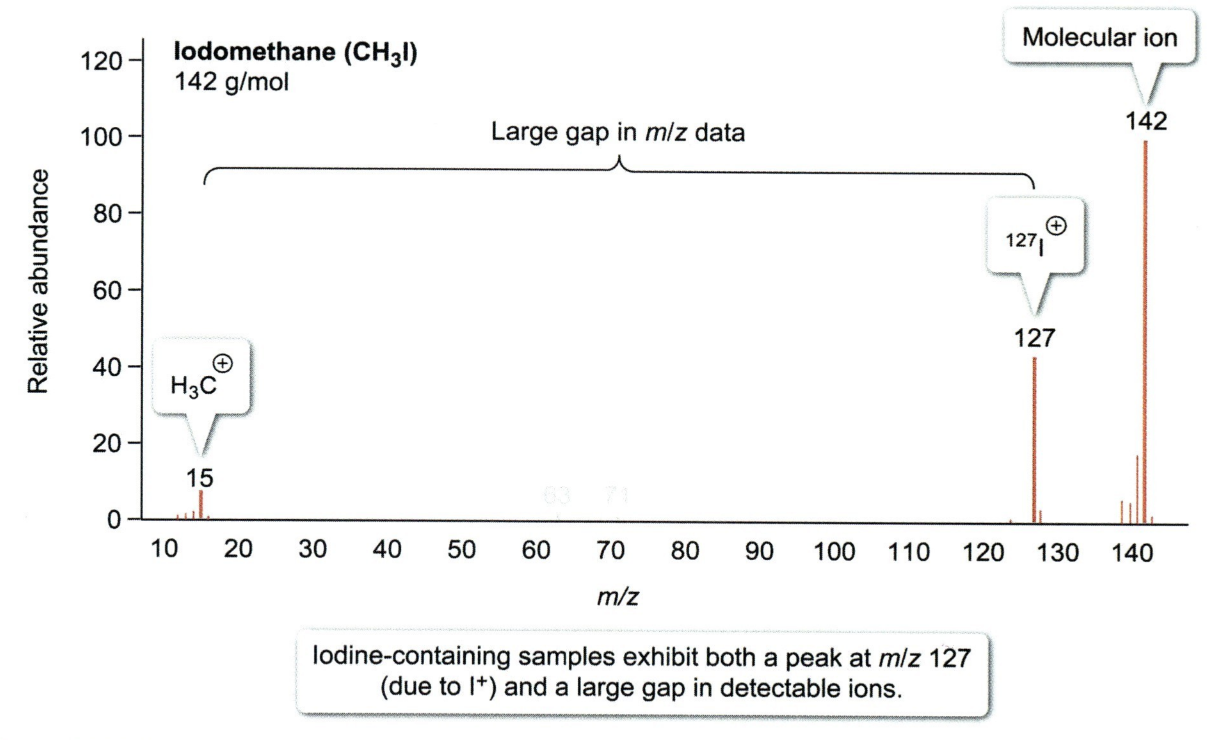

Figure 14.22 Mass spectrometry isotope signature of an iodine atom.

Concept Check 14.3

If the molecular ion for the mass spectrum of an organic compound appears at *m*/*z* 122 and the abundance of the *m*/*z* 124 peak is less than 1% the abundance of the *m*/*z* 122 peak, what conclusions can be drawn about the presence of chlorine, bromine, sulfur, or iodine atoms in the structure of the organic compound?

Solution

Note: The appendix contains the answer.

14.4.06 Fragmentation of Radical Cations

Although electron ionization (EI) always results in the loss of an electron and the formation of a radical cation (Concept 14.4.02), the abundance of *specific* radical cations formed through EI varies. EI is nonspecific, but it is not entirely random within a population of molecules. Electrons that are *easiest* to remove (ie, require less energy to remove) are removed with the greatest *frequency* within the population.

Radical cations that have lost an electron from a sigma bond are particularly unstable and may undergo **fragmentation** (ie, break into two or more smaller chemical structures) *prior* to ion detection. A radical cation that has an unpaired electron in a sigma bond may fragment at the weakened sigma bond to generate a neutral radical *and* a *separate* cation (Figure 14.23).

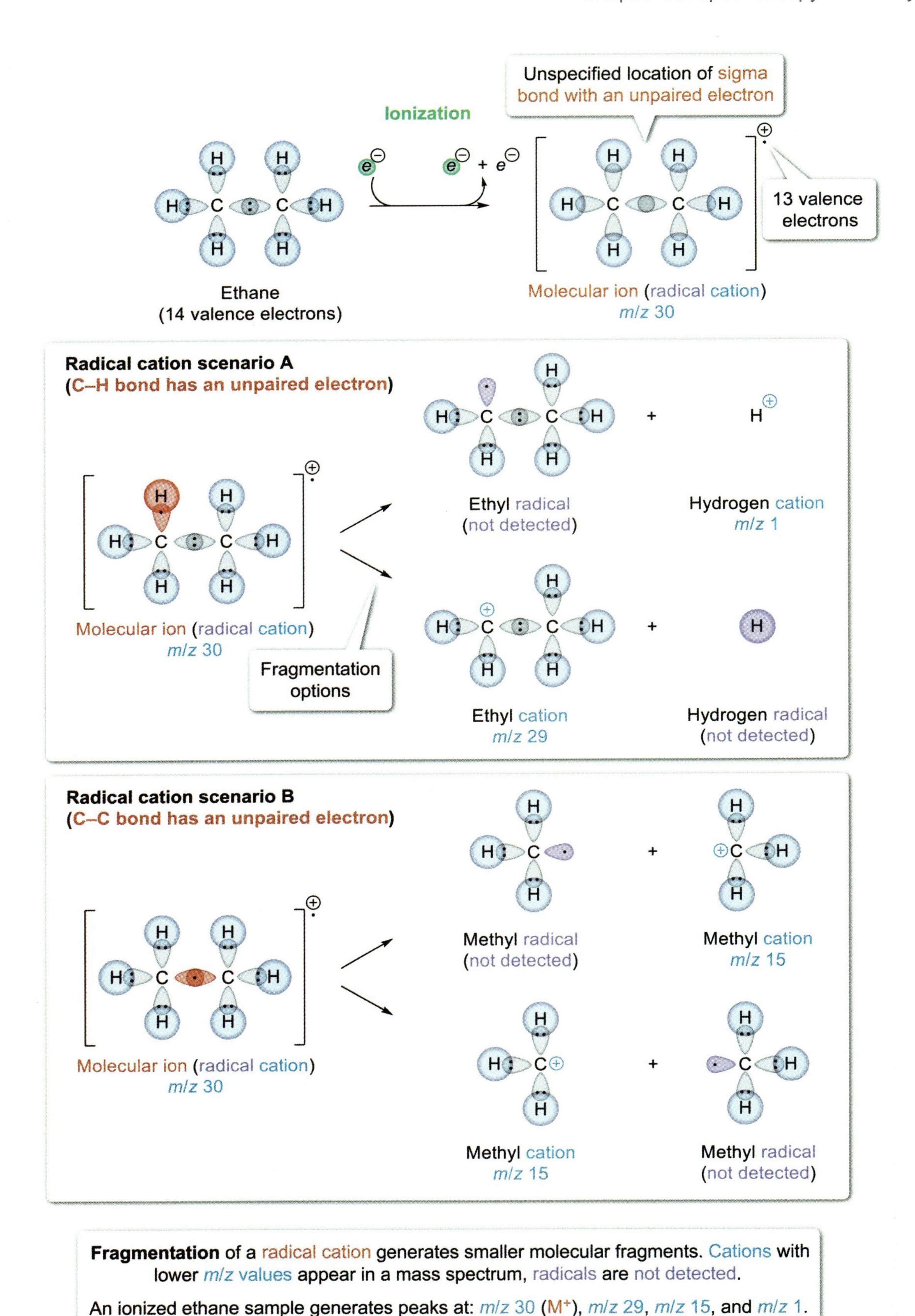

Figure 14.23 Fragmentation of a radical cation.

Because carbon radicals (Concept 5.2.03) and carbocations (Concept 5.2.01) have the same trends in stability, the *directionality* of fragmentation (ie, which fragment becomes the radical and which the carbocation) is random in a population of identical radical cations. However, *only* the cationic (ie, positively charged) fragments are detected because they curve in the magnetic field of the mass spectrometer.

The combination of nonspecific ionization and fragmentation is responsible for the *population* of ions detected by a mass spectrometer and represented as *m*/*z* peaks in the mass spectrum. Analysis of the mass spectrum of a sample to identify patterns can help in drawing conclusions about the structure of the sample compound.

Fragmentation Notation

When evaluating the population fragmentation patterns of radical cations, shorthand notation is often used (Figure 14.24). The sigma bond broken during a fragmentation event is annotated with a line perpendicular to the bond. This annotation typically has a tail oriented toward the fragment half bearing the cation (ie, the fragment detected by mass spectrometry) and a number representing the *m*/*z* value for the fragment. This type of notation can be provided at *multiple* fragmentation points within the structure of the parent radical cation to denote the *population* of fragments that may be observed in a mass spectrum.

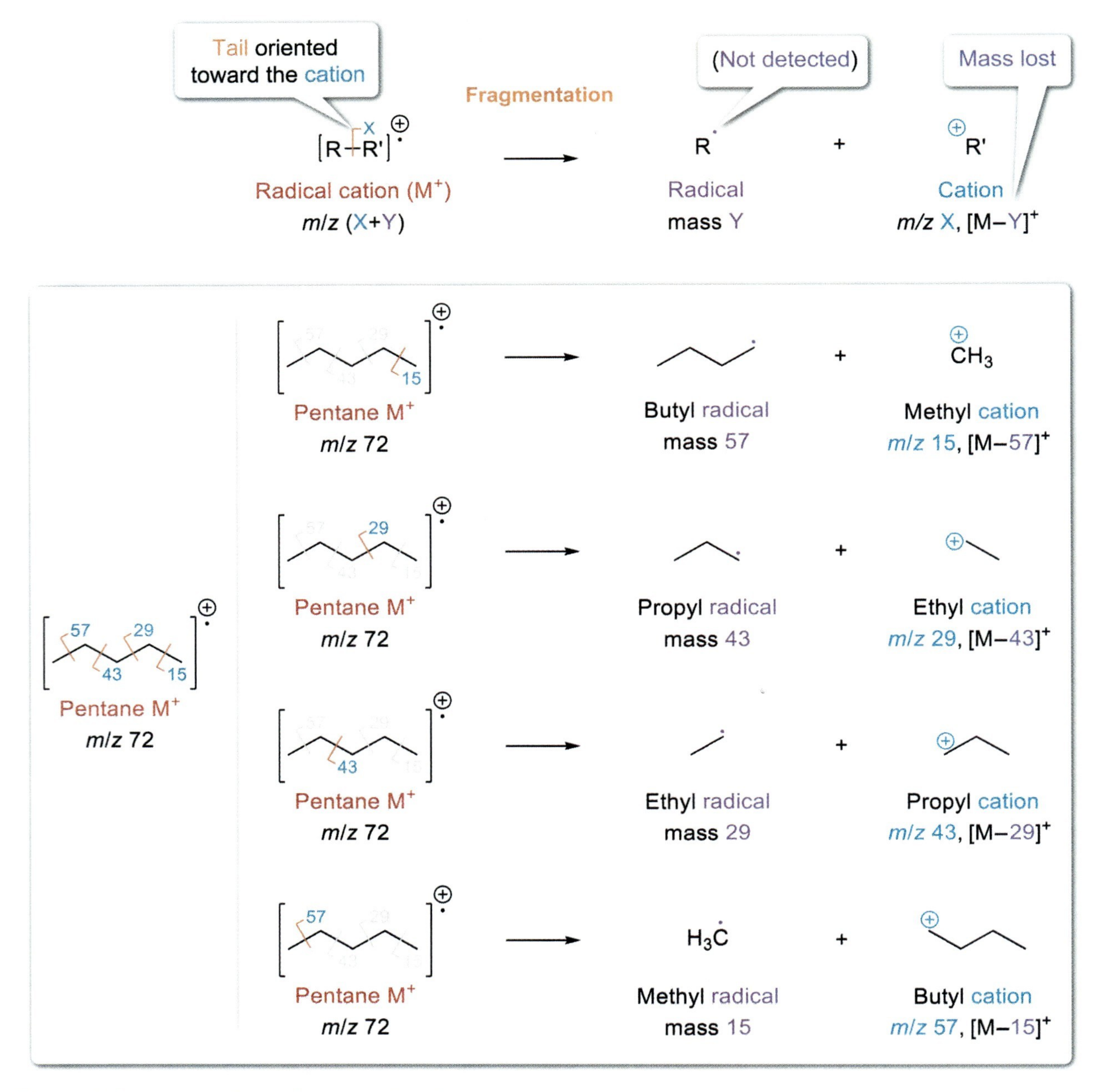

Figure 14.24 Notation for the fragmentation of a radical cation.

Fragments can also be described using **$[M-X]^+$ notation**, where X indicates either the:

- Atoms lost during fragmentation (eg, $[M-CH_3]^+$ or $[M-CH_2CH_3]^+$)
- *Numerical* mass (in amu) of the atoms lost during fragmentation (eg, $[M-15]^+$ or $[M-29]^+$)

Fragmentation of Alkyl Groups

The fragmentation patterns of alkanes and alkyl chains are particularly useful given the prevalence of these groups in organic molecules (Table 14.3).

Table 14.3 Alkyl fragmentation in mass spectrometry.

Alkyl fragment	Molecular formula of fragment	*m/z* remaining or mass lost
Methyl	CH_3	15
Ethyl	C_2H_5	29
Propyl	C_3H_7	43
Butyl	C_4H_9	57

Each additional C is a difference of 14

The smallest alkyl group that can be lost through fragmentation is a methyl group (CH_3), which has a mass of 15 amu and leads to an $[M-15]^+$ product. Each *additional* carbon atom in a linear alkyl chain adds an additional 14 amu to the mass of the lost group (eg, $[M-29]^+$ for an ethyl group, $[M-43]^+$ for an *n*-propyl group). Consequently, organic molecules containing alkyl chains typically have this pattern of fragments present in their mass spectrum (Figure 14.25).

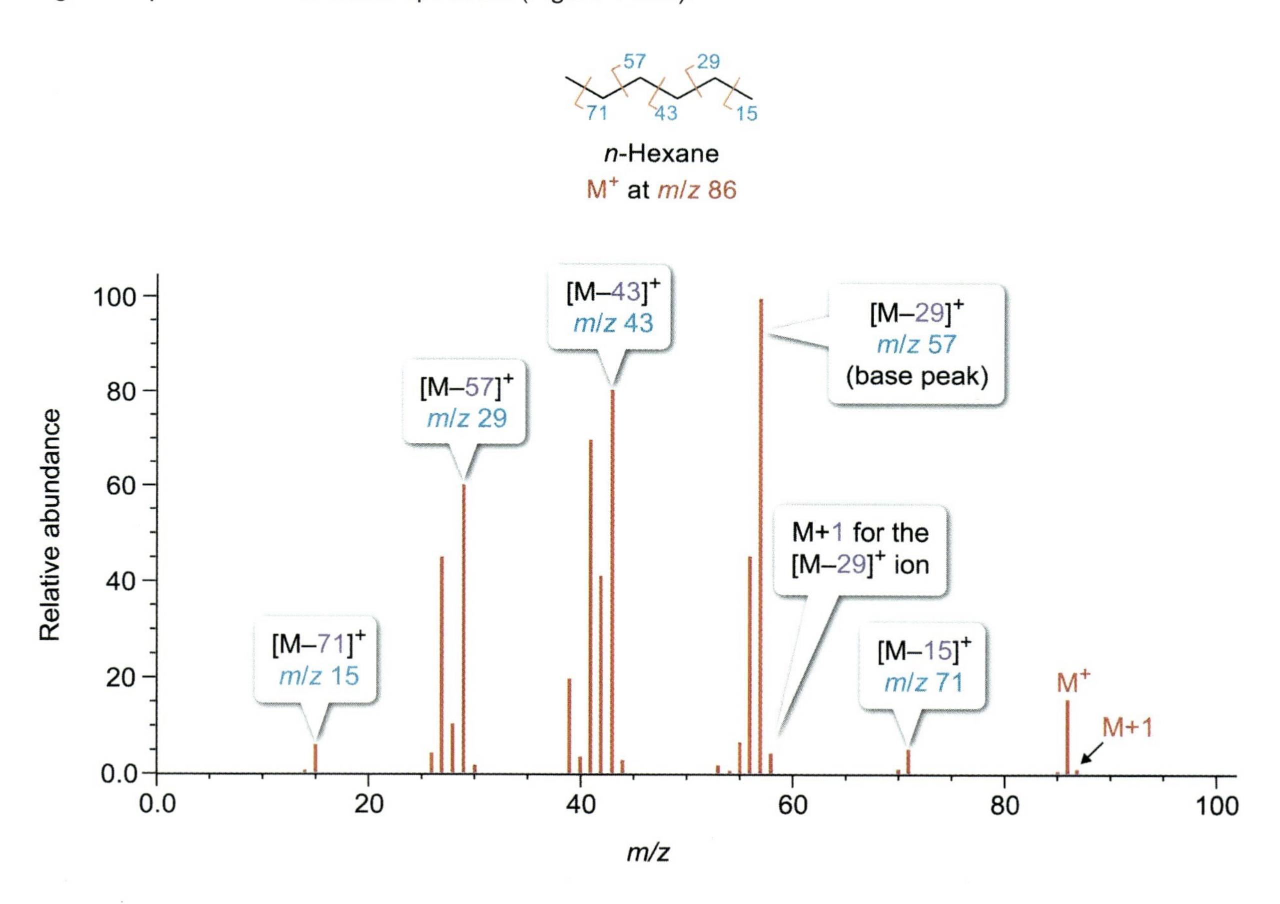

Note: Other *m/z* values in the mass spectrum arise from ions that extend beyond the scope of radical cation fragmentation.

Figure 14.25 The mass spectrum of *n*-hexane.

Fragmentation of Other Functional Groups

Details about the fragmentation of most other functional groups are likely beyond the scope of the exam. However, three common types of fragmentation events involving π electrons are useful to recognize. The first two types are **allylic fragmentation** and **benzylic fragmentation**, which generate a resonance-stabilized allylic and benzylic cation, respectively (Figure 14.26).

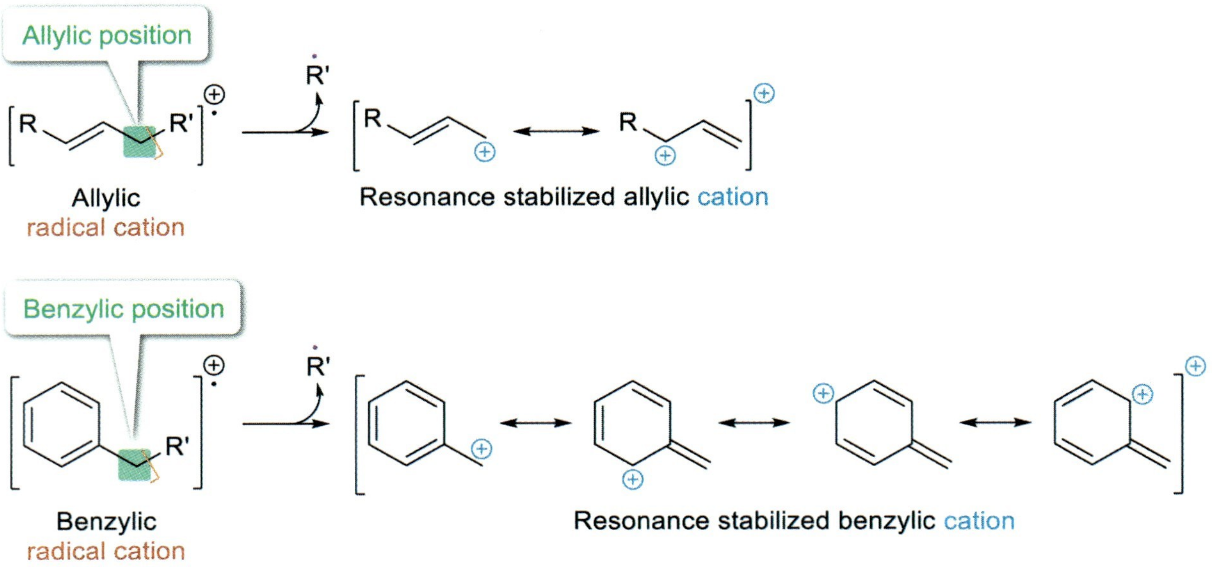

Figure 14.26 Allylic and benzylic fragmentation.

The third type of fragmentation involving π electrons is the formation of a **phenyl cation** through loss of an electron from the sigma bond connecting the rest of a molecule to a monosubstituted phenyl ring (Figure 14.27). The presence of an *m*/*z* 77 peak is a significant clue that the original sample molecule contains a phenyl group.

A phenyl cation is *not* stabilized by resonance because the cation is located in an sp^2 orbital perpendicular to the *p* orbitals of the aromatic ring. In contrast, a **benzyl cation** is *especially* stable due to the resonance delocalization of the carbocation through the pi bonds of the aromatic ring. Consequently, molecules that contain a benzyl group tend to have an *m*/*z* 91 peak with high percent abundance in their mass spectrum.

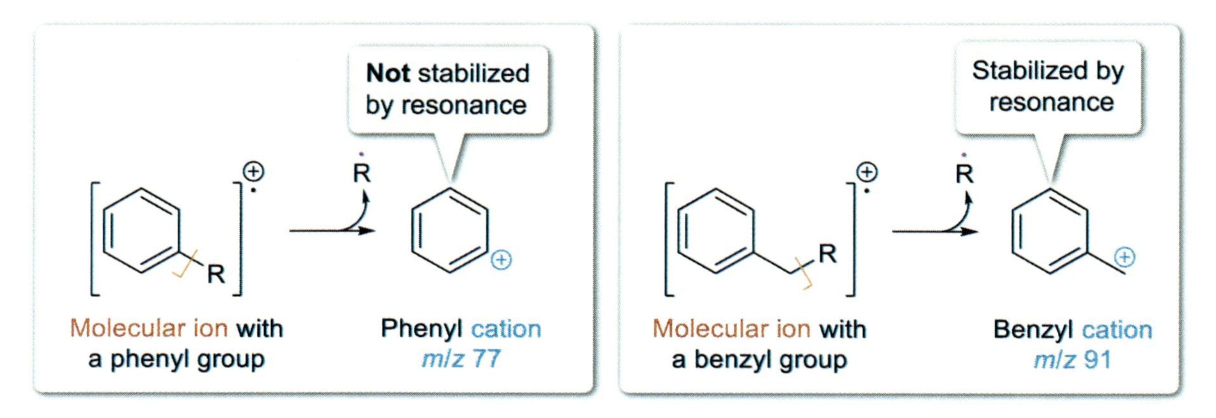

Figure 14.27 Fragmentation of aromatic groups.

Lesson 14.5

UV-Vis Spectroscopy

Introduction

Spectroscopy is the study of the interactions between electromagnetic radiation and matter. Different regions of the electromagnetic spectrum have photons with the correct energy to promote different molecular changes, several of which correlate to the four types of organic spectroscopy covered in this book.

This lesson focuses on the following concepts related to **ultraviolet-visible (UV-Vis) spectroscopy**:

- The transition of electrons between orbitals
- The Beer-Lambert law
- The UV-Vis spectrophotometer
- The UV spectrum
- The visible spectrum

14.5.01 Electronic Transitions

Many types of electronic transitions are possible within an atom or molecule. These transitions are facilitated by photons that have an energy that corresponds to the energy difference between the original electron orbital (the **ground state** of the electron) and the final electron orbital (the **excited state** of the electron). Photons in the ultraviolet (UV) and visible (Vis) regions of the electromagnetic spectrum have the correct energy range to **excite** an electron from a *lower-energy* π molecular orbital to a vacant *higher-energy* π molecular orbital (Figure 14.28).

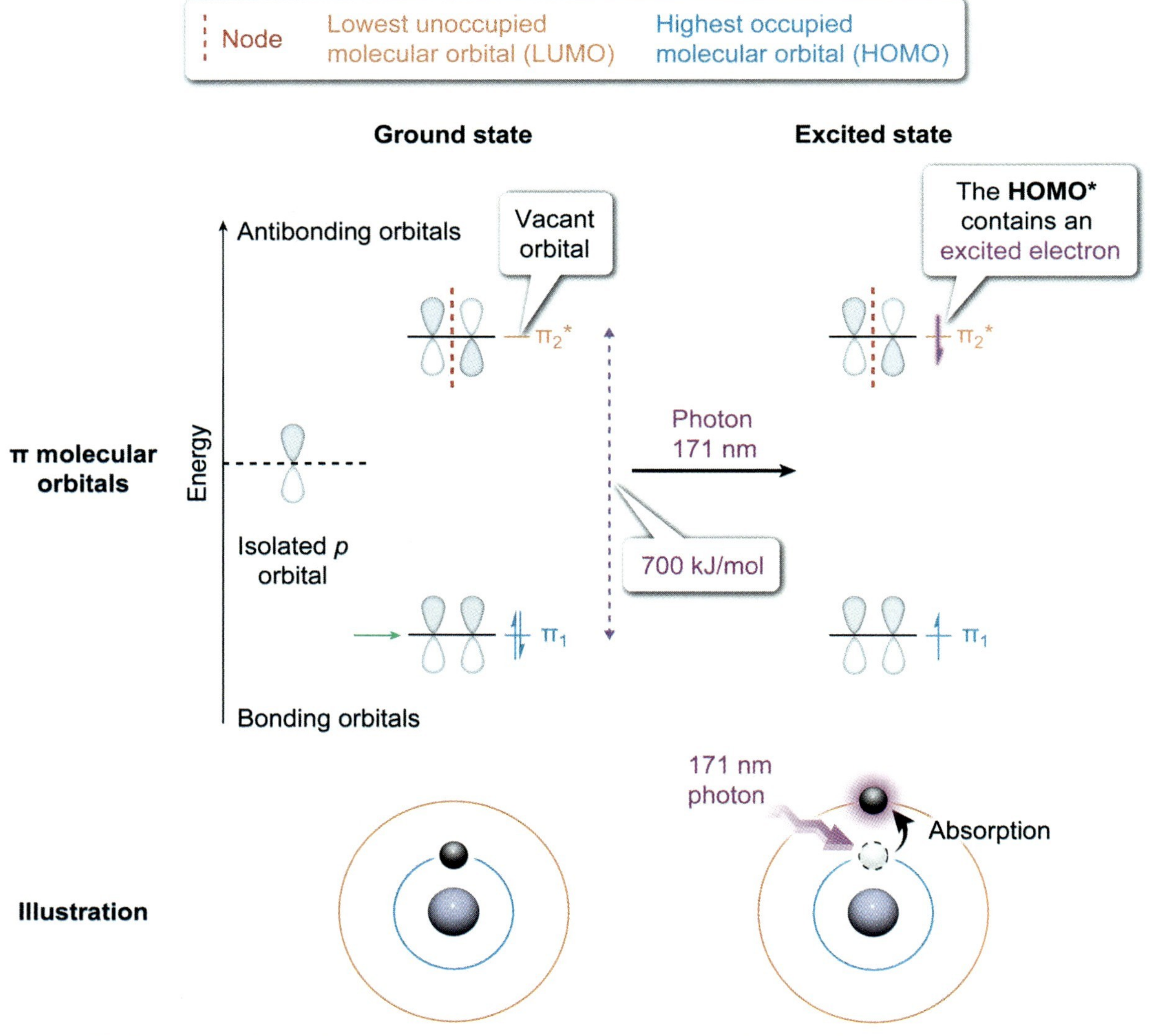

Figure 14.28 Transition of a π electron from the ground state to the excited state of ethene.

The **highest occupied molecular orbital (HOMO)** is defined as the highest-energy molecular orbital containing at least one electron; the **lowest unoccupied molecular orbital (LUMO)** is the lowest-energy *vacant* molecular orbital. For π electrons in molecular orbitals, the electronic transition requiring the *smallest* amount of energy (ie, easiest to achieve) occurs when an electron from the HOMO is promoted to the LUMO. Because the LUMO contains an electron in the *excited* molecule, the LUMO is sometimes called the **HOMO*** in excited atoms or molecules.

Figure 14.28 shows that the HOMO is a bonding molecular orbital (π) and the LUMO is an antibonding molecular orbital (π*) when they are in the ground state. Consequently, the promotion of a π electron from the HOMO to the LUMO is broadly called a **π→π* transition**.

The energy for a π→π* transition is variable and changes with molecular structure. As described in Concept 6.5.01, the energy difference between π molecular orbitals is *smaller* for molecules that have *greater* conjugation. As such, conjugation *decreases* the energy between the HOMO and LUMO, allowing *lower*-energy (ie, *longer*-wavelength) photons to promote a π→π* transition. The correlation between molecular structure and photon energy is discussed further in Concepts 14.5.04 and 14.5.05.

14.5.02 The Beer-Lambert Law

Photons in the UV-Vis range cause π→π* transitions and are **absorbed** by molecules, and this absorption is detectible by **UV-Vis spectroscopy**.

Transmittance and Absorbance

Laboratory measurements of photon absorption involve two related quantities: transmittance and absorbance. The **transmittance (*T*)** is the fractional intensity of electromagnetic radiation detected *after* passing through a sample (I) relative to the intensity of light detected with no sample present (I_0).

$$T = \frac{I}{I_0}$$

Although transmittance is unitless, it is usually represented as a percentage (ie, $T \times 100\%$); experimental transmittance values range from 0% (*all* photons are absorbed, *no* photons pass through the sample) to 100% (*no* photons are absorbed, *all* photons pass through the sample).

By comparison, **absorbance (*A*)** is calculated as the negative logarithm of the transmittance (Figure 14.29). Like transmittance, absorbance is also mathematically unitless; however, absorbance is sometimes reported in terms of absorbance units (AU).

$$A = -\log(T) = -\log\left(\frac{I}{I_0}\right) = \log\left(\frac{I_0}{I}\right)$$

Most instruments have a *maximum* detectable absorbance of approximately 4.00, but the *greatest* analytical sensitivity is obtained for absorbance values between 0.10 and 1.00. Absorption spectroscopy data can be reported as *either* percent transmittance or absorbance, depending on the conventions for the particular method.

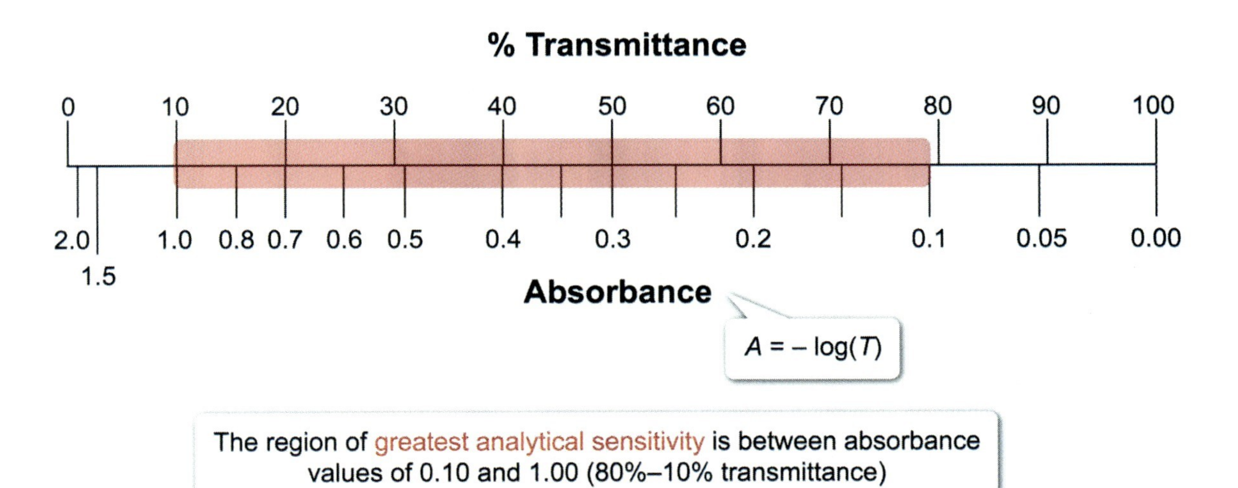

Figure 14.29 Mathematical relationship between percent transmittance and absorbance.

The Beer-Lambert Law

The mathematical relationships between the observed absorbance (A) of a solution, the molar absorptivity (ε) of the solute, the concentration of the solution (c), and the optical path length (ℓ) are described by the **Beer-Lambert law**:

$$A = \varepsilon c \ell$$

Molar absorptivity (ie, extinction coefficient, molar absorption coefficient) essentially describes the *intensity* of photon absorption by a substance at a given wavelength. Molar absorptivity is typically expressed with the standard units of $L \cdot mol^{-1} \cdot cm^{-1}$ when concentration is expressed in $mol \cdot L^{-1}$ and optical path length is expressed in cm. *Intensely* absorbing compounds (ε = 10,000–100,000 $L \cdot mol^{-1} \cdot cm^{-1}$) provide the *greatest* detection sensitivity because *low* solute concentrations can still provide measurable absorbance. Compounds with molar absorptivity values below 1,000 $L \cdot mol^{-1} \cdot cm^{-1}$ are typically more challenging to quantify.

Experimental Determination of Molar Absorptivity

The simplest method of experimentally determining the molar absorptivity of a substance uses a *single* data point. The molar absorptivity is calculated by dividing a sample's absorbance measurement by the known concentration ($mol \cdot L^{-1}$) and optical path length (in cm) parameters.

Alternatively, the molar absorptivity can be experimentally determined using *multiple* data points (Figure 14.30). Usually, the absorbance for a *series* of solutions of known concentration (eg, **standard solutions**) is determined in a sample cell of a constant and known optical path length. A plot of absorbance (*y*-axis) versus concentration (*x*-axis) provides a linear relationship (ie, a **calibration curve**) between data points, where the slope (*m*) of the line of best fit is proportional to the molar absorptivity. If concentration is expressed in $mol \cdot L^{-1}$, the slope needs only to be divided by the optical path length in cm. This approach provides a molar absorptivity value with *greater* precision and reduced error.

$$\varepsilon = \frac{A}{c\ell} = \left(\frac{A}{c}\right)\left(\frac{1}{\ell}\right) = m\left(\frac{1}{\ell}\right) = \frac{m}{\ell}$$

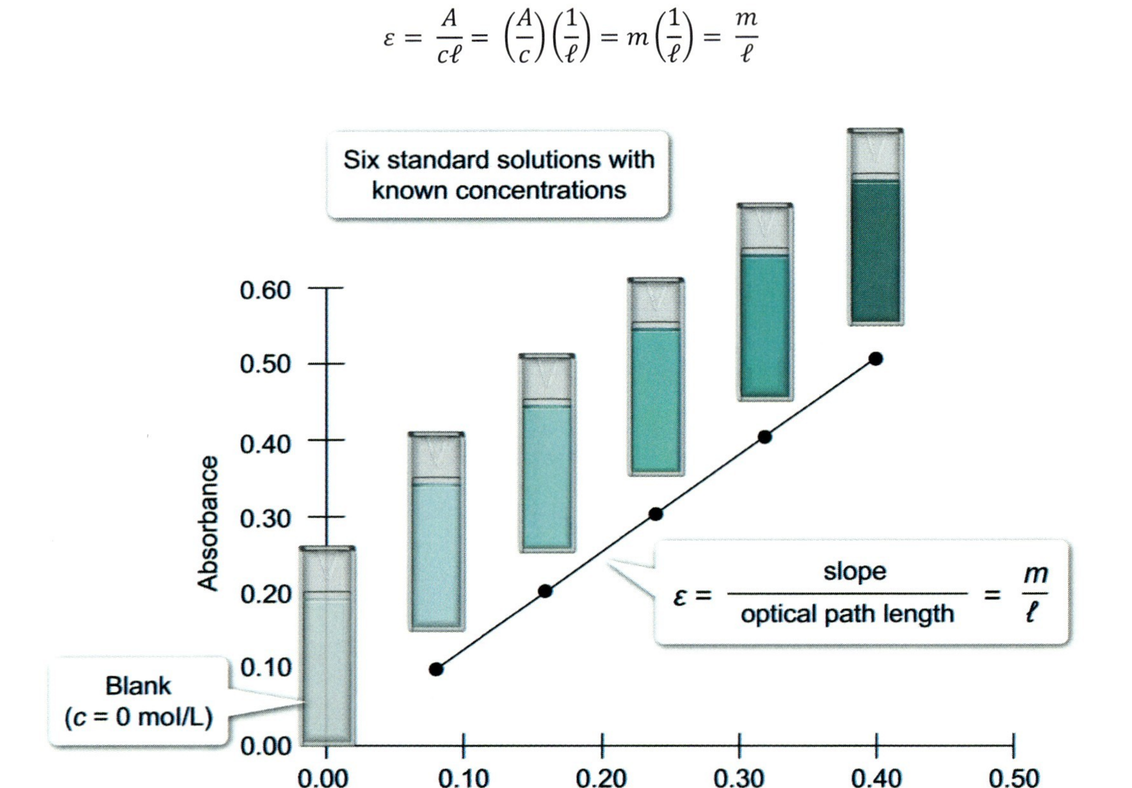

Figure 14.30 Calculating molar absorptivity using a calibration curve.

14.5.03 The UV-Vis Spectrophotometer

The instrument used to measure the absorption of photons in the UV-Vis regions of the electromagnetic spectrum is called a **UV-Vis spectrophotometer**. There are various models of UV-Vis spectrophotometers, but the essential components (Figure 14.31) include a:

- UV and/or visible **light source**
- Monochromator, which filters the emitted light to a narrow wavelength range
- Cell to hold a sample cuvette
- **Detector**, which measures the intensity of light

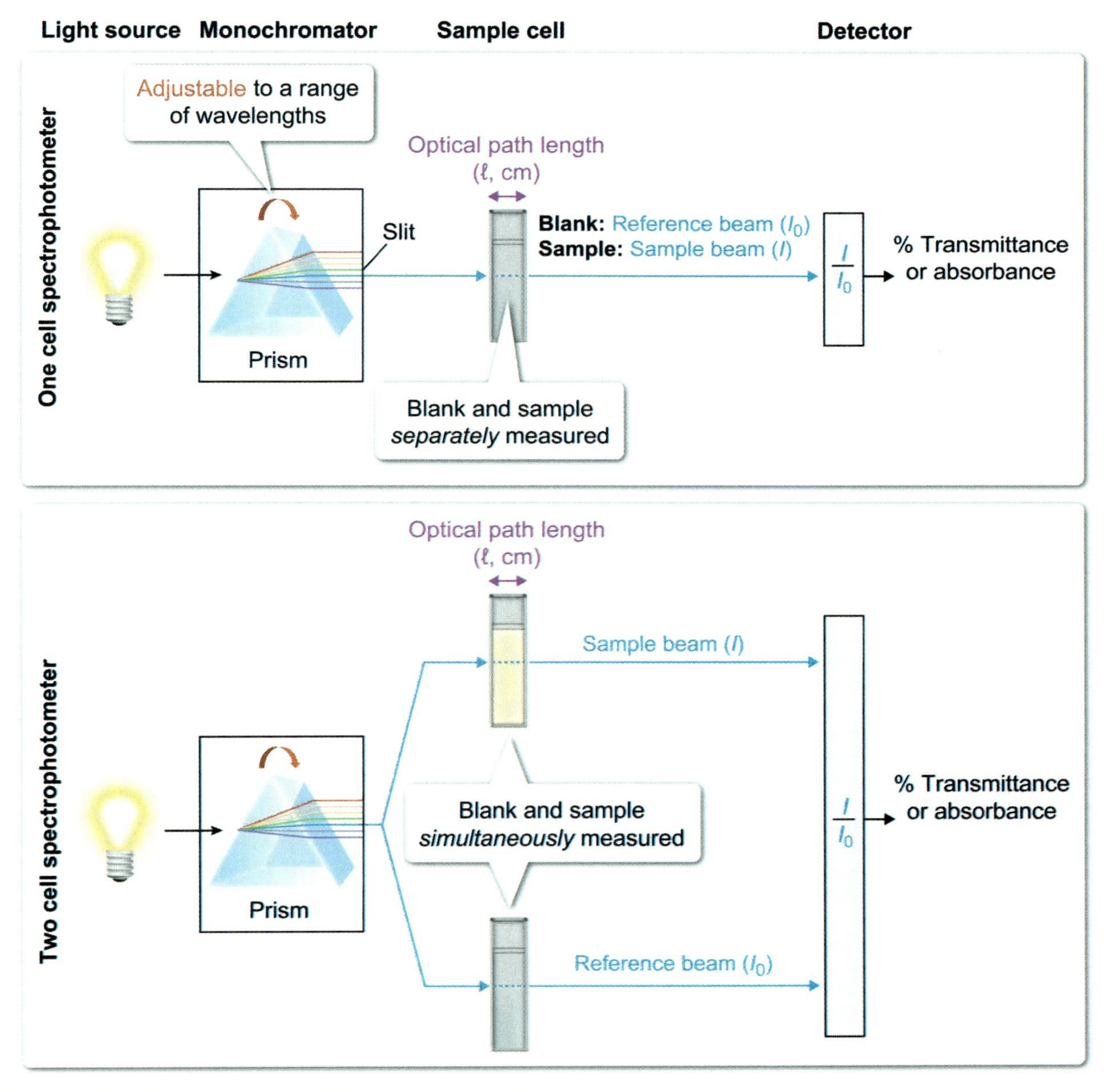

Figure 14.31 The essential components of a UV-Vis spectrophotometer.

Because most UV or visible light sources emit a *range* of wavelengths of electromagnetic radiation, UV-Vis spectrophotometers use a **monochromator**, which allows only a narrow wavelength range to illuminate the sample. Some experiments measure only a *single* wavelength of light; others scan a *range* of wavelengths and produce a UV-Vis spectrum (see Concept 14.5.04).

Most UV-Vis spectrophotometers have one or two **sample cells**, which accept a cuvette holding the material to be measured. Models with *one* sample cell require the spectrophotometer to be initially **blanked** using a solvent sample that does *not contain* the solute to be analyzed. This process calibrates the reference beam (I_0) to 100% transmittance (zero absorbance). After blanking, the sample is added to the sample cell and the intensity of the sample beam (I) is measured.

Spectrophotometer models with *two* sample cells split the light emitted by the monochromator into two equivalent beams. The reference beam passes through the blank solution, and the sample beam passes through the sample being analyzed. In this type of model, the detector *simultaneously* measures the intensity of *both* light beams and uses this ratio to calculate the percent transmittance or absorbance.

14.5.04 The UV Spectrum

A spectrometer scanning a *range* of wavelengths in the ultraviolet (UV) region of the electromagnetic spectrum (200–380 nm) records the measured absorbance of the sample (*y*-axis) as a function of wavelength (*x*-axis) in a **UV spectrum** (Figure 14.32). In some cases, the UV spectrum is scanned along with the visible spectrum (380–780 nm, see Concept 14.5.05), producing a combined **UV-Vis spectrum**.

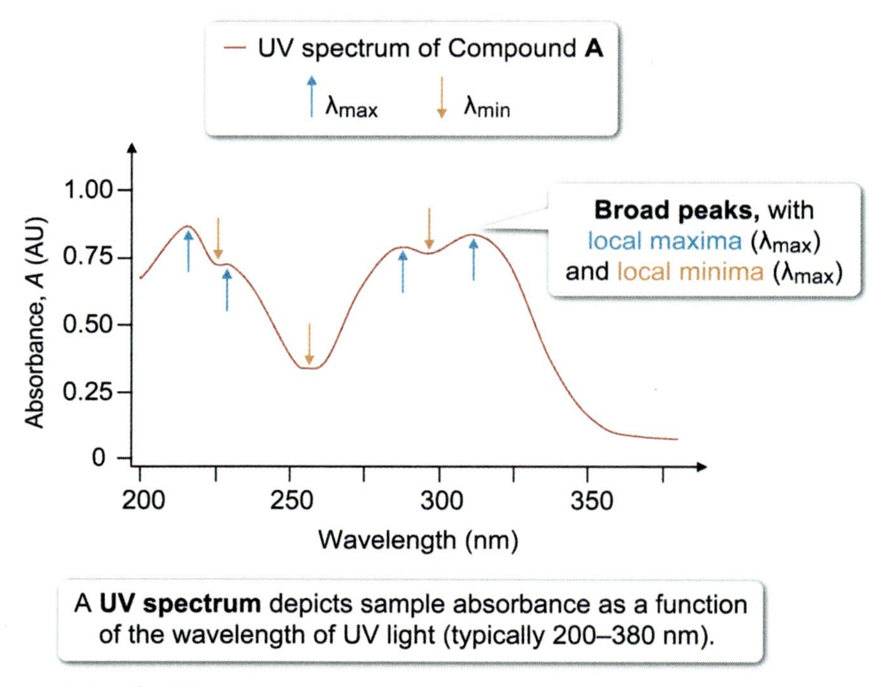

Figure 14.32 The anatomy of a UV spectrum.

Unlike the data from some other forms of spectroscopy, the data in a UV spectrum are typically composed of *broad* peaks and valleys in absorbance as a function of wavelength (λ). *Peaks* (ie, maxima) in absorbance indicate wavelengths that are *highly* absorbed by a sample; these peaks are designated as λ_{max} values. In contrast, *valleys* (ie, minima) in absorbance are designated as λ_{min} values and indicate wavelengths that are *minimally* absorbed by a sample.

Although UV spectroscopy experiments can be conducted using *any* wavelength of radiation in the UV region, experiments tend to focus on the λ_{max} values, as they correspond to wavelengths for which the molar absorptivity is the *largest*. Measuring at a λ_{max} provides the *greatest* analytical sensitivity for detecting small changes in concentration.

The Relationship between Organic Structure and UV Spectroscopy

As described in Concept 14.5.01, organic molecules absorb UV light when a pi electron in a bonding molecular orbital (π) is excited and transitions into an antibonding molecular orbital (π*). The energy required to promote a π→π* transition in ethene (the simplest organic molecule containing π electrons) is 700 kJ/mol, which corresponds to photons with a wavelength of 171 nm—a value just outside the UV spectrophotometer range of 200–380 nm. Consequently, simple alkenes cannot typically be analyzed using UV spectroscopy.

However, **conjugation** *decreases* the energy difference between π molecular orbitals (see Concept 6.5.01), lowering the energy required for a π→π* transition. Because of the inverse relationship between the energy of a photon and its wavelength (see Concept 14.3.01), the π→π* transition for a conjugated alkene (eg, buta-1,3-diene) requires photons with a *lower* energy (eg, 551 kJ/mol) and a *longer* wavelength. Accordingly, the UV spectrum of buta-1,3-diene is *shifted* to the right and has a λ_{max} at 217 nm (Figure 14.33).

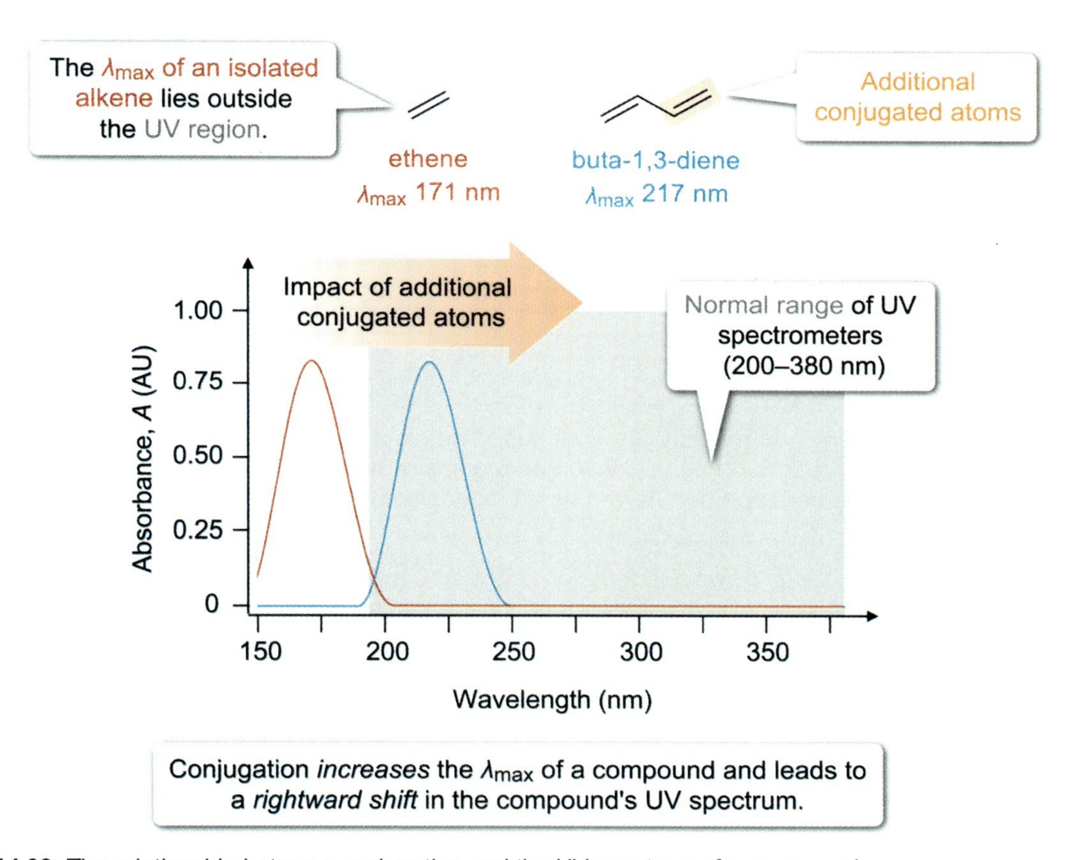

Figure 14.33 The relationship between conjugation and the UV spectrum of a compound.

The effects of conjugation are cumulative, such that *larger* delocalized π systems have increasingly *higher* λ_{max} values. *Highly* conjugated compounds can exhibit *very high* λ_{max} values that extend into the visible region of the electromagnetic spectrum. (The features of visible spectroscopy are discussed in Concept 14.5.05.) The ability of a conjugated or aromatic molecule to absorb UV light is further leveraged in thin-layer chromatography as one method of detecting the location of UV-active materials on a TLC plate (see Concept 13.3.05).

Concept Check 14.4

The following is a UV spectrum of Compound **A** generated using a 6.00×10^{-5} mol/L solution and an optical path length of 1.00 cm. If the absorbance at 247 nm is 0.446, what is the molar absorptivity of Compound **A** at 247 nm?

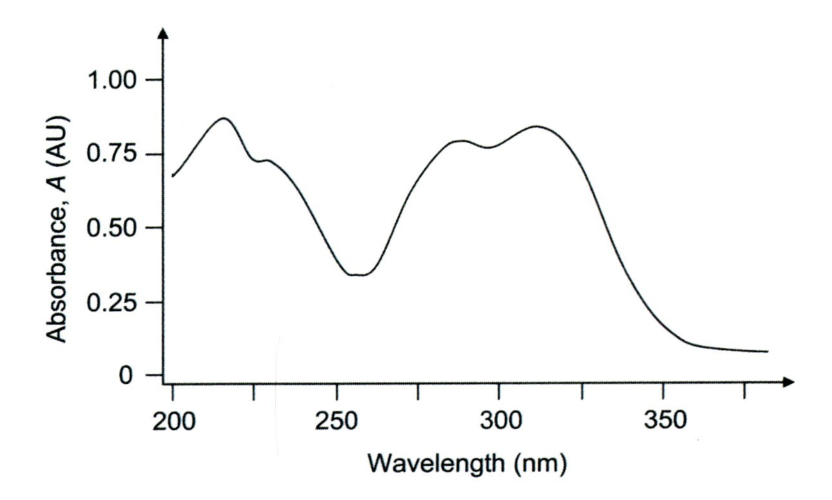

Solution

Note: The appendix contains the answer.

14.5.05 The Visible Spectrum

UV spectroscopy and visible spectroscopy are *both* based on electronic transitions (eg, $\pi \rightarrow \pi^*$ transitions) that occur when molecules absorb photons from these regions of the electromagnetic spectrum. As such, visible spectroscopy can be conceptually viewed as an *extension* of UV spectroscopy using visible light instead of UV light.

A spectrometer scanning a *range* of wavelengths of visible light (approximately 380–780 nm) records the measured absorption by the sample (*y*-axis) as a function of wavelength (*x*-axis), yielding a **visible spectrum** (Figure 14.34). It is common practice for a visible spectrum to display the *full range* of visible wavelengths, and the UV spectrum (200–380 nm) is sometimes also included to give a combined **UV-Vis spectrum**.

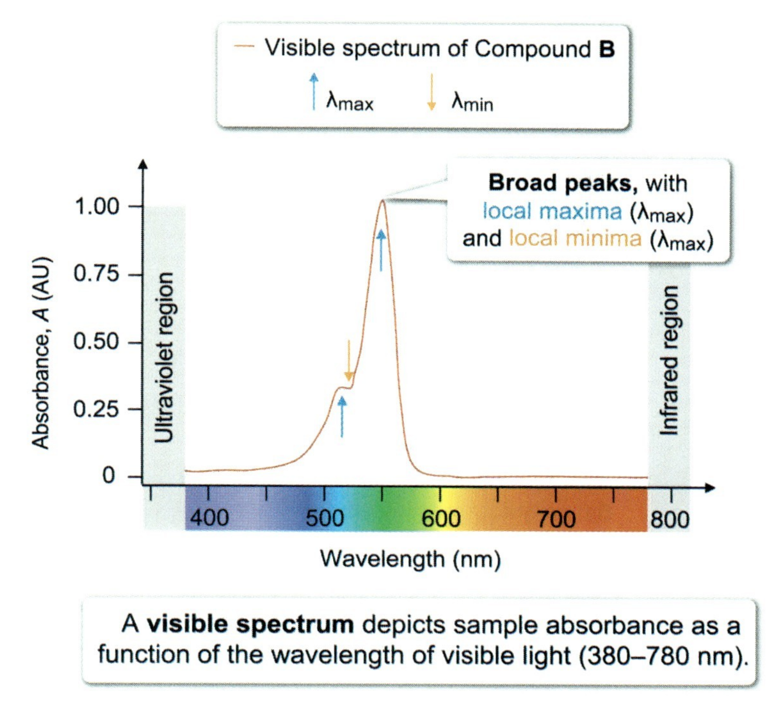

Figure 14.34 The anatomy of a visible spectrum.

Like UV spectrum data, the data of a *visible* spectrum are typically shown as a function of wavelength (λ) and exhibit *broad* peaks and valleys in absorbance, designated as λ_{max} and λ_{min}, respectively. *Unlike* the UV spectrum, the wavelength range of visible light corresponds to *colors* ranging from violet (shorter wavelength) to red (longer wavelength).

As such, a key difference between a UV spectrum and a visible spectrum is the impact of spectral features on the apparent *color* of a compound. However, the **color** of an object is not determined by the visible light absorbed by the object but is instead determined by the visible light *reflected* or *transmitted* by the object and *detected* by the human eye (Figure 14.35).

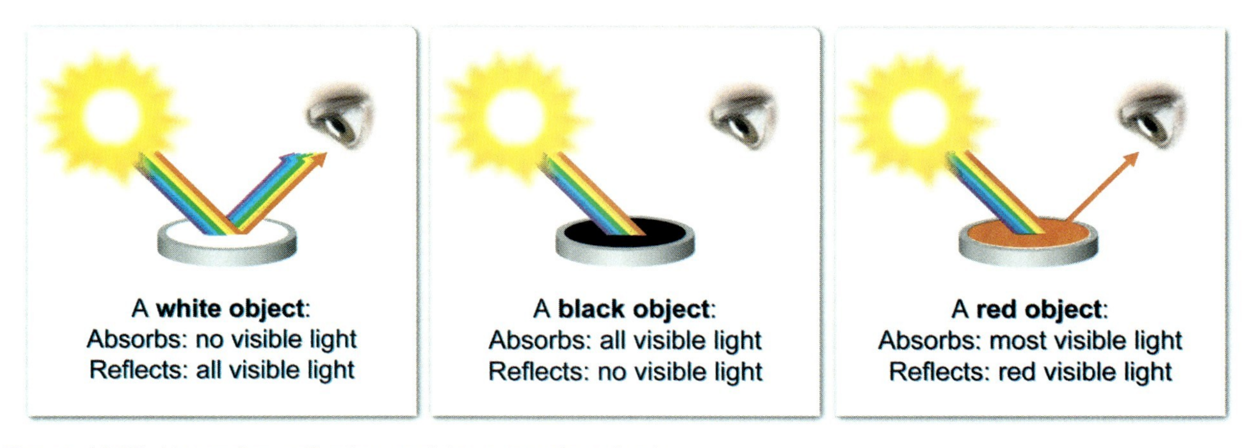

Figure 14.35 Absorption, reflection, and the perception of color.

The relationship between photon *absorption* (ie, color of the λ_{max}) and reflection or transmission (ie, *perceived* color) can be described in the context of complementary colors. Given the **color wheel** shown in Figure 14.36, the **complementary color** for any given color is the color *opposite* it on the color wheel.

When a compound preferentially absorbs one color, the compound is *perceived* as its complementary color. For example, blue and orange are complementary colors, so a compound that absorbs primarily orange light is perceived to have a blue color (and vice versa).

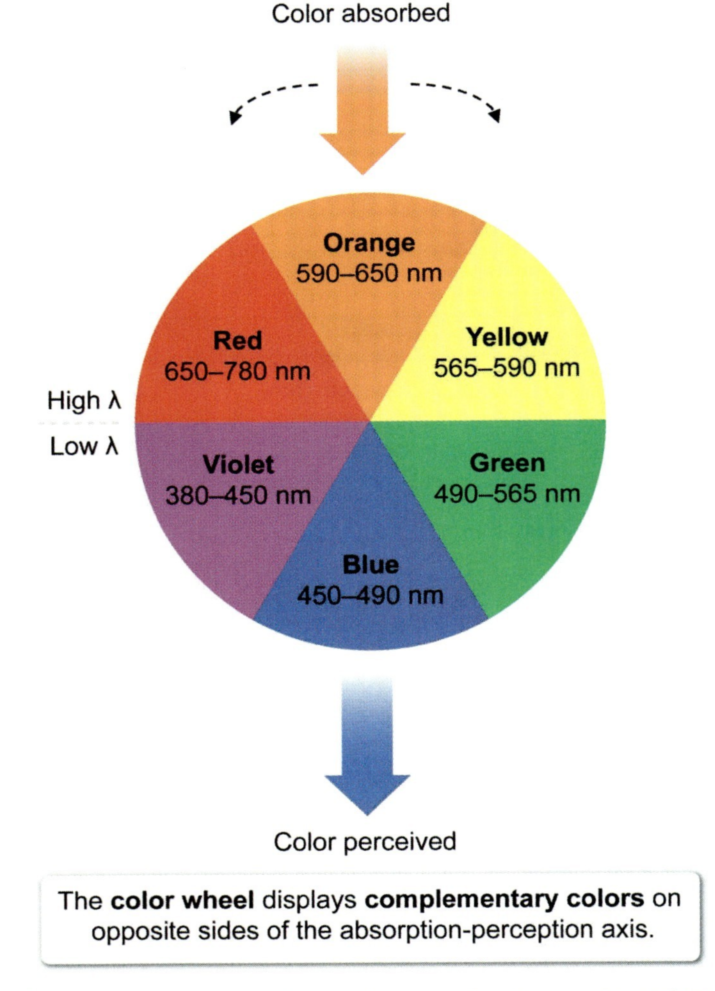

Figure 14.36 The color wheel, complementary colors, and the relationship between color absorption and perception.

The Relationship between Organic Structure and Visible Spectroscopy

As discussed in Concept 14.5.04, conjugation *lowers* the energy of $\pi \rightarrow \pi^*$ transitions, causing conjugated molecules to absorb light of *longer* wavelengths and lower energy. Because the effects of conjugation are cumulative, compounds that are more extensively conjugated absorb light at *greater* λ_{max} values.

Most organic solids have either no conjugated π systems or lightly conjugated π systems, which absorb *only* in the UV region (200–380 nm) of the electromagnetic spectrum. Because such compounds *do not* absorb visible light, all wavelengths of visible light are reflected or transmitted, and the compounds are perceived as being white or, more technically, **colorless**. For this reason, most organic solids are described as *either* white or colorless.

Organic compounds that contain *highly* conjugated π systems can have λ_{max} values within the visible region of the electromagnetic spectrum. Because such compounds absorb a *portion* of visible light and reflect or transmit the wavelengths that are *not absorbed*, they are perceived as having the complementary color of the wavelengths primarily absorbed (ie, λ_{max}).

Figure 14.37 provides three examples of colored organic compounds and their corresponding visible spectra. β-Carotene, a conjugated hydrocarbon present in carrots and other plants, has a λ_{max} of 450 nm (violet light) and is perceived as an orange color. Chlorophyll *a*, a pigment present in the leaves of many plants, has λ_{max} values of 430 nm (violet light) and 662 nm (orange light) and is perceived as a green color. Indigo, a pigment found in the indigo plant, has a λ_{max} of 616 nm (yellow light) and is perceived as a violet color.

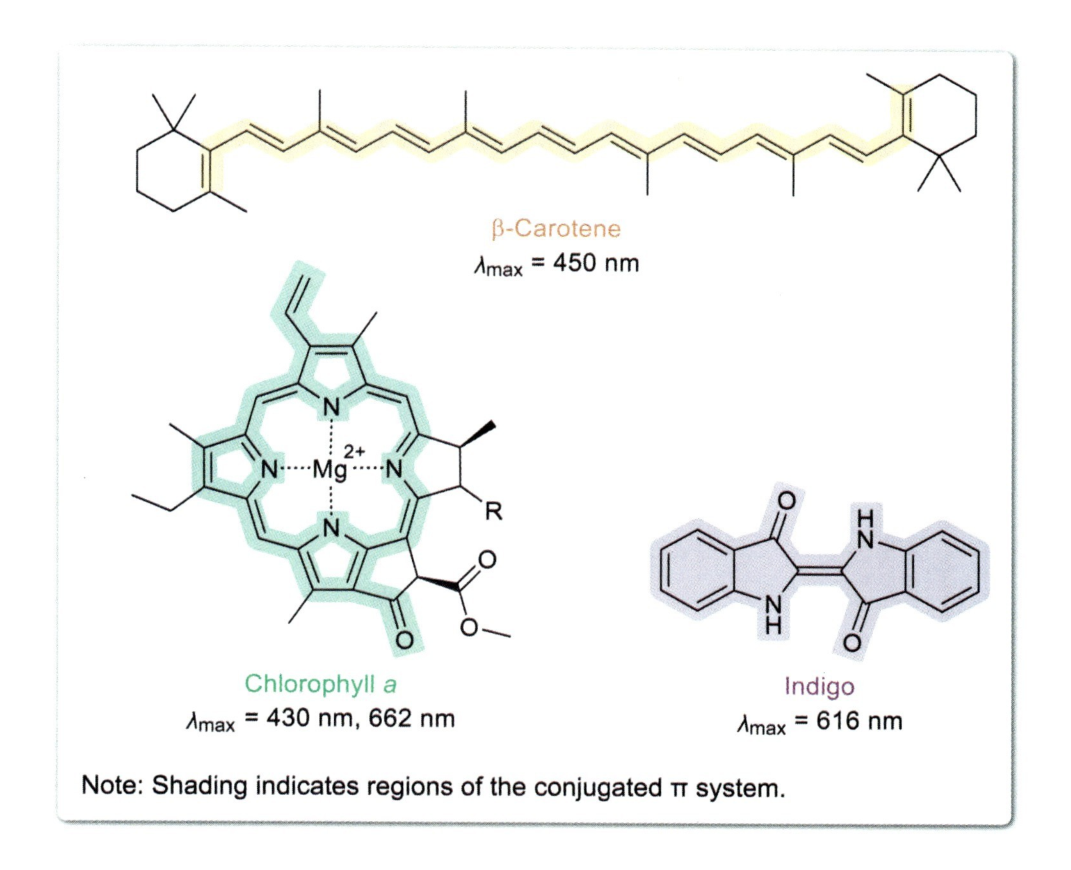

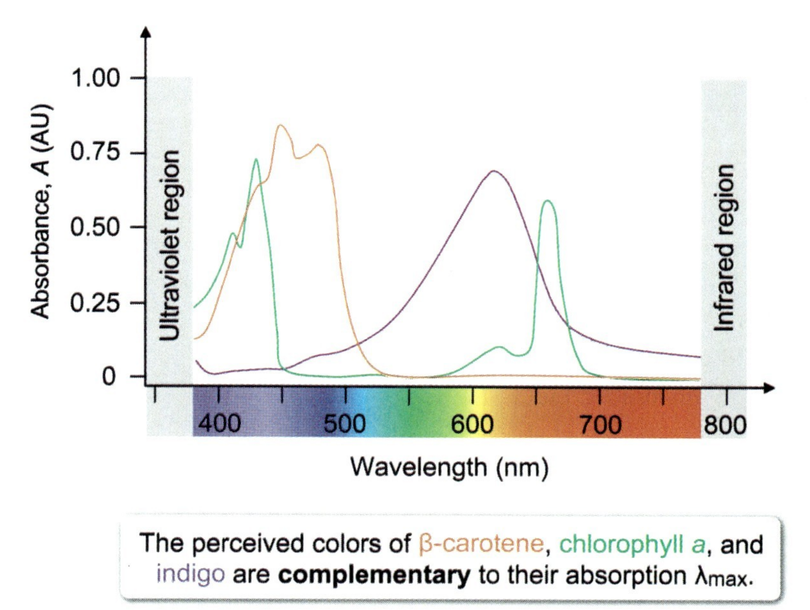

Figure 14.37 Structures and visible spectra of three highly conjugated organic compounds.

Lesson 14.6

Infrared Spectroscopy

Introduction

The interactions between molecules and photons from the infrared (IR) region of the electromagnetic spectrum cause molecular bond vibrations that can be modeled by Hooke's law, which describes oscillations in a spring. This lesson provides an overview of IR spectroscopy, which uses an IR spectrophotometer to detect the absorbance of IR light by molecules across a range of frequencies (generating an IR spectrum) and correlates molecular vibrations with specific organic functional groups within the molecular structure.

14.6.01 Molecular Vibrations

Photons in the **infrared (IR) region** of the electromagnetic spectrum have the correct energy range to cause vibrations of the bonds in a molecule. Although chemical bonds have an optimum (ie, lowest energy) length and angle with respect to other atoms in the molecule (see Concept 1.1.05), absorption of the energy from an IR photon can lead to a *bond distortion*, as illustrated in Figure 14.38.

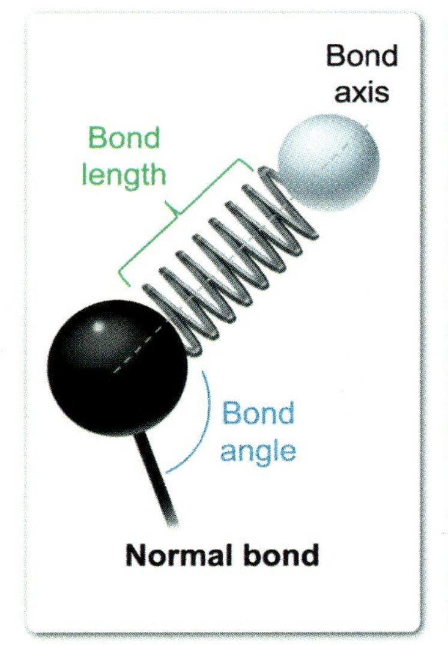

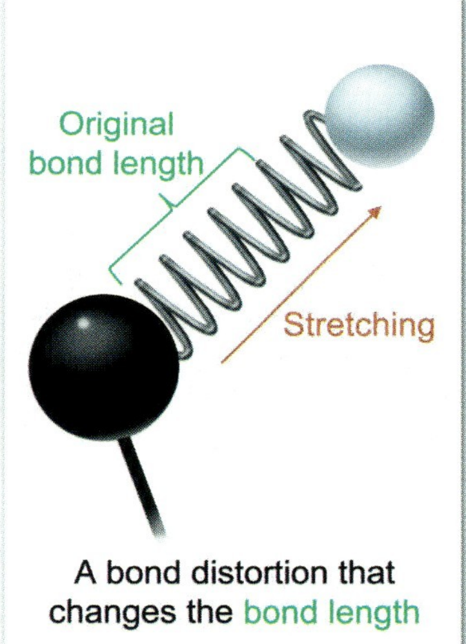

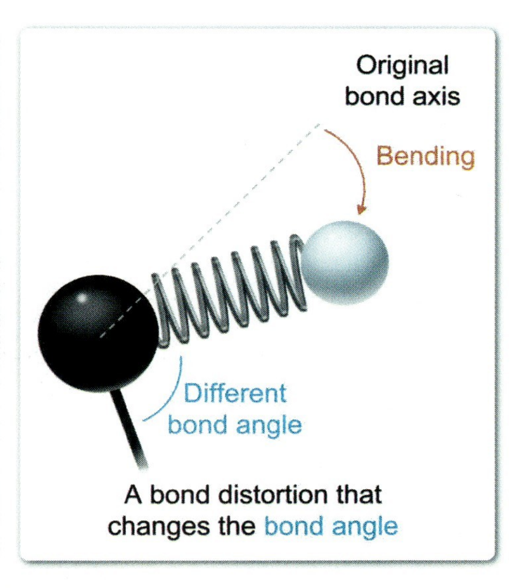

Figure 14.38 Bond distortions that occur following absorption of IR photons.

After a photon induces a bond distortion, the *restoration* of the bond back toward (and past) its optimum length and angle can lead to a regular oscillation of the bond (ie, a **molecular vibration**), which can be described by the *frequency* of the repeating motion. By viewing a chemical bond like a vibrating spring, the physics of the molecular vibrations can be described by Hooke's law (see Concept 14.6.02).

Some chemical bonds in a molecule absorb IR photons more efficiently than others. In general, the absorption intensity decreases as the polarity of a bond decreases. Consequently, the most intense absorptions occur for highly polar bonds, whereas truly nonpolar bonds with zero difference in electronegativity and certain bonds in *symmetrical* molecules may not readily absorb IR radiation.

Vibrational Modes

Molecular vibrations occur in relation to all *other* bonds in a molecule. A single type of molecular vibration is called a **vibrational mode**, and a nonlinear molecule with n atoms has $(3n - 6)$ vibrational modes. For example, a molecule of water (H_2O, $n = 3$) has 3 vibrational modes ($[3 \times 3] - 6 = 3$). Larger organic molecules have a *tremendous* number of possible vibrational modes. For example, a molecule of hexane (C_6H_{14}) has 54 vibrational modes ($[3 \times 20] - 6 = 54$).

IR spectroscopy most commonly studies six common types of vibrational modes, as summarized in Figure 14.39. For the sake of consistency, these vibrational modes are described in the context of a methylene unit ($-CH_2-$).

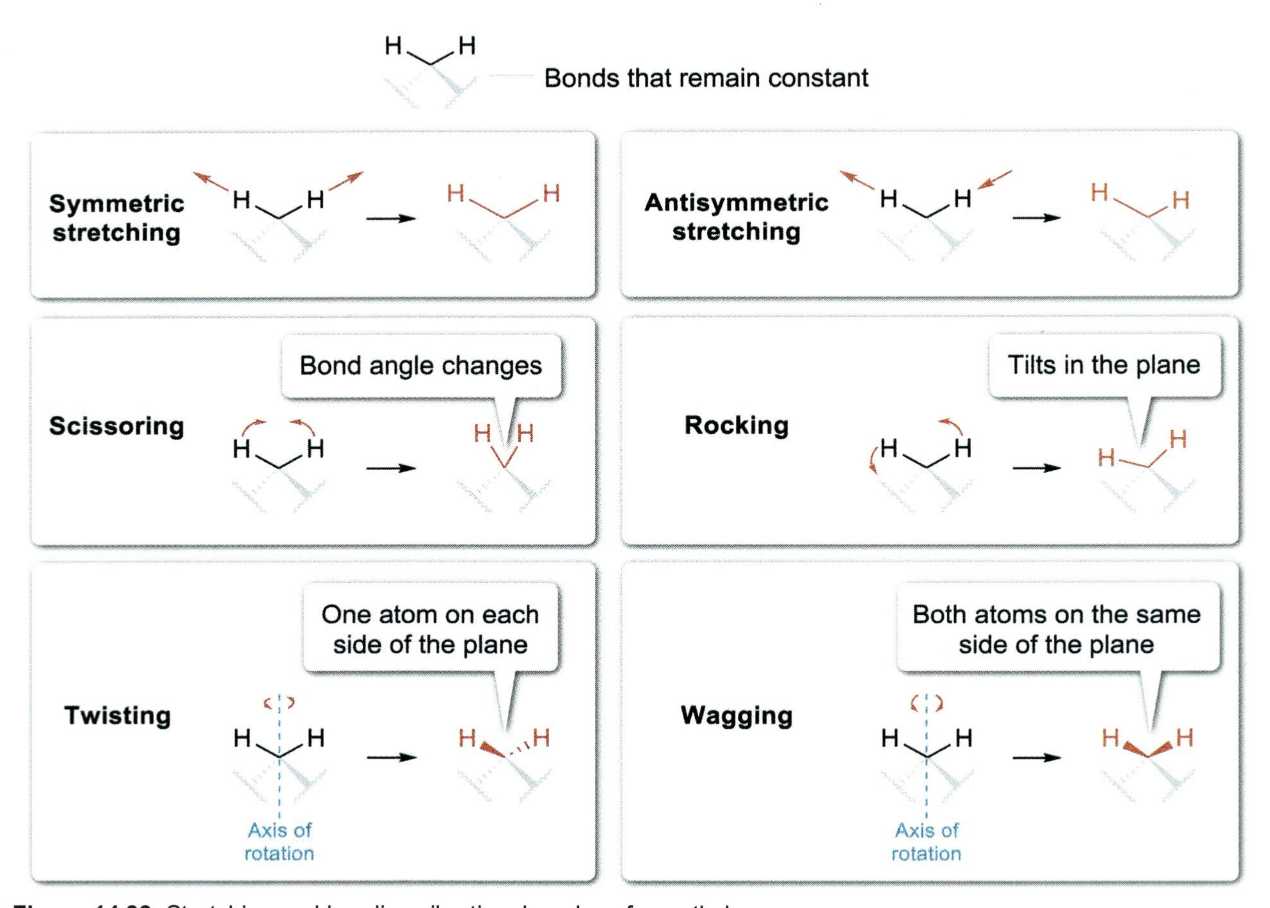

Figure 14.39 Stretching and bending vibrational modes of a methylene group.

Most bending vibrational modes (eg, scissoring, rocking, twisting, wagging) for molecules are present within the fingerprint region of the IR spectrum (see Concept 14.6.04) and are unlikely to be emphasized on the exam. As such, the rest of this lesson focuses on *stretching* vibrational modes.

14.6.02 Hooke's Law

Unlike UV-Vis spectroscopy, which describes electromagnetic radiation by the *wavelength* of the photon, IR spectroscopy (see Concept 14.6.04) describes electromagnetic radiation by the *inverse* of the photon wavelength in centimeters (ie, cm^{-1})—a unit of measure called a **wavenumber**. Because wavelength and frequency are *inversely proportional* to one another (see Concept 14.3.01), *higher*-wavenumber photons have a *higher* frequency and a *higher* energy. The equation for the wavenumber of a photon is:

$$\text{wavenumber}\ (\text{cm}^{-1}) = \frac{1}{\lambda\ (\text{cm})} = \frac{\nu}{c} = \frac{E}{h \cdot c}$$

where λ is the wavelength, v is the frequency, c is the speed of light, E is the energy, and h is Planck's constant.

The physical basis of a molecular vibration is derived from modeling a chemical bond as a spring between two masses. **Hooke's law** (Figure 14.40) describes the relationship of a spring oscillation (ie, the frequency or wavenumber of the spring's motion) with respect to the mass of the atoms (m_1 and m_2) on each side of the spring and a force constant (k), which describes the strength and length of the spring. Because a chemical bond is treated as a spring in this model, the force constant (k) also describes the nature of the chemical bond. In this context, bond strength is proportional to bond order (see Concept 6.5.01).

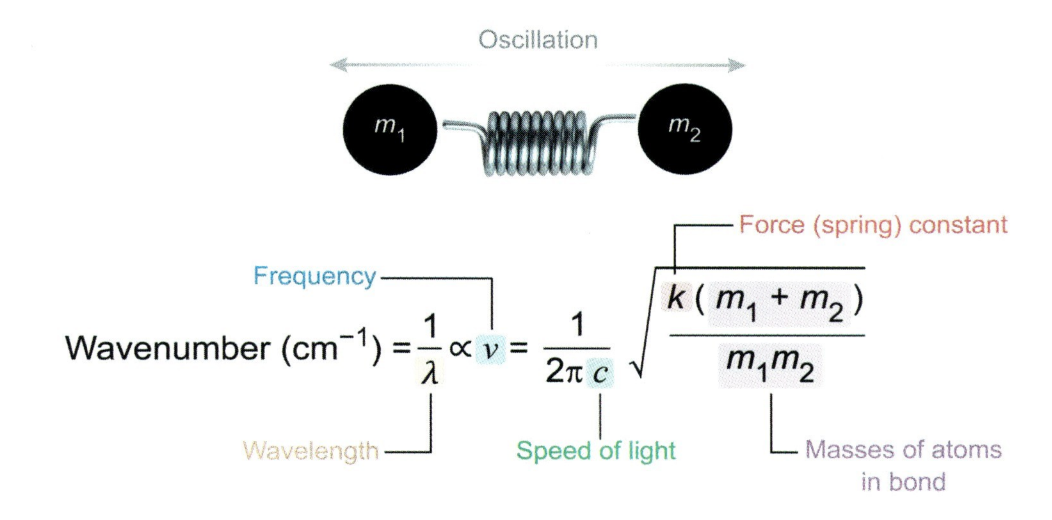

Figure 14.40 Hooke's law and IR spectroscopy.

The wavenumber ranges corresponding to the stretching vibrations for select isolated bonds are listed in Table 14.4. *Increasing* the atomic mass of an atom with a single bond to carbon *decreases* the expected wavenumber of the bond's stretching vibration. This effect is seen with as little as a 1 neutron (ie, 1 amu) mass difference in the atom, as demonstrated by the decrease in the wavenumbers of a C–H bond (3300–2700 cm^{-1}) compared to a C–D bond (2250 cm^{-1}). The symbol D stands for deuterium, the ^{2}H isotope of hydrogen (see Concept 14.4.05).

Table 14.4 Relationships between atomic mass, bond order, and the wavenumber of a stretching vibration.

Bond	Atomic mass (amu)	Bond order	Wavenumber range of a stretching vibration (cm^{-1})
C–H	1	1	3330–2700
C–D	2	1	2250
C–O	16	1	1150–1085
C–Cl	35.5	1	850–550
C–C	12	1	1130
C=C	12	2	1678–1626
C≡C	12	3	2260–2100

Atomic mass and wavenumber have an **inverse** relationship.

Bond order and wavenumber have a **proportional** relationship.

For a carbon–carbon bond, *increasing* the bond order from 1 (C–C) to 2 (C=C) to 3 (C≡C) *increases* the strength of the bond and *increases* the wavenumber of the stretching vibration. A similar trend is observed for most other second row heteroatoms (eg, N, O, F). The wavenumber absorption ranges of chemical bonds in the major functional groups are discussed in Concept 14.6.05.

14.6.03 The Infrared Spectrophotometer

The instrument used to measure the absorption of photons in the IR region of the electromagnetic spectrum is an **IR spectrophotometer**. Although various types of modern IR spectrophotometers operate in slightly different ways (as discussed later in this concept), the essential components of a *traditional* IR spectrophotometer are represented in Figure 14.41.

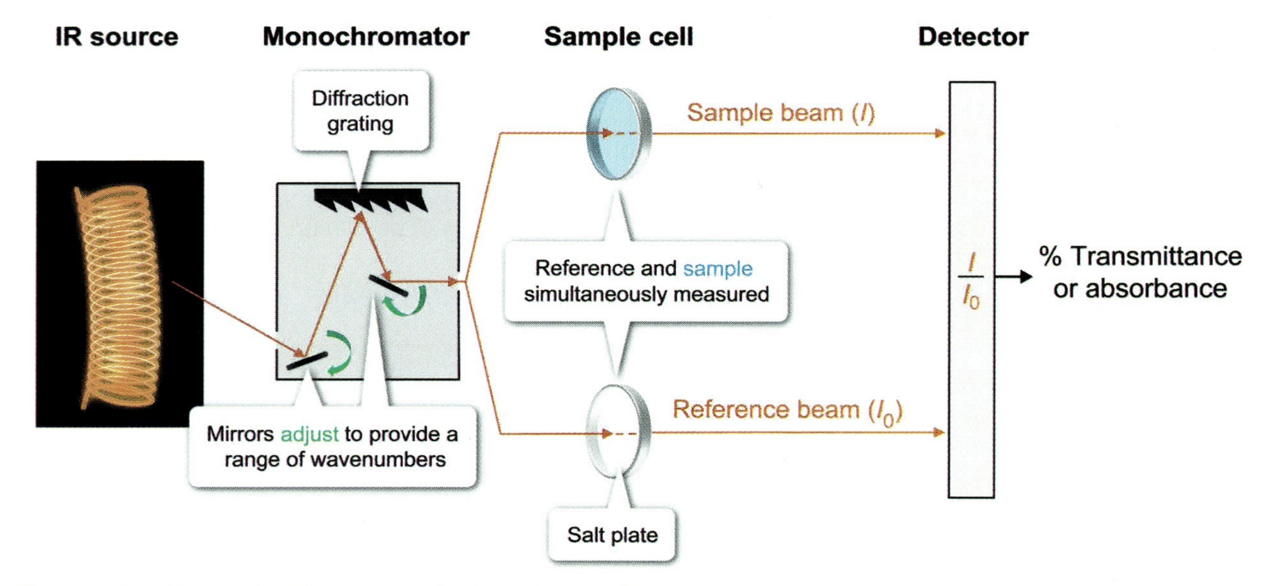

Figure 14.41 The essential components of a traditional IR spectrophotometer.

The source of IR radiation in an IR spectrophotometer is frequently a glowing metal coil, similar to a heating element. Because this type of source emits a *range* of wavenumbers of electromagnetic radiation, a monochromator (often a diffraction grating) is positioned to function like a prism and separate the light into narrower wavenumber ranges of radiation. Rotatable mirrors before and after the diffraction grating allow the selected range of radiation to be adjusted, enabling the instrument to **scan** the range of wavenumbers used by a traditional IR spectroscopy experiment (4000–400 cm^{-1}).

IR spectroscopy of organic molecules is typically performed on a thin layer of a *pure* liquid or solid sample (ie, *not* in a solution). Liquid IR samples are usually sandwiched between two transparent, polished plates composed of an IR-inactive salt (eg, potassium bromide), and solid IR samples are often embedded (ie, dispersed) within a freshly made, transparent salt pellet.

A reference beam passes through an empty salt plate and the air (eg, CO_2, H_2O) within the sample chamber (ie, the background), which together act as the blank. The intensity of this reference beam (I_0) is compared to the intensity of the sample beam (I), which passes through another salt plate holding the organic sample. The percent transmittance is then calculated as $I \div I_0 \times 100\%$ for each wavenumber in the analyzed range, and the resulting data are compiled and displayed as an IR spectrum.

14.6.04 The IR Spectrum

The data generated by an IR spectroscopy experiment are recorded in an **IR spectrum** (Figure 14.42), which displays the measured percent transmittance of the sample (*y*-axis) as a function of wavenumber (*x*-axis). By convention, the *x*-axis is typically depicted with *higher* wavenumbers (eg, 4000 cm^{-1}) on the left and *lower* wavenumbers (eg, 400 cm^{-1}) on the right (ie, wavenumbers *decrease* from left to right). This orientation places high-energy (low-wavelength) data on the left and low-energy (high-wavelength) data on the right.

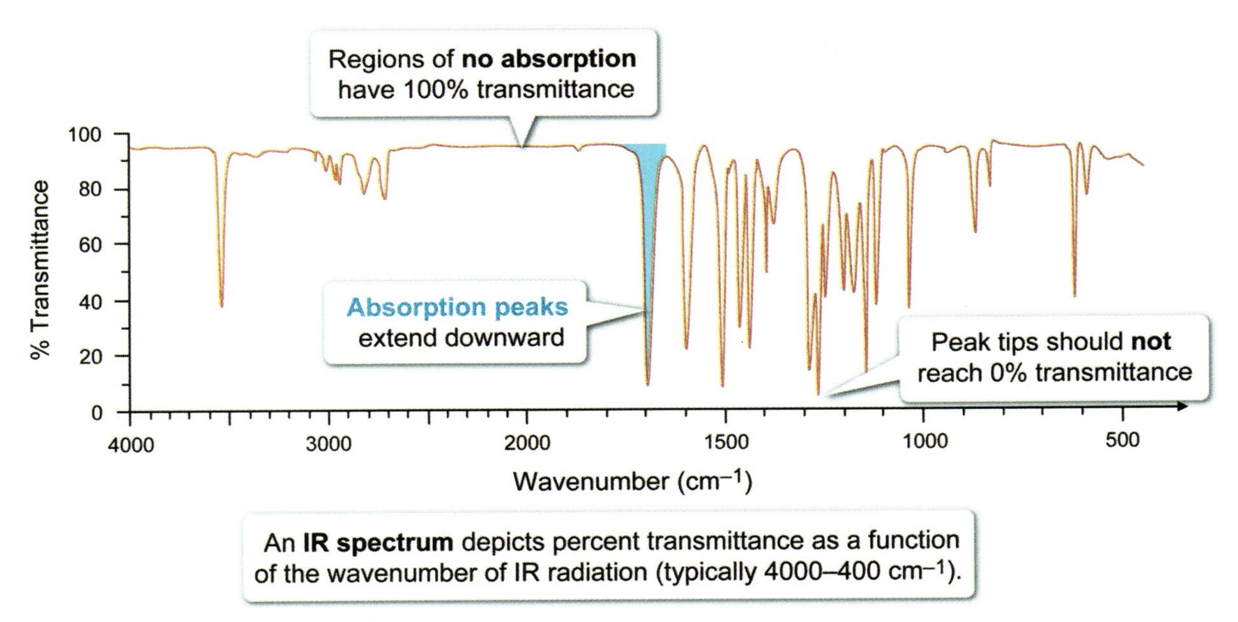

Figure 14.42 The anatomy of an IR spectrum.

The baseline for an IR spectrum runs along the *top* of the spectrum (ie, no absorption, 100% transmittance). Consequently, **absorption peaks** extend *downward* from the baseline. The intensity of an IR absorption peak is described using the relative terms strong, moderate, and weak (Figure 14.43). A **strong (*s*) peak** typically extends along most of the *y*-axis, a **moderate (*m*) peak** extends approximately half of the *y*-axis, and a **weak (*w*) peak** extends less than a third of the *y*-axis. Peak intensity is usually related to the polarity of a bond, such that *more* polar bonds have *more* intense peaks.

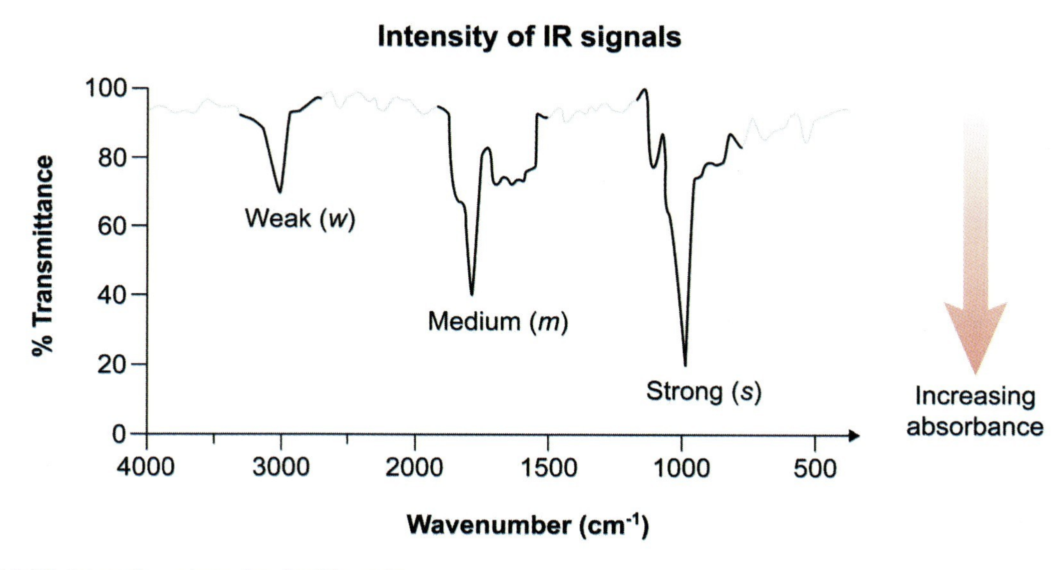

Figure 14.43 Intensity categories for IR peaks.

Although most IR absorption peaks are **sharp** (ie, appear over a narrow range of wavenumbers), some IR peaks are very **broad** (ie, appear over a wide range of wavenumbers). Broad IR peaks are most commonly observed due to variable intermolecular interactions that affect vibrational frequencies (eg, hydrogen bonding).

An IR spectrum depicts a series of peaks representing specific vibrational modes, with each mode resulting from bonding patterns within the structure of a molecule. Given that two *unique* substances inherently have *different* structures, every *unique* compound generates a *unique* IR spectrum, much like a human fingerprint. The sole exception involves enantiomers, which have *identical* IR spectra to one another despite being unique substances.

As such, the region of the IR spectrum in the range of 1600–400 cm^{-1} is sometimes called the **fingerprint region** because the complex peaks normally present in this region form a distinct pattern for every unique compound. Significant information can be gained through the analysis of the fingerprint region of an IR spectrum, but this type of information extends beyond what is likely to be encountered on the exam.

Concept 14.6.05 focuses on the distinct IR spectral features of functional groups that appear in the range of 4000–1600 cm^{-1}.

14.6.05 Overview of IR Spectral Features for Functional Groups

Because IR spectroscopy provides the ability to detect vibrational modes for different types of chemical bonds in a molecule, it is particularly useful for analyzing a sample to determine the presence or absence of particular **functional groups** (see Concept 2.8.01). The specific bonding patterns in a functional group can be identified by considering *both* the absorption peaks that are *present* and those that are *absent* in the IR spectrum of a sample.

General Trends within the IR Spectrum

Hooke's law (Concept 14.6.02) describes how the wavenumber of an IR absorption is related to the mass of the bonded atoms and the bond order. Accordingly, there are particular regions of the IR spectrum where certain types of chemical bonds are typically observed, as summarized in Figure 14.44.

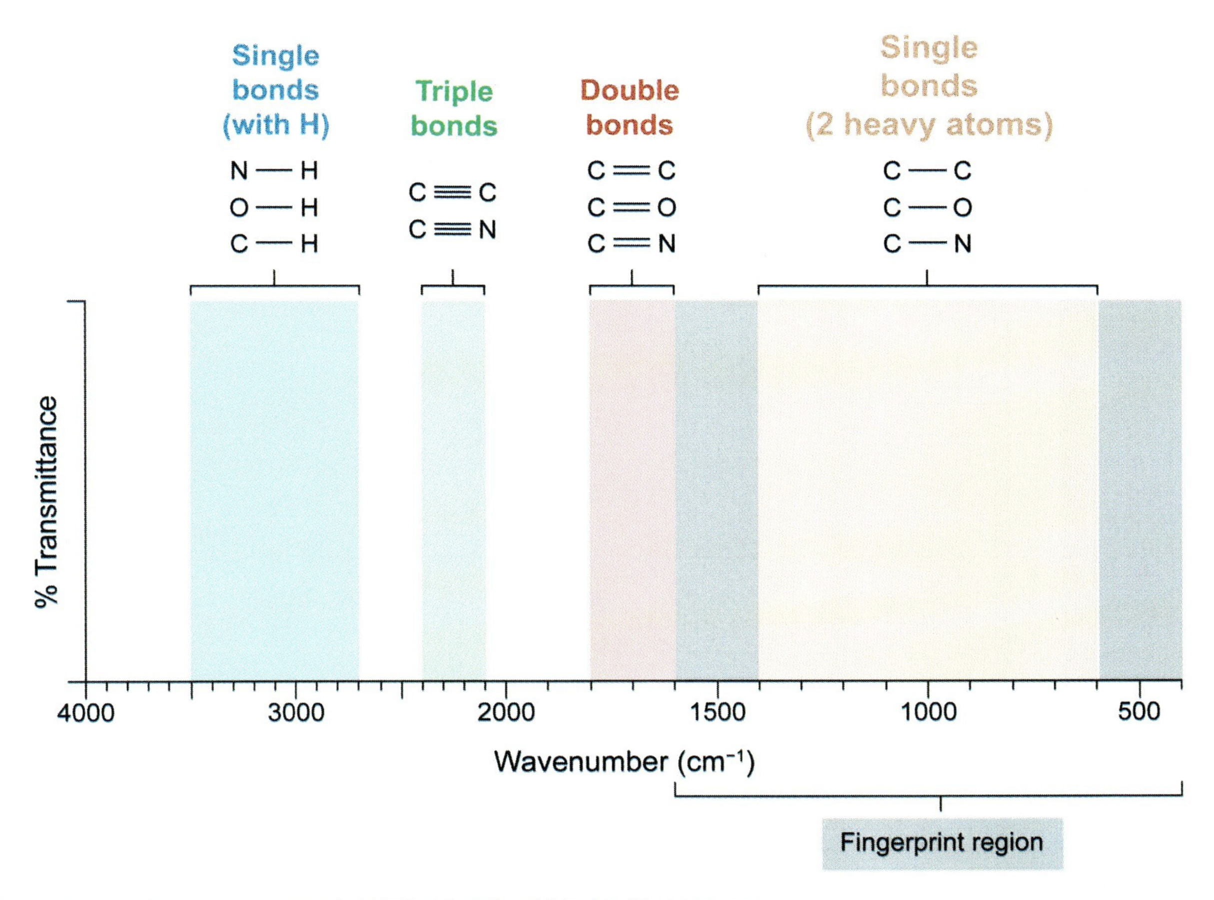

Figure 14.44 General regions of chemical bonds within the IR spectrum.

As discussed in Concept 6.5.01, the bond order of a conjugated π system is *less* than the bond order of an isolated π system (ie, simple alkene or alkyne). Consequently, IR absorption peaks for *conjugated* chemical bonds are shifted rightward to a *lower* wavenumber in an IR spectrum relative to their isolated counterparts (see Concept 6.5.03).

The wavenumber of a single bond stretch also varies with hybridization. Because an *s* orbital is located closer to the nucleus than a *p* orbital, a *higher* percent *s* character in a hybrid orbital leads to a *stronger* bond and an *increase* in the wavenumber of the corresponding IR absorption peak (Figure 14.45).

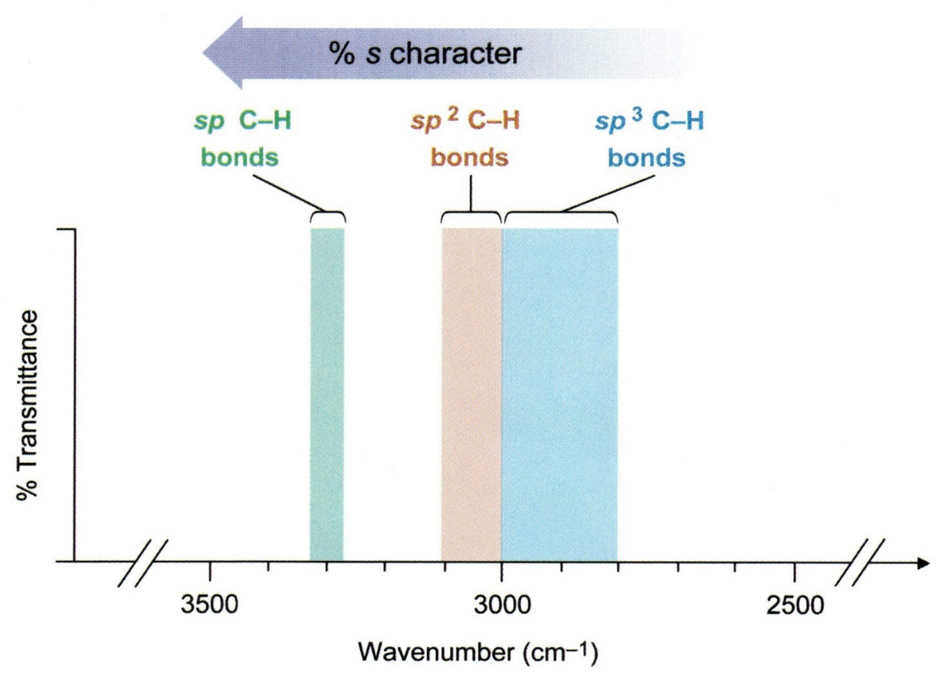

Figure 14.45 Relationship between hybridization and IR absorptions for hydrocarbon C–H bonds.

The remaining sections in this concept discuss the expected IR absorption peaks for different classes of functional groups, with a focus on peaks above 1600 cm^{-1} (ie, peaks *not* in the fingerprint region).

Hydrocarbons

Because hydrocarbon functional groups consist of *only* carbon and hydrogen atoms, the IR spectrum of a hydrocarbon has absorption peaks only corresponding to some combination of C–H and C–C, C=C, or C≡C bonds. A summary of the characteristic absorption peaks for these bonds is given in Table 14.5.

Table 14.5 IR absorbance peaks for hydrocarbon functional groups.

Functional group	Bond	Vibrational mode	Intensity	Wavenumber range (cm^{-1})
Alkane or alkyl group	sp^3 C–H	stretching	moderate	3000–2800
Alkene	sp^2 C–H*	stretching	moderate	3100–3000
	C=C	stretching	moderate	1680–1640
Alkyne	*sp* C–H*	stretching	moderate	3300
	C≡C (terminal)	stretching	moderate	<2200
	C≡C (internal)	stretching	weak	<2200
Benzene or phenyl ring (aromatic)	sp^2 C–H*	stretching	moderate	3100–3000
	C≡C	stretching	weak	1600

* Observed only if present in the structure of the sample molecule.

Oxygen-Containing Functional Groups

The IR spectrum of an oxygen-containing functional group typically contains *specific combinations* of absorption peaks corresponding to C–O, O–H, or C–H bonds. The stretching vibration for a C–O *single* bond typically absorbs below 1600 cm^{-1}. Therefore, an **ether** does not show functional group absorptions in the 4000–1600 cm^{-1} range outside the fingerprint region.

Although an alcohol and a carboxylic acid both contain an O–H bond, the location and width of the broad O–H stretching vibration is *different* for the two functional groups (Figure 14.46). The broadness of an O–H stretch is primarily due to hydrogen bonding. The O–H stretch for an **alcohol** is typically *centered* around 3300 cm^{-1} and extends from about 3600 cm^{-1} to 3000 cm^{-1}. Consequently, an alcohol O–H stretch is usually *separated* from the alkyl group sp^3 C–H stretching peaks spanning 3000–2800 cm^{-1}.

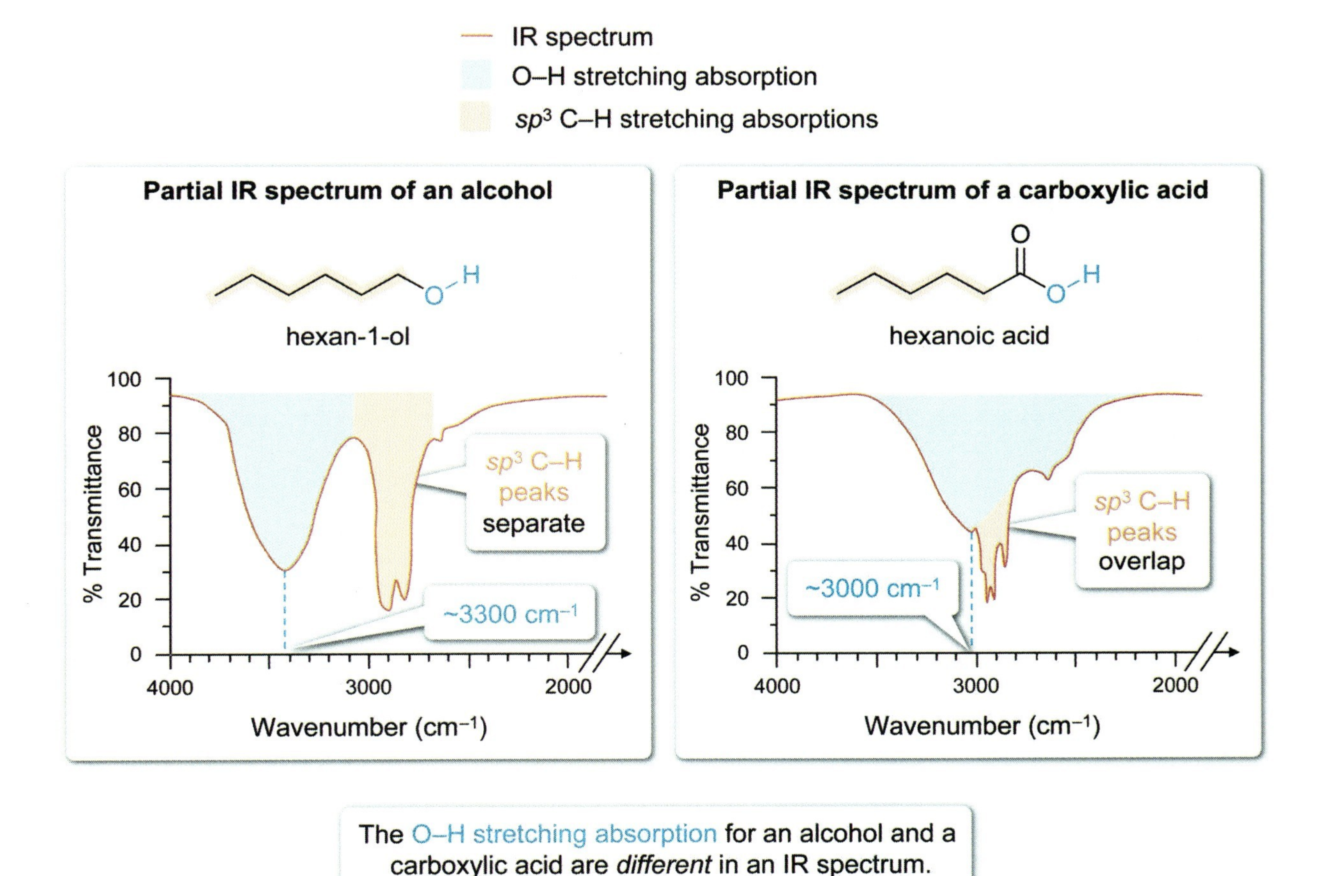

Figure 14.46 Comparison of O–H stretches for an alcohol and a carboxylic acid.

The O–H stretch for a carboxylic acid is *much broader* than the O–H stretch for an alcohol and is typically *centered* around 3000 cm^{-1} (range: 3500–2500 cm^{-1}). The increased broadness of an O–H stretch for a carboxylic acid is partially due to the formation of dimers between carboxylic acid molecules (see Concept 10.1.02). Often, a carboxylic acid O–H stretch overlaps with the alkyl group sp^3 C–H stretching peaks spanning 3000–2800 cm^{-1}.

Most of the oxygen-containing functional groups contain a carbonyl (C=O) group. Carbonyl stretching absorptions are observed over a relatively small wavenumber range (1800–1640 cm^{-1}), and the *specific* wavenumber can often be used to determine or narrow down the type of carbonyl-containing group in a molecule. This is especially true when evidence of a C=O bond is combined with evidence of other bond absorptions (eg, C–H, O–H). Table 14.6 provides a summary of the significant IR absorptions for the oxygen-containing functional groups.

Table 14.6 IR absorbance peaks for oxygen-containing functional groups.

Functional group	Bond	Vibrational mode	Intensity	Wavenumber range (cm^{-1})
Alcohol	O–H	stretching	strong (broad)	3600–3000
Ether	C–O	stretching	moderate	<1600
Aldehyde	*sp*^{2} C–H	stretching	moderate	2700 2800
	C=O	stretching	strong	1725
Ketone	C=O	stretching	strong	1710
Carboxylic acid	O–H	stretching	strong (broad)	3500–2500
	C=O	stretching	strong	1710
Acid halide	C=O	stretching	strong	1800
Anhydride	C=O	stretching	strong	1800 (antisymmetric) 1750 (symmetric)
Ester	C=O	stretching	strong	1735
Amide	N–H*	stretching	moderate	3500–3200**
	C=O	stretching	strong	1680–1640

* Observed only if present in the structure of the sample molecule.
** A 1° amide has two peaks (symmetric and antisymmetric modes) and a 2° amide has one peak.

Nitrogen-Containing Functional Groups

The IR spectrum of a nitrogen-containing functional group typically contains *specific combinations* of absorption peaks corresponding to C=N, C–N, N–H, or C–H bonds. Like C–C and C–O *single* bonds, the stretching vibration for a C–N *single* bond typically absorbs below 1600 cm^{-1}.

Amine and amide functional groups demonstrate nitrogen-hydrogen stretching absorptions in an IR spectrum *if* the amine or amide contains at least one N–H bond (Figure 14.47). N–H stretching absorptions appear in a similar region as an O–H peak, around 3400–3300 cm^{-1}, and are also broadened due to hydrogen bonding. However, the *intensity* of an N–H stretch (ie, moderate) is typically lower than the strong intensity of an O–H stretch because an N–H bond is *less polar* than an O–H bond.

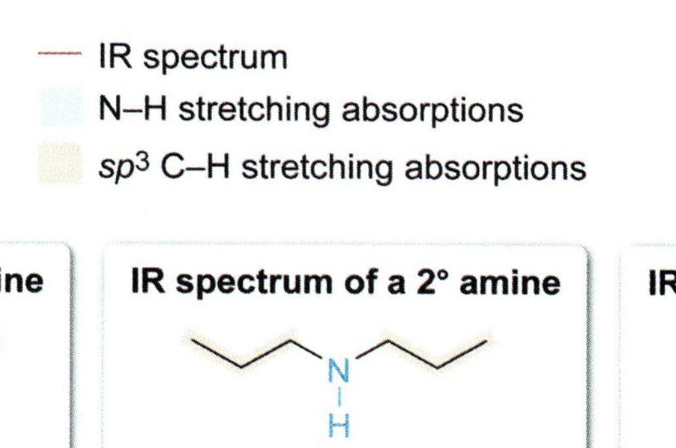

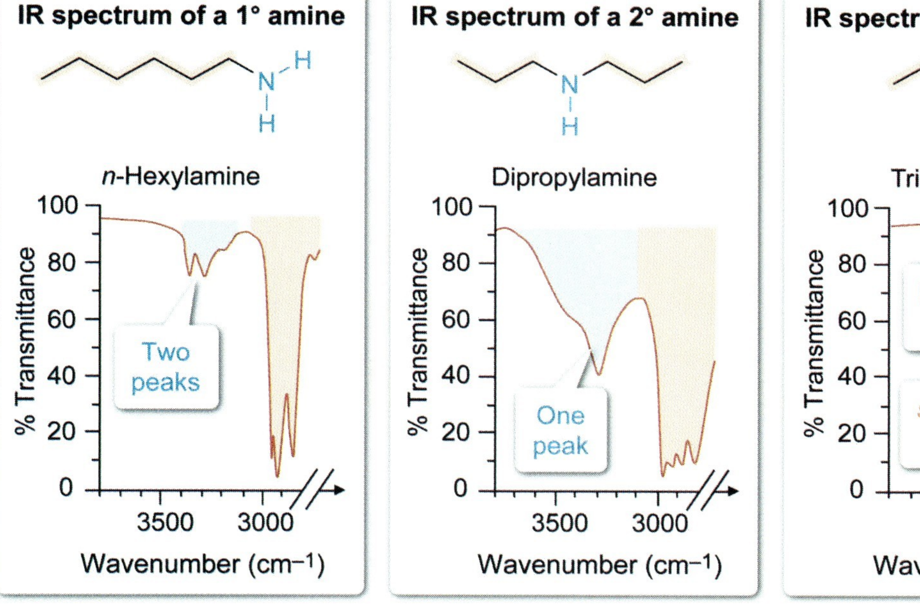

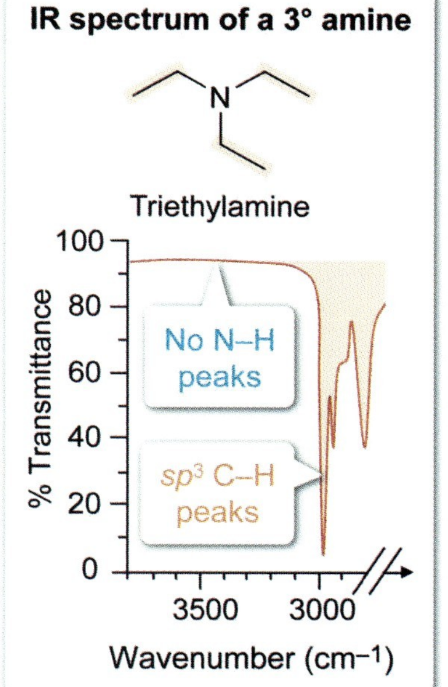

The N–H stretching absorptions for amine classifications are *different* in an IR spectrum.

Figure 14.47 Comparison of N–H stretches for 1°, 2°, and 3° amines.

A **nitrile** is indicated by a moderate C≡N stretch just above 2200 cm^{-1}. Given that *polar* chemical bonds are more efficient at absorbing IR radiation (see Concept 14.6.01), a polar nitrile generates a more intense absorption peak than an alkyne.

Table 14.7 provides a summary of the significant IR absorptions for the nitrogen-containing functional groups.

Table 14.7 IR absorbance peaks for nitrogen-containing functional groups.

Functional group	Bond	Vibrational mode	Intensity	Wavenumber range (cm^{-1})
Amine	N–H*	stretching	moderate	3500–3200**
Amide	N–H*	stretching	moderate	3500–3200**
	C=O	stretching	strong	1680–1640
Nitrile	C≡N	stretching	moderate	>2200

* Observed only if present in the structure of the sample molecule.
** A 1° amine or 1° amide has two peaks (symmetric and antisymmetric modes) and a 2° amine or 2° amide has one peak.

Concept Check 14.5

The IR spectrum of an organic compound spanning 4000–1600 cm^{-1} has *only* moderate absorption peaks in the range of 3000–2800 cm^{-1} and a strong peak at 1712 cm^{-1}. Based on this information, what are the most probable functional groups in the structure of this compound?

Solution

Note: The appendix contains the answer.

Lesson 14.7

^{1}H NMR Spectroscopy

Introduction

Certain atomic nuclei have a quantum mechanical property called nuclear spin, which causes them to respond to magnetic fields and behave like miniature bar magnets that align within a magnetic field. Photons in the radio wave region of the electromagnetic spectrum have the correct energy range to cause transitions in the spin state (alignment) of nuclei. Detection and analysis of these transitions can be used to determine the relative arrangement of the nuclei (ie, the molecular structure) and are the basis of nuclear magnetic resonance (NMR) spectroscopy. This lesson provides an overview of NMR spectroscopy involving the detection of ^{1}H atoms within molecules.

14.7.01 Nuclear Spin

An atomic nucleus with an *odd* atomic number (ie, number of protons) or an *odd* mass number (ie, sum of the numbers of protons and neutrons) has a non-zero **nuclear spin**, which is a quantum mechanical property that relates to the ability of the nucleus to respond to magnetic fields. Fundamentally, a nucleus with a non-zero nuclear spin can be conceptualized as a rotating body of charge that generates a magnetic field (Figure 14.48). In this way, such a nucleus behaves like a small bar magnet with a magnetic dipole moment (field) that can be expressed as a vector quantity (ie, an arrow representing the orientation of the bar magnet from the south pole to the north pole).

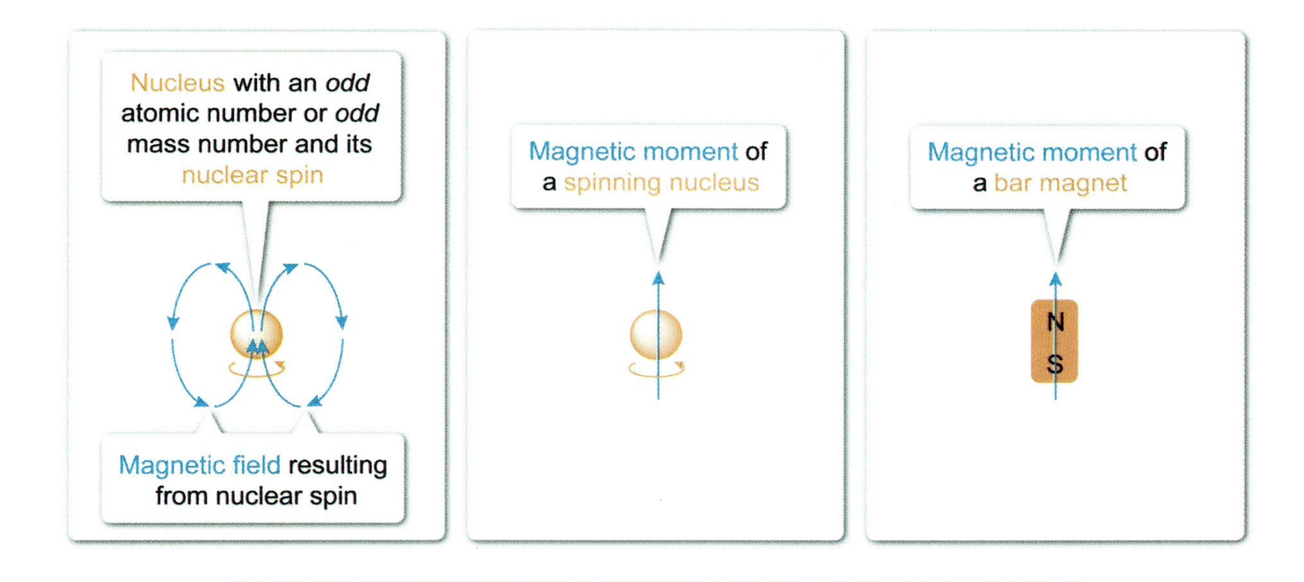

Figure 14.48 Nuclear spin causes a nucleus to act like a bar magnet.

Because a nucleus of **protium** (ie, the major isotope of hydrogen, ^{1}H) contains one proton and zero neutrons, ^{1}H has a nuclear spin and behaves like a miniature bar magnet. Consequently, ^{1}H is one of the *many* types of nuclei that can be analyzed by NMR spectroscopy and is an **NMR-active nucleus**. In the absence of any external influence, the magnetic moments of individual NMR-active nuclei are *randomly* oriented in a sample; however, applying an external magnetic field causes the magnetic moments to align with the field.

14.7.02 Nuclear Spin States and Resonance Energy

In the absence of external forces, the magnetic dipole moment of individual nuclei in a sample is *random* (ie, oriented in all possible directions), with *no net orientation*. However, when an NMR-active nucleus interacts with an **external magnetic field (B_0)**, the magnetic dipole moments of the individual nuclei *align* to the magnetic field, much like how magnetic iron filings orient with respect to a nearby magnet (Figure 14.49).

Most NMR-active nuclei interacting with B_0 adopt one of *two* different **nuclear spin states**. The *lower*-energy spin state is the **α spin state**, in which the nuclear magnetic moment is aligned *with* the magnetic field in the *same* orientation as B_0. The *higher*-energy spin state is the **β spin state**, in which the nuclear magnetic moment is aligned *against* the field in the *opposite* orientation as B_0.

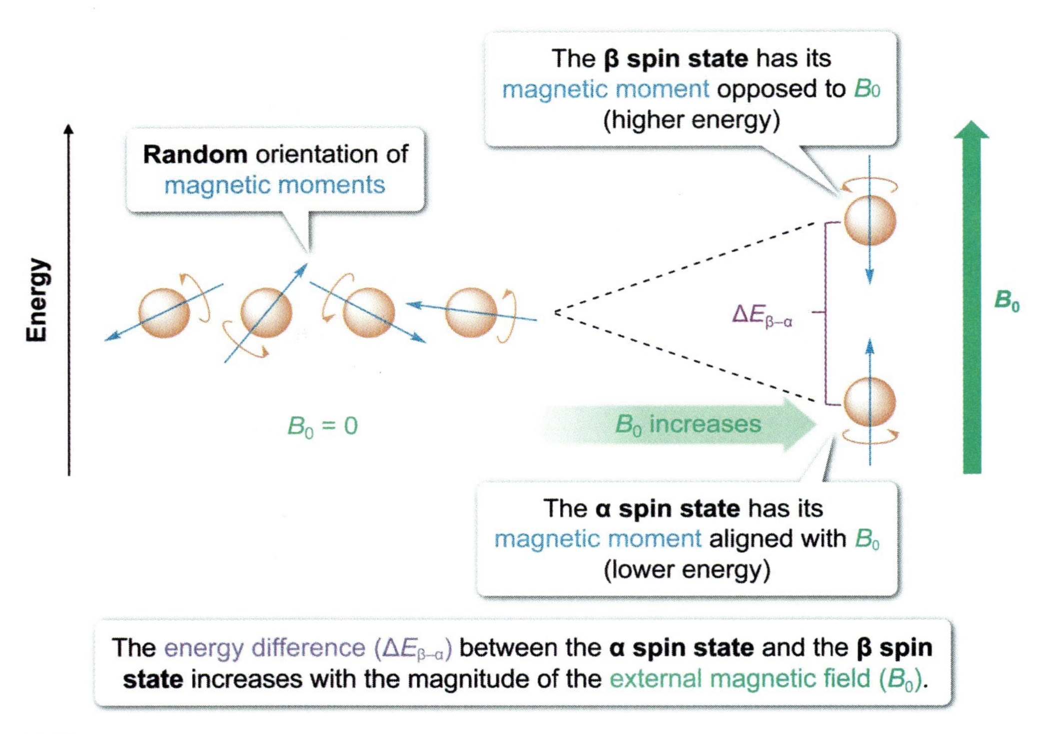

Figure 14.49 The α and β spin states.

In a low B_0, the population of molecules in the α spin state and the β spin state exist in a nearly 1:1 ratio. *Increasing* the magnitude of B_0 *increases* the energy difference between the α spin state and the β spin state ($\Delta E_{\beta-\alpha}$), leading to a larger population of molecules in the lower-energy α spin state.

Photons in the **radio wave** region of the electromagnetic spectrum have the correct energy range to cause transitions between the α and the β spin states, a process called **resonance** (Figure 14.50). Note that *resonance* in this context is *different* from the delocalization of electrons between valid Lewis structures. Resonance between the nuclear spin states forms the basis of **nuclear magnetic resonance (NMR) spectroscopy**.

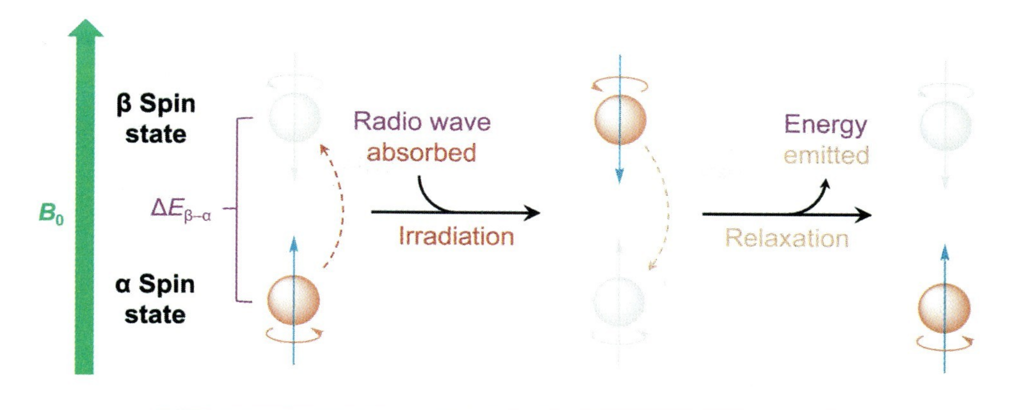

Figure 14.50 Resonance between nuclear spin states.

When nuclei in the lower-energy α spin state are subjected to radio waves of the correct energy or frequency (a process called **irradiation**), the α-state nuclei can each absorb a photon and transition to the higher-energy β spin state. If the radio waves are removed, nuclei in the β spin state *equilibrate* back to the α spin state (a process called **relaxation**). The relaxation process produces a signal that can be detected by a **detector** in an NMR spectrometer (see Concept 14.7.04).

14.7.03 Magnetic Shielding

To this point, the discussion of nuclear spin states is focused solely on the effect of the external field B_0 on the nuclear subatomic particles (ie, protons, neutrons). However, B_0 also affects the surrounding electrons, inducing their movement (ie, circulation), which then generates an **induced magnetic field ($B_{induced}$)** that has an orientation *opposite* to the external magnetic field B_0 (Figure 14.51).

Electron density **shields** the nucleus from the external magnetic field

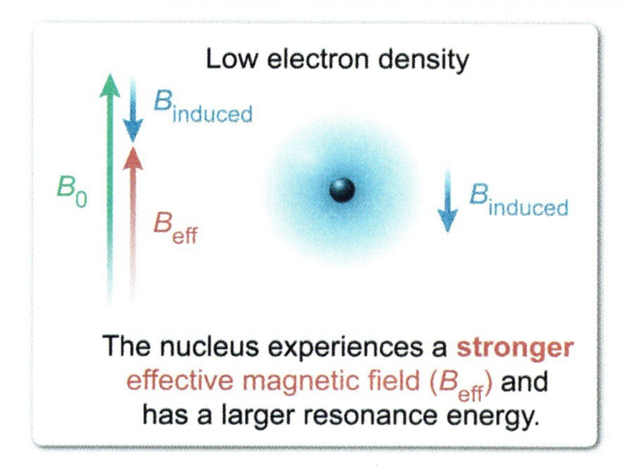

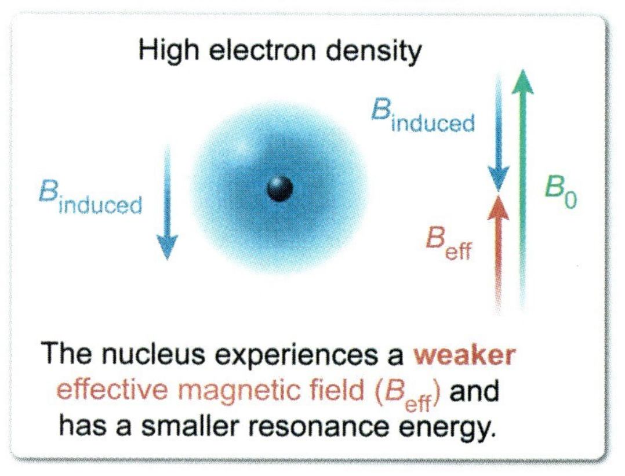

Figure 14.51 Electron shielding.

Given that magnetic fields are vector quantities (ie, they can be added together) and that a nucleus is located *inside* a surrounding electron cloud, the magnetic field experienced by a nucleus is the vector sum of B_0 + $B_{induced}$, a quantity called the **effective magnetic field (B_{eff})**. Because the orientation of the electron cloud $B_{induced}$ is *opposite* to B_0, B_{eff} is always *smaller* than B_0, and the electron density surrounding a nucleus is said to partially **shield** the nucleus from the full effects of B_0. Consequently, electron shielding leads to a *decrease* in the resonance energy between the nuclear spin states.

The magnitude of electron shielding for a given nucleus is proportional to the electron density surrounding it. Therefore, ^{1}H nuclei located at *different* places within the structure of a molecule experience *different* amounts of electron shielding and have *different* resonance energies. Table 14.8 depicts how bond dipole moments can influence electron density and shielding and how they can affect the resonance energy of a ^{1}H nucleus.

Table 14.8 Relationship between bond dipole moment, electron shielding, and resonance energy.

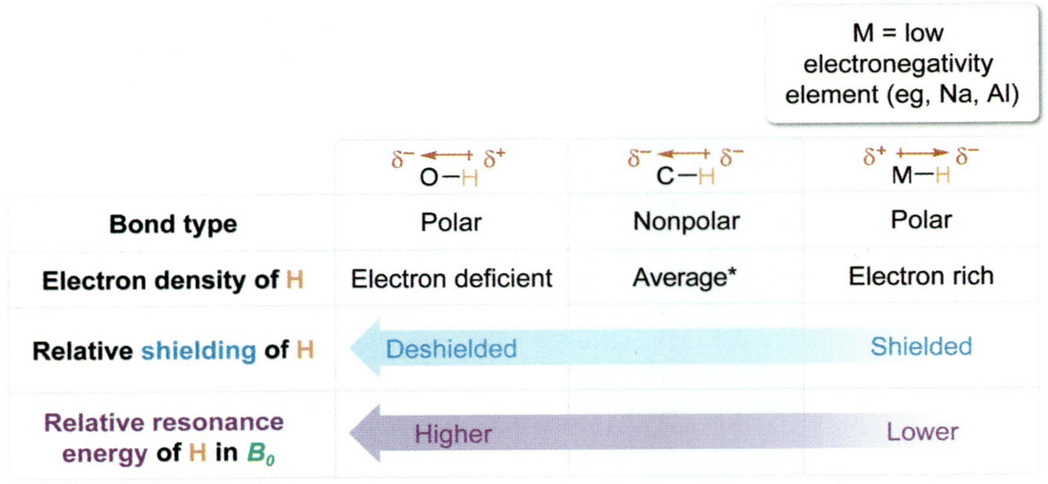

M = low electronegativity element (eg, Na, Al)

	δ^- ← δ^+ O–H	δ^- ← δ^- C–H	δ^+ → δ^- M–H
Bond type	Polar	Nonpolar	Polar
Electron density of H	Electron deficient	Average*	Electron rich
Relative shielding of H	Deshielded		Shielded
Relative resonance energy of H in B_0	Higher		Lower

* In most organic molecules, the ^{1}H in a nonpolar C–H bond is among the most shielded (and lowest resonance energy) ^{1}H nuclei.

The terms magnetic *shielding* or *deshielding* are regularly used in NMR spectroscopy to describe the relative effects of a structural feature on the B_{eff} (and therefore, resonance energy) for a given ^{1}H nucleus. **Magnetic shielding** refers to a feature that *decreases* both B_{eff} and the resonance energy. In contrast, **magnetic deshielding** refers to a feature that *increases* both B_{eff} and the resonance energy. Later in this lesson, the effects of magnetic shielding and deshielding are discussed further in the context of the ^{1}H NMR spectrum (Concept 14.7.05), chemical shift (Concept 14.7.07), and spin-spin splitting (Concept 14.7.09).

14.7.04 The NMR Spectrometer

The instrument used to measure the resonance energies (or frequencies) of NMR-active nuclei (eg, ^{1}H) in a sample is a **nuclear magnetic resonance (NMR) spectrometer** (Figure 14.52).

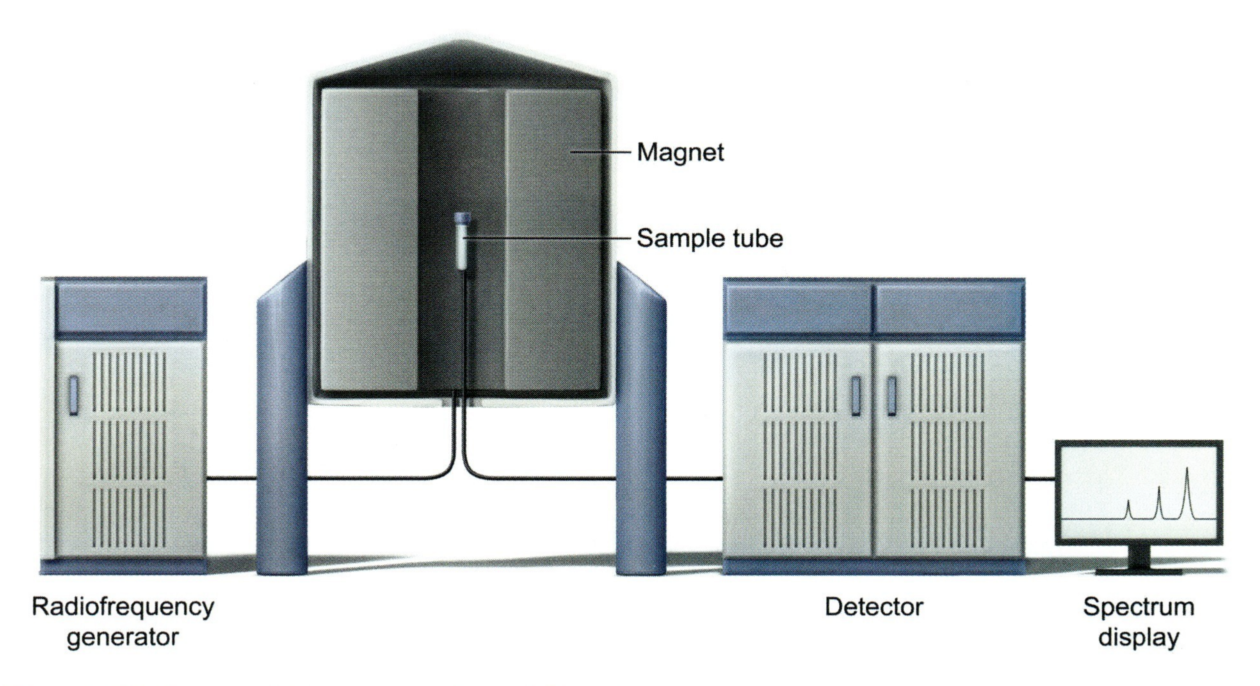

Figure 14.52 The essential components of an NMR spectrometer.

In an NMR experiment, the sample tube is placed into an external magnetic field (often the inner bore of a large magnet). Most ^1H NMR samples are dissolved in a **deuterated solvent**—a solvent in which a large majority of the ^1H atoms are replaced by deuterium (D) (ie, ^2H), which has different NMR-active properties that can be distinguished from the ^1H signals. Use of a deuterated solvent (eg, deutero-chloroform [$CDCl_3$], or deutero-methanol [CD_3OD]) ensures that the sample being analyzed is the primary source of ^1H nuclei.

Modern NMR spectrometers typically use a *constant* magnetic field generated by a superconducting magnet, which is itself cooled by cryogens such as liquid helium and liquid N_2. The sample is irradiated with broadband (ie, range of wavelengths) radio waves created by a radiofrequency generator for a short period of time (ie, a radio wave **pulse**) to promote nuclear resonance and increase accumulation of nuclei in the β spin state.

Then, the radio wave pulse is stopped, and the detector measures the signal frequencies produced by the sample as the nuclei relax back to the α spin state. The pulse-relaxation cycle is completed *many* times to accumulate sufficient signal from the relaxing ^1H nuclei. Computer processing of the detected signals is usually presented as an NMR spectrum (see Concept 14.7.05).

Medical Applications of NMR Spectroscopy

Although NMR spectroscopy is regularly employed in a chemical laboratory to assist in the analysis of molecular structure, the technology has also become a cornerstone of modern medicine. In a clinical setting, **magnetic resonance imaging (MRI)** is a powerful diagnostic technique that relies on the principles of NMR spectroscopy.

Rather than placing a chemical sample into the field of a superconducting magnet to measure resonance frequencies, an MRI places a human being into a magnetic field and then applies pulses of radio waves to induce resonance in the water molecules in tissues (Figure 14.53). The resultant signal detected during nuclei relaxation is used alongside computer technology to create detailed images of tissue to aid physicians in diagnosis.

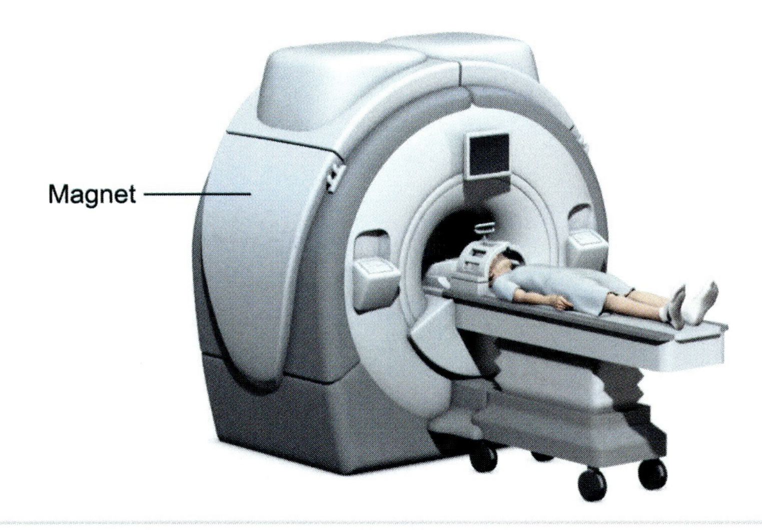

Magnetic resonance imaging (MRI) uses similar technology as a nuclear magnetic resonance (NMR) spectrometer to clinically analyze tissues in patients.

Figure 14.53 Magnetic resonance imaging (MRI).

14.7.05 The ^{1}H NMR Spectrum

The data generated by a ^{1}H NMR spectroscopy experiment are recorded in a **^{1}H NMR spectrum**, which displays the measured intensity of signals from a sample (*y*-axis) as a function of resonance frequency (*x*-axis, Figure 14.54). By convention, the *x*-axis is typically depicted with frequency *decreasing* from left to right. Consequently, regions to the left of a ^{1}H NMR spectrum represent ^{1}H nuclei with a *higher* resonance frequency (ie, deshielded nuclei) whereas regions to the right represent ^{1}H nuclei with a *lower* resonance frequency (ie, shielded nuclei). The **baseline** is indicated by flat regions where no resonance frequencies are detected.

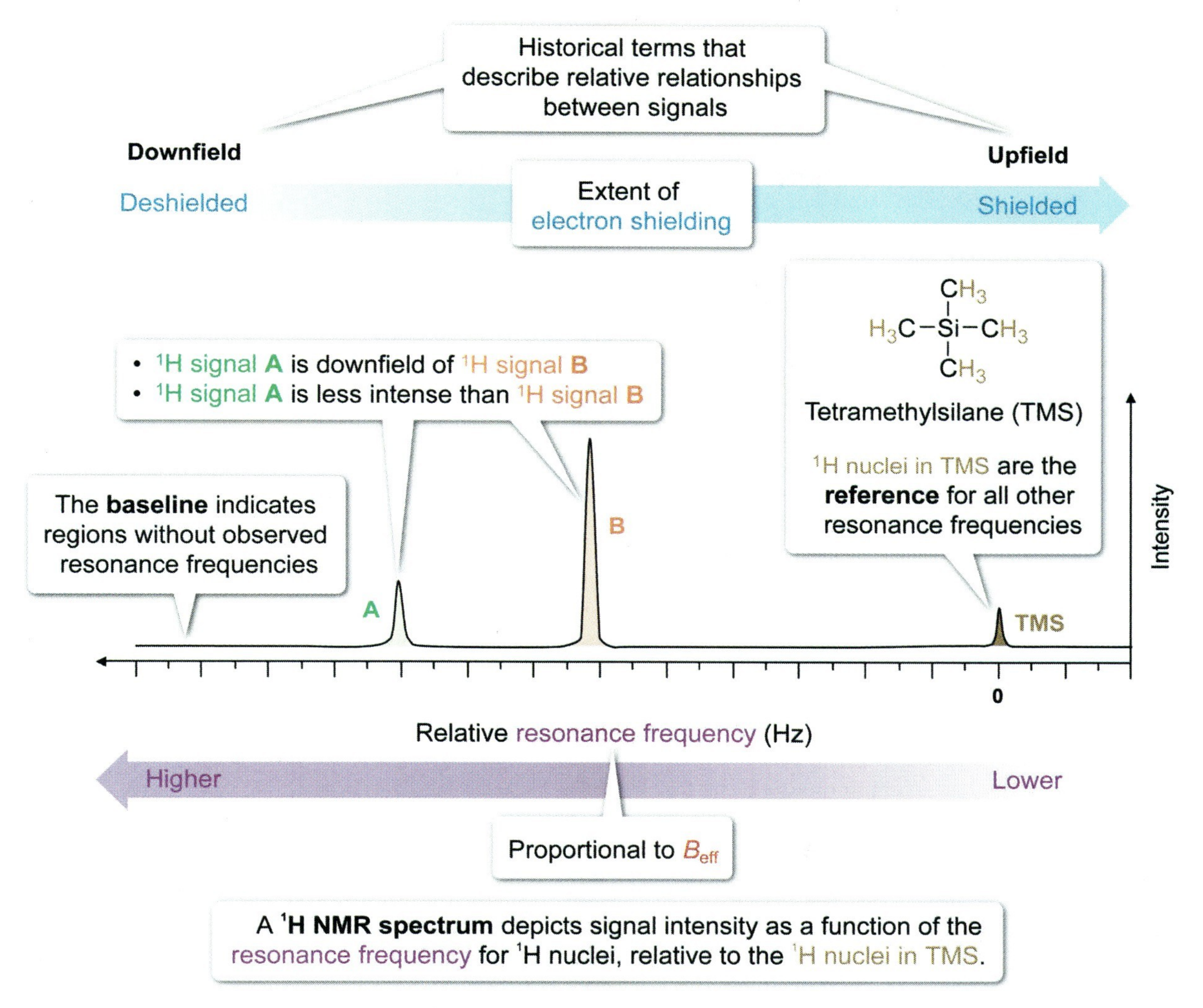

Figure 14.54 The anatomy of a ^{1}H NMR spectrum.

The terms *upfield* and *downfield* are sometimes used to describe the relative position of a signal in a ^{1}H NMR spectrum, where **downfield signals** are synonymous with *higher* resonance frequencies and **upfield signals** are synonymous with *lower* resonance frequencies.

Most forms of NMR spectroscopy use an **external reference** (ie, a standard ^{1}H signal from a molecule *other* than the sample being analyzed) to determine the zero-point for the resonance frequency axis. In ^{1}H NMR spectroscopy, the ^{1}H signal for tetramethylsilane (TMS) is used as the external reference. All other resonance frequencies are reported relative to the reference point. In most situations likely to be encountered on the exam, the resonance frequencies of ^{1}H nuclei in a sample are *downfield* of (ie, have a higher resonance frequency than) TMS.

Summary of Information Gained from a ^{1}H NMR Spectrum

A ^{1}H NMR spectrum can provide a *tremendous* amount of information about the structure of the analyzed compound, including the:

- **Total number of ^{1}H signals**, which indicates the number of *unique* ^{1}H electron environments in the structure of the compound (see *chemical equivalence* in Concept 14.7.06)
- **Locations of each ^{1}H signal** (ie, resonance frequency), which indicate the relative amount of shielding or deshielding experienced by the nuclei (see *chemical shift* in Concept 14.7.07)

- **Relative intensity of ^1H signals**, which indicates the number of ^1H nuclei that have the same resonance frequency and electron environment (see *integration* in Concept 14.7.08)
- **Splitting of ^1H signals** into more than one peak, which indicates the number of ^1H atoms on adjacent carbon atoms (see *spin-spin splitting* in Concept 14.7.09)

Together, these pieces of information may allow the *complete structure* (ie, connectivity of all atoms) of an organic molecule to be determined.

14.7.06 Chemical Equivalence

Concept 14.7.03 describes how a nucleus's electron cloud partially shields it from the full magnitude of the external magnetic field. ^1H nuclei that experience an *identical* magnitude of electron shielding have an *identical* resonance frequency. For this reason, the *number* of unique ^1H resonance frequencies observed in a ^1H NMR spectrum represents the total number of *unique* ^1H chemical environments in the structure of the sample.

In NMR spectroscopy, two atoms have the same **chemical environment** when they both have the same relative proximity to all other groups and experience the same effects to their electron density. In other words, ^1H nuclei in the same chemical environment are chemically *indistinguishable* from one another and are **chemically equivalent**. In most cases, ^1H nuclei attached to the *same* atom (eg, usually carbon) are chemically equivalent due to **rotational symmetry** (Figure 14.55), which results when a single sigma bond rotates with respect to its atoms to generate different conformations for a molecule (see Lesson 3.2).

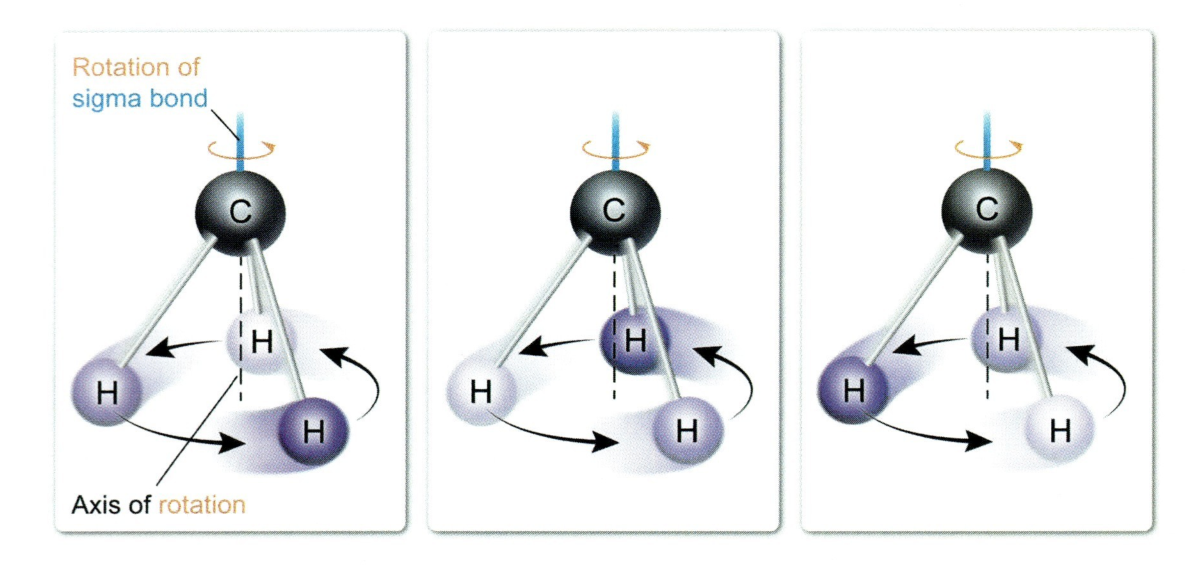

^1H nuclei on a methyl group rapidly rotate about the sigma bond and are **chemically equivalent** to one another due to **rotational symmetry**.

Figure 14.55 Rotational symmetry of ^1H nuclei on a methyl group.

Due to rotational symmetry, the three ^1H nuclei within a methyl group ($-CH_3$) are chemically equivalent and have identical resonance frequencies. Similarly, the two ^1H nuclei within a single methylene group ($-CH_2-$) are *usually* chemically equivalent and have identical resonance frequencies. However, it is important to note that rotational symmetry does *not* imply that *all* methyl groups or *all* methylene groups in a molecule are chemically equivalent with one another—the structural context of *each* group must be considered individually.

Rotational symmetry can also lead to chemical equivalence when the structure of a molecule has an **axis of symmetry**, which is an imaginary line that separates two *identical* portions of a molecular structure. Figure 14.56 provides examples of organic molecules with their unique ^{1}H chemical environments labeled using a letter. Any relevant axes of symmetry that lead to chemically equivalent ^{1}H nuclei are noted.

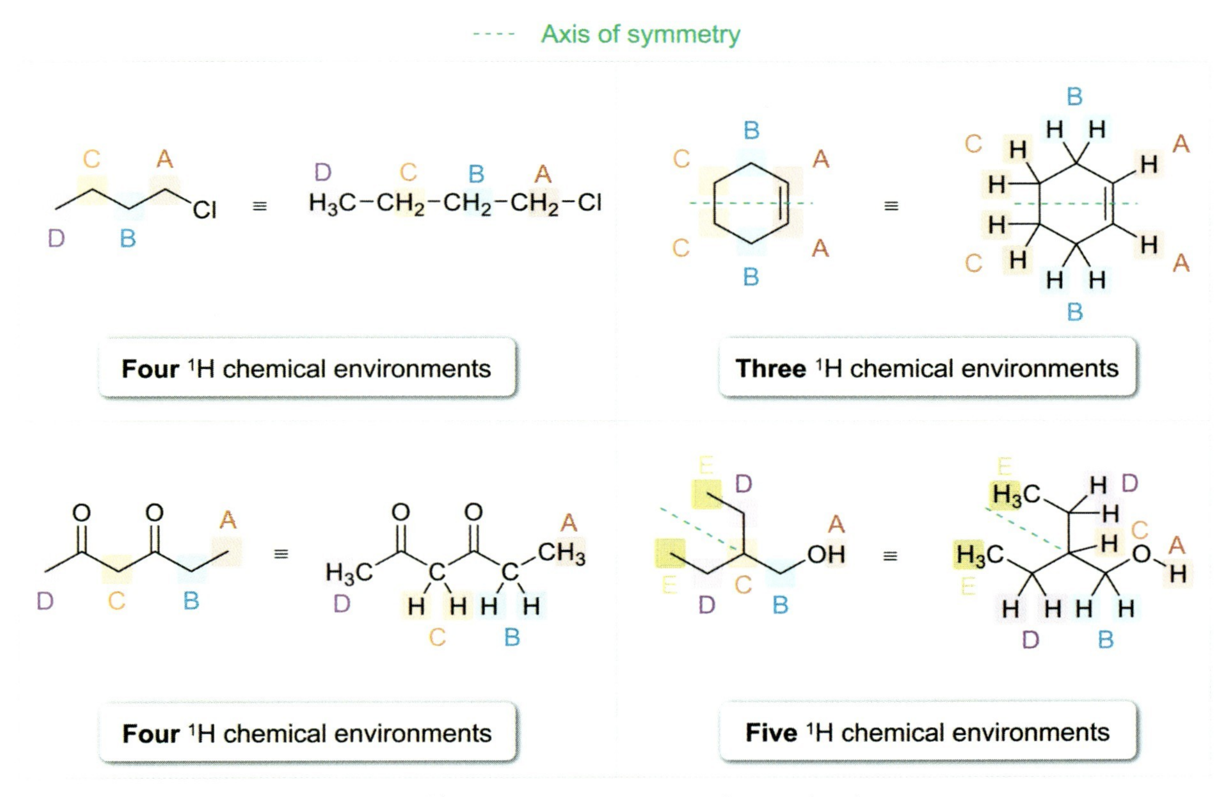

Figure 14.56 Determining the number of ^{1}H chemical environments from molecular structure.

In 1-chlorobutane (top left) and hexan-2,4-dione (bottom left), shown in Figure 14.56, protons on any individual carbon atom are chemically equivalent. However, because there is no additional axis of symmetry observed, the protons of each methyl and methylene group are *chemically inequivalent* from the protons of the other groups.

Cyclohexene (top right) has an axis of symmetry that *crosses* the double bond and the single bond connecting carbons 4 and 5 (C protons). In this case, the *whole molecule* can rotate 180° about that axis, resulting in an *identical structure* in which the *top* A, B, and C protons have swapped places with the *bottom* A, B, and C protons, respectively. Therefore, the two A protons are chemically equivalent to each other, the two B protons are chemically equivalent to each other, and the two C protons are chemically equivalent to each other.

2-Ethylbutan-1-ol (bottom right) has an axis of symmetry that goes *through* the sigma bond between carbons 1 (B) and 2 (C). Instead of rotating the whole molecule, the individual C1–C2 *bond* can rotate. A 180° rotation of this bond would swap the positions of the two sets of D and E protons. Consequently, these protons are chemically equivalent to each other.

Concept Check 14.6

How many unique ${}^1H$ chemical environments are expected in the ${}^1H$ NMR spectrum for the following compound?

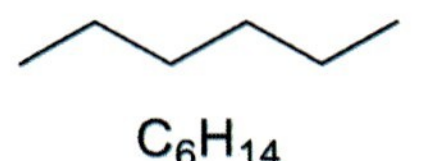

C_6H_{14}

Solution

Note: The appendix contains the answer.

14.7.07 Chemical Shift

Although the superconducting magnet in an NMR spectrometer can be described by the *strength* of its magnetic field (B_0) measured in tesla, spectrometers are more often described by the average *frequency* of the broadband radio waves used to induce nuclear resonance in protons, measured in megahertz (MHz, 1 MHz = 1×10^6 Hz). Because the magnitude of the absolute resonance frequency of a given ${}^1H$ nucleus is proportional to B_0 (ie, a property of the *specific* instrument used to obtain the spectrum), ${}^1H$ resonance frequencies generated by *different* NMR spectrometers are *different*.

To standardize NMR data generated by *different* NMR instruments, chemists report resonance frequencies as a **chemical shift** (Figure 14.57), which is calculated as the ratio of the detected resonance frequency relative to TMS protons (measured in Hz) to the *absolute* resonance frequency of TMS protons (measured in MHz).

$$^{1}\text{H chemical shift } (\delta, \text{ppm}) = \frac{\text{Resonance frequency}_{\text{sample}} - \text{Resonance frequency}_{\text{TMS}} \text{ (Hz)}}{\text{Resonance frequency}_{\text{TMS}} \text{ (MHz)}}$$

As such, chemical shift is similar to a percent change (ie, parts per one hundred units). However, because chemical shift is a ratio of Hz to MHz (ie, 1 million Hz), chemical shift is reported in units of parts per *million* (ppm) on a scale often referred to as the **delta (δ) scale**.

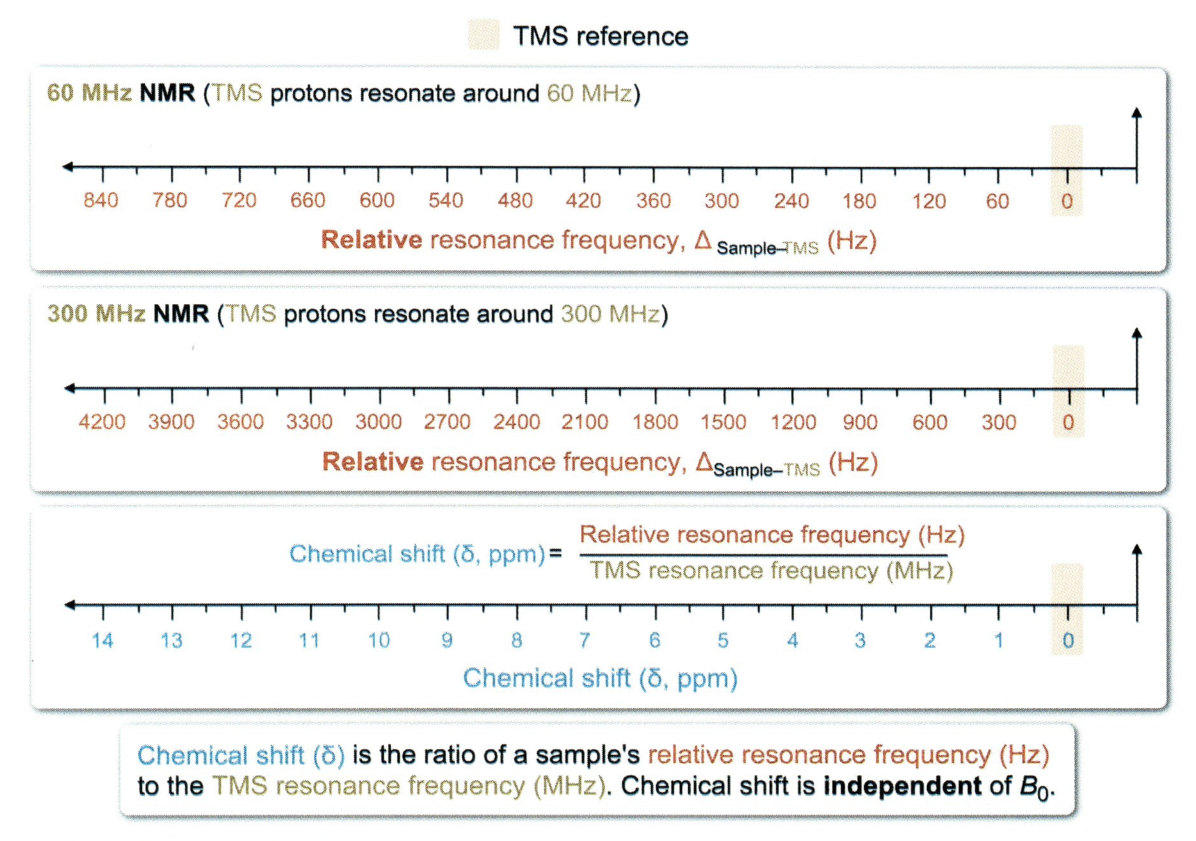

Figure 14.57 Chemical shift and the delta scale.

The chemical shift (δ) of ^{1}H NMR signals in a spectrum is *independent* of the strength of B_0 and the instrument used to acquire the data. Chemical shift values represent the *midpoint* of a ^{1}H NMR signal—a criterion especially important if a ^{1}H signal is *split* (see Concept 14.7.09).

The rest of this concept focuses on the structural and electronic basis of ^{1}H chemical shift values for functional groups, which are influenced in two major ways:

- The inductive effect
- Anisotropy of pi bonds

Chemical Shift and the Inductive Effect

Concept 6.5.03 describes how the inductive effect is one of the major ways to influence the electron density of atoms in a molecule. In NMR spectroscopy, the inductive effect is most often observed in relation to sigma electron–withdrawing groups, which remove electron density from a region through the sigma bond network (Figure 14.58). The presence of a sigma electron–withdrawing group (eg, highly electronegative atoms) in a molecule inductively removes electron density from nearby ^{1}H nuclei and *increases* the resultant chemical shift in a ^{1}H NMR spectrum.

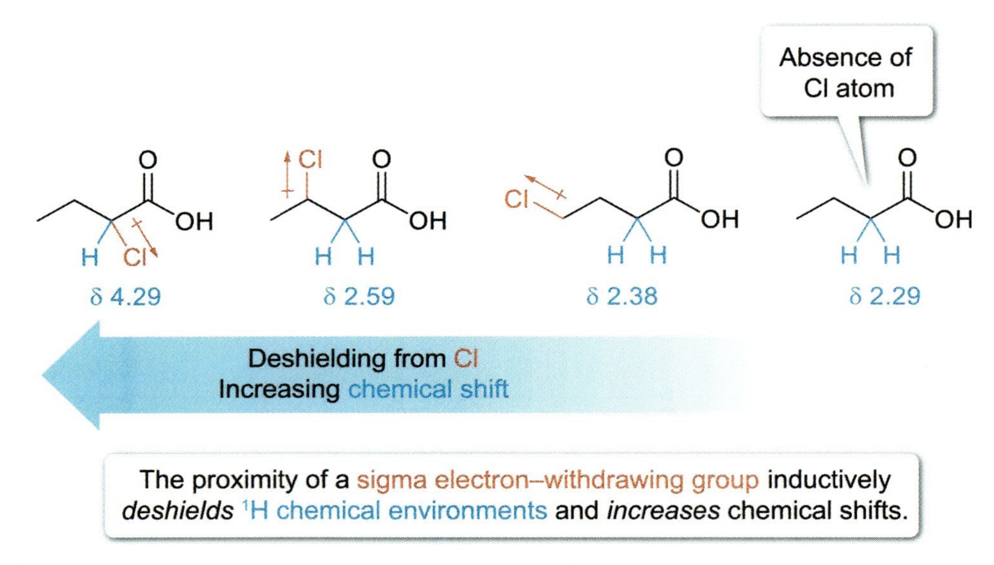

Figure 14.58 The impact of the inductive effect on chemical shift.

The impact of the inductive effect is most pronounced when the electronegative atom is closer to a ^{1}H nucleus. For example, the chemical shift of the α protons in an alkyl carboxylic acid is normally around δ 2.29. Adding a single electronegative chlorine atom to the molecule deshields the α proton environment through the inductive effect, with the most pronounced effect (δ 4.29) observed when the chlorine atom is closest to the α proton environment.

Chemical Shift and the Anisotropy of Pi Bonds

Concept 14.7.03 describes how placing electrons in a magnetic field leads to their movement (ie, a current) and the generation of an induced magnetic field ($B_{induced}$). A similar process to generate $B_{induced}$ also occurs for electrons in π bonds. Because a ^{1}H nucleus is *always* situated within its individual electron cloud, its $B_{induced}$ can be treated as a simple vector to illustrate the relationship between B_0 and B_{eff}.

However, ^{1}H nuclei can be at many locations relative to a source of π electron density, where the $B_{induced}$ from π electron field lines (ie, arrows) *vary* in direction depending on their location. This type of variation is called **anisotropy**. Although the details of π bond anisotropy are unlikely to be evaluated on the exam, this concept is *essential* to explain the observed chemical shifts of ^{1}H nuclei in functional groups containing π bonds.

Near the *center* of the π bond containing the circulating electrons, the field lines for its $B_{induced}$ are oriented *opposite* B_0 (Figure 14.59). A hypothetical ^{1}H nucleus located at this position would experience a $B_{induced}$ that would lead to a *smaller* B_{eff} (ie, the $B_{induced}$ shields the nucleus and leads to a *lower* chemical shift). However, on the *outside* of the π bond, the field lines for its $B_{induced}$ are *aligned with* and *reinforce* B_0. A hypothetical ^{1}H nucleus located at this position would experience a $B_{induced}$ that would lead to a *larger* B_{eff} (ie, the $B_{induced}$ deshields the nucleus and leads to a *higher* chemical shift).

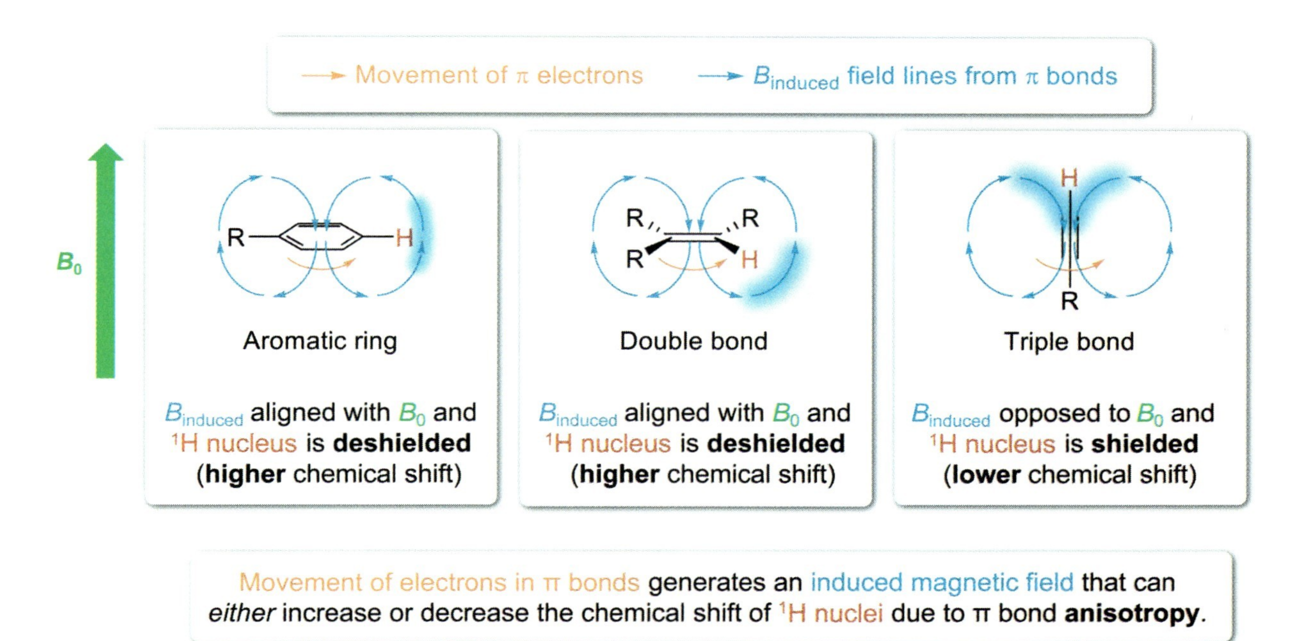

Figure 14.59 Relationship between chemical shift and π bond anisotropy.

For *most* organic functional groups containing π bonds (eg, alkenes, aromatic rings, ketones, aldehydes, carboxylic acids, acid halides, anhydrides, esters, amides), the anisotropy of the $B_{induced}$ field lines generated by π electron currents has a *deshielding* effect on nearby ^{1}H nuclei and leads to an *increase* in chemical shift. The magnitude of this anisotropic effect is more pronounced for ^{1}H nuclei *closer* to the π system that generates the $B_{induced}$, and this effect is especially pronounced for aromatic rings due to the formation of ring currents.

Alkynes and nitriles (ie, functional groups with a *triple* bond) are the sole exceptions to this trend. Whereas double-bonded π bonds align perpendicular to B_0, the two π bonds of a triple bond cause these functional groups to align *parallel* to B_0. Therefore, nearby protons (eg, *sp* C–H of a terminal alkyne, α protons to the triple bond) are *shielded* by $B_{induced}$ and observed at a *lower* chemical shift.

Summary of ^{1}H NMR Chemical Shifts for Functional Groups

The chemical shifts of ^{1}H nuclei contained within and adjacent to functional groups are influenced by a combination of both the inductive effect and π bond anisotropy. Consequently, chemical shift regions of a ^{1}H NMR spectrum often suggest the presence of certain functional groups. The chemical shift ranges for each functional group are graphically represented in Figure 14.60. Importantly, these values represent the chemical shift range when a ^{1}H environment is *solely* influenced by one functional group; *combinations* of two or more functional groups may lead to chemical shifts that lie outside the listed ranges.

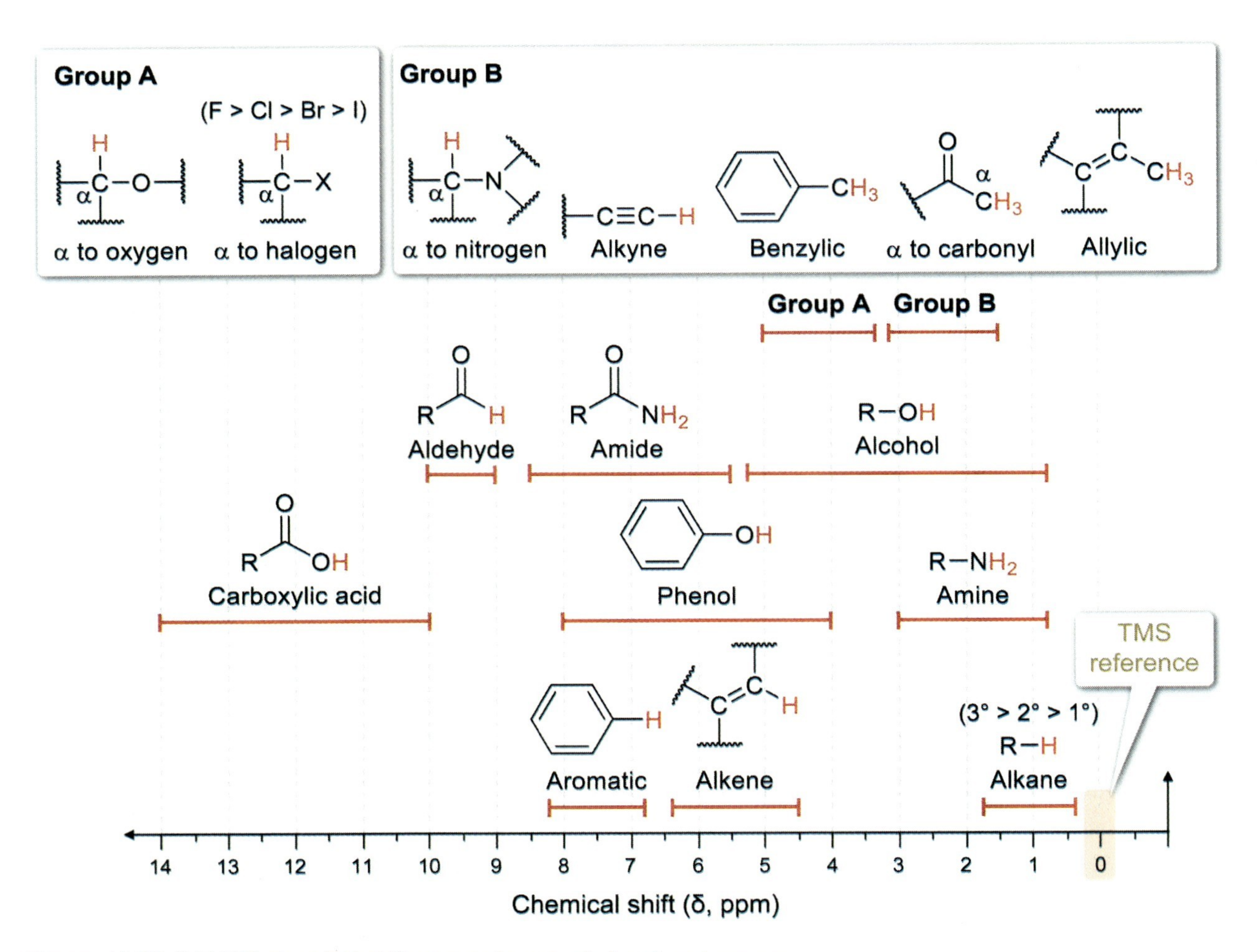

Figure 14.60 ^{1}H NMR chemical shift ranges for organic functional groups.

Many different types of functional groups have overlapping ^{1}H chemical shift ranges that appear in the range of δ 1.5–5.0. To streamline the presentation of data in Figure 14.60, the range δ 3.2–5.0 is designated as Group A and the range δ 1.5–3.2 is designated as Group B.

14.7.08 Integration of NMR Spectra

Understanding chemical equivalence (Concept 14.7.06) is a necessary step toward determining the total number of unique ^{1}H NMR signals that appear in a compound's ^{1}H NMR spectrum. The amount (ie, intensity) of signal generated for *each* unique resonance frequency in an NMR experiment is proportional to the *number* of ^{1}H nuclei present in that chemical environment. For example, the signal generated for a chemical environment consisting of one methyl ($-CH_3$) group is *three times greater* than the signal generated for a chemical environment consisting of one methine (–CH–) group.

The process of quantifying the amount of signal produced for ^{1}H resonance frequencies is called **integration** because the process involves measuring the area underneath the ^{1}H signal peaks. ^{1}H NMR spectra may display integration data in one of three ways, as summarized in Figure 14.61.

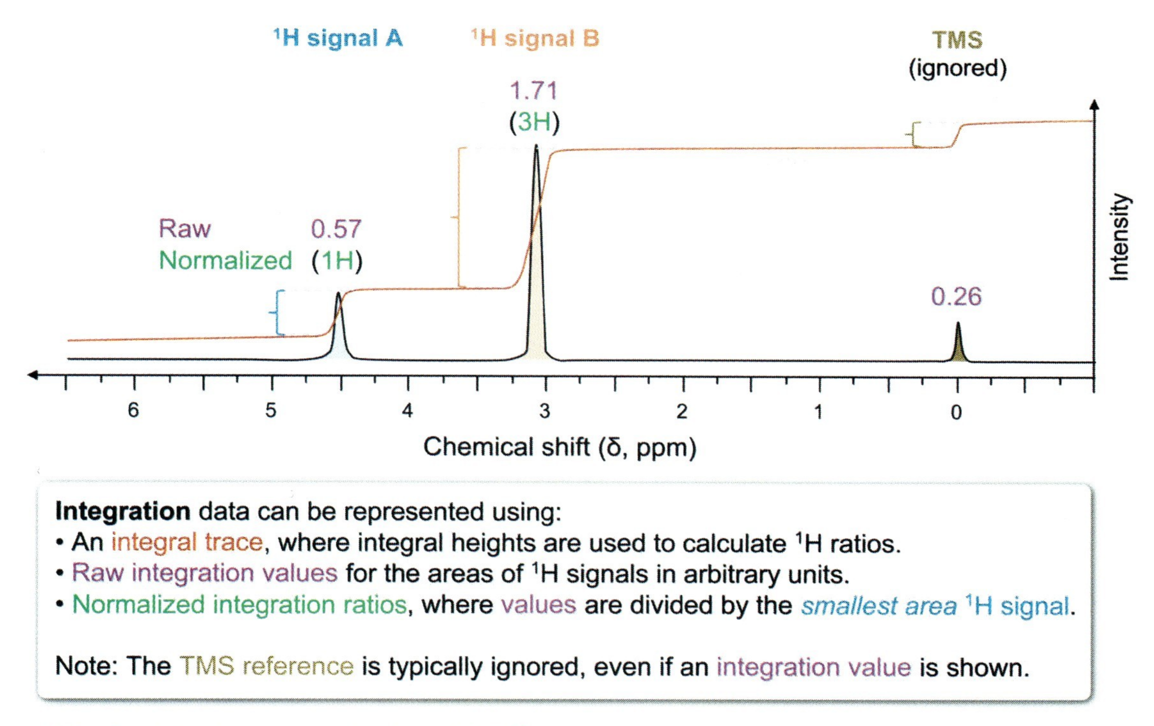

Integration data can be represented using:

- An integral trace, where integral heights are used to calculate ^{1}H ratios.
- Raw integration values for the areas of ^{1}H signals in arbitrary units.
- Normalized integration ratios, where values are divided by the *smallest area* ^{1}H signal.

Note: The TMS reference is typically ignored, even if an integration value is shown.

Figure 14.61 Depicting integration data for a ^{1}H NMR spectrum.

Although modern NMR instruments typically retain the capability of displaying integral trace data, most provide peak areas as raw integration values in arbitrary units. Ultimately, integration data are used to develop **ratios** between ^{1}H signal areas. Because a compound's molecular formula contains a whole number of hydrogen atoms (ie, it is not possible to have 0.5 hydrogen atoms in a molecule), raw integration values are typically divided by the value for the ^{1}H signal with the *smallest area*. This process **normalizes** the integration data to a 1 hydrogen (ie, 1H) signal.

Sometimes this type of normalization process results in integration data in which *other* peaks contain a fractional number of hydrogen atoms (eg, 3.5H). Should this occur, *all* normalized integration values are multiplied by the same factor (eg, 2, 3) to achieve whole number normalized data for all ^{1}H signals.

General Considerations for Integration of ^{1}H NMR Spectra

The following tips can be used to help navigate the analysis of integration data in a ^{1}H NMR spectrum:

- Deductive reasoning is a *powerful* tool when working with integration data. Integration data should indicate structural features that would be expected for an organic molecule—if the data do not, an error may have occurred during analysis.
- Adding up the normalized integration data for a ^{1}H spectrum should provide a value that equals the *total* number of hydrogen atoms in the molecular formula of the sample (eg, 1H + 3H + 2H + 3H = 9H when the molecular formula is C_5H_9Cl).
- Except in the case of the ^{1}H NMR spectrum of methane (CH_4), it is not possible for a *single* carbon atom to have more than three hydrogen atoms attached to it. Consequently, normalized integration data *greater* than 3H is likely an indication of *either* overlapping ^{1}H signals (ie, coincidentally similar chemical shifts for two or more chemical environments) or the presence of symmetry within the molecular structure (ie, more than one carbon atom and its protons have the *same* chemical environment).
- When a ^{1}H signal is split into more than one peak (see Concept 14.7.09), it is important to include the areas of *all* of the signal's peaks during integration.

14.7.09 Spin-Spin Splitting

In addition to being influenced by inductive effects and the anisotropy of π bonds, ^1H chemical environments are *also* influenced by the presence of nearby *nonequivalent* ^1H nuclei (ie, nuclei in a *different* chemical environment and with a *different* chemical shift) within the structure of the molecule. Concept 14.7.01 describes how an NMR-active nucleus (eg, ^1H) can be viewed as a miniature bar magnet with a nuclear spin (ie, miniature magnetic field, B_{spin}) *aligning with* or *opposing* the external magnetic field B_0 (Figure 14.62).

Consider two ^1H nuclei: a nucleus of interest H_A, and a nearby, but *non*-resonating, H_B nucleus. The alignment of H_B with or against the external field would create a *slight* difference in the B_{eff} experienced by H_A and would result in *slightly* different resonance frequencies within its ^1H NMR spectrum.

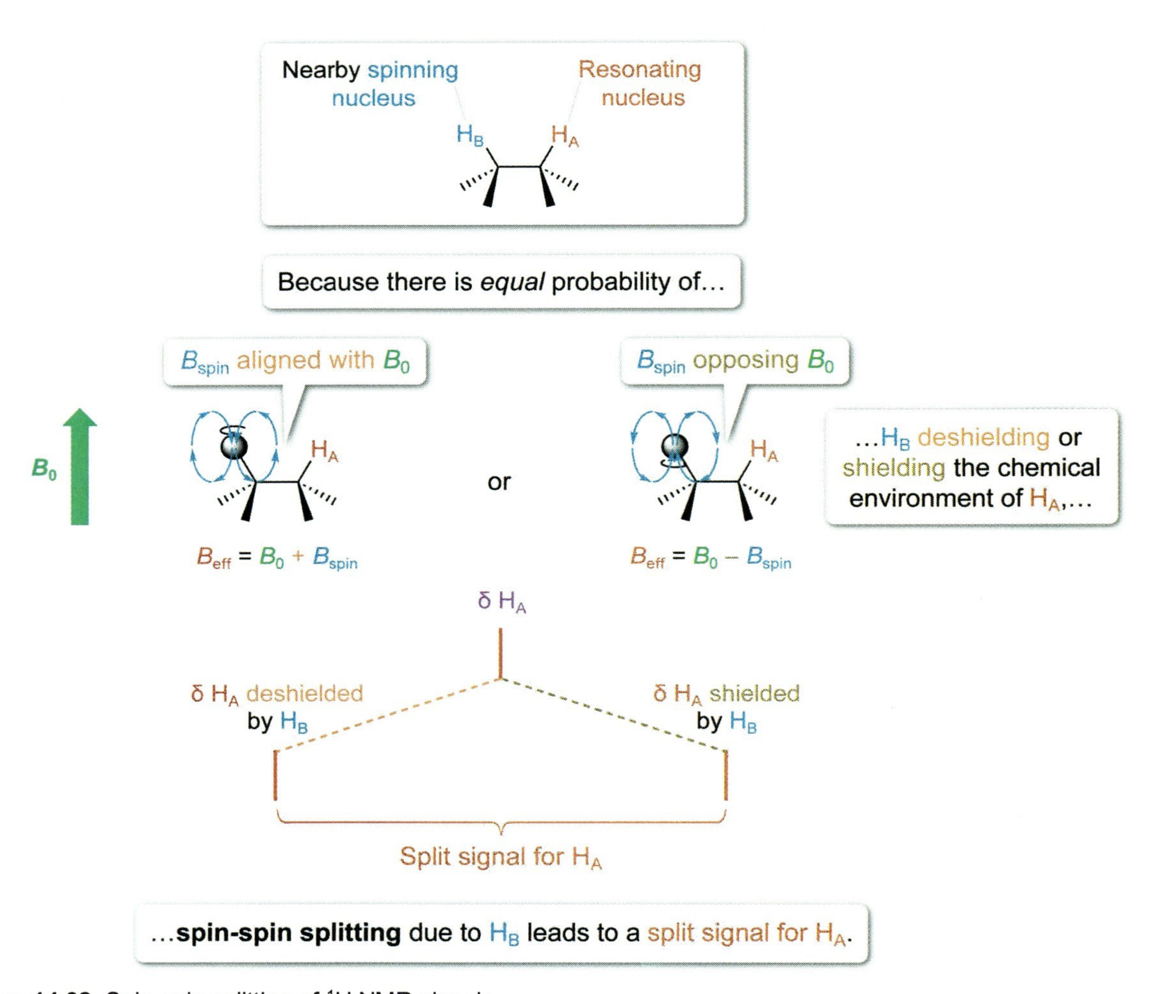

Figure 14.62 Spin-spin splitting of ^1H NMR signals.

In a population of these example molecules, when the nearby H_B nucleus has its magnetic field *aligned with* B_0, the resonating H_A nucleus experiences a *slightly larger* B_{eff} and has a *slightly higher* chemical shift. In contrast, when the nearby H_B nucleus has its magnetic field *opposing* B_0, the resonating H_A nucleus experiences a *slightly smaller* B_{eff} and has a *slightly lower* chemical shift. Consequently, the resonating ^1H nucleus (H_A) no longer exhibits one peak in its ^1H NMR spectrum, but rather *two* peaks (ie, the original ^1H signal is *split*).

The phenomenon in which nearby, *nonequivalent* ^1H nuclei either *increase* or *decrease* the chemical shift for a resonating ^1H nucleus is known as **spin-spin splitting**. Spin-spin splitting is a *bidirectional* process—if two ^{1}H nuclei, H_A and H_B, are close enough that the spin of H_B splits the signal for H_A, then the spin of H_A *also* splits the signal for H_B. Spin-spin splitting is sometimes described as **coupling** (ie, **magnetic coupling**) because the magnetic features of ^{1}H nuclei are influenced by other coupled ^{1}H nuclei.

Spin-spin splitting is typically observed when ^{1}H nuclei are separated by 2–3 sigma bonds. Consequently, ^{1}H nuclei attached to *adjacent* carbon atoms (ie, a vicinal relationship) are typically magnetically coupled to each other. Spin-spin coupling between ^{1}H nuclei separated by 4+ sigma bonds (ie, **long-range spin-spin splitting**) is possible, but such situations usually require other factors, including specific spatial relationships or the presence of π bonds. Importantly, equivalent protons (eg, most geminal protons) do *not* split each other; splitting occurs only between *nonequivalent* nuclei.

The *magnitude* of the magnetic influence of spin-spin splitting is measured by the **coupling constant** (ie, ***J*** **coupling constant** or ***J***), which is the *distance* (in Hz) between the peaks in a split ^{1}H signal (Figure 14.63). Although the underlying theory explaining *J* coupling magnitudes is beyond the scope of content likely to be evaluated on the exam, *different* bonding and conformational relationships have *different* characteristic magnitudes of *J*. For example, the typical *J* coupling constant for ^{1}H nuclei in a linear alkyl chain is 7 Hz.

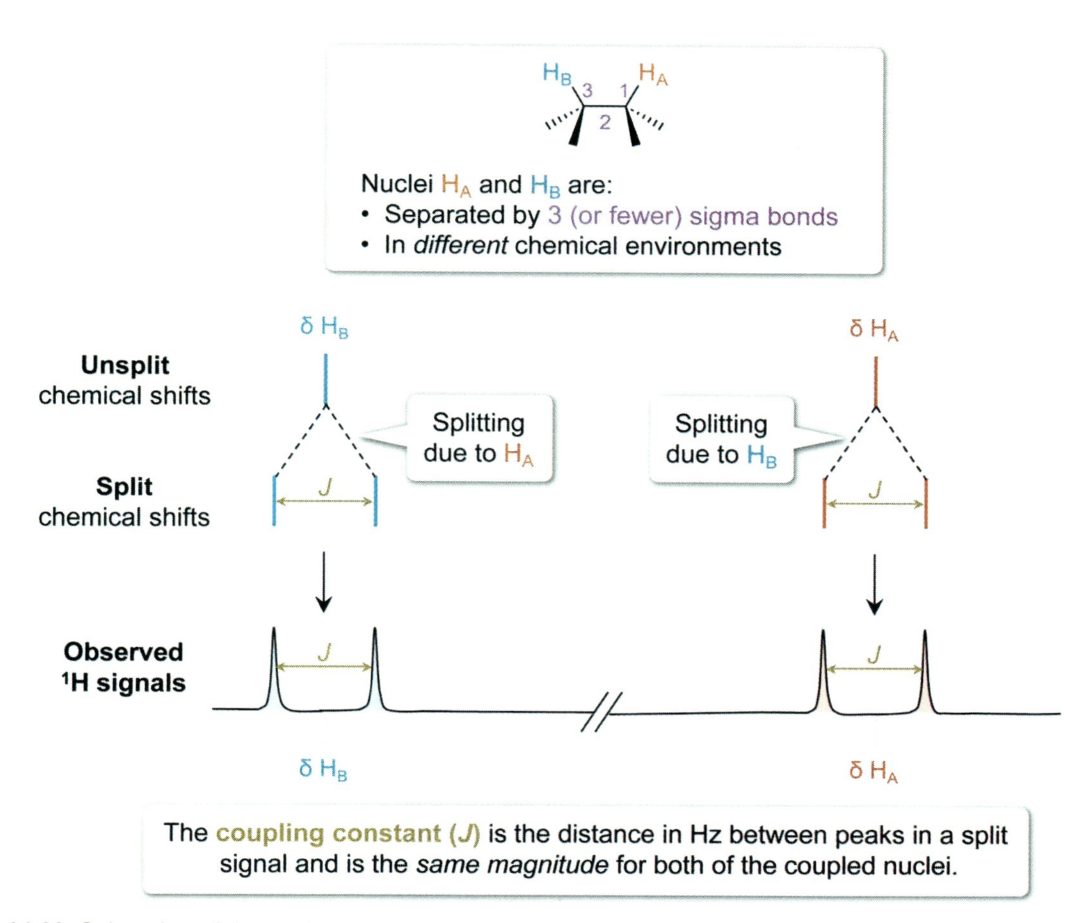

Figure 14.63 Spin-spin splitting and the coupling constant (*J*).

Because spin-spin splitting is *bidirectional*, the magnitude of a *J* coupling constant for a magnetic coupling is *identical* for both coupled ^{1}H nuclei. For example, if the nuclear spin of H_B splits the signal for H_A with a magnitude of 7 Hz, the nuclear spin of H_A *also* splits the signal for H_B with the *same* magnitude of 7 Hz.

Although it may seem that spin-spin splitting adds unnecessary complexity to the analysis of ^{1}H NMR spectra, spin-spin splitting provides a *tremendous* amount of information about the structural details of a molecule. The bidirectional nature of spin-spin splitting between nearby, nonequivalent ^{1}H nuclei allows chemists to use magnetic coupling interactions to obtain information about the *surrounding* ^{1}H nuclei that influence a particular signal. Consequently, spin-spin splitting provides information about the **connectivity** between ^{1}H nuclei in different chemical environments and, ultimately, the *entire* molecular structure of a sample.

Pascal's Triangle and the *N*+1 Rule

To this point, discussion of spin-spin splitting is limited to coupling events between two ^{1}H nuclei—one in each coupled chemical environment. In this scenario, *one* ^{1}H signal is split into *two* signals based on the magnetic influence of *one* nearby nonequivalent ^{1}H nucleus. However, organic molecules frequently have more than one ^{1}H atom in the same chemical environment (eg, ^{1}H attached to the same carbon atom) within 3 sigma bonds of other *dissimilar* ^{1}H nuclei.

The *N*+1 rule provides the ability to quickly identify the splitting pattern for a ^{1}H signal when all the individual coupling constants are *identical* or *nearly identical*. According to the ***N*+1 rule**, when a ^{1}H signal is split by *N* nearby, nonequivalent ^{1}H nuclei, the resultant split signal has *N*+1 peaks, as summarized in Table 14.9.

Table 14.9 Using the *N*+1 rule and Pascal's triangle to determine the peaks in a multiplet.

	Characteristics of the split ^{1}H signal (ie, the multiplet)				
Nearby ^{1}H nuclei (*N*)	**Peaks (*N*+1)**	**Name**	**Ratio of peak sizes (from Pascal's triangle)**	**Appearance in ^{1}H spectrum**	**Structural example**
0	1	Singlet	1		
1	2	Doublet	1 1		
2	3	Triplet	1 2 1		
3	4	Quartet	1 3 3 1		
4	5	Quintet	1 4 6 4 1		
5	6	Sextet	1 5 10 10 5 1		
6	7	Septet	1 6 15 20 15 6 1		
7	8	Octet	1 7 21 35 35 21 7 1		
8	9	Nonet	1 8 28 56 70 56 28 8 1		

The distance (in Hz) between each of the peaks in an *N*+1 multiplet is the *same* value—the coupling constant, *J*. However, the *ratio* of peak heights for an *N*+1 multiplet varies. **Pascal's triangle** is a triangular relationship of numbers in which the value of each number is determined by the sum of the two numbers directly above it. Each level of Pascal's triangle represents a ratio of peak sizes (ie, integration, height) for an *N*+1 multiplet. Consequently, a doublet (two peaks) has a 1:1 ratio of peaks, a triplet (three peaks) has a 1:2:1 ratio of peaks, and a quartet (four peaks) has a 1:3:3:1 ratio of peaks.

Most ^{1}H NMR signals within a linear (ie, not cyclic) carbon chain have *very* similar coupling constants (ie, J = 7 Hz) and multiplets that abide by the *N*+1 rule. Therefore, determining the splitting pattern (ie, multiplet) for these types of ^{1}H signals requires counting the number of *nonequivalent* ^{1}H nuclei on adjacent carbon atoms (*N* nuclei), adding 1 to the total, and using Pascal's triangle to determine the ratio of peak sizes within the multiplet. A structural example that would result in each *N*+1 multiplet is depicted in Table 14.9.

END-OF-UNIT MCAT PRACTICE

Congratulations on completing **Unit 3: Separation Techniques, Spectroscopy, and Analytical Methods**.

Now you are ready to dive into MCAT-level practice tests. At UWorld, we believe students will be fully prepared to ace the MCAT when they practice with high-quality questions in a realistic testing environment.

The UWorld Qbank will test you on questions that are fully representative of the AAMC MCAT syllabus. In addition, our MCAT-like questions are accompanied by in-depth explanations with exceptional visual aids that will help you better retain difficult MCAT concepts.

TO START YOUR MCAT PRACTICE, PROCEED AS FOLLOWS:

1) Sign up to purchase the UWorld MCAT Qbank
   IMPORTANT: You already have access if you purchased a bundled subscription.
2) Log in to your UWorld MCAT account
3) Access the MCAT Qbank section
4) Select this unit in the Qbank
5) Create a custom practice test

Appendix

Concept Check Solutions

You will find detailed, illustrated, step-by-step solutions for each concept check in the digital version of this book.

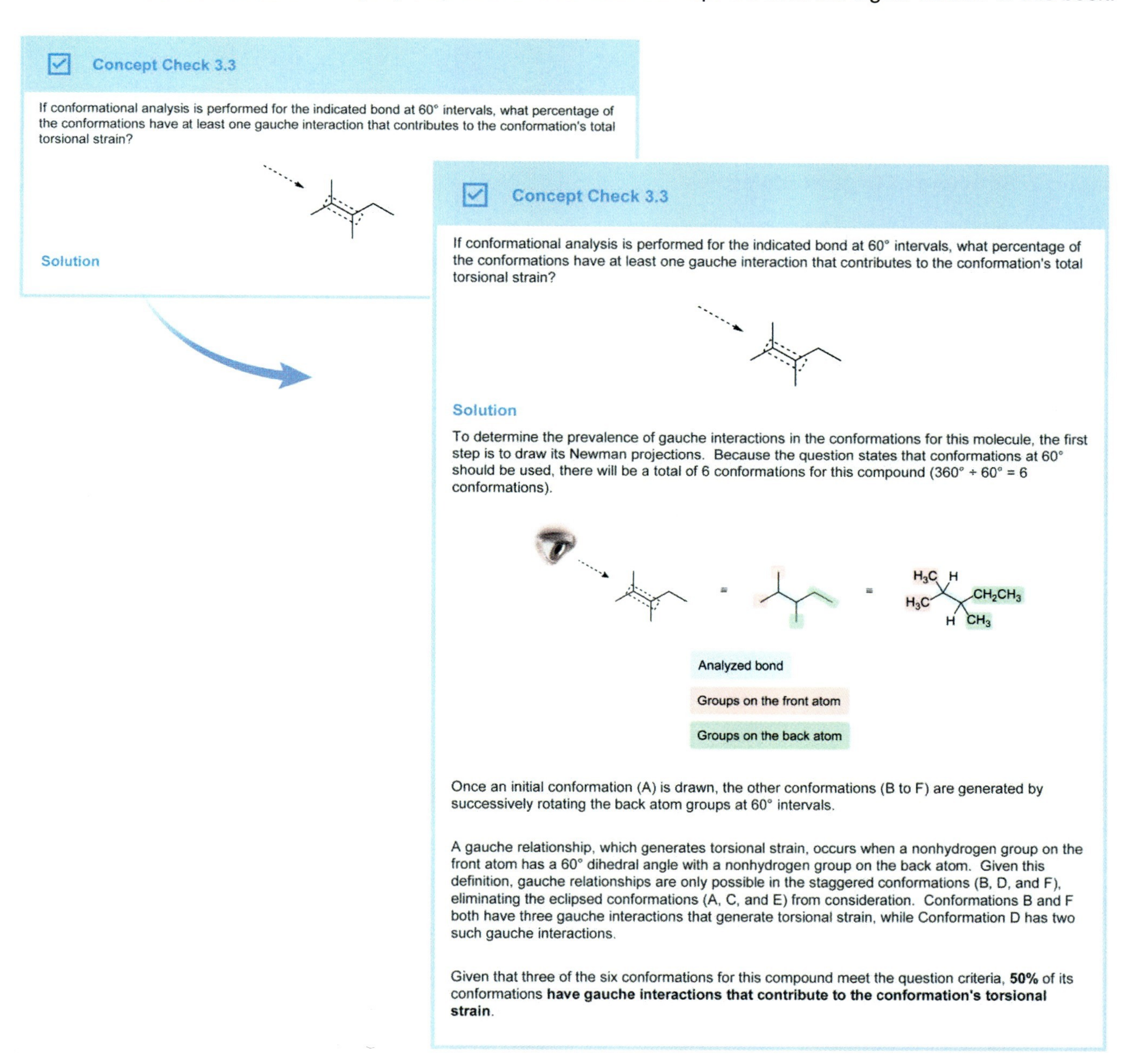

Concept Check 3.3

If conformational analysis is performed for the indicated bond at 60° intervals, what percentage of the conformations have at least one gauche interaction that contributes to the conformation's total torsional strain?

Solution

Concept Check 3.3

If conformational analysis is performed for the indicated bond at 60° intervals, what percentage of the conformations have at least one gauche interaction that contributes to the conformation's total torsional strain?

Solution

To determine the prevalence of gauche interactions in the conformations for this molecule, the first step is to draw its Newman projections. Because the question states that conformations at 60° should be used, there will be a total of 6 conformations for this compound (360° ÷ 60° = 6 conformations).

Once an initial conformation (A) is drawn, the other conformations (B to F) are generated by successively rotating the back atom groups at 60° intervals.

A gauche relationship, which generates torsional strain, occurs when a nonhydrogen group on the front atom has a 60° dihedral angle with a nonhydrogen group on the back atom. Given this definition, gauche relationships are only possible in the staggered conformations (B, D, and F), eliminating the eclipsed conformations (A, C, and E) from consideration. Conformations B and F both have three gauche interactions that generate torsional strain, while Conformation D has two such gauche interactions.

Given that three of the six conformations for this compound meet the question criteria, **50%** of its conformations **have gauche interactions that contribute to the conformation's torsional strain**.

In this section of the print book, you will only find short answers to the concept checks included in each chapter. Please go online for an interactive and enhanced learning experience with visual aids.

Unit 1. Introduction to Organic Chemistry

Chapter 1. Chemical Bonding

Lesson 1.1

1.1 Left to right: N = +1, C = 0, C = 0, O (top) = 0, O (right) = −1.

Lesson 1.2

1.2 1) sp^3, 2) sp^2, 3) sp, 4) sp^3

Lesson 1.3

1.3 1) covalent polar, 2) ionic, 3) covalent nonpolar

Chapter 2. Structure of Organic Molecules

Lesson 2.1

2.1 Left to right: N = tetrahedral, C = tetrahedral, C = trigonal planar, O (top) = trigonal planar, O (right) = tetrahedral.

2.2 O = bent, <109.5°, C = trigonal planar, 120°, N = bent, <120°.

Lesson 2.3

2.3 Yes.

2.4 ion-ion, dipole-dipole, hydrogen bonding, London dispersion.

Lesson 2.5

2.5 3 lone pairs.

2.6 Molecule A, Molecule C.

Lesson 2.6

2.7 C_7H_{10}.

2.8 There is 1 C=C, 0 rings, and N and O functional groups containing only single bonds.

Lesson 2.7

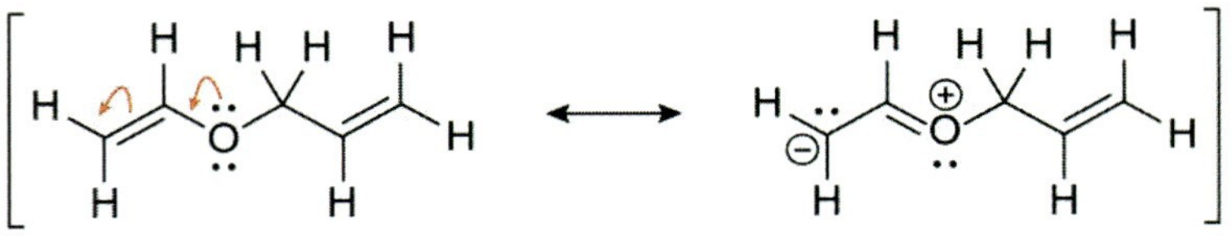

2.9

Lesson 2.8

2.10 Formulas 1 and 6 contain a ketone. Formulas 3 and 6 contain an amine.

Chapter 3. Isomers

Lesson 3.1

3.1 1) same molecule, 2) functional group isomers, 3) skeletal isomers, 4) positional isomers, 5) non-isomers

Lesson 3.2

3.2

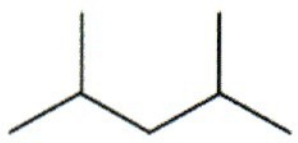

3.3 50%.

3.4

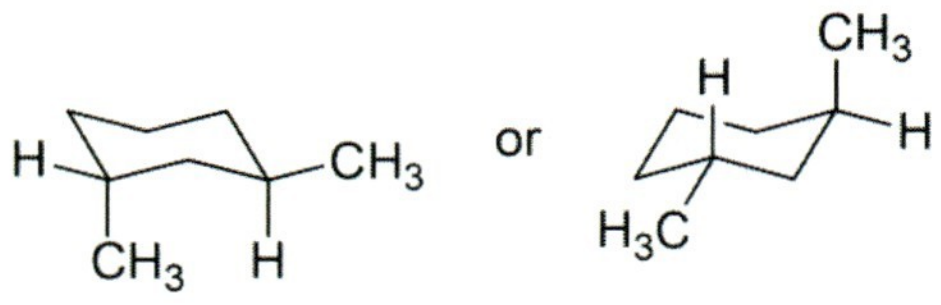

Lesson 3.3

3.5 Carbon A is not chiral. Carbon B and Carbon C are chiral.

3.6 *S*

3.7 1) diastereomers, 2) anomers, 3) enantiomers, 4) epimers

3.8

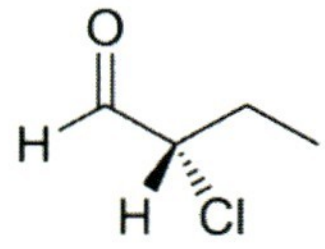

3.9 *Z*

Chapter 4. Organic Nomenclature

Lesson 4.1

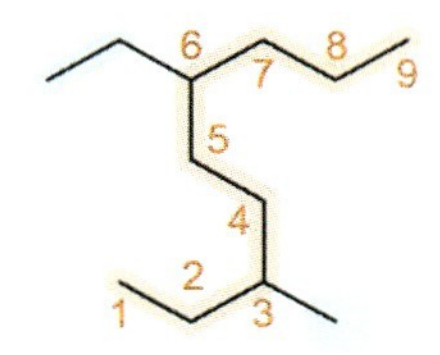

Nonane

4.1

4.2

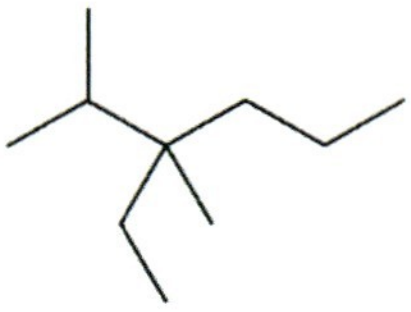

Lesson 4.2

4.3 3-fluoro-2,2-dimethylhexane.

4.4 When the parent chain contains both alkene groups, the IUPAC name is 4-ethyl-2-methylhexa-1,3-diene.

4.5 2-(5,5-dibromocyclopent-1-enyl)ethyl.

Lesson 4.3

4.6

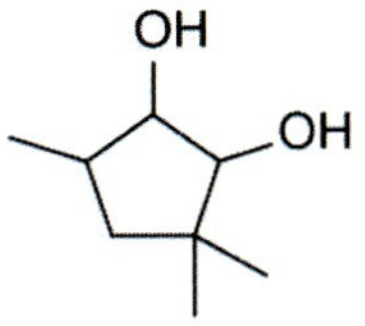

4.7 No, it violates the octet rule.

4.8 An ester, propyl 2-methylcyclopentanecarboxylate.

Lesson 4.4

4.9 6.

4.10 3-(cyanomethyl)cyclopentanecarbonitrile.

Lesson 4.5

4.11

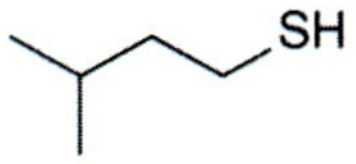

Lesson 4.6

4.12 (3*R*)-3,5-dihydroxy-3-methylpentanoate.

Lesson 4.7

4.13

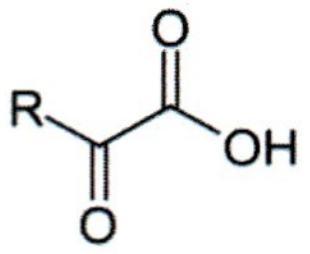

Chapter 5. Overview of Organic Reactions

Lesson 5.1

5.1 Brønsted-Lowry.

5.2

5.3 No.

Lesson 5.2

5.4 1) 1,2-methyl shift

H_3C H H_3C CH_3 CH_3

CH_3 CH_3 H H

2) 1,2-hydride shift

3) No rearrangement likely to occur.

5.5 B, C, A

Lesson 5.3

5.6 1) CH_3SH 2) NH_2 3) $Br^{\ominus}$ in O N H

Electrophilic carbon

5.7 1) Br I Stronger electrophile

Type 2: I is less electronegative and more polarizable than Br.

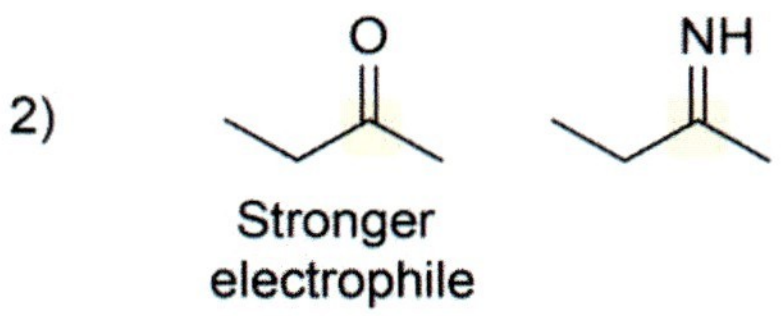

Type 2: O is more electronegative than N.

Lesson 5.4

5.8 H_3C Ö: H C H_3C + H H_3C C Ö: H_3C

5.9 CH_3 $H_3C\ddot{O}-C$ H CH_2CH_3

Lesson 5.5

5.10 Two groups, $^{-}C{\equiv}N$ and H^+, add across a pi bond, resulting in a mixture of enantiomers.

Lesson 5.7

5.11 Reduction.

5.12 The sum of the carbon oxidation numbers is unchanged.

Unit 2. Functional Groups and Their Reactions

Chapter 6. Hydrocarbons

Lesson 6.2

6.1

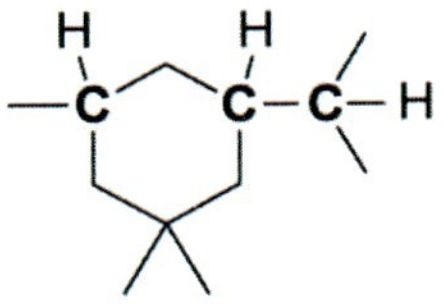

Lesson 6.5

6.2 Molecule A.

6.3 1) antiaromatic, 2) aromatic, 3) nonaromatic, 4) aromatic, 5) aromatic.

Chapter 7. Alkyl Halides, Ethers, and Sulfur-Containing Groups

Lesson 7.1

7.1 1) alkyl iodide, 2) alkyl chloride.

Lesson 7.2

7.2

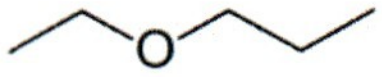

Chapter 8. Alcohols

Lesson 8.1

8.1 1)  2)

Lesson 8.2

8.2

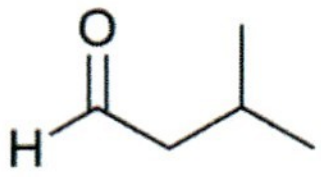

Lesson 8.3

8.3

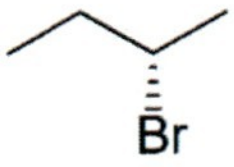

Chapter 9. Aldehydes and Ketones

Lesson 9.2

9.1 6 moles of PCC

Lesson 9.3

9.2 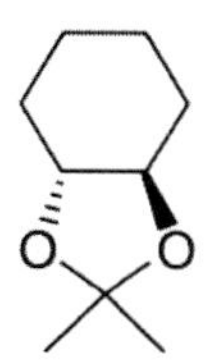

Lesson 9.4

9.3 1) Yes, 2) Yes, 3) No.

9.4 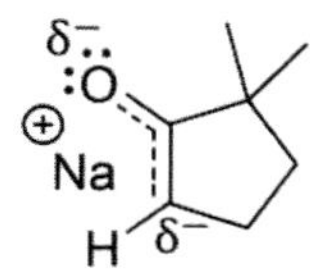

9.5 Slowly add acetone to a stirring solution of benzaldehyde and sodium hydroxide.

Chapter 10. Carboxylic Acids

Lesson 10.1

10.1 1) because it has a higher number of electron-withdrawing fluorine atoms and experiences a greater inductive stabilization of its conjugate base.

2) because it does not have any electron-donating groups to destabilize its conjugate base.

Lesson 10.2

10.2 It is not possible to prepare this carboxylic acid using solely the malonic ester synthesis, since an S_N2 reaction is not possible on an sp^2 carbon atom.

Lesson 10.3

10.3 Heating a solution of the 2-methylbutanoic acid in ethanol with a sulfuric acid catalyst.

10.4

Chapter 11. Carboxylic Acid Derivatives

Lesson 11.2

11.1 No, a methoxide ion nucleophile is not strong enough to react with the electrophilic carbonyl carbon atom of an amide in the absence of heat.

Lesson 11.3

11.2 Molecule A, Molecule C.

Chapter 12. Amines and Amides

Lesson 12.2

12.1

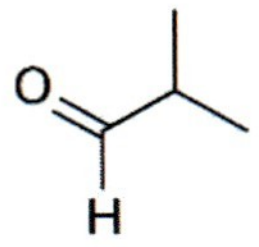

Unit 3. Separation Techniques, Spectroscopy, and Analytical Methods

Chapter 13. Separation and Purification Methods

Lesson 13.1

13.1 Dissolving the mixture in a water-immiscible organic solvent (eg, ethyl acetate) and treating it with an aqueous solution of $NaHCO_3$ extracts 4-methylbenzoate into the aqueous layer. Acidification of the aqueous layer reforms 4-methylbenzoic acid.

Lesson 13.2

13.2 Vacuum distillation.

Lesson 13.3

13.3 The peak at 2.3 minutes is Compound 2. The peak at 3.0 minutes is Compound 1. The peak at 5.1 minutes is Compound 3.

Chapter 14. Spectroscopy and Analysis

Lesson 14.2

14.1 $-8.00°$

Lesson 14.3

14.2 3.7×10^{-22} kJ.

Lesson 14.4

14.3 The molecule does not contain any of these atoms.

Lesson 14.5

14.4 7,430 $L \cdot mol^{-1} \cdot cm^{-1}$.

Lesson 14.6

14.5 The IR spectrum is consistent with the structure of a ketone.

Lesson 14.7

14.6 3.

Index

B

C

D

E

F

G

H

J

K

L

M

O

P

R

S

V

W

Z